Chemistry of
Natural Products

Chemistry of Natural Products

A Unified Approach

SECOND EDITION

N R Krishnaswamy

Former Professor and Head, Department of Chemistry,
Sri Sathya Sai Institute of Higher Learning (Deemed University),
Prasanthi Nilayam

CRC Press

Taylor & Francis Group
Boca Raton London New York

CRC Press is an imprint of the
Taylor & Francis Group, an **informa** business

Cover and book design
© Universities Press (India) Private Limited 2010

First published in paperback 2024

First published 2010 by CRC Press
2385 NW Executive Center Drive, Suite 320, Boca Raton FL 33431

and by CRC Press
4 Park Square, Milton Park, Abingdon, Oxon, OX14 4RN

First issued in hardback 2019

CRC Press is an imprint of Taylor & Francis Group, LLC

© 2010, 2019, 2024 Taylor & Francis Group, LLC

Publisher's Note
The publisher has gone to great lengths to ensure the quality of this reprint but points out that some imperfections in the original copies may be apparent.

ISBN: 978-1-4398-4965-1 (hbk)
ISBN: 978-1-03-291996-6 (pbk)
ISBN: 978-0-429-06580-4 (ebk)

DOI: 10.1201/9780429065804

Typeset in Minion Pro 11/13 *by*
MacroTex Solutions, Chennai 600 088

Visit the Taylor & Francis Web site at
http://www.taylorandfrancis.com

and the CRC Press Web site at
http://www.crcpress.com

Foreword to the First Edition

The *Handbook of Information* of the Massachusetts Institute of Technology, USA, contains this thoughtful statement:

> *A person who teaches in the classroom should be concurrently engaged in creative explorative work. Equally, a research scientist should seek the opportunity to teach students. Higher Education thrives best in an environment in which the performance of the dual tasks receives good encouragement.*

The author of this book, Prof. Dr NR Krishnaswamy, is an outstanding example of a teacher-cum-researcher. This volume carries the imprint of his genius and the role he has played, and continues to play, as an academician par excellence.

In the plethora of practical sciences, organic chemistry has a relatively young history, having emerged as a well-defined discipline less than two centuries ago. Its growth has been phenomenal and it has given birth to numerous subdivisions. The known organic compounds include not only those that possess a natural origin but also a large number synthesised in laboratories all over the world. Despite the synthetic chemist's remarkable contributions to the enrichment of organic chemistry, the subject remains 'organically' linked to Nature. It should be borne in mind that it was from studies on Nature that organic chemistry originated and evolved, and it continues to receive from Nature the inspiration for new and exciting discoveries. This is the reason why the adjective 'organic' remains relevant and appropriate to this day.

Nature produces a large number of chemical compounds whose structures and properties have fascinated organic chemists who have evolved a language of their own to describe the chemistry of these compounds. It has turned out that it is no easy task to present the resulting vast stock of information as a part of the curriculum within a prescribed time-frame. There are other factors which add to the complexity of the situation, resulting from the explosive growth of the subject: the range and diversity encountered in the structures and properties of organic compounds of natural origin, and the imperative to acquire an understanding of their biological functions in terms of chemical structure.

The manner the subject is traditionally taught under headings such as alkaloids, terpenoids, flavonoids, and the like, fails in the attempt to illumine the common fundamental chemical principles which help to unite the diverse structures into one single chemical family in which the diversities do not stand out as so many aberrations but reveal themselves as multiple reflections of the intrinsic chemical principles. In the traditional approach that is commonly followed, emphasis is placed on the chemistry of individual compounds, which results in the students not being adequately trained and assisted in looking beyond specific structures and recognising the beauty and utility of the unifying principles.

The author joined the Sri Sathya Sai Institute of Higher Learning (Deemed University) at Prasanthi Nilayam, at its inception in 1981, as Professor and Head of the Department of Chemistry.

He utilised the golden opportunity that he had, while framing the curriculum in Chemistry at this new institution, to set aside the deficiency perceived by him in the prevalent mode of teaching the subject 'Chemistry of Natural Products'. The result was the emergence of a unique syllabus, with the distinctive feature that this subject would no longer be presented in terms of the enumeration of a number of alkaloids, terpenes, etc., but in terms of representative chemical compounds produced by Nature, apparently with an eye on the structure–biological activity relationships. The author's design was to present the subject to the students under heads such as Structure, Stereochemistry, Dynamic Aspects (reactions), Synthesis, Biosynthesis and Biological Properties, facilitating the development of the subject in the format of organic chemistry and advancing it logically to the doorstep of—if not into— biology. When teaching based on this approach was put into practice, it quickly became evident that the learners were able to assimilate significant knowledge and not merely receive chunks of data which would fade from their memories in no time. One handicap encountered in implementing this approach was the non-availability of a suitable book on which this style of teaching could be based. The student needs, besides a good teacher, a good book that supports the teaching, especially when a new ethos is sought to be created. It is this genuine requirement that induced the author to write this book, to benefit not only the students of the Sri Sathya Sai Institute but other institutions as well. During my years as Vice-Chancellor of the University (1991–93), I had the opportunity to watch Prof. Krishnaswamy at work in developing his vision in the field of chemistry of natural products, into reality.

This book contains the strong imprint of the author's wide-ranging experience in teaching chemistry at a number of institutions, and of the heart-warming manner in which this exercise was conducted by him at the Sri Sathya Sai Institute over a twelve-year period (1981–93). Geared for a new goal to serve a high purpose, and enriched by lucidity of expression, this book is certain to gain a pre-eminent place among books written by Indian authors in the domain of chemistry. In the words of Confucius, Chinese Savant of yesteryears,

The essence of Knowledge is that you have it and you share it with others.

The author has responded to this dictum in his characteristic way, and he proceeds to illuminate the minds of students as well as young teachers who wish to gain proficiency in this branch of chemistry.

S Sampath
Former Vice-Chancellor
Sri Sathya Sai University
Prasanthi Nilayam
Andhra Pradesh

Preface

Since the publication of the first edition of this book, there have been several significant and important developments in the subject, particularly with regard to the biological aspects of natural products. Determination of structures of organic compounds isolated from natural sources has now become streamlined due to the extensive and systematic use of spectral methods, which have more or less displaced time-consuming chemical methods of analysis. However, a student of the subject has to be aware of the classical methods, not only to get a historical perspective of the subject, but also to appreciate the role of experimentation in the evolution of the subject. Therefore, the format adopted in the first edition has been retained in this enlarged, revised edition. The unique style of presentation of a complex subject has also, by and large, been appreciated and accepted by the teaching fraternity and students.

A new chapter, General Introduction, has been added, wherein the different types of secondary metabolites are briefly discussed with the help of examples. It provides students new to the subject with the much needed background for understanding the succeeding chapters. A new introductory section has been added to each chapter to provide basic information regarding the methods and strategies described in the chapter. The number of case studies under Structure, Stereochemistry, Reactions and Rearrangements, and Synthesis, has been increased from ten to fourteen to enlarge the scope of the book, and to cover a wider range of secondary metabolites. In so doing, the requirements as per the prescribed syllabi of different universities in India have been taken into account, at the same time retaining the objectives delineated in the introduction to the first edition. The chapters on Biosynthesis and Biological Aspects have been revised and expanded to incorporate some new discoveries. The Problems section has also been expanded to include a few recently discovered compounds. While making the book more useful to a wider section of students and teachers, care has been taken to keep the book as compact as possible.

Acknowledgements

I had the good fortune to learn the chemistry of natural products under a great master of the subject, the late Prof. TR Seshadri, FRS. His mastery of the subject and inspiring teaching have left an indelible imprint on me. My career as a researcher and teacher has been moulded by his teaching and training. I shall ever remain grateful to him for these gifts.

I would also like to recall the training I obtained in experimental methods from Prof. VVS Murti. It was a pleasure to watch him set up and execute an experiment and I have greatly benefited from the lessons I learnt from him.

I have also had a number of good students whose enthusiastic response made my work interesting and worthwhile. This book is the end-result of a long and satisfying teaching career.

In the preparation of the manuscript, my daughter, Dr Komala Krishnaswamy, and my student and colleague at the Sri Sathya Sai Institute of Higher Learning (SSSIHL), Dr SSR Kumar gave me invaluable help. The manuscript was read by Prof. Charles D Hufford of the University of Mississippi, USA and Prof. PV Raman of Madurai Kamaraj University. Their responses were encouraging and their comments useful and valuable.

My heartfelt thanks are due to Prof. S Sampath, former Vice-Chancellor of the SSSIHL, Prasanthi Nilayam, and my erstwhile colleagues in that institute for their support and help. As Prof. Sampath has mentioned in his foreword, the idea for this book was conceived while I was teaching at the SSSIHL, where the environment was conducive for exploring fresh and new modes of teaching. I could get instant feedback from my graduate students, which went a long way in giving shape to the book.

In the preparation of the Second edition the suggestions made by Prof. PSN Reddy of Osmania University, Hyderabad, have been very useful and valuable. Prof. PV Raman of Madurai generously loaned me his notes on oxygen heterocyclic compounds, and Dr JD Ramanathan and Dr BV Ramani sent me the reprints of their publications on mangiferin, ougenin and anthraquinones. My daughter, Dr Komala Krishnaswamy, meticulously went through the second proofs of all the chapters and pointed out a few corrections. She also prepared a comprehensive list of compound, plant and animal names to be included in the index. I am grateful to all of them.

I am grateful to the publishers, Universities Press, for their encouragement and support. I thank Ms Javanthi Singaram for her careful editing, and the typesetters, MacroTex Solutions, Chennai, for accurately reproducing the structural formulae of a large number of compounds from the sketches provided by me.

Finally, I wish to thank my wife, Mrs Usha Krishnaswamy, for her encouragement and constant goading to get on with the work.

NR Krishnaswamy
krishnaswamynr@gmail.com

Introduction to the First Edition

To write a book on the chemistry of natural products is no easy task. How could one condense the beauty, colour and grandeur of a rainbow stretching across the sky into a tiny dew-drop? The wide range of variety in structures and shapes, and the many reflections of these in reactivities pose a problem to one trying to capture the essence of the structure-shape-reactivity conglomerate within the confines of a book. No account of natural products would be complete without reference to their possible biological functions. In attempting to reason out the connection between chemistry and biological functions, one enters a gray area, full of pitfalls and potholes at the level of our present-day knowledge, but this exercise should not be avoided even if it results only in suggestions lacking, as yet, firm experimental support. Thus, the canvas is wide and the subject complex but one redeeming feature is the constant recurrence of the reflections of a few fundamental principles. Therefore, it should be possible to reduce the large mass of descriptive information to a manageable size of material without sacrificing any of the essentials. One way to do this is to remove the artificial barriers which now separate one group of natural products from another, and to consider them under unifying heads such as structure, stereochemistry, dynamic behaviour, synthesis, and biological activities rather than as so many alkaloids, terpenoids, etc. Another is to eliminate repetitive examples of the same phenomenon and instead use a variety of different examples to illustrate varied aspects so that the kaleidoscopic range of structure and behaviour exhibited by these compounds is retained within the set framework. In this book, such an approach has been adopted in the first part. The second part represents another approach, namely problem-solving as a logical sequel to the first part. Problem-solving, apart from stimulating further interest in these compounds, helps to develop an analytical and critical evaluation of the data presented. Problems also provide, to a limited extent, additional 'meat' in the form of information which might have been left out in the first part.

Confines of the term 'Natural Products'

'Natural Products', in the broadest sense, should connote all the chemical compounds which occur in nature. However, by convention and practice, the term is now used to refer only to the organic compounds occurring in nature. The boundaries are further defined by restricting the term to the secondary metabolites, leaving out the amino acids, proteins, carbohydrates, nucleic acids and lipids, whose biochemical functions are more or less well-known and therefore covered under Biochemistry. By thus restricting the connotation of the term 'natural products', it was possible, till recently, to keep the focus on the chemistry of these compounds (since, at any rate, their biological roles were ill-defined), and thus keep the subject under the umbrella of Organic Chemistry. However, with the growing knowledge about the biological functions of these compounds (which were, not long ago, dismissed as waste products of metabolism!), natural products are emerging from the cocoon of Organic Chemistry. This development was, indeed, predicted several years ago by Lord Todd. Sir Robert Robinson had also anticipated or advocated the emergence of this trend as is clear from his following words: 'The structures are points of reference on a chart, or milestones on a road, but it is the surrounding country which is of greatest interest. Not what the molecules are, but what they will do, how they may be formed and what transpositions they may undergo or be induced to undergo—these are the significant problems". Viewed thus, the large number of different compounds prominently exhibit their common traits while the differences among them bring out the diversity.

Organisation of the Book

The first part, as already mentioned, provides the framework for displaying the essentials of the make-up, despite their variety, of naturally occurring organic compounds, other than the primary metabolites. This has been done under five heads: structure, stereochemistry, dynamic behaviour (reactions and rearrangements), synthesis, and biosynthesis. A brief excursion is also made into biology by discussing some of the biological aspects of these compounds viewed from a chemist's perspective.

The book is primarily directed towards students majoring in Chemistry. Since learning will be more effective if it is enjoyed, the book endeavours to present the material in as easy a style as possible. From the instructional point of view, it is best to approach a problem from different angles so that the final picture which emerges is as complete as possible. For example, structures of naturally occurring compounds have been solved exclusively by the so-called classical method, exclusively by physical methods, and in most cases in recent years, by a judicious combination of the two. A student can get a complete picture of structure-solving, including a historical perspective, by studying the structures which have been investigated and deduced by all the three approaches. This is what has been done in this book. The examples chosen are illustrative—other examples could have been chosen—and the focus is on methods and principles rather than on the individual compounds. A couple of examples have been chosen for the simple reason that the author was personally involved in solving their structures, and therefore the information is first-hand and authentic. The book begins with 'Structure' and with a 'classical' example, namely strychnine. We feel that this would be a good beginning.

The problem-solving section of the book has twenty-five problems chosen from published work of the past twenty-five years; several of them have been taken from recent publications. Each exercise has three sections, namely statement of the problem, analysis, and notes. Students are advised to analyse the problems on their own before reading the last two sections of each problem. They should also make every attempt to go through the original publications, as many of the problems, as presented in the book, are 'touched up' to make them more easily analysable at the instructional level. In the original papers, the data, in some cases, could be inadequate for a complete analysis by someone not experienced enough to substitute intuition for hard facts!

The author hopes that the readers will enjoy the portraits, which follow, of some interesting 'personalities' among molecules of nature. The structural formula of a compound is, after all, a picture which is drawn on the basis of interpreted responses of the compound to a number of stimuli. Yes, the stimuli should be properly chosen and administered, and the responses correctly read and interpreted. The completeness or otherwise of the final picture would therefore depend on the range and accuracy of the stimuli and the 'reading' of the responses. The picture, in turn, should therefore be able to 'speak' about its subject—whether it would be colourless or coloured, and how it could be expected to behave in a specific reaction, and so on. So when one walks into a gallery of a wide variety of structural formulae, one gets an opportunity to study an equally wide range of molecular behaviour. Nature has, indeed, provided ample variety in the structures of its molecules to reflect a plethora of behavioural patterns.

Contents

1

Introduction

As mentioned in the preface to the first edition, the term 'Natural Products' refers, by convention, to secondary metabolites which are distinctly different in their biological functions as compared to the primary metabolites (proteins and carbohydrates) and the semantides (nucleic acids). These compounds are mainly of plant origin though recent research has revealed that secondary metabolites are also produced by insects and marine organisms. Contrary to the earlier view that these compounds are waste products of metabolism, recent research work has shown that they do have well-defined biological functions though they are not essential to life. These activities are discussed at greater length in Chapter 7 of this book.

The major types of secondary metabolites are the terpenes, the alkaloids, the flavonoids and related compound, such as the anthocyanins, isoflavones, neoflavonoids and condensed tannins and the polyketides. The primary building blocks involved in the biosynthesis of these compounds are respectively isopentenyl pyrophosphate, different amino acids, shikimic acid and a combination of acetyl coenzyme-A and malonyl coenzyme-A. Shikimic acid is also the precursor for the aromatic amino acid, phenylalanine, which is the primary building block for the benzylisoquinoline alkaloids.

Each of these groups of secondary metabolites is a large family comprising of a great number of compounds representing a wide variety of skeletal structures, peripheral features and molecular size. In this chapter the salient features of each type are discussed in order to provide background material to the individual compounds described in the succeeding chapters.

1.1 THE TERPENES

Introduction

The terpenes form a large group of secondary metabolites, chiefly of plant origin, though several of these compounds are being increasingly discovered in insects and marine flora and fauna. In plants, the lower terpenes, and particularly the monoterpenes, are generally found as constituents of essential oils which are usually obtained by the steam distillation of the plant parts in which they occur. The presence of these compounds is easily detected because of their characteristic (by and large pleasant) and distinctive odours. Some of the familiar and common plant sources of these compounds are the various species of the genera Eucalyptus, Cymbopogon (for example, lemon grass), Geranium, Pinus and related conifers, Citrus and the spices. Several of these compounds are heat sensitive and are likely to decompose during steam distillation. Such compounds are best isolated using a non-polar solvent for extraction of the plant material at a low temperature. The essential oils and extractives thus obtained are invariably mixtures of several compounds, many of which are closely related to each other. They can be further fractionated using chromatographic methods among which the most ideally suited for the isolation of individual terpene compounds is gas chromatography. Separation and simultaneous identification of the components of an essential oil can be effectively carried out by GC-MS, wherein a gas chromatograph (GC) operates in tandem with a mass spectrometer (MS). High pressure liquid chromatography and critical temperature fluid extraction techniques are also being increasingly used for this purpose.

Classification

Terpenes can be considered to be formally built up by the combination of isoprene units which are 5-carbon hydrocarbon moieties. Accordingly, they are classified into the following six types, depending on the number of isoprene units involved in each type.

1. *Monoterpenes* contain 10 carbon atoms and are apparently formed by the combination of two isoprene units.
2. *Sesquiterpenes* contain 15 carbon atoms are made up of three isoprene moieties.
3. *Diterpenes* are formed from four C-5 units and possess 20 carbon atoms.
4. *Sesterterpenes* which are of rarer occurrence, compared to the other types, are C-25 compounds arising from five isoprene units.
5. *Triterpenes* are C-30 compounds and are derived from six isoprene moieties.
6. *Tetraterpenoids*, which as also known as the carotenoids, are formed by the combination of two diterpene units and hence contain 40 carbon atoms.

In each type, there are hydrocarbons, alcohols, aldehydes, ketones, carboxylic acids and epoxides. The skeletal structure could be acyclic or possess one or more rings. The consequence is an enormous range of compounds which is continually increasing as a result of new discoveries. Even as this brief account is being written up some research group or the other would be isolating a new terpene! In spite of this bewildering variety, all these compounds can be biosynthetically traced to one compound, namely, isopentenyl pyrophosphate. Biosynthesis of terpenes is discussed in detail in Chapter 6. However, to maintain a contextual continuity, a brief account is given below.

Biosynthesis of terpenes

Though a terpene structure can be formally dissected to give isoprene units, the smallest building block is the acetate unit, three of which combine to give R (–) Mevalonic acid (1.1.1). This compound undergoes phosphorylation followed by dehydrative decarboxylation to yield isopentenylpyrophosphate (IPP; 1.1.2). IPP is isomerised by the action of an enzyme, isomerase, to form dimethylallyl pyrophosphate (DMAPP; 1.1.3). The next step is a combination of one IPP molecule with a DMAPP molecule, with the elimination of a phosphate unit resulting in the formation of geranyl pyrophosphate (GPP) (1.1.4). This compound is the first member of the terpene famiy and is the progenitor of all monoterpenes. The biochemical transformation of GPP into other acyclic, monocyclic and bicyclic monoterpenes is discussed in detail in Chapter 6. The primary building block, namely, the IPP molecule, is also produced by a different, non-mevalonic acid, biosynthetic pathway which operates in the chloroplast, wherein the mono- and di-terpenes are produced (see Chapter 6 for further details).

Further elongation of the terpene chain by linear combination of a geranyl pyrophosphate unit with a IPP molecule, accompanied by the loss of a phosphate unit, yields farnesyl pyrophosphate (FPP; 1.1.5), which is the precursor of all sesquiterpenes. The longer chain, as compared to that of monoterpenes, and consequent conformational flexibility (amenable to intramolecular interactions) enables these compounds to exhibit a wider range in skeletal framework (acyclic to tricyclic). A further addition of another IPP unit to a FPP molecule, with the elimination of a phosphate unit, gives the first member of the diterpene family, geranylgeranyl pyrophosphate (GGPP; 1.1.6). One more addition of a IPP molecule to GGPP results in the first member of the sesterterpene group. Till this stage, the chain elongation is by head-to-tail combination of the reacting units. Therefore, the primary member in each of the above-mentioned four types of terpenes, is the pyrophosphate of a primary alcohol. In contrast to these four types, the tri- and tetra-terpenes are formed by a head-to-head combination of the reacting moieties. Thus, a head-to-head reaction of two FPP units, accompanied by the loss of both phosphate units, results in a hydrocarbon, squalene (1.1.7), which is the first member of the triterpene group. Similarly, a head-to-head combination of two GGPP units yields a tetraterpenoid hydrocarbon which is the progenitor of carotenoids. The biosynthesis of squalene and its subsequent transformation into other triterpenes are described in detail in Chapter 6. It should, however, be noted that a few monoterpenes, designated as irregular monoterpenes, though of rarer occurrence, are also formed by a method other than head-to-tail combination of the two isoprenoid units. Their biosynthesis is also briefly described in Chapter 6.

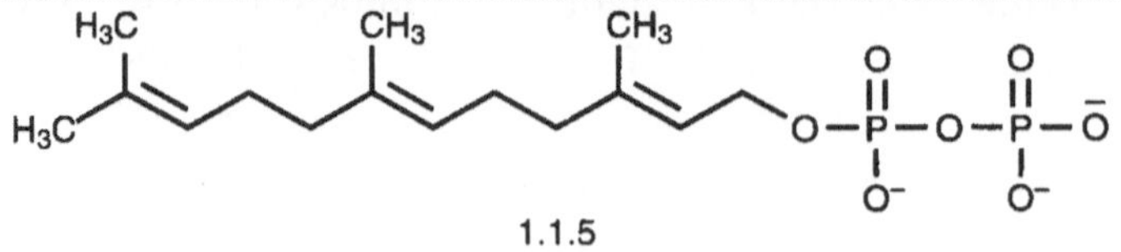

1.1.5

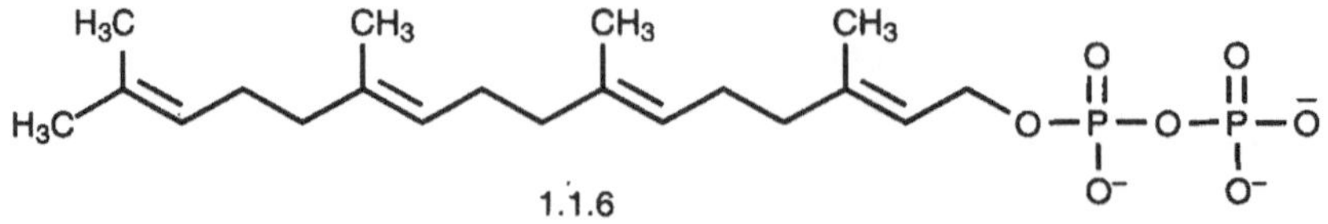

1.1.6

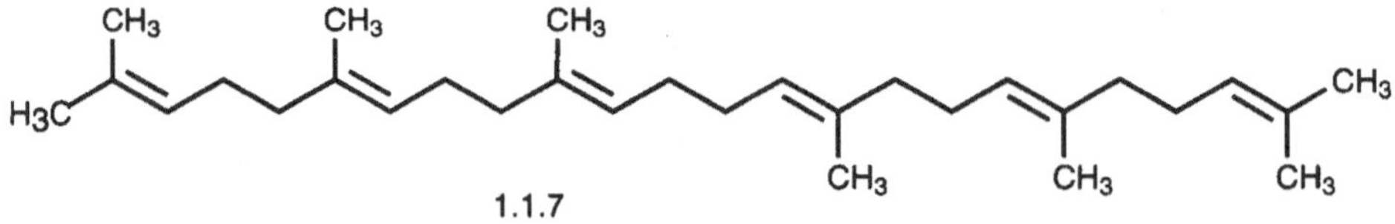

1.1.7

Some Representative Monoterpenes

Geraniol (1.1.8) and nerol (1.1.9) are E- and Z-isomers and are primary alcohols. Geraniol occurs in several essential oils including that of geranium, from which it gets its name, but the major source is rose oil. As a matter of fact the characteristic scent of rose flowers is due to geraniol.

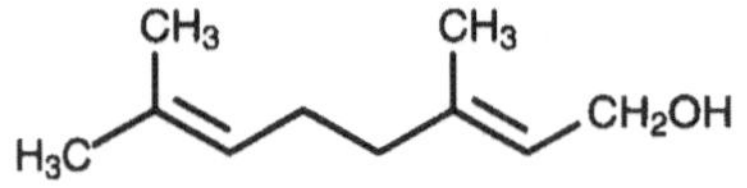

1.1.8

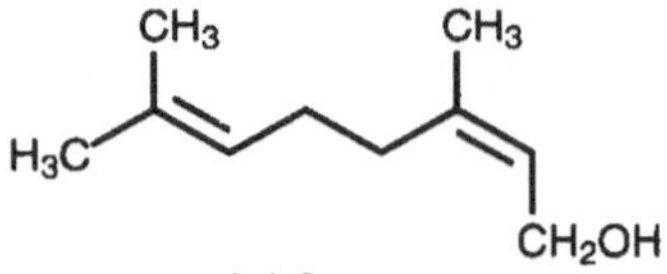

1.1.9

Oxidation of either of these compounds results in citral which is roughly a 3:1 inseparable mixture of the corresponding aldehydes, geranial (1.1.10) and neral (1.1.11). The reason why they are inseparable can be understood if one looks at their canonical forms (1.1.12), and their interconversion as shown beside. Citral, as the name indicates, is a component of the essential oils of citrus fruits, but it is commercially obtained from the oil of lemon grass, wherein it is present to the extent of 60%—80%, particularly in cultivated varieties.

Linalool (1.1.13) is a structural isomer of geraniol and nerol and is a tertiary alcohol. Since it possesse a chiral centre it is optically active. Both the enantiomorphs occur in nature: rose oil is a source for the (−) isomer whereas the (+) form is found in orange oil. It is a characteristic component of the essential oil of linaloe, from which its name is derived. It also occurs in the esterified form in the oil of *Mentha citrata* from which it can be obtained in about 60% yield. It is widely used in the perfumery industry because of its exotic odour. Other commonly occurring acyclic monoterpenes are the hydrocarbons, myrcene (1.1.14) and ocimene (1.1.15), the aldehyde citronellal (1.1.16), and the corresponding alcohol, citronellol (1.1.17).

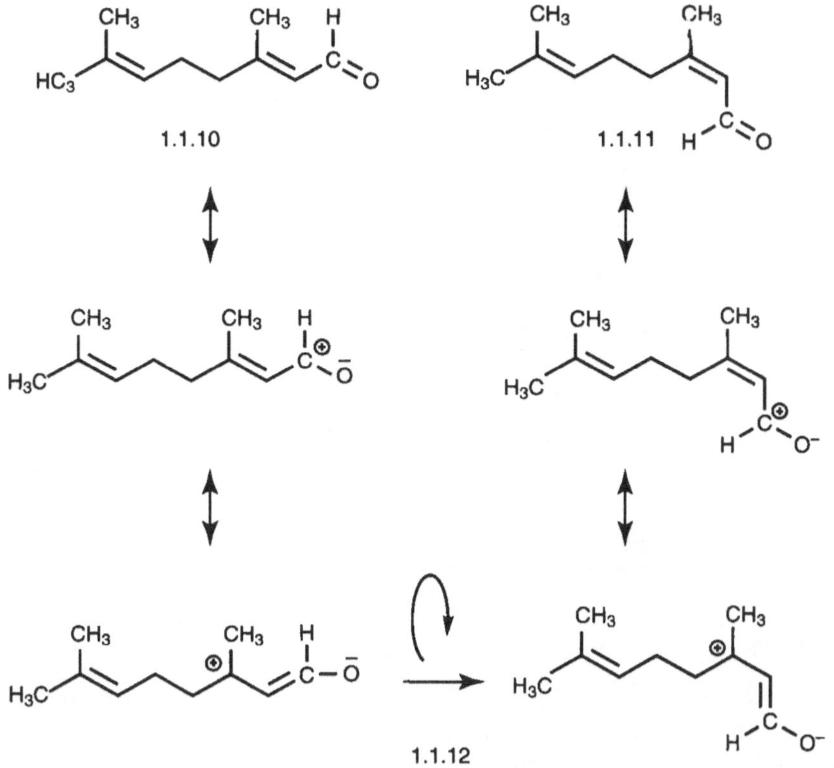

1.1.10

1.1.11

1.1.12

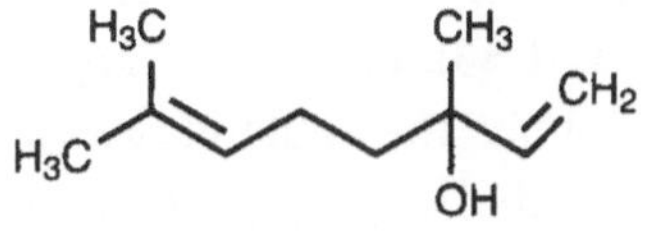

1.1.13

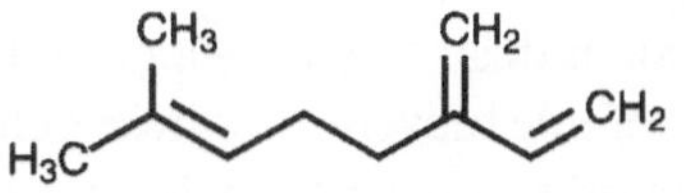

1.1.14

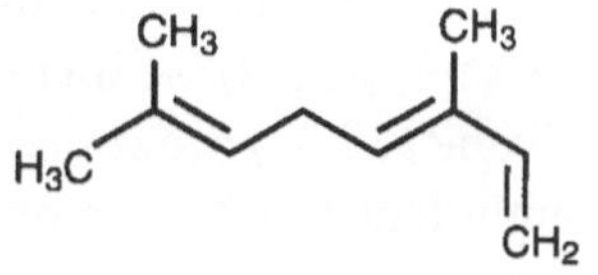

1.1.15

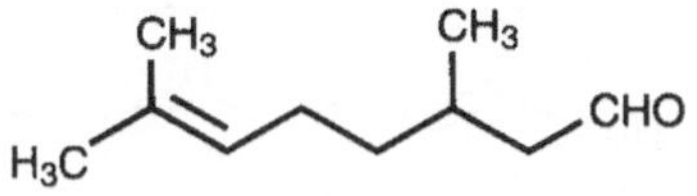

1.1.16

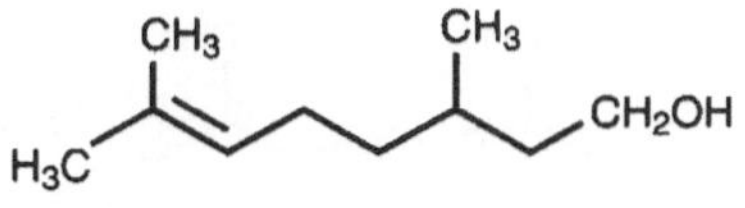

1.1.17

A number of monocyclic monoterpenes find use in the perfumery industry. These include the hydrocarbon, limonene (1.1.18), the tertiary alcohol, α-terpineol (1.1.19), the secondary alcohol, (–) menthol (1.1.20) and the ketones, (–) menthone (1.1.21) and carvone (1.1.22). An important bicyclic monoterpene is the ketone, camphor (1.1.23). Equally important is α-pinene (1.1.24) which is a component of turpentine oil. It serves as a starting material for the preparation of santalaire (1.1.25) which has the smell of sandalwood. Other important bicyclic monoterpenes are car-3-ene (1.1.26) and camphene (1.1.27), both of which are hydrocarbons.

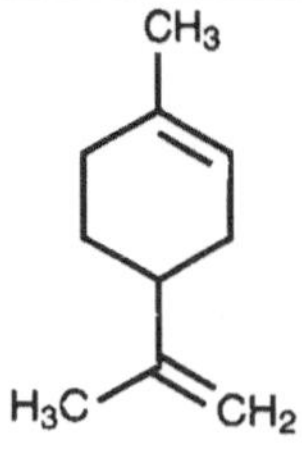

1.1.18	1.1.19	1.1.20	1.1.21	1.1.22

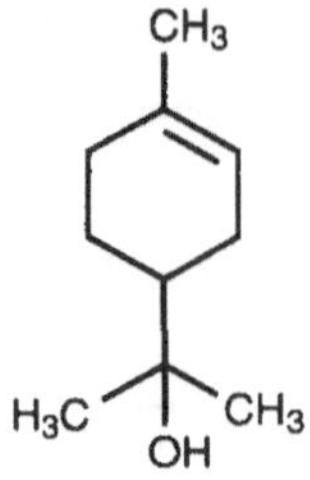

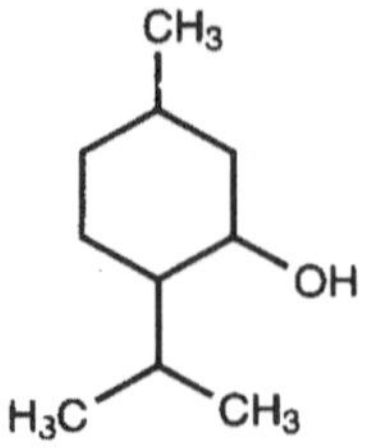

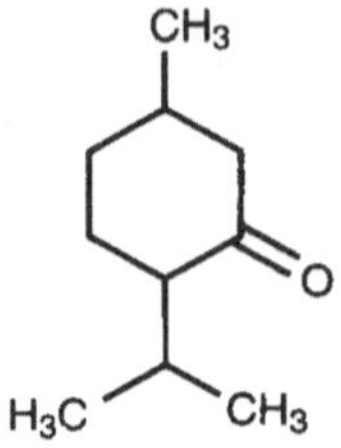

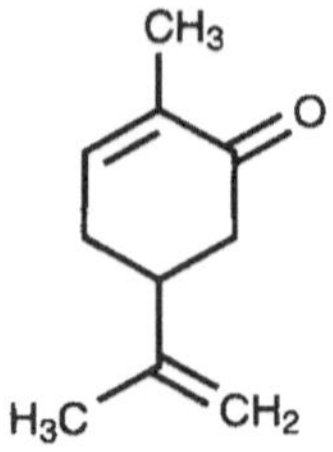

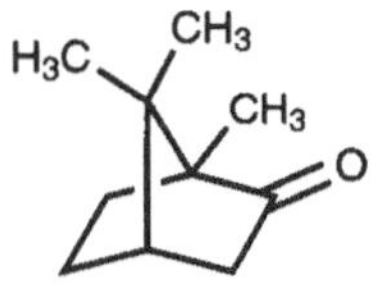

1.1.23	1.1.24	1.1.25	1.1.26	1.1.27

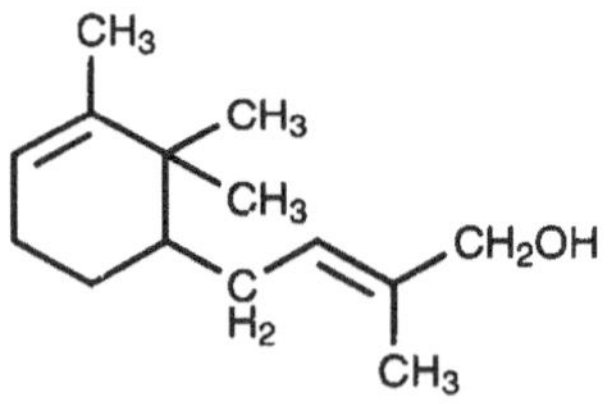

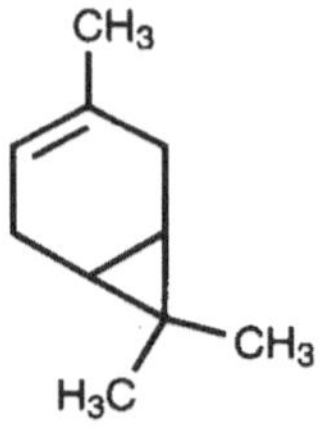

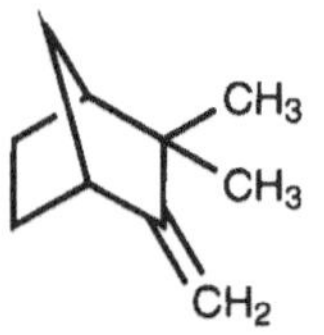

Sesquiterpenes

Farnesol (1.1.28) and nerolidol (1.1.29) are related to each other in the same way as geraniol and linalool are related. The former is a primary alcohol present in the essential oils of ambrette seeds, lily of the valley and lime flowers. Nerolidol, a tertiary alcohol, is a component of Peru balsam.

Bisabolene (1.1.30), and zingiberene (1.1.31) are well known monocyclic sesquiterpenes. Cadinene (1.1.32), selinene (1.1.33), eudesmol (1.1.34), caryophyllene (1.1.35), guaiol (1.1.36) and vetivone (1.1.37) are some of the notable bicyclic sesquiterpenes. Guaiol has the azulene skeleton. Longifolene (1.1.38) is a much studied sesquiterpene. Its synthesis was a challenging problem and was successfully accomplished by Corey, WS Johnson and others (see Chapter 5).

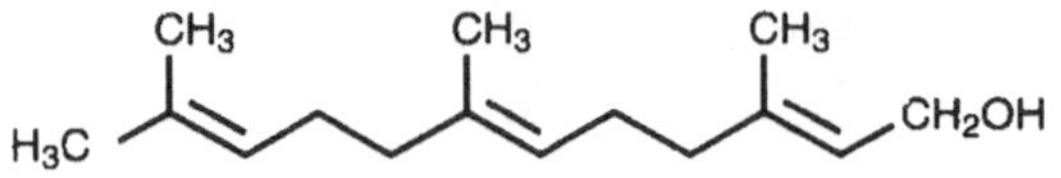

1.1.28	1.1.29

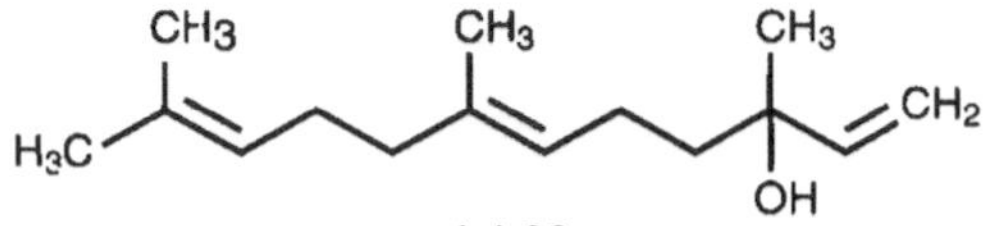

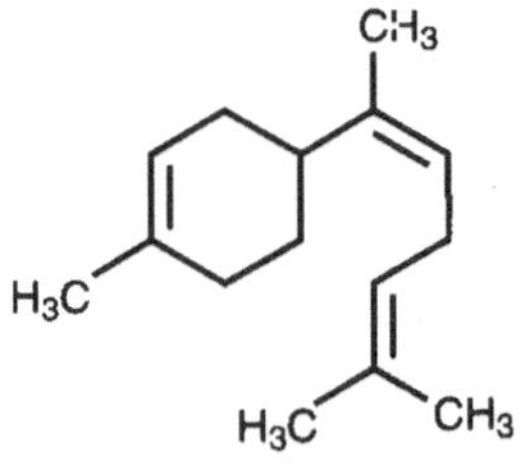

1.1.30	1.1.31	1.1.32

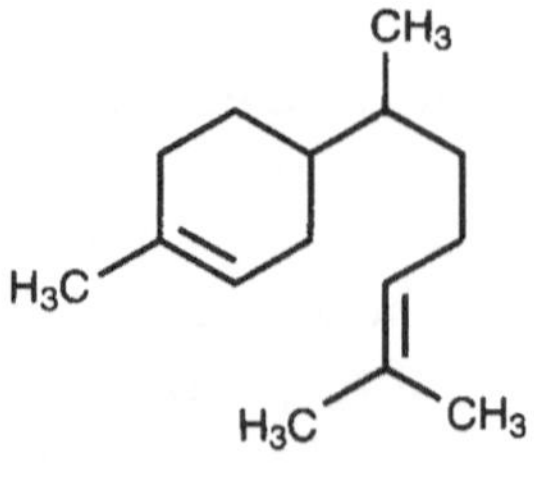

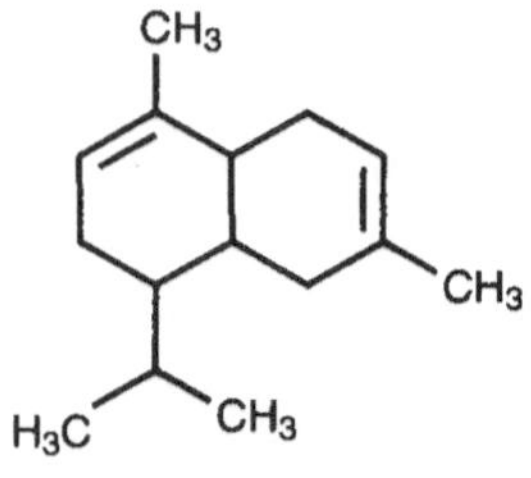

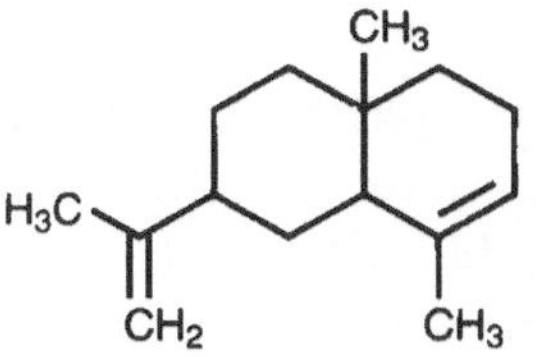

Compounds 1.1.33, 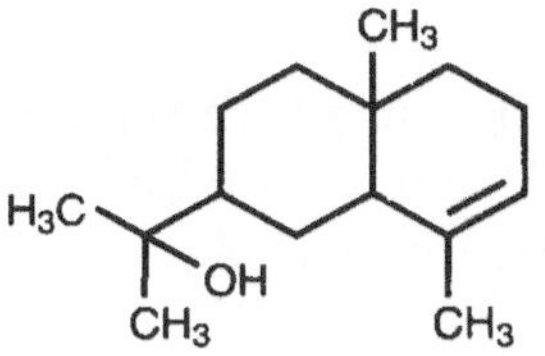 1.1.34, 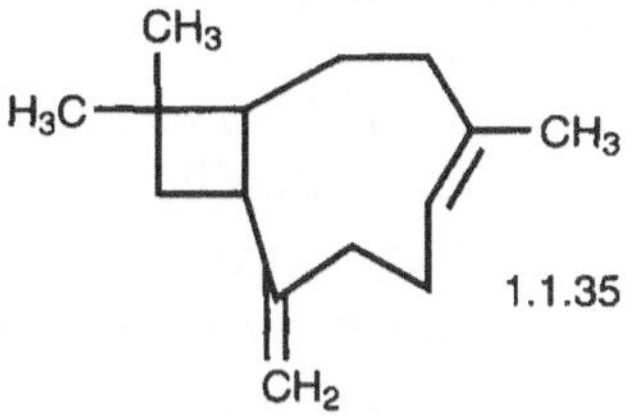 1.1.35, 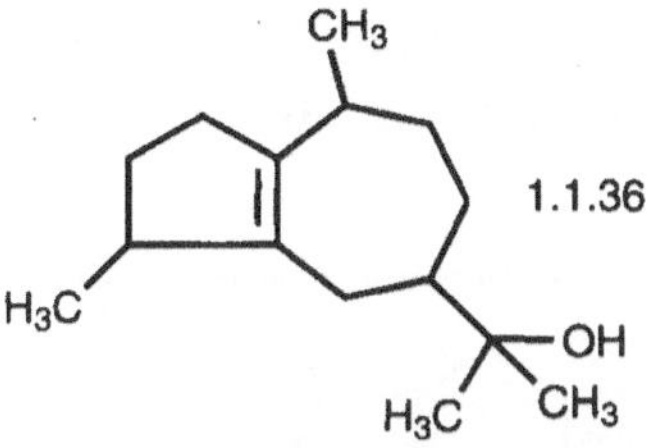 1.1.36, 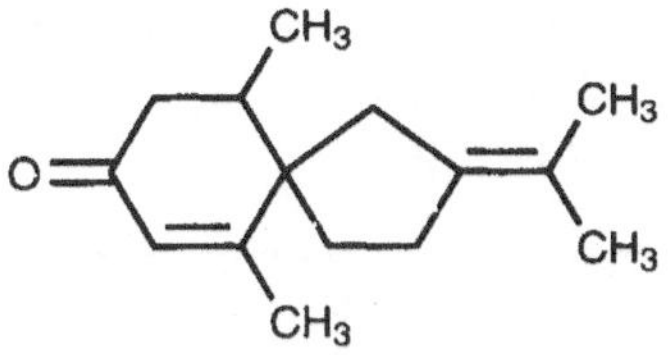 1.1.37, 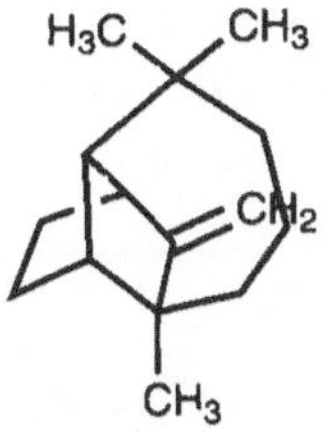 1.1.38

Sesquiterpene lactones occur widely in several plants of the Compositae family. One of the earliest compounds of this group to be isolated and characterised was α-santonin (1.1.39). Germacranolides, as for example, costunolide (1.1.40) and isabelin (1.1.41), have a 10-membered carbocyclic ring fused to a 5-membered lactone; isabelin is a dilactone. Several sesquiterpene lactones possess the naphthalene skeleton and others have the azulene framework. Two examples of the former type are encelin (1.1.42) and Dihydropicridine (1.1.43). Examples of the latter type are ambrosin (1.1.44) and coronopilin (1.1.45). Artemisin (1.1.46) is a novel type of sesquiterpene lactone.

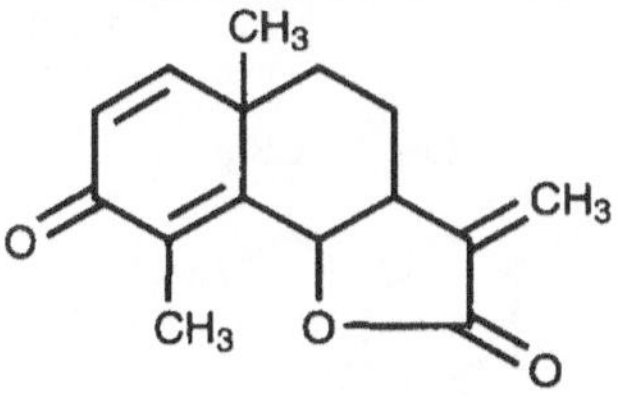

Compounds 1.1.39, 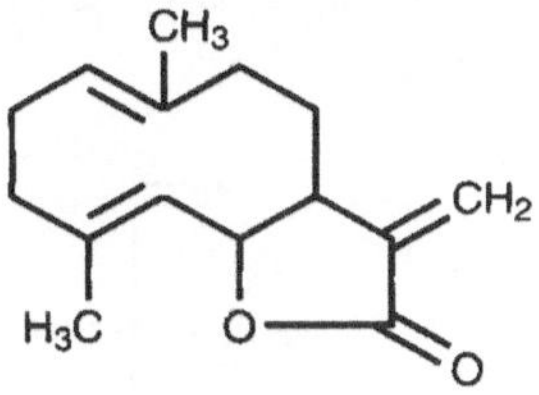 1.1.40, 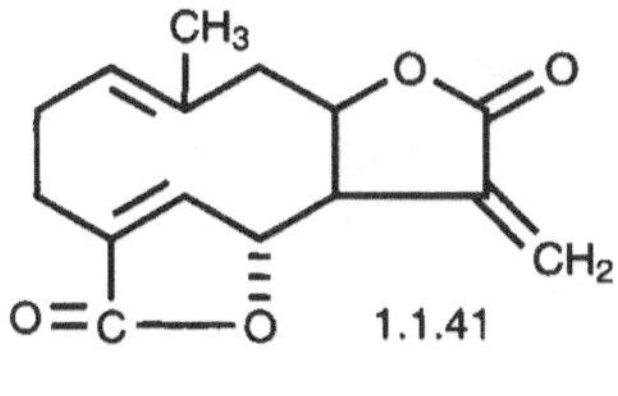 1.1.41, 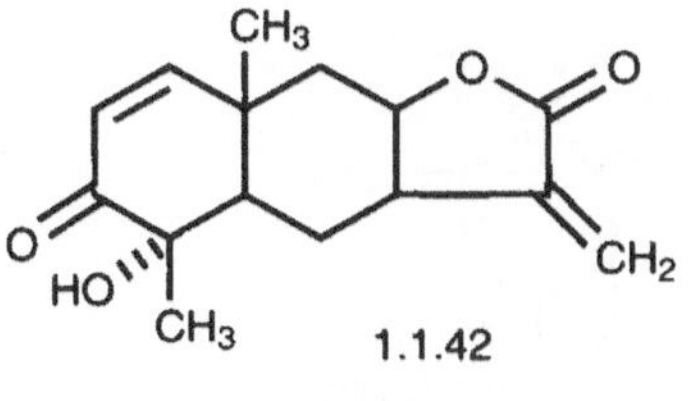 1.1.42, 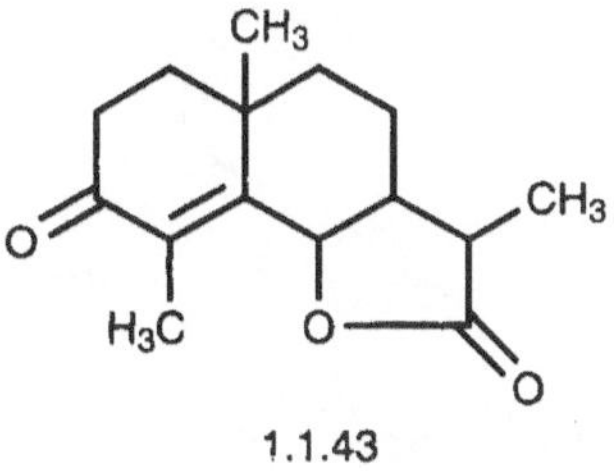 1.1.43, 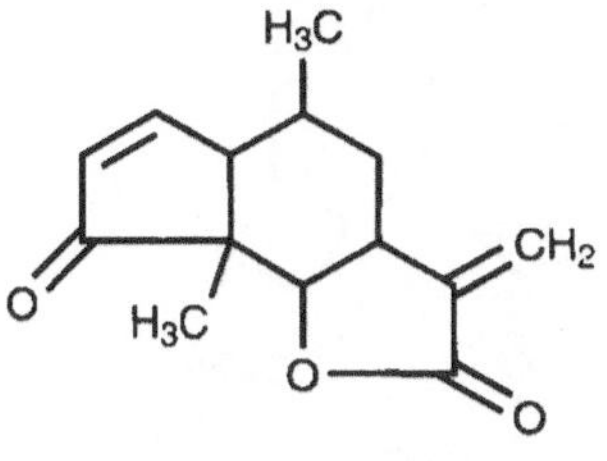 1.1.44, 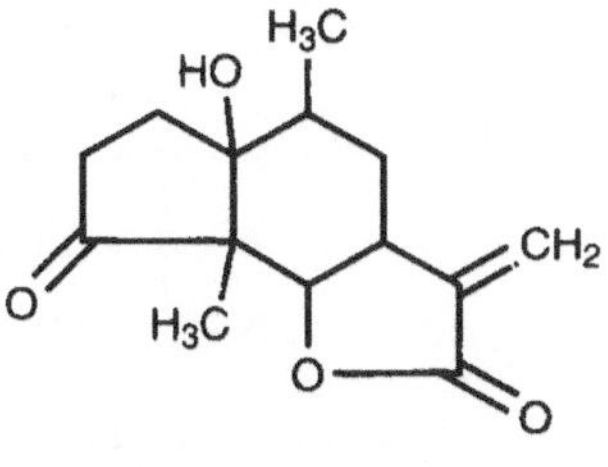 1.1.45, 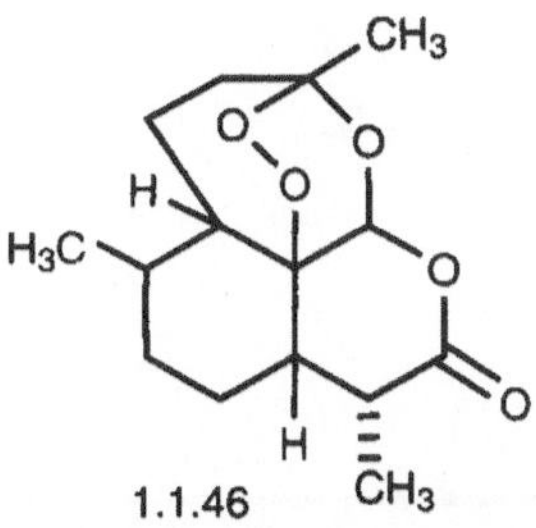 1.1.46

Diterpenes

Phytol (1.1.47) is an acyclic diterpene alcohol which is a constituent of chlorophyll. Abietic acid (1.1.48) is a tricyclic compound isolated from rosin, which is the residue remaining after the steam volatile components of turpentine are removed. Gibberellins, for example, gibberellin-A3 (1.1.49), are complex norditerpenes, with one carbon atom less than a normal diterpene. An example of a tetracyclic diterpene is kaurene (1.1.50). Several cytotoxic diterpenes have been isolated from marine invertebrates. One such compund is sclereophytin A (1.1.51) isolated from the soft coral, *Sclerophytum capilaris.* Other examples are andrographolide (1.1.52), clerodin (1.1.53), royleanone (1.1.54), which is a quinone, and taxol (1.1.55).

1.1.47

1.1.48

1.1.49

1.1.50

1.1.51

1.1.52

1.1.53

1.1.54

1.1.55

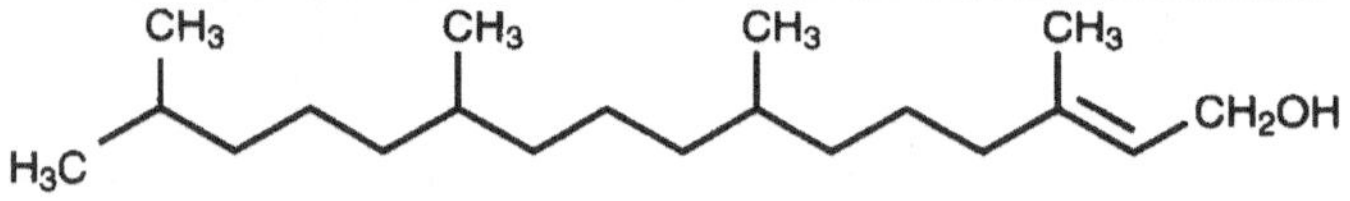

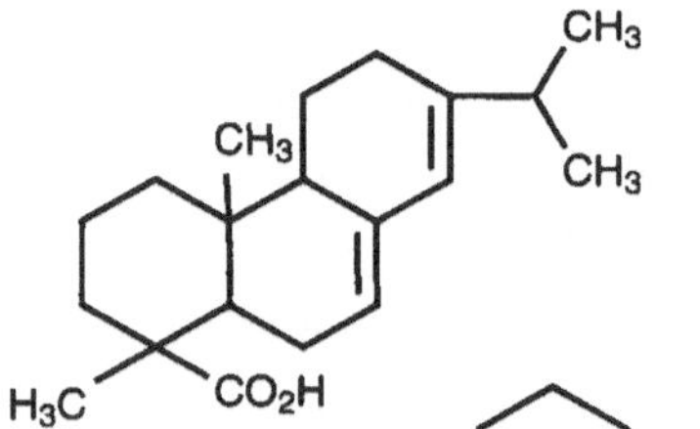

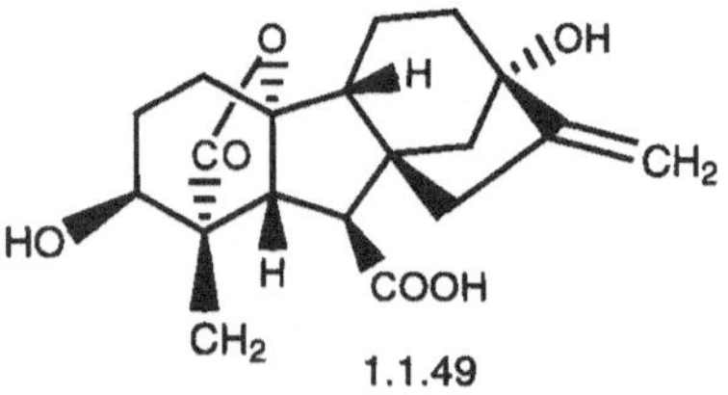

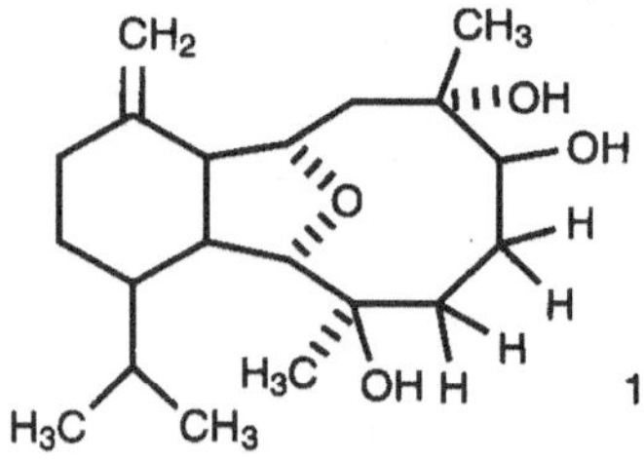

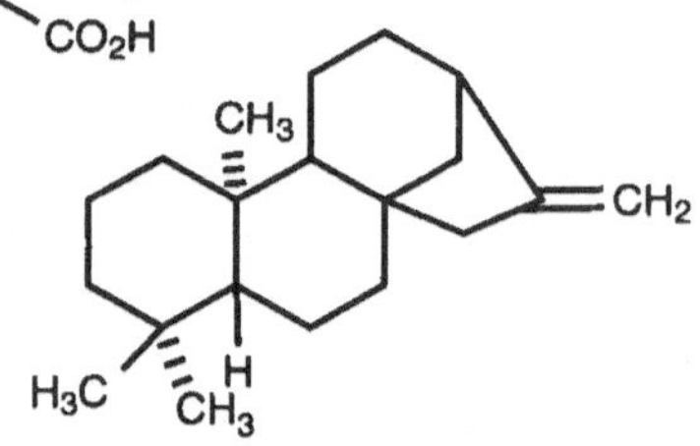

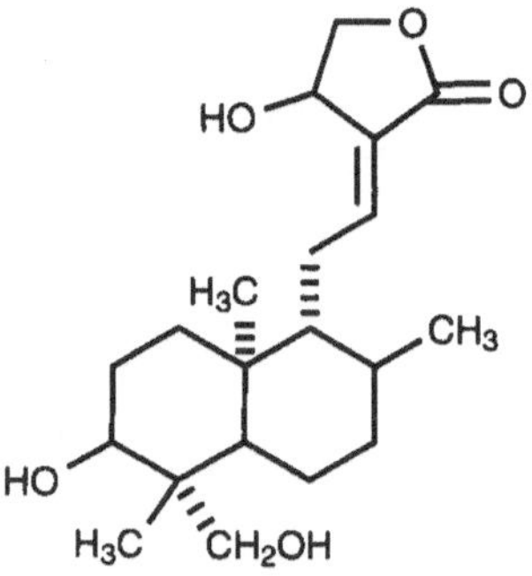

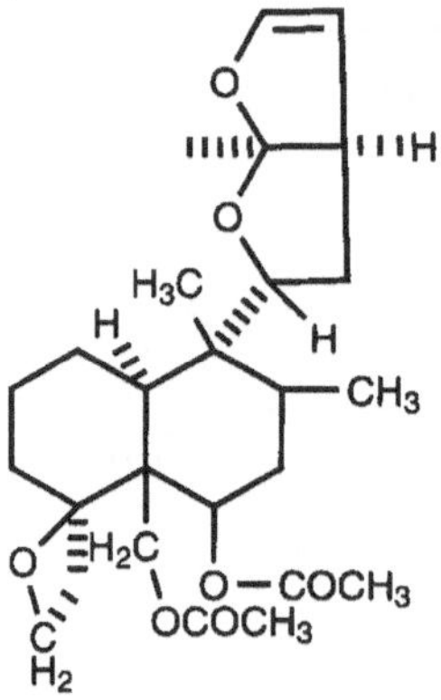

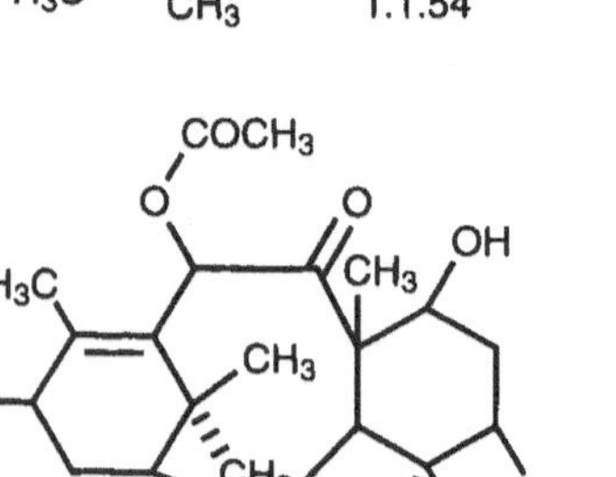

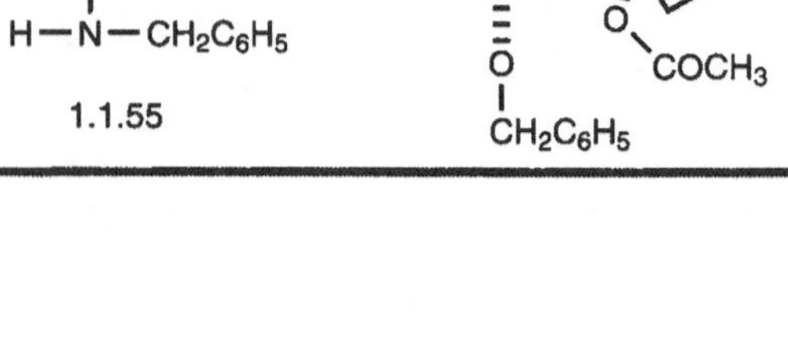

Sesterterpenes

These compounds are products of further transformation of geranylfarnesyl pyrophosphate (1.1.56). As already mentioned, they are of comparatively rarer occurrence. Major sources are micro-flora and marine organisms. The first sesterterpene to be isolated (in 1965) and characterised was ophiobolin-A, also known as cochliobolin-A (1.1.57), which has a tricyclic structure. Since then a number of these compounds have been obtained from sources such as fungi, lichens, ferns, marine organisms and insects. Other examples are cheilanthatriol (1.1.58), which is also tricyclic, and retigeranic acid (1.1.59), which is a pentacyclic compound.

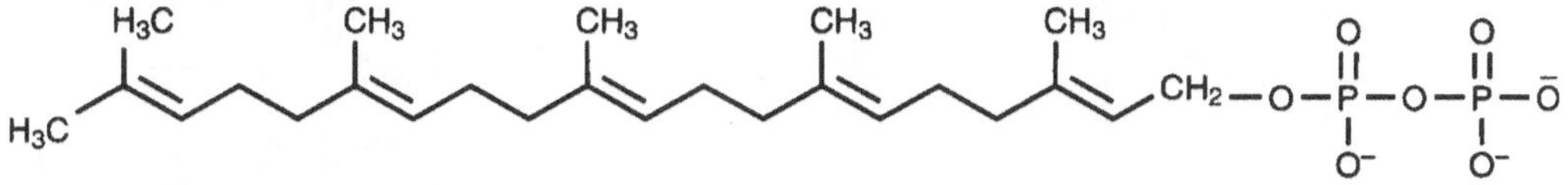

1.1.56

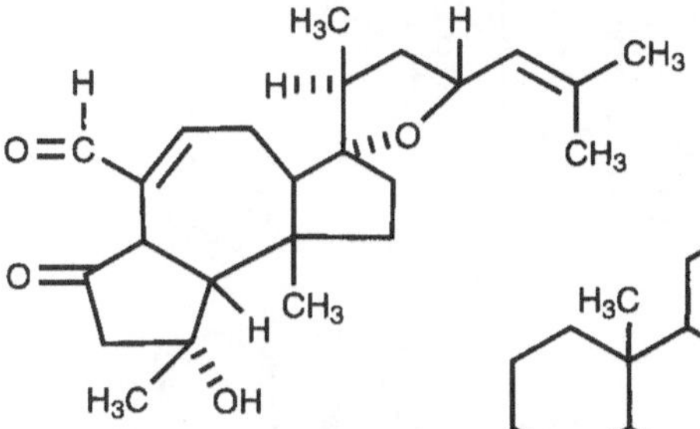

1.1.57

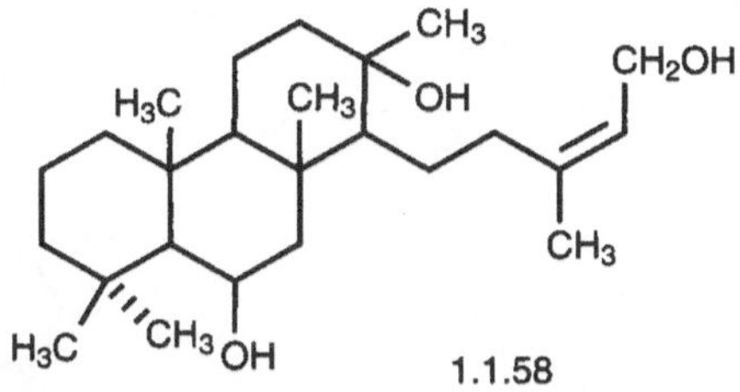

1.1.58

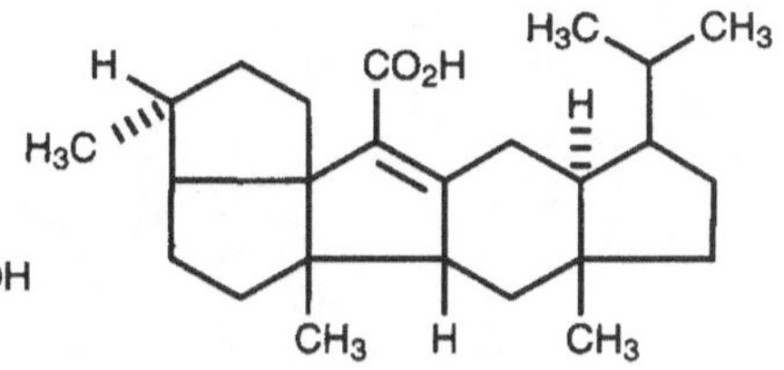

1.1.59

Triterpenes

Squalene (1.1.7), the first member of this large group of compounds, is an acyclic hydrocarbon. Its 2, 3-epoxide (1.1.60), which is enzymatically produced in the liver, is the biosynthetic precursor for the tetracyclic compound, lanosterol (1.1.61), which, in turn, serves as the biosynthetic progenitor of a number of pentacyclic triterpenes. These include α- and β-amyrins (1.1.62. and 1.1.63), oleanolic acid (1.1.64), friedelin (1.1.65) and lupeol (1.1.66). Saponins are glycosides of triterpene alcohols. Unusual triterpenes include the bitter principle of Cucurbitaceae, cucurbitacin B (1.1.67) and limonin

1.1.60

(1.1.68), the bitter component of citrus seeds; the latter has only 28 carbons and has a furan ring in addition to being a dilactone.

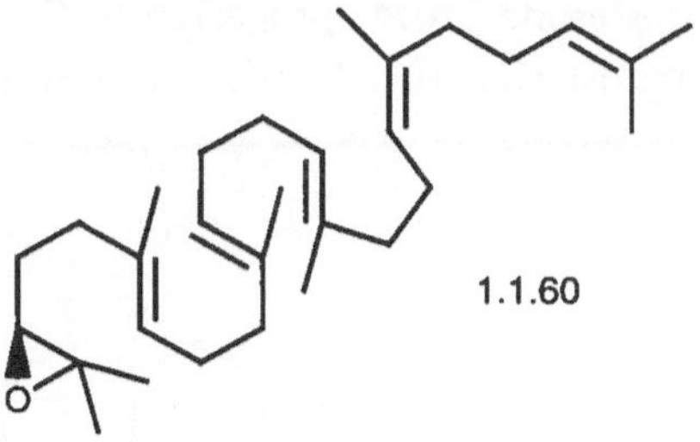

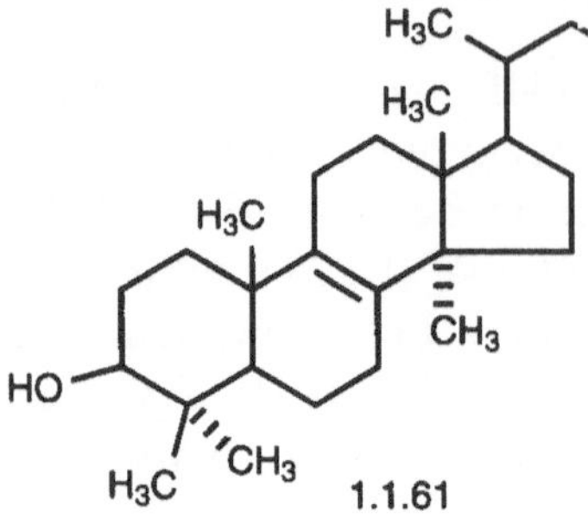

1.1.61

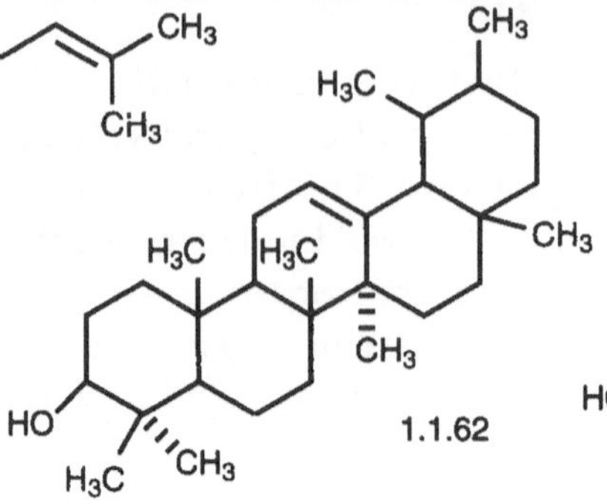

1.1.62

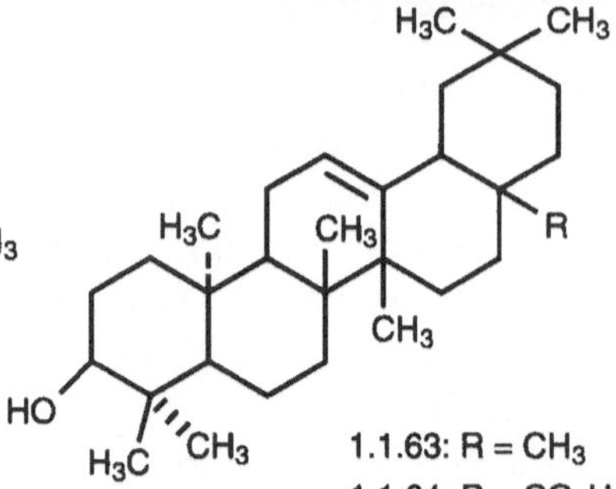

1.1.63: R = CH₃
1.1.64: R = CO₂H

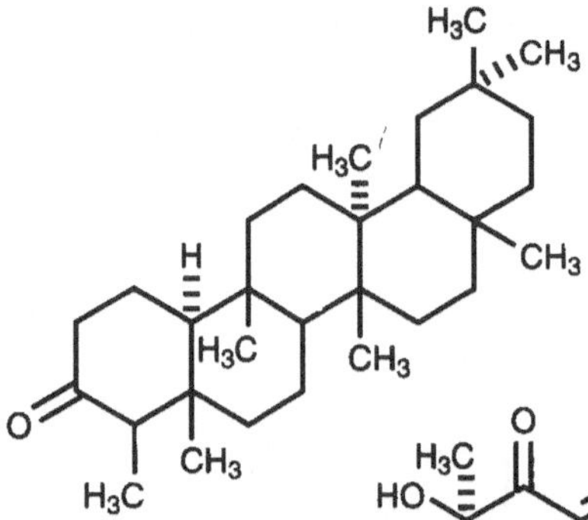

1.1.65

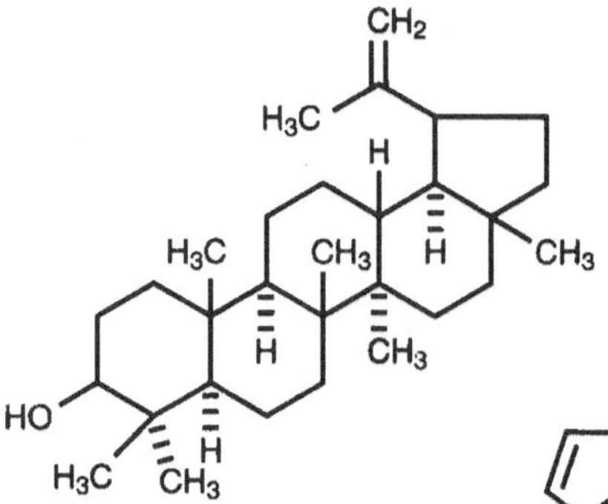

1.1.66

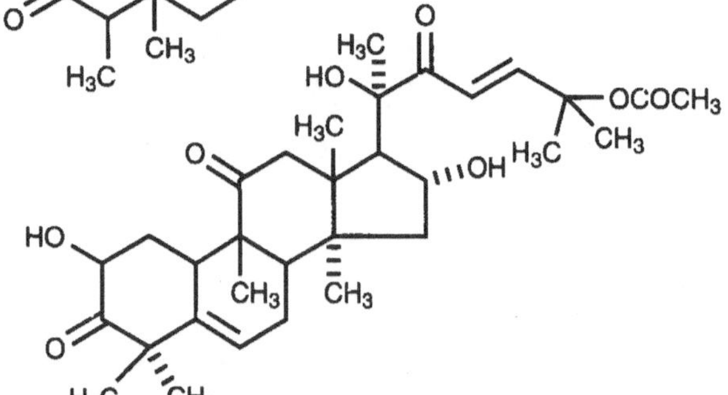

1.1.67

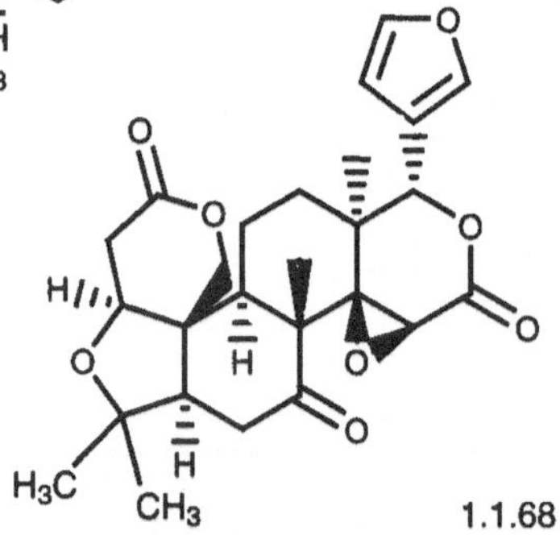

1.1.68

Steroids: This important and large group of compounds, occurring widely in nature, is biosynthetically related to the triterpenes. It includes, besides cholesterol (1.1.69), vitamin-D (calciferol; 1.1.70), ergosterol (1.1.71), the various sex hormones, such as estrone (1.1.72), androsterone (1.1.73), testosterone (1.1.74), progesterone (1.1.75), and the adrenocortical hormones such as cortisone (1.1.76), corticosterone (1.1.77) and aldosterone (1.1.78), all of which are of vital importance in human health.

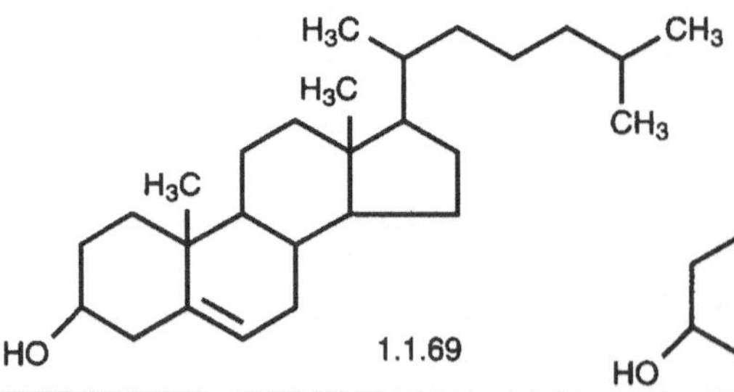

1.1.69

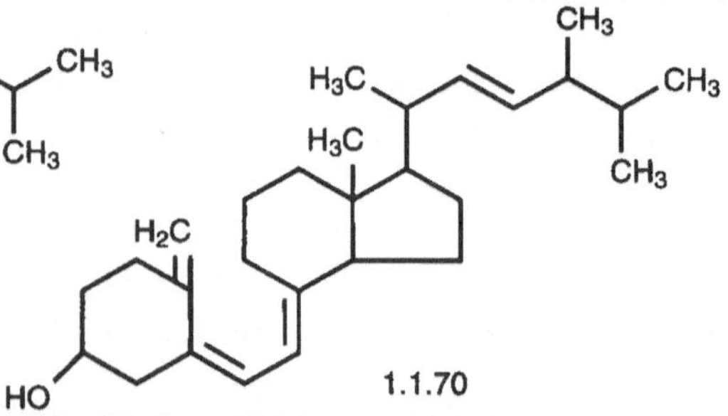

1.1.70

1.1.71

1.1.72

1.1.73

1.1.74

1.1.75

1.1.76

1.1.77

1.1.78

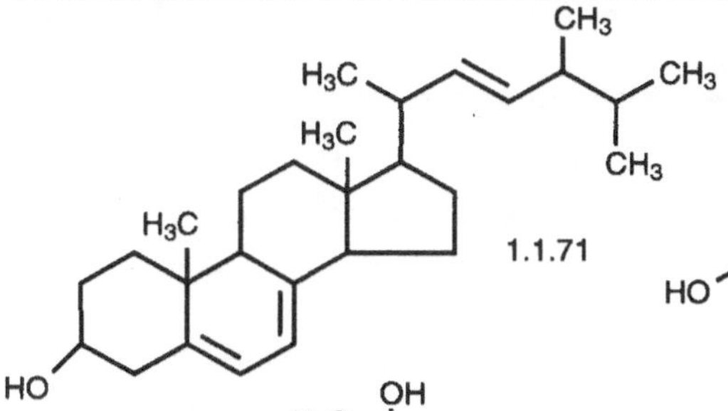

Sterols of plant origin include the cardiac glycosides such as the glycosides of strophanthidin (1.1.79), the sapogenins like diosgenin (1.1.80) and stigmasterol (1.1.81).

1.1.79

1.1.80

1.1.81

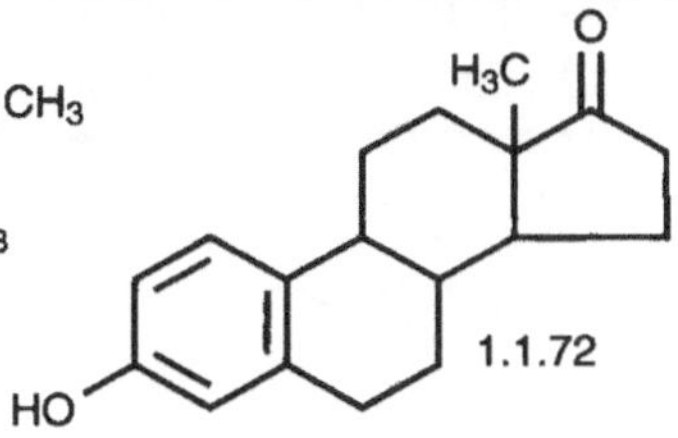

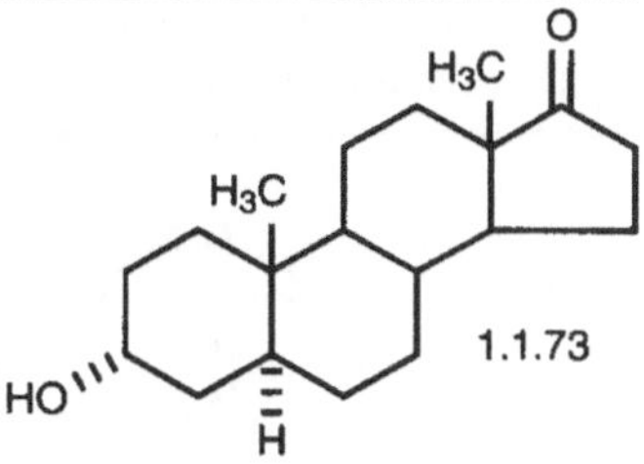

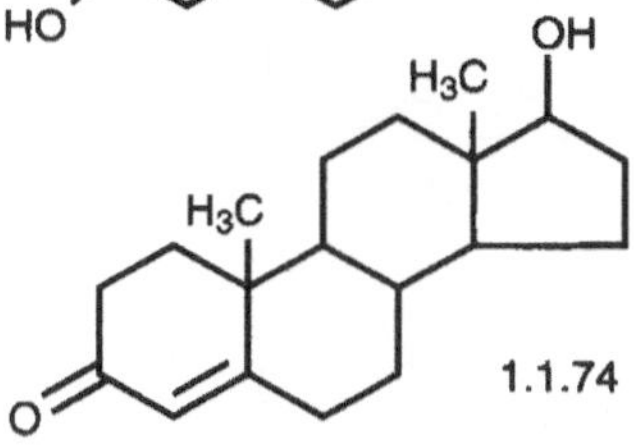

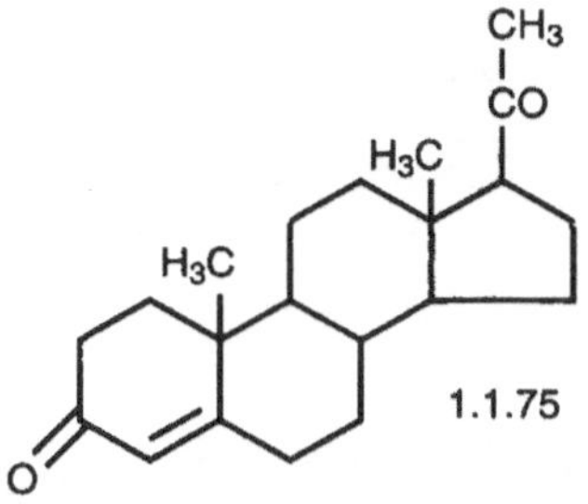

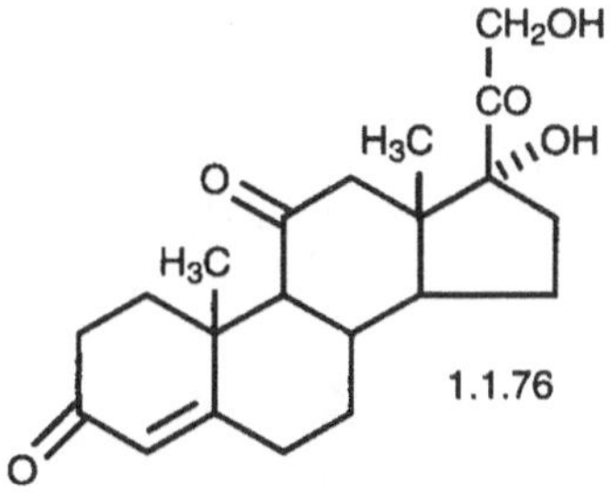

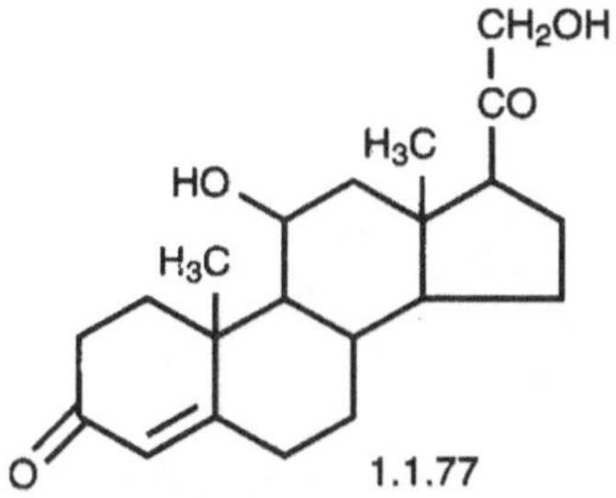

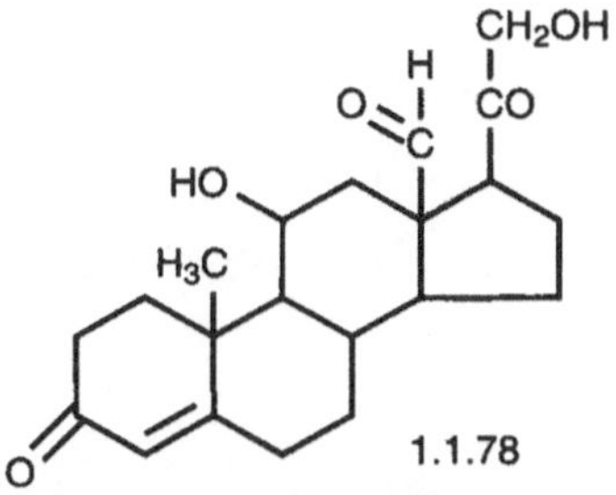

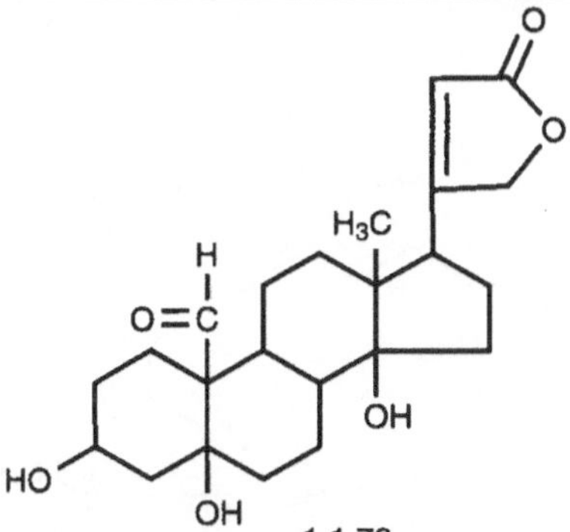

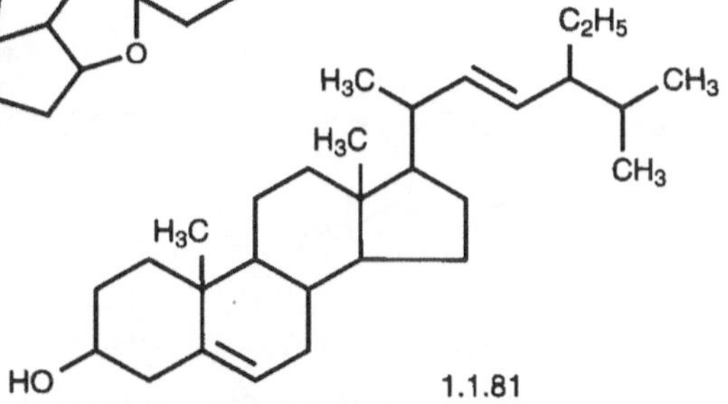

Tetraterpenoids (Carotenoids)

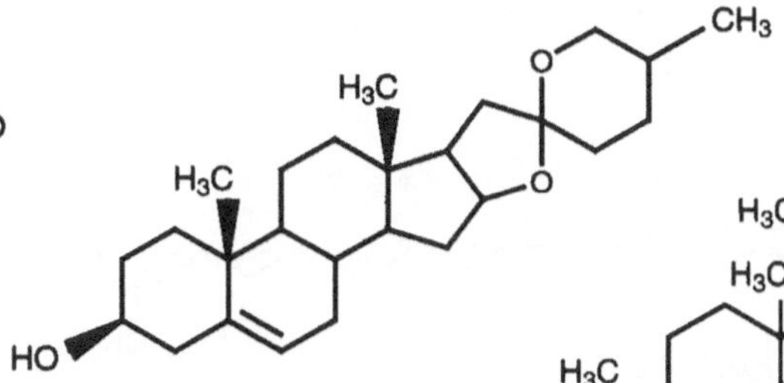

The most well-known acyclic carotenoid is lycopene (1.1.82), which is responsible for the deep red colour of ripe tomatoes. However, the name carotenoid is derived from β-carotene (1.1.83) and α-carotene (1.1.84)—the yellow pigments of carrots. Several oxygenated carotenoids also occur in nature. Two examples are zeaxanthin (1.1.85) and violaxanthin (1.1.86). The pigments of red chillies are capsanthin (1.1.87) and capsorubin (1.1.88).

1.1.82

1.1.83

1.1.84

1.1.85

1.1.86

1.1.87

1.1.88

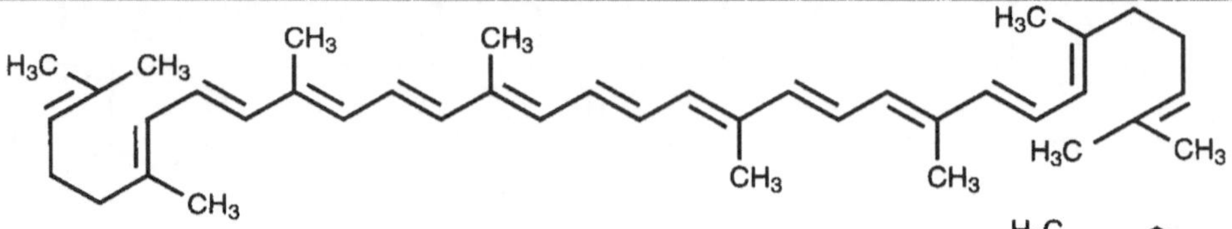

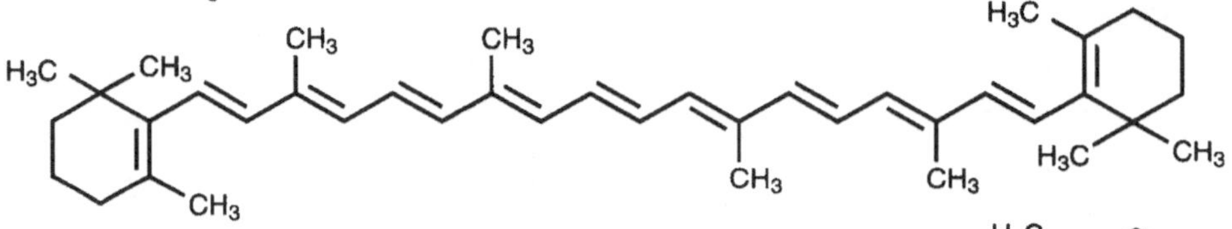

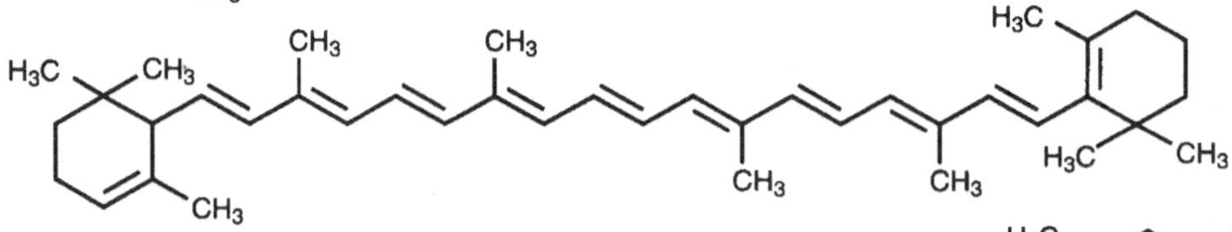

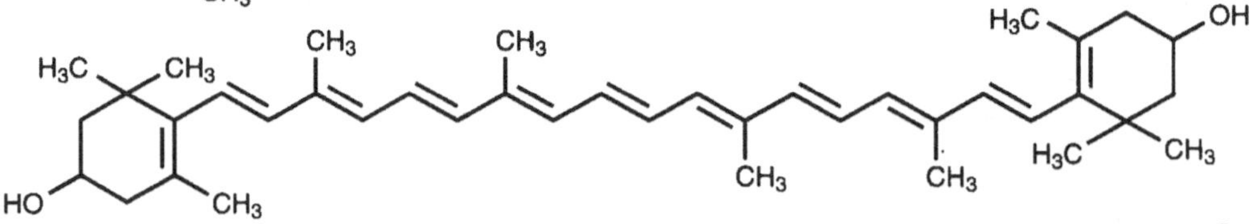

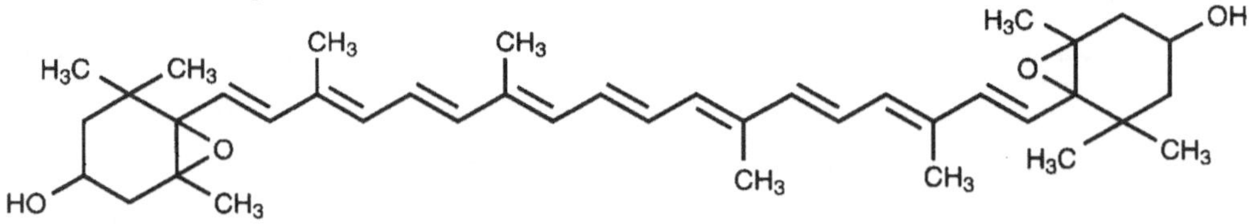

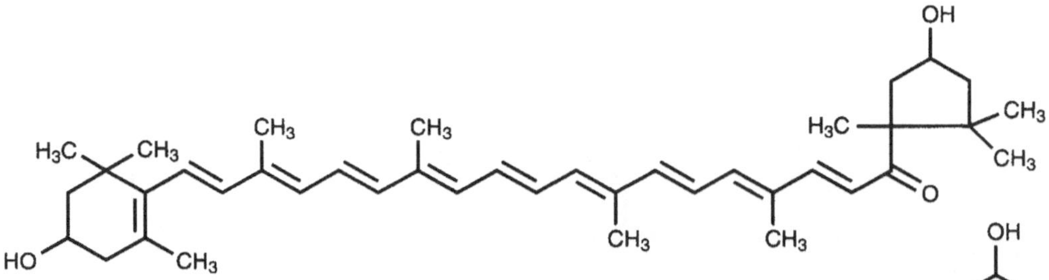

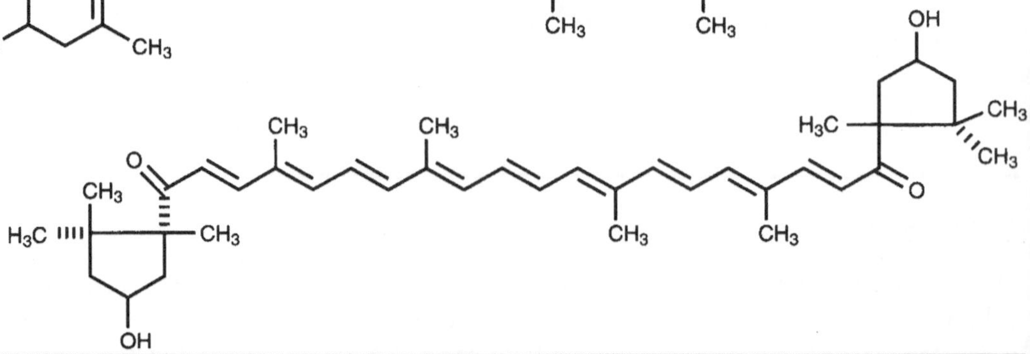

1.2 THE ALKALOIDS

The alkaloids were among the first group of organic compounds to be isolated from plants, particularly those known to be either medicinally useful or toxic. Since these compounds were basic in nature they were designated as 'Alkaloids'. In course of time this definition got narrowed down to a few groups of compounds, now recognised as biosynthetically derived from amino acids. However, all alkaloids are not basic. An important exception is colchicine (1.2.1), which has a N-acetyl group instead of a free amino function. Therefore, a better definition of alkaloids is that they are secondary metabolites derived from amino acids. Colchicine, as well as several other well known alkaloids such as papaverine (1.2.2), morphine (1.2.3), narcotine (1.2.4) and berberine (1.2.5) are biosynthetic products of phenylalanine (1.2.6).

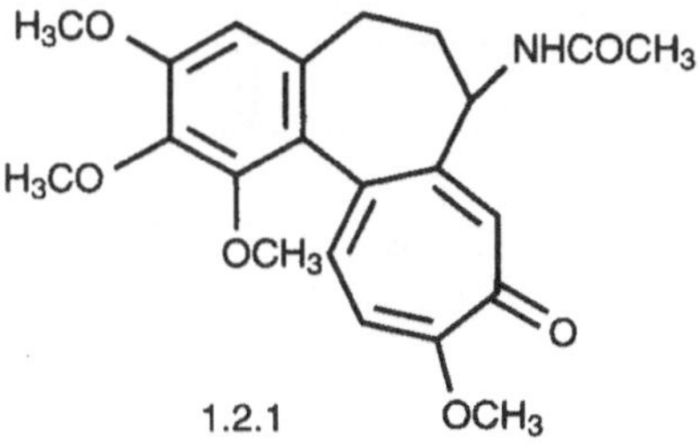

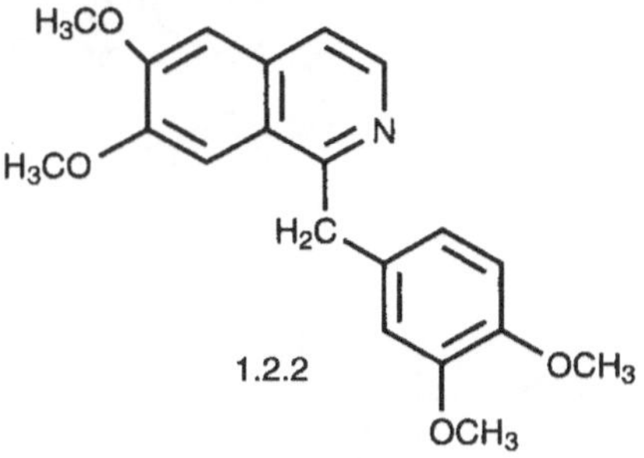

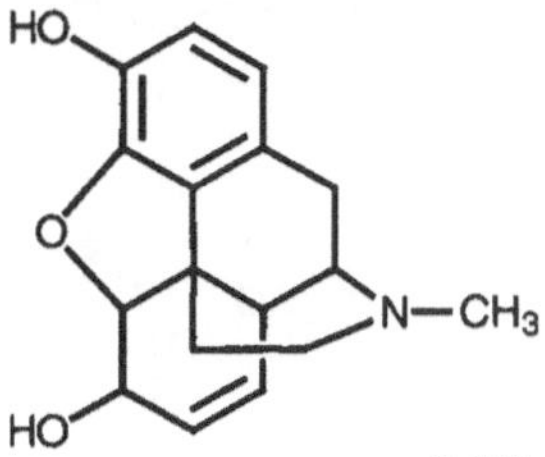

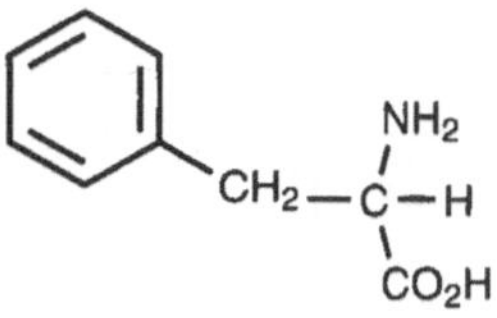

Another aromatic amino acid, namely, tryptophan (1.2.7), is the biosynthetic precursor of the indole alkaloids such as strychnine (1.2.8), reserpine (1.2.9), yohimbine (1.2.10) and the ergot alkaloids which are derivatives of lysergic acid (1.2.11). Some alkaloids are also derived from aliphatic amino acids. For example, hygrine (1.2.12) is biosynthesized from L-ornithine (1.2.13), whereas in the biosynthesis of nicotine (1.2.14), L-aspartic acid (1.2.15) plays a key role.

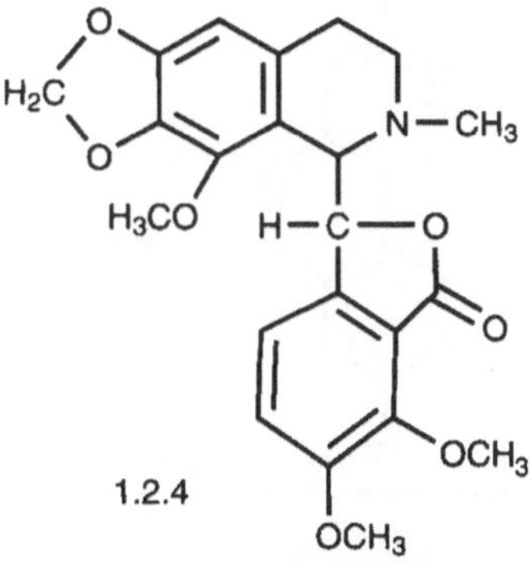

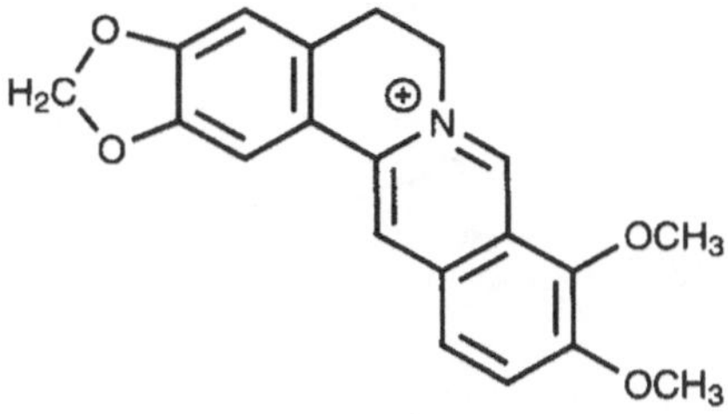

Alkaloids are also classified on the basis of their chemical structures. Thus, papaverine and related compounds are grouped under benzyl isoquinoline alkaloids. Quinine (1.2.16) is a quinoline alkaloid, but is a product of biosynthetic transformation of tryptophan. Some examples of the indole alkaloids were mentioned above. Hygrine is a pyrrolidine alkaloid whereas nicotine is a member of the pyridine group of alkaloids. Thylophorine (1.2.17) has a phenanthrene skeleton.

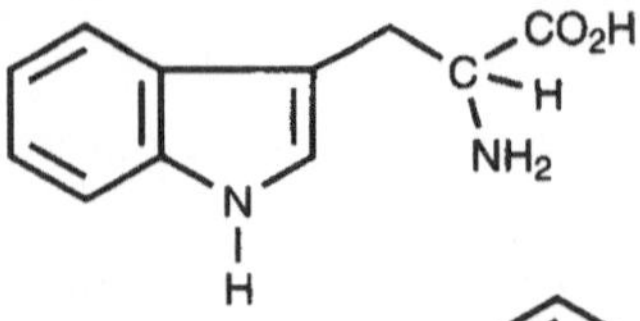

1.2.7

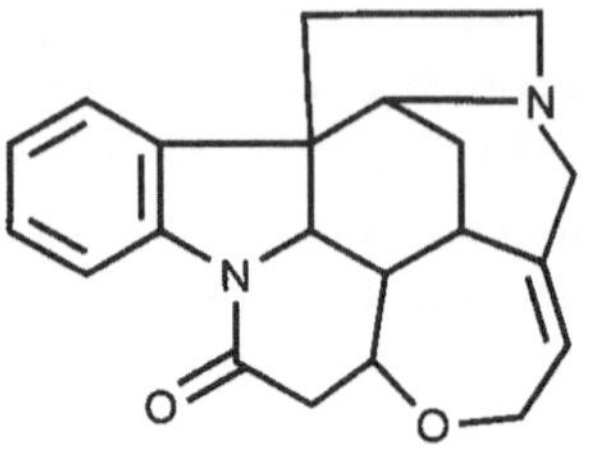

1.2.8

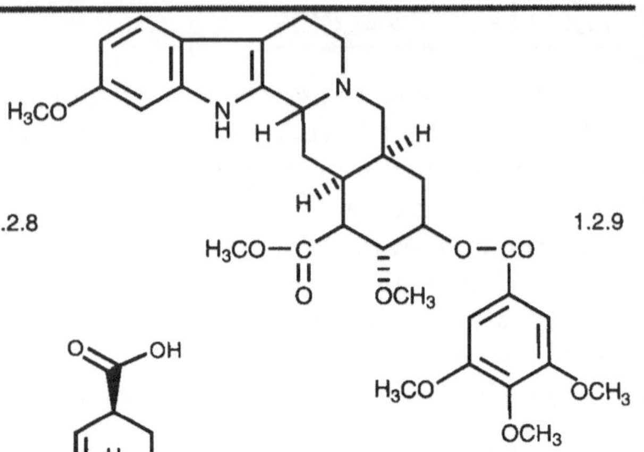

1.2.9

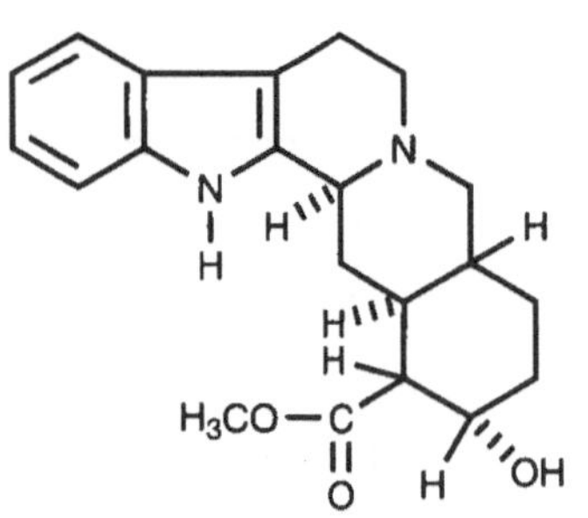

1.2.10

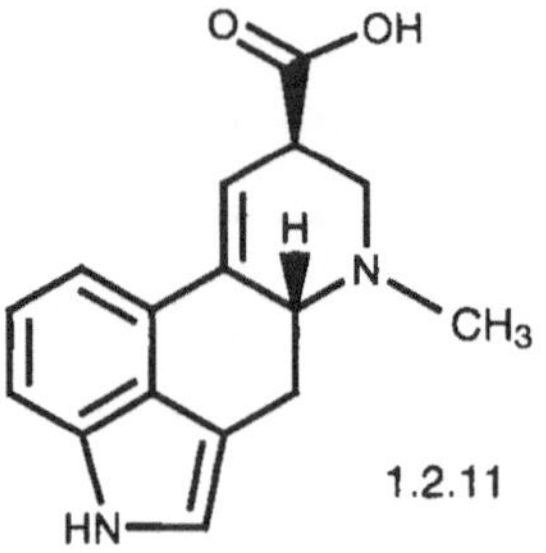

1.2.11

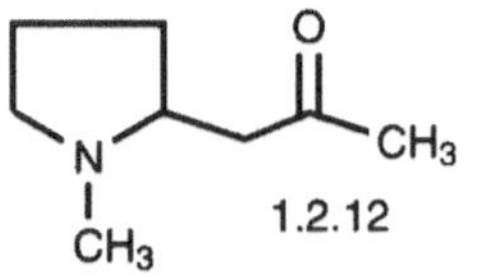

1.2.12

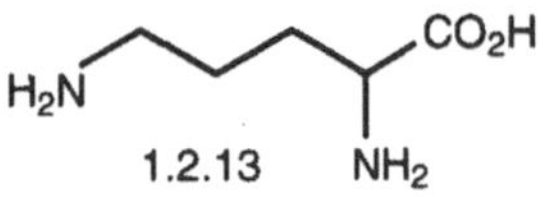

1.2.13

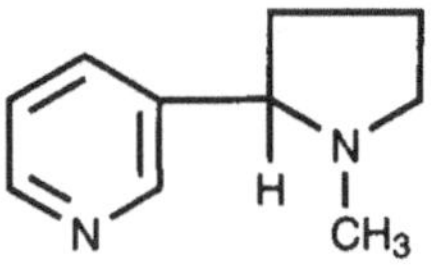

1.2.14

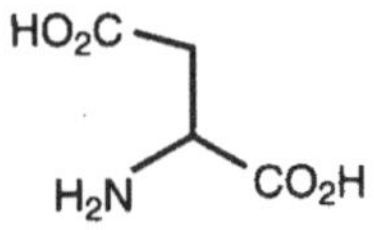

1.2.15

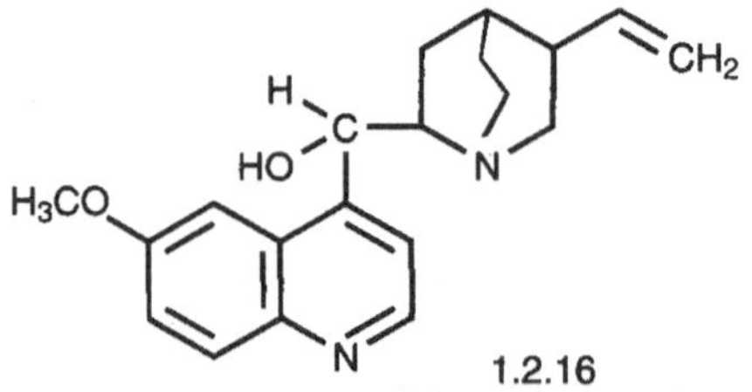

1.2.16

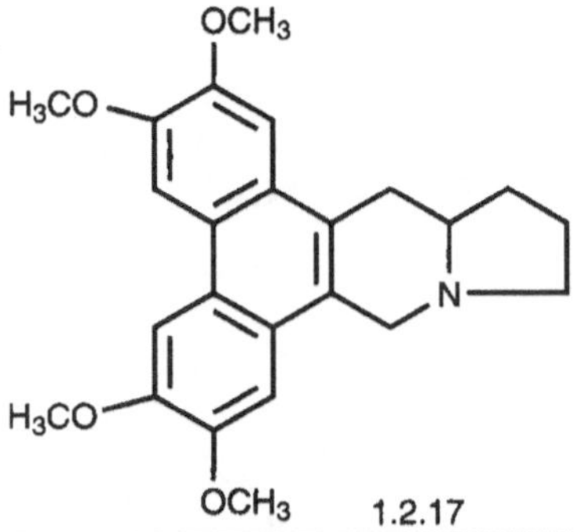

1.2.17

Several alkaloids have also been named after their plant sources. Examples are the senecio alkaloids, the lupine alkaloids and the pomegranate alkaloids. Typical representatives of these types are respectively, senecionine (1.2.18), spartein (1.2.19) and pelletierine (1.2.20). The so-called steroidal alkaloids, such as, solanidine (1.2.21) and conessine (1.2.22), are not true alkaloids in the biosynthetic sense as the nitrogen atom(s) in them are not derived from an amino acid; it is better to classify them as nitrogenous steroids. Along with the several types of diterpene alkaloids, they are termed as pseudoalkaloids.

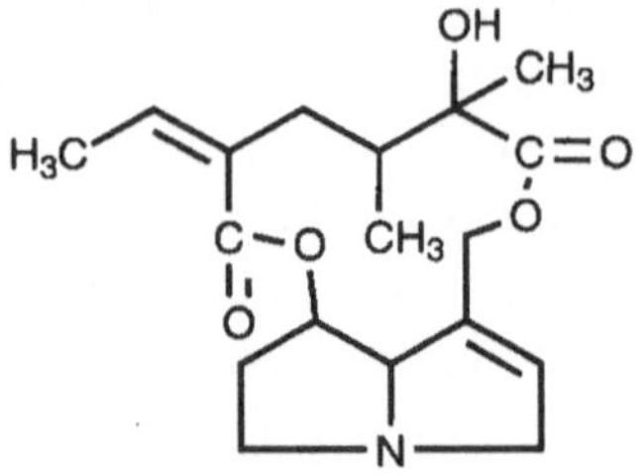

1.2.18

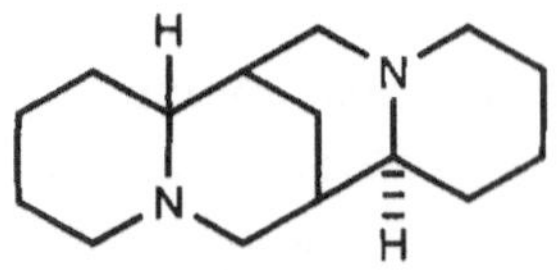

1.2.19

1.2.20   1.2.21   1.2.22

Isolation

Alkaloids which are basic in nature (the large majority of them are, indeed, basic), can be isolated from their plant sources using an acidic solvent such as alcohol containing 2% acetic acid. However, very often these compounds occur in nature in the form of salts with either inorganic or organic acids. In such cases, the free bases have to be liberated first with the help of an alkali before extraction with a solvent. Nicotine can be isolated by steam distillation. Many alkaloids are complex in structure and possess other structural features in addition to the part derived from an amino acid. For instance, the ergot alkaloids have a peptide unit attached to the lysergic acid moiety. The delphinium alkaloids are nitrogenous diterpenes and are predominantly non-polar in character. Therefore, it is not possible to devise an isolation procedure suitable for the extraction of all types of alkaloids. Most often, after extraction with a solvent, it will be necessary to use one or the other of the available chromatographic methods for the isolation of the pure compounds.

Detection

Several specific precipitating and colour reactions are available for the quick detection of alkaloids in plant extracts. These include the Dragendorff's reagent, Mayer's reagent, Wagner's reagent and ammonium reineckate. These yield precipitates with most, if not all, alkaloids. Ultraviolet spectroscopy is useful in detecting the presence of aromatic units in the benzylisoquinoline and indole alkaloids.

The Hoffmann degradation

This technique deserves special mention as it was widely used in classical studies on alkaloids. Also known as Hoffmann's exhaustive methylation, the method involves N-methylation, usually effected with the use of methyl iodide, to the quaternary ammonium stage, followed by treatment with a hot base (usually moist silver oxide) which breaks a C–N bond. If the nitrogen atom is part of an acyclic structure, one methylation (exhaustive) step is enough to convert the compound to a quaternary ammonium derivative. In the next step the nitrogen atom is eliminated. Thus, ephedrine (1.2.23), which is a component of the well-known Chinese medicinal plant, Ma Huang, gives trimethyl amine and a nitrogen free compound (1.2.24) as shown on the next page.

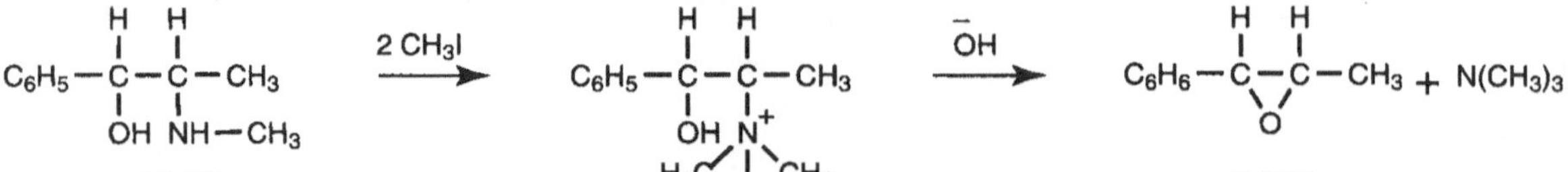

1.2.23 → (2 CH$_3$I) → 1.2.24 + N(CH$_3$)$_3$

If the nitrogen atom is present in a ring as in tropine (1.2.25), which is obtained by the hydrolysis of atropine (1.2.26) (occurring in the poisonous plant, *Atropa belladonna*; deadly nightshade), two Hoffmann methylations are needed to eliminate the nitrogen atom. Tropine is first dehydrated (1.2.27), which after two successive Hoffmann reactions yields trimethylamine and tropylidene (cycloheptatriene) (1.2.28). Three Hoffmann reactions are required to eliminate the nitrogen atom from an alkaloid in which it is common to two rings.

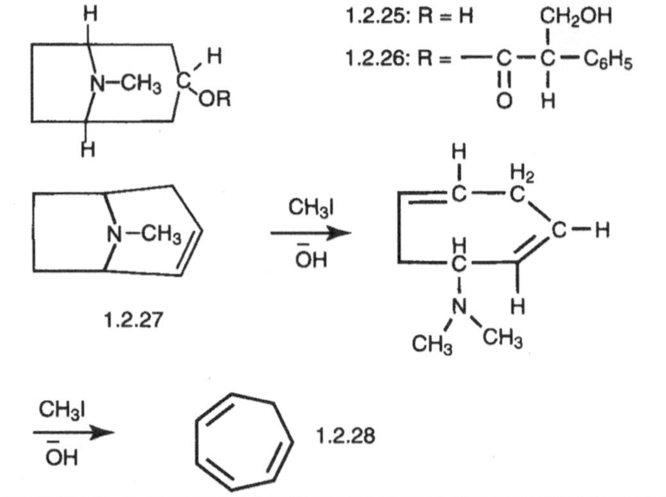

1.2.25: R = H

1.2.26: R = —C—C—C$_6$H$_5$ (with CH$_2$OH, O, H)

1.2.27

1.2.28

Benzylisoquinolines

Apart from the examples already mentioned under 1.2, other important members of this group are, beta-corylidene (1.2.29), nuciferine (1.2.30) and tubocurarine (1.2.31), which is a bisbenzylisoquinoline alkaloid.

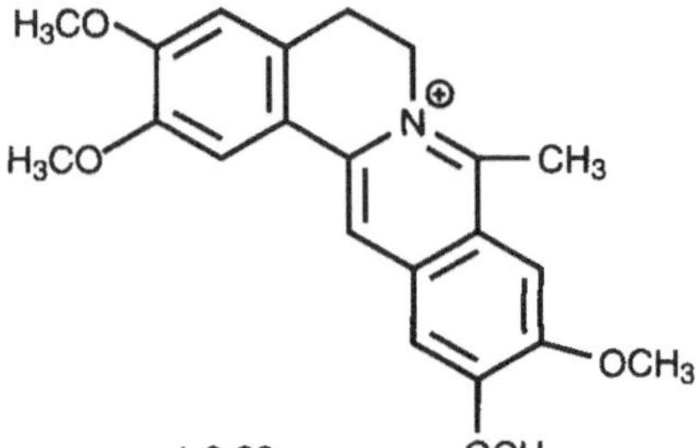

1.2.29

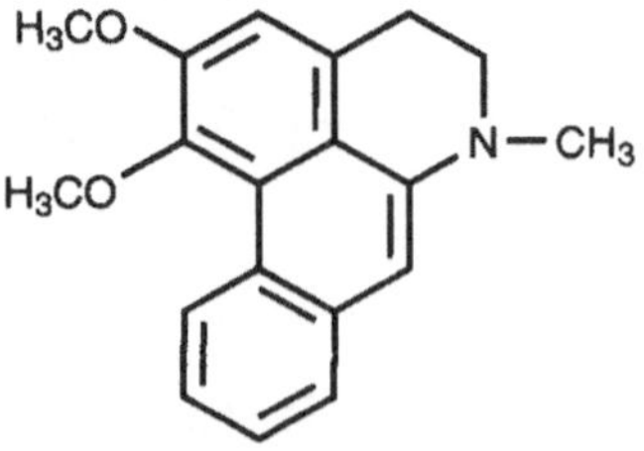

1.2.30

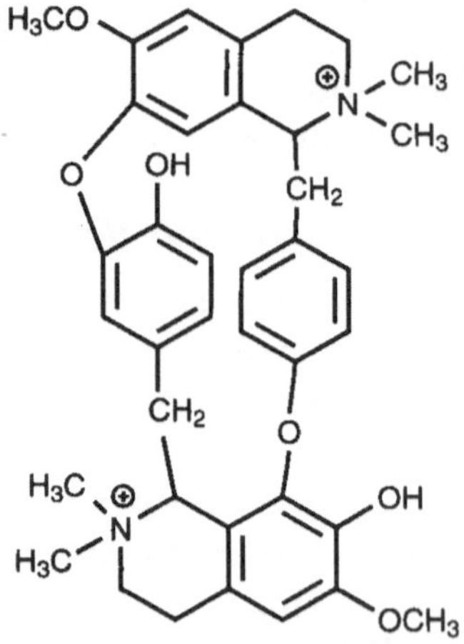

1.2.31

Indole Alkaloids

The other important group of alkaloids is the indole group to which strychnine, reserpine, yohimbine and the ergot alkaloids, which have already been mentioned, belong. Other examples are ajmaline (1.2.32), (a component of *Rauwolfia serpentina* and named after the Indian hakim, Ajmal Khan), catharanthine (1.2.33) (iboga alkaloid), and vinblastine (1.2.34), which is one of the anti-cancer (effective against leukemia) principles of *Catharanthus roseus* (earlier known as *Vinca rosea*).

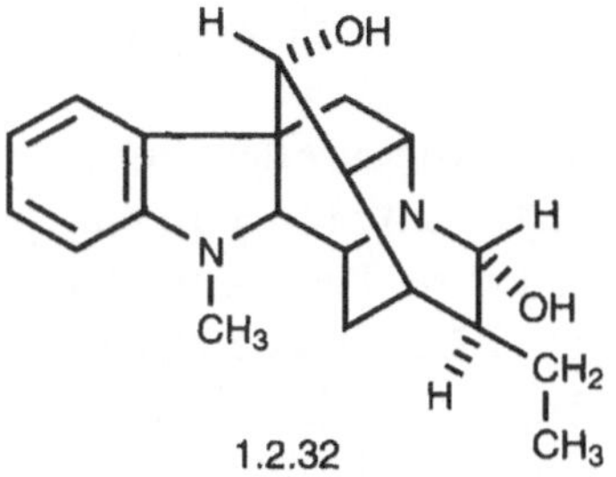

1.2.32

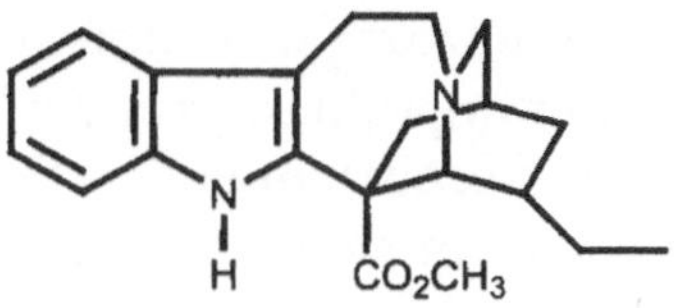

1.2.33

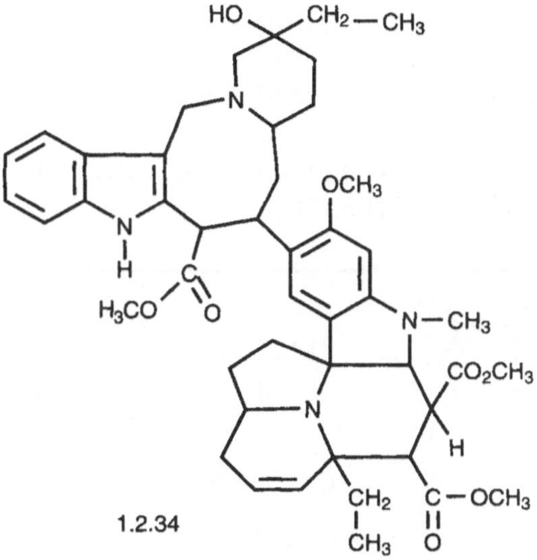

1.2.34

1.3 Flavonoids and Related Compounds

Among plant polyphenols, the flavonoids and related compounds occupy a prime place. They are widely distributed in the plant kingdom and their presence is often obvious because of their bright and striking colours. In recent years, their importance in medicine and human nutrition has been recognised on account of their antioxidant and free radical inhibiting properties. They also have considerable value as chemotaxonomic markers.

The major groups of flavonoids are the anthocyanins, flavones and flavonols, their dihydro derivatives, isoflavonoids and the neoflavonoids. Chalcones, aurones and catechins as well as the proanthocyanidins are also included in this family of compounds. Biflavonoids and C-alkylated flavonoids are of rarer occurrence. Flavonoids occur as free polyphenols as well as O- and C-glycosides; the latter are also of rare occurrence.

Anthocyanins

Anthocyanins are glycosides which are responsible for the majority of floral colours. The colours of fruits such as grapes, apples and several berries are also due to these compounds. However, they often occur along with other flavonoids, thus accounting for the subtle variations of colours we find, for example, in roses and hibiscus flowers.

The sugar-free compounds are the anthocyanidins. Structurally, they are flavylium compounds and being ionic in nature are water soluble. Their colours range from orange to blue primarily, depending on the degree of hydroxylation of the skeletal structure. Pelargonidin (1.3.1), with four phenolic hydroxyl groups, is orange in colour. Cyanidin (1.3.2), whose glycosides are the most abundant among floral anthocyanins, is red in colour. Delphinidin (1.3.3) has a blue colour. Metal complexation of anthocyanins is also known and makes a significant contribution to the modification and deepening of the primary colours. O-Methylation also brings about subtle changes in colour. Peonidin (1.3.4) is the 3'-O-methyl ether of cyanidin. Petunidin (1.3.5) and malvidin (1.3.6) are the mono- and di-methyl ethers, respectively, of delphinidin. Cyanin (1.3.7), the pigment of red roses, is the 3,5-diglucoside of cyanidin. Acylated glycosides also often occur in nature; the acyl part is usually a cinnamic acid derivative, such as p-coumaric acid (1.3.8), caffeic acid (1.3.9) or ferulic acid (1.3.10).

1.3.1: R = R' = H
1.3.2: R = OH; R' = H
1.3.3: R = R' = OH
1.3.4: R = OCH$_3$; R' = H
1.3.5: R = OH; R' = OCH$_3$
1.3.6: R = R'= OCH$_3$

1.3.7

1.3.8: R = H
1.3.9: R = OH
1.3.10: R = OCH$_3$

The blue pigment of *Commelina communis* flowers (Asiatic day flower), commelinin, is a complex molecule in which magnesium metal is linked to six flavocommelin molecules (1.3.11). Hirsutin (1.3.12) is the 3,5-diglucoside of delphinidin 7.3',5'-tri-O-methyl ether (hirsutidin).

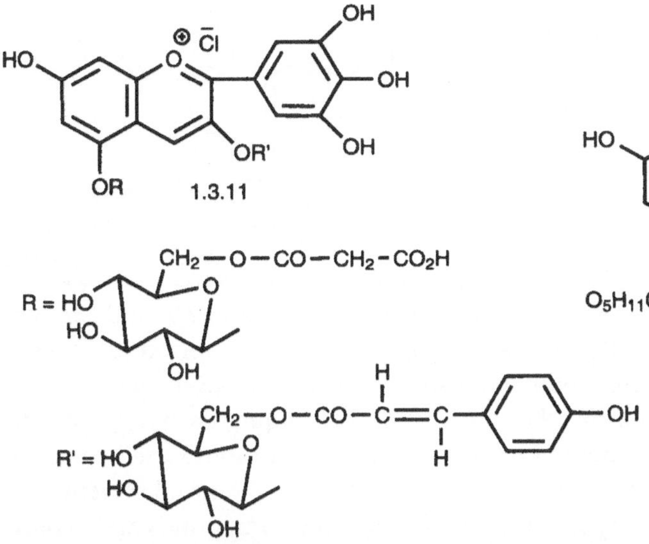

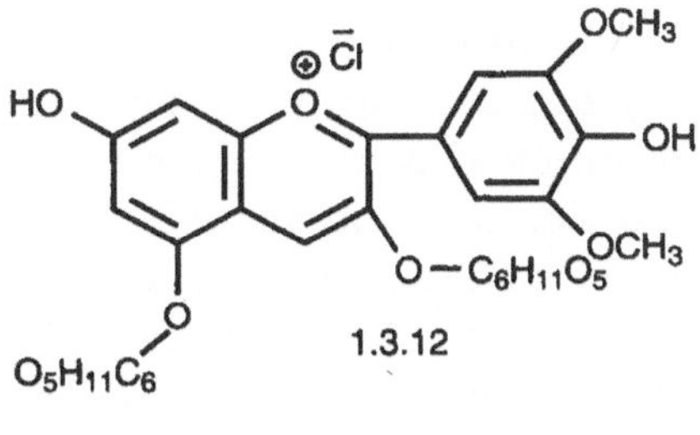

Anthocyanins and anthocyanidins are stable in acidic media but rapidly decompose in alkaline media. As a consequence, their colours depend on the pH of the medium. When the pH is increased beyond 7, the colour first changes to blue and then quickly fades away. This is due to structural changes as shown below, taking cyanin as an example.

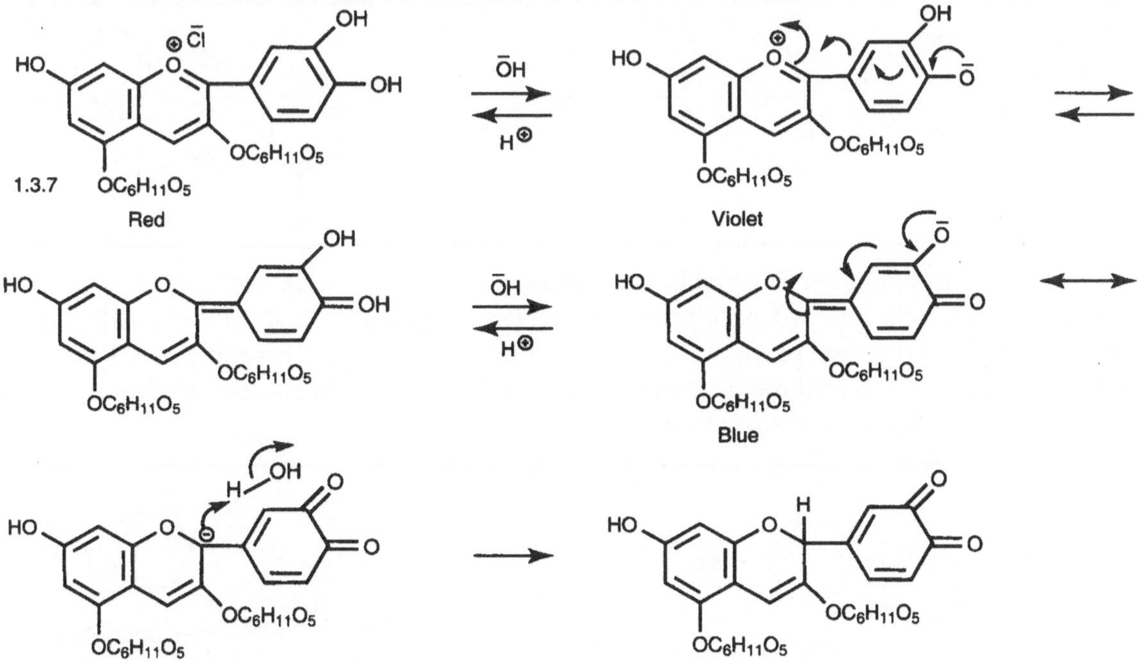

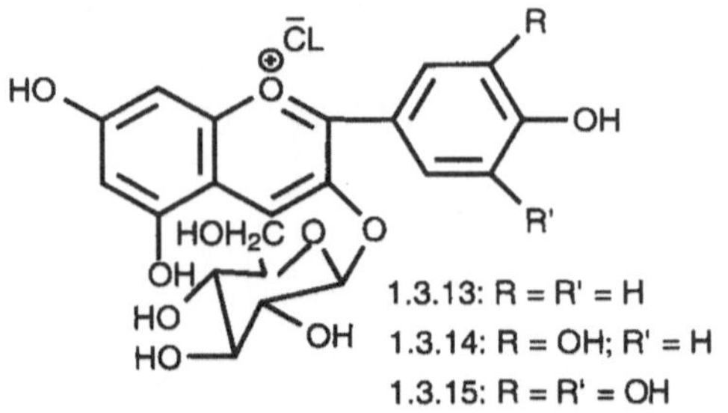

The ultraviolet–visible spectra of anthocyanins and anthocyanidins exhibit two maxima. The maximum in the visible region (475–560 nm) is known as Band-I. This absorption band reflects the colour of the compound. Band II is narrower (275–280 nm) and does not vary from compound to compound. It is instructive to compare the UV maxima of the 3-glucosides of pelargonidin, cyanidin and delphinidin, namely 1.3.13, 1.3.14 and

1.3.13: R = R' = H
1.3.14: R = OH; R' = H
1.3.15: R = R' = OH

1.3.15, respectively. The values are 270 and 506 nm, 274 and 523 nm and 276 and 534 nm, in that order. The maximum of the Band I of the corresponding aglycones (anthocyanidins) are 520, 535 and 546 nm. Thus, glycosidation brings about a hypsochromic shift of 12 to 14 nm.

Paper chromatography—particularly the horizontal technique—is very convenient for the examination of anthocyanin pigments in plant extracts. Commonly used solvents are the upper layer of butanol–acetic acid–water (4:5:1), 15% acetic acid and Forestal solvent (acetic acid, concentrated HCl and water in the proportion 30:3:10). Since the compounds are coloured, there is no need for any spraying agent.

After Willstatter's pioneering work on the isolation of anthocyanin pigments, Karrer and co-workers elucidated their structures by alkaline and oxidative degradations. Robinson and co-workers gave synthetic support to the structures thus deduced. For example, condensation of 2-benzoylphloroglucinaldehyde (1.3.16) with 4,ω-diacetoxy-3-methoxyacetophenone (1.3.17) in the presence of hydrochloric acid yielded the 5-benzoyl derivative of peonidin chloride (1.3.18).

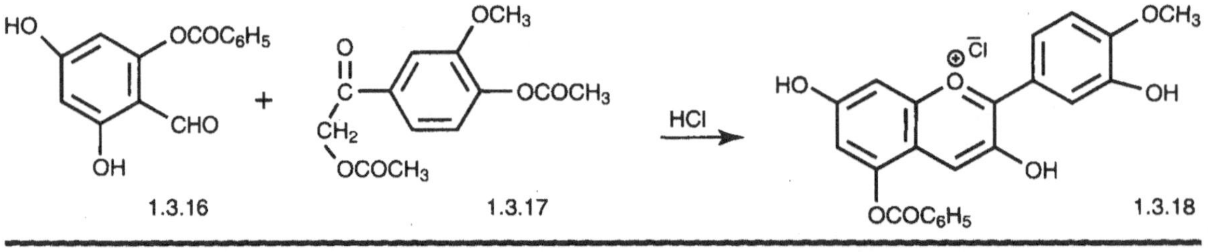

Anthocyanins, like other flavonoids, are naturally occurring antioxidants. As early as 1927, Karrer studied the effect of hydrogen peroxide on these compounds. As shown below, the middle ring opens up and the final product is a diester, which is colourless. This explains why extracts of anthocyanins even in acid media gradually fade in the presence of air, though they retain the colour for longer periods in their natural state.

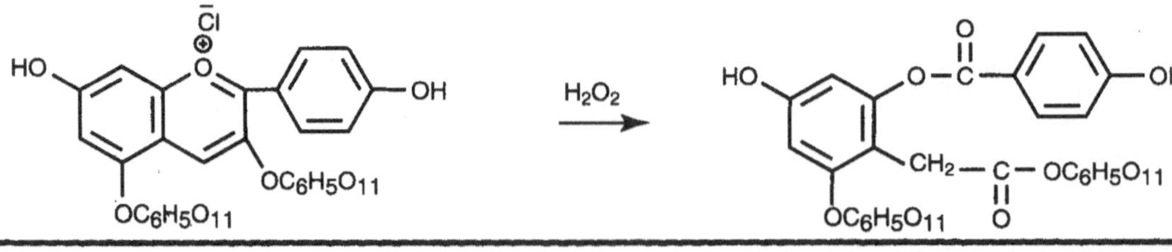

Flavones and Flavonols

Flavones and flavonols form the largest group among flavonoids. Their colour varies from pale yellow to deep yellow, depending on, as in the case of the anthocyanins, the degree of hydroxylation. Their dihydro derivatives are colourless. Methyl ethers are also of common occurrence and so are the glycosides. Besides glucose, rhamnose is often found as a component sugar in many flavone glycosides. Flavone C-glycosides have been isolated from some species of stoneworts, liverworts, mosses and ferns as well as some gymnosperms. Biflavones are of rare occurrence and have been isolated from gymnosperms (for example, species of Cupressus). C-Alkylated flavones (C-methyl and C-isoprenyl) are also known though they are not as abundant as the O-methyl derivatives.

Several specific colour reactions for the detection of flavones and flavonols, even in crude plant extractives, are available. The most important among them is the Shinoda test. In this test, to a methanolic solution of the compound, a small amount of magnesium metal powder is added, followed by the drop-wise addition of concentrated HCl. A red, pink or purple colour develops, primarily due to the conversion of the flavone or flavonol into the corresponding anthocyanidin. In a related colour reaction, specific for flavonols and their glycosides, metallic zinc dust is used in place of magnesium. Circular paper chromatography is a very convenient tool for the detection of these compounds and the Rf values are useful in making identifications. Since these compounds are not as coloured as the anthocyanins, reagents that are sprayed are useful in detecting them. Alcoholic ferric chloride, aqueous ammonia and alcoholic aluminium chloride (anhydrous) are used for this purpose.

Perhaps, the most important single technique used for the characterisation and identification of individual flavones and flavonols is ultraviolet spectroscopy. This technique acquires unique value when specific shift reagents, such as sodium acetate, aluminium chloride, sodium ethoxide and sodium acetate–boric acid are used. Like anthocyanins, these compounds also exhibit two absorption bands in their UV spectra, but in contrast to anthocyanins, both these bands may fall within the ultraviolet range, except where a high degree of hydroxylation makes the compounds deep yellow in colour. The wave length ranges for these bands are: Band I: 310–385 nm and Band II: 250–280 nm. The spectra are recorded in methanol. Addition of sodium acetate to the methanolic solution brings about a bathochromic shift (5 to 20 nm) in Band II if the compound has a free hydroxyl group at position 7 [the most acidic hydroxyl on account of conjugation, through the benzene (A) ring with the carbonyl group at position 4]. Addition of aluminium chloride brings about a large bathochromic shift if the compound has hydroxyl groups at positions 3 (flavonols) and 5. These shifts are unaffected on addition of a few drops of hydrochloric acid if the compound does not have the catechol grouping (3',4') in the side phenyl ring. If such a grouping is present, besides hydroxyls at positions 3/5, the initial bathochromic shift with aluminium chloride alone will be larger and the magnitude of this shift would be reduced on addition of the acid. The reason for these effects is that the aluminium metal-5-hydroxyl chelate complex is stable even in acid, whereas the complex involving the catechol grouping breaks down in acid. This is shown below using luteolin (1.3.19) as a typical example.

Addition of sodium acetate and boric acid causes a bathochromic shift of Band I if the flavone or flavonol has a catechol grouping (3',4'-dihydroxy) in the side phenyl ring. Sodium methoxide brings about a bathochromic shift of Band I, as all the hydroxyl groups ionise in this medium, but if the compound is polyhydroxylated, the spectrum decays, with the intensity of the absorption band continually decreasing.

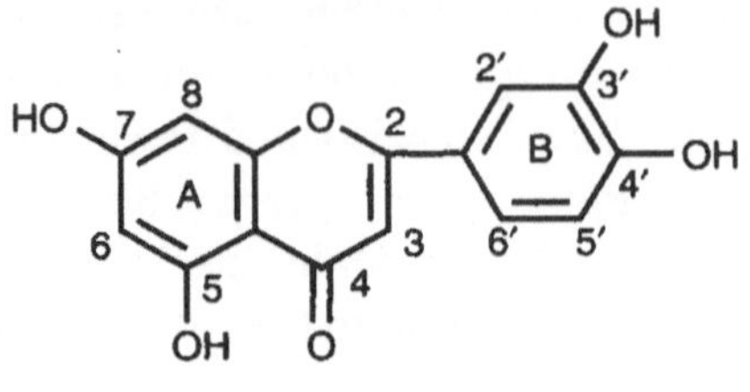

1.3.19

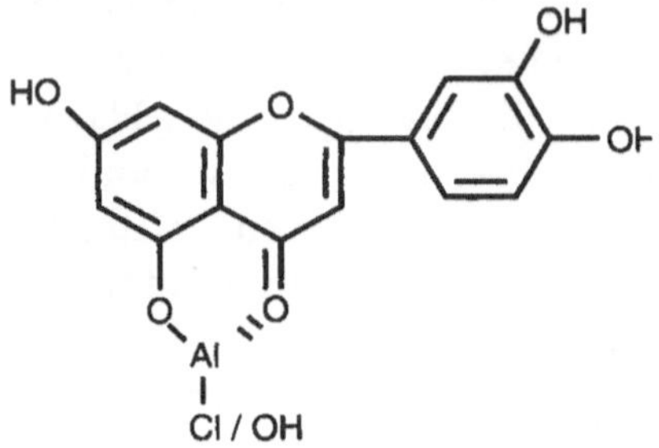

Nuclear magnetic spectra (both proton and carbon) are very useful in the determination of structures of even complex flavones and flavonols. If there are no alkyl or sugar residues, all the protons fall in the aromatic region (6–8 ppm) and from the coupling constants it is possible to deduce the oxygenation pattern. The presence (and number) of methoxyl groups, if any, is readily detected by the signal(s) in the 3.5–4.0 ppm region. For example, in the proton nmr spectrum of apigenin (1.3.20)—recorded as the tetramethylsilane ether in carbon tetrachloride)—the protons at 2',6' and 3',5' positions appear as ortho-coupled doublets at 7.8 and 6.8 ppm, respectively. The hydrogens at positions 6 and 8 give meta-coupled signals at 6.2 and 6.5 ppm. The signal due to H-3 appears as a singlet between these two. Spectra of more complex compounds are discussed elsewhere in this book.

Unsubstituted flavone (1.3.21) occurs in nature but is rare. More common are the hydroxylated compounds. Some examples are chrysin (1.3.22; 5,7-dihydroxyflavone), apigenin (1.3.20), which is 5,7,4'-trihydroxyflavone, luteolin (1.3.19; 5,7,3'4'-tetrahydroxyflavone) and diosmetin (1.3.23), which is the 3'-O-methyl ether of luteolin. The corresponding flavonols (flavonol is 3-hydroxyflavone), are galangin (1.3.24), kaempferol (1.3.25), quercetin (1.3.26) and isorhamnetin

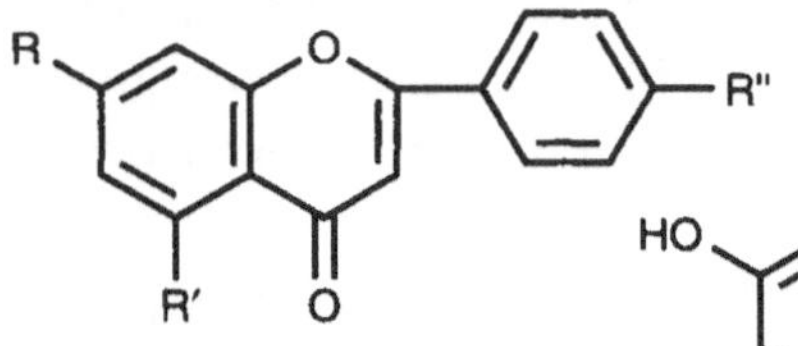

1.3.20: R = R' = R" = OH
1.3.21: R = R' = R" = H
1.3.22: R = R' = OH; R"= H

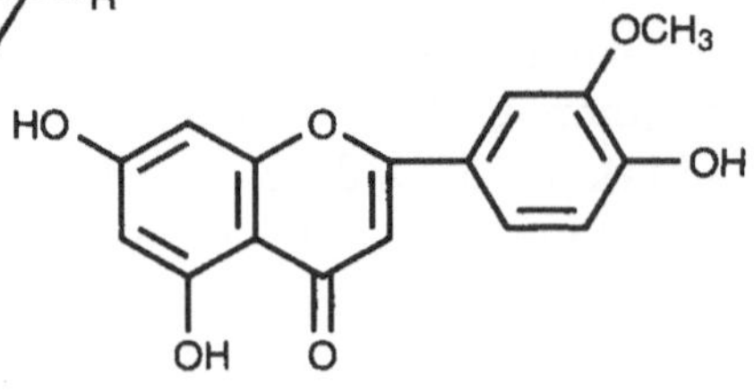

1.3.23

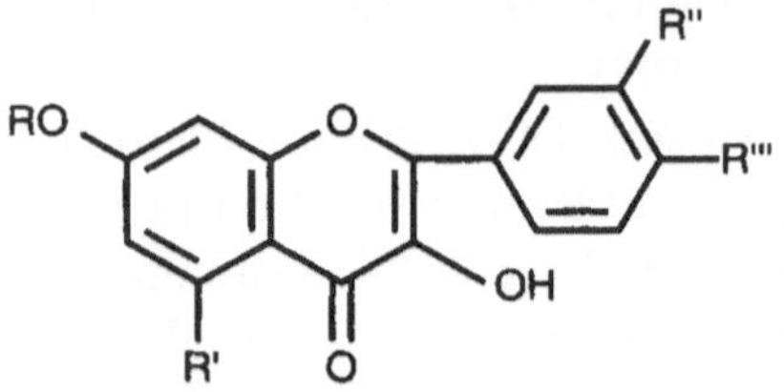

1.3.24: R = R' = OH; R"= R''' = H
1.3.25: R = R' = R''' = OH; R" = H
1.3.26: R = R' = R" = R''' = OH

(1.3.27). Rhamnetin (1.3.28) is the 7-*O*-methyl ether of quercetin. Quercetagetin (1.3.29) and gossypetin (1.3.30) each has an additional hydroxyl in the A ring, whereas myricetin (1.3.31) has an extra hydroxyl in the side phenyl ring. Morin (1.3.32), which is isomeric with quercetin, has the rare 2',4'-dihydroxy (resorcinol) pattern instead of the more common 3',4' (catechol) oxygenation in the 'B' ring.

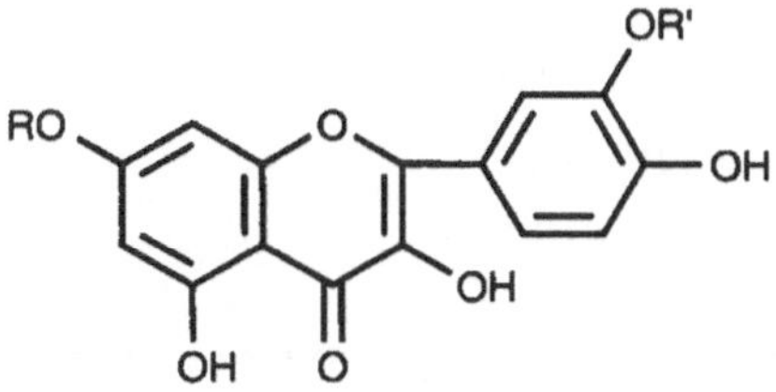

1.3.27: R = H; R' = CH₃
1.3.28: R = CH₃; R' = H

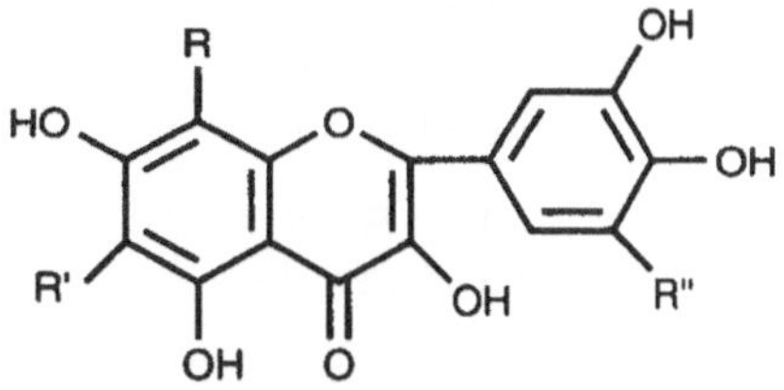

1.3.29: R = R''' = H; R' = OH
1.3.30: R = OH; R' = R'' = H
1.3.31: R = R' = H; R'' = OH

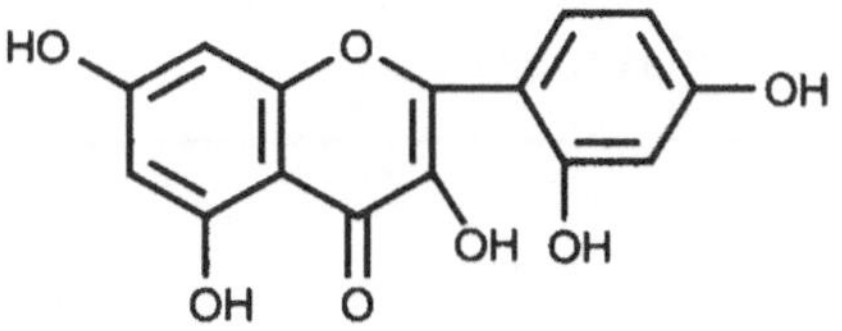

1.3.32

In flavone glycosides, the most common position for the attachment of the sugar residue is position 7. In flavonols, 3-hydroxyl is often involved in glycosidation. Typical examples are quercetrin (1.3.33) which is the 3-rhamnoside of quercetin and rutin (1.3.34), which is a rhamnoglucoside. Several *C*-glycosides have also been isolated and characterized. Some examples are vitexin (1.3.35) and isovitexin (1.3.36), both of which are isomeric monoglucosides of apigenin, and orientin (1.3.37) and isoorientin (1.3.38), which are luteoloin mono *C*-glucosides.

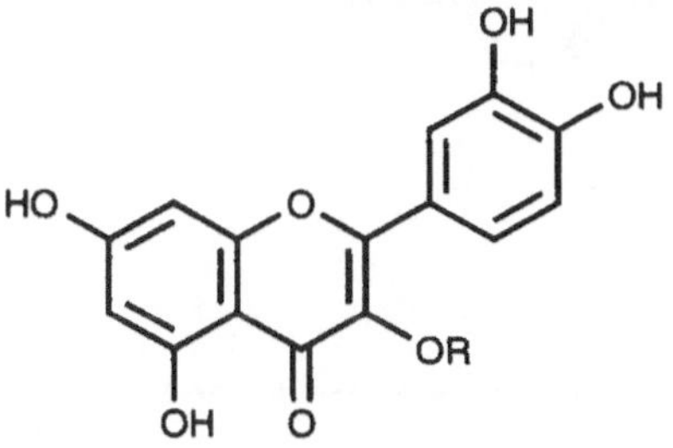

1.3.34: R = HO

1.3.33: R = HO

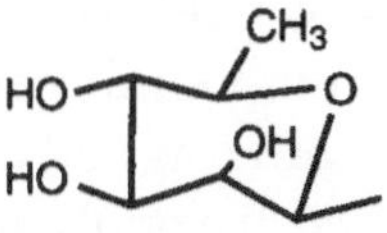

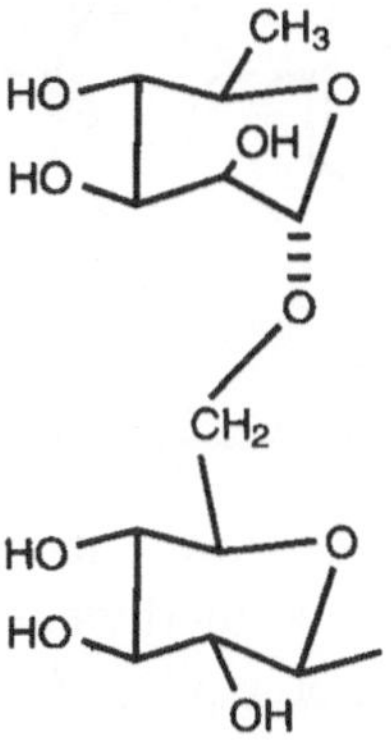

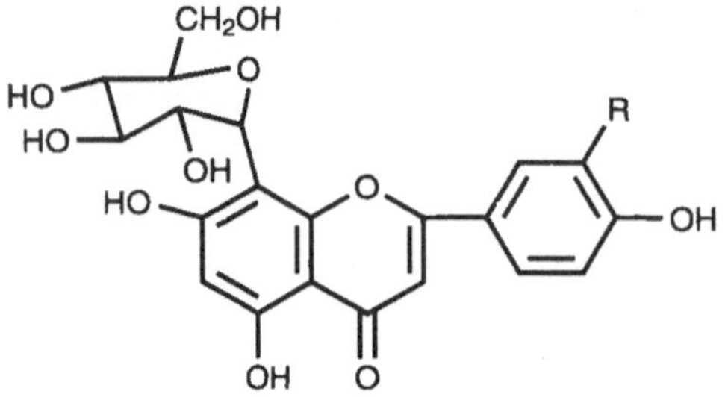

1.3.35 : R = H
1.3.37 : R = OH

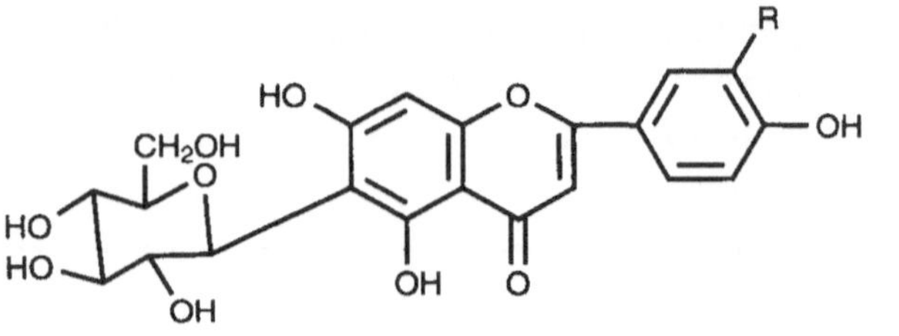

1.3.36 : R = H
1.3.38 : R = OH

Examples of dimeric flavones, called biflavones, are amentoflavone (1.3.39), hinokiflavone (1.3.40), gingketin (1.3.41) and cupressuflavone (1.3.42).

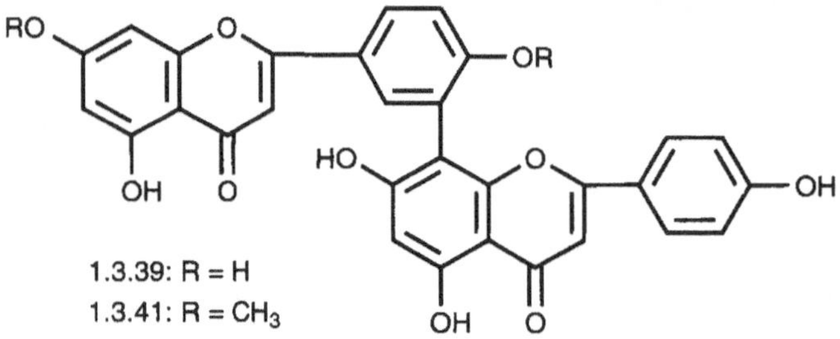

1.3.39: R = H
1.3.41: R = CH₃

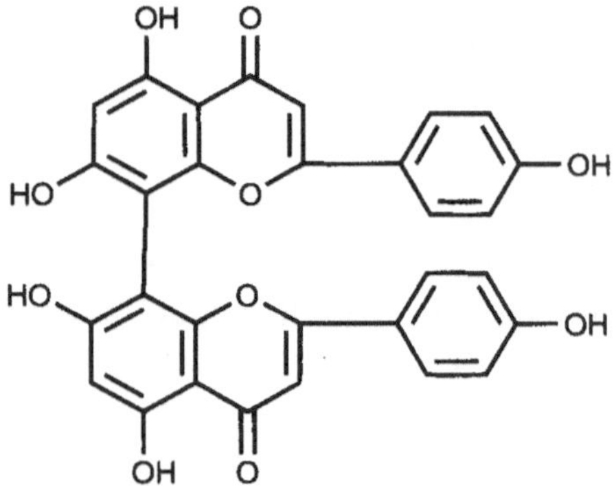

1.3.42

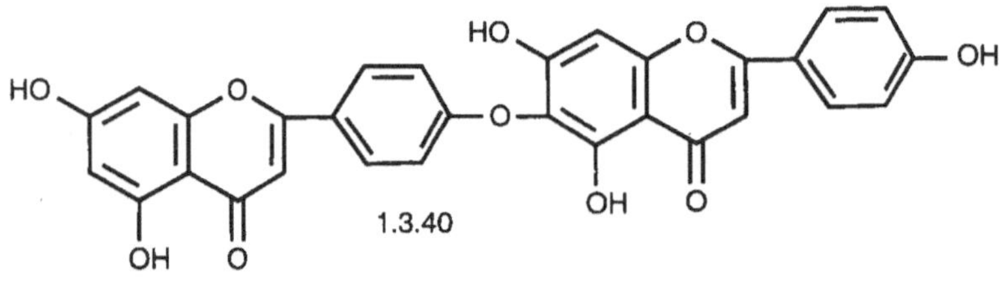

1.3.40

Kanugin (1.3.43) is a flavonol derivative without any free hydroxyl groups; it has three methoxyls and a methylenedioxy group. Another compound having a methylenedioxy functionality is pongapin (1.3.44), which also possesses a furan ring and thus can be considered as a C-alkylated flavonol derivative. Examples of C-methylated flavonols are pinoquercetin (1.3.45) and pinomyricetin (1.3.46). Anhydroicaritin (1.3.47) is a kaempferol derivative with a isoprenyl unit at position 8.

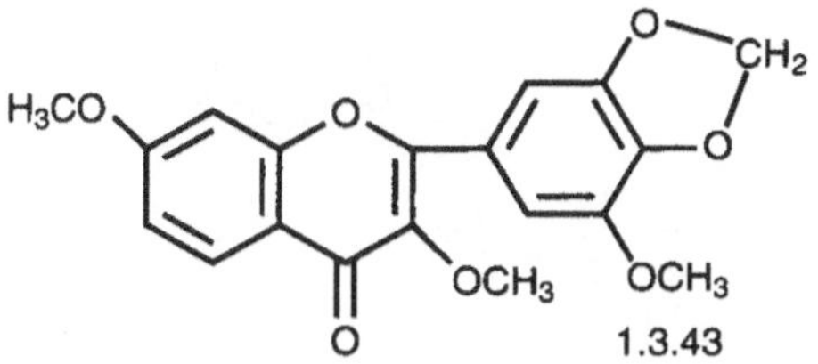

1.3.43

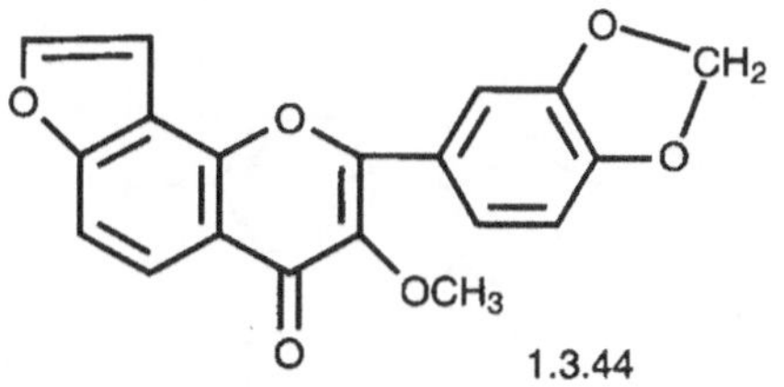

1.3.44

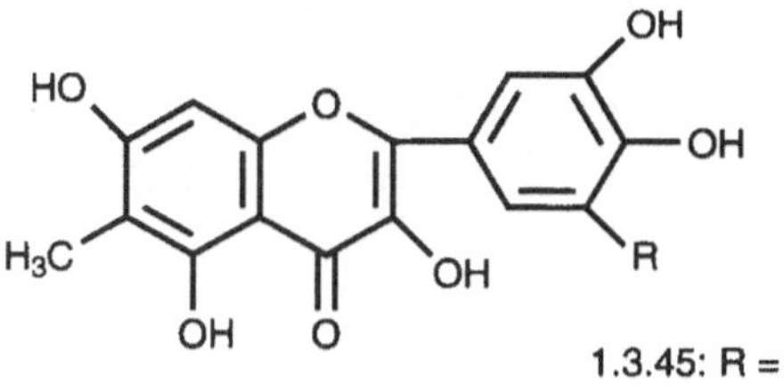

1.3.45: R = H
1.3.46: R = OH

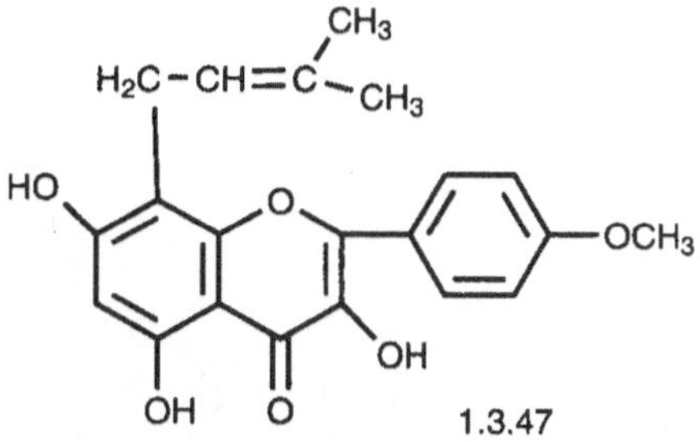

1.3.47

Dihydroflavones and Dihydroflavonols

Flavanones (dihydroflavones) and flavononols (dihydroflavonols) are not as numerous as flavones and flavonols. However, several of them have been isolated from diverse plants, including species of Pinus, Prunus and Eucalyptus. Due to the presence of a chiral carbon atom (position 2), flavanones are optically active. The naturally occurring laevorotatory flavanones possess 2S configuration. Dextro-rotatory dihydroflavonols have 2R, 3R configurations. Some examples of flavanones are pinocembrin (1.3.48) and liquiritigenin (1.3.49), both of which are dihydroxyflavanones. Naringenin (1.3.50) is a trihydroxy compound. Its 7-*O*-methyl ether (1.3.51) is sakuranetin. Hesperetin (1.3.52) is the 4'-*O*-methyl derivative of eriodictyol (1.3.53). Pinobanksin (1.3.54), aromadendrin (1.3.55) and taxifolin (1.3.56) are dihydroflavonols. An unusual flavanone glucoside is poriolide (1.3.57), which has been isolated from a plant of the Ericaceae family. Silybin (1.3.58), is a dihydroflavanolignan, isolated from the medicinal plant *Silybum marianum*; it is anti-hepatotoxic. A number of *C*-methylated flavanones are known. Some examples of this type are strobopinin (1.3.59), cryptostrobin (1.3.60), matteucinol (1.3.61) and farrerol (1.3.62).

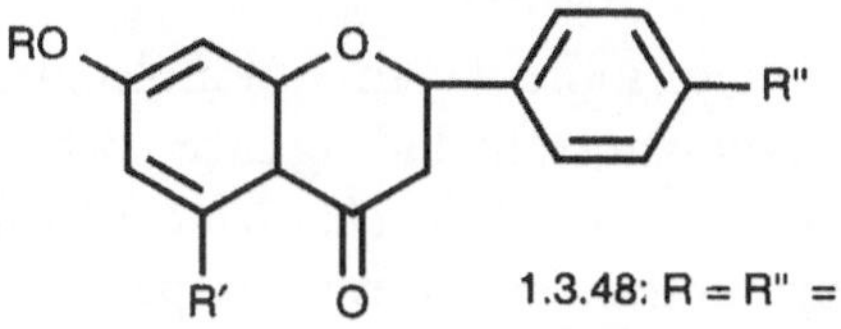

1.3.48: R = R" = H; R'= OH
1.3.49: R = R = H, R" = OH
1.3.50: R = H, R' = R" = OH
1.3.51: R = CH₃, R' = R" = OH

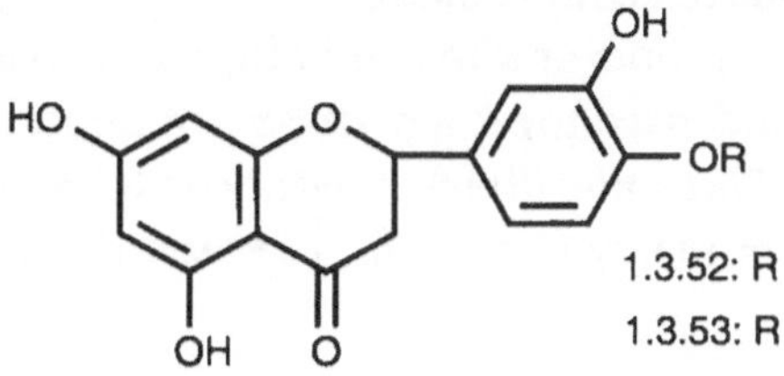

1.3.52: R = CH₃
1.3.53: R = H

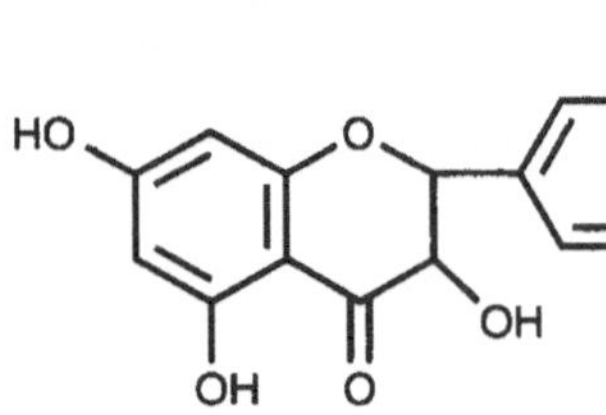

1.3.54: R = R' = H
1.3.55: R = H; R = OH
1.3.56: R = R' = OH

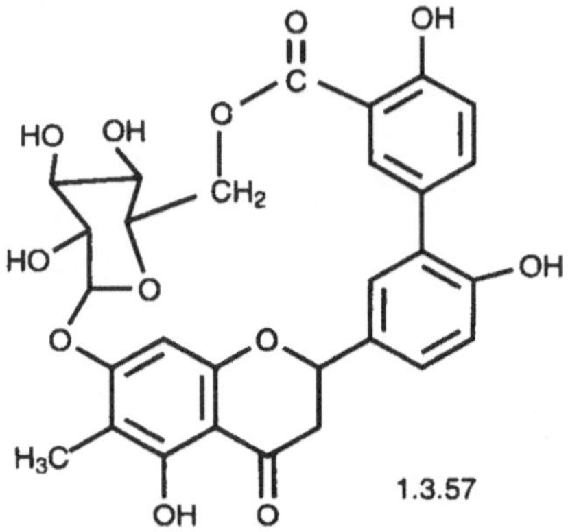

1.3.57

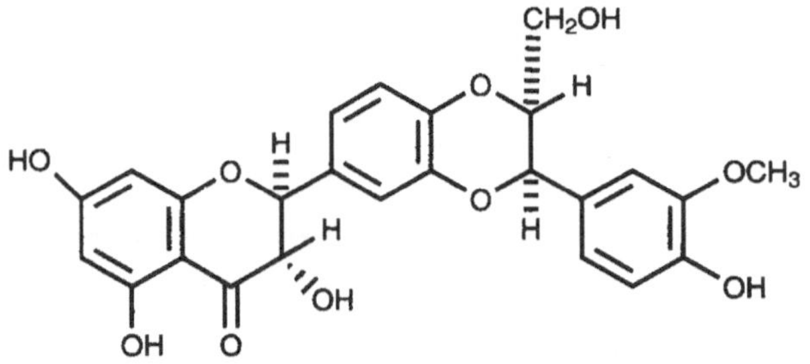

1.3.58

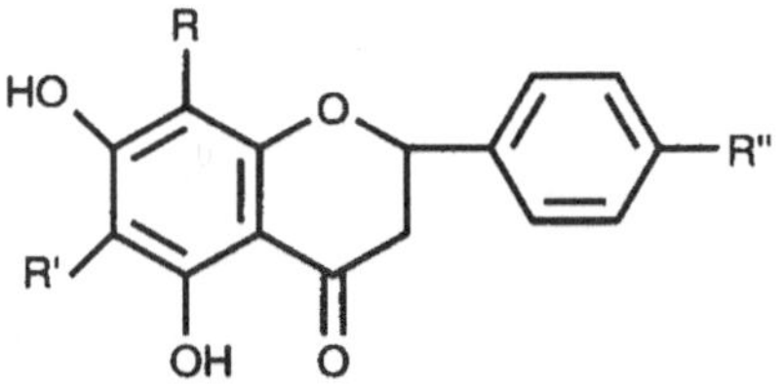

1.3.59: R = CH₃; R' = R" = H
1.3.60: R = R" = H; R' = CH₃
1.3.61: R = R' = CH₃; R" = OCH₃
1.3.62: R = R" = CH₃; R' = OH

Catechins and Proanthocyanidins

Catechins, which are 3-hydroxyflavans, have attracted much attention in recent years because of their beneficial pharmacological properties. Green tea is a rich source of these compounds. However, they were isolated first from *Areca catechu* and hence the name catechin. The most well known members of this group are (–) epicatechin (1.3.63) and (+) catechin (1.3.64), which are diastereoisomers. The former is the astringent principle of *Areca catechu* while the latter has been isolated from *Uncaria gambir*. Green tea catechin is a mixture of epicatechin, its gallate ester (1.3.65), epigallocatechin (1.3.66) and its gallate ester (1.3.67). Another example is (–) epiafzelechin (1.3.68).

Proanthocyanidins, are oligomeric compounds, in which two or more flavan-3-ol molecules are condensed together. These compounds occur widely in several plants, including cacao, white pine (*Pinus maritime*), hawthorne (*Crataegus oxycantha*), honey locust (*Gleditschia triacanthos*) and cranberry (*Vaccinum oxycoccus*) The dimeric proanthocyanidin isolated from grape seeds

(*Vitus vinifera*) has the structure (1.3.69). Polymeric proanthocyanidins are also known as condensed tannins. These compounds are also of widespread occurrence in nature, particularly in astringent plants and exhibit strong interaction with proteins. Much interest in these compounds has arisen in recent years as they possess several useful pharmacological properties such as anti-inflammatory, cardiac- protective and antioxidant activities.

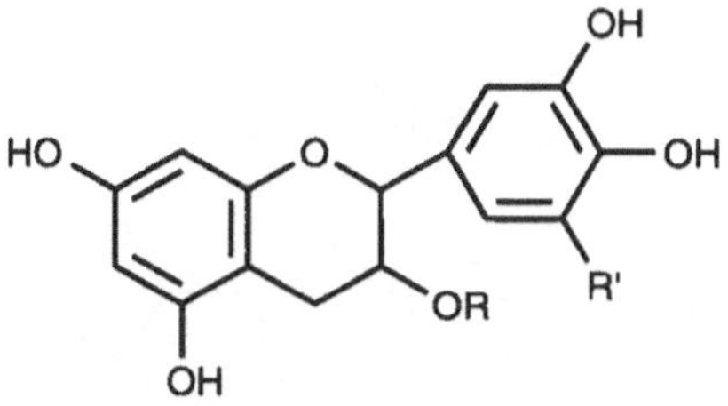

1.3.63: R = R' = H

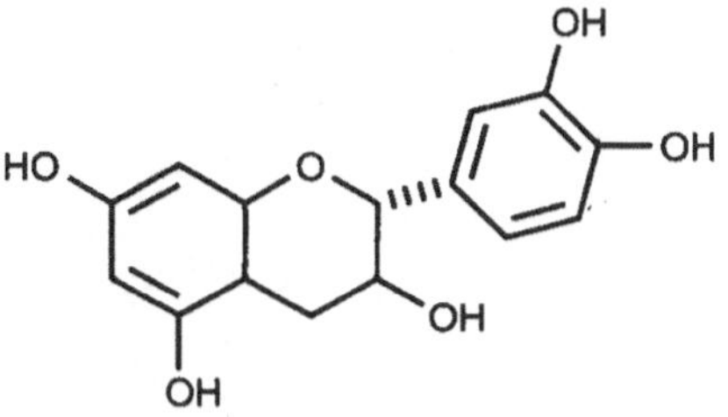

1.3.64

1.3.65: R = —C(=O)— ; R' = H

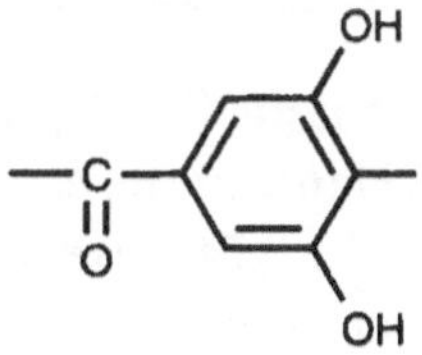

1.3.66: R = H; R' = OH

1.3.67: R = —C(=O)— ; R' = OH

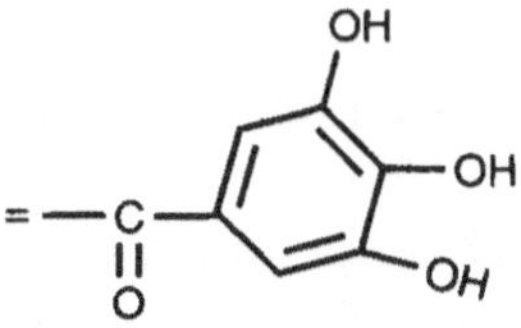

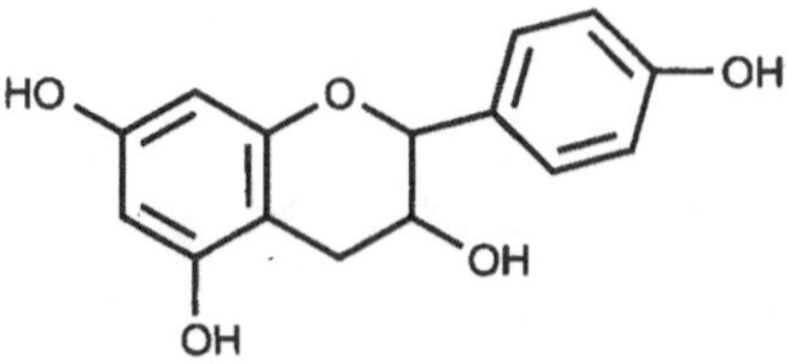

1.3.68

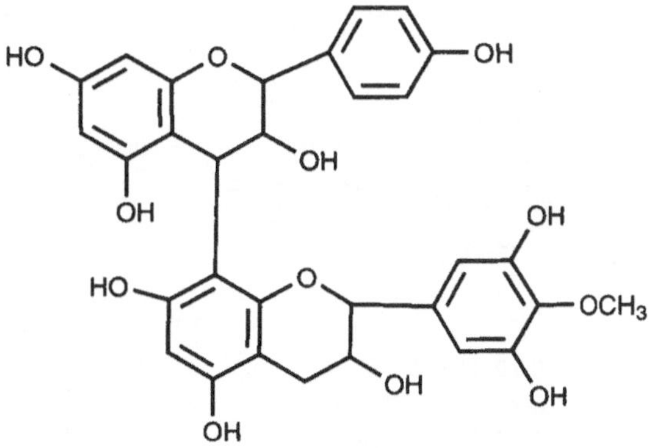

1.3.69

Chalcones

These deeply coloured (yellow to red) compounds are biosynthetic precursors of the flavonoids, though they do not possess the middle pyrone ring. However, they readily cyclise, in the presence of acids, to the corresponding flavanones; this reaction, pictorially shown on the next page, can be considered as an acid-catalysed intramolecular Michael reaction.

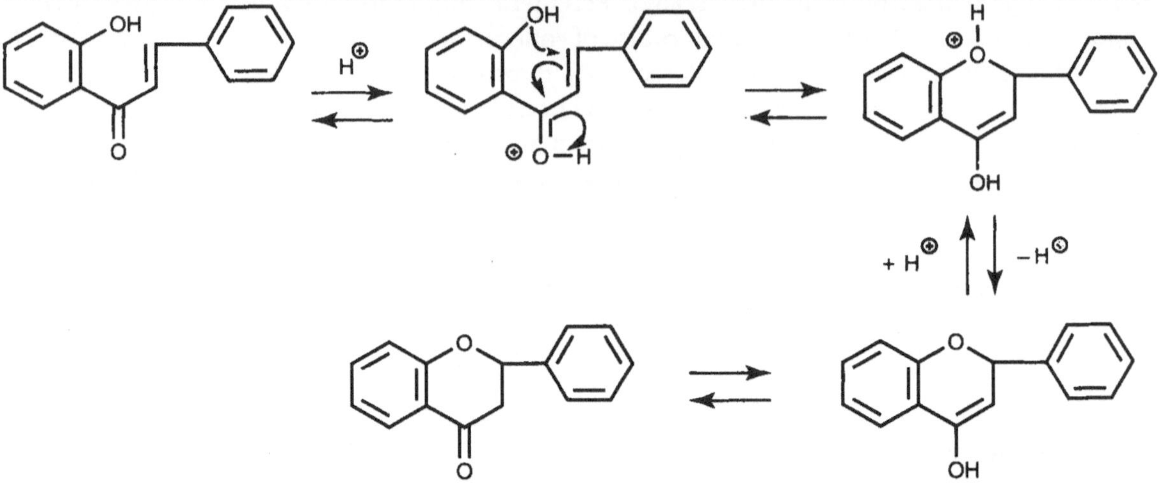

Well-known examples of chalcones are isoliquritigenin (1.3.70), butein (1.3.71) and okanin (1.3.72); butein is the pigment of the brightly colored *Butea frondosa* (Flame of the forest) flowers. Isoprenylated chalcones, such as licochalcone-A (1.3.73) and crotaorixin (1.3.74) have been isolated from licorice roots (*Glycyrrhiza glabra*) and *Crotalaria orixensis*, respectively.

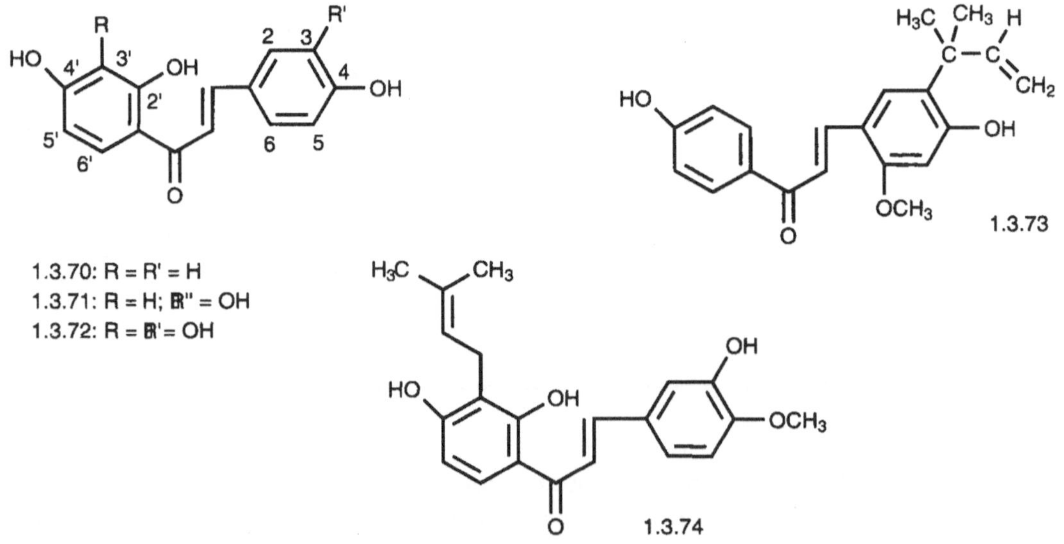

1.3.70: R = R' = H
1.3.71: R = H; R'' = OH
1.3.72: R = R'= OH

Aurones

These golden-yellow coloured compounds, which are of rare occurrence, are biosynthetically derived from chalcone epoxides. Common examples are sulphuretin (1.3.75) and its glucoside, sulphurin (1.3.76) aureusidin (1.3.77), martimetin (1.3.78) and leptosidin (1.3.79).

1.3.75: R = R' = H
1.3.76: R = glucose; R' = H
1.3.78: R = H; R' = OH

1.3.77: R = OH; R' = H
1.3.79: R = H; R' = OCH₃

Isoflavones and Related Compounds

These compounds are derivatives of 3-arylchromone (1.3.80) and
are not as abundant as the flavones and flavonols (derivatives
of 2-arylchromone). However, they exhibit a wider variety of
skeletal structures. This family of compounds comprises of
the isoflavones, the isoflavanones, rotenoids, pterocarpans,
3-phenylcoumarins and coumestans. Due to lack of conjugation
of the side phenyl group with the chromone carbonyl function,
these compounds are not coloured. Their biosynthesis is discussed
in Chapter 6.

1.3.80

Isoflavones

As mentioned earlier, these compounds, unlike the
corresponding flavones, are colourless and their UV
spectra show only one strong absorption band in the
250–260 nm region. However, the spectra exhibit
one or two shoulders and a weaker maximum in
the 300–330 nm region (corresponding to Band I of
flavones). Substituents on the condensed benzene ring
profoundly influence the absorption maximum. As
in the case of flavones, useful information regarding
the pattern of hydroxylation can be obtained with the

1.3.81: R = R' = H
1.3.82: R = OH; R' = H
1.3.83: R = H; R' = CH₃
1.3.84: R = OH; R' = CH₃

use of shift reagents. Thus, the UV spectrum of daidzein (1.3.81) recorded in methanol has
a maximum at 249 nm, with shoulders at 238, 260 and 303 nm. With the addition of sodium
methoxide the maximum shifts to 259 nm. In the case of genistein (1.3.82), the absorption
maximum is seen at 261 nm and this shifts to 276 nm in the presence of sodium methoxide.
Formononetin (1.3.83), biochanin-A (1.3.84), afromosin (1.3.85), tectorigenin (1.3.86) and
tlatlancuain (1.3.87) are some other examples of this group of compounds. From Osaje orange
(*Maclura pomifera*), two isoflavones, namely osajin (1.3.88) and pomiferin (1.3.89) having
isoprenyl and modified isoprenyl groups have been isolated. Munetone (1.3.90), occurring
in *Mundulea suberosa*, has a isopropylfurano ring, which is another modification of a C-
isoprenyl group by interaction with a phenolic hydroxyl *ortho* to it.

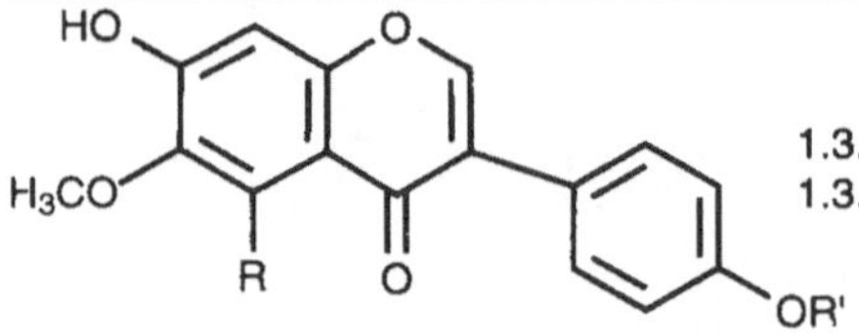

1.3.85: R = H; R' = CH₃
1.3.86: R = OH; R' = H

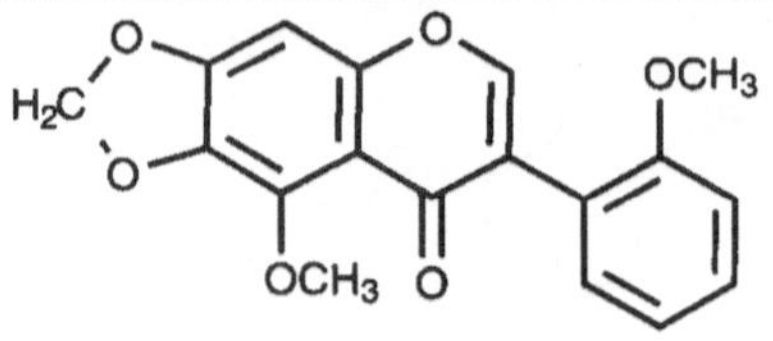

1.3.87

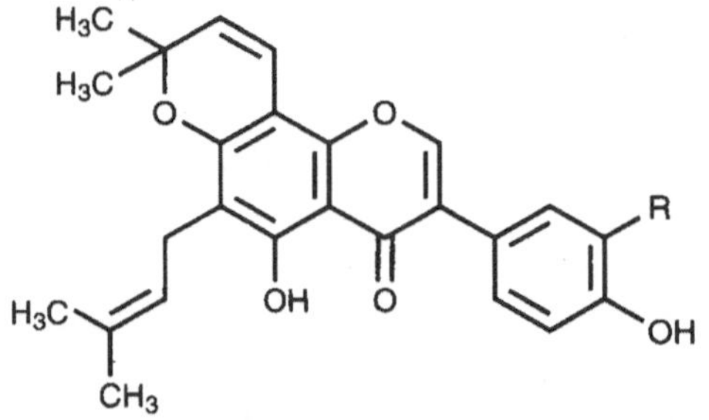

1.3.88: R = H
1.3.89: R = OH

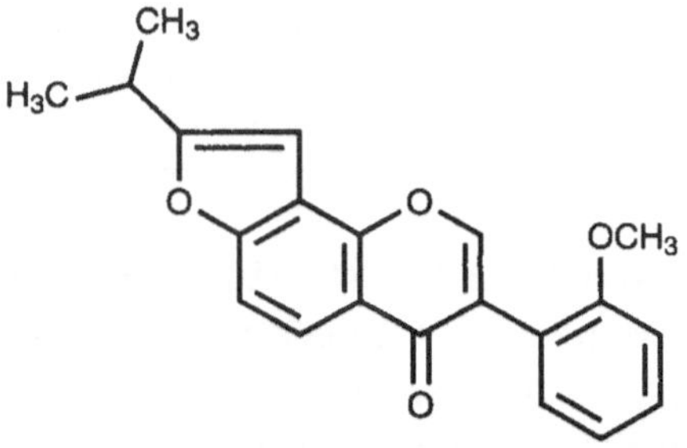

1.3.90

Isoflavanones

These compounds are optically active due to the presence of a chiral centre (position 3; point of attachment of the side phenyl group). They can be dehydrogenated to the corresponding isoflavones. Very often, one finds the 2',4'-dihydroxylation pattern (resorcinol type) in the side phenyl which is rather uncommon among flavones and flavanones. Typical examples are ferreirin (1.3.91), homoferreirin (1.3.92), dalbergioidin (1.3.93) and ougenin (1.3.94). The last three mentioned compounds occur together in the heartwood of *Ougenia dalbergiodes*. Sophorol (1.3.95) has a methylenedioxy group in addition to the 2'-hydroxyl in the side phenyl.

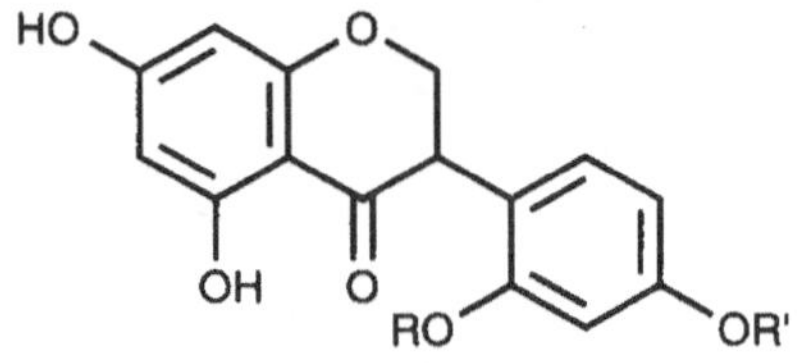

1.3.91: R = H; R' = CH₃
1.3.92: R = R' = CH₃
1.3.93: R = R' = H

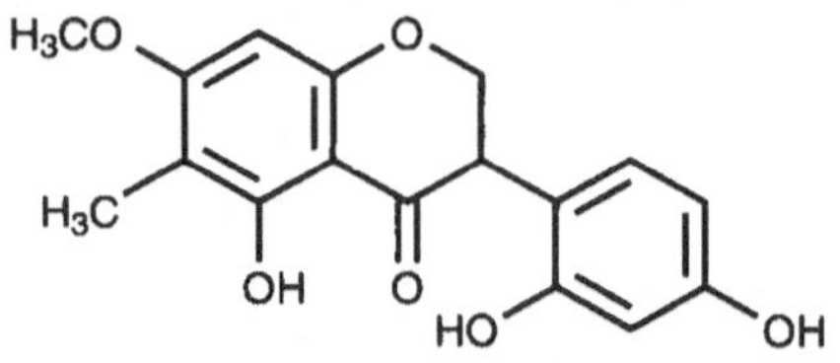

1.3.94

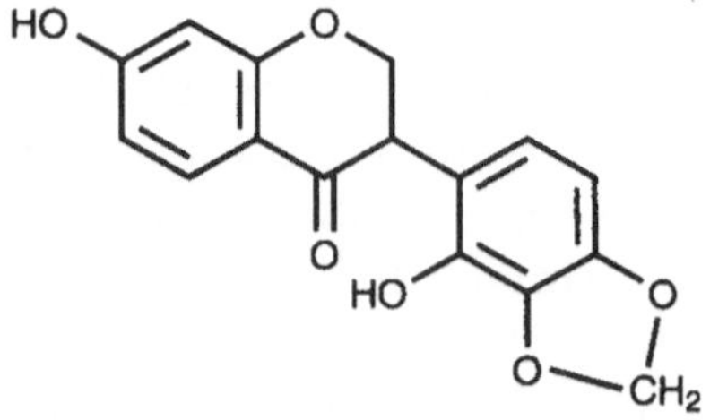

1.3.95

Pterocarpans, coumestans, 3-arylcoumarins and isoflavans

Pterocarpin (1.3.96) and homopterocarpin (1.3.97) were the first pterocarpans to be isolated from the heartwood of *Pterocarpus santalinus* (red sandalwood), where they occur along with the pigments, santalin and santarubin, which are anhydrobases. Pterocarpans posssess a tetracyclic skeletal structure and are biosynthetically derived from appropriately substituted isoflavanones. Wedelolactone (1.3.98) and desmethylwedelolactone (1.3.99) were the first coumestans to be isolated and characterised. Their chemistry is discussed in the next chapter. Coumestol (1.3.100), though a later discovery, is considered as the parent coumestan as it is less substituted. Psoralidin (1.3.101), which occurs in the seeds of *Psoralea corylifolia* is 6-γ,γ-dimethylallylcoumestrol. Erosnin (1.3.102), isolated along with the 3-phenylcoumarin, pachyrrhizin (1.3.103) from yam beans (*Pachyrrhizus erosus*) has an additional furan ring and a methylenedioxy group. The coumestans, as well as the 4-hydroxy-3-phenylcoumarins, described below, are biosynthetically closely related to the isoflavanones and the pterocarpans. Scandenin (1.3.104), lonchocarpic acid (1.3.105) and robustic acid (1.3.106) are 4-hydroxy-3-arylcoumarins isolated from the roots of *Derris scandens* and *Derris robusta*.

1.3.96

1.3.97

1.3.98: R = CH_3
1.3.99: R = H

1.3.100: R = H
1.3.101: R = $-CH_2-C(H)=C(CH_3)_2$

1.3.102

1.3.103

1.3.104

1.3.105: R = $-CH_2-C(H)=C(CH_3)_2$
1.3.106: R = H

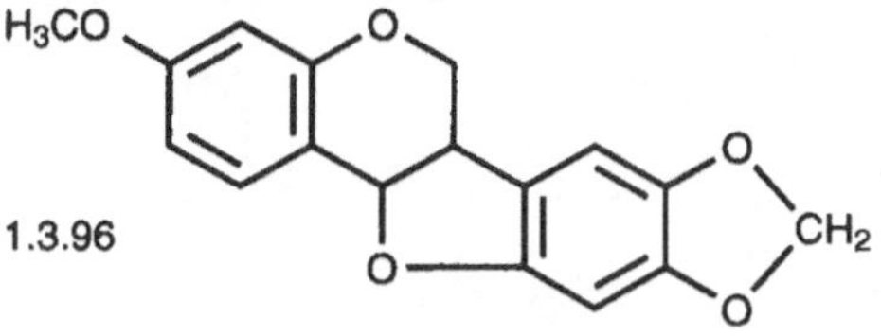

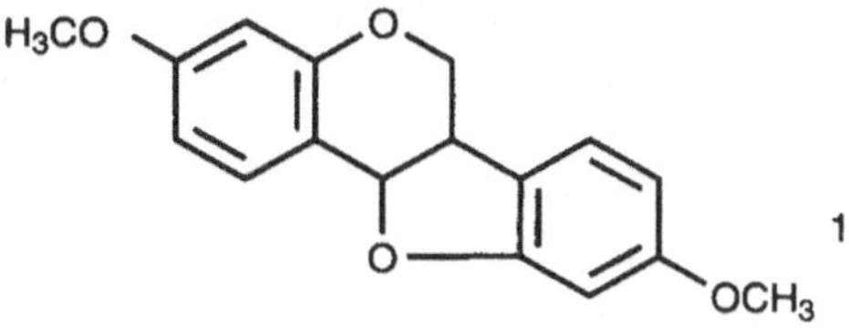

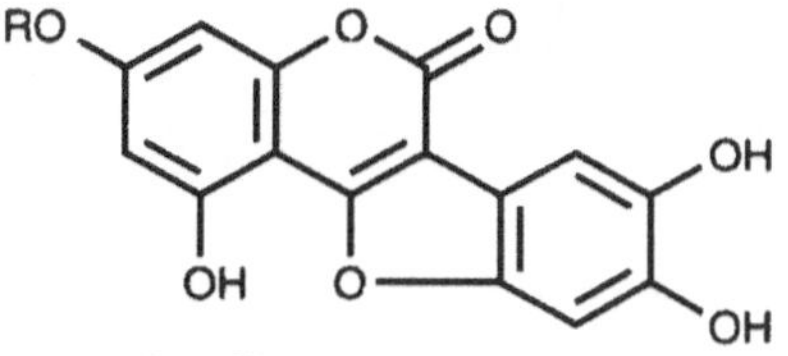

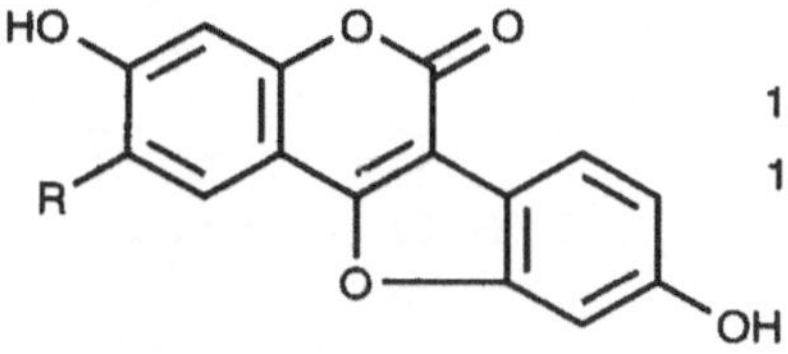

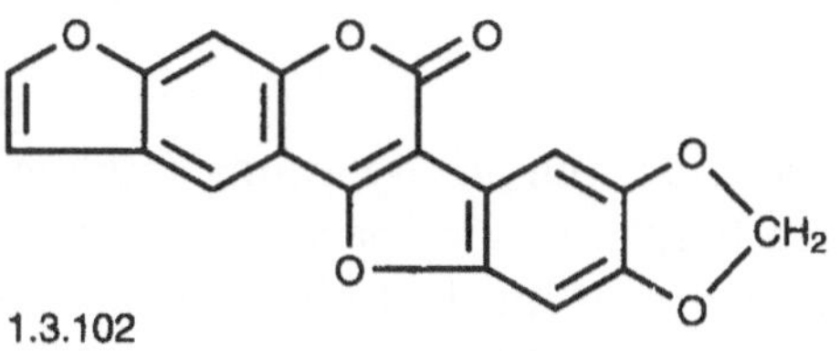

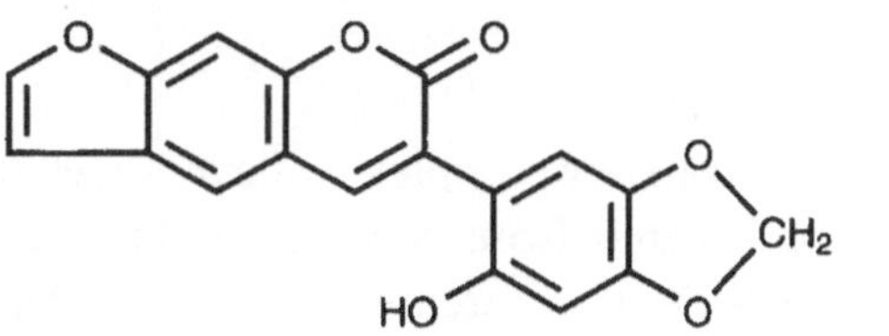

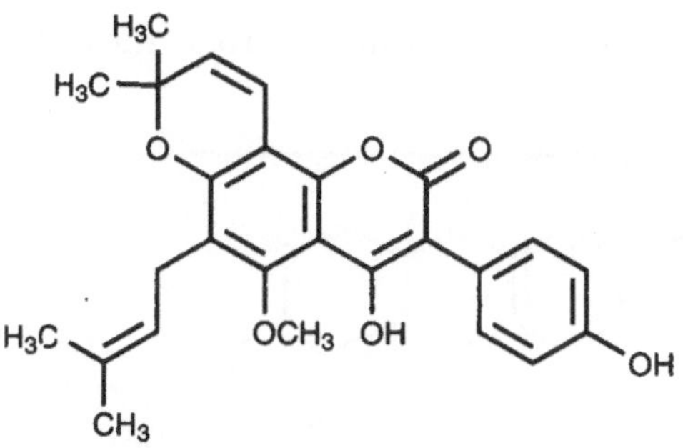

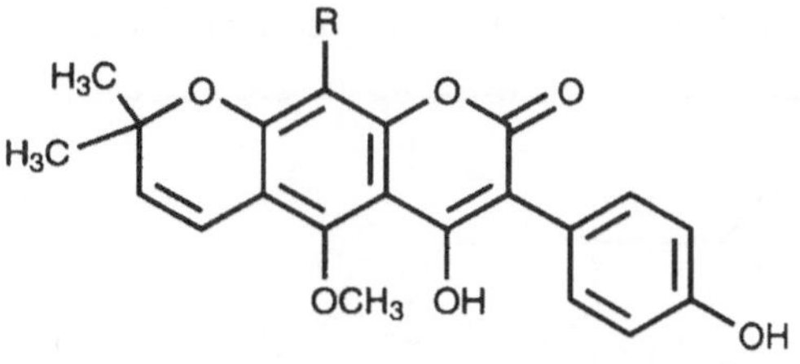

Isoflavans, which are 3-phenylchromans, are of much rarer occurrence. As a matter of fact, for a long time the only known isoflavan was equol (1.3.107), which was isolated from the urine of pregnant mares. The first isoflavan of plant origin to be discovered was lonchocarpin (1.3.108) isolated from a leguminous plant. Another example is vesitol (1.3.109) present in *Machaerium vesitium*. Mucroquinone (1.3.110) is an interesting compound in which the 3-aryl group is oxidised to a benzoquinonoid moiety; it has been isolated from the wood of *Machaerium mucronulatum*. Lisetin (1.3.111), which is a coumaronochromone, can also be considered under this head as it can be readily converted, by treatment with alkali, into the 4-hydroxy-3-phenylcoumarin (1.3.112).

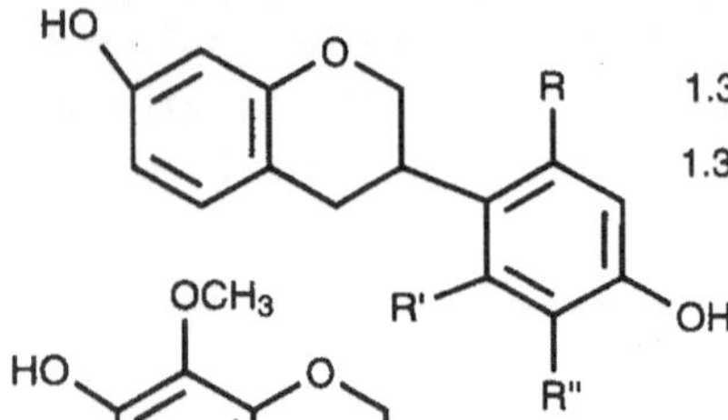

1.3.107: R = R' = R'' = H

1.3.108: R = R' = R'' = OCH_3

1.3.109

1.3.110

1.3.111

1.3.112

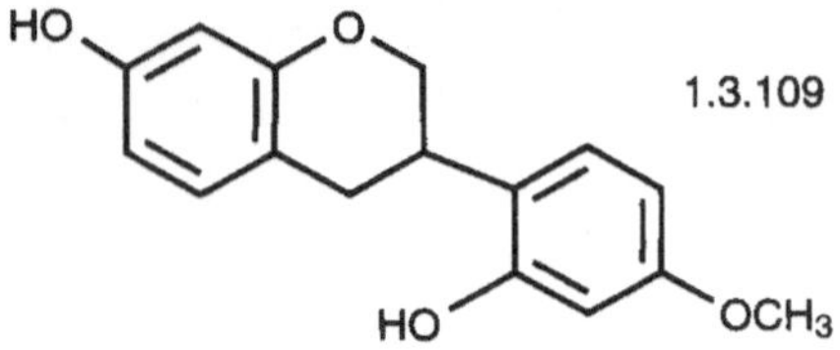

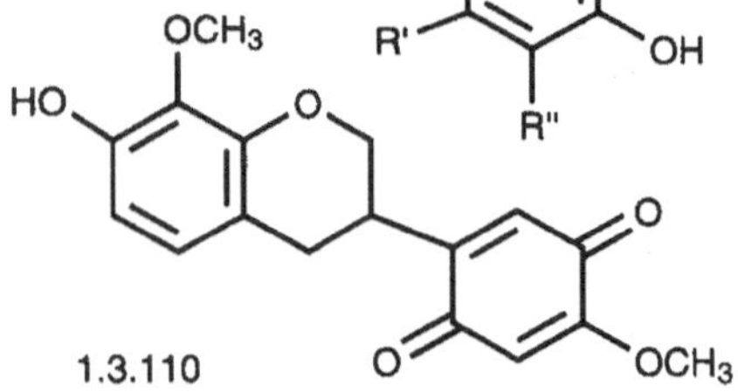

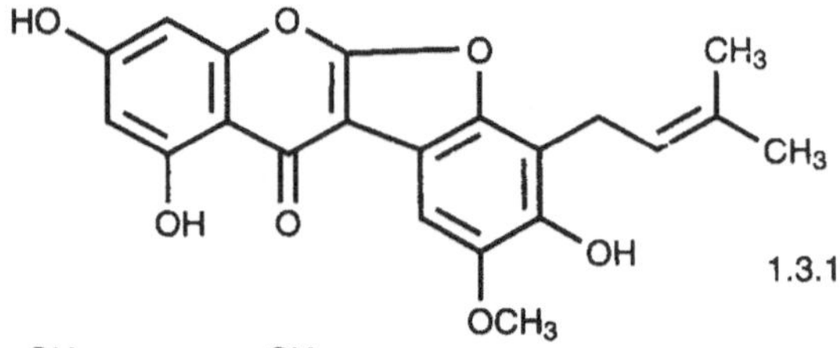

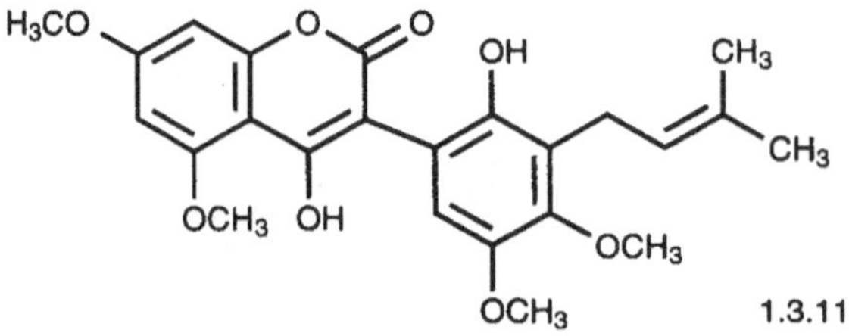

Rotenoids

These comparatively complex isoflavanones, possessing insecticidal and fish poisoning activities have been isolated from several plants of the leguminosae family and, in particular, species of the genus Derris. Rotenone (1.3.113) is the most well-known and much studied compound of this type. Other examples are elliptone (1.3.114), munduserone (1.3.115), deguelin (1.3.116), α-toxicarol (1.3.117) and pachyrrhizone (1.3.118). With the exception of munduserone, the rest of them contain structural features which can be traced to a terpenoid origin.

1.3.113

1.3.114

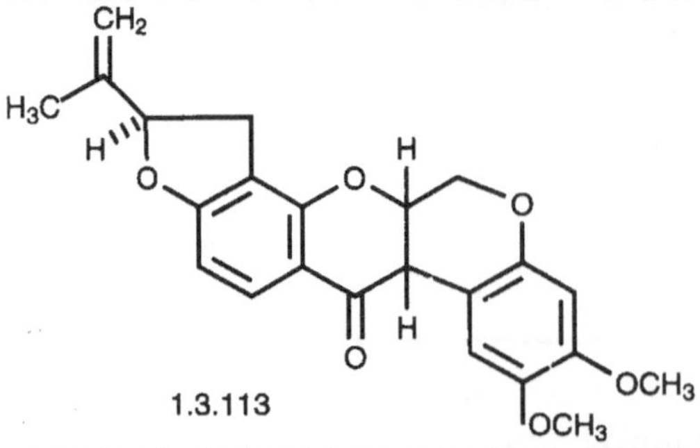

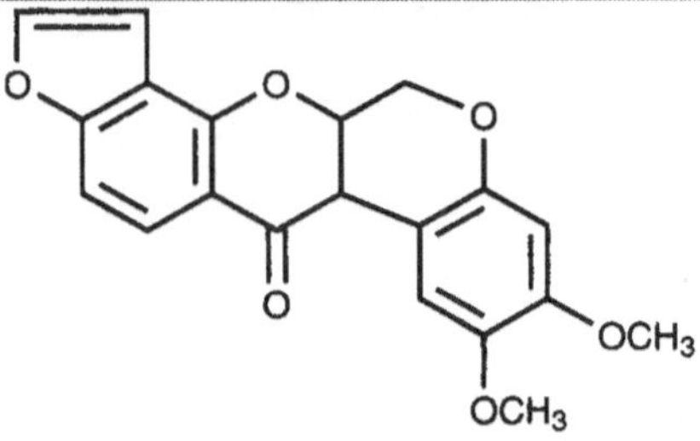

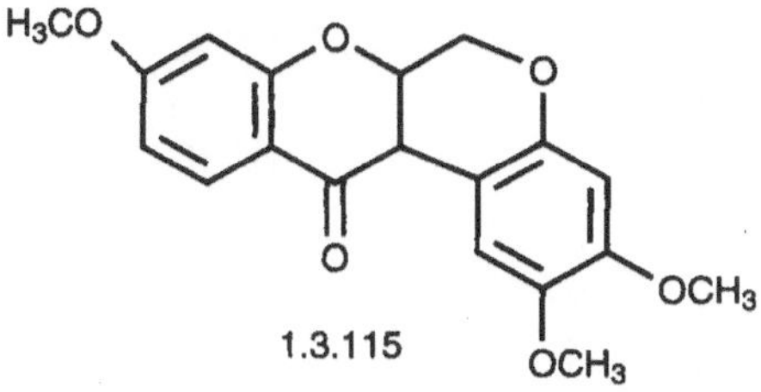

1.3.115

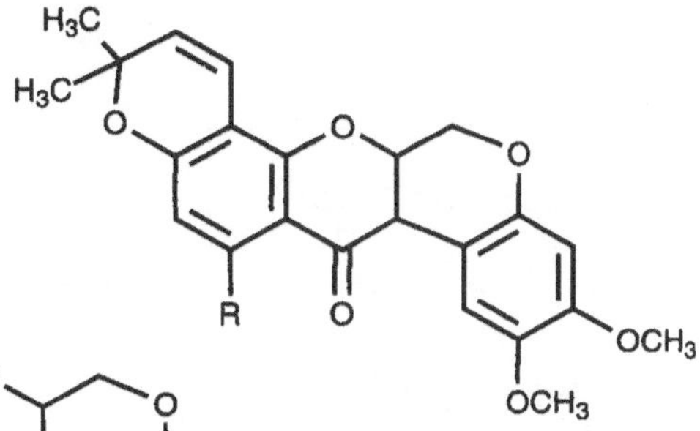

1.3.116: R = H
1.3.117: R = OH

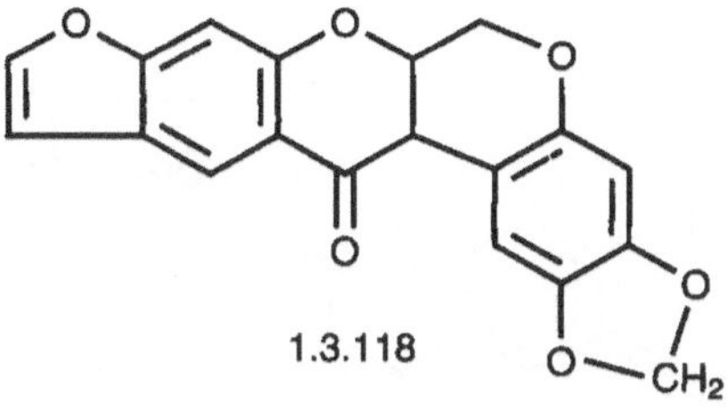

1.3.118

Neoflavonoids

As the name suggests this group of compounds is a later addition to the flavonoid family. Their biosynthesis involves a combination of a C-6 and C-9 (cinnamoyl moiety) by a different mode compared to the formation of flavonoids (and isoflavonoids). They can be considered as 4-arylchroman derivatives and the most common among them are 4-arylcoumarins. Dalbergin (1.3.119) and desmethyldalbergin (1.3.120) were the first neoflavonoids to be characterised. These compounds occur in the heartwood of *Dalbergia sissoo*. Other examples of 4-aryl coumarins are melannein (1.3.121), calophyllolide (1.3.122) and inophyllolide (1.3.123) which occur together in the seeds of *Calophyllum inophyllum* and mammeisin (1.3.124) and related compounds isolated from *Mammea americana*. In dalbergichromene (1.3.125), the lactone carbonyl group is replaced by a methylene. Biosynthetic experiments have shown that it is the immediate biosynthetic precursor of dalbergin. The compound 1.3.125 itself is formed from 4-methoxydalbergione (1.3.126), which is a benzoquinone derivative. In obtusaquinol (1.3.127) and latifolin (1.3.128), the quinone ring is reduced to the corresponding quinol.

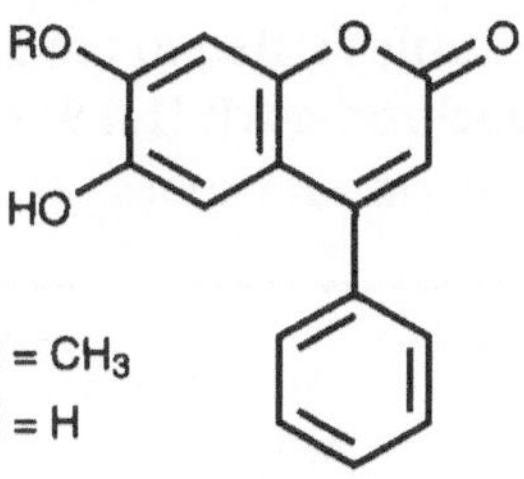

1.3.119: R = CH₃
1.3.120: R = H

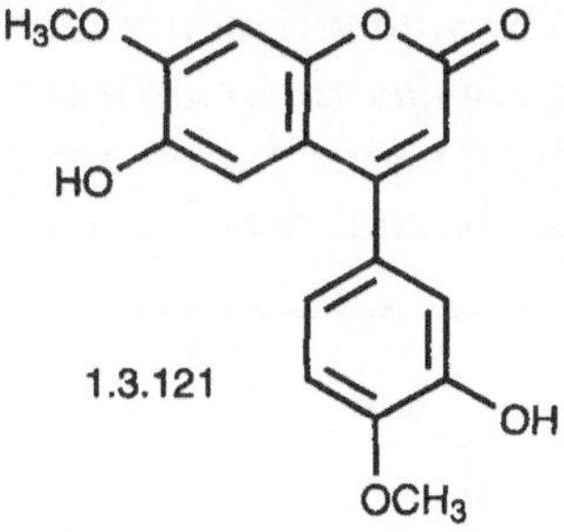

1.3.121

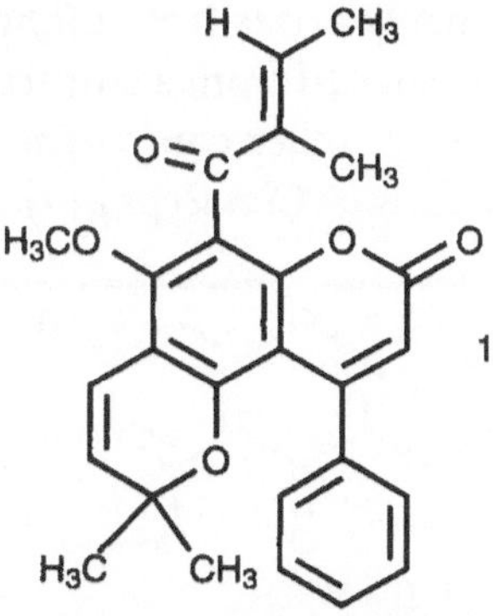

1.3.122

1.3.123

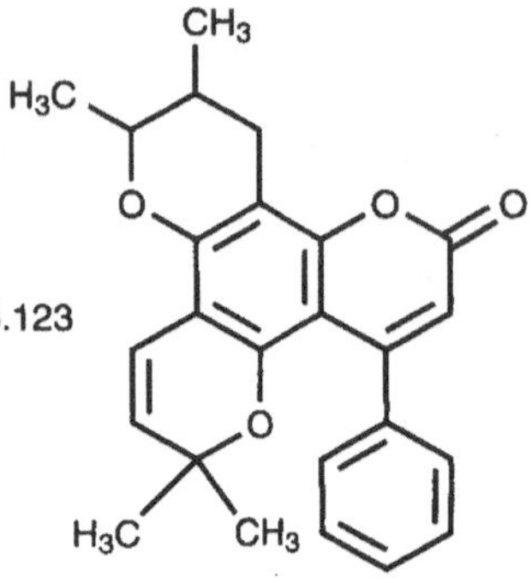

1.3.124

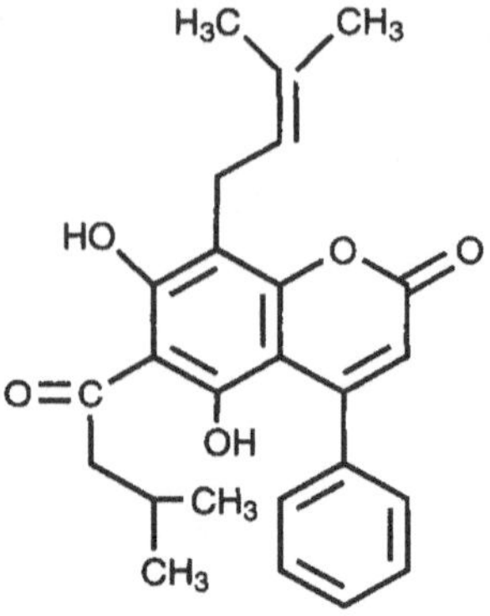

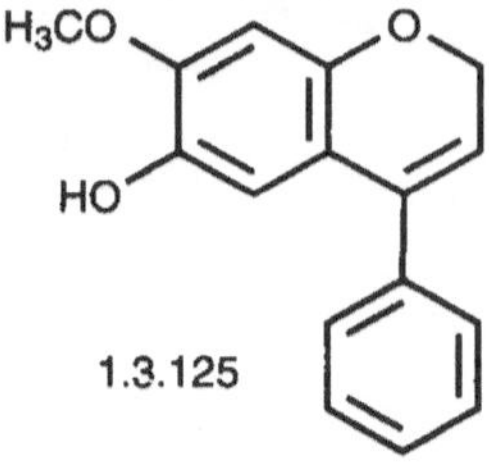

1.3.125

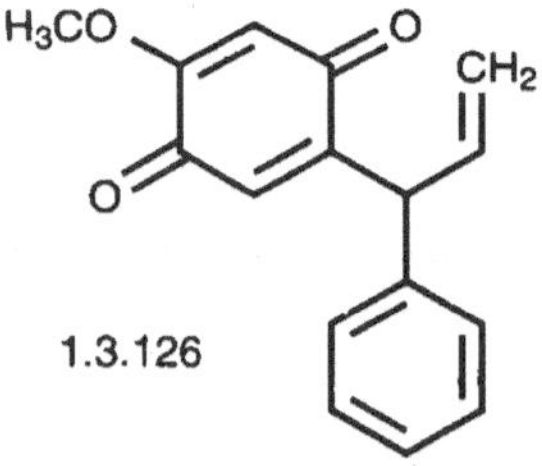

1.3.126

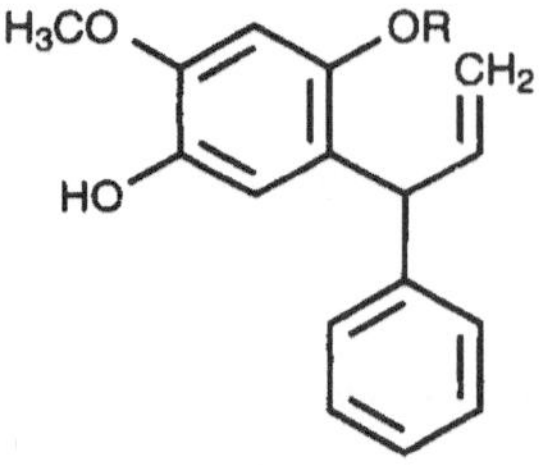

1.3.127: R = H
1.3.128: R = CH₃

1.4 XANTHONES

Structurally, xanthones can be classified as benzo-γ-pyrones and possess a compact tricyclic structure. As the name indicates they exhibit yellow and deeper colours (depending on the degree of hydroxylation). Their occurance is particularly widespread in plants of the Guttiferae family. The carbonyl group, being flanked by the two benzene rings, is not very reactive towards hydroxylamine and phenylhydrazine but can be detected by infrared spectroscopy from the strong carbonyl stretching frequency near 1650 cm^{-1}. A typical example is euxanthin (1.4.1), which is 1,7-dihydroxyxanthone. On fusion with alkali it yields gentisic acid (1.4.2) and resorcinol (1.4.3). Its 7-glucuronide is euxanthic acid (1.4.4). Gentisin (1.4.5), which occurs in the roots of *Gentiana lutea*, mangostin (1.4.6), isolated from *Garcinia mangostana*, and the *C*-glucoside, mangiferin (1.4.7), from *Mangifera indica* (mango) are other examples. The yellow dye, Indian yellow, obtained from the urine of cows fed with mango leaves has been identified as euxanthin, which presumably arises from mangiferin by appropriate metabolic transformations. Sterigmatocystin (1.4.8), isolated from a culture of *Aspergillus versicolor*, is biosynthetically related to the aflatoxins mentioned under coumarins. The ergot pigments, for example, ergochrome AB (1.4.9) produced by the fungus *Claviceps purpurea* can be considered as modified bimeric xanthones.

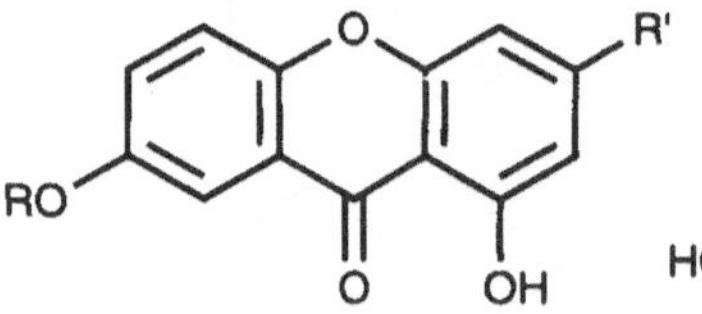

1.4.1: R = R' = H

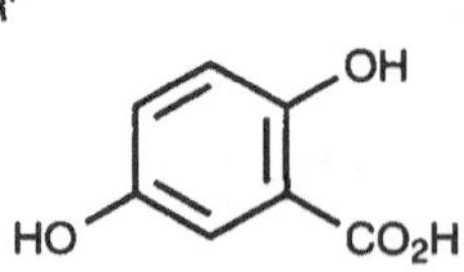

1.4.2

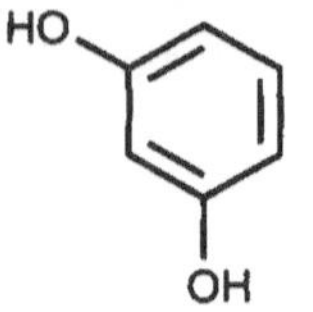

1.4.3

1.4.4: R =   ; R' = H

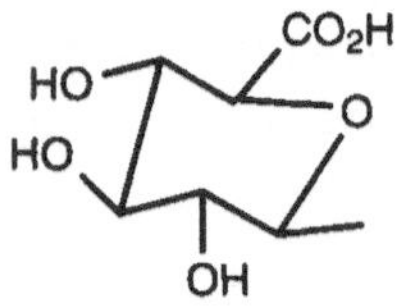

1.4.5: R = H; R' = OCH₃

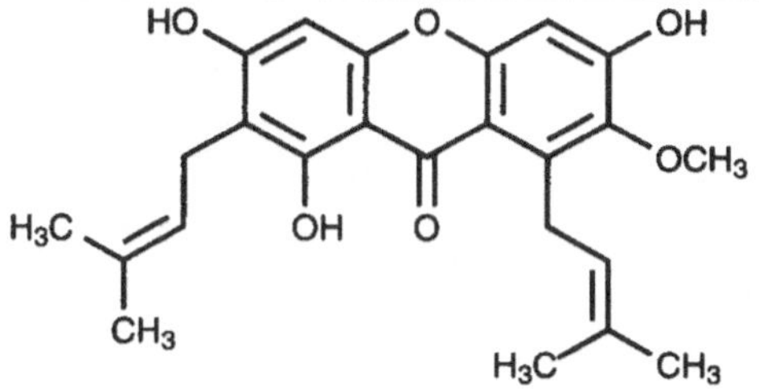

1.4.6

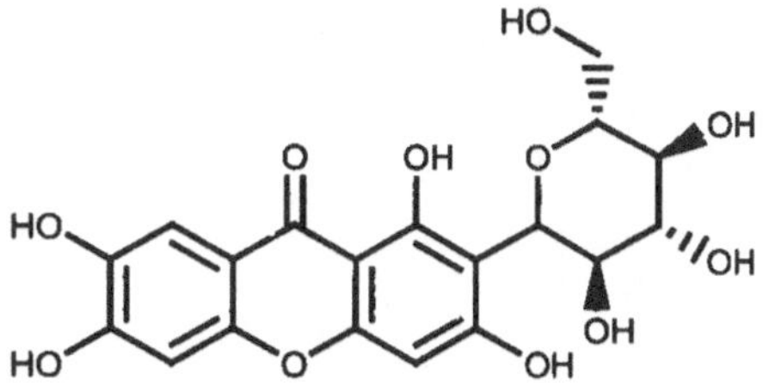

1.4.7

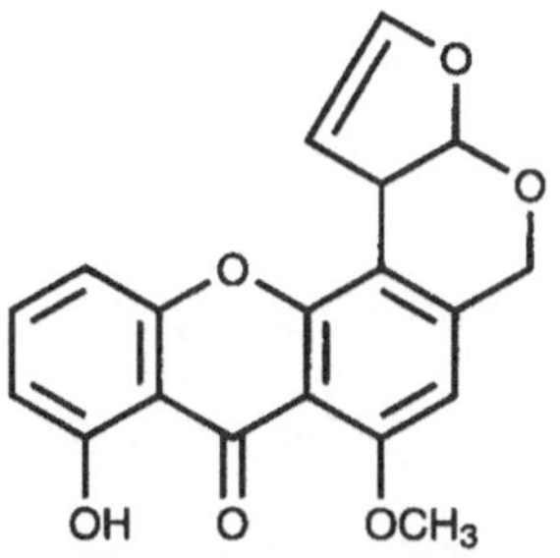

1.4.8

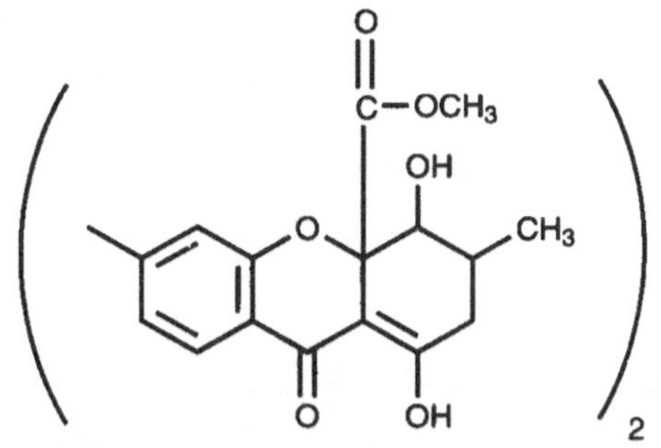

1.4.9

1.5 OTHER OXYGEN HETEROCYCLIC COMPOUNDS

Coumarins and Isocoumarins

Naturally occurring coumarins exhibit a wide variety of structural features. Under this head we will not be discussing the coumestans and 3-arylcomarins which have been dealt with under isoflavonoids. These compounds are also known as benzo-α-pyrones and are biosynthetically derived from cinnamic acids. Unsubstituted coumarin (1.5.1) has been isolated from tonka beans. It has a pleasant smell resembling hay and is used in perfumery. Umbelliferone (1.5.2), which is 7-hydroxycoumarin, is the most well-known compound among hydroxylated coumarins and, as the name suggests, occurs in plants of the Umbelliferae family. Its methyl ether, herniarin (1.5.3) occurs in lavender oil. Other examples of simple, hydroxylated coumarins are esculletin (6,7-dihydroxycoumarin; 1.5.4), daphnetin (7,8-dihydroxycoumarin; 1.5.5) and fraxetin (7,8-dihydroxy-6-methoxycoumarin; 1.5.6). Scopoletin (1.5.7) is the 6-O-methyl ether of esculetin. Both esculetin and scopoletin also occur as their 7-O-glucosides. C-Alkyl, and in particular C-isoprenylcoumarins are of common occurrence. Osthol (1.5.8), ostruthin (1.5.9), aurapten (1.5.10), suberosin (1.5.11), xanthyletin (1.5.12), and mammein (1.5.13) are some typical examples of this type.

Furanocoumarins include psoralen (1.5.14), angelicin (1.5.15), bergapten (1.5.16) and peucedanin (1.5.17). Dicoumarol (1.5.18), a potent anticoagulant, isolated from spoiled sweet clover, is a bis-4-hydroxycoumarin. Ellagic acid (1.5.19) is a benzocoumarin. Coumarins of microbial origin include the antibiotic, novobiocin (1.5.20), from *Streptomyces niveus* and alternariol (1.5.21) produced by the fungus, *Alternaria tenuus*. The aflatoxins, for example aflatoxin B1 (1.5.22), possess the coumarin skeletal structure. These poisonous and carcinogenic compounds are metabolites of species of the fungi Aspergillus and Penicillium.

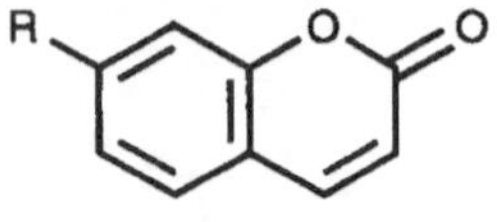

1.5.1: R = H
1.5.2: R = OH
1.5.3: R = OCH₃

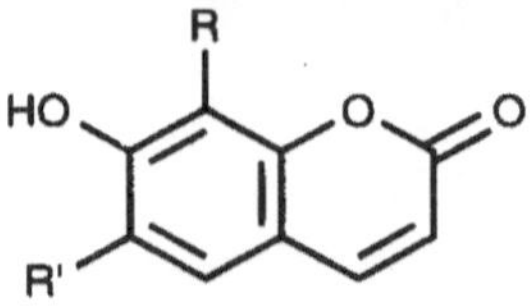

1.5.4: R = H
        R' = OH
1.5.5: R = OH; R' = H
1.5.6: R = OH; R' = OCH₃

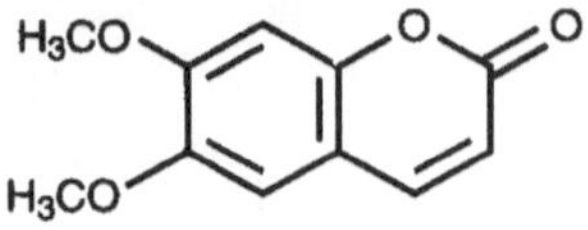

1.5.7

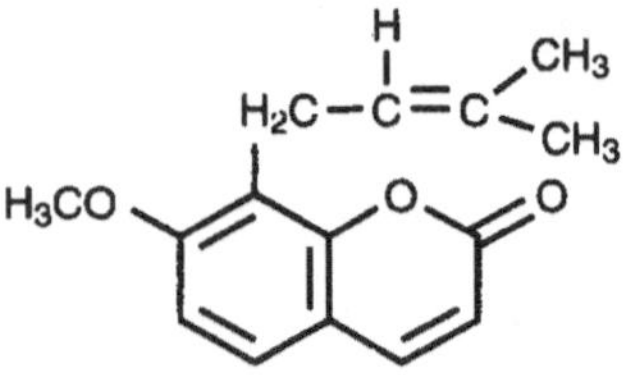

1.5.8

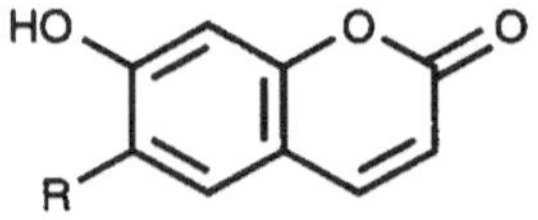

1.5.9 : R=—CH₂

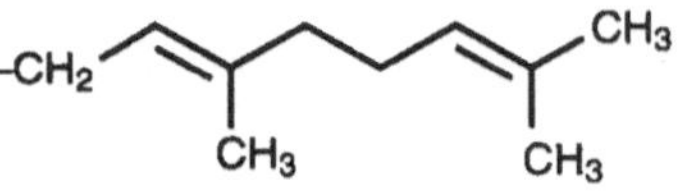

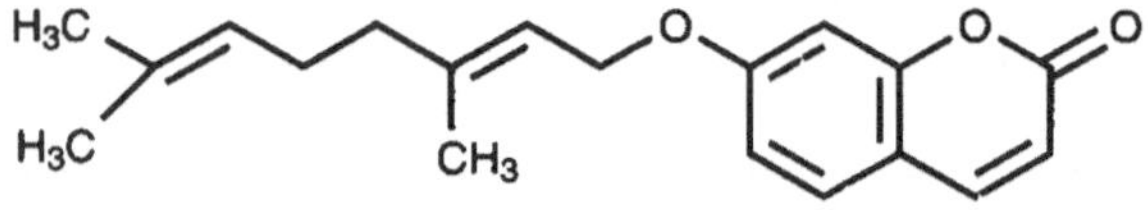

1.5.10

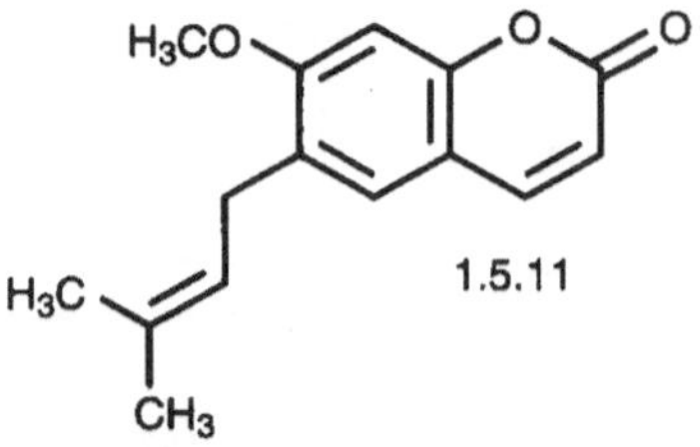

1.5.11

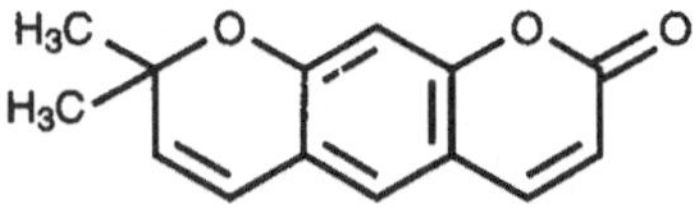

1.5.12

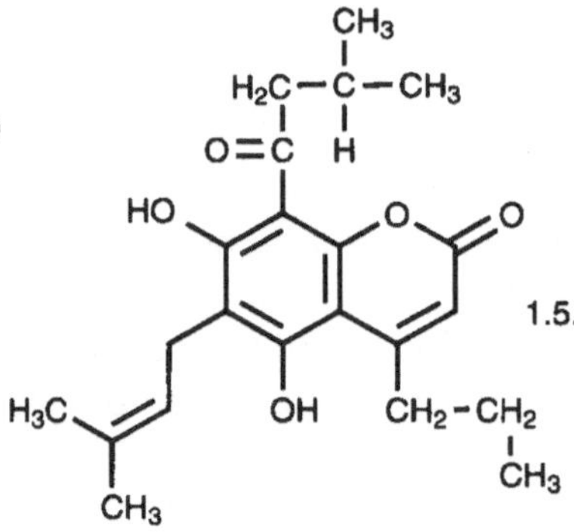

1.5.13

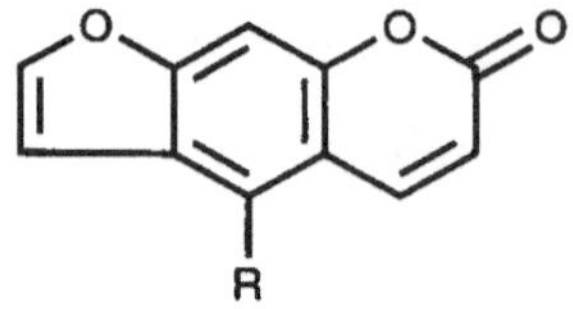

1.5.14: R = H
1.5.16: R = OCH₃

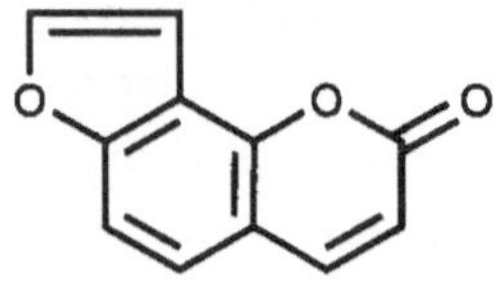

1.5.15

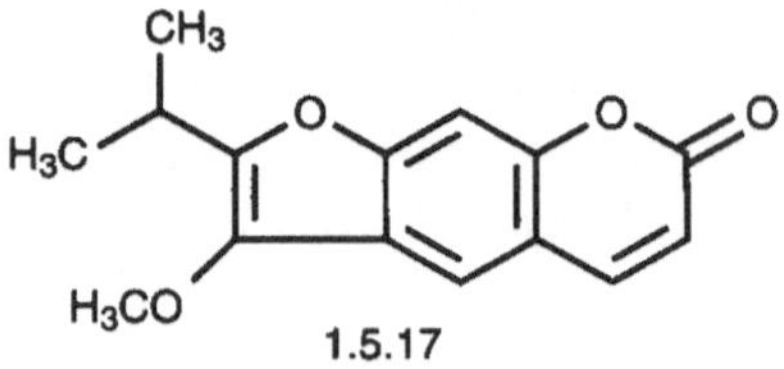

1.5.17

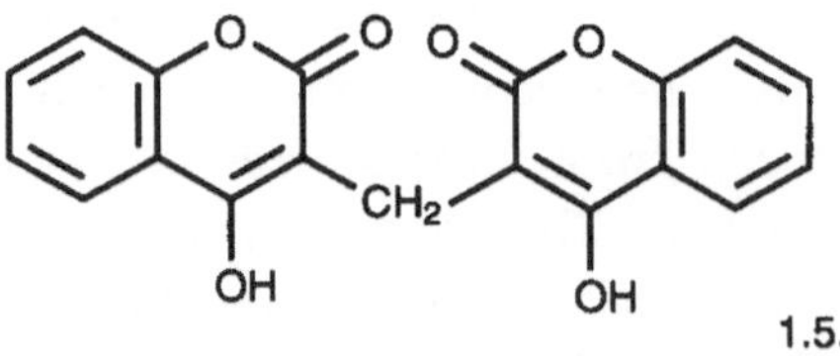

1.5.18

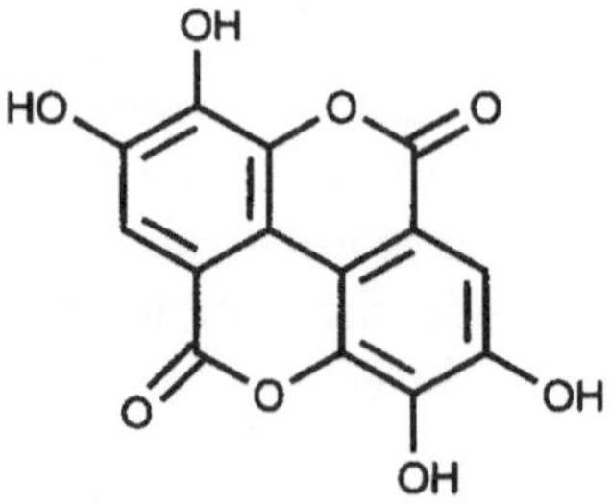

1.5.19

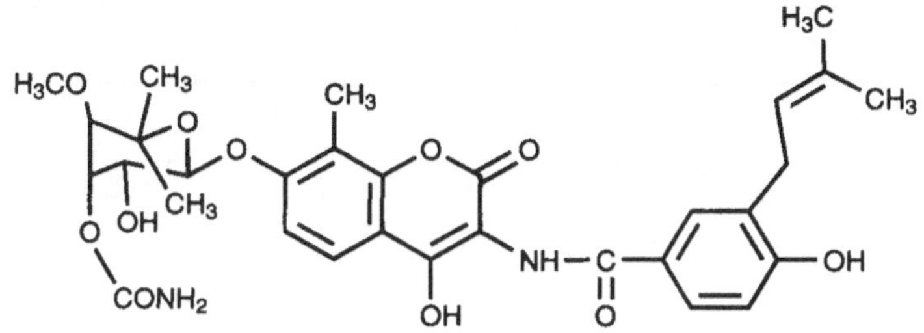

1.5.20

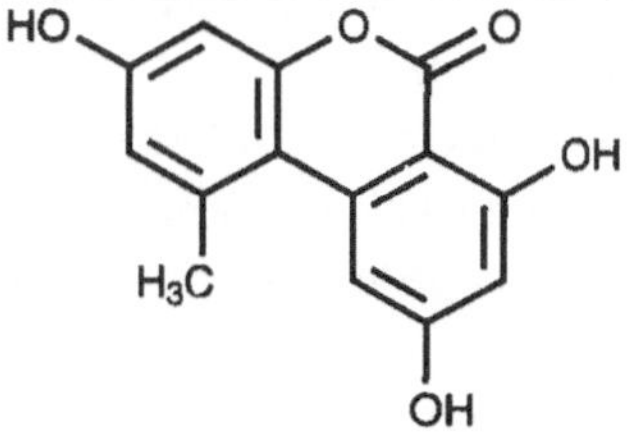

1.5.21

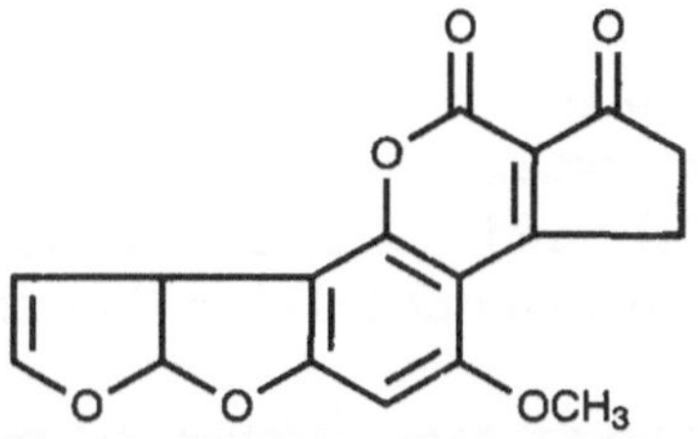

1.5.22

Isocoumarins are of rarer occurrence. 8-Hydroxy-3-methylisocoumarin (1.5.23) has been isolated from a species of Marasimus. Its dihydro derivative, mellein (1.5.24) is a fungal metabolite (from *Aspergillus mellus*). Hydrangeol (1.5.25) and bergenin (1.5.26) are also dihydroisocoumarins. The last mentioned compound has been isolated from *Saxifraga crassifolia* syn., *Bergenia crassifolia* and *Mallotus japonica*.

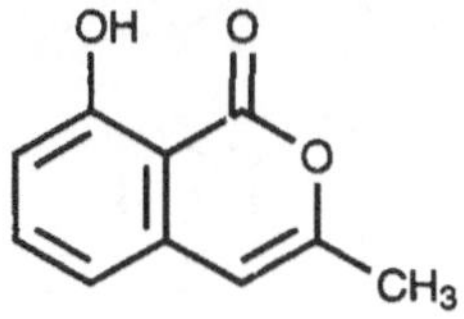

1.5.23

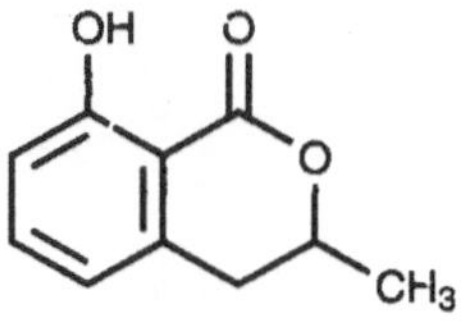

1.5.24

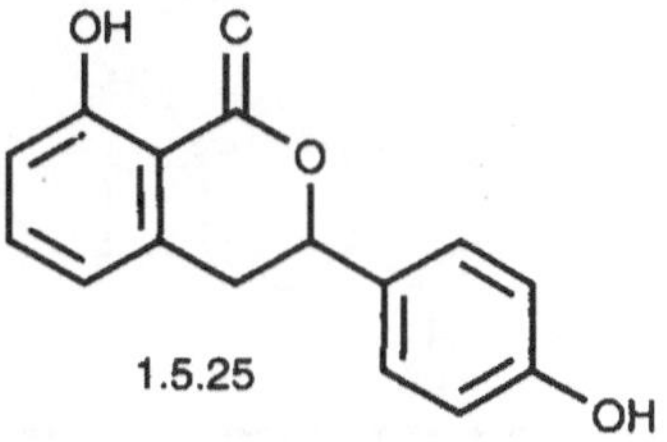

1.5.25

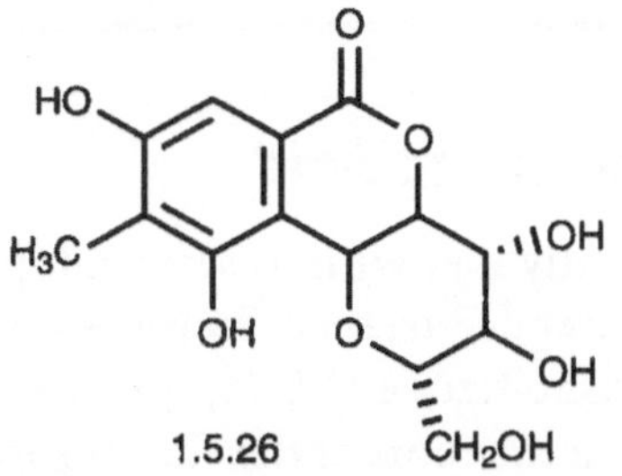

1.5.26

Chromenes and Chromans

These comparatively rare compounds are chemically related to the coumarins. For example, ageratochromene (1.5.27) isolated from some species of the genus Ageratum (Asteraceae family), can be synthesised from esculetin dimethyl ether by a Grignard reaction with methyl magnesium iodide. Other examples are evodionol (1.5.28) and allo evodionol (1.5.29), present in the leaves of *Evodia litoralis.*

Rottlerin (1.5.30), the pigment of 'Kamala dye', *Rottlera tinctora* syn. *Mallotus philipinensis*, is a complex chromene, having additional cinnamoyl and a substituted benzyl groups. Tetrahydrocannabinol (1.5.31), one of the 'controlled substances', is a component of the resin obtained from *Cannabis sativa* (hashish or marihuana). Cannabichromene (1.5.32) and cannabinol (1.5.33) have also been isolated from this resin.

1.5.27

1.5.28: R = H; R' = —C(=O)—CH3

1.5.29: R' = —C(=O)—CH3; R' = H

1.5.30

1.5.31

1.5.32

1.5.33

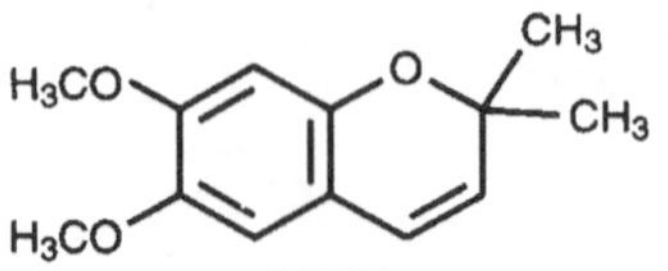

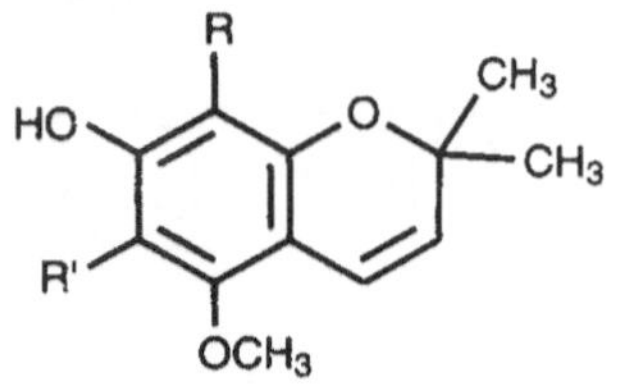

Furans, Benzofurans and Usnic Acid

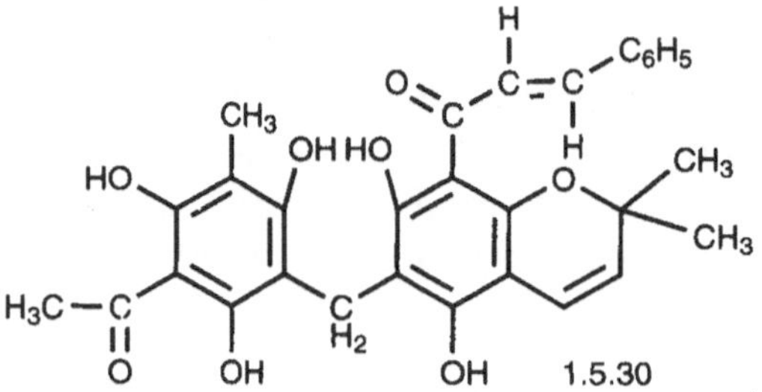

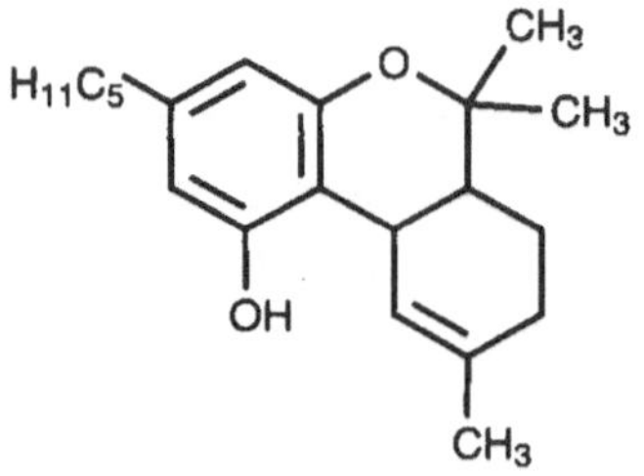

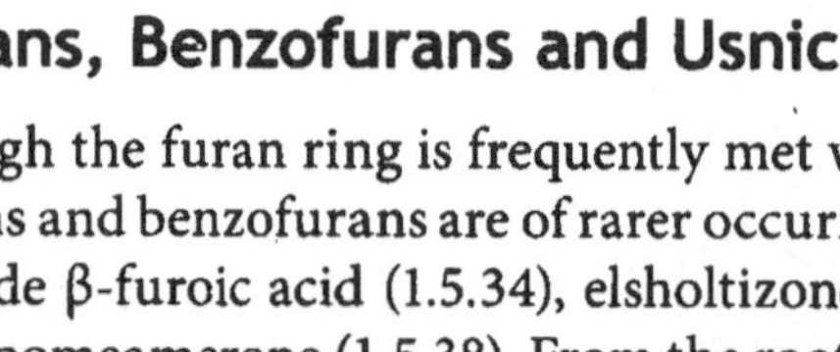

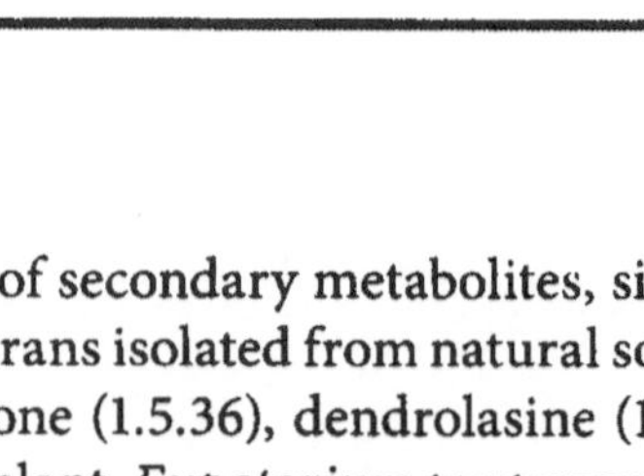

Though the furan ring is frequently met with in several types of secondary metabolites, simpler furans and benzofurans are of rarer occurrence. Examples of furans isolated from natural sources include β-furoic acid (1.5.34), elsholtizone (1.5.35), perilla ketone (1.5.36), dendrolasine (1.5.37) and ipomeamerone (1.5.38). From the roots of the Compositae plant, *Eupatorium purpureum*, the benzofuran, euparin (1.5.39) has been isolated.

Pongamol (1.5.40) can be considered as a benzofuran as well as a diketone. It is present in the seed oil obtained from the tree, *Pongamia glabra.* Complex benzofurans have been identified among lichen metabolites. Two well-known examples are didymic acid (1.5.41) occurring in species of Cladonia and Usnic acid (1.5.42) obtained from species of several lichen genera, including Cladonia, Cetraria, Parmelia, Alectoria and Usnea.

1.5.34: R = CO₂H; R' = H
1.5.35: R = CH₃; R' = ...

1.5.37 1.5.38 1.5.39 1.5.40 1.5.41 1.5.42

α- and γ-pyrones

As we have already seen the pyrone ring (both α and γ) is an integral part of a large number of oxygen heterocyclic compounds of natural origin including, chromones, coumarins and the flavonoids and related compounds. However, there are very few compounds in which the pyrone ring alone forms the skeletal structure. Some examples of naturally occurring α-pyrones are phenylcoumalin (1.5.43), which has been isolated from some plants of the Lauraceae family, yangonin (1.5.44), occurring in the roots of *Piper methysticum*, and alternaric acid (1.5.45), a fungal metabolite. In patulin (1.5.46), the lactone carbonyl is reduced to a carbinol function but it has an additional 5-membered lactone It is a metabolite of *Penicillium urticae*. Examples of γ-pyrone are maltol (1.5.47), isolated from some pine trees and kojic acid (1.5.48), a fungal metabolite.

1.5.43 1.5.44 1.5.45 1.5.46 1.5.47: R = H; R' = OCH₃ 1.5.48: R = CO₂H; R' = H

1.6 QUINONOID COMPOUNDS

The quinonoid chromophore is present in several naturally occurring plant pigments. These can be considered under three categories, namely, benzoquinones, naphthoquinones and anthraquinones.

Benzoquinones

The simplest naturally occurring benzoquinone is 2,6-dimethoxy-1,4-benzoquinone (1.6.1) which occurs in *Adonis vernalis*. The occurance of Ubiquinones (1.6.2) in which an isoprenoid or polyisoprenoid unit is attached to a *para* benzoquinone, occur widespread in the plant kingdom. They are involved in respiratory electron transport and are found predominantly in mitochondria. In their visible spectrum, the absorption maximum at 405 nm is due to the quinone chromophore. In their infra-red spectra, the carbonyl stretching frequency is seen at 1650 cm⁻¹. Embelin (1.6.3) has been isolated from the fruits of *Embelia ribes*. The closely related rapanone (1.6.4) occurs in *Rapanea Maximowiczii* and *Myrsine guianensis*. Fumigatin (1.6.5) and spinulosin (1.6.6) are metabolites of the fungus *Aspergillus fumigatus*; spinulosin has also been isolated from *Penicillium spinulosum* Thom. Perezone (1.6.7) occurs in the roots of *Perezia alamant*. Derivatives of 2,5-diphenyl-*p*-benzoquinone, commonly known as terphenylquinone, have been isolated from some fungi and lichens. For example, atromentin monoacetate (1.6.8) occurs in the fruiting bodies of *Paxillus atrotomentosus*. The fruit bodies of the *Anthracophyllum discolor* contain anthracophyllin (1.6.9), whereas the lichen (bacidomycete), *Thelephora purpurea*, owes its dark purple colour to thelephoric acid (1.6.10).

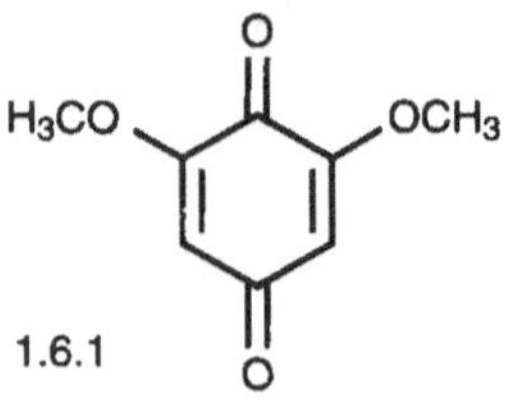

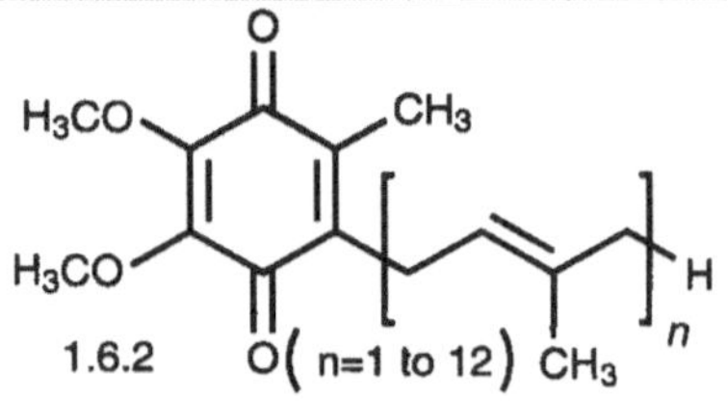

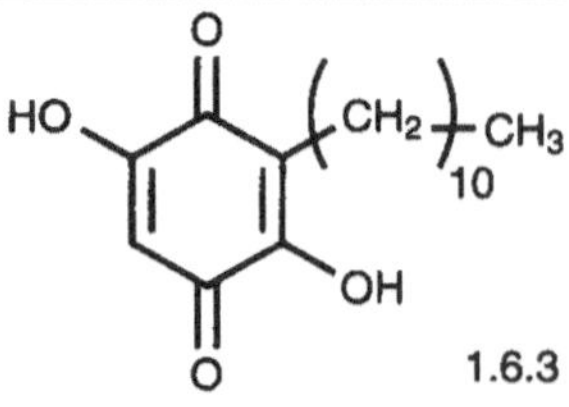

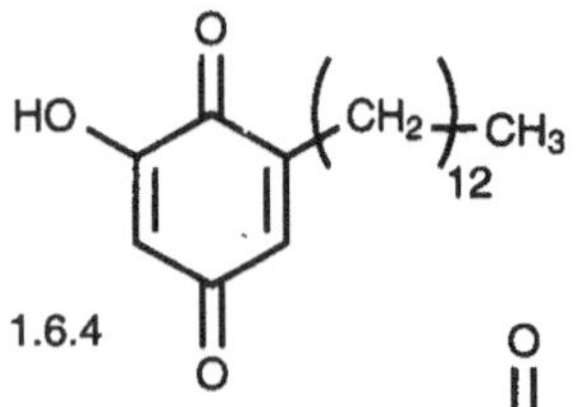

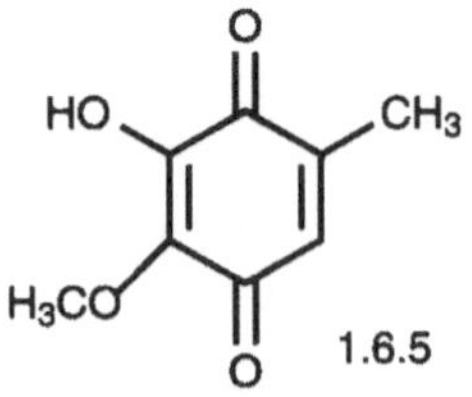

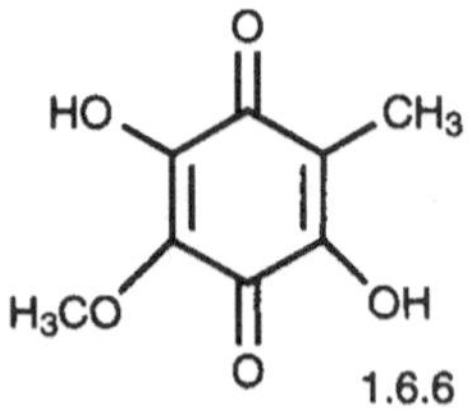

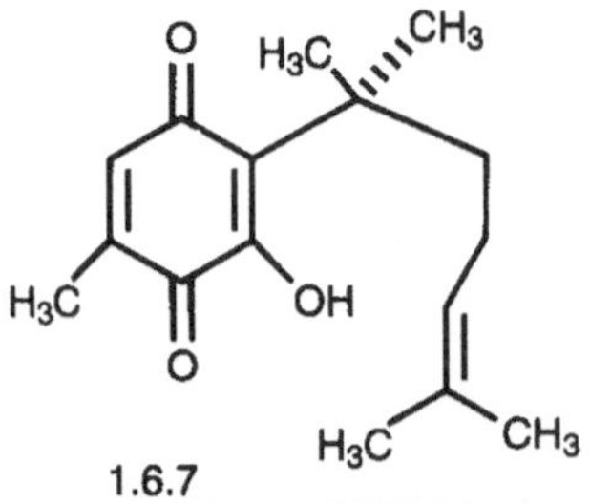

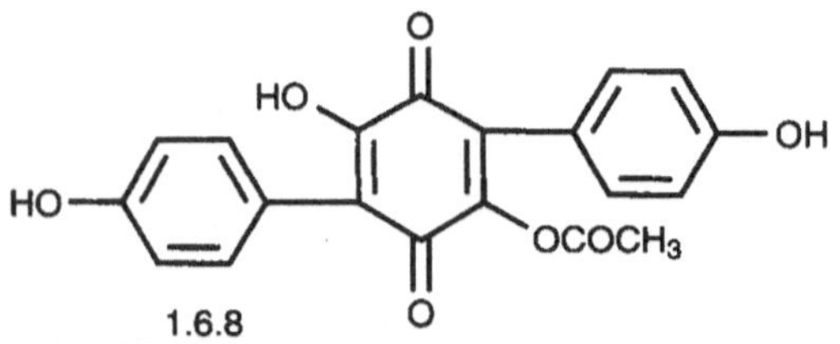

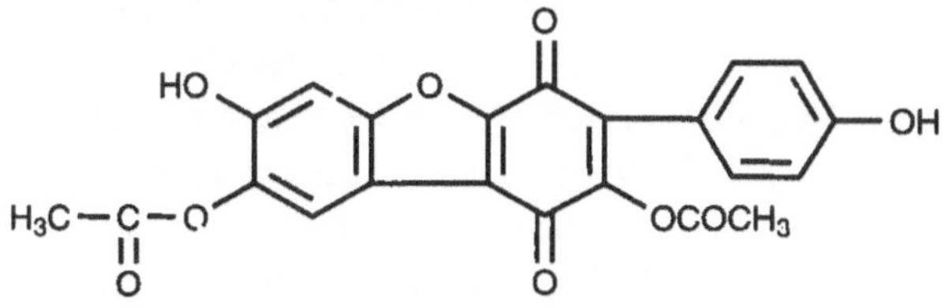

1.6.9 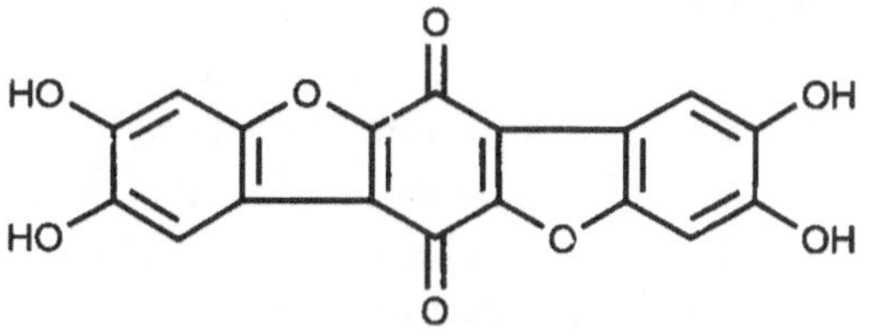   1.6.10

Naphthoquinones

The most important compound among naturally occurring naphthoquinones is phylloquinone, better known as vitamin K1 (1.6.11). It was first isolated from alfalfa and has been subsequently detected in several plants including cabbage, spinach and carrot tops. It plays a vital role in blood coagulation. Plumbagin (1.6.12), isolated from various species of the genus Plumbago, is a much simpler compound. Another well known naphthoquinone pigment of plant origin is lawsone (1.6.13), obtained from the leaves of *Lawsonia alba* (henna). Echinochrome A (1.6.14) is a pigment isolated from the sea urchin (*Arbacia pustulosa*). Several other heavily hydroxylated napthoquinones have also been obtained from the spines of sea urchins (Echinodermata). These include spinochromes B (1.6.15), D (1.6.16) and E (1.6.17).

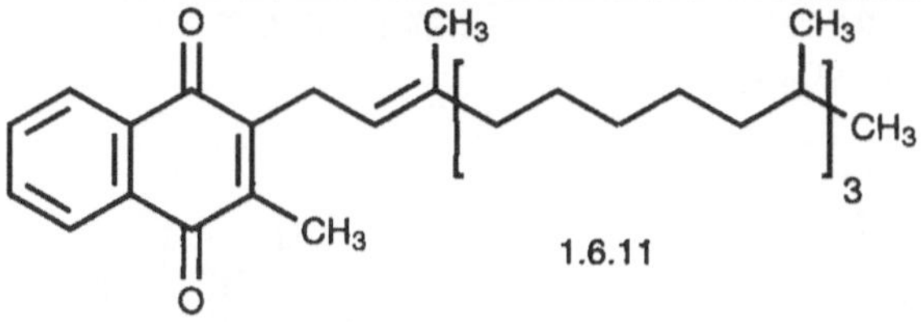

1.6.11 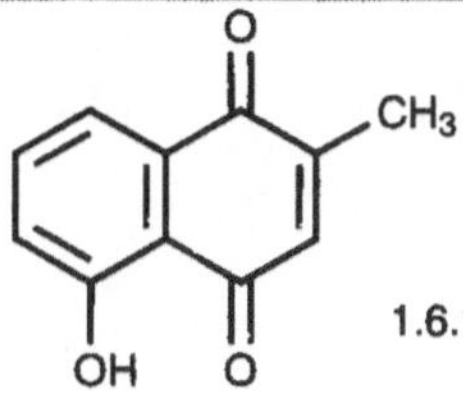   1.6.12

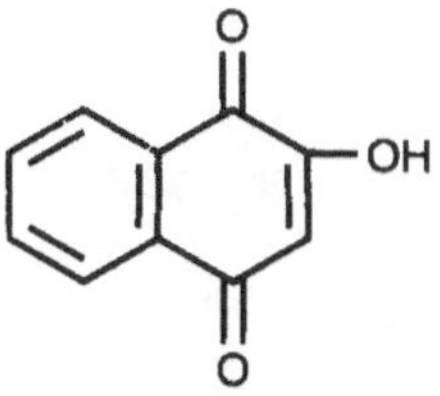

1.6.13 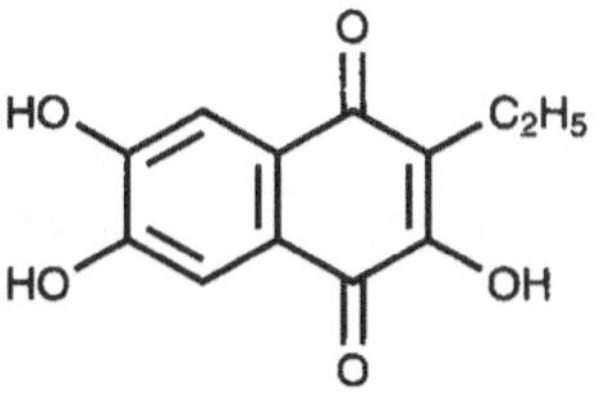   1.6.14 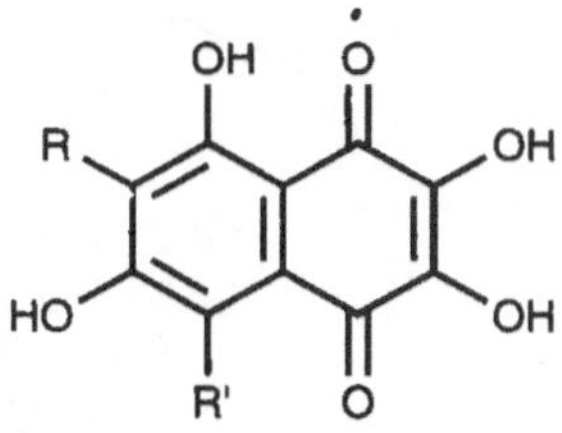   1.6.15: R = R' = H
                               1.6.16: R = H; R' = OH
                               1.6.17: R = R' = OH

Anthraquinones

The most numerous and widely occurring among quinonoids are the anthraquinones. They occur not only in higher plants but also in lichens, fungi and other microorganisms. Among higher plants, those belonging to the families Rubiaceae and Rhamnaceae are particularly rich in anthraquinones. One of the earliest compounds of this group to be isolated was alizarin (1.6.18) which occurs as its primveroside (ruberythric acid; 1.6.19) in the madder root. (*Rubia tinctorum*). along with purpurin (1.6.20). Species of Rubia also yield several other anthraquinones including 2-methylanthraquinone or tectoquinone (1.6.21; also obtained from *Tectona grandis* - teak), 1-hydroxy-2-methylanthraquinone (1.6.22), digiferruginol (1.6.23) and the ethyl ester of 1-hydroxyanthraquinone-2-carboxylic acid (1.6.24). Pachybasin (1.6.25) has been isolated from the rhizomes of *Rheum moorcraftiana* of the family Rhamnaceae. Other examples are emodin (1.6.26), from aloes, chyrsophanol (1.6.27) isolated from species of Rumex, damnacanthal (1.6.28, from species of Damnacanthus and Morinda, morindone (1.6.29) and morindonin (1.6.30) occurring in the heartwood of *Morinda citrifolia*. Endocrocin (1.6.31) and rhodocladonic acid (1.6.32) are two examples of anthraquinones of lichen origin. The pigment of the colchineal insect (*Coccus cacti*) is carminic acid (1.6.33), which is a *C*-glucoside. From *Penicilium islandicum*, islandicin (1.6.34) and catenarin (1.6.35) have been isolated. Skyrin (1.6.36) is a dimeric anthraquinone occurring in several fungi and lichens.

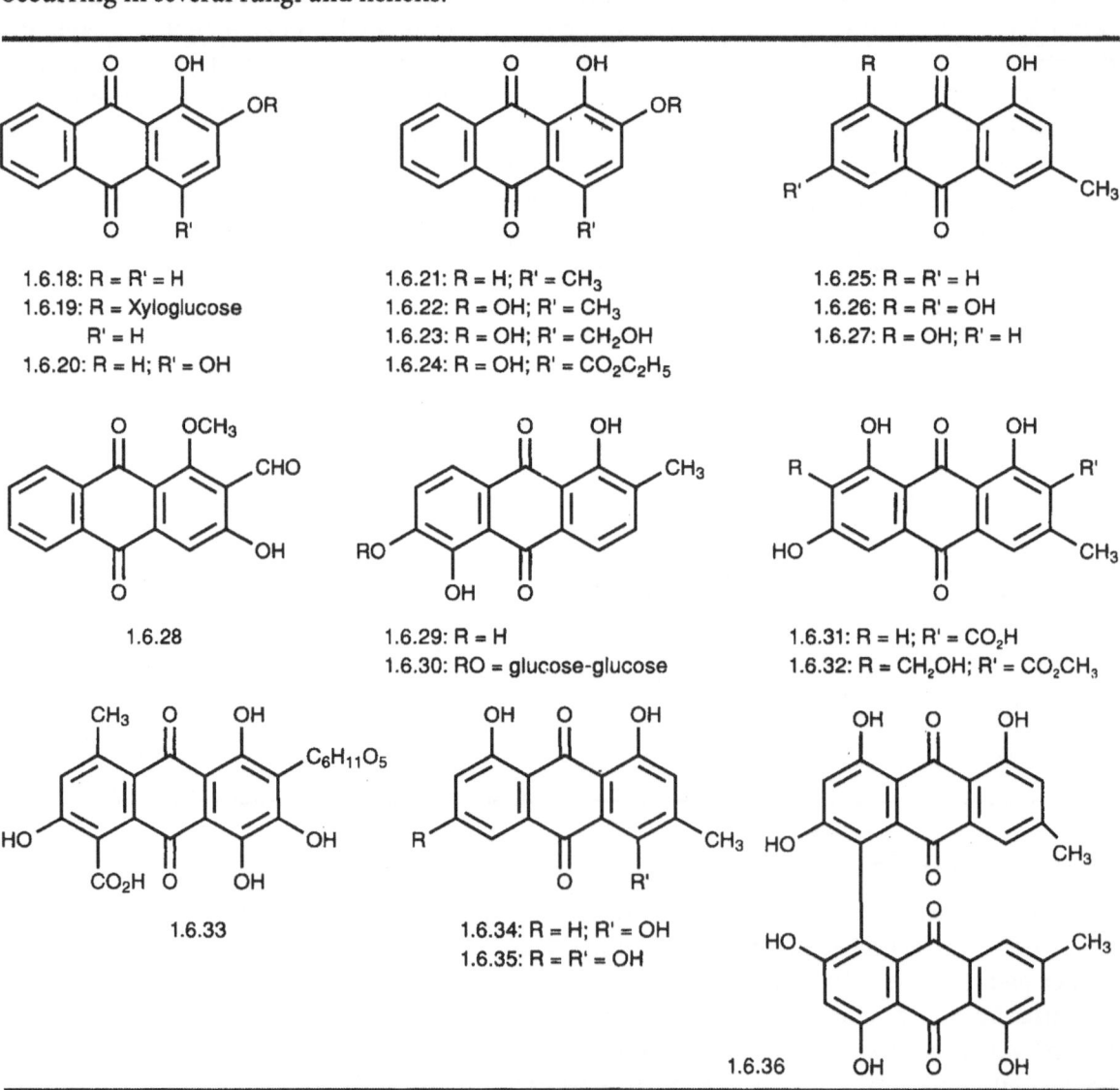

1.6.18: R = R' = H
1.6.19: R = Xyloglucose
         R' = H
1.6.20: R = H; R' = OH

1.6.21: R = H; R' = CH$_3$
1.6.22: R = OH; R' = CH$_3$
1.6.23: R = OH; R' = CH$_2$OH
1.6.24: R = OH; R' = CO$_2$C$_2$H$_5$

1.6.25: R = R' = H
1.6.26: R = R' = OH
1.6.27: R = OH; R' = H

1.6.28

1.6.29: R = H
1.6.30: RO = glucose-glucose

1.6.31: R = H; R' = CO$_2$H
1.6.32: R = CH$_2$OH; R' = CO$_2$CH$_3$

1.6.33

1.6.34: R = H; R' = OH
1.6.35: R = R' = OH

1.6.36

Phenanthrenequinones

Unlike the anthraquinones, only a few quinones with the phenanthrene skeleton have been so far isolated from natural sources. The simplest among them is annoquinone A (1.6.37), which is actually a benzonaphthoquinone. This compound occurs in the stem bark of *Annona montana*. Ochrone B (1.6.38) is closely related; it has been isolated along with ochrone A from *Coclogyne ochracea*, a plant belonging to the Orchidaceae family. Spiranthoquinone (1.6.39), furanospiranthoquinone (1.6.40) and pyranospiranthoquinone (1.6.41) occur together in the roots of *Spiranthes sinensis* which also belongs to Orchidaceae. A true phenanthroquinone is murrayaquinone (1.6.42).

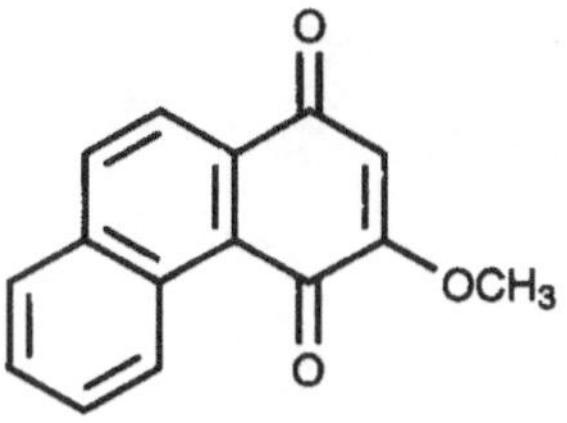

1.6.37

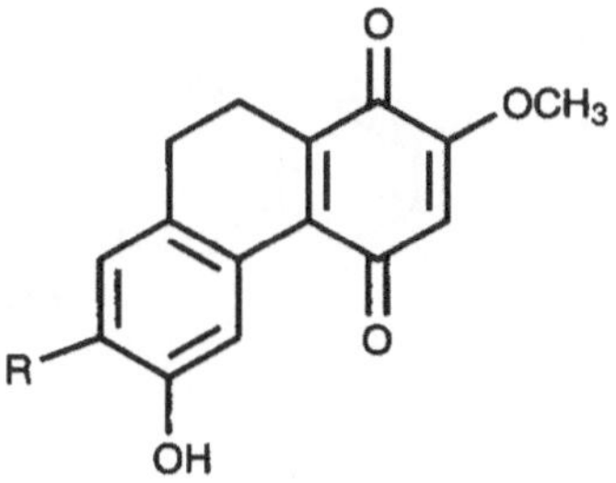

1.6.38: R = H
1.6.39: R = —CH₂-C=C

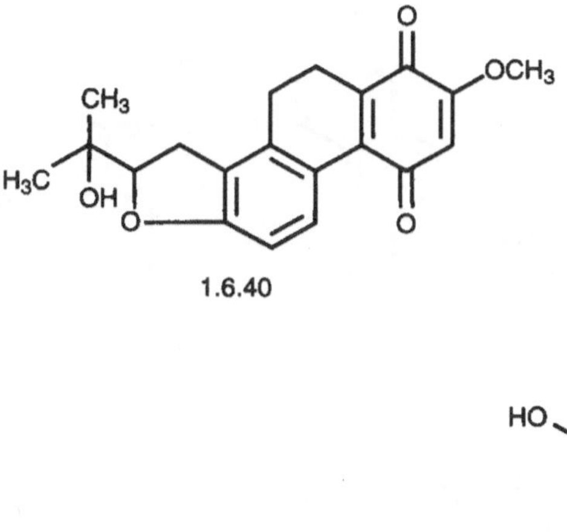

1.6.40

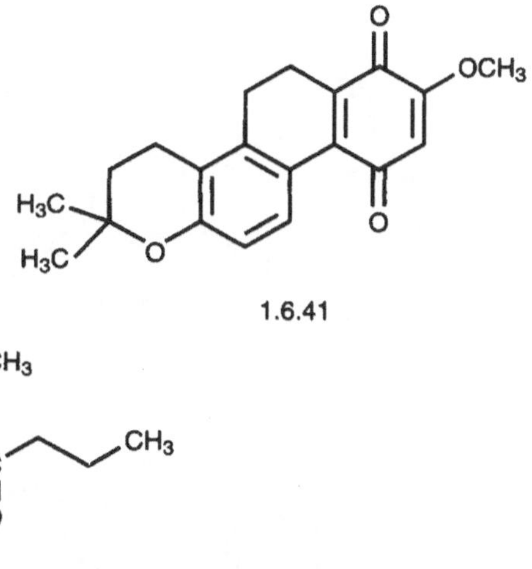

1.6.41

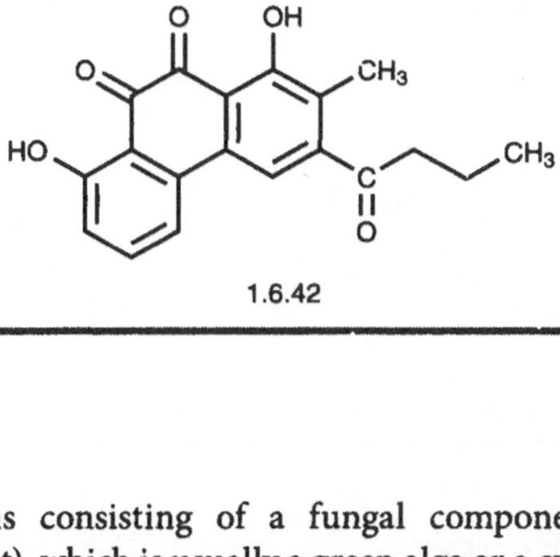

1.6.42

1.7 LICHEN METABOLITES

Lichens are symbiotic organisms consisting of a fungal component (mycobiont) and a photosynthetic partner (photobiont), which is usually a green alga or a cyanobacterium. They are an integral part of every ecosystem and play a vital role in transferring essential nutrients from the atmosphere to their environment. They produce different types of secondary metabolites some of which are unique in the sense that they are not found in higher plants. These compounds are primarily generated by the fungal partners of the lichens but they rarely occur in other fungi, clearly indicating that the symbiotic relationship does play a role in the formation of these secondary metabolites. The majority of these compounds are biosynthesised following the polyketide pathway. Some of these compounds such as usnic acid, didymic acid and a couple of anthraquinones have already been mentioned in the preceding sections. Unique to lichens are the depsides and depsidones. The biosynthetic precursor of these compounds is orsellinic acid (1.7.1). Depsides are aryl esters of (1.7.1). Well known examples are lecanoric acid (1.7.2), which occurs widespread in species of Lecanora, Roccella and Variola, evernic acid (1.7.3), isolated from species of Evernia (for example, oak moss), and atranorin (1.7.4). Depsidones can be considered as diphenyl ethers as well as lactones. Examples are lobaric acid (1.7.5), from Lobaria species, stictic acid (1.7.6) from lichens of the family Strictaceae, cetraric acid (1.7.7) from Cetraria species (for example, *Cetraria islandica* or iceland moss) and salazinic acid (1.7.8).

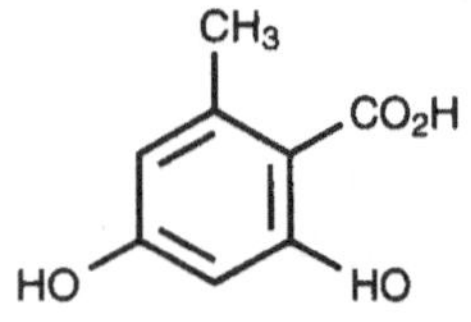

1.7.1

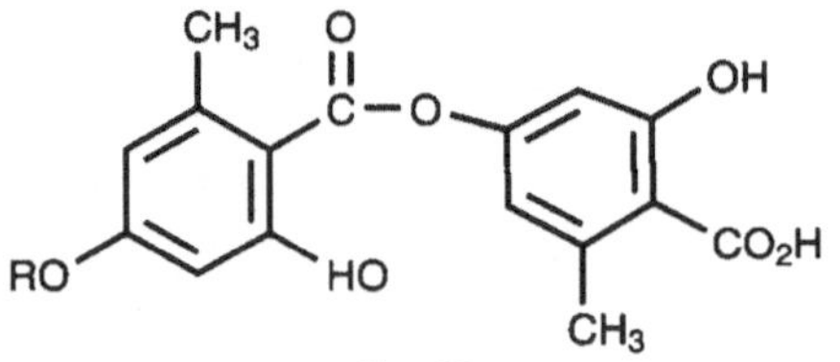

1.7.2: R = H
1.7.2: R = CH₃

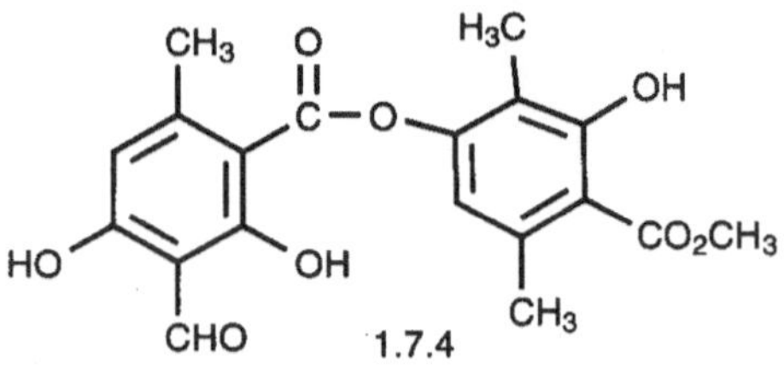

1.7.4

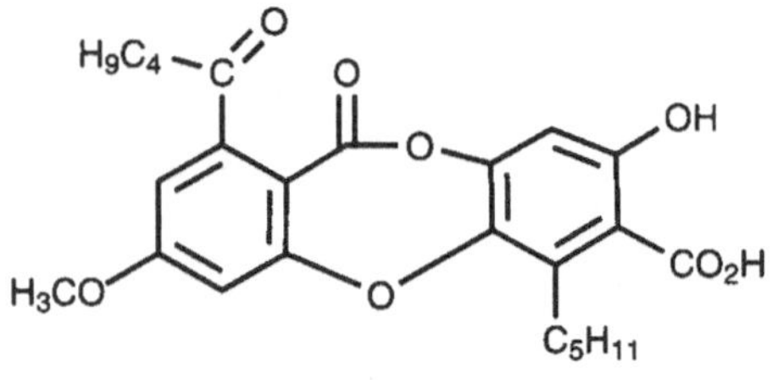

1.7.5

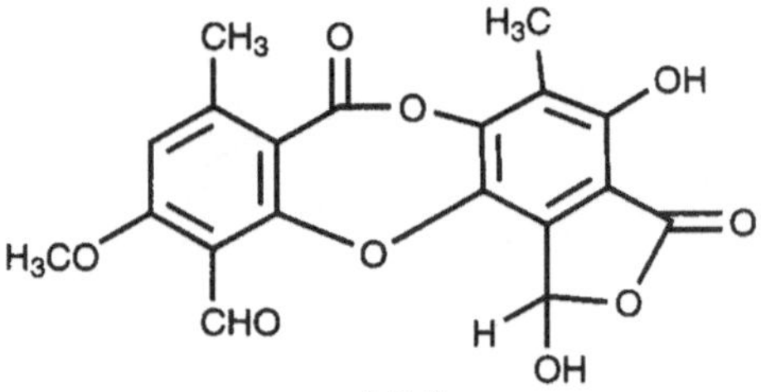

1.7.6

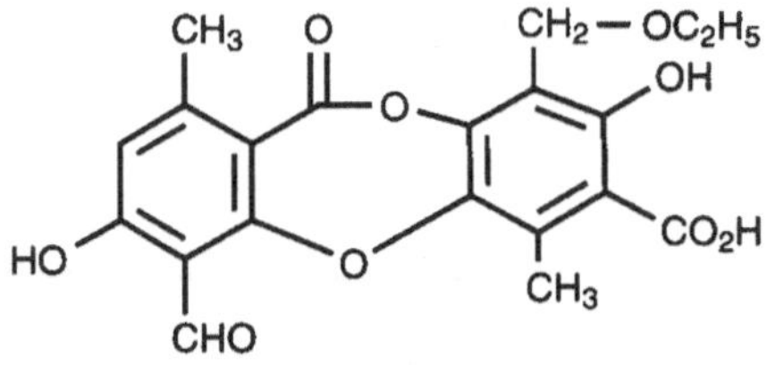

1.7.7

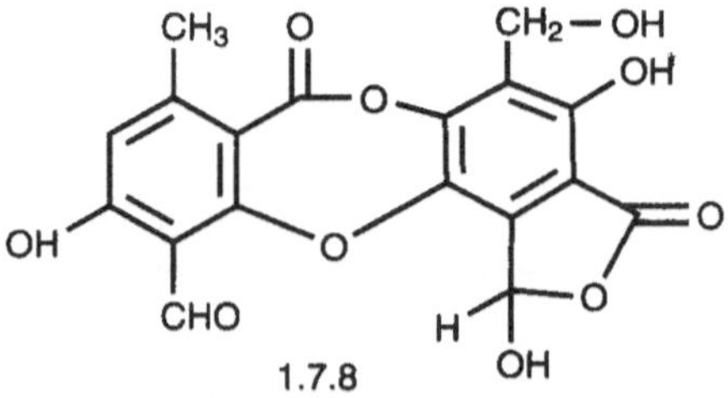

1.7.8

1.8 CONCLUDING REMARKS

In the foregoing sections, some of the major types of secondary metabolites have been briefly discussed with the help of illustrative examples. A few other secondary metabolites, particularly those derived from fatty acids, polyacetylenes and heterocyclic compounds derived from them, and some nitrogenous compounds, other than the alkaloids have been kept out of the purview of this book. Compounds isolated from marine organisms have also not been discussed in detail in spite of the fact that this branch of natural products chemistry is gaining importance. The reason for this omission is two fold:

1. They form a large body of compounds of various types requiring much space for discussion,
2. There are authoritative books on the subject available.

However, some individual compounds belonging to these categories are considered elsewhere in this book in other chapters.

Suggested reading

1. Finar IL, *Organic Chemistry*, Vol. 2, London: Longmans Green, 1975. (A comprehensive and popular textbook).
2. Nakanishi K, Goto T, Ito S, Natori S and Nozoe S, (eds), *Natural Products Chemistry*, Vols 1–3, Tokyo: Kodansha and Mill Valley, California: University Science Books, 1983.
3. Barton DHR, Nakanishi Kand and Meth-Cohn O, (eds), *Comprehensive Natural Products Chemistry*, Vols 1–9, Oxford (UK): Elsevier Science Limited, 1999.
4. Bhat SV, Nagasampagi BA and Sivakumar M, *Chemistry of Natural Products*, New Delhi: Narosa Publishing House, 2004.
5. Kaufman PB, Cseke LJ, Warber S, Duke JA and Brielmann HL, *Natural Products from Plants*, Boca Raton: CRC Press, 1999.

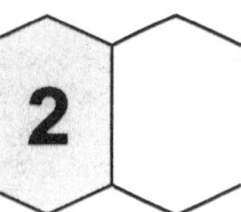

Chapter

2

Structure

2.1 Introduction: A Survey of the Methods used for Determination of Structures

Classical Methods

The determination of the structure of a naturally occurring organic compound is an important milestone in its chemical history. Over the years, various methods and techniques, with procedural variations, have been developed for this purpose. Some of these are of a general nature and some others are specifically designed for a particular type of compound. In the early days of classical organic chemistry, a well-defined sequence of steps was used in structural studies. After determining the molecular formula of a pure compound, it was customary to subject it to various analytical and degradation procedures. Detection of functional groups is the primary step. The presence of basic or acidic functionalities is readily revealed by solubility tests. For example, the presence of a carboxylic acid group is shown by the compound's solubility in aqueous sodium bicarbonate. This could be confirmed by esterification; ethereal diazomethane is a convenient reagent for the preparation of methyl esters. On the other hand, a basic compound, such as an amine, dissolves in aqueous hydrochloric acid. For the detection of other functional groups like the hydroxyl and carbonyl (aldehydic or ketonic), derivatisation is used (acetylation for the hydroxyl and formation of an oxime or 2, 4-dinitrophenyl hydrazone derivative for the carbonyl group).

Quantitative methods are also available to get further information about a compound. Thus, the Zerewittinoff method of active hydrogen estimation, using methyl magnesium bromide and estimating the amount of methane formed, is used to find out the number of hydroxyl groups in an alcoholic compound. Zeisel's method is used for determining the number of methoxyl groups. In this case, the compound under analysis is heated with hydriodic acid and the methyl iodide generated is converted into silver iodide. The latter can be estimated either gravimetrically or volumetrically using Volhard's method. A related method for the estimation of N-methyl groups (for example, in alkaloids), is the Herzig–Meyer method. The degree of unsaturation in a compound (easily detected by decolourisation of bromine water) is found out either by quantitative catalytic hydrogenation or by titration with a peracid. C-Methyl groups, which are of common occurrence, particularly, in terpenes, are estimated by the Kuhn–Roth method. This method, first developed during studies on carotenoids, involves oxidation with chromic acid followed by the estimation of the amount of acetic acid formed. However, these methods are rarely used these days as the same information can be obtained much more quickly from spectroscopic, particularly, nuclear magnetic resonance, data. This aspect will be delineated later on in this section.

For breaking down a compound into smaller, easily identifiable molecules, different methods of oxidation are used. In many of these reactions, the focus of attack is a double bond. In studies on terpenes, ozonolysis is an oft-employed method. For example, the monoterpenes, linalool (2.1.1), geraniol (2.1.2) and pulegone (2.1.3), the sesquiterpenes, farnesol (2.1.4) and the bisabolenes (2.1.5), (2.1.6) and (2.1.7), the triterpene squalene (2.1.8) and the carotenoid, lycopene (2.1.9), all yield acetone among other products when subjected to ozonolysis. This fixes the position of one double bond. Squalene and lycopene yield two moles of acetone, as expected from their structures. β-bisabolene (2.1.6) also gives formaldehyde or formic acid (depending on the mode of work-up) as a product, thus showing the position of another double bond as a terminal group. Ozonolysis is usually followed by either oxidative or reductive work-up. For example, oxidative work-up after ozonolysis of geraniol gives, besides acetone, one mole each of laevulinic acid (2.2.20) and glycolic acid (2.1.11). Thus, all the ten carbon atoms of geraniol could be accounted for and the positions of the two double bonds get fixed. As we shall see later, the same result could be reached with a few milligrams of the compound and without destroying it from its nuclear magnetic resonance spectrum. Ozonolysis has also been usefully employed in studies on the sesquiterpenes longifolene (2.1.12) and caryophyllene (2.1.13). The former gives a ketone (2.1.14) and the latter yields a keto carboxylic acid (2.1.15); in each case one carbon atom is lost as formaldehyde or formic acid.

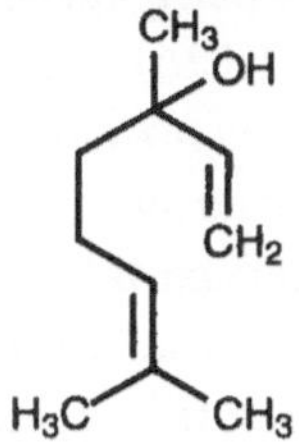

2.1.1 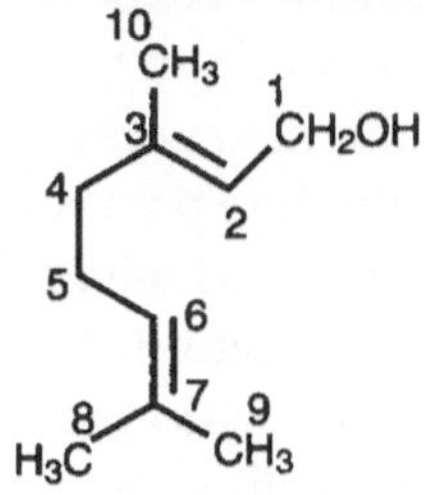   2.1.2 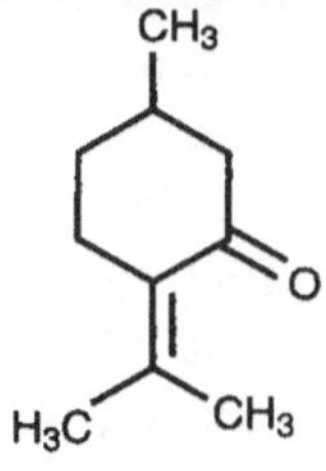   2.1.3 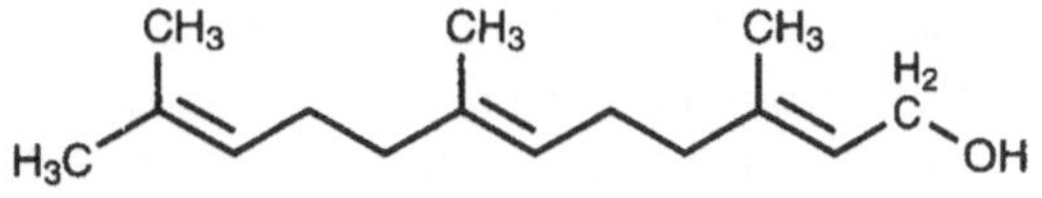   2.1.4

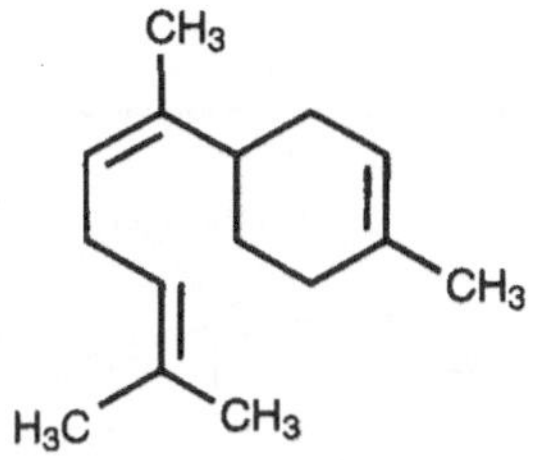

2.1.5 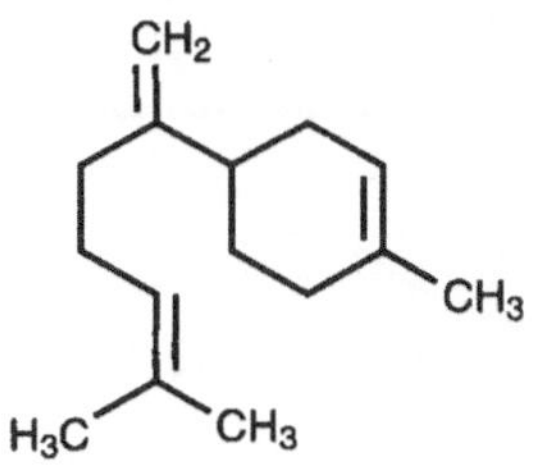   2.1.6 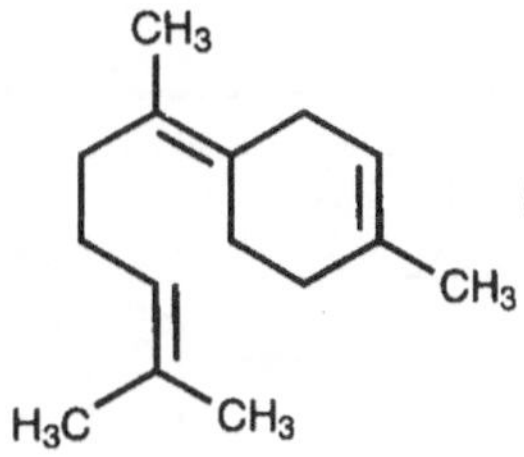   2.1.7

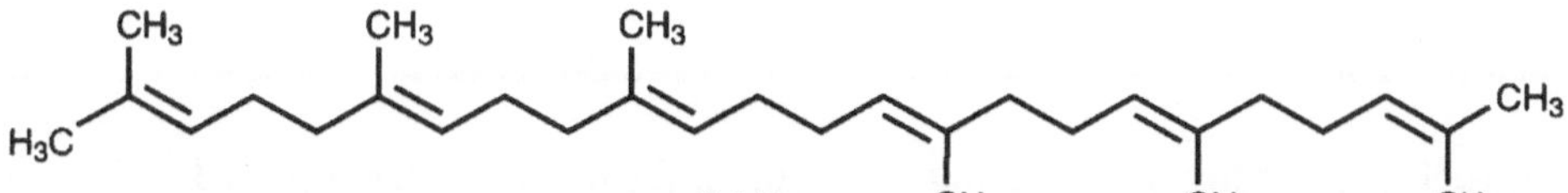

2.1.8

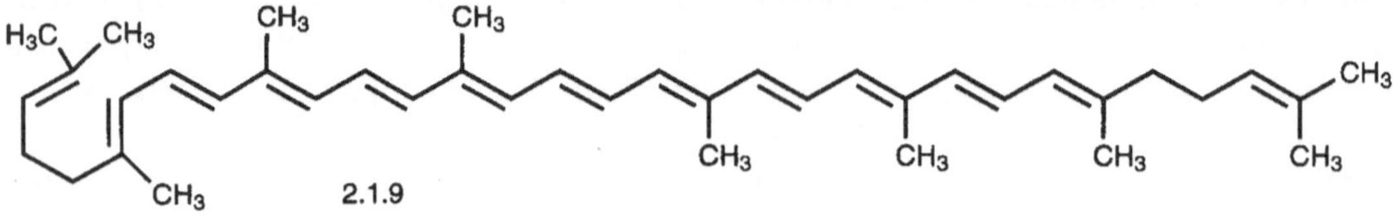

2.1.9

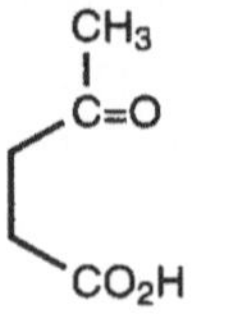

2.1.10 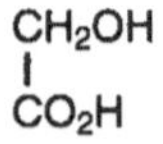   2.1.11 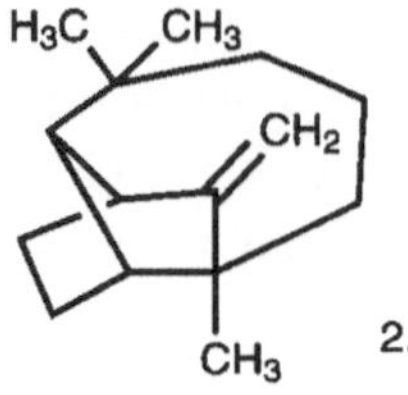   2.1.12 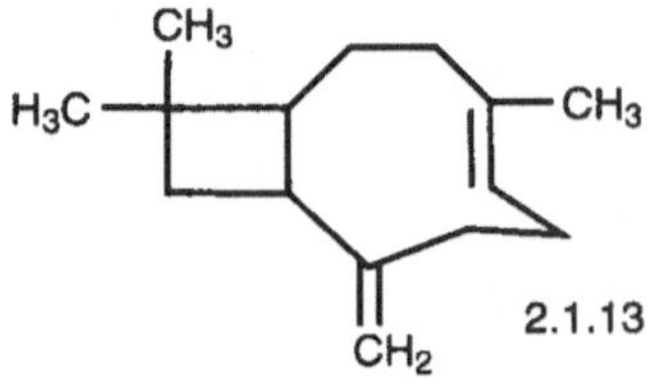   2.1.13

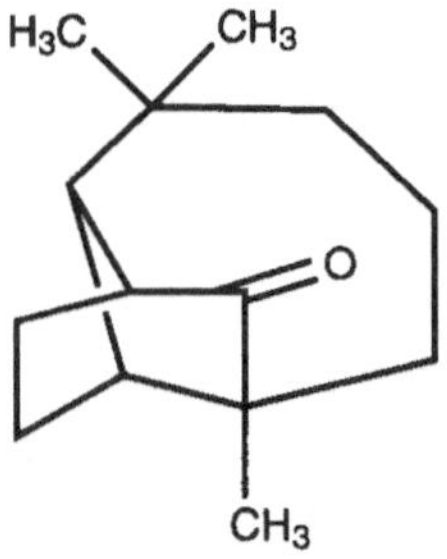

2.1.14 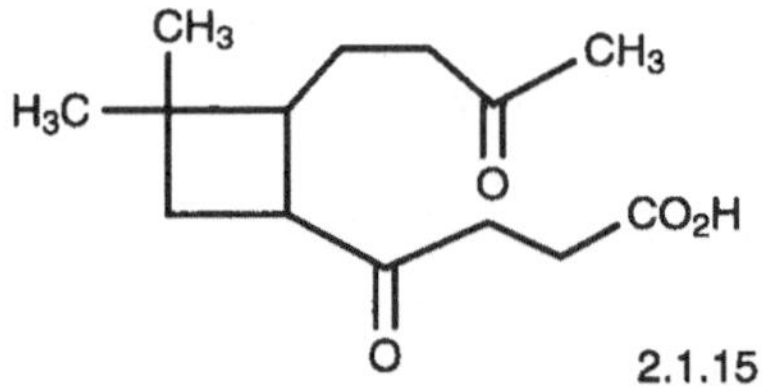   2.1.15

Potassium permanganate—either in neutral or alkaline medium—is another oxidizing agent widely used in structural studies. Baeyer used this reagent in his work on α-pinene (2.1.16). This reaction carried out more than a hundred years ago has considerable instructive value even now and hence merits description here. On treatment with cold, dilute alkaline potassium permanganate solution, 2.1.16 yielded the corresponding diol (pinene glycol, 2.1.17), which on further oxidation with warm potassium permanganate gave pinonic acid (2.1.18). The ketonic function in the latter could be selectively oxidised by sodium hypobromite (haloform reaction) to obtain pinic acid (2.1.19). Pinic acid was treated with bromine followed by barium hydroxide and then oxidised by chromic acid when *cis*-norpinic acid (2.1.20) was obtained. The latter was characterised as *cis*-2, 2-dimethyl-1,3,-dicarboxylic acid by synthesis.

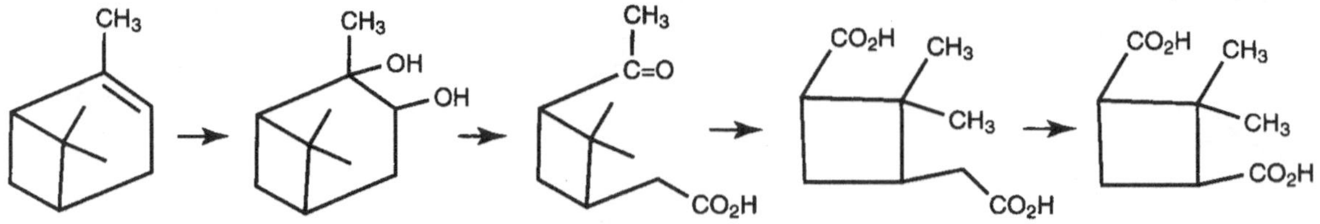

2.1.16   2.1.17   2.1.18   2.1.19   2.1.20

Alkaline fusion (heating with strong aqueous sodium or potassium hydroxide) is another method used in structural studies, particularly on flavonoids and related compounds. For example, under these conditions, cyanidin chloride (2.1.21) breaks down to phloroglucinol (2.1.22) and 3,4-dihydroxybenzoic acid (protocatechuic acid) (2.1.23). The flavonol, quercetin (2.1.24) also yields the same products under similar conditions, thus establishing the close relationship between the two compounds. In this reaction two carbon atoms are lost as the 'C' ring in anthocyanidins and flavonols is vulnerable under these conditions.

However, flavones which lack the 3-hydroxyl group, can be degraded to two fragments without loss of any carbon atom. Thus, chrysin (2.1.25), under milder conditions of alkali fusion, yields 2,4,6-trihydroxyacetophenone (2.1.26) and benzoic acid, presumably via the diketone (2.1.27). Better yields of the degradation products are obtained when methyl ethers of the flavones are used. Thus, when chrysin dimethyl ether (2.1.28) is refluxed with alcoholic potassium hydroxide, 2-hydroxy-4,6-dimethoxyacetophenone (2.1.29) is obtained along with benzoic acid.

2.1.25: R = H
2.1.26: R = CH$_3$

2.1.26: R = H
2.1.29: R = CH$_3$

2.1.27

Isoflavones, on alkaline degradation give the corresponding desoxybenzoins, with the loss of a carbon atom. In this case too, it is better to deal with the complete methyl ethers of the isoflavones under study. For example, 5,7,2',4'-tetramethoxyisoflavone (2.1.30), obtained by the dehydrogenation (using iodine and potassium acetate in acetic acid) of tetra-O-methyldalbergioidin (2.1.31), on boiling with 10% alcoholic potassium hydroxide gives 2-hydroxy-4,6-dimethoxyphenyl-2,4-dimethoxybenzyl ketone (2.1.32). The latter can be synthesised by a Hoesch reaction between phloroglucinol and 2,4-dimethoxyphenylacetonitrile followed by partial methylation. Incidentally, O-methylation is an important preliminary step in structural studies on flavonoids. Dimethyl sulphate in acetone in the presence of anhydrous potassium carbonate is the reagent of choice for this purpose. The desired degree of O-methylation can be achieved by controlling the molar proportion of the methylating agent and the duration of the reaction. Ethereal diazomethane can also be used for O-methylation of hydroxyl groups which are not involved in intramolecular hydrogen bonding.

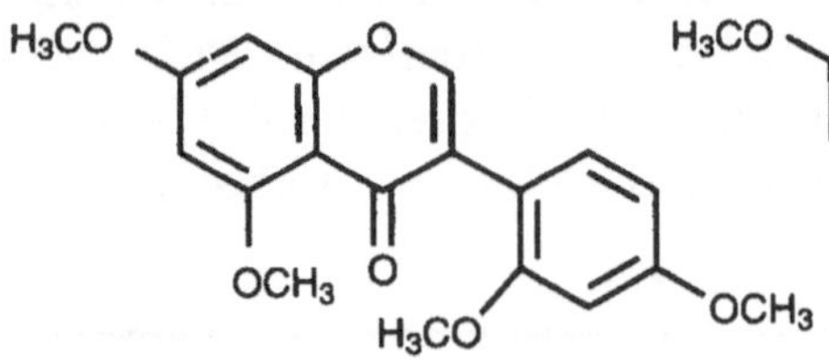

2.1.30 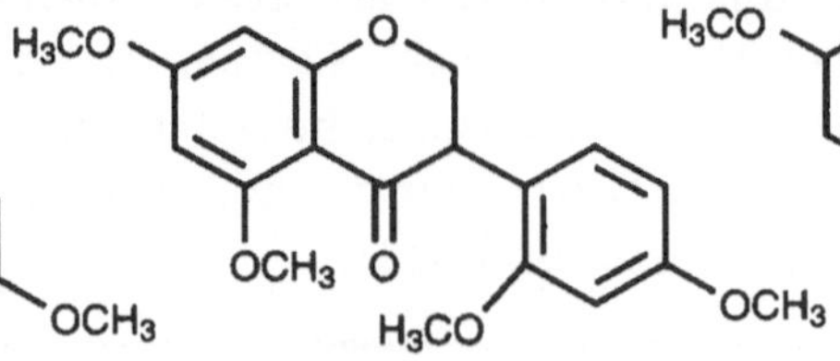   2.1.31 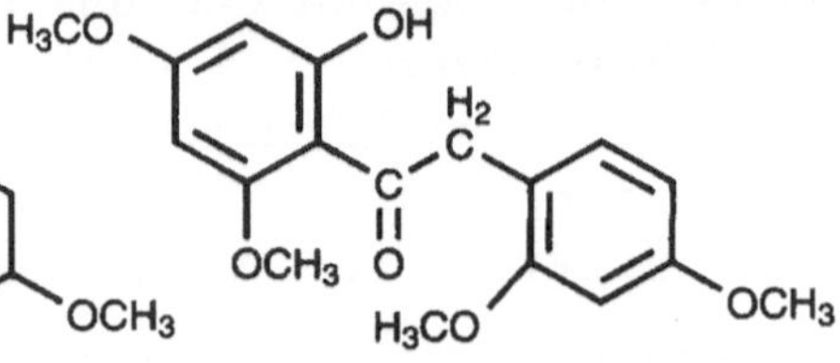   2.1.32

More drastic methods of degradation were sometimes resorted to in earlier days, when simpler methods failed to give the desired results. These were subjecting the compound to dry distillation, alone or in the presence of soda lime, and heating with zinc powder or sulphur or selenium. These reactions are rarely used these days but can give useful information in certain specific cases. For example, strong heating of the bicyclic sesquiterpene, cadinene (2.1.33), with sulphur brings about dehydrogenation to yield the naphthalene derivative, cadalene, which is 1,6-dimethyl-4-isopropylnaphthalene (2.1.34). This reaction, introduced by Ruzicka, helps to establish not only the skeletal structure but also the positions of the alkyl groups. In this reaction sulphur can be replaced by selenium. When the sesquiterpene, guaiol (2.1.35) is subjected to this reaction, the product, gauiazulene (2.1.36) can easily be recognised as an azulene from its blue colour. It is 4,10-dimethyl-7-isopropylazulene. This reaction has also been effectively used in studies on higher terpenes. For instance, the triterpene, lanosterol (2.1.37), on heating with selenium, gives 1,2,8-trimethylphenanthrene (2.1.38). and friedelin (2.1.39) yields, among other products, the pentacyclic aromatic hydrocarbon, 1,8-dimethylpicene (2.1.40). In the last two compounds angular methyl groups are eliminated.

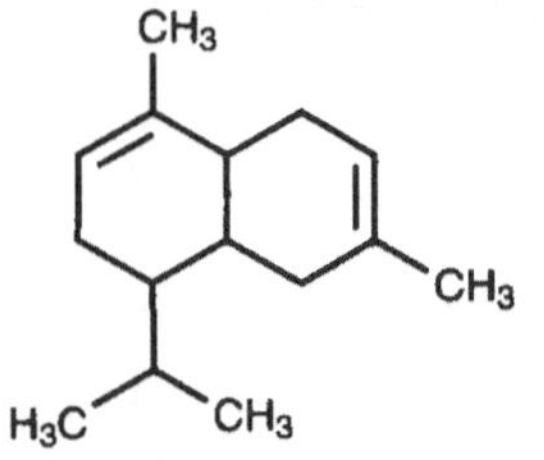

2.1.33 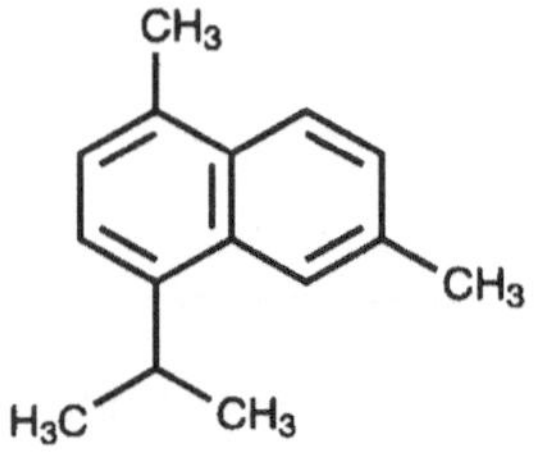   2.1.34 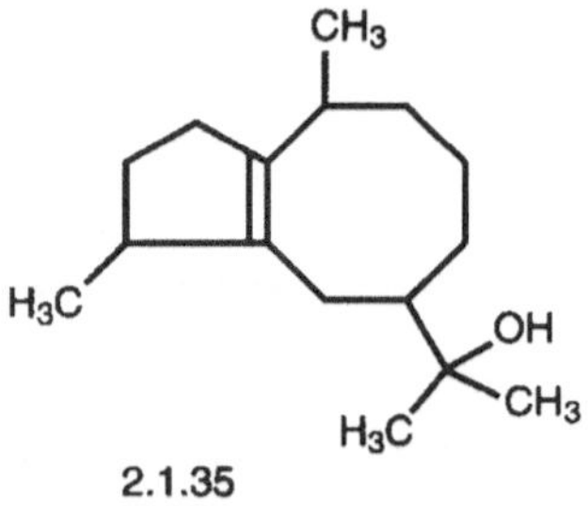   2.1.35

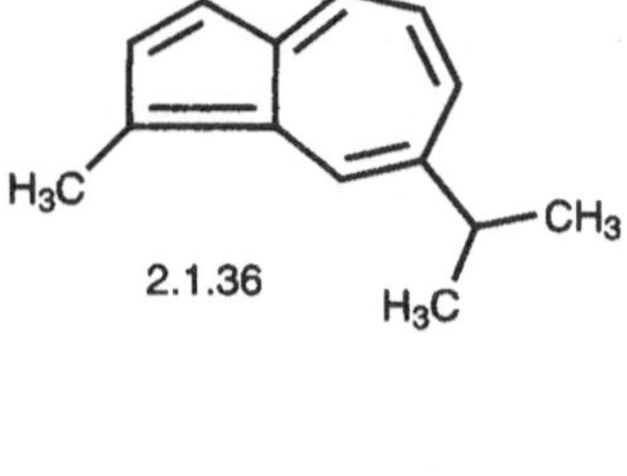

2.1.36 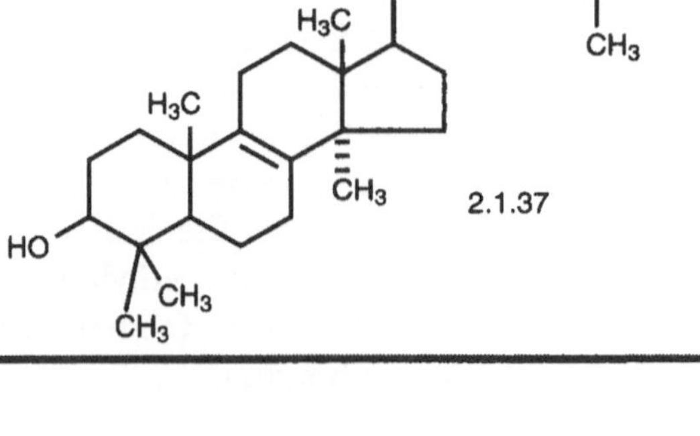   2.1.37 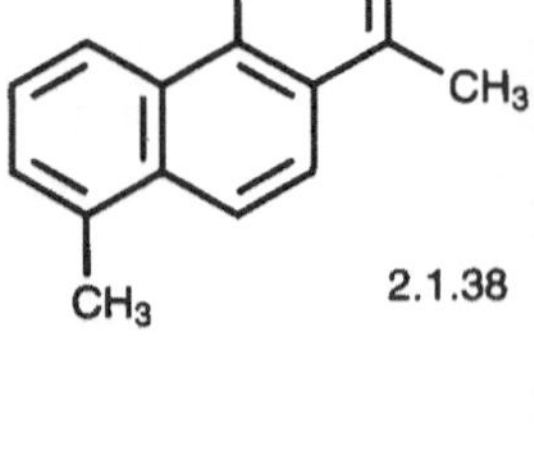   2.1.38

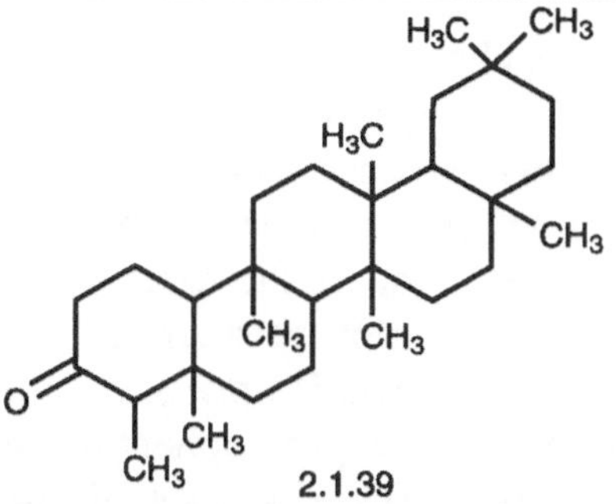

2.1.39 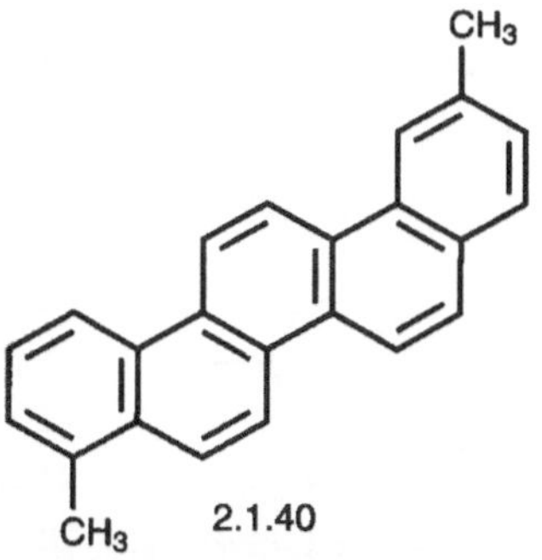   2.1.40

An early indication that morphine (2.1.41) possesses a phenanthrene skeleton was obtained when the alkaloid was distilled with zinc dust; phenanthrene was one of the products. In this respect, the reaction is comparable to the dehydrogenation brought about by sulphur or selenium. The alkaloid coniine (2.1.42) gives 2-propylpyridine (conyrine; 2.1.43) thus unequivocally establishing the skeletal structure of this hemlock alkaloid.

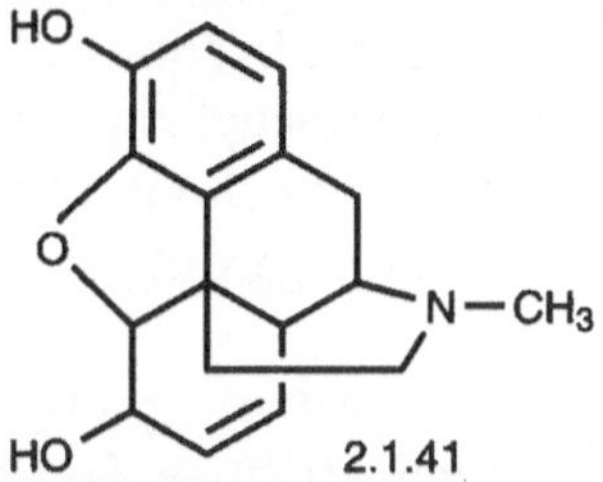

2.1.41 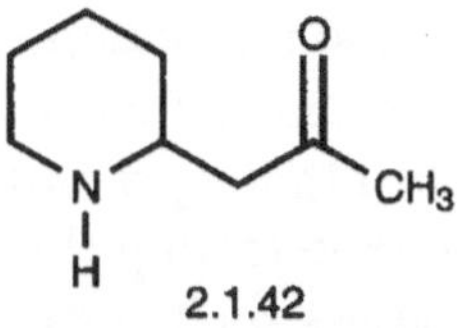   2.1.42 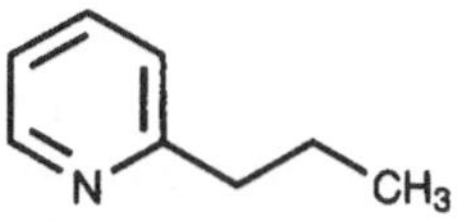   2.1.43

Spectroscopic Methods

Spectroscopic methods have almost completely displaced the above mentioned analytical and degradation techniques. Two chief advantages of spectroscopy are:

1. They, with the exception of mass spectroscopy, are non-destructive and after recording the spectra, the compounds can be recovered.
2. The quantities required for recording a spectrum are a few milligrams and a single spectrum, particularly NMR, can give a plethora of information which can otherwise be obtained only by subjecting the compound to a series of analytical procedures and degradation reactions.

Ultraviolet–visible spectroscopy

Ultraviolet–visible spectra are particularly useful in the study of flavonoids. As mentioned in Chapter 1, flavones and flavonols exhibit two absorption maxima in their ultraviolet spectra and these are designated as Band I and Band II. Substitutions in ring A (condensed benzene ring) affect the position of Band II, whereas substituents on ring B (side phenyl group) influence Band I. Thus, chrysin (2.1.25), absorbs at 268 and 313 nm, apigenin (2.1.44) at 267 and 336 nm and luteolin at 267 and 349 nm. It is obvious, therefore, that increasing hydroxylation in the side phenyl progressively shifts the position of Band I to longer wavelengths, with Band II remaining

2.1.44

2.1.45: R = H
2.1.46: R = OH

2.1.47: R = H
2.1.48: R = OH

unaffected. In flavonols, the extra hydroxyl at position 3 is part of the chromophore responsible for Band I which, therefore, occurs at a longer wavelength, compared to the corresponding flavone. Thus, galangin (2.1.45), absorbs at 267 and 359 nm, whereas kaempferol (2.1.46) absorbs at 266 and 367 nm. In the case of flavanones, wherein there is no conjugation between the side phenyl and the pyrone carbonyl, there is only one absorption maximum corresponding to the propiophenone chromophoric unit present in them. There is a set of rules for predicting the absorption maximum of phenyl alkyl ketones which can be used for predicting the absorption maximum of a flavanone. The base value for a phenyl alkyl ketone (acetophenone or propiophenone) is taken as 246 nm. For *ortho*- and *meta*-hydroxy/methoxy substituents, 7 nm are added and when these groups are in the *para* position, 25 nm are added to the base value. Therefore, for pinocembrin (2.1.47) and naringenin (2.1.48), the calculated UV absorption maximum in methanol is 285 nm, which is in good agreement with the observed value of 289 nm; the extra hydroxyl group on the side phenyl ring in naringenin does not make any contribution. Addition of sodium methoxide brings about a bathochromic shift of 35 nm. The use of other shift reagents for eliciting information regarding the oxygenation pattern in flavonoids has been described in the preceding chapter.

Ultraviolet spectra are also useful in detecting the presence of a carbonyl group, which may not be reactive enough to form derivatives such as oxime or phenylhydrazone. For example, the carbonyl group in the ketone (2.1.14) obtained by the ozonolysis of longifolene is not chemically very reactive as it is in a sterically hindered position. However, the UV spectrum of the compound has an absorption maximum of low intensity (due to n-pi* excitation) at 290 nm which is characteristic of a carbonyl group. Ultraviolet spectra also yield useful information regarding the number and degree of substitution of double bonds in terpenes and related compounds. For example, the UV spectra of β-amyrin (2.1.49) and oleanolic acid (2.1.50) (absorption maximum at about 190 nm) strongly suggest the presence of a tri-substituted double bond in each of these compounds. The Woodward–Fieser rules for polyenes and enones were derived largely from analysis of the UV spectra of terpenes and steroids. For example, the monoterpene ketone, carvone (2.1.51) shows an absorption maximum at 237 nm in perfect agreement with the calculated value (215 base value) + 10 (α substituent) + 12 (β substituent) = 237 nm. Carotenoids generally show three absorption maxima in their visible spectra.

An equation has been evolved for predicting the wavelength of the main absorption maximum (most intense) in their spectra recorded in hexane. The equation is:

$$\text{Absorption maximum} = 114 + 5M + n(48.0 - 1.7\,n) - 16.5R(\text{endo}) - 10R(\text{exo}) \text{ nm,}$$

where n is the number of conjugated double bonds, M is the number of methyl (or alkyl) substituents on the conjugated system, $R(\text{endo})$ is the number of rings with endocyclic double bonds and $R(\text{exo})$ is the number of rings with exocyclic double bonds.

In lycopene (2.1.9), the pigment of red tomatoes, which is an acyclic compound, $n = 11$, $M = 8$ and there are no rings. Therefore, the calculated value is $114 + 40 + 11 (48.0 - 18.7)$nm which is equal to 476.3 nm; the observed maximum is 474 nm. For β-carotene (2.1.52), the calculated maximum is 453.3 nm and the observed value is 452 nm; in this case $n = 11$, $M = 10$ and R(endo) is 2.

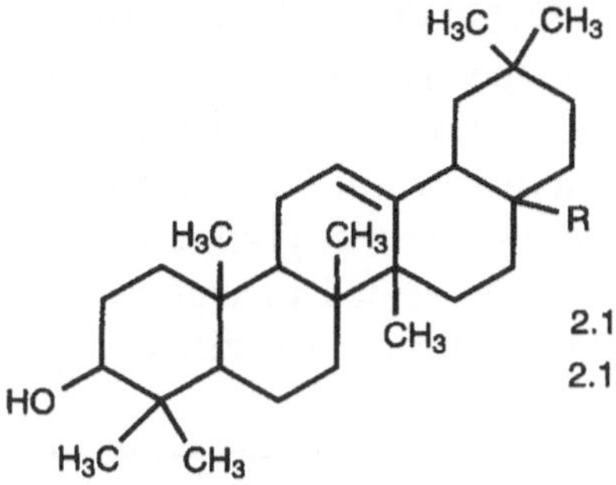

2.1.49: R = CH$_3$
2.1.50: R = CO$_2$H

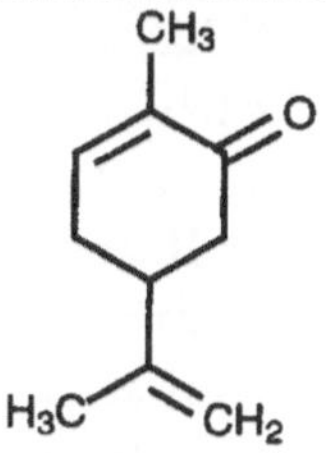

2.1.51

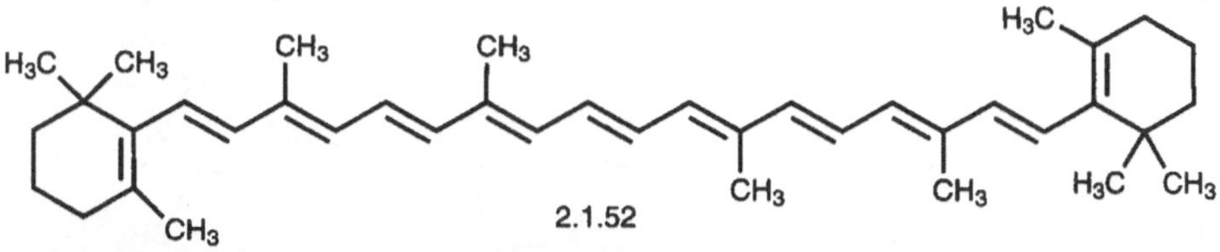

2.1.52

Infrared spectroscopy

Infrared spectra are more useful than ultraviolet spectra as they not only reveal the presence (or absence) of functional groups but also provide valuable information with regard to their environment. The presence of hydroxyl, amino (in alkaloids), and different types of carbonyl groups (aldehydes, ketones, esters and carboxylic acids) is unequivocally shown by the absorptions in the appropriate regions of the IR spectra. For example, the somewhat broad but intense absorption band centred at 3350 cm^{-1} shows the presence of a hydroxyl group in menthol (2.1.53) (spectrum recorded in KBr). In this case, the sharp absorptions at 3000, 2920 and 2860 cm^{-1} further show that the compound has CH$_2$ and CH$_3$ groups as well. The well-defined finger print region (850–1400 cm^{-1}) suggests that the compound is cyclic rather than acyclic; the spectra of acyclic compounds have broad, ill-resolved absorption bands in this region due to conformational mobility. These data, together with the molecular formula of the compound and the fact that it is a monoterpene are sufficient to arrive at the structure shown. However, conclusive evidence is obtained only from NMR spectra (to be described later). The most important application of IR spectra is the critical analysis of the carbonyl stretching frequencies, as carbonyl groups of different types and in different structural environments occur in several types of secondary metabolites. The stretching frequency of an isolated, non-conjugated, ketonic carbonyl group, as found in aliphatic ketones and cyclohexanone, is around 1715 cm^{-1}. In an aldehyde, the frequency is approximately 10 cm^{-1} higher. When the carbonyl function is part of a five-membered ring (cyclopentanone), the stretching frequency increases to 1745 cm^{-1}. On the other hand, conjugation as in α-, β-unsaturated ketones and aromatic ketones (acetophenone, for example), lowers the frequency to around 1680 cm^{-1}. Normal ester carbonyl groups, as well as six-membered lactones, absorb at 1735 cm^{-1}. In five-membered lactones, due to steric strain, the frequency increases to about 1770 cm^{-1} or higher. The presence of a carboxylic acid group is easily recognised from the broad O–H stretching frequency band between 2500 and 3300 cm^{-1}, and the sharper carbonyl stretching frequency around 1710 cm^{-1}. These generalisations can be used effectively in the analysis of the IR spectra of different types of natural products. Thus, the IR spectrum of menthone (2.1.54) has a strong absorption band at 1715–1720 cm^{-1} due to the carbonyl group in a six-membered carbocyclic ring. In contrast, in the spectrum of camphor (2.1.55), wherein the carbonyl group can be considered as a part of a five-membered ring, the absorption band is seen at 1740 cm^{-1}. In the IR spectrum of the sesquiterpene, β-vetivone (2.1.56), the carbonyl stretching frequency is 1690 cm^{-1}. Incidentally, this compound was earlier given a wrong structure based on the observation that on selenium dehydrogenation it yielded an azulene derivative. This fact emphasises that drastic chemical reactions can sometimes give misleading results, whereas spectral data are dependable.

| 2.1.53 | 2.1.54 | 2.1.55 | 2.1.56 |

In the IR spectra of flavones, flavanones and the isoflavonoids, the position of the carbonyl stretching frequency is influenced by conjugation with the condensed benzene ring, conjugation with the side phenyl ring through the intervening double bond (in flavones) and intramolecular hydrogen bonding with neighbouring hydroxyl groups as in 5-hydroxy flavones and flavonols. In complete methyl ethers, as for example, 5,7,4'-trimethoxyflavanone (2.1.57), the carbonyl group absorbs at 1700 cm^{-1}, like the corresponding acetophenone derivative. In the spectrum of silybin (2.1.58), on the other hand, this absorption band is seen at 1640 cm^{-1} due to intramolecular hydrogen bonding with the neighbouring hydroxyl groups.

| 2.1.57 | 2.1.58 |

The carbonyl group in the isoflavone, auriculatin (2.1.59), absorbs at 1650 cm^{-1}, whereas in the isoflavanone, rotenone (2.1.60), this absorption occurs at 1680 cm^{-1}. The lactone carbonyl group in coumarins, as for example, coumurrayin (2.1.61) and auraptenol (2.1.62), gives rise to an absorption band at 1710 cm^{-1}. In the spectrum of aflatoxin (2.1.63), this functionality absorbs at 1684 cm^{-1}. In podophyllotoxin (2.1.64), which has a five-membered lactone, the carbonyl frequency is higher and is seen at 1785 cm^{-1}. However, if the γ-lactone is α-, β-unsaturated as in several sesquiterpene lactones which occur frequently in plants of the Compositae family, the carbonyl stretching frequency is lower. For example, in the IR spectrum of heliangine (2.1.65), this frequency is seen at 1750 cm^{-1}.

2.1.59

2.1.60

2.1.61

2.1.62

2.1.63

2.1.64

2.1.65

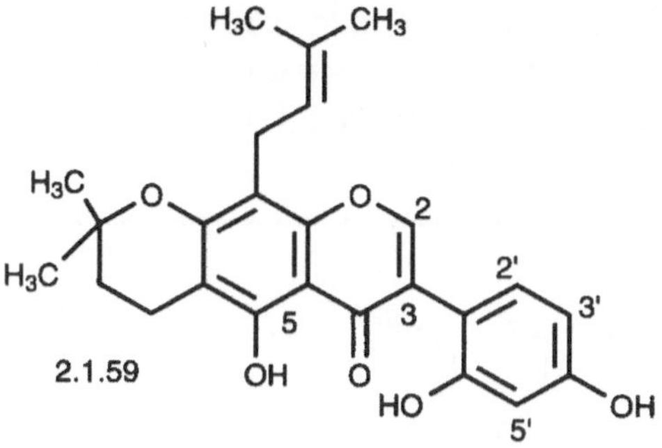

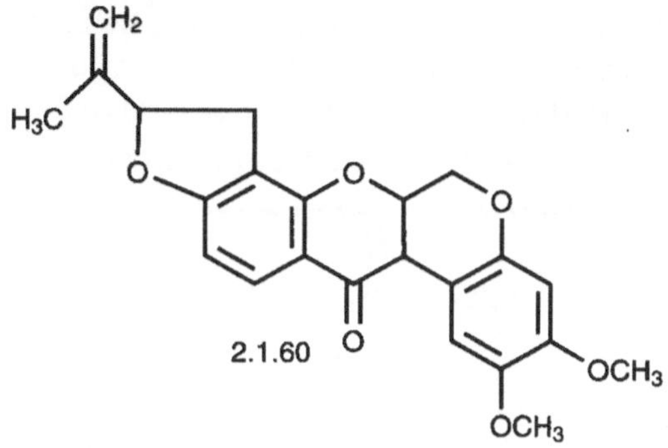

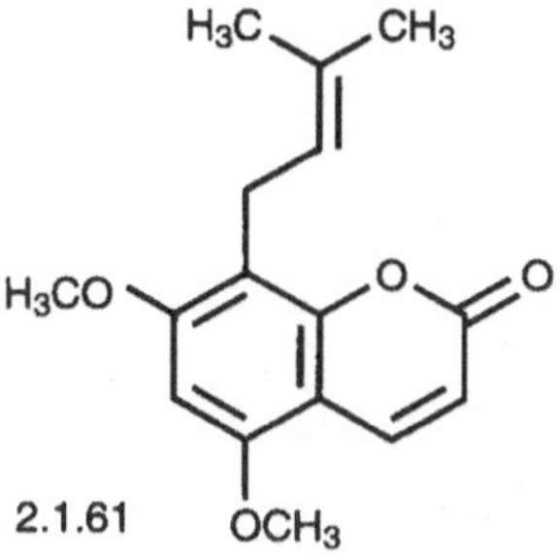

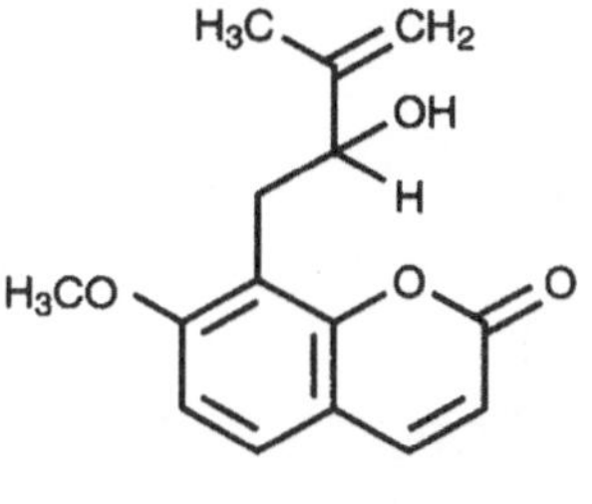

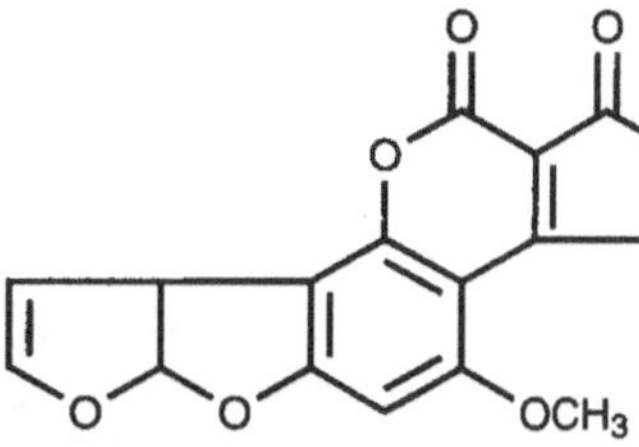

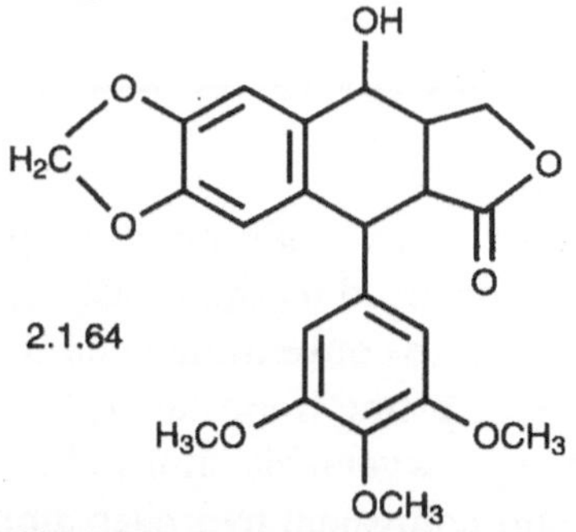

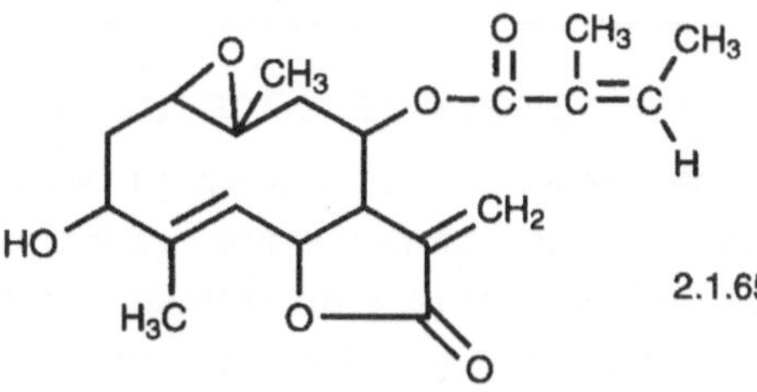

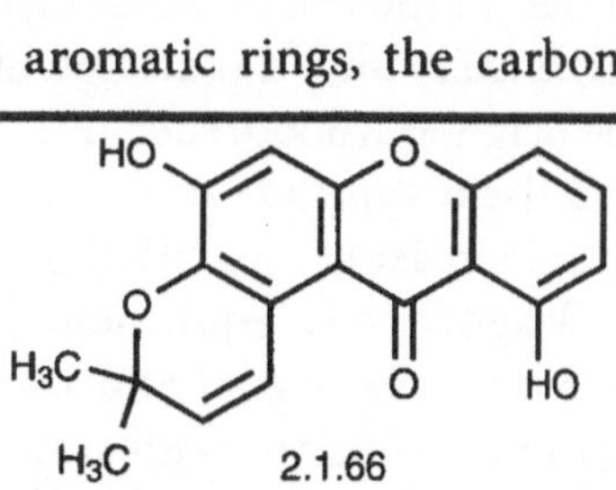

In xanthones, where the carbonyl group is flanked by two aromatic rings, the carbonyl stretching frequency is much lower than the normal value. Thus, in the spectrum of tovoxanthone (2.1.66), this group absorbs at 1640 cm^{-1}; in this case intramolecular hydrogen bonding between the carbonyl and the hydroxyl peri to it is also a contributing factor for the low frequency. In quinones, as in the ubiquinones (2.1.67), which are benzoquinones, and lawsone (2.1.68) which is a naphthoquinone, the

2.1.66

stretching frequency of the carbonyl group is 1650 cm^{-1} or near about. In the anthrquinones, pachybasin (2.1.69) and phomarin (2.1.70), there are two carbonyl absorption bands: one at 1677 cm^{-1} due to the non-chelated carbonyl group and the other at 1640 cm^{-1} which can be assigned to the other carbonyl which is hydrogen bonded to the hydroxyl group at position 1.

The presence of ester groups is readily recognized from the strong absorption band in the frequency range 1730–1740 cm^{-1}. If the ester is α-, β-unsaturated, the frequency is lowered to 1700–710 cm^{-1}. Thus, in the spectrum of heliangine (1.65), besides the band at 1750 cm^{-1}, there is an additional absorption band at 1710 cm^{-1}, in the carbonyl region. In the spectrum of helianginol (2.1.71), this absorption band is absent. Helianginol acetate, on the other hand, absorbs at 1750 and 1740 cm^{-1} due to the lactone and normal ester carbonyls respectively.

2.1.67 n = 1 to 12

2.1.68

2.1.69: R = H
2.1.70: R = OH

2.1.71

Nuclear magnetic resonance spectroscopy

NMR spectroscopy is the most powerful technique in structure analysis. Both proton and carbon-13 NMR techniques are routinely used along with advanced methods such as 2D NMR these days in structural studies on all types of natural products. In conventional one-dimensional proton magnetic resonance spectrum two parameters are of prime importance. These are the chemical shifts and the multiplicity of the signals. The former gives valuable information regarding the chemical environment of the different, magnetically non-equivalent hydrogen atoms present in the compound under analysis. The latter tells us about the interaction of the hydrogen atom(s) responsible for a signal showing multiplicity with other, magnetically non-equivalent protons in the neighbourhood; such interactions are considerable between protons on adjacent carbon atoms and the magnitude of the interaction indicated by the coupling constant (J) rapidly decreases with increasing separation between the interacting protons.

Magnetically equivalent protons such as the three hydrogen atoms in a methyl group give only one signal and therefore the intensity of a signal is also an important input. The chemical shifts are expressed in parts per million on the delta scale with reference to an internal reference, namely, tetramethylsilane (TMS), whose chemical shift value is taken as zero. All hydrogen atoms in an organic compound (except hydrogen atom(s) directly attached to a metal atom) resonate downfield of the tetramethylsilane signal with the delta value increasing with decreasing electronic shielding. The usual range of a proton nmr spectrum is 0–10 ppm, though in several cases there are signals beyond 10 ppm. We shall now consider the chemical shift values of the different types of hydrogen atoms commonly seen in natural products of different types. The signals due to methyl and methylene protons in an alkane and alkyl groups are nearest to the TMS signal. Their range is 0.5–1.5 ppm. Methyl and methylene groups next to a double bond or a carbonyl group or a nitrogen atom or a phenyl group give rise to signals in the region 1.6–3 ppm. Methoxyl protons are readily recognised from the singlet signal at 3.5–4.0 ppm. Olefinic protons resonate from 4.5 ppm onwards depending on whether they are isolated or conjugated; conjugation results in lowered shielding and hence a larger delta value. Aromatic protons give rise to signals from 6.5 to 8.5 ppm; in rare cases, they may be seen below 6.0 ppm. Aldehydic protons are also easily recognised from the signals in the 9–10 ppm region. Carboxylic acid protons resonate further downfield (10–12 ppm). The position of the signals due to hydroxyls groups vary depending on the solvent and concentration but they can also be recognised as they disappear when the solution containing the sample is shaken with a few drops of D_2O, due to deuterium exchange. Intramolecularly hydrogen-bonded phenolic hydroxyls, commonly seen in flavonoid compounds, also resonate far downfield (12 ppm and beyond). Splitting patterns (multiplicity) and coupling constants are invaluable in the finer analysis of a spectrum and are often the key factors in arriving at an unambiguous structure for a compound under examination. With this brief introduction we shall now illustrate the utilitarian value of this technique with a few examples chosen from the different types of secondary metabolites.

Earlier, in this chapter, it was mentioned that the structure of geraniol (2.1.2) could be detemined by subjecting it to a series of analytical and degradative experiments, at the expense of a considerable amount of the compound. The same result could be obtained from a PMR spectrum of the compound, using just a few milligrams which can even be recovered after recording the spectrum. A 500 MHz- spectrum of the compound dissolved in $CDCl_3$ has signals at 1.65 (3H, s), 1.70 (3H, s), 1.75 (3H, s), 2.0 (2H, m), 2.1(2H, m), 2.8 (1H, s, disappears on addition of D_2O), 4.1 (2H, d, $J = 6.8$ Hz), 5.1 (1H, t, $J = 6.8$ Hz) and 5.4 (1H, t, $J = 6.8$ Hz). Thus, all the 18 hydrogen atoms can be accounted for and it is possible to recognise the three methyl groups on double bonds, two methylene groups next to double bonds, a hydroxyl proton, a methylene group flanked by a double bond and the hydroxyl and two olefinic protons, each coupled to a methylene group. The assignments are confirmed by a 2D NMR spectrum. In a homonuclear correlation study (COSY), two normal proton NMR spectra are recorded perpendicular to each other in a square plot. The proton spectrum is seen as contours of varying proportional intensities along the diagonal cutting through the grid. Several off-diagonal contours are also seen and these are called cross peaks. From each cross peak, if a horizontal line is drawn it will meet a contour on the diagonal, while a vertical line from the same cross peak will meet with another contour on the diagonal. This suggests that the two contours on the diagonal are the signals of two spin–spin interacting protons. Thus, in the 2D NMR spectrum of geraniol, on the vertical line above the signal at 5.4 ppm there are two cross peaks corressponding to signals at 4.1 and 1.7 ppm clearly showing that this signal is due to H-2. Similarly, the signal at 5.1 ppm has cross peaks corresponding to the diagonal signals at 2.1, 1.75 and 1.65 ppm and could be assigned to H-6. For numbering of the carbon atoms, see fig 2.1.2 given earlier.

A more complex example from among the terpenoids is santonin (2.1.72). A well-resolved 500 MHz proton NMR spectrum of this compound (a 60 MHz- or even a 100 MHz-spectrum of the same compound has poorly resolved, overlapping signals, which are difficult to interpret) shows signals at 1.25 (3H, d; H15), 1.3 (3H, s; H11), 1.52 (1H, dd; H9a), 1.7 (1H, dq; H8b), 1.82 (1H, dq; H7), 1.9 (1H, td; H9b), 2.02 (1H, qd; H8a), 2.12 (3H, s H12), 2.42 (1H, dq; H13),

2.1.72

4.8 (1H, d; H6), 6.25 (1H, d; H2) and 6.7 (1H, d; H1). (s = singlet; d = doublet; t = triplet; q = quartet; dd = doublet of a doublet; dt = doublet of a triplet; dq = doublet of a quartet; td = triplet of a doublet and qd = quartet of a doublet). The complex spin–spin interactions between the methylene protons at C-8 and C-9, all of which are magnetically non-equivalent and the proton at C-7 are clearly brought out in this spectrum recorded at a high magnetic field (Instruments operating at even higher magnetic fields such as 600 and 800 MHz, are now available). A critical analysis of the multiplicities of the signals between 1.5 and 2.1 ppm allows one to arrive at the stereochemical structure shown in. Figure 2.1.72.

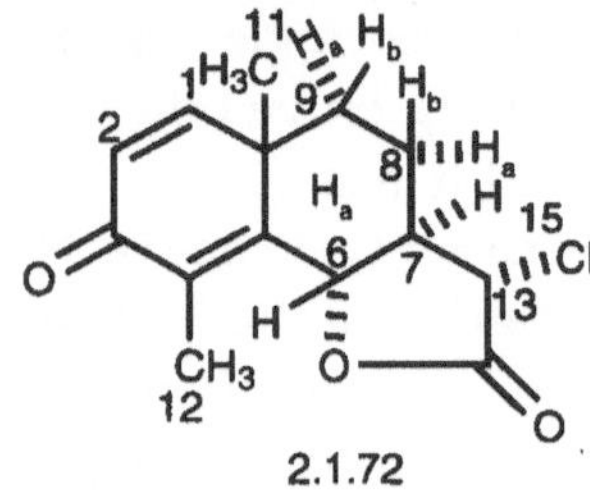

Another illustrative example is the sesterterpene, ophiobolin A (2.1.73), which has been isolated from Helminthosporium species as well as from a marine sponge (*Phyllospongia madagascarensis*).

2.1.73

In the proton nmr spectrum of this compound, the aldehydic proton (H21) gives rise to a signal at 9.26 ppm. The protons of the geminal dimethyl allyl group (H24 and 25) appear as a six proton singlet at 1.70 ppm. The three-proton singlet signal at 0.82 ppm can be attributed to methyl group (H22) attached to carbon 11. The methyl group at position 15 (H23) is responsible for the 3H doublet at 1.08., whereas the one at C-3 (H20) appears as a singlet at 1.34 ppm. The olefinic proton at C-18 appears as a doublet at 5.51 ppm (typical value for a trisubstitued olefinic proton) while the one at C-8 is seen as a triplet at 7.13 ppm; the higher delta value for this signal is due to conjugation with the aldehyde grouping. The methylene protons at C-4 give rise to an AB quartet (pair of doublets due to geminal coupling) at 2.46 and 2.77 ppm.

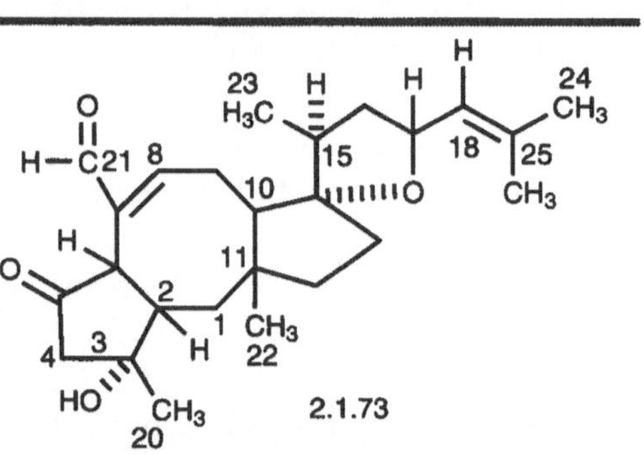

We shall now consider a few examples from other types of secondary metabolites. Proton NMR spectra of flavonoids are easy to interpret. Some reference to them was made in the previous chapter. As mentioned therein, the oxygenation patterns can be discerned quite readily from the splitting patterns and coupling constants of the aromatic protons of the condensed benzene ring and the side phenyl. Intramolecularly hydrogen-bonded hydroxyl groups, such as the hydroxyl in 5-hydroxyflavones and flavonols, are also easily detected by the signals beyond 10 ppm. We shall now take up a few specific examples chosen from the different types of flavonoids and related compounds.

The proton NMR spectrum of luteolin 7-glucoside (2.1.74), as its TMS ether, has signals between 3 and 5 ppm, besides those in the aromatic region, clearly indicating the presence of a non-aromatic substituent. A closer look reveals that the 1 proton doublet signal at 5.0 ppm can be attributed to the anomeric carbon of the sugar residue

2.1.74

and further the position of this signal clearly shows that the compound is a O-glucoside and not a C-glucoside. The remaining sugar protons are responsible for the multiplets between 3.5 and 4.0 ppm. At higher magnetic fields, this region can be resolved and appropriately interpreted. The one-proton singlet at 6.2 ppm shows that the compound is a flavone and not a flavonol (H-3) though it does overlap with the signal due to H-6. H-6', appearing at 7.3 ppm shows *ortho*-coupling with H-5' (signal at 6.8 ppm). The doublet with a small coupling constant (ca 2 Hz) at 6.6 ppm is clearly due to H-8. The H-2' signal overlaps with that of H-6'.

In the proton NMR spectrum of the isoflavone, auriculatin (2.1.75) in CDCl$_3$, the far downfield signal at 12.45 can be attributed to the 5-hydroxyl proton. H-2 appears as a singlet at 7.95 ppm. Of the side phenyl protons, the doublet at 6.96 (J = 8Hz) is due to H-2'. The double doublet (J = 8, 2 Hz) at 6.42 and the doublet (J = 2 Hz) at 6.50 ppm can be attributed, respectively, to H-3' and H-5'. The somewhat downfield signal (singlet) at 8.53 ppm of the hydroxyl proton at position 6' (normal value for a phenolic proton is about 6 ppm (indeed, the signal due to the 4'hydroxyl proton appears at 6.64 ppm), shows that this hydroxyl is hydrogen-bonded to the carbonyl group at position 4, though this bonding is not as strong as that between the carbonyl and the 5-hydroxyl. The two one-proton doublets (each with a

2.1.75

J value of 10 Hz) at 5.59 and 6.69 ppm are due to protons (Hb and Ha) of the dimethylchromene ring, the methyl protons of which resonate at 1.42 ppm (6 H, s). The broad triplet (broadening due to allylic coupling with the methyl protons) at 5.1 ppm, the two-proton doublet at 3.5 ppm and the the two three-proton singlets at 1.68 and 1.78 ppm are due to the hydrogens of the isoprenyl group at position 8.

The proton NMR spectra of eucomin (2.1.76) and eucomol (2.1.77) are instructive, particularly with regard to the signal attributed to the hydroxyl proton at position 7. In the spectrum of eucomin it appears at 10.60 ppm, whereas in eucomol, it takes a more normal value (for a phenolic hydroxyl) at 6.03 ppm. This fact, in conjunction with the difference in the signal positions of the 5-hydroxyl proton (12.7 ppm in eucomin and 11.2 ppm in eucomol), suggests that the carbonyl group in eucomol may not be as perfectly co-planar with the condensed benzene ring as in eucomin; consequently the conjugative interaction between the 7-hydroxyl and the carbonyl group and the hydrogen bonding between the 5-hydroxyl and the carbonyl are stronger in eucomin than in eucomol. The methoxyl group can be readily recognised from the three-proton singlet at 3.73–3.76 ppm. The side phenyl hydrogens in eucomin give rise to two *ortho*-coupled ($J = 9$ Hz) two-proton signals at 7.34 (2',6' H) and 6.98 (3',5' H) ppm. In eucomin the methylene protons at position 2 appear as a doublet ($J = 2$ Hz) signal at 5.30 ppm. That the splitting is due to allylic coupling with the proton at position 9 is evident from the one-proton triplet signal ($J = 2$Hz) at 7.65 ppm, due to H-9. H-6 and H-8 appear somewhat upfield (for aromatic protons) at 5.83 and 5.88 ppm (in eucomin) and 5.94 and 5.98 ppm (in eucomol). The benzylic methylene protons in eucomol give rise to a signal at 2.90 ppm, whereas the hydroxyl proton at C-3 appears at 3.32 ppm; obviously, though next to the carbonyl group, it is not hydrogen-bonded to the latter.

2.1.76   2.1.77

Another illustrative example is the 4-hydroxy-3-aryl coumarin, scandenin (2.1.78). In the

2.1.78

PMR spectrum of this compound, the proton of the acidic hydroxyl at position 4 resonates at 10.4 ppm, whereas the signal due to phenolic hydroxyl proton at position 4' is seen at 6.11 ppm. The side phenyl protons exhibit signals of a typical *para* disubstitued benzene ring; they appear as two two-proton doublets ($J = 8.4$Hz) at 6.88 and 7.43 ppm. The protons of the methoxyl group at position 5 resonate at 3.95 ppm. The hydrogens of the isoprenyl group at position 6 give rise to signals, as expected, at 1.68 (3H, s), 1.86 (3H, s), 3.43 (1H, d), 3.55 (1H, d) and 5.26 (1H, t). The protons of the dimethylpyrene ring bridging positions 7 and 8 are responsible for the signals at 1.46 (6H, s), 5.58 (1H, d, $J = 10.2$ Hz) and 6.58 (1H, d, $J = 10.2$ Hz).

As mentioned earlier, with the advent of high field NMR spectroscopy, it is now possible to completely interpret multiple spin–spin coupling interactions in complex compounds. This was illustrated with the example of the spectrum of santonin recorded with a 500 MHz instrument. The main advantages of using such instruments are increased signal sensitivity, improved signal separation and, more importantly, increased symmetry of splitting patterns. In other words, the pattern of splitting approaches the first order and thereby becomes easier to interpret. However, instruments operating at lower fields (60, 90 or 100 MHz) have not become obsolete and are functional in many laboratories, since they are much cheaper and easier to maintain. Even with such instruments, it is possible to derive extra information regarding the compound under examination by resorting to a few additional techniques other than a routine recording in commonly used solvents such as CCl_4 or $CDCl_3$. The simplest of these is to change the solvent. For example, when the sample size is small (2–3 mg), even a trace of moisture in the solvent

(CDCl$_3$) shows up as a sharp singlet at 1.5 ppm. This is the region where signals due to methyl and methylene groups—if present in the compound—are expected and therefore the water signal becomes a serious interference. However, when the spectrum is recorded in C$_6$D$_6$, the water peak appears at 0.5 ppm and any signal appearing at 1.5 ppm can be unambiguously interpreted. Benzene (or hexadeuterated benzene) also specifically interacts with certain functionalities such as the carbonyl group and thereby shield any protons attached to the carbonyl function. Thus, while the methyl protons of an acetyl group resonate at 2.0 or 2.1 ppm in CDCl$_3$, their signal is shifted slightly upfield (1.7 or 1.8 ppm) in C$_6$D$_6$.

Paramagnetic shift reagents are also used to resolve overlapping signals seen particularly in the spectra of long chain aliphatic alcohols and similar oxygenated compounds. Such shift reagents which are organic complexes of a rare earth metal such as europium coordinate with the oxygen atom of the hydroxyl group (or for that matter that of a carboxyl or ester function) and thereby cause a considerable downfield shift of the protons on the carbon next to the hydroxyl function. This effect is transmitted down the chain, with decreasing magnitude and the result is a much better resolved specrum which is easier to interpret. The shift reagents commonly used are Eu(fod)3, which is tris-1,1,1,2,2,3,3-heptafluoro-7,7′-dimethyl-3,5-octanedionatoeuropium (2.1.79) and Eu(dpm)3, which is tris(dipivalomethanato)europium (2.1.80). For example, in the normal PMR spectrum of the methyl ester of 9,12-octadienoic acid (2.1.81), the methoxyl protons resonate, as expected, at 3.7 ppm. On addition of Eu(fod)3, this signal shifts to 4.7 ppm and the protons of the methylene group next to the ester function shift from 2.3 to 3.5 ppm. The signal due to the C-3 methylene protons appears as a clear quintet at 2.4 ppm, whereas in the normal spectrum it partly overlaps with the water signal at 1.6 ppm.

2.1.79   2.1.80   2.1.81

Another method to simplify complex spectra is to bring about selective spin–spin decoupling using the double resonance technique. In an A–X system, wherein the proton A is involved in spin–spin interaction through the intervening bonding electrons with the proton X, if the latter is irradiated with a second, strong radio frequency equal to the resonance frequency of X, the rate of exchange between the two spin levels of X increases to such a rate that A is no longer able to differentiate between the two states of X. Depending on the strength of the second frequency, the coupling interaction is transformed into selective population transfer (SFT) or spin tickling or complete decoupling. When decoupling occurs, the signal due to X is not seen, whereas that of A appears no longer as a doublet but as a singlet. In A–X$_3$ system, the signal of A will change from a quartet to a singlet. Since the signal width is considerably reduced, any overlapping signals in that region of the spectrum will separate thus facilitating interpretation, while allowing the unambiguous recognition of the signals due to A and X in the original spectrum. If the original spectrum is subtracted from the decoupled spectrum, the signal due to X (in the A-X$_3$ system), appears as an inverted doublet, whereas that of A shows five lines, with the original quartet inverted and the decoupled singlet appearing above and in the centre of the inverted quartet. Such a spectrum is known as a spin decoupling difference spectrum (SDDS). In the normal proton NMR spectrum of β-ionone (2.1.82), the signals at 7.28 and 6.12 ppm due to the olefinic protons at positions 7 and 8 appear as broadened doublets with a *J* value of 18 Hz (*trans* coupling). The broadening indicates further long range coupling interactions, the nature of which is revealed by a SDD spectrum. When the sample is irradiated with a second strong frequency equal to the resonance frequency of the protons of the methyl group at position 5 (1.76 ppm), each of the signals at 7.28 and 6.12 exhibit four inverted lines clearly showing that both of these olefinic protons are involved in long-range coupling with the methyl group at position 5. The magnitude of this allylic coupling is, however, small, and hence in the normal spectrum only a broadening is seen. When the signal at 7.28 ppm at the top in the SDD spectrum is expanded, it appears as a triplet of a doublet with *J* values of 18 and 1.5 Hz. The obvious inference is that even after decoupling with the C-5 methyl protons, H-7 is involved in spin-spin interaction not only with H-8 but also with two other magnetically equivalent protons. The latter have to be the hydrogens at C-4. This is confirmed by the observation that the signal due to these methylene protons at 2.07 ppm, which appears as a broadened triplet in the normal spectrum, (*J* = 6 Hz) becomes a doublet of a triplet (*J* = 6 and 1.5 Hz) in the same SDD spectrum, with four inverted lines. Thus, it becomes clear that the C-4 hydrogens are involved in vicinal coupling with the C-3 protons as well as long range coupling with the H-7 and C-5 methyl protons.

Nuclear Overhauser Effect (NOE) is another outcome of a double resonance NMR experiment. This phenomenon is seen when one of two spatially proximate protons is irradiated with its resonance frequency. While the signal of the irradiated proton disappears due to saturation, that of the other proton increases in intensity leading to the inference that the two protons are involved in spin-spin interaction through space under conditions used for recording the normal spectrum. With the older CW NMR instruments, it is often difficult to detect NOE, particularly if the magnitude of the effect is small. However, with the modern FT instruments, NOE, difference spectra can be recorded and these give unambiguous information about any finite increase in a signal intensity due to NOE. This can be illustrated taking β-ionone again as an example. Structure (2.1.82) is a conventional representation of the compound. However, to understand its stereochemical nuances it is better to consider the structures (2.1.83) and (2.1.84), which are two conformations of the compound (other conformations are also possible). In (2.1.83), H-7 is seen in close proximity, through space, to one of the methyl groups at C-1 as well as the terminal methyl group (C-10), attached to the carbonyl group. On the other hand, in (2.1.84), H-8 comes close to these two methyl groups. A NOE difference spectrum is recorded in the same way as a SDDS spectrum. The only difference is that we look for increase in intensities of specific signals after a double irradiation experiment. For example, when β-ionone is irradiated with a second frequency equal to the resonance frequency of the 1,1′-geminal dimethyl groups, and the difference spectrum recorded, the intensity of the signal due to H-7 increases by 13%, whereas those of H-8 and H-2 show increases of 9.8 and 5.5% respectively. This result clearly shows that the conformation (2.1.83) is the major contributor to the structure of β-ionone. This is also supported by the observation that irradiation of the C-5 methyl protons brings about a slightly larger NOE in the signal of H-8 (7.2%) compared to that of H-7 (5.0%). On the other hand, irradiation of the C-10 methyl protons results in a 8.7% increase in the intensity of H-7 signal as compared to a 5.2% increase in that of H-8.

13-Carbon NMR spectroscopy is a comparatively newer technique but has now become an invaluable method in structural studies on natural products. In the earlier days, after the advent of this technique, the usual procedure was to completely eliminate all 13-carbon proton spin–spin interactions to make the spectra easy to interpret. The advantages of this technique are two-fold.

1. The spectrum shows only carbon resonances as singlets.
2. Due to the NOE effect, there is considerable increase in the intensities of the signals, particularly of carbon atoms carrying one or more hydrogen atoms.

This is a distinct advantage as the natural abundance of 13-carbon is only 1.1% (the much more abundant 12-carbon is non-magnetic). However, because the NOE depends upon the number of protons attached to a carbon atom (a completely substituted carbon atom will not show any NOE), the intensity of a signal loses its value as a quantitative measure. Further, all information regarding the degree of substitution of a carbon atom is lost. In spite of these disadvantages, this procedure of broadband proton decoupling or noise decoupling continues to be used as it shows the number of magnetically non-equivalent carbon atoms in a compound and the chemical environment of each one of them. For example, in β-ionone the two geminal methyl carbon atoms are equivalent but all the others are in different environments. Therefore, the noise decoupled 13-carbon NMR spectrum of this compound shows twelve signals spread over a region of 10 to 200 ppm from the TMS signal. The least shielded carbon atom is the carbonyl carbon atom and its signal is seen at 198 ppm; it is also one of the less intense peaks. At the other end of the spectrum, at 18 ppm, carbon-3 appears followed closely by the methyl carbon attached to position 5 (21 ppm). The methyl carbon attached to the carbonyl group resonates at 27 ppm and at 28 ppm there is an intense signal due to the carbon atoms of the geminal dimethyl group at position 1. The carbon atoms of the double bonds give rise to signals in the region 130 to 145 ppm (C-8 at 130, C-5 at 135, C-6 at 136 and C-7 at 145 ppm). To get information regarding the degree of substitution on each carbon, the technique of off-resonance decoupling is used. This method eliminates long-range carbon–proton couplings and considerably reduces—but does not completely eliminate—interactions between directly attached carbon and protons. Thus, while tetra-substituted carbon atoms continue to appear as singlets, methyl carbons show up as quartets, methylene carbons as triplets and methine carbons as doublets. Thus, a combination of noise decoupled spectrum and off-resonance decoupled spectrum gives all the information needed about each different carbon atom in a compound. Thus, while the first method gives chemical shift values for the different carbon atoms, the latter tells us about their multiplicities. However, in the case of complex molecules there can be overlap of signals and since the magnitude of NOE is less, signal intensities are also lower in an off-resonance decoupled spectrum. Therefore, two other methods have been developed in recent years to get information lost in the noise decoupled spectrum. These are INEPT (insensitive nuclear enhancement by polarisation transfer) and DEPT (distortionless enhancement by polarisation transfer).

In an INEPT spectrum, signals due to carbon atoms carrying odd number of protons (1 or 3) appear on the positive side (above) whereas methylene carbons show up as inverted signals and those of tertiary carbon atoms are not seen. In this technique the decoupling pulse is applied after a delay time of 6 milliseconds. For example, in the INEPT spectrum of β-ionone, signals of carbon atoms 2, 3 and 4 appear on the negative side, while those of the methyl groups and the olefinic carbons are seen on the positive side. The DEPT technique is an improvement of the INEPT method. In this case the spectra are recorded by varying the pulse width by 135 degrees and the result is the same as in the INEPT spectrum described above. If the pulse width is 90 degrees, only signals of C–H carbons (methine) are seen and they appear on the positive side. For example, in the 13-carbon NMR spectra (combination of noise decoupled and DEPT 90 and 135), of the diterpene viridinol (2.1.85), isolated from the roots of *Salvia viridis*, 21 signals are seen, out of which eight signals are not seen in the DEPT spectra. These are, therefore, due to quaternary carbon atoms. The chemical shifts of these signals are 33.5 (C-4), 40.4 (C-10), 125.2 (C-8), 138.4 (C-13), 139.8 (C-9), 146.6 (C-14), 156.4 (C-11) and 184.3 (C-12). In the DEPT 135-spectrum three signals appear on the negative side and these are identified as due to C-1 (41.2 ppm), C-2 (18.3) and C-3 (35.7). Of the ten signals appearing on the positive side in the DEPT 135-spectrum, six are due to methyl groups and they appear at 18.0 (C-19), 21.5 (C-20), 23.3 (C-17), 23.5 (C-16), 26.7 (C-18), and 61.9 (methoxyl carbon). The remaining four signals can be identified as due to methine carbons as they were also seen in the DEPT 90-spectrum. Their chemical shift values are 33.5 (C-15), 50.3 (C-5), 73.3 (C-6) and 117.5 (C-7). In the off-resonance decoupled spectrum, some of these signals overlap making the interpretation difficult.

13-Carbon NMR spectra are also useful in structural studies on flavonoids, and in particular flavonoid *O*- and *C*-glycosides. Signals due to the skeletal carbon atoms appear in the region 90 to 165 ppm. The carbonyl carbon signal in flavones and flavonols is seen at about 180 ppm. This signal shifts nearer to 200 ppm in dihydroflavones and dihydroflavonols. Tabulated data are available for the identification of different flavonoid aglycones and glycosides from their 13-carbon NMR spectra. The presence of methoxyl groups is revealed by signals in the 55–63 ppm region and that of methylenedioxy group by a signal near 100 ppm. The latter, of course, could overlap with some signals in the aromatic region but can be differentiated from them by the

DEPT technique. Carbon atoms of any sugar residue give rise to signals in the 'aliphatic' region, namely, 61–80 ppm, with the anomeric carbon appearing at about 100 ppm in O-glycosides. A couple of specific examples are discussed in detail in the Problems section of this book.

Reference was made to the 2D NMR spectrum of geraniol earlier. The spectrum discussed was result of a homonuclear (proton–proton) correlation spectroscopy. A much more complex example is the diterpene alkaloid, delphisine (2.1.86), whose COSY spectrum is analysed in depth later in this chapter. We shall now consider the homo COSY spectrum of β-ionone (2.1.82) whose normal proton NMR spectrum has already been discussed. As explained in the case of geraniol, chemical shift values of the different (magnetically non-equivalent) protons can be read directly from the peaks on the diagonal of the 2D spectrum. Then, the various off-diagonal cross peaks are noted. The signal having the highest delta value, at 7.28 ppm in the spectrum of β-ionone is due to H-7. There are three cross peaks correlating with this signal. The most intense of which corresponds with the diagonal peak at 6.12 (H-8). The other two, weaker, cross peaks correspond with the signals at 2.07 and 1.76 ppm given by H-4 (methylene) and H-5 (methyl) protons respectively. Thus, this COSY spectrum confirms the conclusions drawn from the SDDS spectra described earlier.

In a hetero COSY spectrum, one axis represents the proton NMR spectrum of the compound under study and the other its 13-carbon spectrum. It is common practice to plot the proton spectrum on the vertical axis on the right hand side and the carbon spectrum on the horizontal axis at the top. In contrast to the homo COSY spectrum, there is no diagonal in this case but cross peaks are seen. Taking β-ionone as an example again, if we draw a horizontal line from the 7.28 ppm signal on the vertical line to the left, a cross peak is reached which corresponds with the signal at 142 ppm on the carbon spectrum. Therefore, C-7 can be unambiguously assigned this chemical shift value. Similarly, the signals at 6.12 on the proton spectrum and 131 on the carbon spectrum can be correlated and hence 131 ppm is taken as the chemical shift value of C-8. The main advantage of this technique is that using the proton spectrum for correlation, 13 carbon assignments can be made with a degree of certainty which is not possible from normal 13 CMR spectra. There are several refinements of these basic techniques but they are not indispensable in routine structural studies. However, these newer techniques become useful when the quantity available of a natural product is very small, as in the case of compounds isolated from many marine organisms. Two such techniques are HMQC (1H-detected heteronuclear multiple quantum coherence) and HMBC (1H-detected mutiple-bond heteronuclear multiple-quantum coherence) spectra. For further information, the books listed under suggested reading may be consulted.

Mass Spectra

The chief advantage of mass spectroscopy is that the sample size needed is very small (about a milligram) and it is often possible to determine the molecular weight of the compound just using a milligram of it. The one limitation, however, is that the compound should be sufficiently volatile to be injected into the ionisation chamber in the vapour form under the operating conditions of temperature and high vacuum. Heavily hydroxylated compounds, such as carbohydrate derivatives and high molecular weight compounds, therefore, are not readily amenable to be examined directly by mass spectroscopy.

The most widely used method for generating the mass spectrum of a compound is to bombard it with a beam of high energy electrons (70 eV). Such an impact removes an electron from the target molecule generating its molecular ion, which is a positive ion radical. This high energy species further disintegrates into various fragment ions and ion radicals which are separated and detected according to their mass-to-charge ratios (m/z values) by means of a variable magnetic field. In the following discussion we will assume that the readers are familiar with the instrumentation, the basic facts and the nomenclature used in mass spectroscopy. In structural studies on natural products, mass spectral data help to substantiate inferences drawn from other spectra and in particular NMR spectra. The recognition of the molecular ion peak wherever possible, is an important first step. Its relative abundance with regard to other peaks and in particular the base peak is also a useful input. For example, in te case of natural products which are predominantly aromatic in character, such as the flavonoid aglycones, the molecular ion peak is a major peak, if not the base peak. On the other hand, in the spectra of compounds which are largely aliphatic in nature, such as the acyclic terpenoids, its abundance will be relatively low. We shall now discuss the salient features of the mass spectra of the major types of secondary metabolites, starting with the flavonoids.

The majority of flavones and flavonols are volatile enough to be studied without any modifications. However, flavonoid glycosides, biflavonoids and anthocyanidins require suitable derivatisation before they can be volatilised. Permethylation (complete O-methylation) and acetylation are two convenient methods for improving the volatility of polyhydroxy flavones and their glycosides. Trimethylsilyl ethers have also been used. As mentioned earlier, flavone and flavonol aglycones and, in particular, their methyl ethers, give recognisable and fairly abundant molecular ion peaks. If the compound has a methoxyl group, the loss of a methyl

radical from the molecular ion is a common occurrence. However, the most important and informative fragmentation pathway involves a retro Diels–Alder type reaction which splits the molecule into two fragments, one representing the A ring and the other the side phenyl. For example, in the mass spectrum of 5,7,3′,4′, 5′-pentahydroxyflavone, tricetin (2.1.86), the base peak corresponds to that of the molecular ion (m/z 302), while the fragment ions (2.1.87; m/z 153) and (2.1.88; m/z 159) are also prominent (44% and 24% respectively). In the mass spectrum of the flavonol derivative, jaceidin, which is myricetin 3,6,3′-tri-O-methyl ether (2.1.89), too, the molecular ion peak at m/z 360 is the base

2.1.86

2.1.87   2.1.88

peak. In this case, the M+-15 peak at m/z 345 has 65% relative abundance. It probably arises due to the loss of a methyl group from the methoxyl at position 6. The resulting ion is stabilised by resonance as shown in (2.1.90). The loss of a COCH₃ radical results in (2.1.91) (m/z 317, 20%).

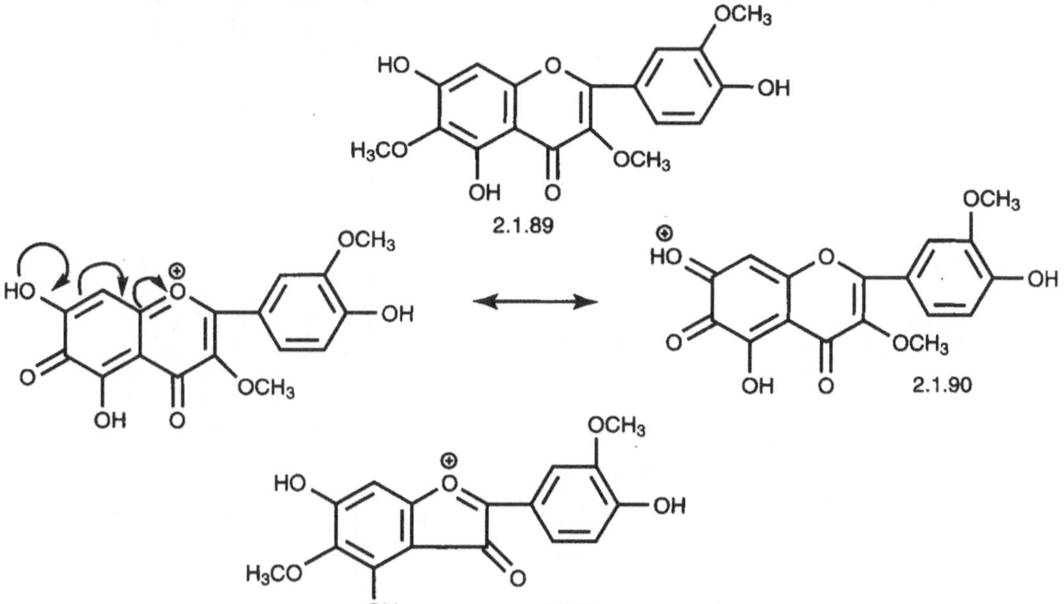

2.1.89

2.1.90

2.1.91

In the mass spectrum of lonchocarpan (2.1.92), an isoflavan isolated from *Lonchocarpus laxiflorus*, the molecular ion signal at m/z 332 is the base peak. The benzylic carbocation (2.1.93), with a m/z value of 197, consisting of the entire B ring and C-3 of the pyran ring is also abundant (98%). The ion radical (2.1.94) with m/z 210 is responsible for the third most intense peak (88%) in the spectrum shown below.

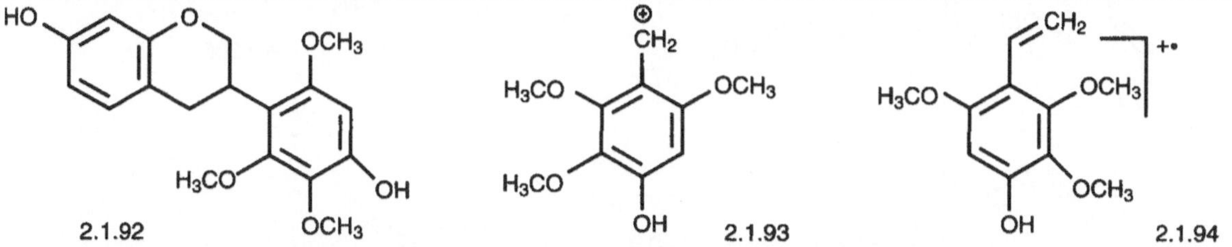

2.1.92

2.1.93

2.1.94

In the mass spectrum of scandenin (2.1.95), the molecular ion peak at m/z 434 is the base peak. It readily loses one of the allylic C-methyl groups in the chromen ring part to yield the ion with m/z 419 (2.1.96). Other prominent peaks are seen at m/z 285 (2.1.97) and 257 (2.1.98).

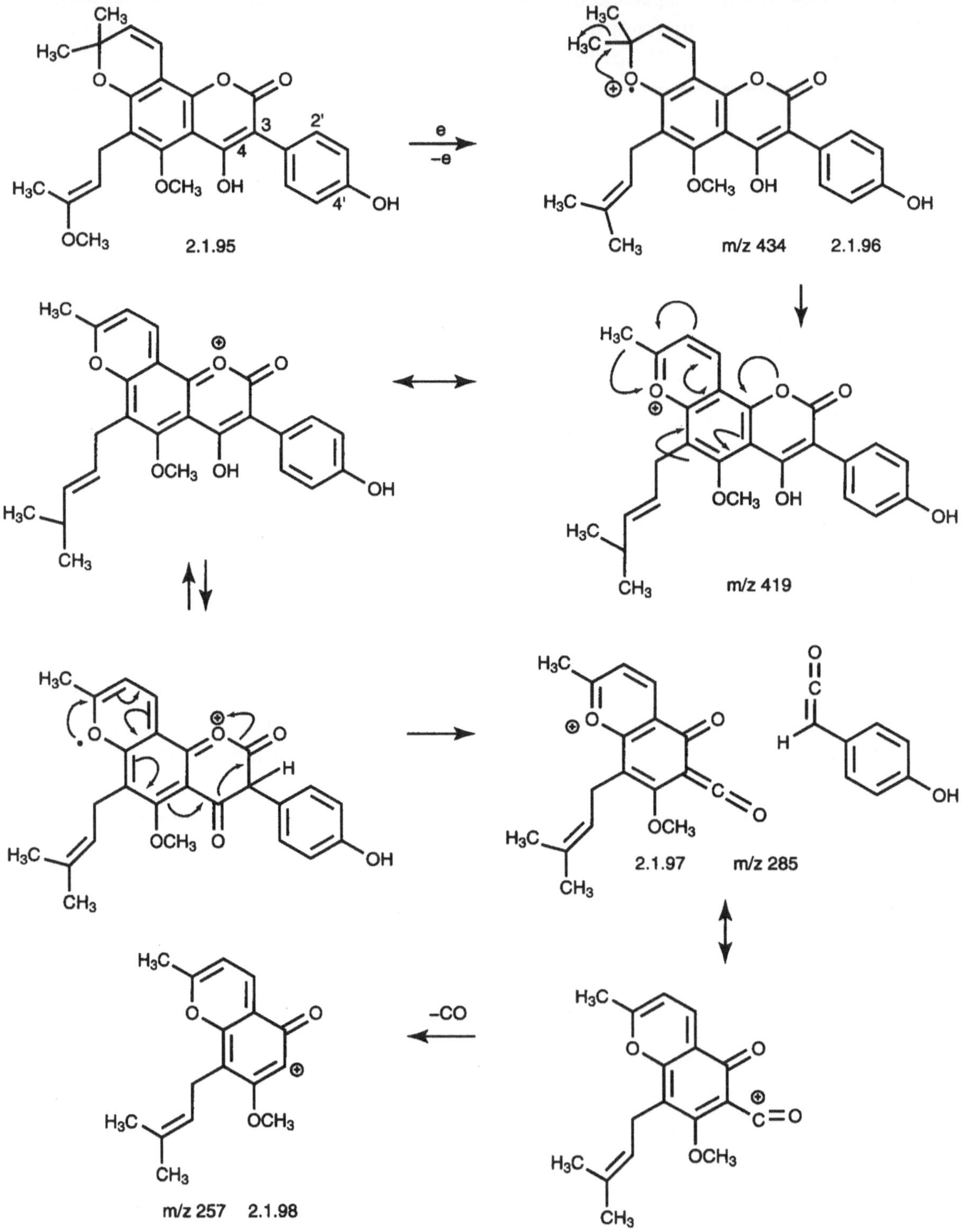

CH3
H3C
O
O
O
3
2'
4
OCH3 OH
4'
OH
H3C
OCH3
2.1.95
e
−e
m/z 434
2.1.96
m/z 419
2.1.97
m/z 285
H
−CO
m/z 257
2.1.98

Two homoisoflavones, (2.1.99) and (2.1.100) were isolated from the roots of a Chinese medicinal plant, *Ophiphogon japonicus*. In each case, the molecular ion (m/z 354.0768 and m/z 340.0965) was responsible for the base peak. In the case of 2.1.99, other peaks were seen at m/z 194 and m/z 160. The spectrum of 2.1.100 had peaks at m/z 194 and m/z 146.

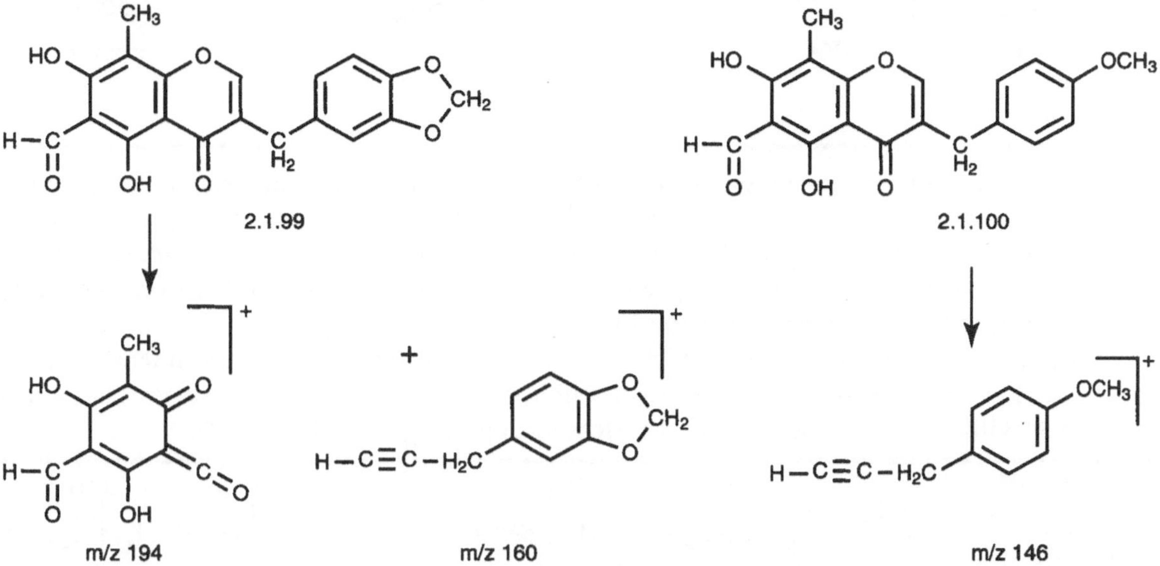

2.1.99   2.1.100

m/z 194   m/z 160   m/z 146

In the mass spectrum of the lichen metabolite, usnic acid (2.1.101), the molecular ion has 60% relative abundance; the base peak is due to the ion with m/z 233 (2.1.102). The fragment (2.1.103) with m/z 260 has 70% relative abundance. In the mass spectrum of the lignan, pinoresinol (2.1.104). the relative intensity of the molecular ion is 48.5%. The base peak occurs at m/z 151.

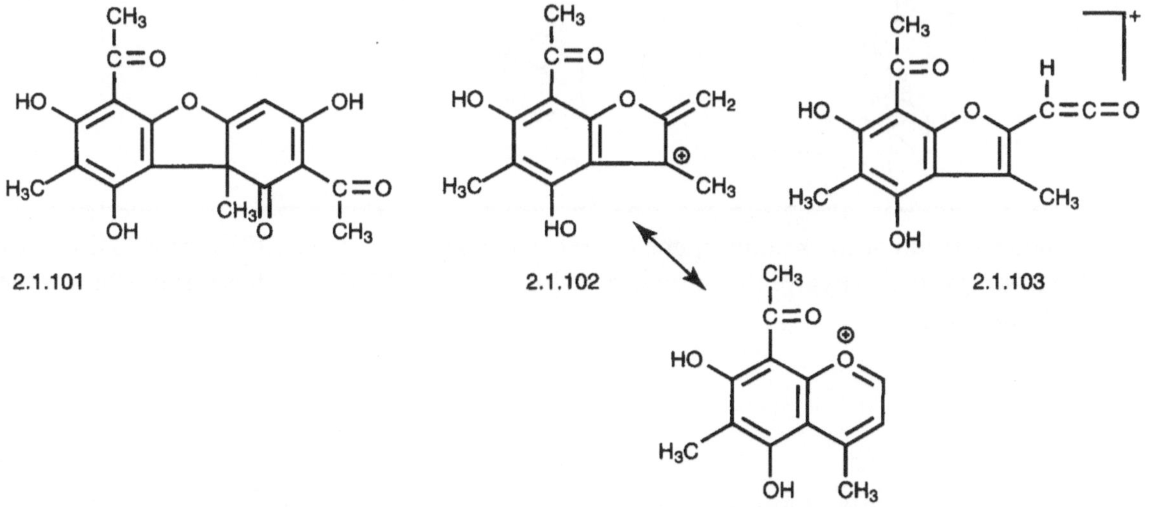

2.1.101   2.1.102   2.1.103

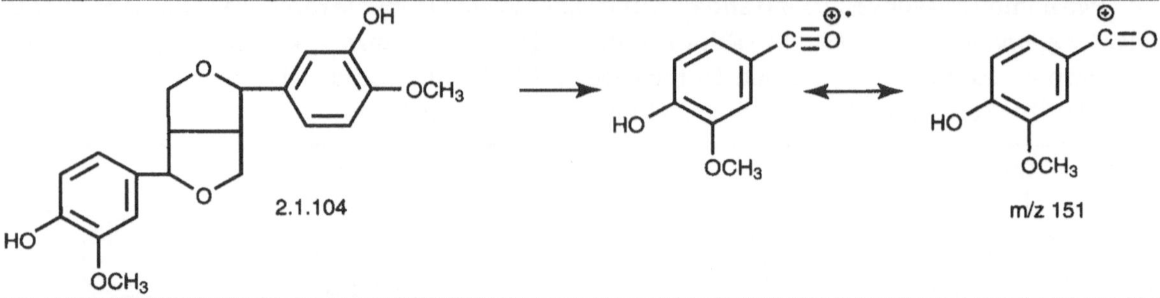

2.1.104

m/z 151

We shall now examine the mass spectra of a few representative examples from the terpenoids. As can be expected and in contrast to the compounds mentioned above, which are predominantly aromatic, the acyclic monoterpene, linalool (2.1.1) gives rise to its molecular ion (m/z 154), whose relative intensity is just 0.5%. The base peak is seen at m/z 93. The loss of the hydroxyl group, as water, is responsible for the peak with m/z 136.

In the case of α-pinene (2.1.16), which has a more compact structure, the intensity of the molecular ion (m/z 136) is greater (10%), but still low. In this case too, the base peak has a m/z value of 93. Other peaks are seen at m/z 121 (loss of a methyl group), 106 and 77.

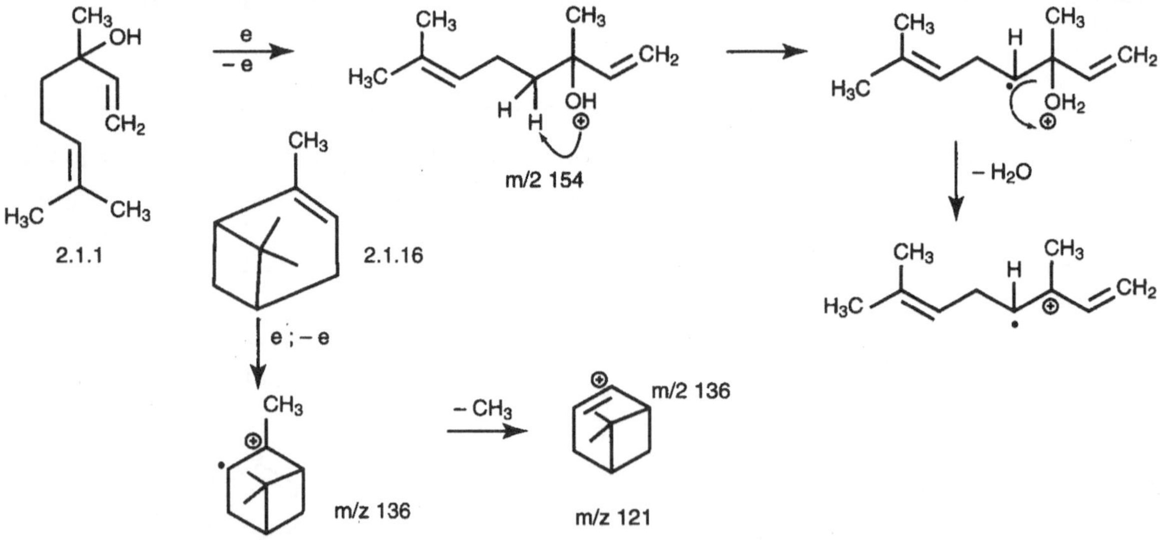

2.1.1

m/2 154

2.1.16

− H$_2$O

e ; − e

− CH$_3$

m/z 136

m/z 121

m/2 136

The molecular ion of the sesquiterpene hydrocarbon, germacrene (2.1.105), (m/z 204) readily loses the isopropyl group to give the resonance stabilised ion (2.1.106) which is responsible for the base peak (m/z 161).

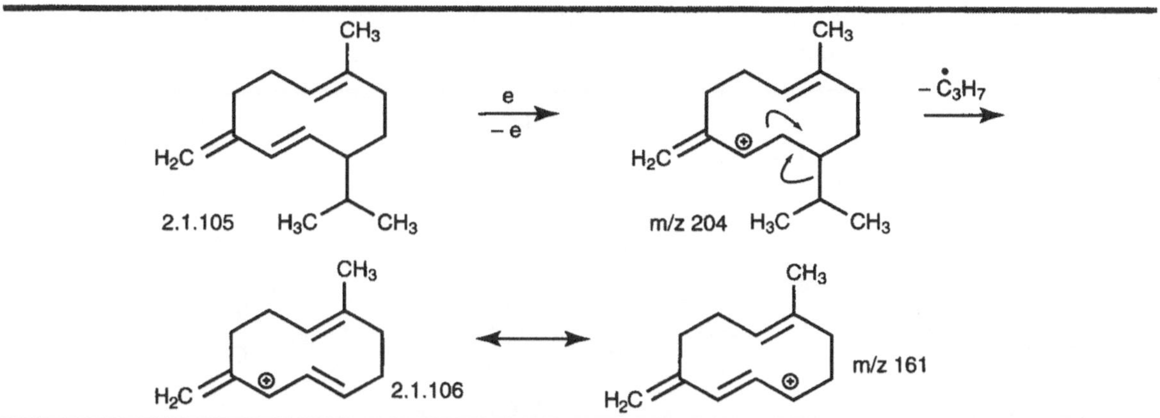

In the mass spectrum of γ-cadinene (2.1.107), the relative intensity of the molecular ion at m/z 204 is 20%. The base peak is seen at m/z 161 due to the ion 2.1.108. Other peaks are seen at m/z 119 and 91. In this case, prior to the loss of the isopropyl radical, apparently, there is a shift of a hydrogen atom in the original molecular ion as shown below.

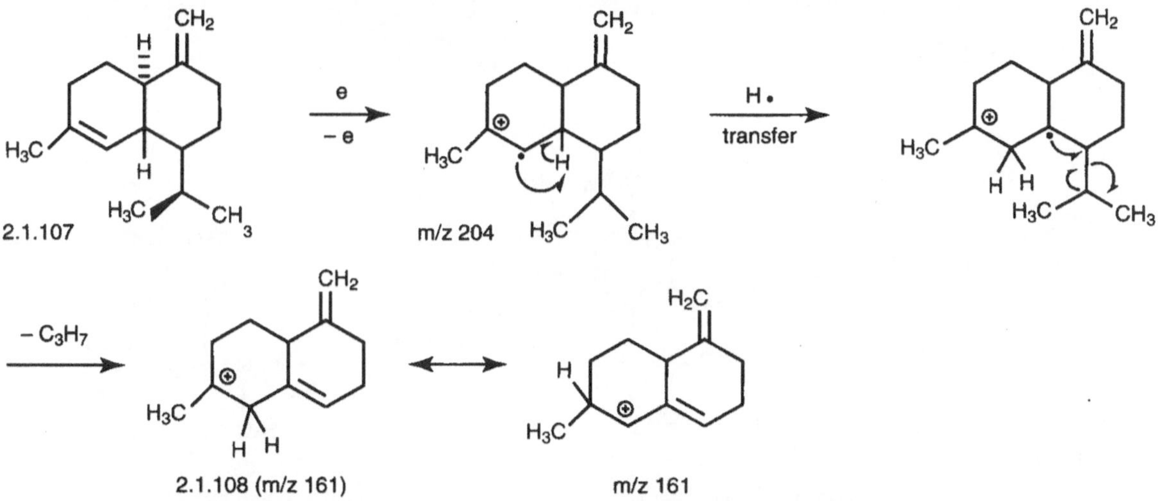

In the case of the diterpene, salviviridinol (2.1.109), the presence of a benzene ring and the compact structure are probably responsible for ensuring that the base peak is due to the molecular ion (m/z 348); this is in spite of the fact that the compound has two hydroxyl groups. Other prominent peaks are those having m/z values of 330 (60%), 315 (83%), 287 (70%), and 245 (88%). The formation of the first three fragmentation ions can be rationalised as shown. It must, however, be noted that it is not always possible to explain, in mechanistic terms, each and every fragmentation pathway in most cases.

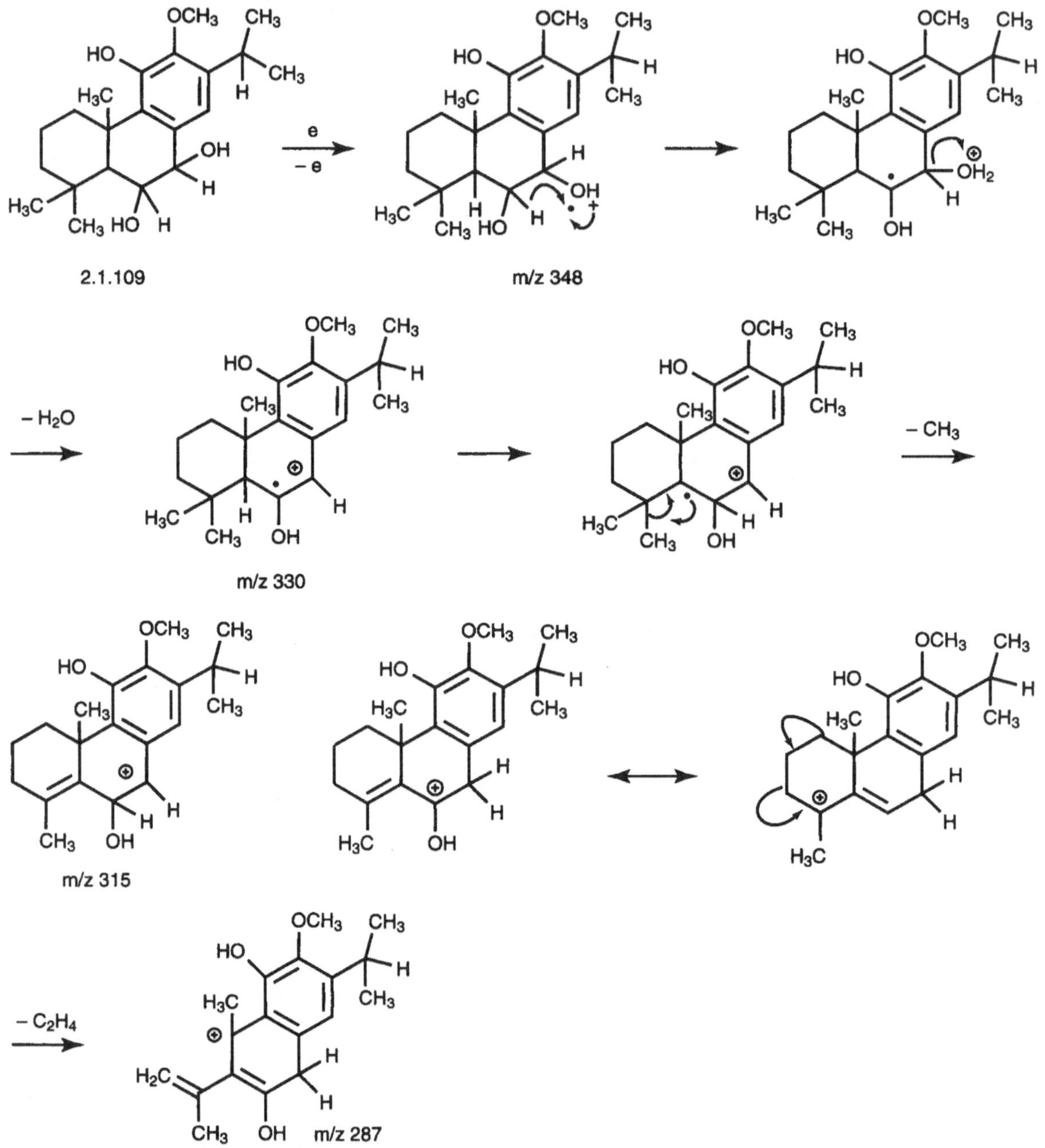

The mass spectral fragmentation of β-amyrin (2.1.110) is also instructive. In this case, the relative intensity of the molecular ion (m/a 426) is 17%. The base peak occurs at m/z 218 and the ion radical responsible for it can be assigned the structure 2.1.111.

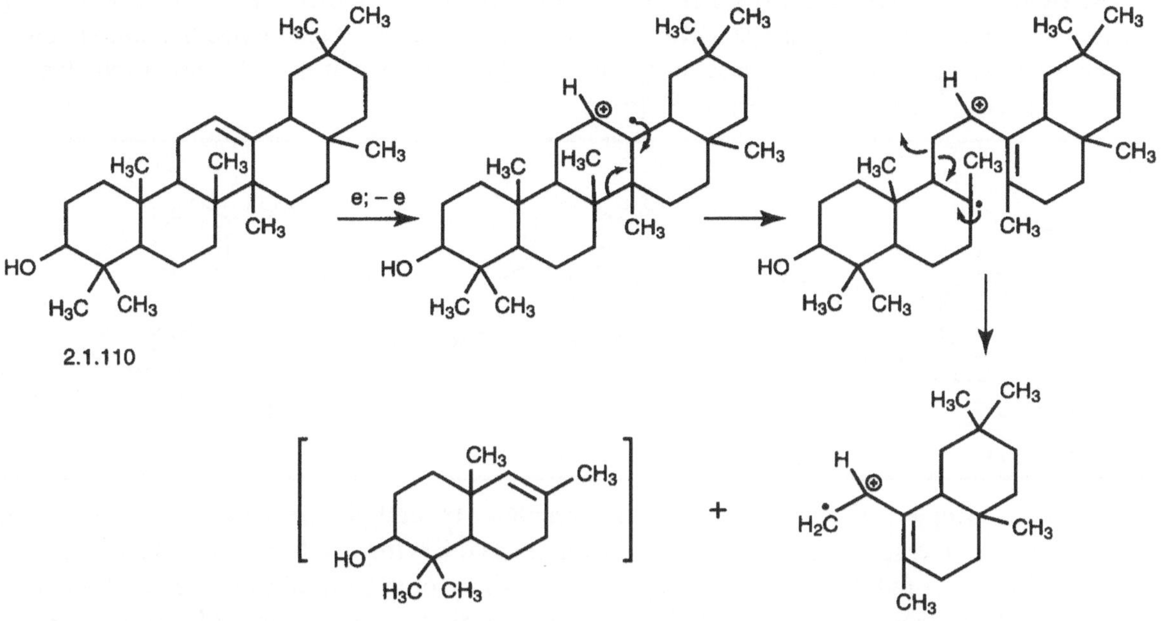

2.1.110

2.1.111   m/z 218

The carotenoid pigment, capsorubin (2.1.112) on electron impact yields its molecular ion which undergoes fragmentation to give two ions having m/z values of 127 (20% relative intensity) and 109 (base peak). These can be assigned the structures 2.1.113 and 2.1.114 respectively.

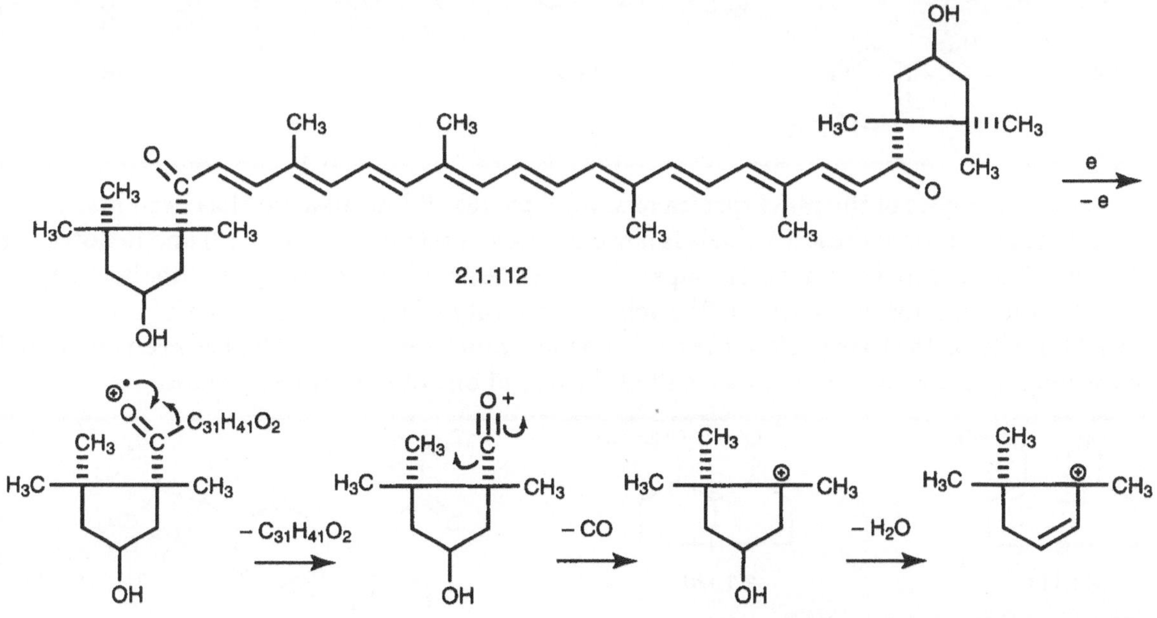

2.1.112

2.1.113   m/z 127

2.1.114   m/z 109

We shall now look at the mass spectrum of the opium alkaloid, thebaine (2.1.115). Reflecting its compact and partly aromatic structure, the relative intensity of the molecular ion is 100%. Another prominent peak in the mass spectrum of thebaine is seen at m/z 296, due to the loss of the methyl radical of the enolic methoxyl group.

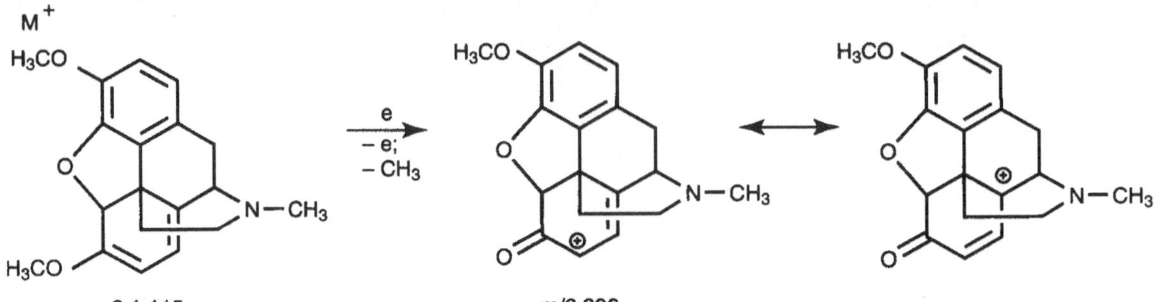

2.1.115   m/2 296

Another example is the benzylisoquinoline alkaloid, trimethylcoclaurine (2.1.116) in which case the base peak is seen at m/z 306 due to the ion 2.1.117; it is formed from the molecular ion by the loss of a benzyl radical. In the mass spectrum of the proton form of the piperidine alkaloid (2.1.118), isolated from *Senna spectabilis*, the relative intensity of the molecular ion (M+H) is only 10%. The base peak at m/z 280 is obviously due to the loss of a molecule of water from the molecular ion.

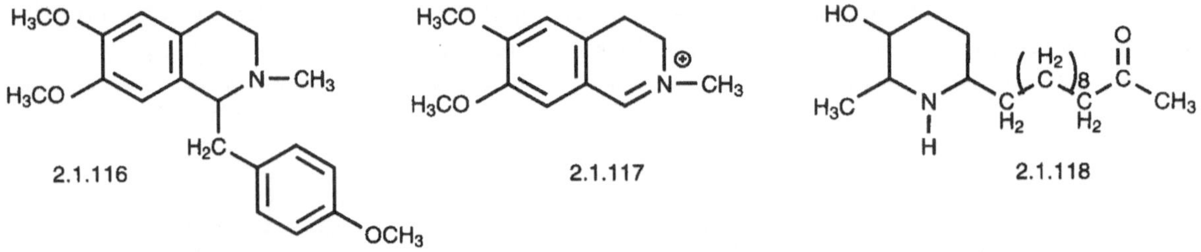

2.1.116   2.1.117   2.1.118

Detailed analysis of the mass spectra of several pyrrazolidine alkaloids has been made. The basic structural unit of these alkaloids is a necine such as retronecine (2.1.119) and heliotridine (2.1.120). One or both the hydroxyl groups is/are esterified to give compounds such as heliosupine (2.1.121) and senecionine (2.1.122). The relative intensity of the molecular ion of senecionine (m/z 335) is 20%. The base peak in the case of heliosupine , seen at m/z 220, has been attributed to the ion 2.1.123, which arises as a result of the loss of one of the acyloxy groups.

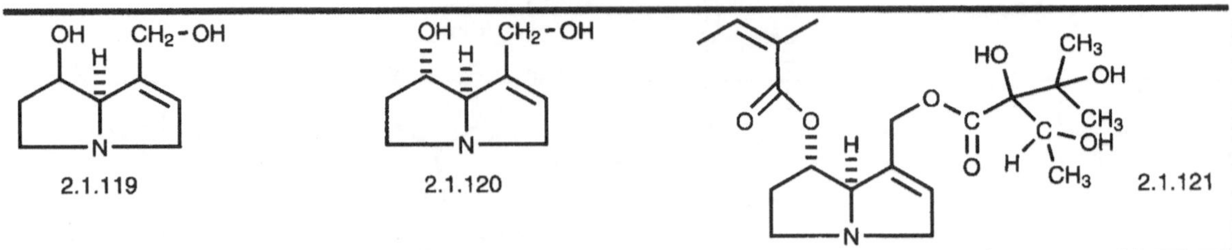

2.1.119   2.1.120   2.1.121

2.1.122   2.1.123

In conclusion we may briefly note that besides electron impact, other methods such as electron spray, chemical ionisation, field sorption, field ionisation and fast atom bombardment are also used to generate mass spectra of natural products. The electron-spray technique is a milder method which does not lead to much fragmentation; the molecular ion is readily identified, most often as M+1peak. In chemical ionisation, a reagent gas such as methane or ammonia or isobutene is subjected to electron impact and the ions thus produced allowed to interact with the compound under study. The advantage of this method is that the spectrum has fewer peaks due to less fragmentation thus enabling easier identification of the molecular ion, for example, as M +H.

Field desorption and field ionisation techniques are also softer than direct electron impact and involve electron tunneling from an emitter which is subjected to a high potential. The compound under examination is deposited on the emitter.

In fast atom bombardment (FAB), the substrate is dissolved in a liquid matrix such as glycerol and placed on a target which is bombarded by a beam of fast (6 keV) xenon atoms. As a result, molecular ions and fragments are generated from the substrate. The use of this technique in the study of natural products is becoming increasingly popular.

For finding out the exact molecular formula the method of choice is high resolution mass spectroscopy (HRMS). Examples in which these techniques have been used can be seen in the problems section of this book.

Suggested reading

1. Nakanishi K, Goto T, Ito S, Natori S and Nozoe S, *Natural Products Chemistry,* (eds), Vols 1–3, Tokyo: Kodansha and Mill Valley, California: University Science Books, 1983.
2. Paech K and Tracey MV, (eds), *Modern Methods of Plant Analysis,* Vols 1–7, Berlin: Springer-Verlag, 1956–64.
3. Markham KR, *Techniques of Flavonoid Identification,* London: Academic Press, 1982.
4. Nakanishi K and Solomon PH, *Infrared Absorption Spectroscopy-Practical,* 2nd edn, San Francisco: Holden-Day, 1977.
5. Kemp W, *NMR* in *Chemistry: A Multinuclear Introduction,* London: Macmillan, 1986.
6. Wehrli FW, Marchand AP and Wehrli S, *Interpretation of Carbon-13 NMR Spectra,* John Wiley, 1988.
7. Nakanishi K, (ed), *One-dimensional and Two-dimensional NMR Spectra by Modern Pulse Rechniques,* Tokyo: Kodansha and Mill Valley, California: University Science Books, 1990.
8. McLafferty FW, *Registry of Mass Spectral Data,* 5th edn, New York: Wiley, 1989.
9. Ulubelen A, Oksuz S, Kolak U, Bozok Johansson C, Celik C, Voelter W, *Planta Med.,* 2000, 66(5), 458–62.

10. Pivatto M, Crotti AEM, Lopes NP, Castro-Gamboa I, Amanda de Rezende, Viegas C, Young MCM, Furlan M, Bolzani VS, *J. Braz. Chem. Soc.*, 2005; 16, 1431–38.

We will now discuss a few specific examples chosen from different types of secondary metabolites as case studies. The first of these is strychnine.

2.2 Strychnine: A Hard Nut to Crack!

Among secondary metabolites of plant origin, the alkaloids were among the first to attract the attention of organic chemists as these compounds were well-known for their pronounced

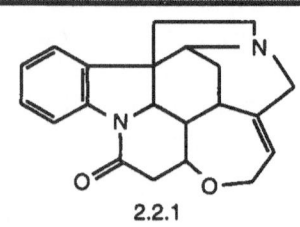

2.2.1

pharmacological and toxic properties. Several of these compounds were first isolated by the French chemists Pelletier and Caventou (Pelletier was later honoured when the pomegranate alkaloids were named after him). One of the compounds isolated by them was strychnine, the toxic component of *Strychnos nux-vomica*. The isolation of this compound did not pose any special problem and it could be obtained in a pure, crystalline condition, and in considerable quantities. In spite of this, its structure remained a mystery for nearly 130 years after its isolation. Robinson, who finally arrived at the correct structure, has told the story of the unravelling of the complex structure in his inimitable style. The following is a condensed version of the story which outlines only those features which have implications for structure–reactivity relationships in general, as exemplified by the experiences with strychnine (2.2.1).

The remarkable degree of reticence of strychnine in responding to chemical stimuli was ultimately traced to its compact, fused, heptacyclic skeletal framework. Mild methods evoked only routine responses which revealed just the peripheral features such as a lactam function, a basic tertiary nitrogen and a double bond, and not much more. On the other hand, under drastic conditions such as fusion with an alkali or distillation with zinc dust, the compound suffered extensive degradation and yielded small molecules like indole (2.2.2), tryptamine (2.2.3), carbazole (2.2.4) and 3-ethyl-4-methylpyridine (2.2.5).

2.2.2	2.2.3	2.2.4	2.2.5

However, these observations were useful, as they gave valuable information regarding parts of the skeletal structure. Such drastic methods, which were routinely employed for determining structures of organic compounds, are not always dependable as they can lead to rearrangements of sensitive structures under the impact of high temperature and strong alkalis. However, in the case of strychnine, the results of these reactions could be accepted as all of them gave either the same or similar products, indicating that no skeletal rearrangement had occurred. Further, the lack of reactivity of the compound under milder conditions showed its stability and resistance to breakdown. Of the 21 carbon atoms constituting the skeletal framework, along with the 2 nitrogen atoms, only 15 could be accounted for on the basis of the observations stated above. Thus, approximately a third of the molecule had been burnt beyond recognition under the drastic methods, whereas under milder conditions of probing, they remained concealed.

Tafel, and later Leuchs, had carried out some experiments which were remarkable for their fidelity. (Indeed, German chemical work of that period consistently had this mark of dependability.) Their observations were correctly interpreted later by Robinson, whose interest in strychnine had originated during his association with WH Perkin, Jr. Some outstanding work was also done by Wieland and co-workers [for example, the preparation of the Wieland–Gumlich aldehyde (2.2.6)], which

2.2.6

necessitated the revision of an earlier structure put forward by Robinson. With this preamble, we will consider the reactions which touched a chord, so to say, in the molecule and which brought out responses which could be understood in terms of fundamental chemistry.

Strychnine has two nitrogen atoms and two oxygen atoms. Of the two nitrogen atoms, only one is basic enough to enable the compound to form a monohydrochloride. The lack of basicity of the other nitrogen atom is due to its being part of a lactam group which includes one oxygen atom. The second oxygen atom was inferred to be a part of a cyclic ether function, mainly on the basis of negative responses to tests for carbonyl, hydroxyl and methoxyl groups. Since strychnine readily reacted with benzaldehyde (in the presence of a base) to form a benzal derivative (2.2.7) and with nitrous acid to yield an oximino derivative (2.2.8), the presence of an active methylene group could be inferred. The only activating group being the carbonyl of the lactam unit, this methylene could be placed next to it.

2.2.7   2.2.8

The presence of a double bond was readily revealed by the formation of a dihydro derivative (2.2.9) on catalytic hydrogenation. Its location somewhere in the neighbourhood of the basic nitrogen was indicated by a Raney nickel-induced isomerisation of strychnine to the less basic neostrychnine (2.2.10); in the latter, the double bond is next to the basic nitrogen as in an enamine and the resulting conjugative interaction between the lone pair of electrons on the nitrogen and the π electrons of the double bond is responsible for the reduction in base strength.

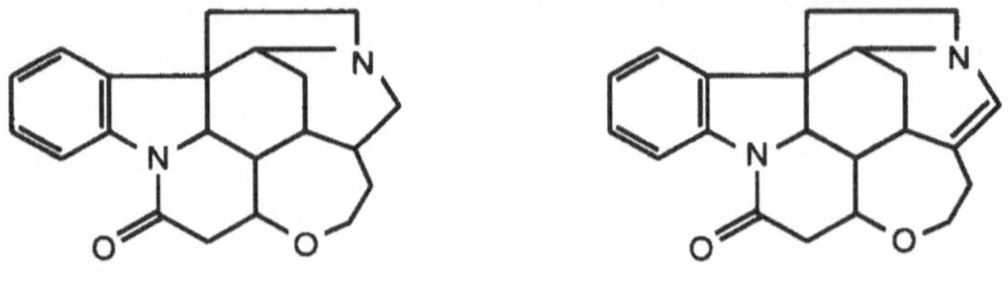

2.2.9   2.2.10

For a Raney nickel-induced isomerisation (which apparently involves hydrogenation and dehydrogenation) to have occurred, the double bond in strychnine should have been close enough to its new position (in neostrychnine) but farther away from the basic nitrogen. These deductions are perhaps not far-reaching as far as the structure of strychnine itself is concerned, but they are nevertheless remarkable as they link the compound to some fundamental concepts in chemistry. Such observations and their logical interpretations shift the emphasis from an individual compound to something more comprehensive and substantive. These part transformations are shown below.

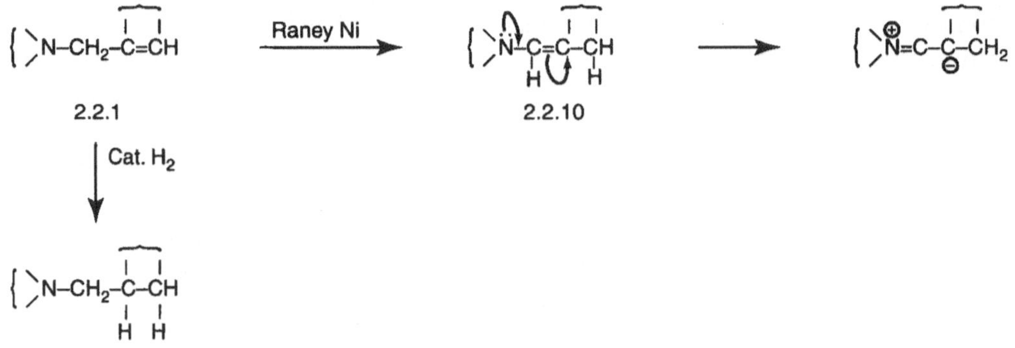

The only chink in the armour of strychnine is the double bond and this was the point of attack for a remarkable oxidative reaction which opened up the structure and paved the way for further degradation. What was remarkable was that this oxidation did not remove any of the carbon or nitrogen atoms of strychnine but added four more oxygens to the two already present. The reaction brought about by the action of potassium permanganate on strychnine in acetone–chloroform resulted in the formation of strychninonic acid (2.2.11). Strychnine has the molecular formula $C_{21}H_{22}N_2O_2$ whereas the formula of strychninonic acid is $C_{21}H_{20}N_2O_6$. The latter is a monocarboxylic acid but is devoid of any basic property indicating that the original basic, tertiary nitrogen has also been rendered non-basic, presumably by conversion to an amide. This would have been possible only if, in strychnine, the basic nitrogen was next to a methylene which should have been next to the double bond—the prime target for oxidative attack. Strychninonic acid, indeed, behaves as an α-ketoamide as shown by the following observation. When heated with barium hydroxide and hydrogen peroxide, strychninonic acid gave barium carbonate and an amino dicarboxylic acid (2.2.12). These transformations could be explained as shown below, and thus led to the recognition of the part structure shown therein.

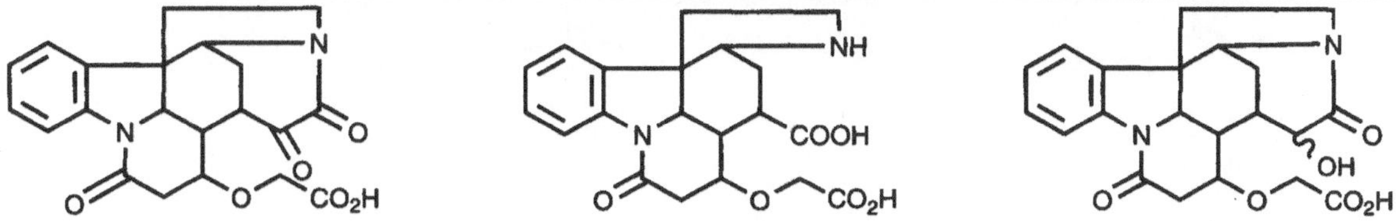

| 2.2.11 | 2.2.12 | 2.2.13 |

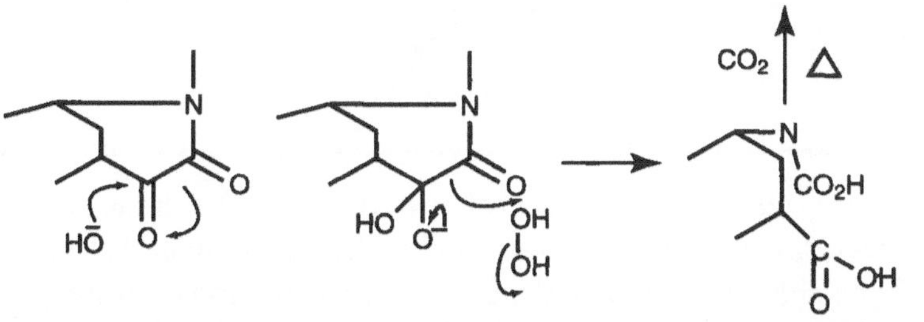

Dihydrostrychninonic acid (2.2.13), obtained by the reduction (with sodium amalgam) of strychninonic acid, lost two carbon atoms and three oxygen atoms in the form of a molecule of glycollic acid on treatment with aqueous alkali. The hydroxyl of the glycollic acid must have been the cyclic ether function of strychnine; in strychninonic acid, it had got transformed to -O-CH$_2$-COOH. The extraordinary rupture of an ether linkage under the influence of a base must be the result of a β-elimination—the activating group for which is the carbonyl group attached to nitrogen as shown below.

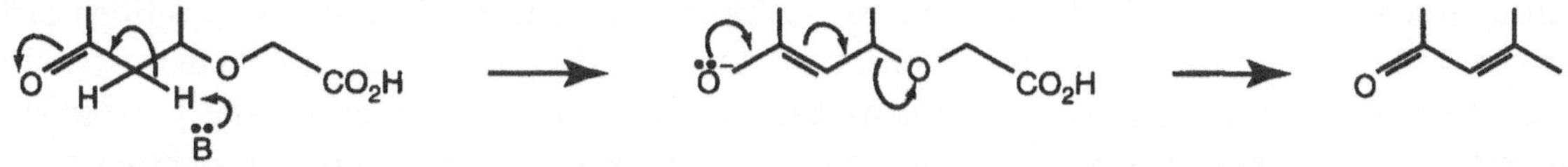

The resulting product, strychninolone A (2.2.14), was indeed shown to possess a double bond in agreement with the above interpretation. Further action of alkali brought about a reversible rearrangement of strychninolone A into strychninolone B (2.2.15) and then on to strychninolone C (2.2.16). Strychninolones A and C are diastereoisomers whereas strychninolone B is a structural isomer of either of them. Thus, while strychninolones A and C yielded the corresponding dihydro derivatives, which were diastereoisomers, strychninolone B yielded a mixture of dihydro derivatives of strychninolone A and C. These results are also shown below. The part structures of strychninolones A, B and C were further supported by the results of oxidative and subsequent hydrolytic degradation of the three isomers.

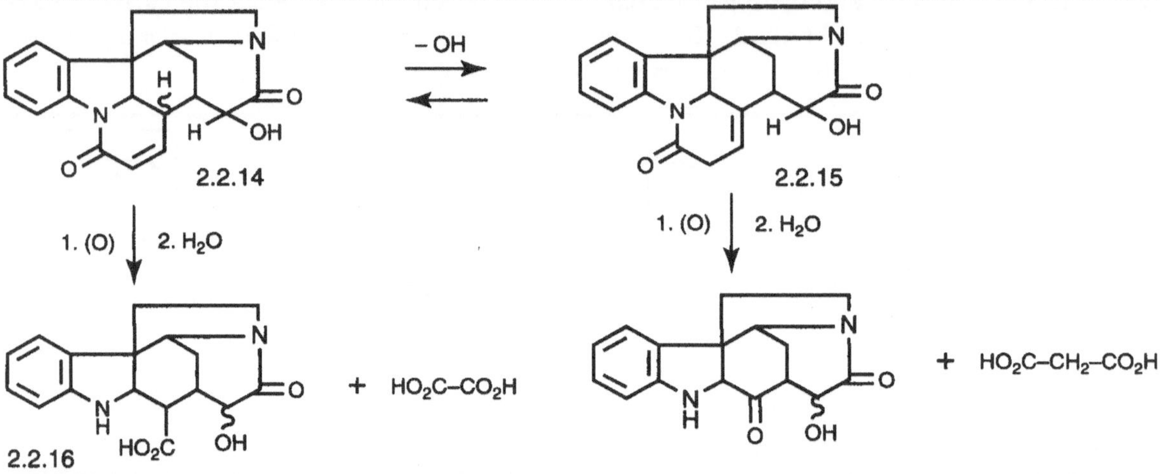

2.2.14

2.2.15

2.2.16   HO₂C   OH   +   HO₂C–CO₂H

+   HO₂C–CH₂–CO₂H

This is as far as one could go using chemical methods of probing into the structure of strychnine. Summarising what has been said so far, one may note that the experimental strategy involved the systematic opening of one side of the molecule, beginning at the most vulnerable point, namely the double bond, thereby generating data which were amenable to a logical interpretation. In this respect, oxidative and hydrolytic reactions, which can be understood in terms of mechanistic principles, are the most suitable (if they work, in the particular case under investigation) as chemical tools for structure elucidation.

Among the drastic methods used in the case of strychnine, perhaps, the mildest and therefore more readily acceptable than the others, was the use of nitric acid as an oxidising agent. The product was dinitrostrychol carboxylic acid (2.2.17), which was characterised, by further degradation and synthesis, as 5,7-dinitroindole-2,3-dicarboxylic acid. It is tempting to assume that in this case, nitration presumably precedes oxidative degradation; strychnine can be considered as an acetanilide derivative and therefore nitration would occur at positions *ortho* and *para* to the acetamido function, as observed. On the other hand, if an indole derivative had been the primary product, subsequent nitration ought to have occurred on the pyrrole ring. In certain colour reactions, such as the Otto reaction (where blue colour is generated with a trace of potassium dichromate in the presence of sulphuric acid), strychnine, indeed, resembles acetanilide. Incidentally, colour reactions have been used as diagnostic tools for the detection of specific functional groups or even structural units and types, though a majority of them are too empirical to be accepted on par with other methods based on firmer mechanistic grounds.

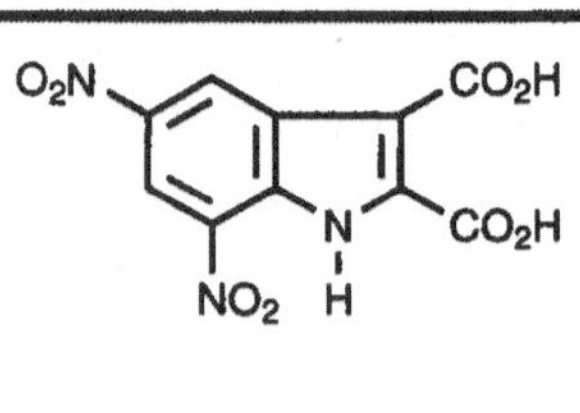

2.2.17 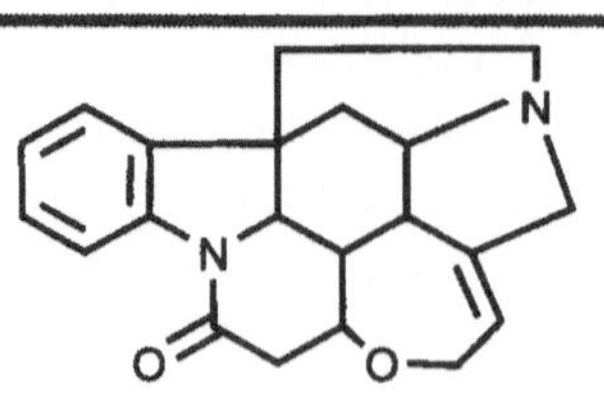   2.2.18 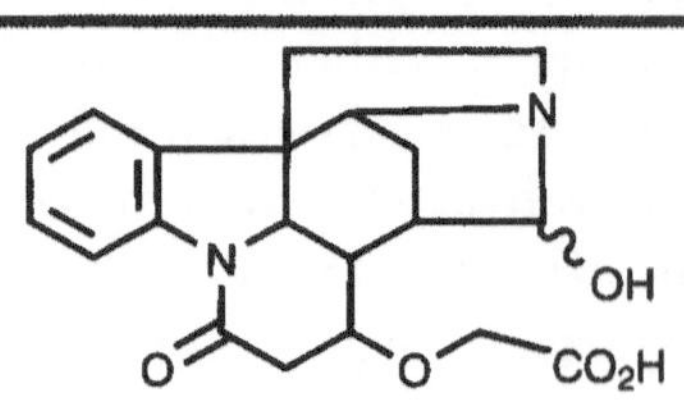   2.2.19

Mainly on the basis of the above observations, and supported by a few others, strychnine was first formulated as 2.2.18. The points of linkage, for which there was no direct experimental evidence, were merely reasonable conjectures. The structure was later revised to that shown in Fig. 2.2.1 to account for the formation of 3-ethyl-4-methylpyridine as one of the products of alkali fusion; by no stretch of imagination can one think of its formation from 2.2.18. Structure 2.2.1 was also in agreement with the fact that the Wieland–Gumlich aldehyde (2.2.6) formed a stable, cyclic aza-acetal (2.2.19) which should therefore be five-membered rather than four-membered. That the structure thus arrived at entirely by a chemical approach has been confirmed by x-ray diffraction analysis is proof enough for the reliability of the chemical method. It was undoubtedly time- and material-consuming, but the method, being a composite of a number of different types of reactions, generated a good deal of new chemistry while illustrating several known principles.

One may recall what Robinson said about the excessive use of modern spectroscopic methods to the exclusion of chemical methods in structure elucidations: "Would it have been good for the development of Organic Chemistry if Baeyer had been able to ascertain the molecular structure of indigotin by applying physical techniques to a few milligrams?". From this statement it should not be misconstrued that chemical methods are superior to physical techniques. Indeed, they are not; but they provide information which the instrumental methods do not give. The best results are obtained from a judicious combination of the two approaches as illustrated by some subsequent examples in this chapter.

Suggested reading

1.   Robinson R, In Cook JW, (ed.), *Progress in Organic Chemistry*, Vol. 1, London: Butterworths, 1952.
2.   Cordell GA, *Introduction to Alkaloids—A Biogenetic Approach*, New York: John Wiley & Sons, 1981.

2.3 NEPITRIN AND PEDALIIN: A CASE OF MISTAKEN IDENTITY!

Polyhydric phenols occur widely in nature. The most widespread and well-studied compounds among plant polyphenols are the flavonoids.

It is now generally accepted that the main physiological functions of these compounds in plants are associated with two properties, namely their strong capacity for absorbing ultraviolet (UV) light and their susceptibility to oxidation. An important function is to act as inhibitors of electron transport in mitochondria. Leaf flavonoids probably function as UV screening agents and also in conjunction with the photosynthetic pigments for the effective utilisation of incident sunlight.

The determination of structures of flavonoids is a comparatively straightforward and routine affair but to grasp the nuances in their structures would require an intuitive approach. Besides a number of phenolic hydroxyls, flavonoids (with the exception of flavans and flavan 3,4-diols and their derivatives) also possess a carbonyl group which plays a key role in differentiating the hydroxyls at different positions. To be more specific, the ionisabilities, and therefore, their reactivities towards, for example, alkylating agents, are significantly controlled by the carbonyl group which is a constituent of the pyrone ring. This feature is the underlying principle for a very reliable spectroscopic method routinely used these days for finding out the disposition of free hydroxyls in a flavonoid. By and large devoid of stereochemical complexities, these compounds are also readily amenable to chemical manipulations which yield unambiguous responses.

The structures thus determined can also be confirmed, in most cases, by syntheses which do not require a Woodward, an Eschenmoser or a Corey for execution—you and I can do it! Thus, a flavonoid structure can be determined with the help of a 'tool box', containing almost every common tool in the organic chemist's workshop but nothing which is extraordinary. Someone aspiring to become a structural chemist could probably begin with a flavonoid as the first test case to learn the tricks of the trade. One does stumble occasionally, even on such firm ground, particularly when the data are based heavily on one set of experiments (say, spectroscopy) or when the identification of a key degradation product is made on the basis of chromatographic data. In this section we will be discussing one such example, where there was a confusion of identity due to a minor aberration in the experimental strategy.

Sesamum indicum is widely cultivated in India for the oil obtained from its seeds. This plant belongs to the family Pedaliaceae. About fifty years ago, Morita isolated a flavone glucoside which he formulated as 7-glucosyloxy-6-methoxy-5,3',4'-trihydroxyflavone (2.3.1). The reactions which enabled Morita to arrive at this structure are summarised as follows. The analytical evidence was in perfect agreement with the molecular formula, which includes a methoxyl group, three free phenolic hydroxyls and a glucose unit in glycosidic linkage. A few years later, Krishnaswamy, Seshadri and Tahir isolated a flavone glucoside with the same molecular formula from *Nepeta hindustana* of the family Labiateae. Their systematic studies, as summarised below, led to this compound, named as nepitrin, being assigned the same structure that was earlier given to pedaliin. A direct comparison between the two compounds, however, showed that they were different. Therefore, the structure of pedaliin was re-examined and on the basis of the observations summarised below, the compound was reformulated as 6-glucosyloxy-7-methoxy-5,3',4'-trihydroxyflavone (2.3.2). No detailed analysis of the results summarised in these diagrams is included here, as most of the observations are readily understood and do not require any commentary. However, there are a few points which deserve a little explanation.

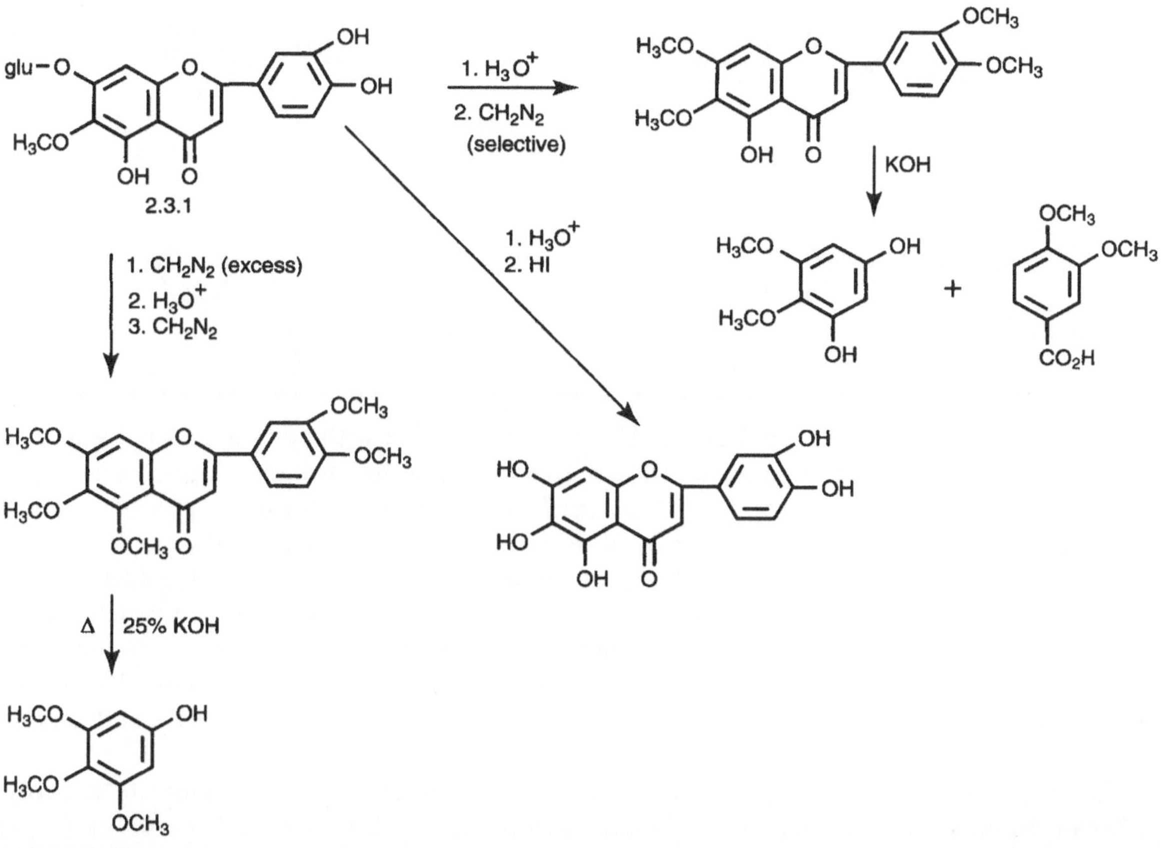

For example, if one carefully examines the data in the diagrams, one can see that the only flaw in Morita's work was the identification of a degradation product on the basis of R_f value as determined by a paper chromatographic analysis. Chromatographic data are useful but they have to be supported by other data for identification purposes.

The UV spectra of pedaliin and nepitrin (the two isomeric glucosides) are practically indistinguishable from each other. On the other hand, there is one significant difference in the spectra of the corresponding aglucones, namely pedalitin (2.3.3) and nepetin (2.3.4). Thus, in the spectrum of the former, the shorter wavelength absorption band at 285 nm is 10 nm higher than the corresponding band (at 275 nm) in the spectrum of nepetin. Both the compounds have the longer wavelength absorption band at 345 nm. The shorter wavelength absorption band is due to the chromophoric part present in the A ring, extended by the pyrone carbonyl. The differences in the disposition of the free hydroxyls and the lone methoxyl group in these two isomeric aglucones are thus reflected in this region of the spectra. There is supporting evidence for this in the following observation. The spectrum of pectolinarigenin, which is 6,4'-dimethoxy-5,7-dihydroxyflavone (2.3.5), has absorption maxima at 277.5 nm and 334 nm, whereas scutellarein 7,4'-di-O-methyl ether, that is, 7,4'-dimethoxy-5,6-dihydroxyflavone (2.3.6), absorbs at 285 nm and 330 nm.

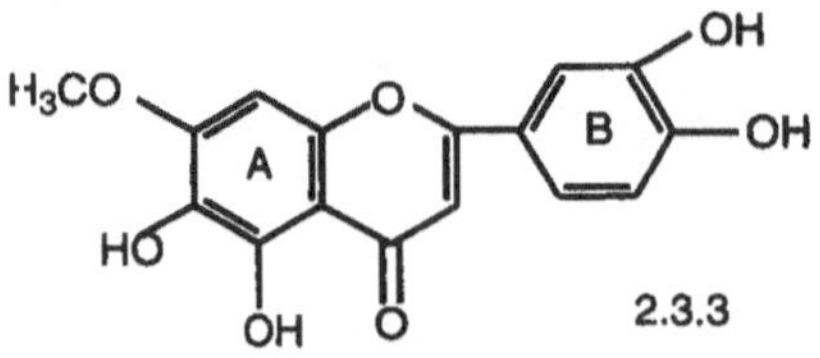

Apart from this empirical application, UV spectra recorded with certain shift reagents give definitive information regarding the location of free hydroxyls in flavonoids. Thus, for example, the presence of a free hydroxyl at position 5, which is hydrogen-bonded to the pyrone carbonyl in pedaliin, nepitrin and their aglucones, is revealed by bathochromic shifts, particularly in the longer wavelength band, on the addition of aluminium chloride as a chelating agent. In the absence of HCl these shifts range from 15–18 nm, as in the case of nepitrin and nepetin, to as high as 75 nm as observed with pedaliin and pedalitin. Similarly, the presence of a catechol group (1,2-dihydroxybenzene unit) in the B ring is shown by the 15–35 nm bathochromic shift on the addition of sodium acetate–boric acid. In this case, the shift is greater with nepitrin and nepetin than with pedaliin and pedalitin.

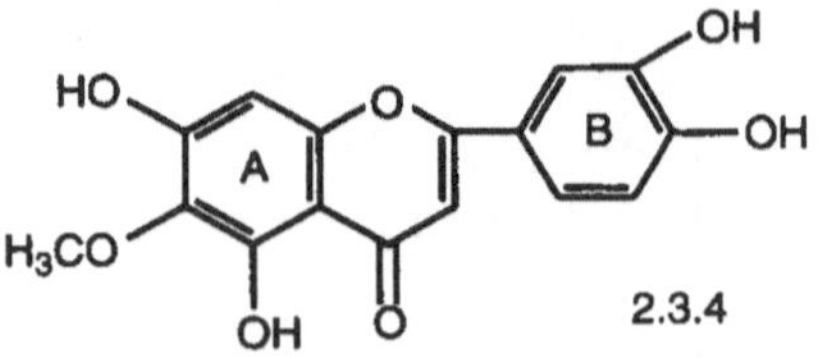

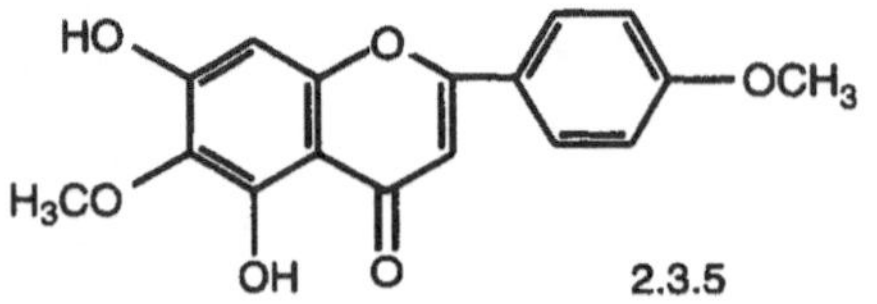

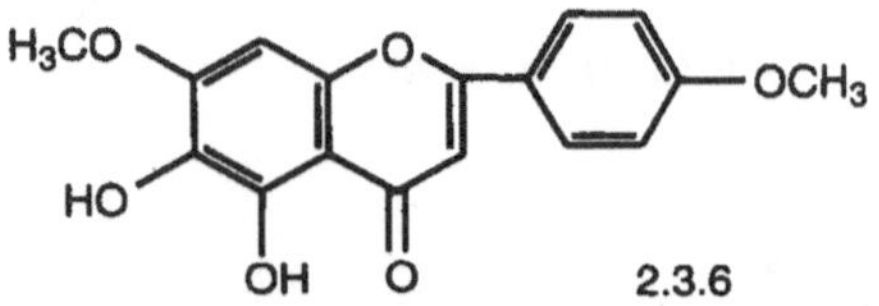

The position of the methoxyl group in nepetin was confirmed by converting it into its tetraethyl ether and by establishing the identity of the latter as 6-methoxy-5,7,3',4'-tetraethoxyflavone (2.3.7) by direct comparison with a synthetic sample.

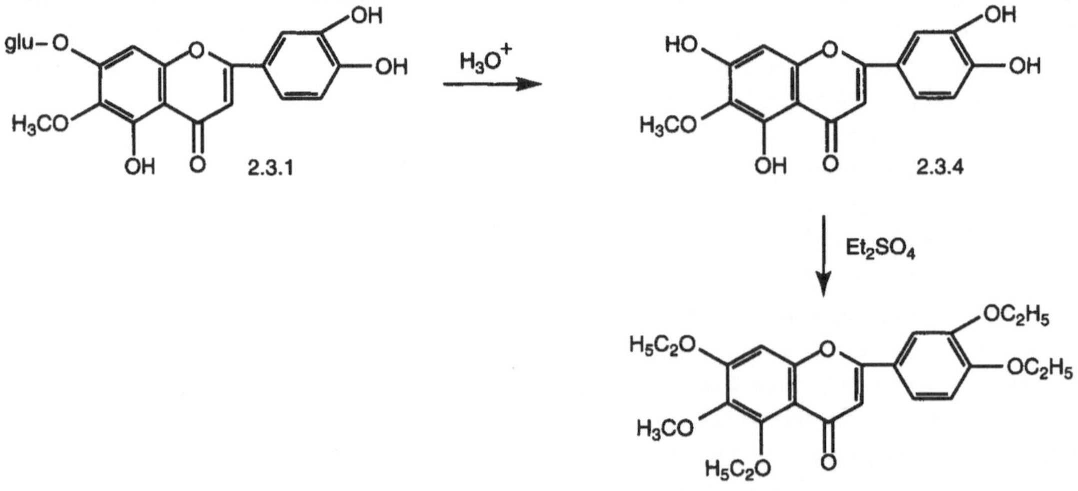

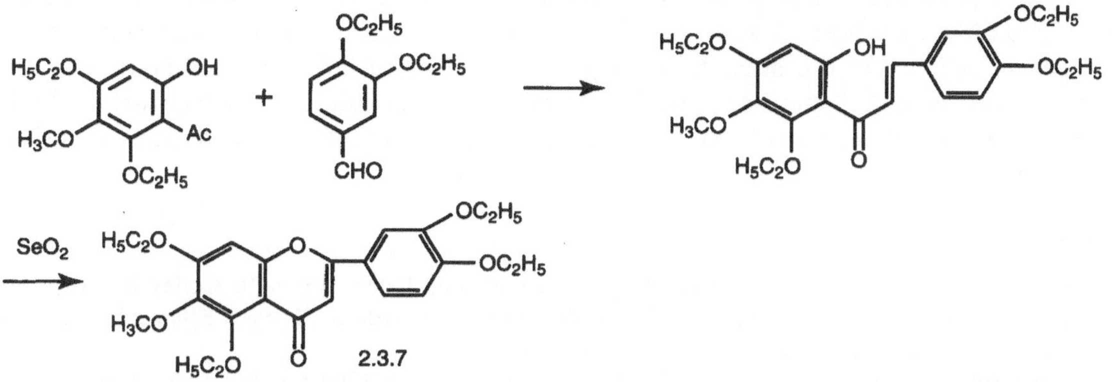

This is an example of a classical approach in which structural studies include a synthetic component as an integral part of analysis. For locating the position of the sugar residue in pedaliin, the permethyl ether of the glucoside was hydrolysed and the product converted into its O-ethyl ether. This compound was found to be 6-ethoxy-5,7,3′,4′-tetramethoxyflavone (2.3.8), thus proving that the glucose is attached to position 6 in pedaliin.

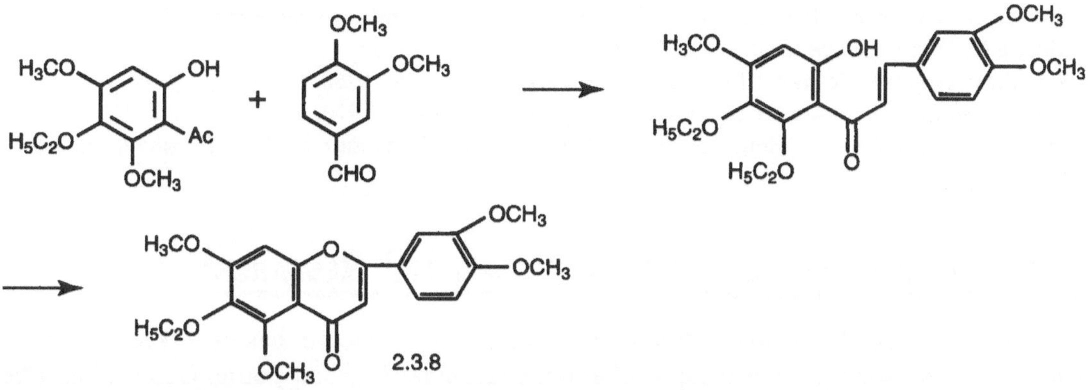

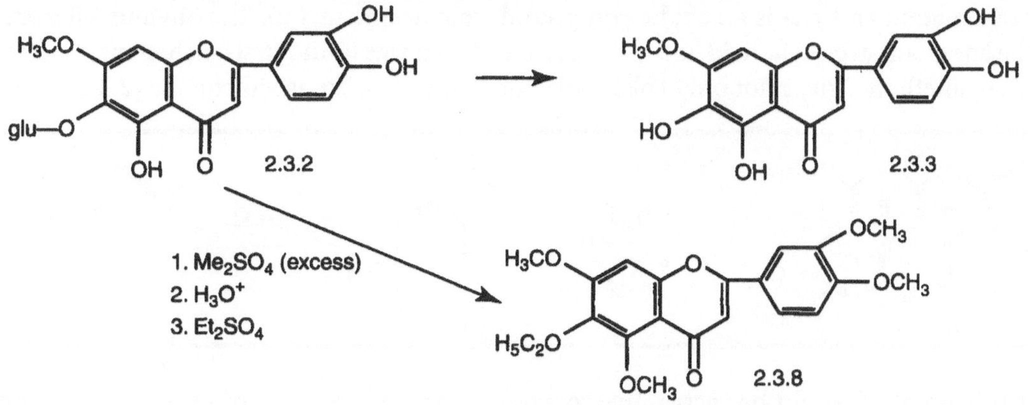

Another point worth noting about pedalitin is its sensitivity towards alkalis, unlike nepetin. Thus, pedalitin dissolved in aqueous alkali giving a transient green colour which rapidly changed to yellow. On the other hand, nepetin gave a stable solution (that is, no change of colour was observed). Apparently, pedalitin undergoes oxidation (by dissolved air) in alkaline medium due to the presence of a catechol group, in the form of the free hydroxyls at positions 5 and 6 in ring A.

Suggested reading

1. Livingstone R, Six-membered ring compounds with one hetero atom, In Coffey S, (ed), Rodd's *Chemistry of Carbon Compounds*, 2nd edn, Vol. 4 (E), Amsterdam: Elsevier Science Limited, 1977 (for a general account of flavonoids and isoflavonoids).
2. Harborne JB, (ed.), *The Flavonoids—Advances in Research since 1980*, London: Pergamon Press, 1988 (for recent advances in the chemistry of flavonoids, isoflavonoids and neoflavonoids).
3. Morita N, *Chem. Pharm. Bull.*, 1960, 8: 59.
4. Krishnaswamy NR, Seshadri TR and Tahir PJ, *Indian J. Chem.*, 1968, 6: 676.
5. Krishnaswamy NR, Seshadri TR and Tahir PJ, *Indian J. Chem.*, 1970, 8: 1074.
6. Kupchan SM, Sigel CW, Hemingway RJ, Knox JR and Udayamurthy MS, *Tetrahedron*, 1969, 25: 1603.

2.4 COLCHICINE: STRUCTURE FROM 'ARMCHAIR' RESEARCH!

The meadow saffron, *Colchicum autumnale*, of the family Liliaceae, has been known for more than a century as a source of the unique alkaloid, colchicine. This compound is indeed unique in more than one sense. Unlike almost every other alkaloid, it is not basic, but like all alkaloids it has an amino acid (phenylalanine) as its primary biosynthetic precursor. It has a tropolone ring and is photosensitive. As a matter of fact, the concept and chemistry of tropolones originated from the puzzling (at that time) chemical behaviour of colchicine. This was perhaps the first case where an interpretation of experimental data based on first principles—that is, a theoretical approach—gave a satisfactory solution to the problem.

Colchicine (2.4.1) has the molecular formula $C_{22}H_{25}NO_6$. On controlled acidic hydrolysis it loses one carbon atom and yields an acidic compound, colchiceine (2.4.2). The obvious interpretation that the latter is a carboxylic acid had to be revised when it was found that colchiceine, on treatment with diazomethane, gives not only colchicine but also an isomer, isocolchicine (2.4.3).

At this point, it would be instructive to examine the mechanism of *O*-methylation brought about by diazomethane. This reagent, being dipolar ionic in nature, has two canonical forms which render it a weak carbanionic nucleophile/base. To act as a methylating agent, it has to abstract a proton from an acidic-enough hydroxyl group, such as the -OH of a carboxyl group or a phenol or an enol. Being a weak base, diazomethane by itself is incapable of removing a proton from an alcoholic hydroxyl group; in such cases the hydroxyl can be made acidic enough by prior coordination with a Lewis acid, such as boron trifluoride.

With this background, when we examine the colchiceine to colchicine/isocolchicine transformation it becomes apparent that the hydroxyl group undergoing methylation (in colchiceine) is an enolic -OH in conjugation with a carbonyl group. In the light of earlier observations suggesting a tricyclic skeletal structure for colchiceine, the enolic hydroxyl and the carbonyl group, connected by a conjugated system of double bonds, could only be accommodated in a unique seven-membered ring structure as conceived by Dewar; he called it the tropolone ring. Of the six oxygen atoms present in colchiceine, one could be accounted for as part of an *N*-acetyl function, three as methoxyl groups and one, as pointed out above, as an enolic-OH. The sixth oxygen atom was eventually characterised as that of an inert carbonyl group. This is the carbonyl group of the tropolone ring. Not surprisingly, earlier researchers interpreted the available data in terms of a tricyclic structure in which all the three rings are six-membered. The

correct structure could be deduced only after a reinterpretation of the data, particularly with regard to the reactions concerning ring C. Dewar also drew attention to the fact that a few other naturally occurring compounds, for example the mould metabolite, stipitatic acid (2.4.4), exhibited similar properties which could be traced to the presence of a tropolone ring.

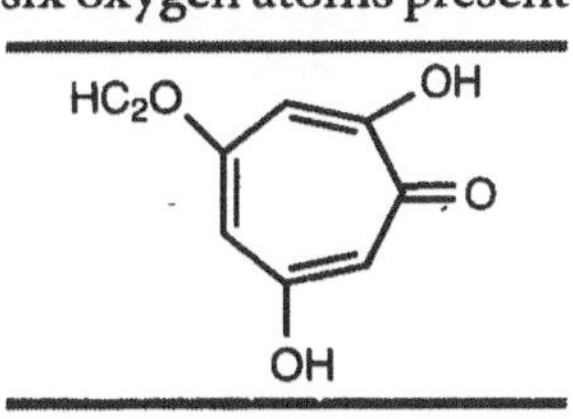

The tropolone ring in colchicine, though aromatic, is susceptible to rearrangements and shows a remarkable tendency for transformation into a benzenoid ring. Thus, for example, when treated with sodium methoxide in methanol, colchicine suffered a C ring contraction to yield allocolchicine (2.4.5), which possesses a carbomethoxy group. The rearrangement apparently occurred due to an initial attack of the methoxide ion on the carbonyl carbon followed by a skeletal rearrangement and simultaneous elimination of the original methoxyl group as shown below.

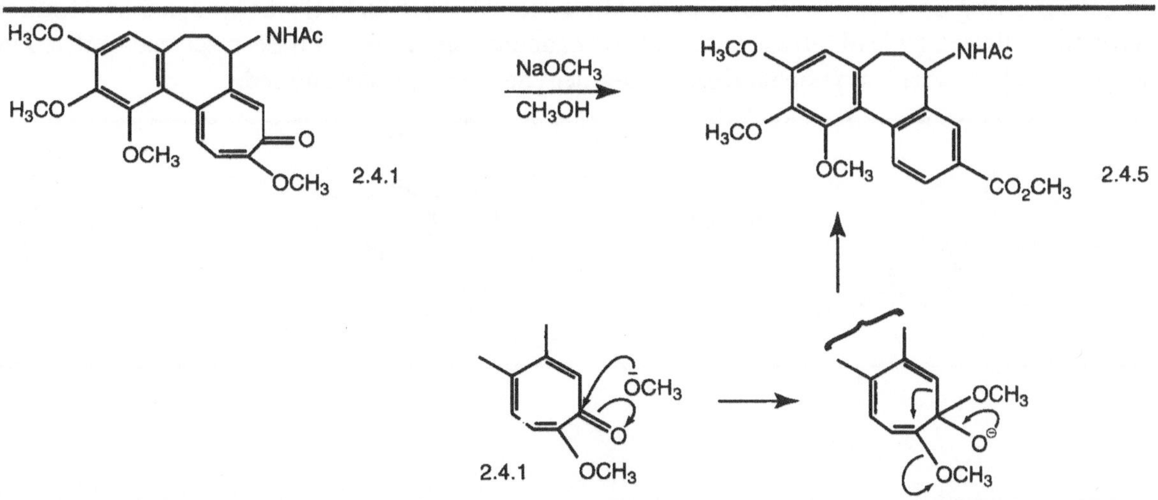

The other reactions which led to the elaboration of the structure of colchiceine are as follows. On rigorous oxidation, colchiceine yielded 3,4,5-trimethoxyphthalic acid (2.4.6). On acid-catalysed hydrolysis colchiceine (2.4.2) gave one mole of acetic acid and trimethylcolchiceinic acid (2.4.7) which answered tests for a primary amino group. Colchiceine could be converted into N-acetylcolchinol methyl ether (2.4.8) by a two-step process involving treatment with alkaline hydrogen peroxide followed by methylation with diazomethane. Hydrolysis of 2.4.8 gave colchinol methyl ether (2.4.9) which could be oxidised to 4-methoxyphthalic acid (2.4.10).

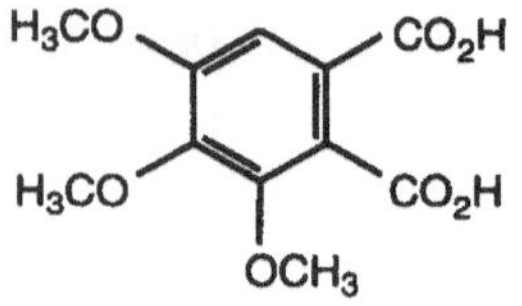

2.4.6 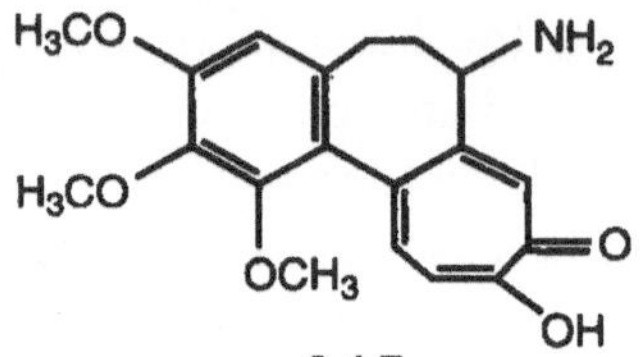   2.4.7 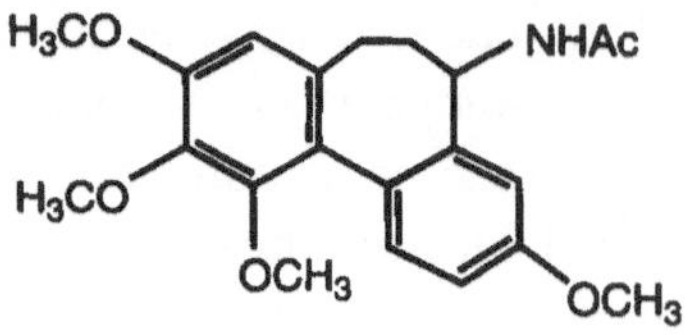   2.4.8

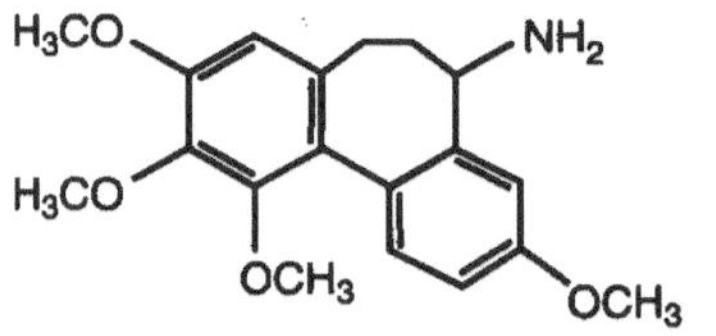

2.4.9 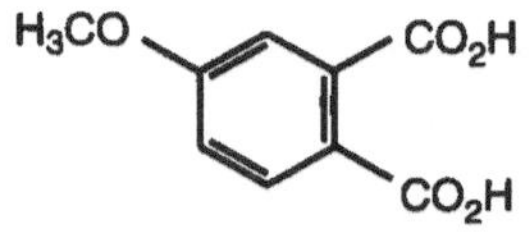   2.4.10

The methiodide of 2.4.9 on treatment with a base (Hoffmann degradation) gave deaminocolchinol methyl ether (2.4.11). The latter on oxidation with potassium dichromate yielded a mixture of the phenanthrene quinone (2.4.12) and a ketone (2.4.13). The structures of these compounds were confirmed by synthesis. Thus, 2.4.13 was prepared from 2,3,4,7-tetramethoxy-10-methylphenanthrene (2.4.14) by a three-step sequence, proceeding through 2.4.15 and 2.4.16 as shown overleaf. The formation of the ketone 2.4.13 from deaminocolchinol methyl ether proved that the middle ring in colchicine is seven-membered. This inference was also supported by the following observation. Deaminocolchinol methyl ether (2.4.11) was oxidised by osmium tetroxide followed by lead tetraacetate and the product subjected to a base-catalysed cyclisation when 2,3,4,7-tetramethoxyphenanthrene-10-aldehyde (2.4.17) was obtained.

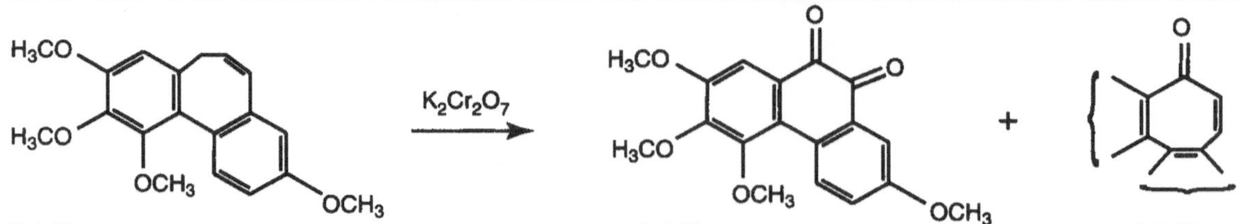

2.4.11   2.4.12   2.4.13

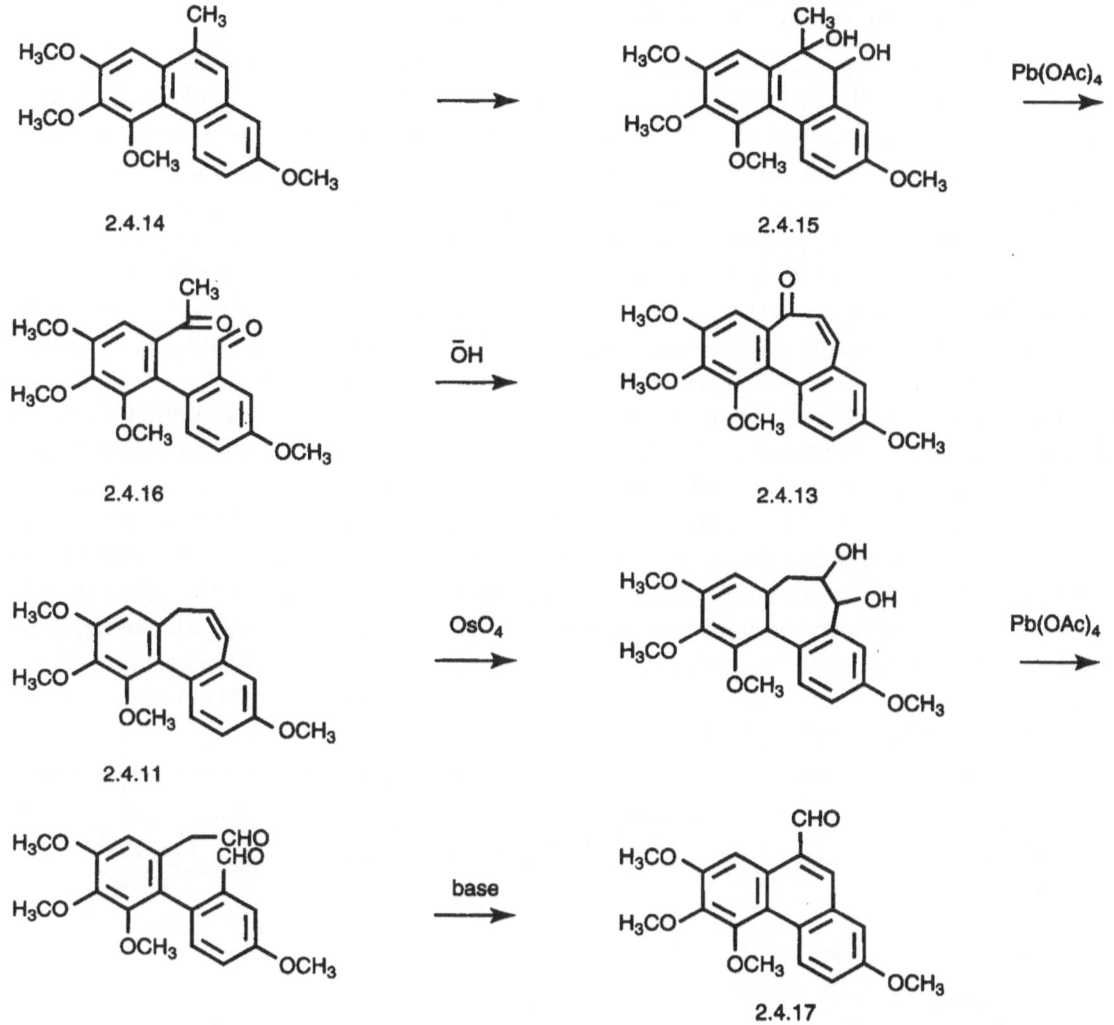

2.4.14   2.4.15

2.4.16   2.4.13

2.4.11

2.4.17

Suggested reading

1. Dyke SF, The Isoquinoline Alkaloids, In Coffey S, (ed), Rodd's *Chemistry of Carbon Compounds*, 2nd edn, Vol. 4 (H), Amsterdam: Elsevier Science Limited, 1978.
2. Dewar MJS, *Nature*, 1945, 155: 50–141.
3. Ito S, in Nakanishi K, Goto T, Ito S, Natori S and Nozoe S, (eds), *Natural Products Chemistry*, Vol. 2, Tokyo: Kodansha Ltd, 1973,.
4. Chapman OL, Smith GH and King RW, *J. Amer. Chem. Soc.*, 1963, 85: 806.

2.5 LONGIFOLENE: A MOLECULAR ACROBAT!

Sir Robert Robinson once called morphine 'a star performer among molecular acrobats'. We are not sure whether the sesquiterpene hydrocarbon, longifolene (2.5.1), could qualify for this unique title but it is an acrobat all right, just like many other terpenoids. Its tendency to undergo rearrangements even under mild conditions, coupled with its unique tricyclic skeletal structure made it an enigmatic molecule whose structure could ultimately be solved only by x-ray diffraction analysis. Longifolene is a component of the oleo resins obtained from *Pinus longifolia* and *Pinus maritima*, and was first studied by the pioneer in terpene chemistry—Sir John Simonson.

Longifolene is a comparatively small hydrocarbon with molecular formula $C_{15}H_{24}$. The presence of a double bond was readily shown by its reactions with perbenzoic acid or OsO_4, followed by lead tetraacetate or ozone, all of which resulted in the liberation of a molecule of formaldehyde and a compound (2.5.2) with the molecular formula $C_{14}H_{22}O$. That the latter was a ketone was indicated only by the strong absorption band at 1740 cm^{-1} in its infrared (IR) spectrum. Otherwise, the compound failed to react with most of the common and characteristic diagnostic reagents for carbonyl groups, even under rigorous conditions. It was also unaffected by selenium dioxide or bromine or cold alkaline potassium permanganate. These observations suggested that the carbonyl group was most probably present in a five-membered ring and that the adjacent carbons were either fully substituted or were in bridge-head positions. The ketone was attacked by sodium amide in boiling xylene and an amide $C_{14}H_{25}ON$ (2.5.3), was obtained. The presence of the exo-cyclic double bond in longifolene was also shown by the facile formation of the dihydro derivative (longifolane; 2.5.4) on catalytic hydrogenation.

Even the addition of HCl to the double bond brought about a rearrangement. This was shown by the following observation. Longifolene hydrochloride on reduction with sodium and alcohol gave a compound $C_{15}H_{26}$ (2.5.5) which was different from longifolane.

Summing up the discussion up to this point, we find, from the data presented above, that longifolene is a mono-unsaturated tricyclic hydrocarbon which is highly susceptible to skeletal rearrangements and that it contains a vinylidene group which is, most probably, situated between two bridge-head positions. The molecular formula indicates that it is a sesquiterpene, and the susceptibility to rearrangements brings to mind the chemistry of camphene. Indeed, the optical rotatory dispersion (ORD) curves of (+)camphene (2.5.6) and (+)longifolene are very similar to each other. That longifolene does possess the camphene skeleton with an additional isoprene 'cap' was conclusively proved by x-ray diffraction analysis of longifolene hydrochloride. The reactions of longifolene are shown below. One interesting reaction is the formation of a secondary alcohol by successive treatment of longifolene with trifluoroacetic acid and lithium aluminium hydride. Apparently, the rearrangement involves a transannular hydride shift in the intermediate longibornyl carbenium ion.

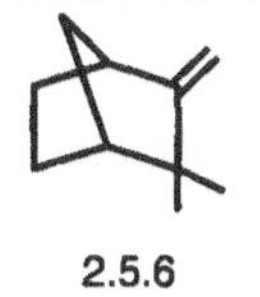

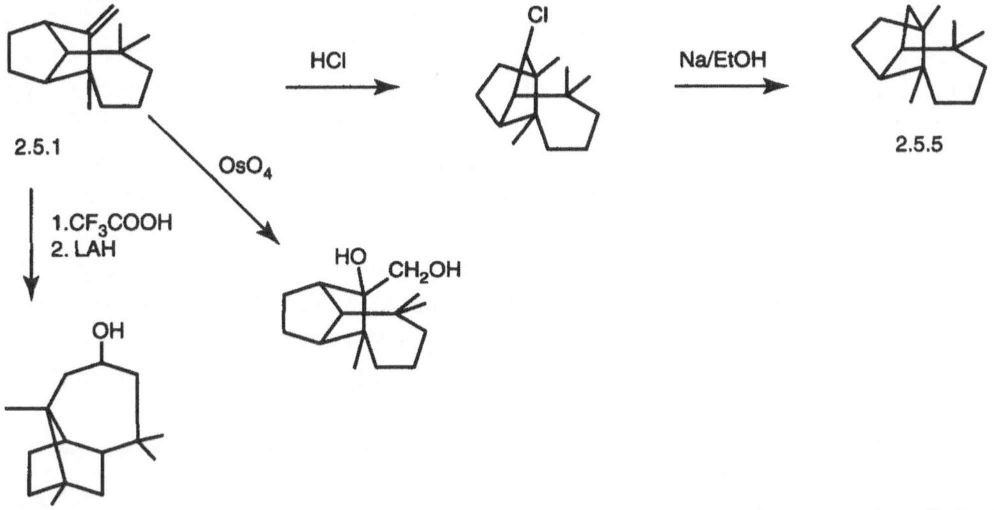

2.5.6

Suggested reading

1. Bryant R, in *Rodd's Chemistry of Carbon Compounds*, In Coffey S, (ed), Rodd's *Chemistry of Carbon Compounds*, 2nd edn, Vol. 2 (C), Amsterdam: Elsevier Science Limited, 1969.
2. Simonson JL, *J. Chem. Soc.*, 1920, 117: 578.
3. Moffett RH and Rogers D, *Chem. & Ind.*, 1953, 916.
4. Nozoe S, In Nakanishi K, Goto T, Ito S, Natori S and Nozoe S, (eds), *Natural Products Chemistry*, Vol. 1, Tokyo: Kodansha Ltd, 1974.

2.6 β-AMYRIN: A 'STRAIGHT' MOLECULE

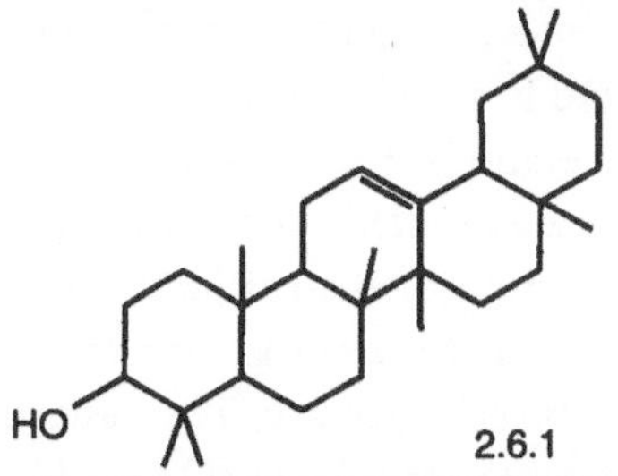

β-Amyrin (2.6.1) is a pentacyclic triterpene derived from six mevalonate units via the acyclic hydrocarbon squalene (2.6.2). The latter is obtained by the tail-to-tail combination of two farnesyl pyrophosphate units as described in Chapter 6. The conversion of squalene epoxide into amyrin and other triterpenes and sterols will also be discussed therein. We may,

2.6.1

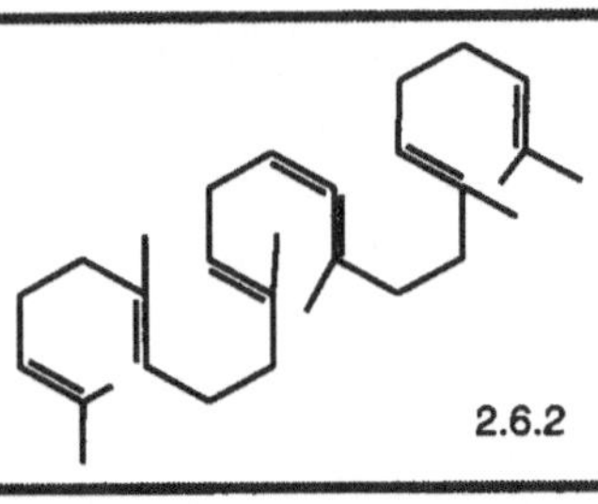

however, note here that the positions of the C-methyl groups in β-amyrin are pre-determined by its biosynthetic origin from squalene.

β-Amyrin, with molecular formula $C_{30}H_{50}O$, is more than twice as large as longifolene, but unlike the latter it is not a 'wayward' molecule! Its responses to various probes, physical and chemical, are straightforward and, therefore, it can be taken as a model compound, parts of which could be taken

2.6.2

apart and put back again. It occurs widely in the plant kingdom and is a major component of the Manila elemi resin. It occurs in nature both in the free state and in the form of esters such as the acetate, palmitate and myristate. The molecular formula indicates that it is a triterpene and this is supported by the characteristic play of colours given by the compound on treatment with the Liebermann–Burchard reagent (acetic anhydride–sulphuric acid). The presence of a secondary hydroxyl group is revealed by the formation of acyl derivatives as well as by the facile (Oppenauer) oxidation to the corresponding ketone. That the compound is mono-unsaturated is shown by titration with per-acids and by the colour reaction with tetranitromethane. However, β-amyrin resists catalytic hydrogenation, indicating that the double bond is sterically unapproachable to the catalyst surface due to a high order of substitution. The UV spectrum of the compound indicates that it has a trisubstituted ethylene chromophore which implies the presence of an olefinic hydrogen; this was confirmed by the NMR spectrum. On selenium dehydrogenation—a technique which was introduced by Ruzicka in his extensive studies on terpenoids—β-amyrin yielded a number of naphthalene derivatives and, more significantly, two pentacyclic aromatic compounds (2.6.3 and 2.6.4) as shown below.

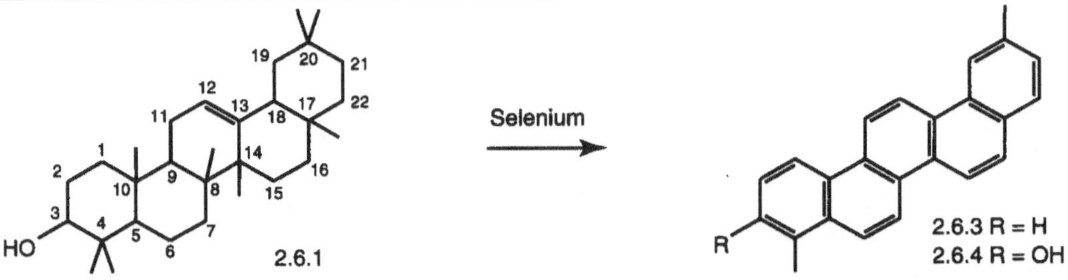

Selenium

2.6.1

2.6.3 R = H
2.6.4 R = OH

The reactions which were designed to find out the location of the double bond in β-amyrin are very instructive. As mentioned in the discussion on strychnine, a double bond in a compound is an easy target for attack by a suitable oxidising agent. In the case of β-amyrin, the hydroxyl group had first to be protected by acetylation before the double bond could be targeted. Thus, β-amyrin acetate (2.6.5) was subjected to oxidation with potassium persulphate or hydrogen peroxide in acetic acid. The product was a ketone (2.6.6) which was apparently formed by the oxidative opening of the initially formed epoxide (2.6.7). This conjecture is supported by the fact that the same ketone could also be obtained by the action of perbenzoic acid or monoperphthalic acid on β-amyrin acetate. The ketone (2.6.6) on Wolff–Kishner reduction yielded the saturated alcohol, β-amyranol, $C_{30}H_{25}O$ (2.6.8).

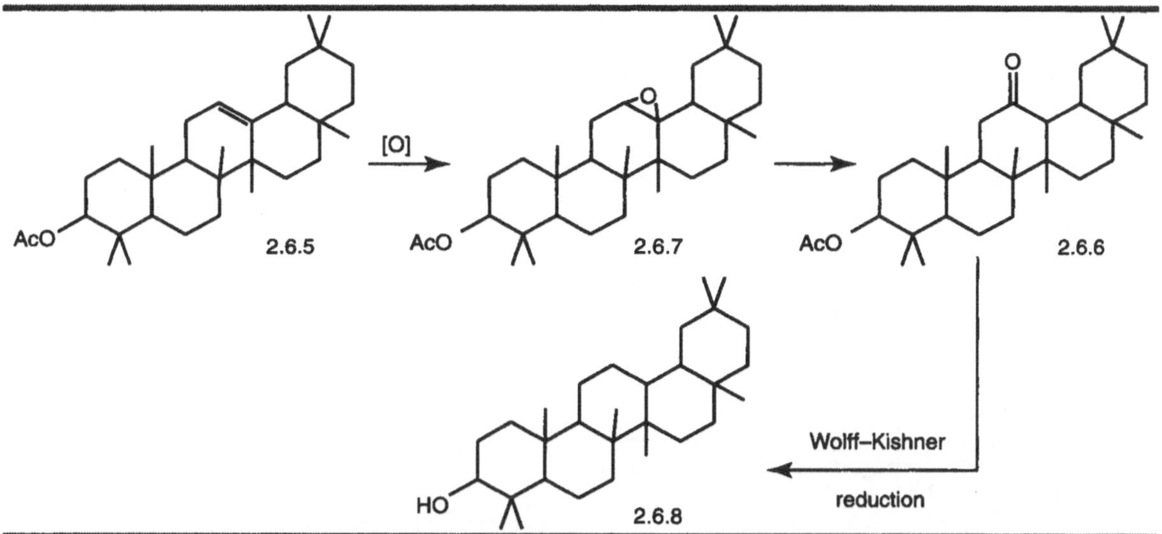

If one constructs a skeletal structure for β-amyrin on the basis of the results of selenium dehydrogenation and biogenetic considerations, it would become obvious that there are only four possible locations for a trisubstituted double bond in it. These are the 5,6-, 9,11-, 12,13-, and 18,19-positions. Therefore, in the ketone mentioned above, namely β-amyranone acetate (2.6.6), the carbonyl group should be at one of the following positions—6, 11, 12 and 19. β-Amyranone acetate gave, on oxidation with fuming nitric acid, a dicarboxylic acid (2.6.9) which on heating readily yielded a five-membered ketone (2.6.10; $\bar{\nu}_{max}$ 1735 cm^{-1}). This was an anticipated result since the dicarboxylic acid had four carbon atoms separating the two carboxyl groups. Position 19 for the carbonyl group in β-amyranone could be ruled out at this stage because such a ketone could not have given rise to a dicarboxylic acid without the loss of any carbon atoms.

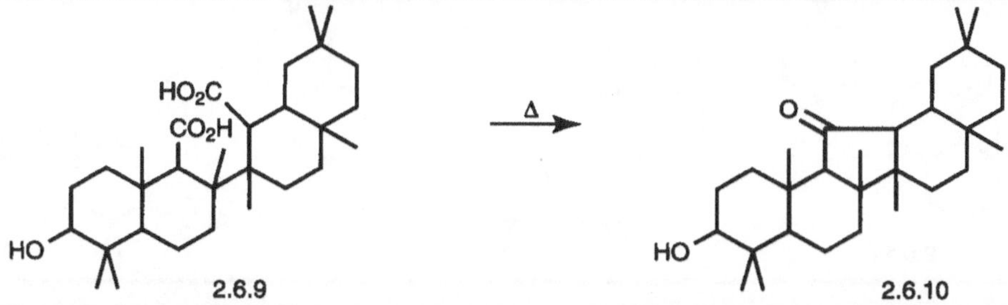

Another approach based on a different strategy of oxidation was also useful. This involved an allylic oxidation of β-amyrin acetate brought about by the action of chromium trioxide. The product was β-amyrenone acetate (2.6.11) which possessed a β, β-disubstituted enone chromophore as shown by its absorption in the UV region at 249 nm. Such a ketone could have arisen with the double bond being located between 5,6- or 9,11- or 11,12-positions. On reduction with sodium and alcohol and subsequent thermolysis with sodium acetate and acetic anhydride, it yielded a diene (2.6.12) which had an absorption maximum at 281 nm, thus indicating its homoannular diene character.

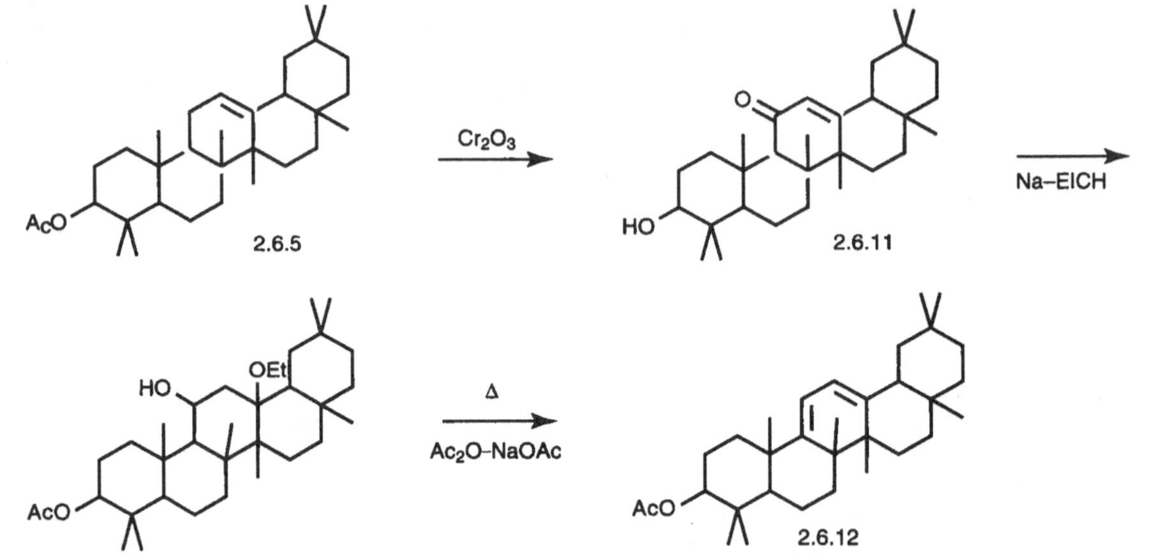

From the figure, it can be seen that these results can be rationalised by locating the double bond either at the 9,11- or 11,12-positions but not in terms of a double bond between positions 5 and 6. The same diene was also obtained as one of the products of reduction (using lithium aluminium hydride) followed by dehydration of β-amyrenone acetate. The other product of this reaction sequence was a heteroannular diene (2.6.13) absorbing in the 240–245 nm region. This result led to a unique solution to the problem of locating the double bond—it could now be firmly placed between positions 11 and 12 as indicated.

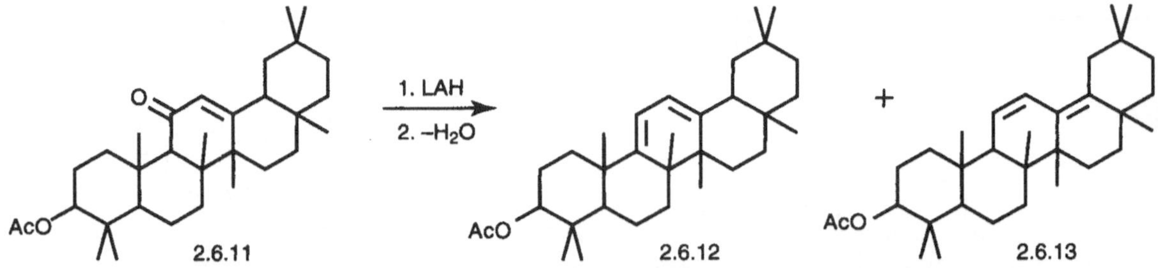

Supporting evidence was provided by the following observations. The benzoate of β-amyrenone (2.6.14) on treatment with bromine in glacial acetic acid gave a dienone (2.6.15), apparently by bromination followed by dehydrobromination as shown below. This dienone had an absorption maximum at 305 nm in agreement with a heteroannular dienone chromophore with appropriate substituents. Such a ketone could not have arisen from a 9,11-ene-12-one structure for β-amyrenone. Further, β-amyrin acetate underwent acid-catalysed rearrangement to give the isomeric 13,18-ene in which the double bond could be detected only by oxidation by hydrogen peroxide in acetic acid to the corresponding epoxide. This epoxide suffered ring opening and subsequent dehydration on treatment with an acid to yield the heteroannular diene (2.6.13) mentioned earlier.

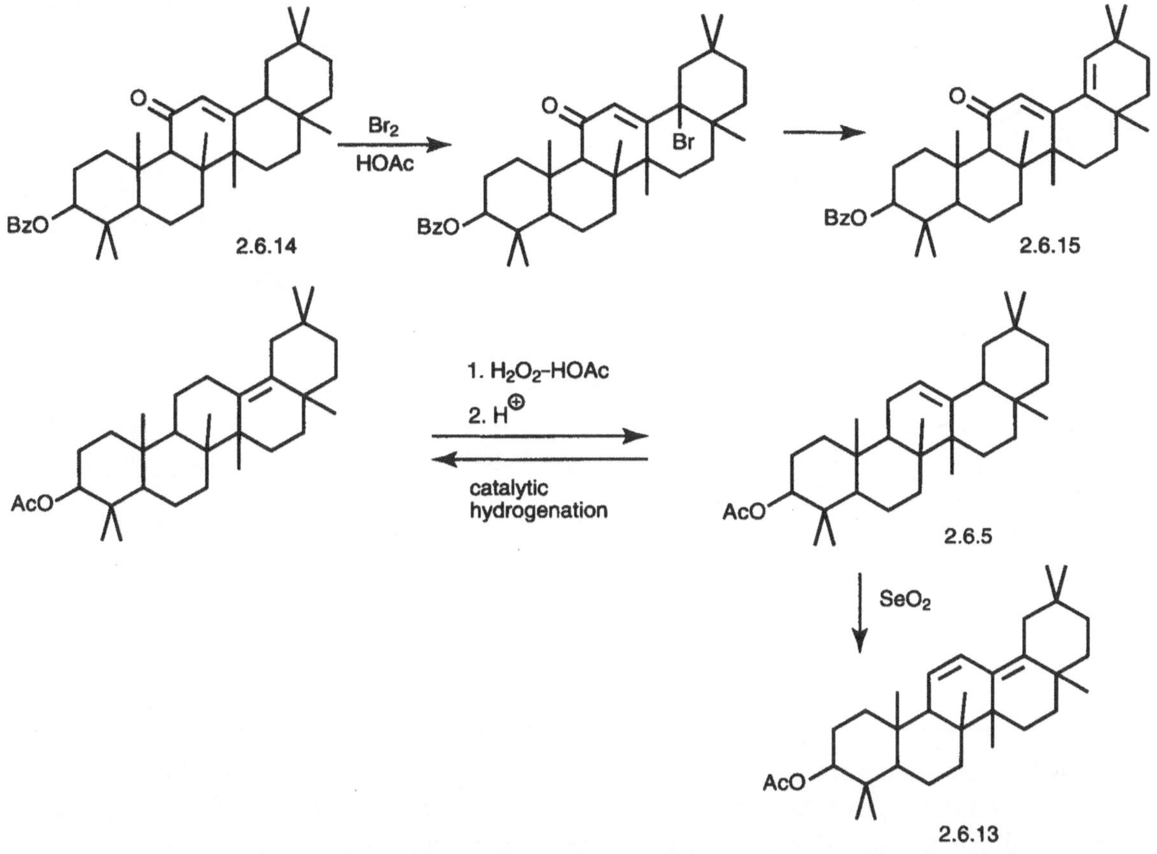

Oleanolic acid (2.6.16), which has the same skeletal structure as β-amyrin, also occurs widely in plants. Oleanolic acid could indeed be converted into β-amyrin by a sequence of unambiguous reactions as shown below. The carboxyl group could be readily esterified by reaction with diazomethane but not with methanol and acid. The mechanism of O-methylation by diazomethane was referred to earlier in Section 1.3. This reaction can be classified, in Ingold terminology, as following a B_{Al2} mechanism. On the other hand, Fischer esterification comes under the A_{Ac2} pathway. Therefore, one can conclude that the carboxyl group in oleanolic acid is sterically hindered and is hence unaffected under the A_{Ac2} esterification conditions but can undergo ionisation, and thereafter O-methylation with diazomethane. In other words, the carboxyl group should occupy an angular position, replacing a methyl group. The possible positions are 8, 10, 14 and 17. A few reactions of oleanolic acid which prove the position of the carboxyl as 17 are shown below. This is also supported by the mass spectrum of acetyl methyloleanolate (2.6.17) in which a prominent peak appears at m/z 203. This can be attributed to an ion arising from the top portion of the molecule by a retro-Diels–Alder fragmentation of the molecular ion.

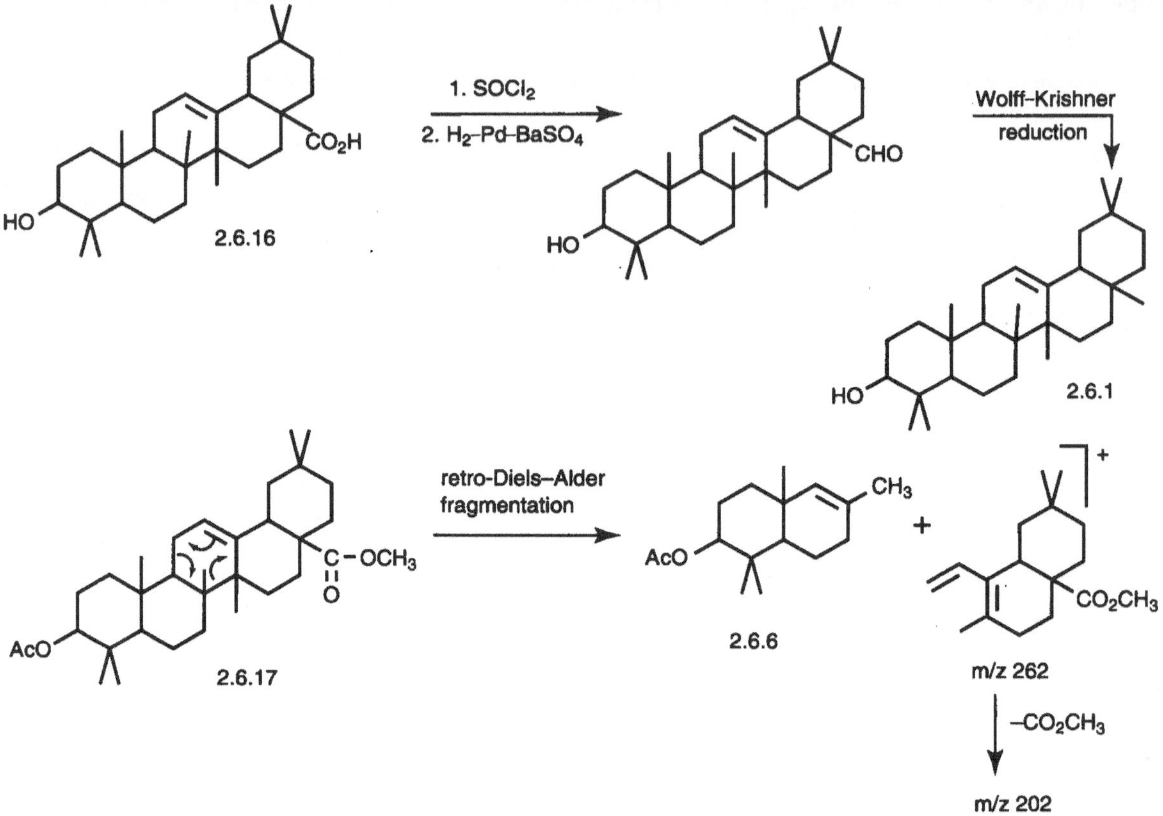

We will be discussing stereochemical problems in Chapter 3. However, it is difficult to ignore the slim, lathe-like shape of the amyrin skeleton. Rings A and B are *trans*-fused but rings D and E are *cis*-fused. The absolute stereochemistry of the compound has been determined.

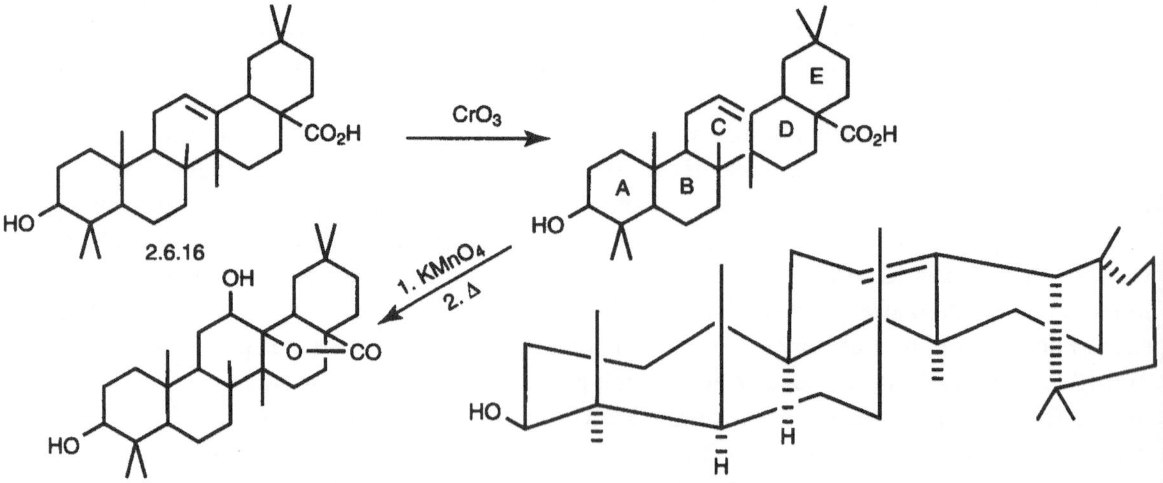

Suggested reading

1.   Barton DHR, In Cook JW, (ed.), *Progress in Organic Chemistry,* Vol. 2, London: Butterworth, 1953.
2.   Ito S, In Nakanishi K, Goto T, Ito S, Natori S and Nozoe S, (eds), *Natural Products Chemistry,*Vol. 1, Tokyo: Kodansha Ltd, 1974.

2.7 WEDELOLACTONE: THE FIRST OF THE COUMESTANS

Wedelolactone (2.7.1) has a simpler architecture than any of the other compounds discussed in this chapter. However, it is also unique in certain respects. It possesses a compact tetracyclic structure and can be viewed either as a 3-phenylcoumarin derivative or as an isoflavonoid. Biosynthetically, it is indeed related to other isoflavonoids. Its compact, fused, tetracyclic heteroaromatic structure is responsible for its brilliant blue fluorescence which can be seen even in neutral solvents. However, the compound is not photodynamic, unlike psoralen (2.7.2), which is a furanocoumarin. When first isolated and characterised in 1956, wedelolactone represented a new skeletal type among isoflavonoids, though the closely related pterocarpan structure (2.7.3) was known. Subsequently, the simpler coumestrol (2.7.4) was discovered. The skeletal structure possessed by these compounds has been designated as the coumestan structure.

2.7.1   2.7.2   2.7.3

2.7.4   2.7.5

The elucidation of the structure of wedelolactone is instructive since it was based on unambiguous and well-established methods. Indeed, the strategy used to find out its structure conformed to a textbook pattern. Thus, routine analytical studies gave the molecular formula as $C_{16}H_{10}O_7$, and showed the presence of a methoxyl (Zeisel) and three phenolic hydroxyl groups [formation of a triacetate (2.7.5) and a tri-O-methyl ether (2.7.6)].

The presence of a six-membered, α, β-unsaturated lactone ring was indicated by the absorption band at 1705 cm^{-1} in the IR spectrum of wedelolactone. This was also supported by the observation that while the tri-O-methyl ether was insoluble in cold alkali, it dissolved on heating the solution. Acidification of the cold alkaline solution resulted in the re-formation of tri-O-methylwedelolactone. On the other hand, when the alkaline solution was treated with dimethyl sulphate, the resulting product was a carboxylic acid (2.7.7), which had four methoxyl groups as shown by a Zeisel estimation. This was a strategy which was successfully used by A Robertson and his co-workers in their studies on coumarins. Tetra-O-methylwedelic acid thus obtained was then subjected to decarboxylation, followed by ozonolysis and a final alkaline hydrolysis. The final products were characterised as 2,4,6-trimethoxybenzoic acid (2.7.8) and 2-hydroxy-4,5-dimethoxybenzaldehyde (2.7.9). Working backwards, the structure of tri-O-methylwedelolactone (2.7.6) could, therefore, be deduced as shown.

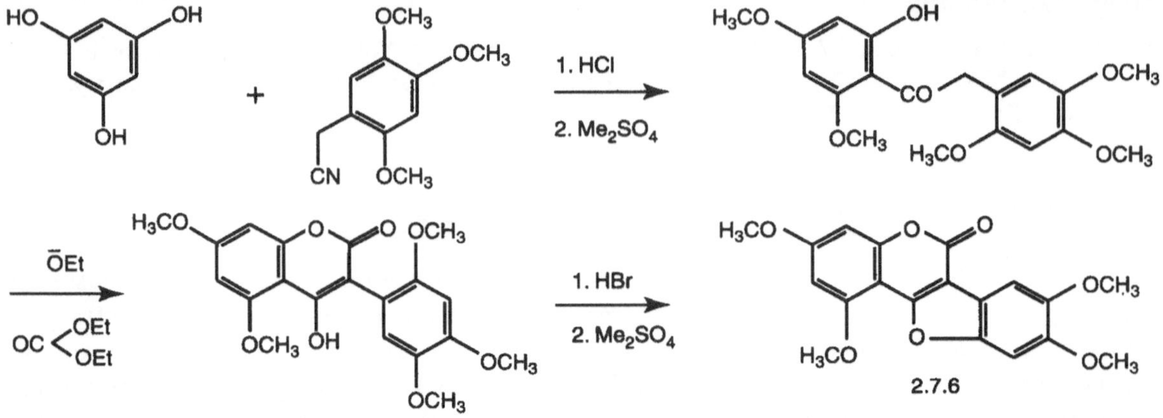

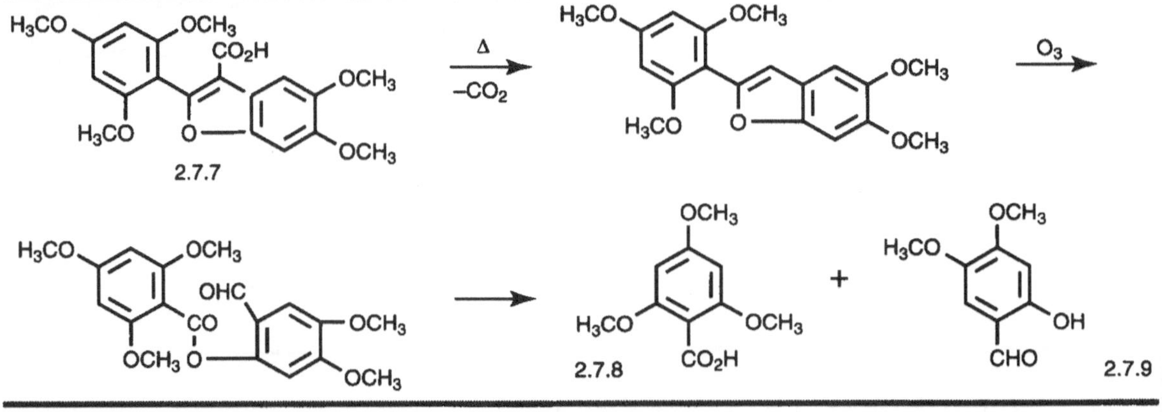

The position of the methoxyl group in wedelolactone was determined by subjecting the tri-O-ethyl derivative (2.7.10) of the compound to a similar series of reactions, whereby 2,6-diethoxy-4-methoxybenzoic acid (2.7.11) and 2-hydroxy-4,5-diethoxybenzaldehyde (2.7.12) were obtained. The skeletal structure and the oxygenation pattern were confirmed by two independent syntheses of the tri-O-methyl ether following a synthetic strategy developed by Robertson and co-workers. The structure of wedelolactone was confirmed by a partial synthesis involving a selective demethylation of the tri-O-methyl ether. This result could be taken as confirmatory evidence for the location of the methoxyl group in wedelolactone because it is supported by theory as explained as follows.

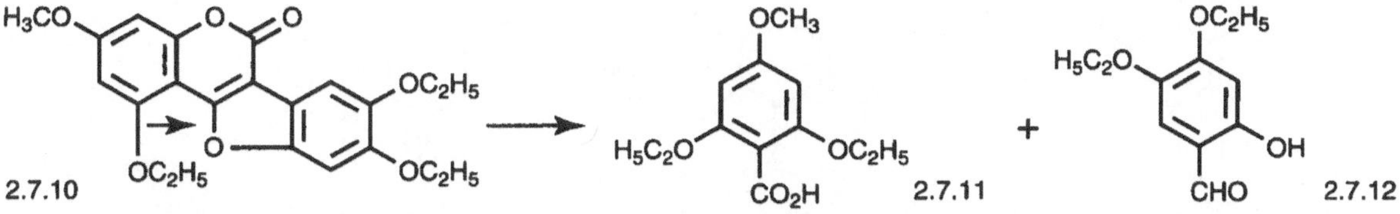

2.7.10   2.7.11   2.7.12

A good look at the tri-O-methylwedelolactone structure (2.7.6) would show that, of the four methoxyl groups, only the two at positions 5 and 7, that is, in the coumarin part of the structure, are in effective conjugation with the lactone carbonyl group. Therefore, their basicities would be less than those of the other two methoxyls, and hence, they could be expected to resist acid-catalysed demethylation. These two could also be differentiated from each other, because of the proximity of one of them, namely the one at position 5, to the oxygen of the furan ring. Its demethylation would create a hydroxyl which can be H-bonded to the furan oxygen. Therefore, on theoretical grounds, it is possible to predict the order of demethylation of the four methoxyls as $3', 4' > 5 > 7$, as is confirmed by experimentation. By adopting milder conditions of demethylation, it was possible to obtain a monomethyl ether of wedelolactone, which was conclusively shown to be the 5-O-methyl derivative (2.7.13) of wedelolactone by comparison with a synthetic sample. Incidentally, the most elegant and unambiguous method of synthesis of wedelolactone is the one developed by Wanzlick. The strategy is based on the reactivity of 4-hydroxycoumarin as a carbanionic nucleophile (earlier studies by Anschutz, Link and their co-workers had shown the versatility of this compound as a nucleophile in aldol and related reactions) in a Michael reaction on o-benzoquinone, produced in situ. Thus, an oxidative coupling in alkaline medium of 4,5-dihydroxy-7-methoxycoumarin (2.7.14) with catechol (2.7.15) yielded wedelolactone.

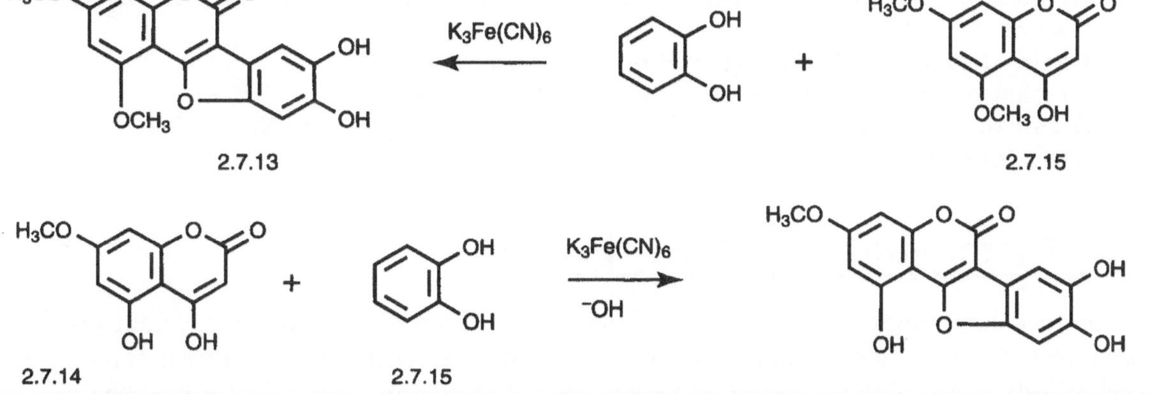

2.7.1   2.7.6

2.7.13   2.7.15

2.7.14   2.7.15

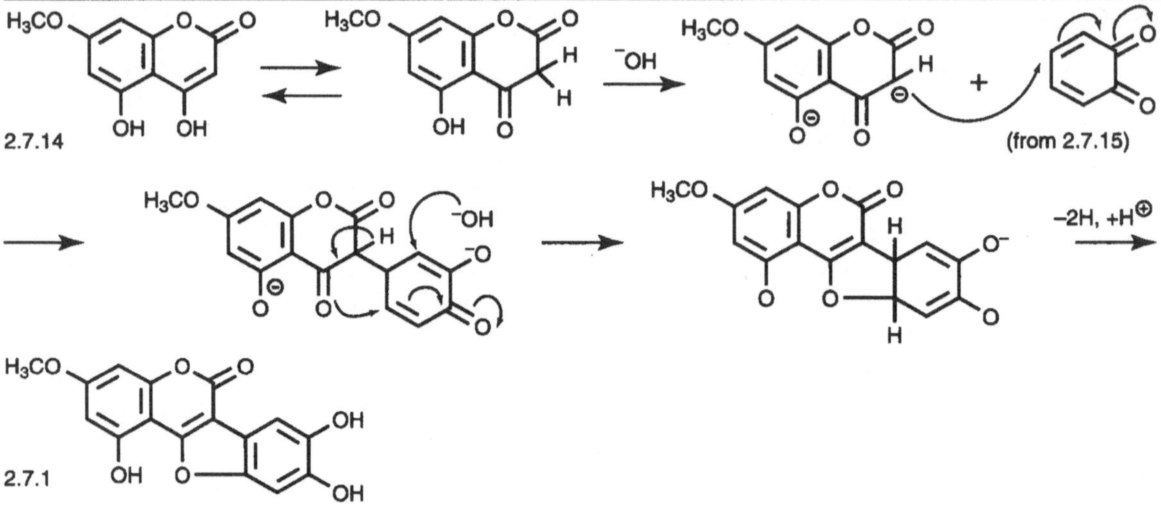

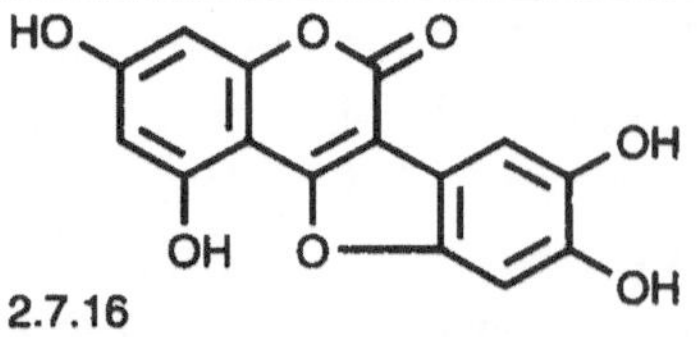

2.7.14

(from 2.7.15)

−2H, +H⊕

2.7.1

Wedelolactone occurs in *Eclipta alba* and *Wedelia calendulaceae*, both of which belong to the family Compositae, and in *Hypericum erectum* (Hypericaceae). In all three plants, it occurs along with desmethylwedelolactone (2.7.16). The two compounds exhibit certain interesting colour reactions which indicate their sensitivity towards even mild oxidising agents such

2.7.16

as silver oxide, benzoquinone, nitric acid and dissolved oxygen in a weakly alkaline medium. It is possible that their functions in plants are associated with these properties.

Suggested reading

1. Krishnaswamy NR and Seshadri TR, In Gore TS, Joshi BS, Sunthankar SV and Tilak BD, (eds), *Recent Progress in the Chemistry of Natural and Synthetic Colouring Matters and Related Fields*, New York & London: Academic Press, 1962.
2. Ingham JL, in Herz W, Grisebach H and Kirby GW, (eds), *Progress in the Chemistry of Organic Natural Products*, Vol. 43, Wien New York: Springer–Verlag, 1983.
3. Govindachari TR, Nagarajan K, Pai BR and Parthasarathy PC, *J. Chem. Soc.*, 1957, 548 (and references therein).
4. Bower WJ, Robertson A and Whalley WB, *J. Chem. Soc.*, 1957, 542.
5. Emerson OH and Bickoff EM, *J. Amer. Chem. Soc.*, 1958, 80: 4381.
6. Dean FM, in Apsimon J, (ed.), *The Total Synthesis of Natural Products*, Vol. 1, New York: Wiley Interscience, 1973.

2.8 PROTOAPHIN-fb: AN INSECT PIGMENT

Colour is a dominant aspect of nature and almost all kinds of life forms sport colour either for ornamentation or for a more fundamental reason connected with the very process of living. The compounds responsible for colour in nature vary as widely in their structures as the colours do. Their chemistry is also equally varied and fascinating. In this section, we will discuss the structure of a quinonoid pigment produced by the tiny insect *Aphis fabae*. Our knowledge of the chemistry of this interesting group of natural pigments is due to the systematic studies, spread over two decades and more, of Lord Todd and his co-workers.

Aphis fabae is the common black fly which thrives on cultivated beans. Its haemolymph, which has a reddish purple colour, contains a pigment, protoaphin-fb, with the molecular formula $C_{36}H_{38}O_{16}$. The insect also has an enzyme which acts on protoaphin-fb (2.8.1) after the insect's death, breaking it down to a mixture of glucose and a fluorescent, yellow-coloured aglucone called xanthoaphin, $C_{30}H_{26}O_{10}$. Therefore, for the isolation of protoaphin-fb, the insects had to be killed by plunging them into hot water to deactivate the enzyme. The aglucone, xanthoaphin (2.8.2), is itself unstable and with time, slowly loses a molecule of water to yield chrysoaphin (2.8.3), which is equally unstable. The result is the loss of another molecule of water and the formation of the stable red-coloured erythroaphin (2.8.4), $C_{30}H_{2}O_{8}$. These changes are catalysed either by acids or by alkalis.

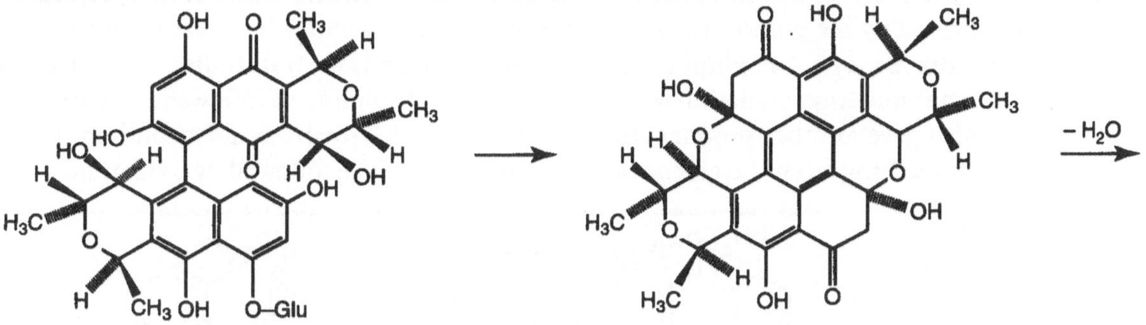

2.8.1   2.8.2

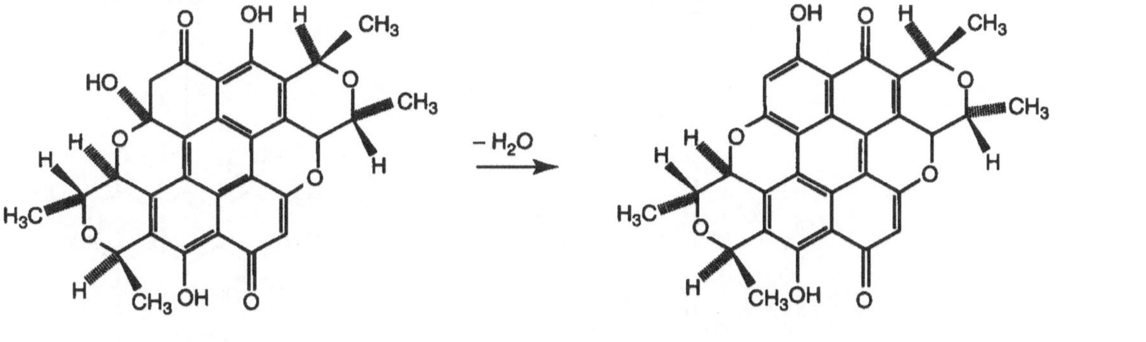

2.8.3   2.8.4

Protoaphin-fb is a brownish-yellow acidic compound, owing its acidic nature to the presence of three free phenolic hydroxyl groups. It may be instructive to note that the pK_a of the compound is 6.2. Therefore, it is sufficiently ionised at the physiological pH to exhibit the deep violet-red colour characteristic of the anion. The main chromophore is a quinonoid group whose presence is shown by reduction to the leuco-compound by means of sodium dithionite. The quinone is regenerated by aerial oxidation of the colourless dihydro derivative (2.8.5).

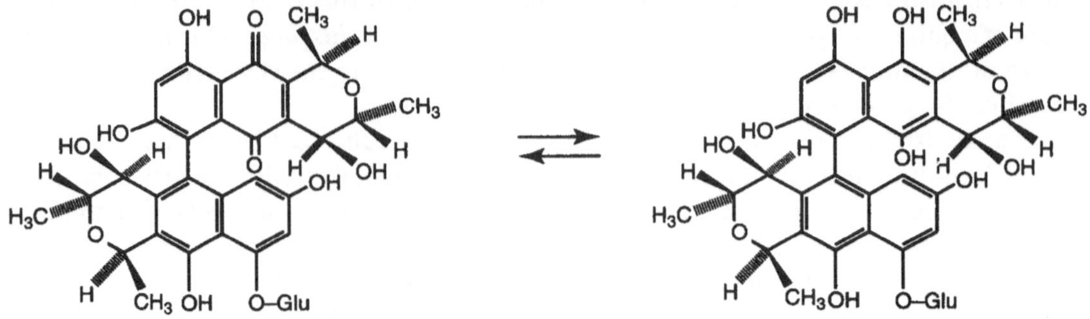

2.8.1   2.8.5

In an aqueous medium, protoaphin-fb is acted upon by sodium dithionite resulting in the formation of two products which can be isolated, after aerial oxidation, as a quinone (2.8.6) and a glucoside (2.8.7). The former has two phenolic hydroxyls, one of which is H-bonded to a quinone carbonyl as indicated by the purple colour that it gives with ferric chloride. It was shown to be a derivative of 5,7-dihydroxy-1,4-naphthoquinone by analytical and spectral studies. The glucoside yielded the same quinone after oxidation with potassium nitrodisulphonate followed by hydrolysis. The oxidation also gave another product characterised as a 2,5-dihydroxy-1,4-naphthoquinone (2.8.8) presumably arising from the glucoside by glucosidic cleavage followed by oxidation.

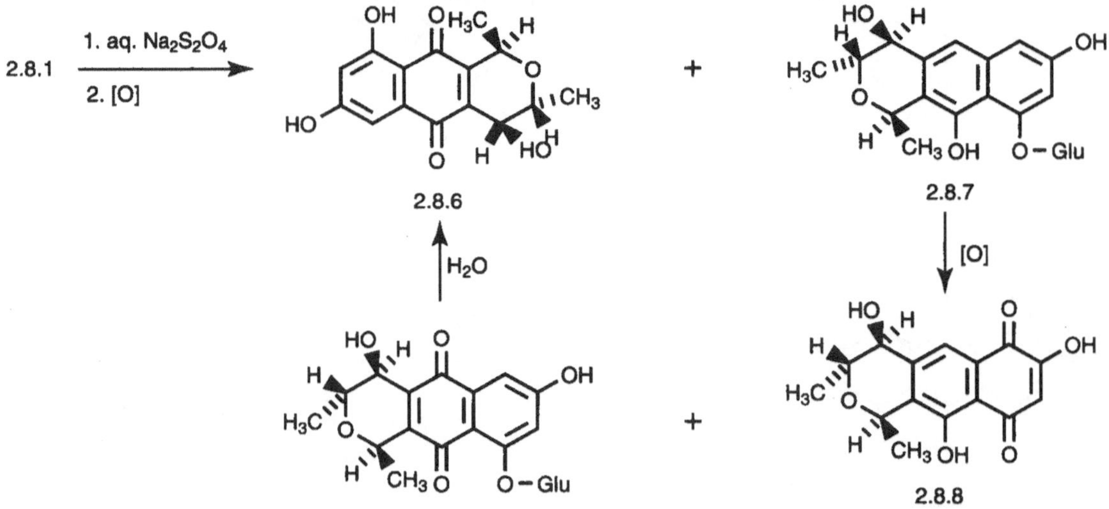

The UV spectrum of protoaphin-fb is an additive of the spectra of the quinone and the glucoside mentioned above, thus indicating that it is a dimer. Obviously, sodium dithionite brings about a reductive cleavage of the dinaphthyl structure. The transformation of protoaphin-fb (2.8.1) into xanthoaphin (2.8.2), apparently involving the loss of the glucose unit and two water molecules is complex, since the UV spectrum of the end-product is very different from that of the parent compound. The spectrum of xanthoaphin, which is the initial product, shows that it is an anthracene derivative. For its formation, the removal of the glucose unit is a necessary first step. The subsequent acid-catalysed condensation and acetal formation can be understood in terms of mechanistic principles though the order of these reactions is not known.

An interesting and significant fact is that while the natural enzymes efficiently convert protoaphin into erythroaphin (2.8.3) via xanthoaphin, in the acid-catalysed reaction, only traces of xanthoaphin and erythroaphin are obtained. It has been suggested that the enzymes hold protoaphin in a conformation-cum-configuration amenable to the formation of the fused pentacyclic perylene structure present in xanthoaphins and erythroaphins. Though there is no experimental evidence in support of this suggestion, it is plausible since protoaphin, being a sterically hindered dinaphthyl derivative, is unlikely to assume a planar conformation. Indeed, the total optical activity of the compound is the resultant of contributions from the individual chiral centres as well as the molecule as a whole since it possesses a chiral axis.

Erythroaphin readily reacts with primary and secondary amines to yield an adduct which undergoes facile aerial oxidation to give a nitrogen-containing quinone (2.8.9) as shown below. Implicit here is a fundamental reaction characteristic of fully methylated benzoquinones, such as duroquinone (2.8.10). The course of the reaction can be explained in more general terms by considering the reactive part of the molecule. This is a β-methylated enone and what is observed is its anticipated behaviour towards a base-cum-nucleophile.

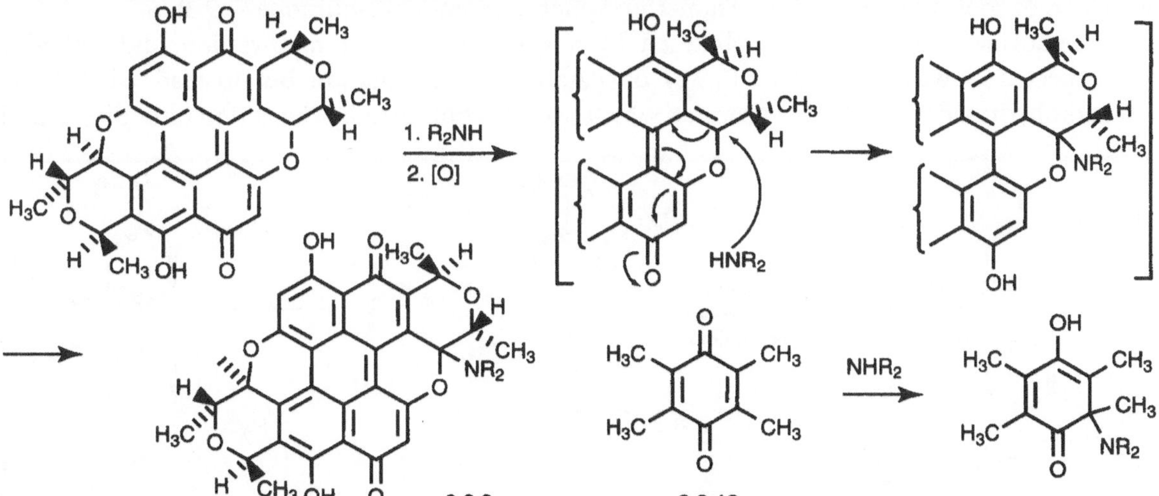

It should however be noted that the incorporation of the reactive part in an appropriate skeletal framework is often necessary for the proper articulation of predictable behaviour. In this specific context, the other carbonyl group of the quinonoid unit is also needed for making the reaction proceed.

Suggested reading

1.　Todd AR, *Chemistry in Britain*, 1966, 2: 428, The Robert Robinson Lecture of the Chemical Society, London.

2.   Weiss U, Merlini L and Nasini G, in Herz W, Grisebach H and Kirby GW, (eds), *Progress in the Chemistry of Organic Natural Products*, Vol. 52, Wien New York: Springer–Verlag, 1987.

2.9 TYLOPHORINE: ANOTHER 'STRAIGHT' MOLECULE WITH A KINK OR TWO!

Tylophora asthmatica, as the species name indicates, is a plant whose leaves are used in traditional Indian medicine as a cure for asthma. It is a perennial climber belonging to the family Asclepiadaceae. Govindachari and co-workers isolated tylophorine and a couple of other related alkaloids from this plant. The structure of tylophorine was determined by a combination of the classical method of Hoffmann exhaustive methylation and spectral data. One of the products of degradation also exhibited unusual behaviour due to a transannular interaction as will be explained later. Tylophorine (2.9.1) has the molecular formula $C_{24}H_{27}NO_4$. Zeisel estimation showed the presence of four methoxyl groups. The compound reacted with one equivalent of methyl chloride to form a methochloride, revealing the tertiary nature of the nitrogen atom. Absence of an *N*-methyl group further indicated that the nitrogen atom was common to two rings. The UV spectrum of tylophorine had the characteristic contour exhibited by phenanthrene derivatives. Three successive Hoffmann exhaustive methylations yielded a nitrogen-free compound (2.9.2), which on ozonolysis gave 2,3,6,7-tetramethoxyphenanthrene-9,10-quinone (2.9.3).

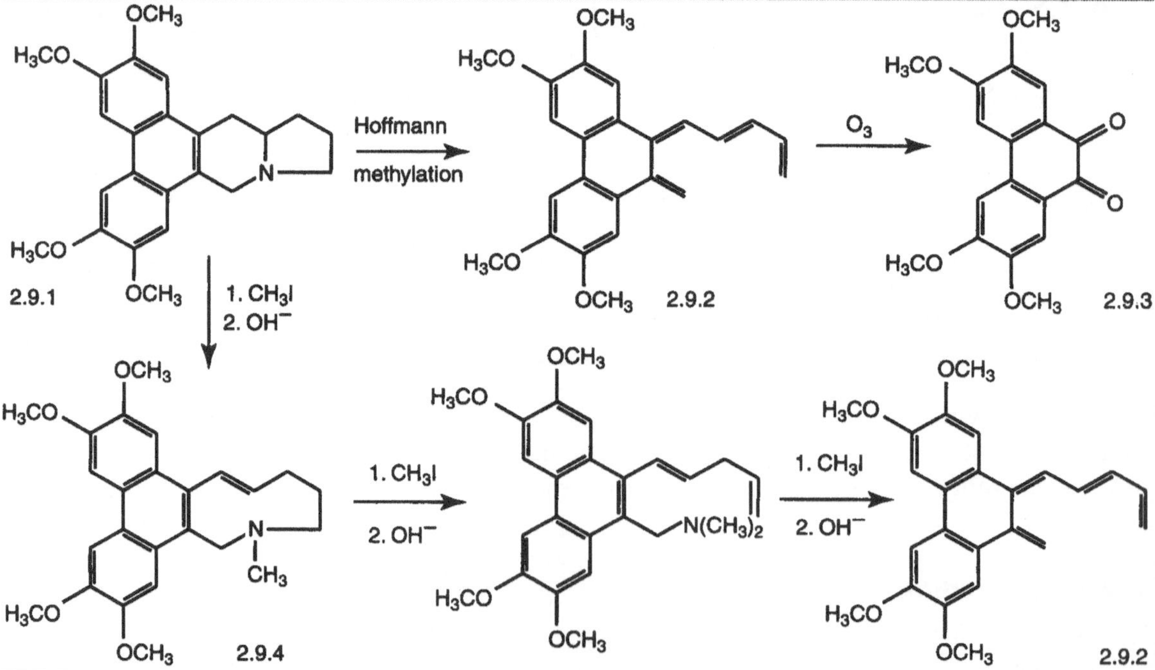

Tylophorine methine (2.9.4), obtained by a single Hoffmann methylation, did not give the expected dihydro derivative (2.9.5) on catalytic hydrogenation but instead yielded the isomethohydroxide (2.9.6). This product, which was also obtained by the action of boiling alkali on tylophorine methiodide, apparently arises from the methine by a transannular reaction followed by ring fission and ring closure to yield the configurational isomer as shown.

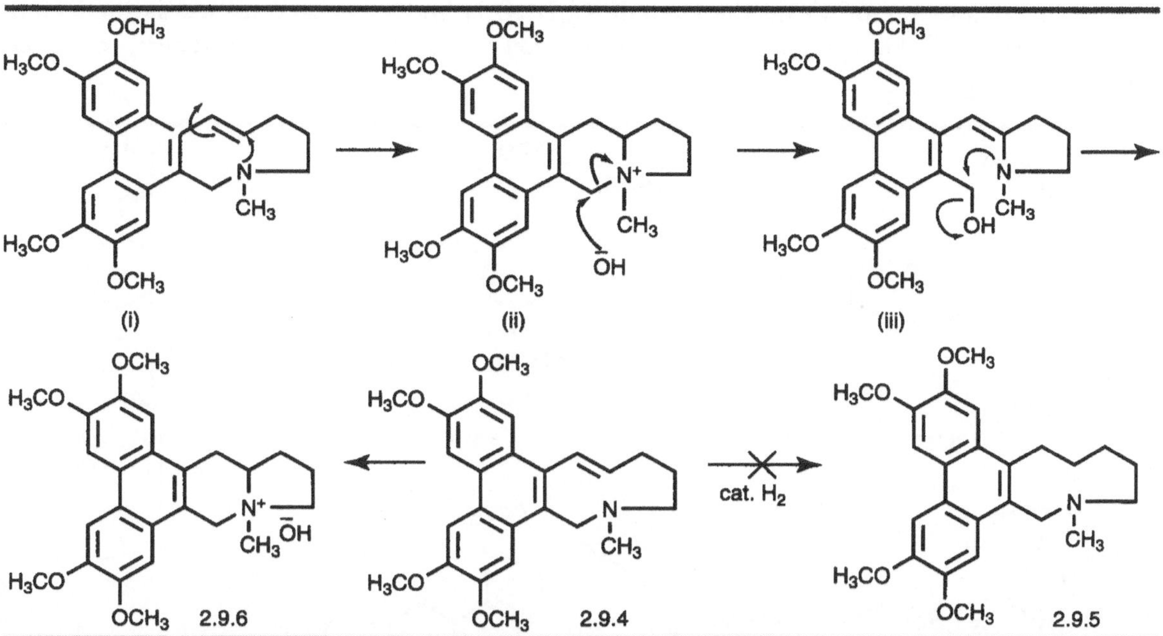

(i) (ii) (iii)

2.9.6 2.9.4 2.9.5

Transannular interaction is also obviously involved in the Emde degradation of tylophorine methiodide or isomethiodide. The Emde reaction is an alternative strategy to the Hoffmann exhaustive methylation and involves the reaction of a quaternary ammonium compound with sodium and alcohol or sodium in liquid ammonia or with hydrogen in the presence of a catalyst. In this case, the product was isodihydrohomotylophorine (2.9.7). This compound underwent dehydrogenation on treatment with palladised charcoal to yield the tetradehydro-derivative (2.9.8) which answered the Ehrlich test for pyrroles.

After two successive Hoffmann exhaustive methylations, isodihydrohomotylophorine (2.9.7) lost its nitrogen atom to give a product (2.9.9) which on oxidation yielded 2,3,6,7-tetramethoxy-10-methylphenanthrene-9-carboxylic acid (2.9.10) and 2,3,6,7-tetramethoxyphenanthrene-9,10-dicarboxylic acid (2.9.11), which were isolated as their esters. These and a few other reactions of tylophorine and its derivatives are shown.

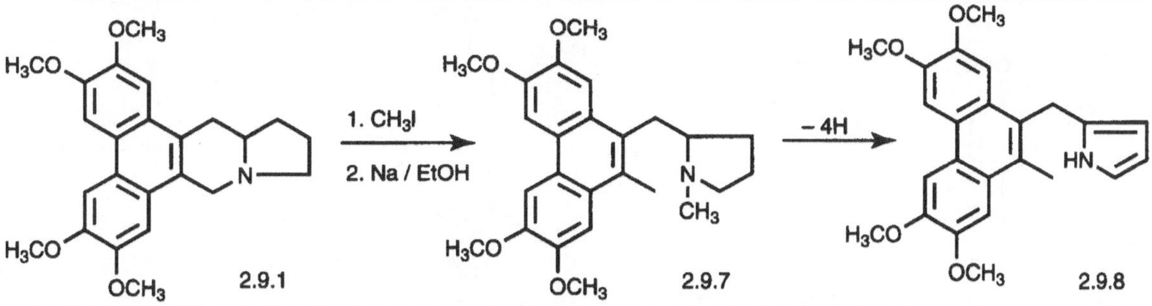

2.9.1

1. CH$_3$I
2. Na / EtOH

2.9.7

− 4H

2.9.8

An interesting reaction, which has implications beyond the compound under discussion, is the oxidation of tylophorine methine with potassium permanganate in pyridine. The product was a neutral compound having the formula $C_{25}H_{28}O_5N$ and was characterised as the lactam (2.9.12), in which the benzyl group of tylophorine, next to the nitrogen atom, had been converted into a carbonyl group.

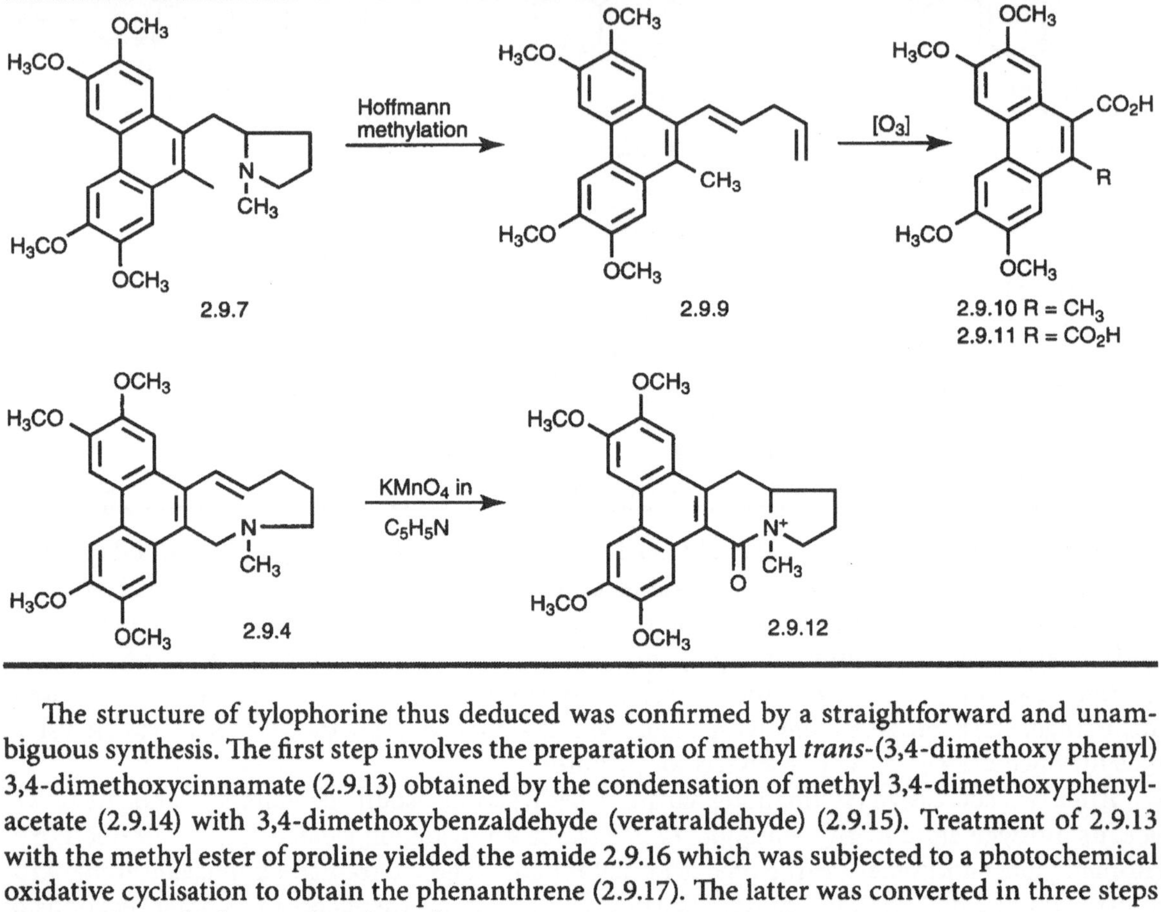

2.9.7

2.9.9

2.9.10 R = CH₃
2.9.11 R = CO₂H

2.9.4

2.9.12

The structure of tylophorine thus deduced was confirmed by a straightforward and unambiguous synthesis. The first step involves the preparation of methyl trans-(3,4-dimethoxy phenyl) 3,4-dimethoxycinnamate (2.9.13) obtained by the condensation of methyl 3,4-dimethoxyphenyl-acetate (2.9.14) with 3,4-dimethoxybenzaldehyde (veratraldehyde) (2.9.15). Treatment of 2.9.13 with the methyl ester of proline yielded the amide 2.9.16 which was subjected to a photochemical oxidative cyclisation to obtain the phenanthrene (2.9.17). The latter was converted in three steps into racemic tylophorine (2.9.1), as shown.

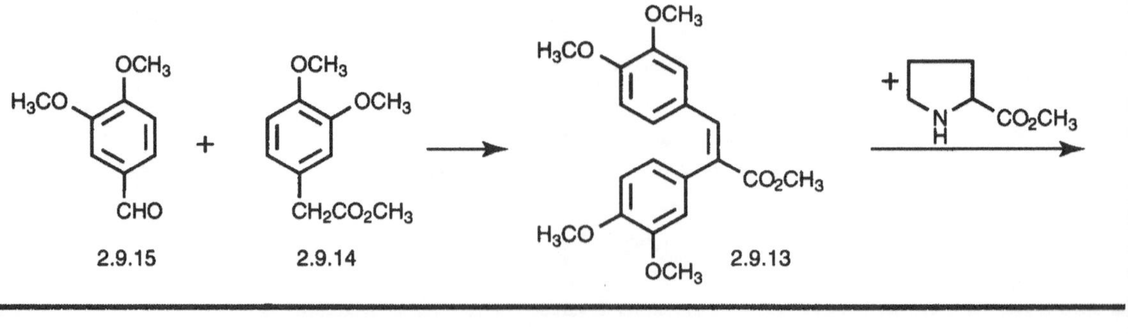

2.9.15

2.9.14

2.9.13

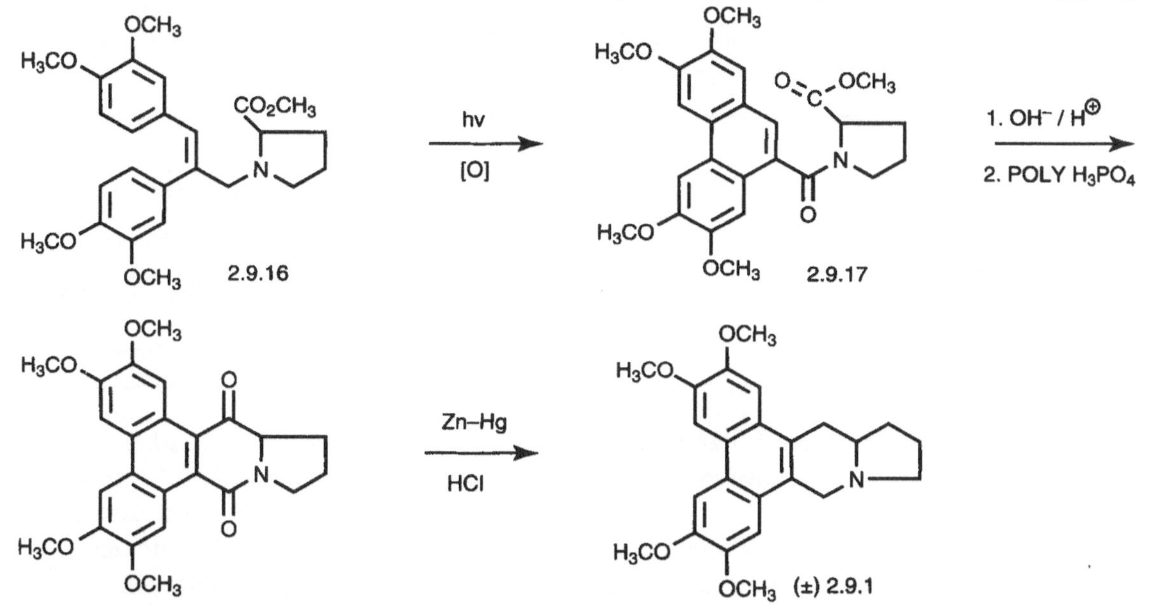

2.9.16 → 2.9.17 → (±) 2.9.1

Suggested reading

1. Govindachari TR In *The Alkaloids* (Ed. R.H.F. Manske), Vol. 9, 517 (1967), Academic Press, New York.
2. Govindachari TR, Lakshmikantham MV, Pai BR and Rajappa S, *Tetrahedron*, 1960: 9: 53.
3. Nakanishi K, In Nakanishi K, Goto T, Ito S, Natori S and Nozoe S, (eds), *Natural Products Chemistry*, Vol. 3, Tokyo: Kodansha Ltd, 1983.

2.10 HELIANGINE: AN ANTI-AUXIN

So far we have discussed structures which were deduced largely, if not exclusively, on the basis of classical chemical studies. Spectral data, where used, were employed only to confirm a structural feature suggested by chemical data. We will now consider an example where spectral data made a substantial and significant contribution to the determination of the structure. This can be considered as a model to show the advantages of a judicious combination of chemical and spectral approaches.

Sesquiterpene lactones are fairly widespread in higher plants. Members of the family Compositae are particularly capable of producing these compounds as end-products of secondary metabolism. Germacranolides are monocyclic lactones containing a ten-membered ring. Heliangine has the gross skeletal structure shown in 2.10.1 and is present in the leaves of *Helianthus tuberosus* L. It is an anti-auxin, that is, it inhibits the growth of plants. It is assumed to serve as an allelopathic compound, though there is no conclusive proof.

Heliangine has the molecular formula $C_{20}H_{26}O_6$, is optically active and forms a monoacetate, indicating the presence of a free hydroxyl group. It consumes two moles of alkali and undergoes saponification to yield tiglic acid (2.10.2) and a neutral compound, helianginol, $C_{15}H_{20}O_5$ (2.10.3), which forms a diacetate. Helianginol also consumes one mole of alkali—though on acidification it is recovered unchanged—indicating the presence of a lactone ring.

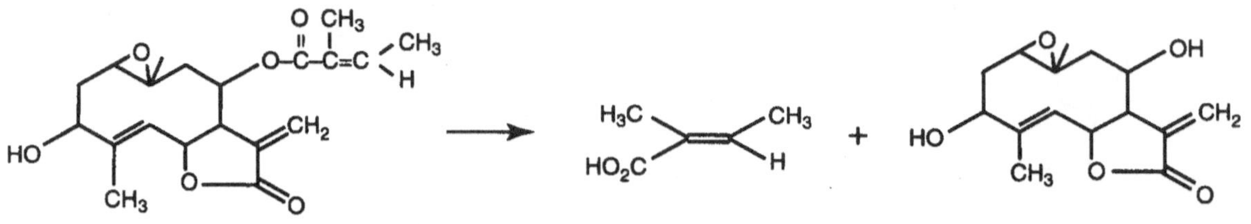

2.10.1   2.10.2   2.10.3

On catalytic hydrogenation, heliangine readily forms a dihydro derivative (2.10.4) and under more rigorous conditions, tetrahydro- and hexahydro- products (2.10.5 and 2.10.6 respectively), thus revealing the presence of three double bonds, one of which is easily hydrogenated, and therefore less substituted than the other two. Dihydroheliangine (2.10.4) retains the tigloyl unit, thus showing that the readily hydrogenated double bond is elsewhere in the molecule.

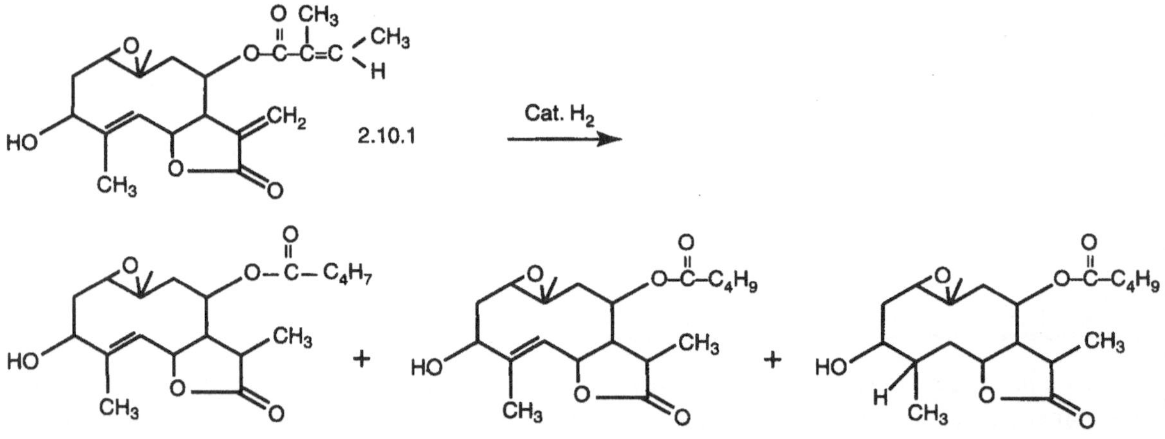

2.10.4   2.10.5   2.10.6

Its nature was revealed by the observation that heliangine yields formaldehyde and acetaldehyde on ozonolysis, while helianginol gives only formaldehyde. These observations indicated the presence of a vinylidene group which was supported by the formation of pyrazoline derivatives (2.10.7 and 2.10.8) by both heliangine and helianginol. The acetaldehyde formed by ozonolysis of heliangine obviously arose by the oxidative cleavage of the double bond in the tiglic acid moiety. At this stage, an objective analysis of some spectral characteristics of heliangine and its derivatives in the light of the chemical data described above will be helpful.

The UV spectrum of heliangine has an absorption band at 204 nm with a large molar absorption coefficient. In the spectrum of helianginol, the absorption maximum occurs at

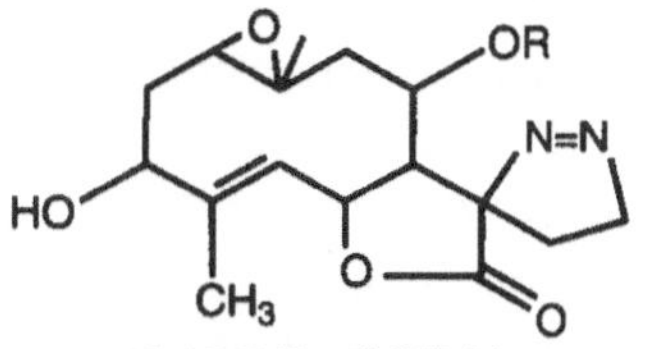

210 nm but with an ε value which is approximately half that of helianginе. Tiglic acid itself absorbs at 210 nm with an ε value almost equal to that of helianginol. These data lead to the obvious conclusion that in heliangine there is another α, β-unsaturated ester (or lactone) chromophore in addition to the tigloyl unit. This conjecture is further supported by the observation that dihydroheliangine also absorbs at 210 nm with an ε value very close to that of helianginol. It may be noted here that dihydroheliangine

2.10.7 R = COC$_4$H$_7$
2.10.8 R = H

retains the tigloyl unit but does not give formaldehyde on ozonolysis. Indeed, subtraction of the curve of dihydroheliangine from the UV spectrum of heliangine gave a spectral curve which was identical to that of helianginol. Thus, it could be concluded that heliangine and helianginol possess an α, β-unsaturated lactone chromophore in which the double bond is exocyclic to the lactone ring. The IR spectrum of heliangine has absorptions at 1750, 1700 and 1650 cm^{-1}, of which the one at 1700 cm^{-1} is absent in the spectrum of helianginol. Therefore, this absorption could be attributed to the ester carbonyl of the tigloyl unit. The absorption at 1750 cm^{-1} should, therefore, be due to the lactone moiety, which should be five-membered as indicated by the position of the absorption band. Thus, five of the six oxygen atoms of heliangine could be accounted for. The sixth oxygen was subsequently characterised as a cyclic ether (epoxide) linkage by the NMR spectral analysis of two products obtained as follows. The dihydro and tetrahydro derivatives of heliangine were oxidised to the corresponding ketones (2.10.9 and 2.10.10) which were then treated with acids to obtain two dienones (2.10.11 and 2.10.12), each of which contained a tertiary hydroxyl which was absent in the original compounds. Their NMR spectra could be satisfactorily interpreted in terms of the structures shown below.

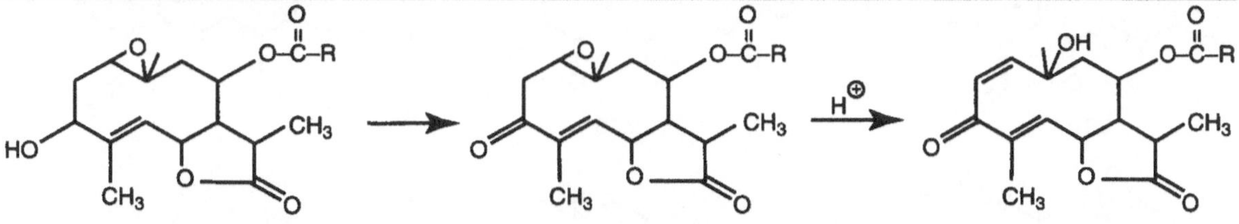

2.10.4 R = C$_4$H$_7$
2.10.5 R = C$_4$H$_9$

2.10.9 R = C$_4$H$_7$
2.10.10 R = C$_4$H$_9$

2.10.11 R = C$_4$H$_7$
2.10.12 R = C$_4$H$_9$

The presence of four C-methyl groups in heliangine and of two in helianginol was also indicated by Kuhn–Roth estimations. The part structures and functional groups thus shown by the above data were fully supported by the NMR spectra of heliangine and its derivatives. The skeletal structure of heliangine as a germacranolide could be deduced by considering the molecular formula of the fully saturated derivatives, as well as the NMR data, and in particular, information regarding spin–spin interactions. Further, the transformation of dihydroheliangine (2.10.4) into the hydroxydienone (2.10.11) mentioned earlier accounted for a continuous chain of six carbon atoms which could only be accommodated by a ten-membered ring structure.

The disposition of the tigloyl unit, the lone hydroxyl and the lactone group in heliangine, relative to each other, was indicated by the observation that hexahydroheliangine (2.10.6) yielded, on rigorous oxidation, methylsuccinic acid. This could have arisen from two parts of the molecule, the correct structure of which was shown by an x-ray diffraction analysis. In this structure, the free hydroxyl group in heliangine is allylic in nature. Therefore, the corresponding ketone, helianginone (2.10.13), should be expected to behave like an α, β-unsaturated ketone. However, its spectral properties were anomalous. A possible explanation of this anomaly may be found in the original publication (Ref. 2), but the reader is advised to construct a framework model of the compound and check it against the explanation given by the authors.

A few further reactions of heliangine and some of its derivatives are shown below. Particular mention may be made of the meticulous care with which the oxidation of deoxyhexahydroheliangine (2.10.14) was carried out. This reaction yielded, besides methylsuccinic acid, methylglutaric and methyladipic acids in the ratio 2:2:1. Under the influence of an alkali, hexahydroheliangine also underwent skeletal rearrangement. Such rearrangements, involving alternate lactone opening and reclosure, are not uncommon among sesquiterpene lactones.

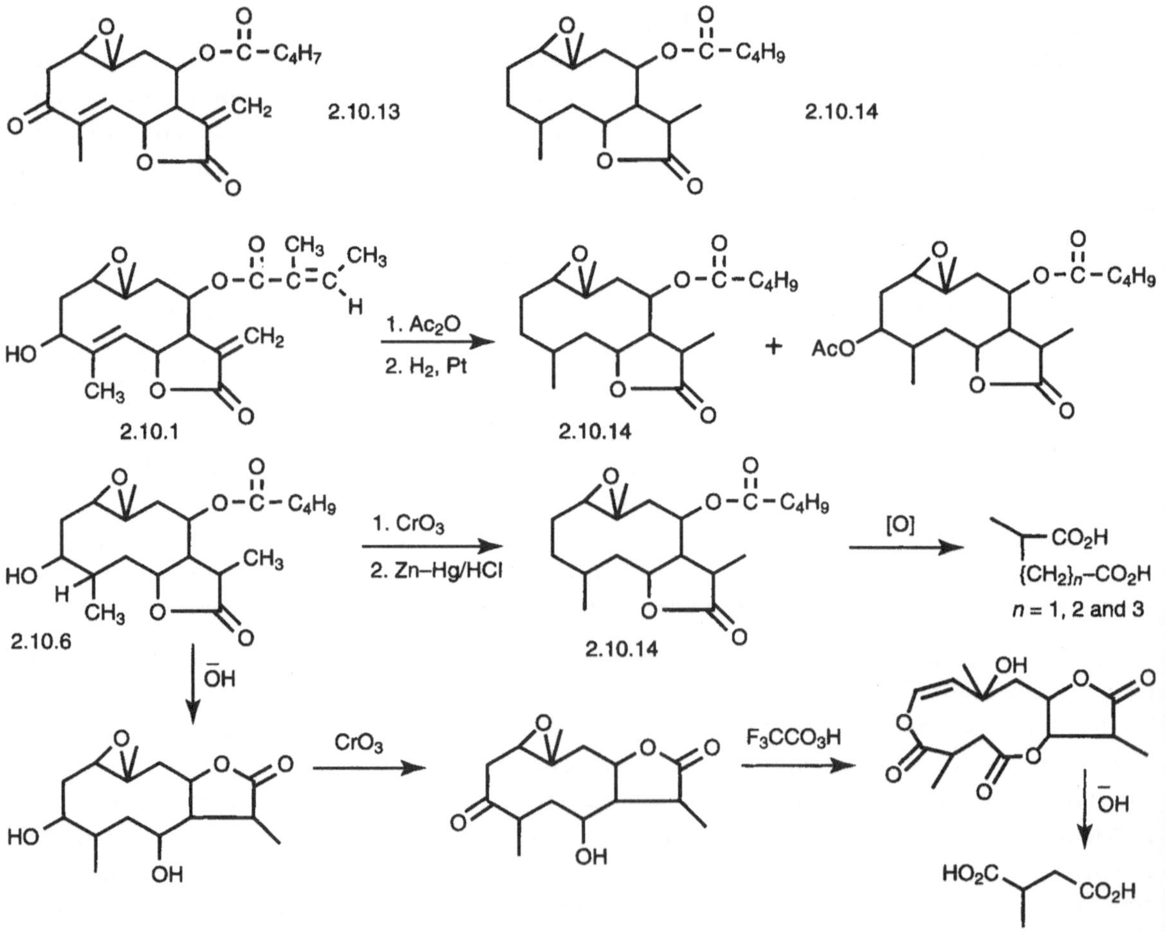

Suggested reading

1.   Sorm F and Dolejs L, *Guianolides and Germacranolides*, San Francisco: Holden-Day, 1966.
2.   Iruichijima S, Kuyama S, Takahashi N and Tamura S, *Agr. Biol. Chem.*, 1960, 30: 1152.

2.11 DELPHISINE: CONFIRMATION OF A STRUCTURE BY NMR SPECTRA

We will now discuss a structure which could be interpreted satisfactorily by NMR spectral data, including some of the very recent techniques.

The example chosen to illustrate this methodology is a diterpene alkaloid, delphisine. Diterpene alkaloids are fairly widespread in higher plants but the majority of the known compounds have been isolated from species of the genera Aconitum, Consolida and Delphinium of the family Ranunculaceae.

The structure of delphisine (2.11.1), a minor constituent of the seeds of *Delphinium staphisagria,* was deduced—on the basis of chemical and degradative experiments—as a pentacyclic diterpene alkaloid. It is a tertiary base and carries a *N*-ethyl group, besides possessing three methoxyls, two acetoxyls and a secondary alcoholic hydroxyl as shown in the structure given. The structure has been confirmed by a combination of ^{1}H NMR, ^{13}C NMR and correlation studies (such as COSY) which bring out information of instructional value.

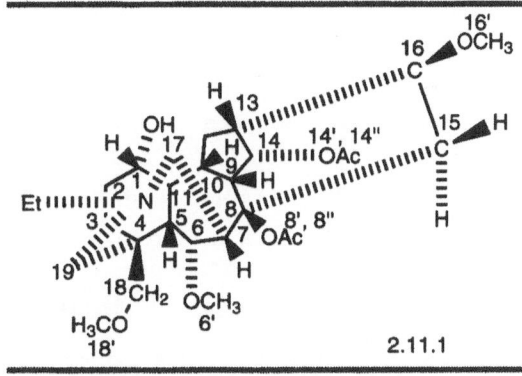

2.11.1

Conventional ^{1}H NMR spectral studies confirm the presence of the various functional groups mentioned above. Their dispositions in relation to the skeletal structure were confirmed by a homonuclear shift correlated 2D NMR spectrum (COSY). The data thus generated are shown in Table 2.1. Anyone well-versed in the interpretation of NMR spectra of organic compounds can see that the assignments shown in the table are, by and large, unambiguous. The grey areas in the interpretation of the conventional ^{1}H NMR spectrum were removed by the 2D spectrum. As a consequence of the presence of as many as six different functional groups, all the carbon atoms in delphisine are differentiated from each other and give rise to separate signals. Most of these signals are readily assigned by a one-bond heteronuclear shift correlated 2D NMR spectrum (HETCOR). The data are shown in Table 2.2. This technique, however, is not useful for determining the chemical shift values of quarternary carbons as their resonances do not show up in a HETCOR spectrum. For this purpose, techniques like the Distortionless enhancement by polarisation transfer (DEPT) sequence, long-range HETCOR and Difference nuclear Overhauser enhancement (DIFNOE) are useful. We may note here that a similar methodology was used to find out the correct structure of the complex insect anti-feedant component of neem (*Azadirachta indica*) seeds, azadirachtin. In this case, x-ray diffraction could not be used due to the microcrystalline nature of thecompound.

The longrange HETCOR spectrum is useful in differentiating the three methoxyl groups in delphisine as shown by the data given in Table 2.3. The DIFNOE technique could be used to differentiate the two acetoxyl groups and to locate them correctly. These data are also shown in Table 2.3.

Table 2.1 COSY experiment on delphisine

δ (ppm)	Multiplicity	H	Off-diagonal elements	Multiplicity	H
4.81	t	14	3.32	d	16
			2.50	dd	9
			2.42	ddd	13
4.03	d	6	3.05	br, s	7
			2.23	s	5
3.65	br, s	1	1.87	m	2
			1.59	m	2
3.57	d	18	3.10	d	18
3.32	d	16	4.81	t	14
			2.85	dd	15
			2.08	dd	15
3.05	br, s	7	4.03	d	6
2.85	dd	15	3.32	d	16
			2.08	dd	15
2.59	d	19	2.26	d	19
2.50	dd	9	4.81	t	14
			1.99	m	10
2.42	ddd	13	4.81	t	14
			1.84	m	12
2.23	s	5	4.03	d	6
1.99	m	10	2.50	dd	9
1.90	m	12	1.84	m	12
1.59	m	2	3.65	br, s	1
			1.79	m	2
			1.55	m	3
1.55	m	3	1.79	m	2

The 2D-COSY spectrum was recorded in CDCl₃ at 300 MHz; chemical shifts in ppm (δ) downfield from TMS; s: singlet; d: doublet; t: triplet; br: broad; m: multiplet; dd: double doublet; ddd: doublet of double doublet.

Table 2.2 C/H Correlation of delphisine

Carbon (ppm)	Multiplicity	Assignment	Proton correlation
84.0	d	6	4.03, t
82.6	d	16	3.32, d
79.7	t	18	3.57, 3.10, d
75.5	d	14	4.81, t
72.0	d	1	3.65, br, s
62.6	d	17	2.62, s
59.1	q	18	3.30, s
58.0	q	6	3.23, s
56.6	q	16	3.31, s
56.6	t	19	2.59, 2.26, d
48.2	t	N-CH$_2$-	2.54, 2.49, m
47.8	d	7	3.05, br, s
43.8	d	5	2.23, s
43.1	d	9	2.50, dd
43.1	d	10	1.99, m
38.4	d	13	2.42, ddd
38.3	t	15	2.85, 2.08, dd
30.0	t	3	1.55, m
29.4	t	2	1.87, 1.59, m
29.4	t	12	1.90, 1.84, m
22.2	q	8''	1.96, s
21.2	q	14''	2.03, s
12.9	q	N-CH$_2$-CH$_3$	1.11, t

Table 2.3 Long-range HETCOR of delphisine

Carbon (ppm)	H correlation in HETCOR (δ)	Assignment	H correlation in long-range HETCOR (δ)
59.1 (q)	3.30	18'	–
58.0 (q)	3.23	6'	4.03 (H-6)
56.6 (q)	3.31	16'	3.32 (H-16)

Suggested reading

1.   Pelletier SW and Joshi BS, In Pelletier SW, (ed.) *Alkaloids: Chemical and Biological Perspectives*, Vol. 7, (and references therein), New York: Springer–Verlag, 1991.
2.   Pelletier SW and Djarmati Z, *J. Amer. Chem. Soc.*, 1976, 98: 2626.
3.   Pelletier SW, Sudhakar Bhandaru, Desai HK, Ross SA and Sayed HM, *J. Nat. Prod.*, 1992, 55: 736.

2.12 TUBOCURARINE: THE STING IN THE POISONED ARROW

Among the many intriguing discoveries made by European explorers in South America, the arrow poisons used by the natives of Brazil, Peru and Ecuador deserve a place in a book on natural products! These poisons known under the name curare can be further sub-divided into tube curare, calabash curare and pot curare. Tube curare is made from the plant, *Chondodendron tomentosum*, which is a creeper belonging to the family Menispermaceae, along with some other plants which are crushed and cooked, and the resulting dark red resinous mass packed into bamboo tubes. The main ingredient of calabash curare is *Strychnos toxifera* and in this case the preparation is placed in a calabash or gourd.

The main toxic principle of tube curare is the alkaloid, d-tubocurarine, which was isolated by the German chemist, Boehm, in 1897. However, the compound was obtained in the pure crystalline state only in 1935 by the British chemist, H King, who isolated it from a museum sample of curare. He also established its structure as a bisbenzylisoquinoline alkaloid. It is a water soluble quaternary ammonium compound. It is best purified as the chloride salt by crystallisation from hydrochloric acid. It is used as an adjunct in surgical anaesthesia. The following is a concise account of the studies by King and others on the structure elucidaton of tubocurarine.

The alkaloid (2.12.1) has the molecular formula, $C_{38}H_{44}C_{12}N_2O_6$. Zeisel estimation revealed the presence of two methoxyl groups. Methylation with methyl iodide yielded the di-*O*-methyl derivative (2.12.2) showing that the alkalod also has two phenolic hydroxyls. Thus, four of the six oxygen atoms are accounted for. Hoffmann degradation of the chloride of (2.12.2) yielded a mixture of four methine bases which were then converted into their

2.12.1: R = H
2.12.2: R = CH₃

2.12.3

methiodides. Three of the latter were the same as those obtained from the alkaloid beeberine (2.12.3). A second Hoffmann reaction of the methine methochlorides obtained from 2.12.2 gave, besides trimethylamine, a nitrogen- free compound which could be formulated as 2.12.4, as on oxidative degradation it yielded the tricarboxylic acid, 2.12.5. The last mentioned compound, on alkali fusion gave 4-hydroxybenzoic acid (2.12.6). On oxidative degradation, beeberine dimethyl ether also gave another tricarboxylic acid which was assigned the structure (2.12.7) as, on decarboxylation, it yielded 2,2'-dimethoxydiphenyl ether (2.12.8).

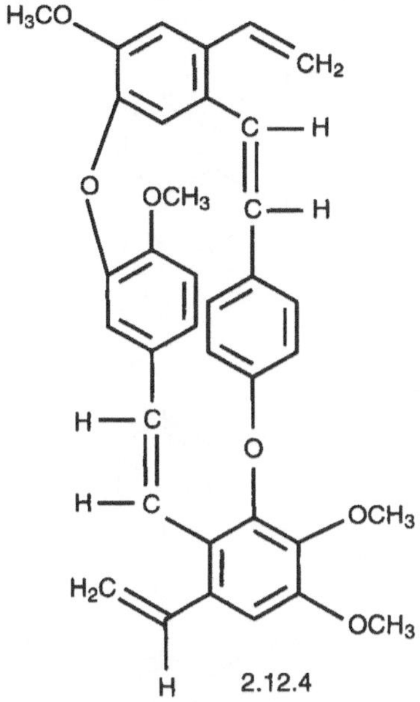

2.12.4

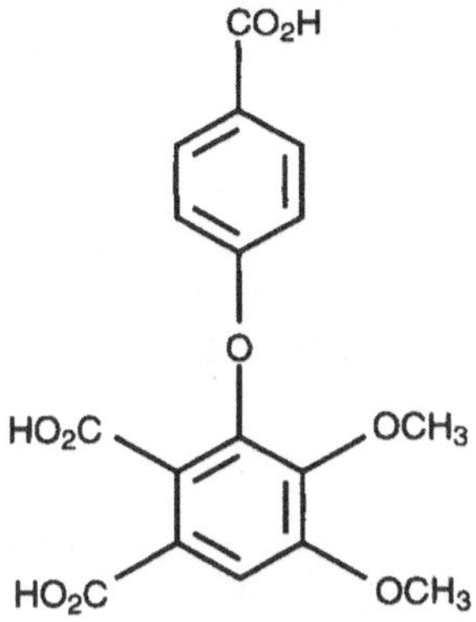

2.12.5

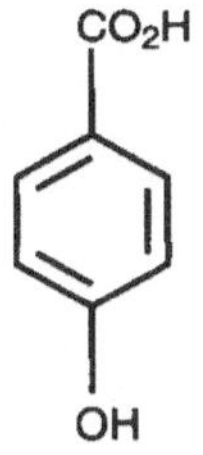

2.12.6

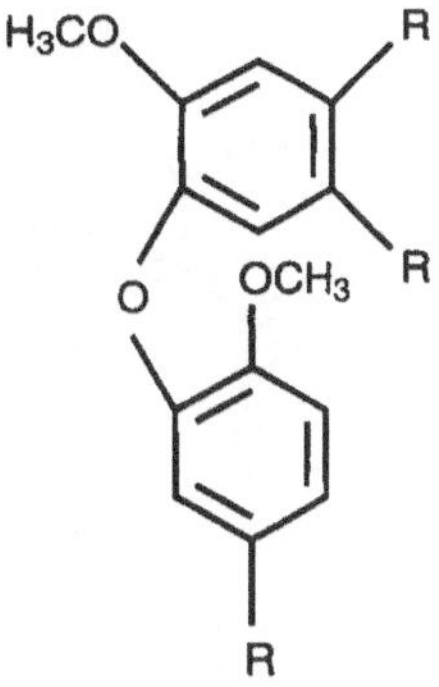

2.12.7: R = CO$_2$H
2.12.8: R = H

After a one-stage Hoffmann reaction, beeberine dimethyl ether yielded a methine base (2.12.9) which on ozonolysis gave a mixture of two dimethylaminodialdehydes. The latter was methylated, oxidised (by potassium permanganate) and boiled with alkali when trimethylamine and a mixture of two vinyl carboxylic acids were obtained. One of these acids was identified as 4',6-dicarboxy-2,3-dimethoxy-5-vinyldiphenyl ether (2.12.10). The other acid was decarboxylated by heating with copper and quinoline and then oxidised by potassium permanganate. The product was identified as 4-carboxy-2,2'-dimethoxydiphenyl ether (2.12.11). It was confirmed by a Ullmann synthesis involving o-bromoanisole (2.12.12) and vanillic acid (2.12.13). Hence, the parent vinyl carboxylic acid could be given the structure (2.12.14). This compound on further oxidation yielded a tricarboxylic acid (2.12.15) whose structure was confirmed by a Ullmann condensation between 4-iodo-5-methoxyphthalic acid (2.12.16) and isovanillic acid (2.12.17).

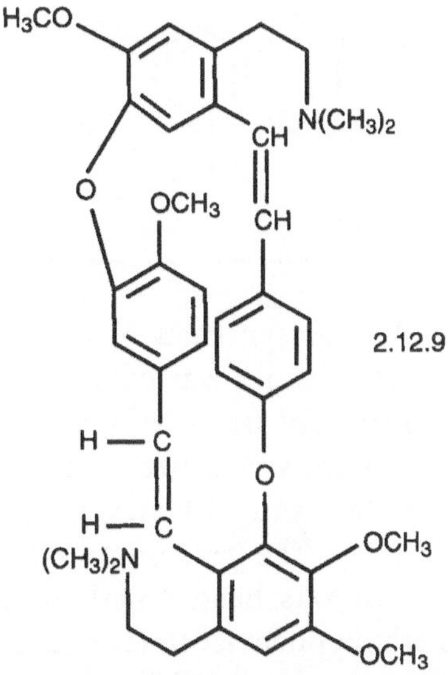

2.12.9

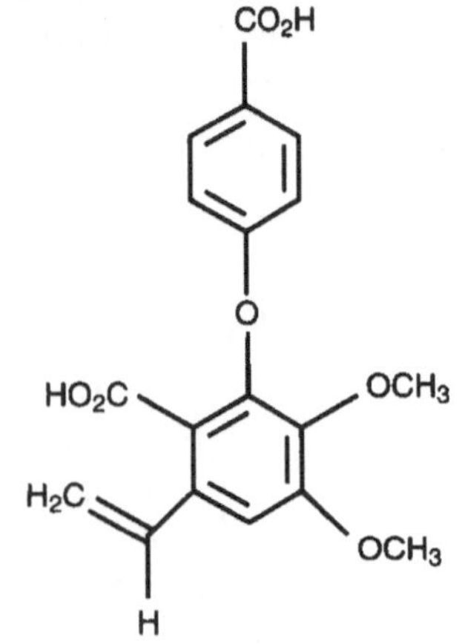

CO$_2$H

2.12.10

H$_3$CO, R, R$_1$, OCH$_3$, O, R$_2$

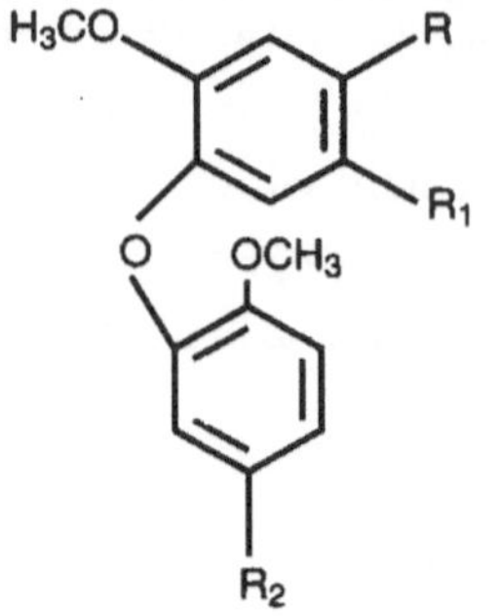

2.12.11: R = CO$_2$H; R$_1$ = R$_2$ = H
2.12.14: R = —CH≡CH; R$_1$ = R$_2$ = CO$_2$H
2.12.15: R = R$_1$ = R$_2$ = CO$_2$H

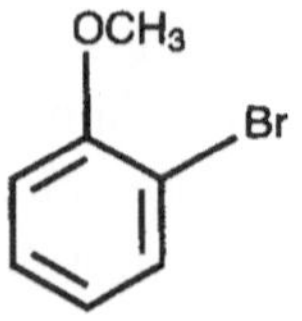

OCH$_3$
Br

2.12.12

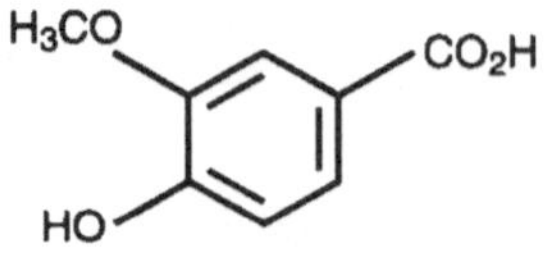

H$_3$CO, CO$_2$H, HO

2.12.13

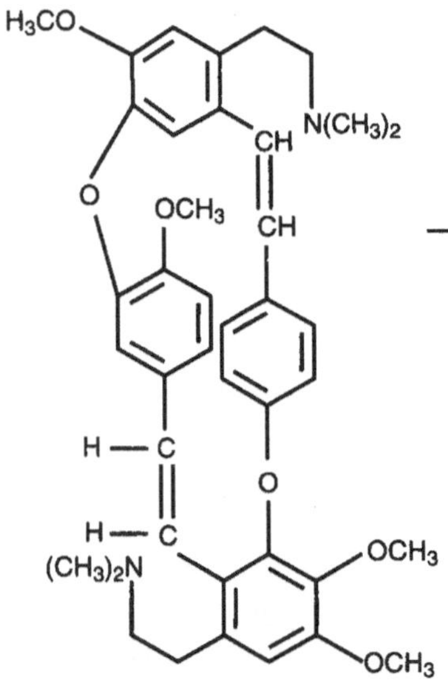

2.12.9

$\xrightarrow{O_3}$

+

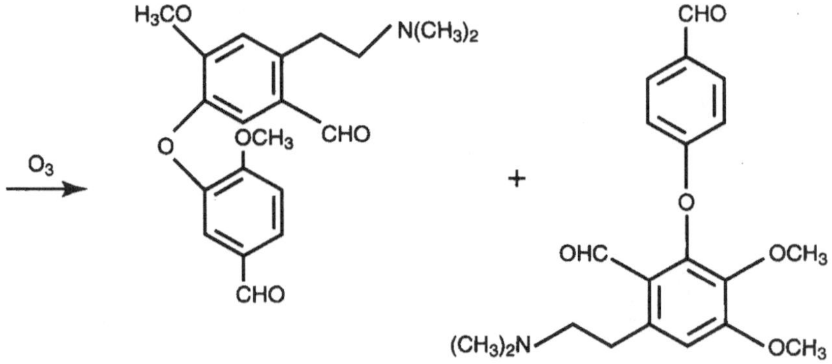

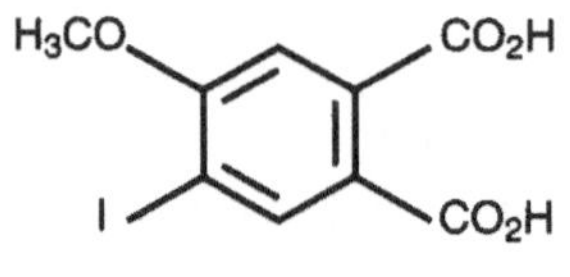

2.12.16

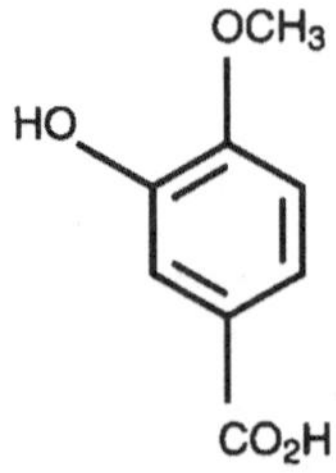

2.12.17

A total synthesis of racemic tubocurarine dimethyl ether was effected by Tokacher et al in 1959. The first step was a reaction between 3-methoxy-4-hydroxyphenylethylamine (2.12.18) and 4-benzyloxyphenylacetic acid (2.12.19). The potassium salt of the resulting 2.12.20 was heated with methyl 3-bromo-4-methoxyphenylacetate (2.12.21) in the presence of copper powder (Ullmann reaction) to obtain

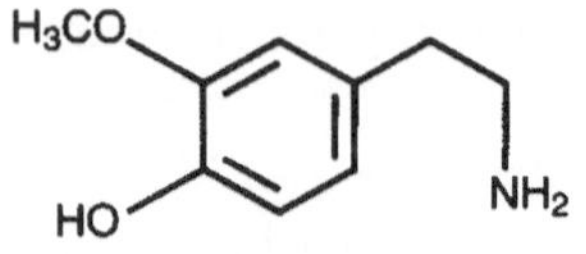

H$_3$CO, HO, NH$_2$

2.12.18

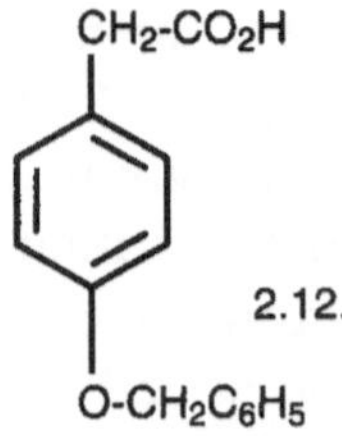

CH$_2$-CO$_2$H

2.12.19

O-CH$_2$C$_6$H$_5$

2.12.22. The latter was condensed with 3-methoxy-4-hydroxy- 5-bromophenylethylamine (2.12.23) and the resulting amide (2.12.24) methylated and subjected to a Bischler–Napieralski reaction (heating with phosphorus oxychloride) when 2.12.25 was obtained. The latter was debenzylated and subjected to an intramolecular Ullmann reaction to get 2.12.26. The final steps involved reduction with zinc and acetic acid followed by a two- stage methylation.

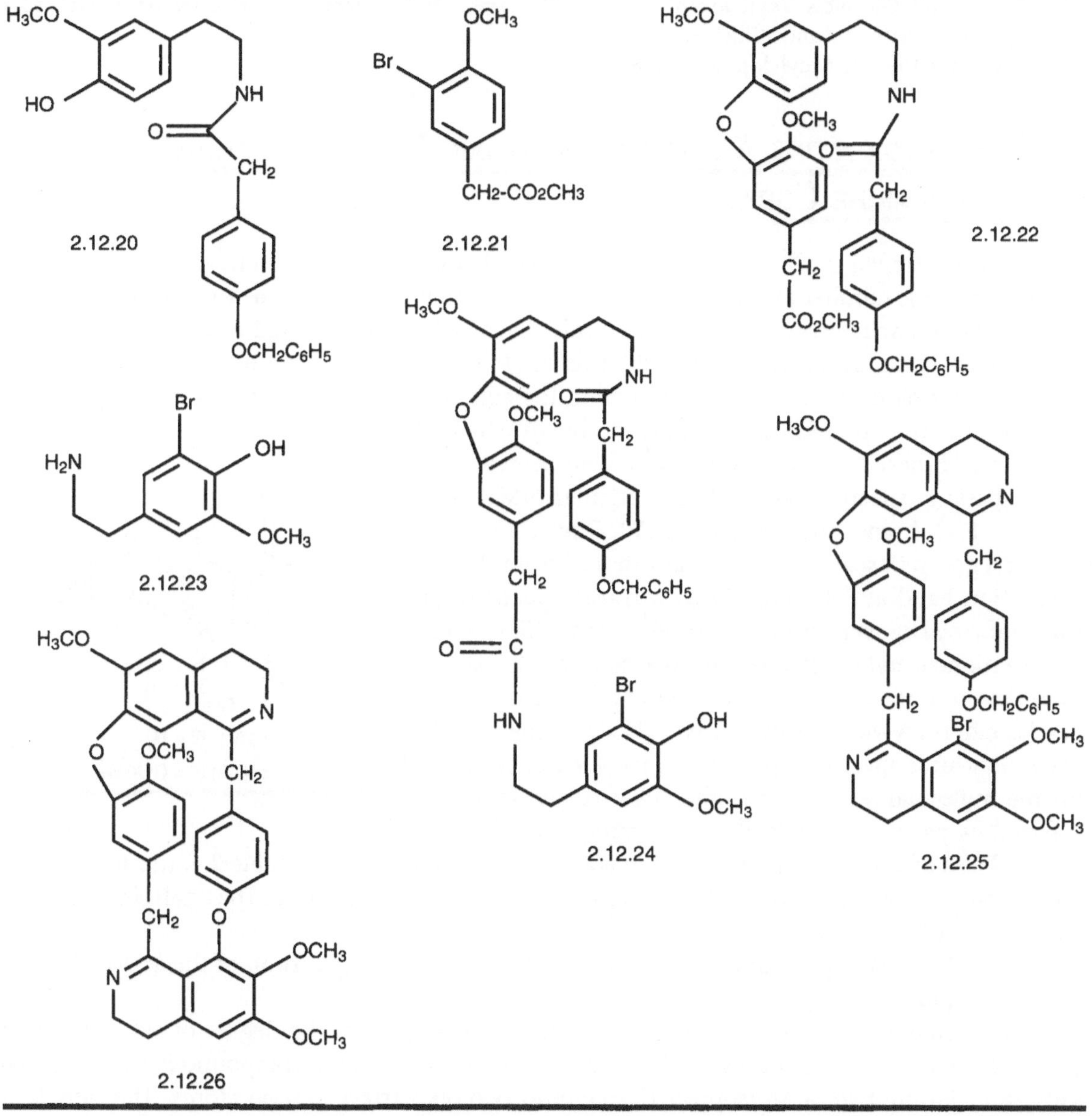

2.12.20

2.12.21

2.12.22

2.12.23

2.12.24

2.12.25

2.12.26

The total package of experimental strategy described above is a typical example of the classical approach to structure determination. It is an accepted norm that a structure deduced from analytical and degradation studies should be confirmed by synthesis—a dissection should be followed by a reconstruction!

Suggested reading

1. King H, *J. Chem. Soc.*, 1935, 1381, 1939, 1157 and 1948, 2645.
2. Tolkacher eta al., *Chem. Abs.*, 1960, 54: 1578.
3. Wilson and Gisvold's *Textbook of Organic Medicinal and Pharmaceutical Chemistry*, 11th edn, Lippincott Williams and Wilkins, 2004.
4. *Wikipedia*, the free encylclopedia, 2008.

2.13 SCLEROPHYTINS A AND B: CYTOTOXIC DITERPENES FROM A MARINE CORAL

As briefly mentioned in Chapter 1, marine flora and fauna have proved in recent years to be rich sources of a large number of secondary metabolites and in particular a wide variety of terpenoids. Among them marine invertebrates (soft corals) have yielded oxygenated diterpenes possessing skeletal structures uncommon among diterpenes of terrestrial plant origin.

Sclerophytins A and B were isolated from the soft coral, *Sclerophytum capitalis*. In distinct contrast to the preceding example (tubocurarine), the structures of these two compounds were determined entirely on the basis of spectral data and in particular NMR spectra (proton, 13C, homo and hetero COSY). The presence of a hydroxyl group in sclerophytin A was inferred from the broad absorption band at 3400 cm^{-1} in its infrared spectrum. This was confirmed by acetylation (with acetic anhydride–pyridine) when a monoacetyl derivative was obtained, which was found to be the same as sclerophytin B. The exact molecular formula of sclerophytin A was determined as $C_{20}H_{32}O_3$ by means of high resolution mass spectroscopy which gave a value of 320.2235 for the molecular weight. The other two oxygen atoms were inferred as parts of chemically inert cyclic ether functions.

2.13.1: R = H
2.13.2: R = COCH₃

On the basis of a detailed analysis of the NMR spectral data given in Table 2.4, the structure of sclerophytin A was deduced as 2.13.1 and that of sclerophytin B as 2.13.2. The analysis of the data is given below in Table 2.4.

In the 13 C NMR spectrum of 2.13.2 the signal of the carbonyl carbon of the acetoxyl group is at 171.8 ppm (s).

From a combination of the proton and carbon NMR spectra, the degree of substitution at each of the twenty carbon atoms can be readily recognised. Thus, the compound has four methyl groups, including two of an isopropyl group and two on tertiary carbon atoms. The signals at 4.64 and 4.67 ppm in the 1H NMR spectrum and the signal at 109.08 in the carbon spectrum are due to a part of a terminal vinal functionality. Carbon–carbon connectivities could be deduced from the COSY spectra. Thus, in the homo COSY (1H–1H) spectrum, the signal at

Table 2.4 1H and 13C NMR data for 2.13.1 and 2.13.2.

Carbon	δ H (J in Hz)	δ C	δ H (J in Hz)	δ C
	2.13.1		**2.13.2**	
1	2.19 (dd, 6.7)	45.16 (d)	2.18 (dd)	45.45 (d)
2	3.63 (br,s)	90.49 (d)	3.63 (br,s)	90.49 (d)
3		74.83 (s)		74.77 (s)
4	1.27 (m)	39.93 (t)		30.82 (t)
5	1.75 and 1.98 (m)	29.37 (t)		28.99 (t)
6	4.55 (d)	78.93 (d)	5.62 (dd)	85.03 (d)
7		77.01 (s)		75.87 (s)
8	1.71 and 2.23 (m)	45.38 (t)	1.70 and 2.23 (m)	45.46 (t)
9	4.12 (m)	78.17 (d)	4.12 (m)	77.97 (d)
10	2.98 (dd, 13.4;6.7)	53.21 (d)	2.98 (dd)	53.2 (d)
11		147.90 (s)		147.87 (s)
12	2.05 and 2.25 (m)	31.58 (t)	2.03 and 2.15 (m)	31.87 (t)
13	1.70 (m)	24.85 (t)		24.79 (t)
14	1.29 (m)	43.69 (d)		43.56 (d)
15	0.80 (d, 6.7)	16.01 (q)	0.79 (d)	16.11 (q)
16	0.96 (d, 6.7)	21.96 (q)	0.95 (d)	21.93 (q)
17	1.78 (m)	29.13 (d)		28.07 (d)
18	1.16 (s)	30.32 (q)	1.14 (s)	30.19 (q)
19	1.20 (s)	23.06 (q)	1.20 (s)	23.70 (q)
20	4.64 and 4.67 (s)	109.08 (t)	4.64 and 4.67	109.23 (t)

4.55 ppm has cross peaks corresponding to the signals at 1.98 and 1.75 ppm. From its chemical shift value and multiplicity, the signal at 4.55 ppm could be attributed to the hydrogen atom on C-6. These assignments are also supported by the corresponding signals in the 13C and hetero COSY (1H–13C) spectra. An interesting observation is that one of the hydrogen atoms of the exocyclic methylene group (signal at 4.67 ppm) shows large interaction with the hydrogen responsible for the signal at 2.05 ppm, indicating allylic spin–spin interaction between the two. The other exocyclic methylene proton (signal at 4.64 ppm) on the other hand, exhibits weak interactions with the hydrogen atoms responsible for the signals at 4.12 and 3.63 ppm. These interactions are, presumably, due to long range interactions possible because of the stereochemical disposition of the concerned hydrogen atoms on the same face of the molecule as shown in 2.13.1. The homo

COSY spectrum also reveals that the hydrogen atom giving the signal at 4.12 ppm is involved in vicinal spin–spin interaction with three hydrogen atoms whose signals are seen at 2.98, 2.23 and 1.71 ppm. These are respectively, the hydrogen atoms on C-9, C-10 and C-8 (methylene group). These assignments are again supported by 13C NMR data. The connectivity between the signals at 2.98 and 2.19 ppm permits the assignment of the latter signal to H-1, which, in turn, exhibits interactions with the signals at 3.63 (H-2) and 1.25 (H-14) ppm.

Subsequently, it has been suggested that the structures originally assigned to sclerophytins A and B, namely 2.13.1 and 2.13.2, have to be revised to 2.13.3 and 2.13.4 respectively. This folowed the observation that synthetic 2.13.1 was different from natural sclerophytin A,. Friedrich and Paquette who have made this suggestion also cite 13C NMR data in support of their contention. They point out that the spectra of sclerophtin A and sclerophytin F, for which the structure 2.13.3 had earlier been established and confirmed by X-ray analysis, are identical except for the differences in the signal due to C-3. Therefore, they have suggested that sclerophytins A and F are C-3 epimers, though they have given the same structure to both the compounds. If their conjecture is correct, sclerophytin A should be 2.13.5 and sclerophytin B 2.13.6. However, the original structures were also supported by high resolution mass spectra which show molecular ion signals at m/z 320 for sclerophytin A and m/z 362 for sclerophytin B. While it is possible that the compounds could have undergone dehydration (involving one of the tertiary hydroxyl groups) prior to electron impact, it should be noted that in the spectrum of sclerophytin A, a peak at m/z 302, due to loss of a water molecule is seen. Further, on acetylation, sclerophytin A has been reported to give only a mono acetyl derivative, though it is possible that the tertiary hydroxyls could have remained unreactive. One

2.13.3: R = H   2.13.5: R = H

2.13.4: R = COCH$_3$ 2.13.6: R = COCH$_3$

may, therefore, conclude that there are supporting data in favour of either structure. Perhaps, if more of the compound was available, a Zerewittinoff estimation could have been done. This highlights the fact that the best approach to a structure determination is to combine spectral data with some analytical, if not degradative evidences as well.

Suggested reading

1. Perveen Sharma and Maktoob Alam, Sclerophytins A and B. Isolation and structures of novel cytotoxic diterpenes from the marine coral *Sclerophytum capitalis, J. Chem. Soc. Perkin Trans.*, 1988, 1: 2537–40.
2. Dirk Friedrich and Leo A Paquette, Structural and stereochemical reassessment of sclerophytin type diterpenes, *J. Nat. Prod.*, 2002, 65: 126–30.
3. Faulkner DJ, *J. Nat. Prod. Rep.*, 1997, 14: 259.

4. Sceuer PJ, (ed.), *Marine Natural Products, Chemical and Biological Perspectives*, Vols 1–5, New York: Academic Press, 1977–81.

2.14 MANGIFERIN: PARENT OF THE INDIAN YELLOW

There is a prevalent story about the pigment, Indian yellow. Chemically, it is euxanthic acid (2.14.1), which is the glucuronide of euxanthone. The story is that it used to be produced in rural India by feeding cows with mango leaves and allowing the urine to dry up. According to Wikipedia this tale was concocted by one Mukerji and may not be true. However, it is a fact that mangiferin, the yellow pigment present in the leaves, heartwood and stem bark of the mango tree does undergo metabolism in experimental animals to turn out as euxanthic acid (2.14.1) in the urine. The yellow pigment used in the Bundi School of painting is made from the urine of cows fed on mangoes. We have also heard that in some villages of Andhra Pradesh the urine from cows fed with mango leaves is mixed with the lime used for whitewashing walls. The authenticity of this has, however, not been verified. Therefore, some more research has to be done before Mukarji's version of the Indian yellow is either accepted or dismissed as a figment of imagination! The magnesium salt of 2.14.1 is a fluorescent dye and was at one time a popular medium in Europe for producing water colours as well as oil paintings.

Mangiferin is best obtained from the stem bark of *Mangifera indica*. It is a C-glucoside of a tetrahydroxyxanthone. Its structure was established as 2.14.2 on the basis of the data outlined below. It has the molecular formula, $C_{19}H_{18}O_{11}$. It is soluble in aqueous ethanol and answers the Molisch test suggesting that it is a glycoside. However, it resists hydrolysis by the action of acids or enzymes indicating that it may not be a *O*-glycoside. It forms an octaacetate, a tri-*O*-methyl ether (2.14.3) (obtained by the action of diazomethane) and a tetra-*O*-methyl ether (2.14.4) (formed by the action of dimethyl sulphate in the presence of potassium carbonate). These observations indicate that the compound possesses eight hydroxyl groups of which four are phenolic in nature; one of the

2.14.1

2.14.2: R =

2.14.3: R = H
2.14.4: R = CH₃

phenolic hydroxyls is less reactive than the other three, perhaps due to intramolecular hydrogen bonding with a carbonyl group. Indeed, mangiferin and its trimethyl ether give deep green colours with alcoholic ferric chloride whereas the tetramethyl ether does not give any colour with this reagent. The presence of a carbonyl group hydrogen bonded to a phenolic hydroxyl is revealed by the presence of a strong absorption band at 1640 cm^{-1} in the infrared spectrum of mangiferin. Other colour reactions (orange red colour with magnesium and hydrochloric acid and red colour with zinc and boiling hydrochloric acid) as well as the ultraviolet spectrum (absorption maxima at 239, 257, 314 and 360 nm) further suggest that the compound is probably a tetrahydroxyxanthone derivative.

When mangiferin is heated with hydriodic acid in phenol, and the product acetylated, a colourless compound, identified as 1,3,6,7-tetraacetoxyxanthone (2.14.5) is obtained. Mangiferin undergoes oxidative degradation when heated with aqueous ferric chloride. The presence of glucose in the aqueous hydrolysate can be detected by paper chromatography and confirmed by conversion into its osazone derivative. From these observations it can be inferred that mangiferin is a C-glucoside of 1,3,6,7-tetrahydroxyxanthone (2.1.4.6). The following experimental strategy was then adopted to determine the position of the glucose moiety. Mangiferin trimethyl ether (2.14.3) consumed 2.2 moles of periodic acid, yielding a mole of formic acid and (2.14.7) in the process. Reduction of 2.14.7 with sodium borohydride followed by acid hydrolysis gave glycerol (2.14.8) and the xanthone (2.14.9). On the other hand, when 2.14.7 was hydrolysed without prior reduction, the product obtained was 2.14.10. The structure of the latter was confirmed by synthesis. 1-Hydroxy-3,6, 7-trimethoxyxanthone (2.14.11) was treated with allyl bromide and the resulting allyl ether (2.14.12) subjected to a Claisen rearrangement. The product (2.14.13) was then reacted with osmium tetroxide and the diol thus obtained (2.14.14) oxidised with periodic acid, when the xanthone acetaldehyde (2.14.15) was obtained. Further oxidation of 2.14.15 with periodic acid yielded a compound identical with 2.14.10. Thus, the position of the glucose unit as 2 in the xanthone structure was established.

2.14.5

2.14.6

2.14.7

2.14.8

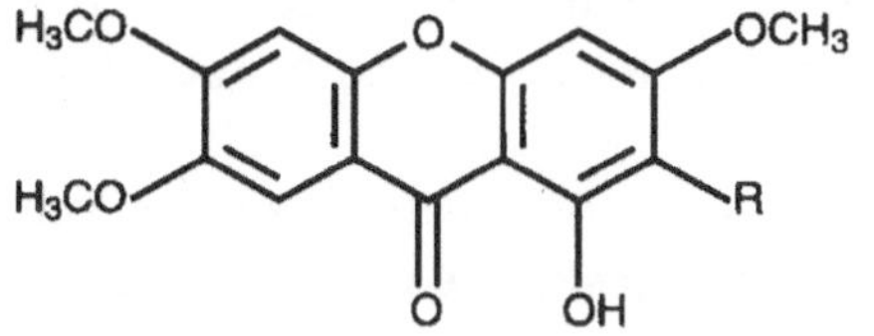

H_3CO ... O ... OCH_3

H_3CO ... R

O   OH

H_3CO ... O ... OCH_3

H_3CO ... OR

O   OR

2.14.9: R = $-\overset{H}{\underset{OH}{C}}-CH_2OH$

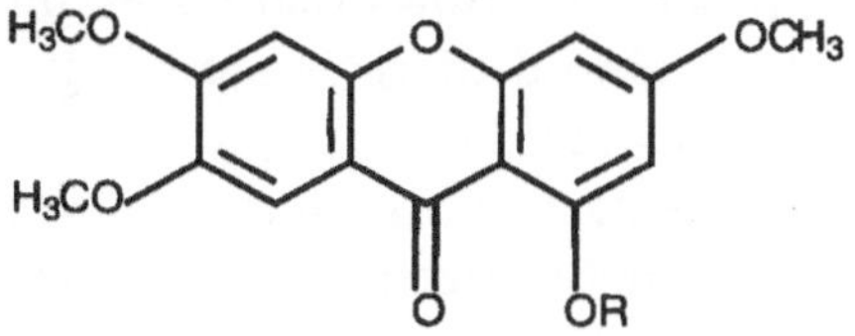

2.14.11: R = H

2.14.12: R = $- CH_2- CH = CH_2$

2.14.10: R = $-\underset{H}{C}\overset{CHO}{\diagdown} \overset{}{OH}$

2.14.13: R = $-CH_2 - CH=CH_2$

2.14.14: R = $- CH_2 - \underset{OH}{CH} - \underset{OH}{CH_2}$

2.14.15: R = $- CH_2 - CHO$

This example also illustrates the classical approach to structure determination with some input of spectral data and has instructive value.

Suggested reading

1. Bhatia VK, Ramanathan JD and Seshadri TR, *Tetrahedron*, 1967, 23: 1363–68.
2. Hawthorne et al., In Gore TS, Joshi BS, Sunthankar SV and Tilak BD, (ed.), *Recent Progress in the Chemistry of Natural and Synthetic Colouring Matters and Related Compounds*, New York: Academic Press, 1962.
3. Horhammer L and Wagner H, In Ollis WD, (ed.). *Recent Developments in the Chemistry of Natural Phenolic Compounds*, London: Pergamon Press, 1961.

2.15 CONESSINE: A PSEUDO ALKALOID

The bark of *Holarrhena antidysenterica* , also known as conessi or kurchi bark, has been used in traditional medicine for the treatment of amoebic dysentery in India and Africa. The main active principle is the steroidal alkaloid, conessine, which was first isolated from the seeds and bark of another plant, *Wrightia antidysenterica*. It has the molecular formula, $C_{24}H_{40}N_2$ and behaves as a diacidic base. On methylaton with two moles of methyl iodide, conessine yields a diquaternary ammonium compound showing thereby that conessine has two tertiary amino functions. On catalytic hydrogenation conessine yields a dihydro derivative (and hence has a double bond), the molecular formula of which indicates five degrees of unsaturation. It can therefore be inferred that the compound is pentacyclic, a conjecture confirmed by the following observation. When conessine dihydrochloride is subjected to dry distillation and the product dehydrogenated with selenium, γ-ethyl-1,2-cyclopentanophenanthrene (2.15.1) is obtained. This result shows that conessine is chemically related to the steroids. The presence of two C-methyl and three N-methyl groups are shown by Kuhn–Roth and Herzig–Meyer estimations respectively.

The diquaternary ammonium derivative of conessine, on heating with alkali, gives a molecule of trimethylamine and another tertiary amine, designated as apoconessine., which could be assigned the structure 2.15.2 based on the data noted below. The ultraviolet spectrum of apoconessine shows an absorption maximum at 235 nm indicating the presence of a trisubstituted (by ring residues) heteroannular diene chromophore. The remaining tertiary amino group in apoconessine is eliminated by the Emde reductive method. The product is pregna-3,5,20-triene (2.15.3). This result establishes not only the skeletal structure of conessine but also the configurations of the chiral centres at 8, 9, 10, 13, 14 and 17 positions. Incidentally, the Emde reaction is used when the Hoffman method fails due to lack of a hydrogen atom beta to the quaternary ammonium group. In such cases, the quaternary base is reduced with sodium amalgam in aqueous ethanol or sodium in liquid ammonia; the double bonds remain unaffected.

Von Braun degradation of conessine (reaction with cyanogen bromide followed by boiling with HBr) results in a nitrile which on hydrolysis and decarboxylation gives isoconessine, which is also a naturally occurring compound. Isoconessine has a secondary amino group in place of a tertiary amino in conessine. The N-acetyl derivative of isoconessine on Hoffman reaction followed by a Emde reduction gives 2.15.4. This structure is assigned to it as it can be prepared from 3-hydroxy-pregna-5,20-diene (2.15.5) by reaction with diethylamine followed by acetylation. Therefore, conessine itself can be assigned the structure 2.15.6.

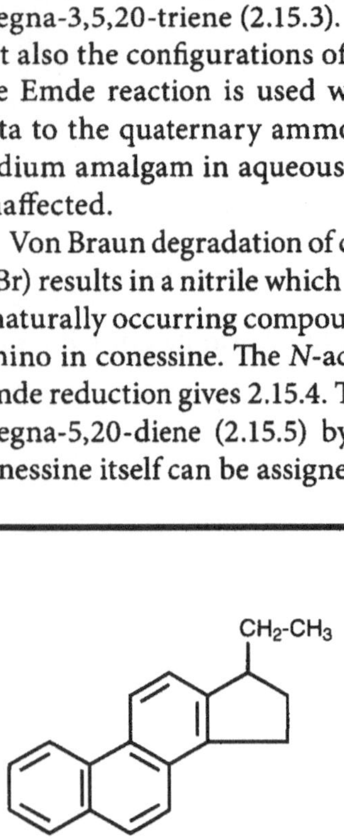

2.15.1 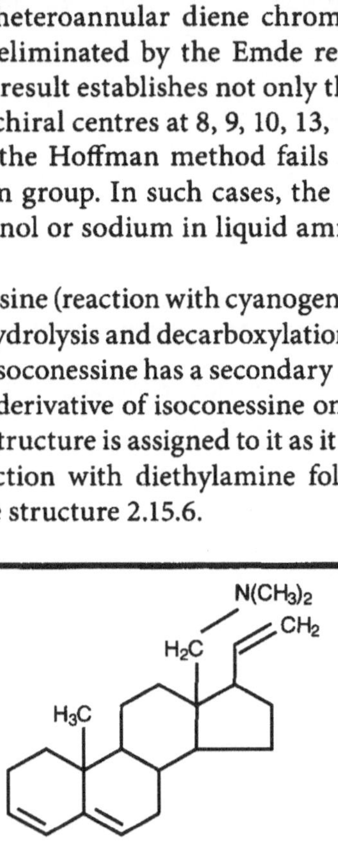   2.15.2 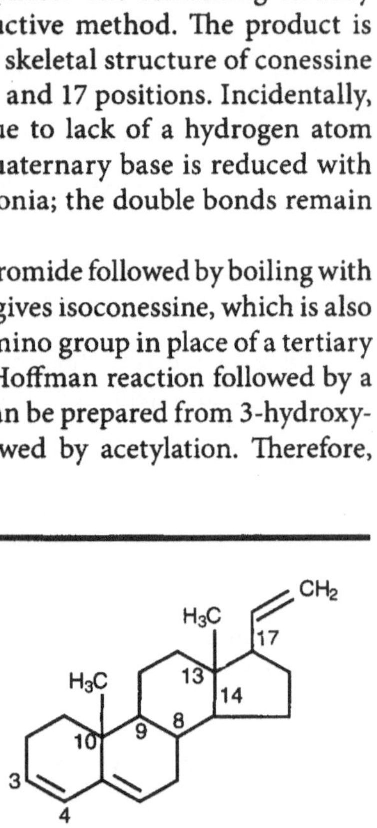   2.15.3

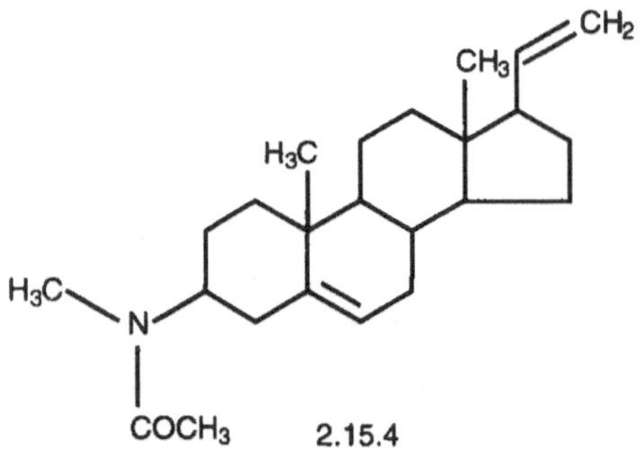

2.15.4 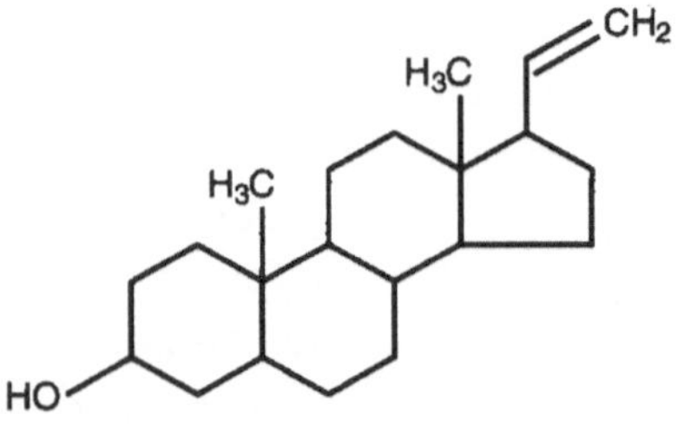   2.15.5

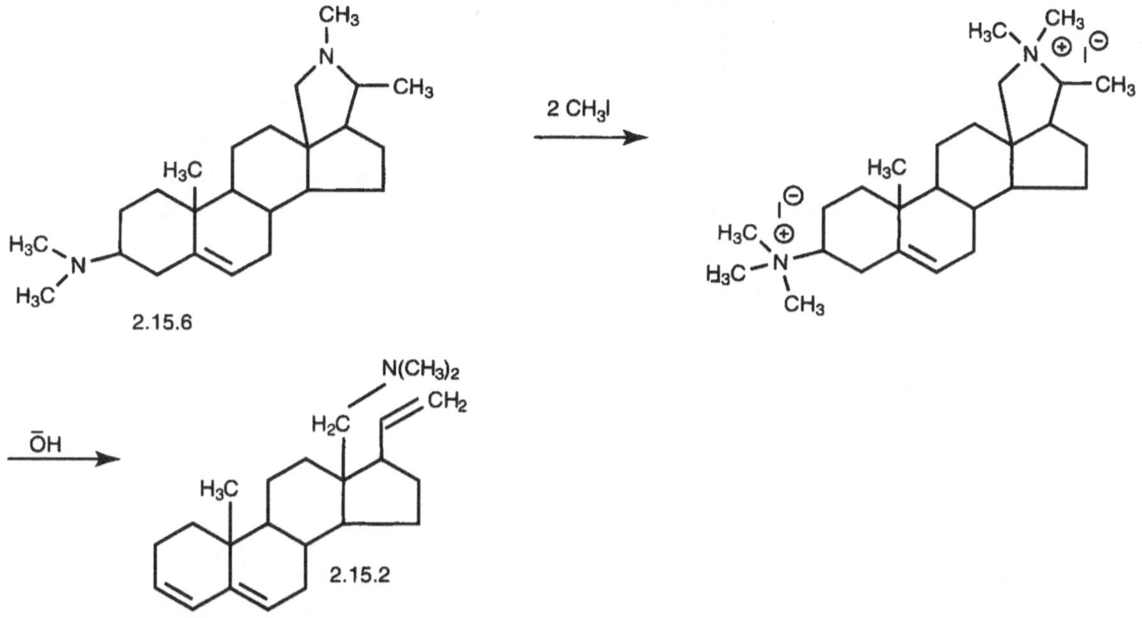

2.15.6

2.15.2

Suggested reading

1. Fieser LF and Fieser M, *Topics in Organic Chemistry*, New York: Reinhold, 1964.
2. Manske RHF and Holmes HL, *The Akaloids, Chemistry and Physiology*, Vol.7, London: Academic Press, 1981.
3. McKenna J, *Quarterly Reviews*, Chemical Society, London, 1972, 231–40.
4. RD Haworth, J Mckenna and Nazar Singh, *Nature*, 1948, 162: 22.

Summary

In this chapter we have reviewed the structures of fourteen naturally occurring compounds, belonging to different chemical types and originating from different biosynthetic pathways and precursors. We have seen how chemists have been able to figure out the structures by conducting ingeniously designed experiments, both chemical and spectroscopic, and logically as well as intuitively interpreting the data thus generated. Every investigation into the secrets of molecular structure is an adventure, and every experiment, however trivial, has its own value. As mentioned in the introduction, the 14 examples discussed here could have been replaced by 14 others and supplemented by fourteen more! The choice is immaterial, but what matters is the collated, systematised and logically interpreted information which the examples provide. We hope that we have succeeded, to some extent at least, in placing the fairly large amount of diverse data in a common theoretical framework, so that one can reason out any particular reaction of any compound in terms of fundamental chemical principles.

Any exploration can succeed only if it is backed by appropriate spadework. Structural studies will provide reliable data only if the compound under investigation is pure. The preliminary work leading to this stage can often be tedious and difficult since the majority of secondary metabolites occur in small quantities and in association with a number of closely related compounds having similar properties. Their separation into individual pure compounds may often involve a combination of methods, including carefully carried out chromatographic fractionations. This part of the total work on a naturally occurring compound may not be intellectually as stimulating as the parts which follow it, but it is no less important. Very often, if one is very observant, the chromatographic behaviour of a compound, its affinities for various solvents during solvent-fractionation, and other characteristics can give very useful clues with regard to its structure.

Chapter

3

Stereochemistry

3.1 INTRODUCTION

All the three types of stereoisomerism—conformational, geometrical and optical—are encountered during studies on natural products. Among these, optical isomerism is the most prevalent. A large number of secondary metabolites exist in nature as configurational isomers exhibiting optical activity. The physiological and pharmacological properties of enantiomers of bioactive compounds often differ. This is not surprising since the biological environments in which they are formed and function are themselves chiral in nature. We assume that users of this book are familiar with the basic concepts of stereochemistry, including the sequence rule for assigning configurations (R and S nomenclature). The following account will, therefore, be confined to the methods used for the determination of relative and absolute configurations of stereoisomers. Specific examples will be taken up later to serve as case studies. We shall begin with chiral compounds.

During the classical period of development of the subject, chemical methods were used to determine the configuration at a chiral centre. This involved chemical correlation of the compound under study with a reference compound whose configuration was known. For historical reasons, D or R (+)glyceraldehyde (3.1.1) was chosen as the reference compound. By a series of chemical reactions (Scheme 3.1), which do not affect the chiral centre, this compound could be correlated with (–)lactic acid (3.1.2),which should, therefore, have the D or R configuration. The reverse in the sign of rotation is intriguing at first sight. However, when the optical rotatory dispersion (ORD) spectra (plotting optical rotation against wavelength of incident light) of the two compounds are compared, it is seen that both of them are positive plain curves (Figure 3.1). Therefore, it is not the sign of optical rotation at the sodium D line (589 nm) but the sign of the ORD curve which should be used for correlating configurations. Figure 3.2 shows negative plain curves exhibited by the enantiomers of 3.1.1. and 3.1.2.

Plain ORD curves are shown by chiral compounds, such as hydrocarbons, non-conjugated olefins and alcohols which do not possess a chromophoric unit in the vicinity of a chiral centre. However, if a compound possesses an inherently symmetric but asymmetrically perturbed

Scheme 3.1

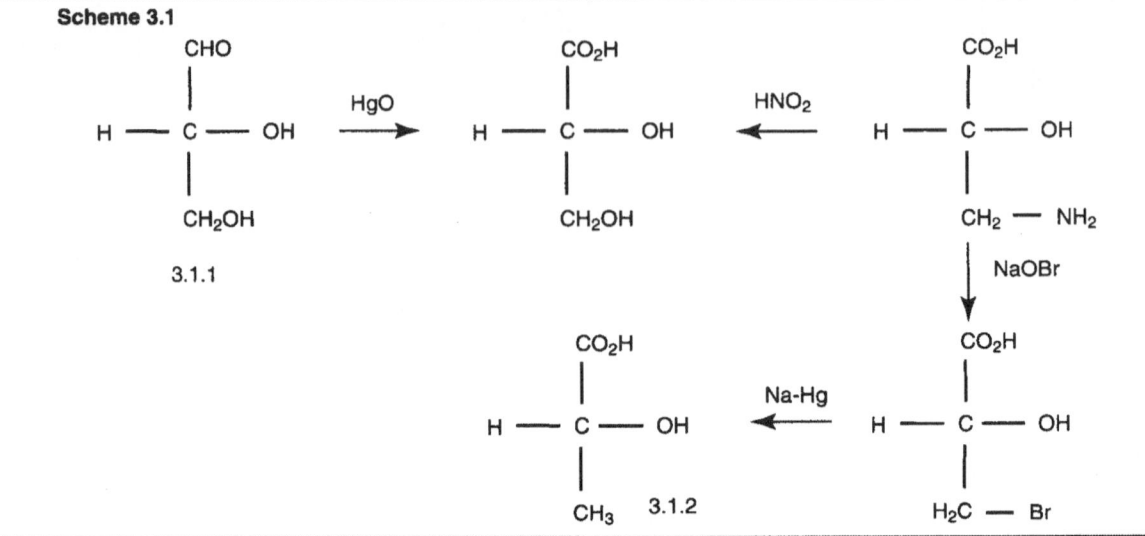

3.1.1

chromophore such as a carbonyl group present in the neighbourhood of a chiral centre, or if the chromophore itself is chiral, such as an optically active biphenyl, the resulting ORD curves will be anomalous. In such cases the optical rotation suddenly increases (becomes more positive or negative as the case may be), as the wavelength of the incident light approaches the absorption maximum of the chromophore (in the case of the carbonyl group this occurs in the region 280–290 nm due to n-π^* excitation), reaches a maximum value and then reverses. Therefore, the resulting ORD spectrum shows a peak and a trough. The difference in specific rotation between these two is known as the amplitude of the Cotton effect and is a measure of the chirality. Figure 3.3 shows an anomalous positive Cotton effect, wherein the peak is at a higher wavelength than the trough. Figure 3.4, in which the trough is at the higher wavelength, is an example of a negative Cotton effect curve.

Chemical correlations and plain ORD curves are of limited application for assigning configurations. The former is tedious, time consuming, and not always feasible, whereas the latter is restricted to hydrocarbons and others. Wider scope is offered by the anomalous

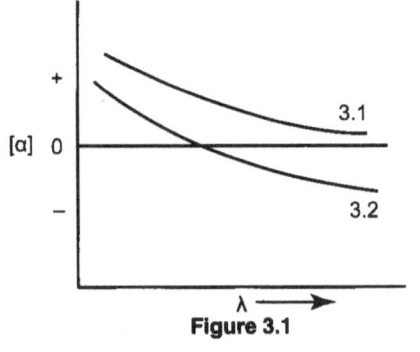

Figure 3.1

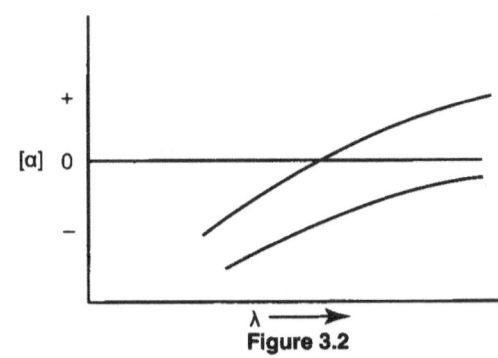

Figure 3.2

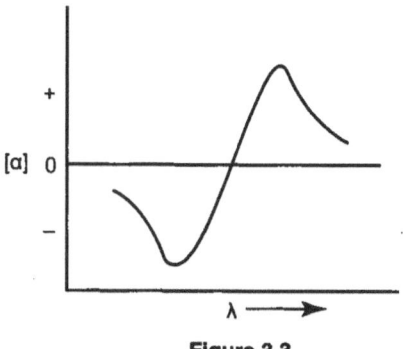

Figure 3.3

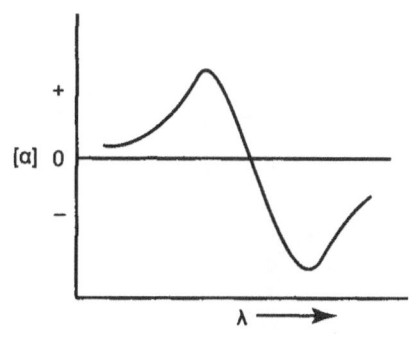

Figure 3.4

ORD spectra and by the related circular dichroism (CD) spectra. The latter are simpler and readily provide the numerical values of the amplitudes of the Cotton effect. In this case, the maximum in optical rotation coincides with the absorption maximum of the chromophore. Indeed, the CD and absorption spectra of a chiral ketone, for example, have identical contours.

A very important and widely used application of the anomalous Cotton effect is the Octant rule, which is semi-empirical in nature. It has been most useful in stereochemical studies on chiral cyclohexanones. The operational part of this rule uses three symmetry planes to divide a cyclohexanone into eight sectors or octants. One horizontal plane disects the molecule into a upper half and a lower half. Another, a vertical one, perpendicular to the first, cuts it into a left half and a right half. The third plane, also perpendicular, divides the molecule into a front part and a back (or rear) part by cutting through the carbonyl group. Of these eight sectors, for most practical purposes, only the rear four sectors need to be considered as the majority of substituents at positions 2 and 6 on the cyclohexanone ring are not long enough to fall into front parts. The cyclohexanone is then viewed in its chair conformation from the carbonyl oxygen side with the alpha substituents on positions 2 and 6 assuming axial orientation. Viewed thus, the rule stipulates that if any substituent falls or nearly falls on any of the symmetry planes, it does not contribute to optical activity. Thus, equatorial substituents at positions 2 and 6 (anti-clockwise numbering) which are almost on the horizontal plane (plane B) and both substituents (equatorial and axial) on carbon 4 (falling on the vertical plane A), do not make any contributions to optical activity and, hence, the Cotton effect. The empirical part of the rule says that substituents occupying the upper left and lower right segments (both substituents on carbon 5 and axial substituent on carbon 2) make positive contributions, whereas substituents occupying the upper right and lower left parts make negative contributions. Axial substituents make larger contributions than equatorial substituents. The rule can be summarised in pictorial form as shown in Figures 3.5 and 3.6. For convenience of expression, the hexagonal form of cyclohexanone is used in Figure 3.5 and in Figure 3.6 only the back ocants are shown. An important application of this rule is the determination of the absolute configuration of (–) menthone (3.1.3) whose ORD and CD spectra show a positive Cotton effect, with an amplitude equal to that of (+) 3-methylcyclohexanone, which has been assigned the stereochemical structure (3.1.4). The value of the amplitude of the Cotton effect is small indicating that the substituent making the contribution

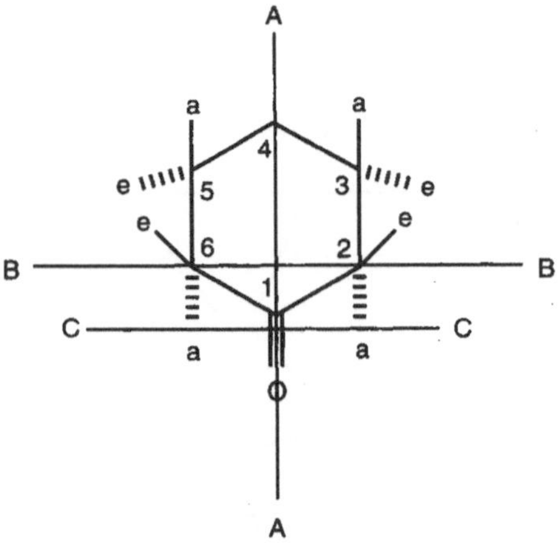

Figure 3.5 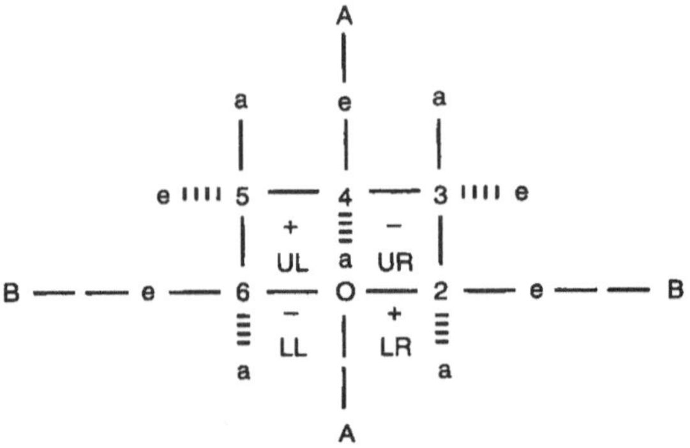   Figure 3.6    Figure 3.7

is equatorially oriented. Using this result and other data, the absolute stereochemistry of (–) menthol has also been determined as described in detail in a later section of this Chapter.

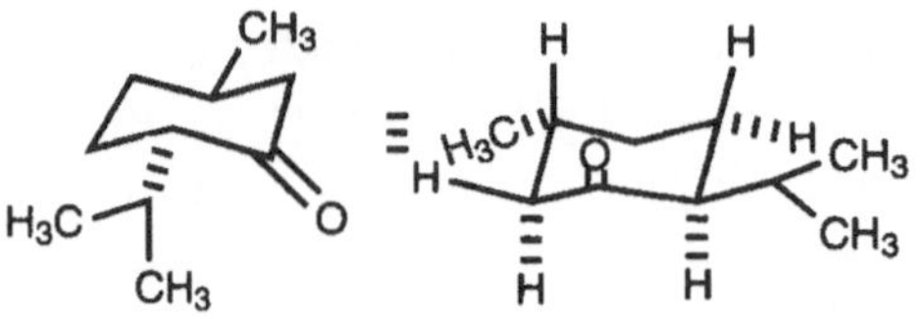

3.1.3 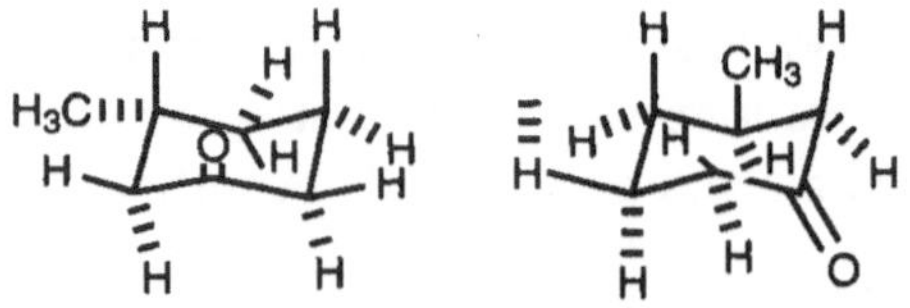   3.1.4

Substituted biphenyls (with bulky substituents at 2,2' and 6,6' positions) represented by the general structure (3.1.5) show optical activity as they possess a chiral axis. The amplitude of the Cotton effect in the ORD and CD spectra of such a compound is much larger than that of a chiral cyclohexanone, as the entire molecule is a chromophore. Among naturally occurring compounds, the dimeric anthraquinone pigments produced by the mould, *Penicillium islandicum*, fall into this category. For example, the CD spectrum of aurantioskyrin (3.1.6) shows a large positive Cotton effect.

The related lactone sector rule is applicable to chiral, saturated lactones wherein the n-π^* excitation maximum occurs near about 220 nm. For example, if we consider a six-membered cyclic lactone (3.1.7), one symmetry plane (horizontal) covers the lactone group. A plane perpendicular to this bisects the lactone group and goes through the carbonyl carbon atom as shown by the arrow in 3.1.7. As in the case of chiral cyclohexanones, these two are the significant planes to be considered; we can ignore the third plane which divides the molecule into a front part and a rear part. Thus, the lactone gets divided into four rear sectors similar to those of a cyclohexanone. However, experimental data show that the signs used in the ketone octant rule have to be reversed for the lactone sectors. This is summarised in Figure 3.7. The rule has also been applied to five-membered lactones such as the eudesmanolide (3.1.8).

3.1.5   3.1.6   3.1.7   3.1.8

The most reliable method for determining absolute configuration and for delineating the complete stereochemical profile of a chiral compound is the x-ray diffraction technique. However, this falls outside the expertise of an organic chemist, requires the skills of a specialist crystallographer, and, therefore, is not within the purview of this book.

Geometrical isomers (cis-, trans- or E-, Z-isomers), being diastereoisomers, differ in their physical as well as chemical properties. By and large, trans-isomers are thermodynamically more stable than their cis counterparts. To determine the configuration at a disubstitued double bond, when the substituents are different, the most reliable physical method is NMR spectroscopy. The coupling constant for a trans-olefinic hydrogen–hydrogen spin interaction is usually between 15 to 20 Hz, whereas for cis-coupling it is less than 10 Hz. An illustrative example is mitrorubin (3.1.9), which is one of the metabolites produced by Penicillium rubrum. Its proton NMR spectrum has six signals in the olefinic-aromatic region. Of relevance to the topic under discussion, are the signals at 6.21 ppm (1H, d, J=16Hz) and 6.55 ppm (1H, dq, J=16 and 6 Hz). These can be assigned respectively to the protons at positions 9 and 10. The large coupling constant between these two protons unequivocally shows that they are trans to each other. In the PMR spectrum of the alkaloid securinine (3.1.10), on the other hand, the coupling constant between H-4 (6.42 ppm) and H-5 (6.67 ppm) is 9 Hz, showing that they are cis to each other. Yet another example is the macrolide anti-cancer compound, phomin (cytochalasin B) (3.1.11) isolated from species of Phoma and Helminthosporium. In its proton NMR spectrum, the coupling constant between the protons H-3 (5.77 ppm) and H-4 (6.77 ppm) as well as that between H-11 (~5 ppm) and H-12 (5.82 ppm) is 16 Hz in agreement with the configurations assigned as in 3.1.11.

Conformations can be crucial in determining the chemical as well as biological properties of several terpenoids and other compounds which are capable of existing as conformational isomers. For delineating preferred conformations, NMR spectroscopy is, once again, the method of choice. As in the case of geometrical isomers, conformational isomers can be distinguished from the coupling constants between appropriately positioned hydrogen atoms. The magnitude of a

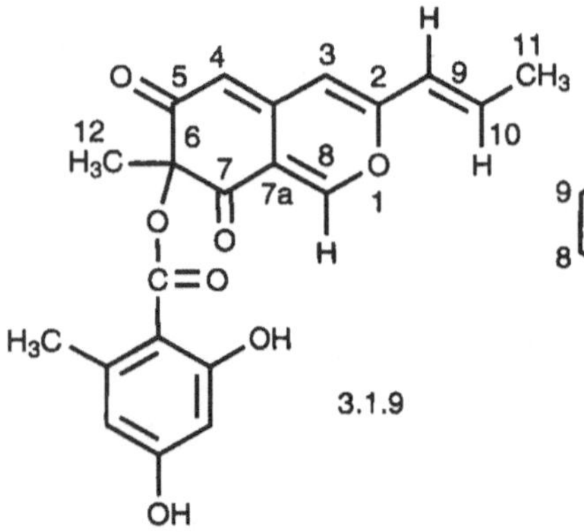

3.1.9

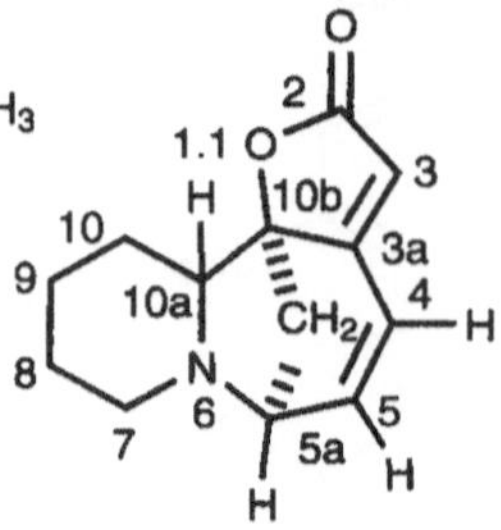

3.1.10

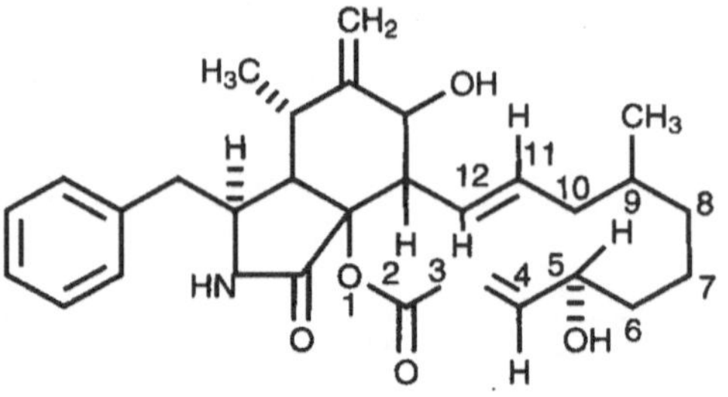

3.1.11

vicinal H–H spin interaction depends on the dihedral angle between the two interacting protons. In alicyclic compounds, particularly those having small rings, these angles are often well defined and, therefore, it has been possible to evolve an empirical relationship connecting J values and dihedral angles. This is the Karplus equation according to which the J value varies with the dihedral angle, reaching a maximum of ~ 10 Hz when the dihedral angle is zero or 180° and a minimum of near zero when the angle is 90°. Thus, the J value for the spin–spin interaction between two axial hydrogen atoms on adjacent carbon atoms in a cyclohexane will be around 10 Hz, whereas that between an equatorial hydrogen and an axial hydrogen will be much less (2–3 Hz). The Karplus equation has also been extensively used in determining the stereochemistry of lactone ring fusion in sesquiterpene lactones such as the eudesmanolides, guianolides and similar compounds wherein a gamma lactone is fused to a conformationally not too very mobile ring system. For example, in the 1H NMR spectrum of the pseudoguianolide, coronopilin (3.1.12), the J value for coupling between H-6 (4.9 ppm) and H-7 (3.4 ppm) is 8Hz indicating that the lactone is *trans*-fused. Another example is enhydrin which is discussed later in this Chapter. Other NMR techniques such as 2D (COSY) and NOE spectra also provide valuable information regarding conformations. The utility of these methods has been discussed in Chapter 2, taking β-ionone as an example.

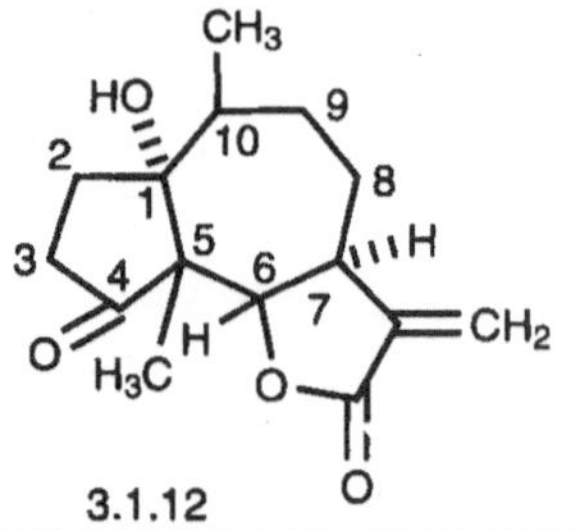

3.1.12

Suggested reading

1. Eliel EL and Wilen SH, *Stereochemistry of Organic Compounds*, Wiley Interscience, 1994.
2. Nasipuri D, *Stereochemistry of Organic Compounds: Principles and Applications*, 2nd edn, New Age Publishers, 1994.
3. Bassindale A, *The Third Dimension in Organic Chemistry*, John Wiley, 1984.
4. Mislow K, *Introduction to Stereochemistry*, Dover Publications, 2003.

3.2 ABSOLUTE STEREOCHEMISTRY OF MORPHINE

This much-studied alkaloid (3.2.1) named after the Greek god of sleep, Morpheus, is a constituent of opium, the dried latex of unripe fruits of the opium poppy, *Papaver somniferum*. Its structure was the subject of several investigations which ultimately led to its formulation as shown in the diagram. It can be seen that there are five chiral centres in this structure, namely positions 5, 6, 9, 13 and 14.

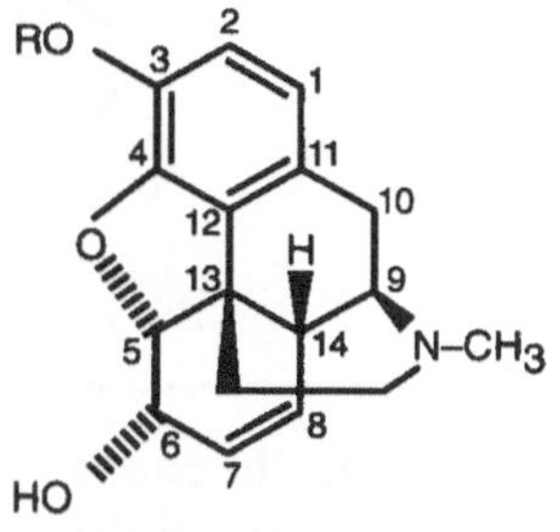

3.2.1: R = H
3.2.2: R = CH$_3$

The most convincing experimental evidence with regard to the relative configurations at C-5 and C-6 was generated by Rapoport and co-workers by selectively 'burning out' the benzene ring (see diagram). The starting material was dihydrocodeine (3.2.3) [codeine (3.2.2), which is the monomethyl ether of morphine is also a constituent of opium], which was subjected to ozonolysis. The ozonide was worked up and subsequently reduced with lithium aluminium hydride (LiAlH$_4$) to obtain a tetrahydroxy compound (3.2.4) which behaved as a 1,2-diol. Since the gross structure of dihydrocodeine was already known, the structure of this tetraol could be figured out as shown below.

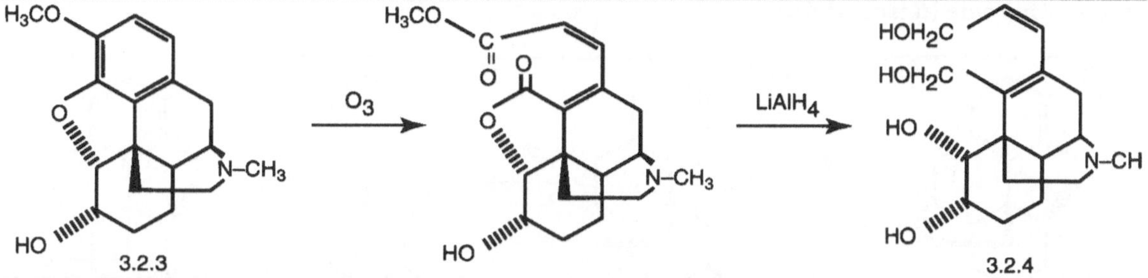

Thus, it could be considered as a cyclohexane 1,2-diol derivative. That the two hydroxyls in this part of the molecule were *cis* to each other was indicated by the facile reaction of the compound with lead tetraacetate. On the other hand, the corresponding tetrahydroxy compound (3.2.6) obtained from dihydroisocodeine (3.2.5) by an identical series of reactions was found to be three times less reactive towards lead tetraacetate. Therefore, in dihydrocodeine, the C-5 oxygen bridge and the C-6 hydroxyl should be *cis* to each other. At this point, it would perhaps be useful to recall the differences in rates of reaction between *cis*- and *trans*-cyclohexane 1,2-diols. The *cis*-compound is also known to react faster with acetone and to increase the conductivity of boric acid to a larger extent. However, reactions of 1,2-diols with lead tetraacetate and periodic acid are more amenable to quantitative analysis.

3.2.5   3.2.6

Relating the configurational centres at C-6 and C-13, which are not adjacent but close enough, was not as straightforward as relating C-5 and C-6. Dihydroisocodeine (3.2.5), on Hoffmann exhaustive methylation, followed by catalytic hydrogenation and a second Hoffmann reaction, yielded, as expected, an unsaturated, nitrogen-free compound (3.2.7) which underwent a facile acid-catalysed rearrangement to give a cyclic ether (3.2.8). For the formation of such an ether, the hydroxyl at position C-6 and the newly formed vinyl group at C-13 in the end-product should be on the same side of the ring C. On the other hand, in agreement with this logic, the corresponding nitrogen-free, unsaturated compound (3.2.9) similarly obtained from dihydrocodeine was not affected by acids; it failed to give any cyclic ether. Therefore, in codeine, the C-6 hydroxyl and the C-13–C-15 bond should be *trans* to each other as they are on the same side in isocodeine.

3.2.5   3.2.7   3.2.8

The chiral centres C-9 and C-13 could be readily related to each other as follows. The C-13–C-15 bond and the C-9–N bond should be on the same side of the middle ring as they are an integral part of a bridged bicyclic system which could only be *cis*-fused. A similar situation exists, for example, in camphor.

Perhaps the most difficult part of the chemical approach towards delineating the stereochemistry of morphine and its derivatives was to devise a suitable strategy to relate C-13 and C-14. This correlation depended on arguments which make sense to an organic chemist but are not derived from 'first principles'. All the same, the strategy used was ingenious and instructive. Thebaine (3.2.10), which is not pharmacologically active and which is also a constituent of opium, contains only three chiral centres—C-5, C-9 and C-13. It is known that it undergoes oxidation with hydrogen peroxide to give stereospecifically 14-hydroxycodeinone (3.2.11). In this reaction, hydrogen peroxide obviously adds on to the diene system in thebaine in a Diels–Alder type of reaction.

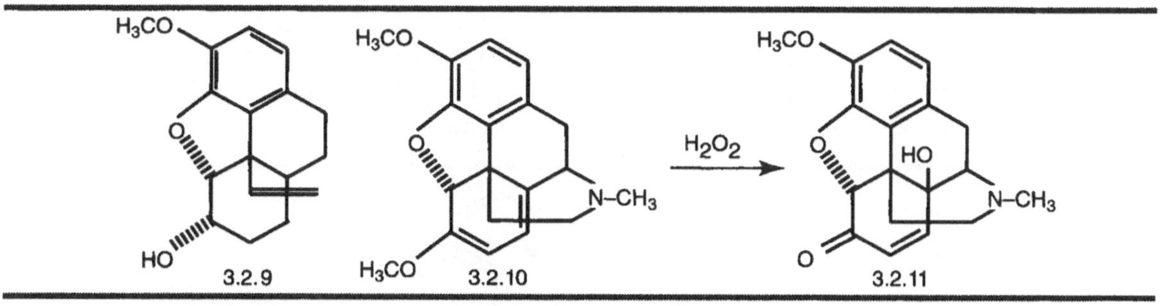

3.2.9   3.2.10   3.2.11

The product (3.2.11) on catalytic hydrogenation gave the dihydro derivative (3.2.12) which on Hoffmann exhaustive methylation yielded a compound (3.2.13) which readily gave a cyclic ether (3.2.14). In this case, the groups interacting to form the ether bridge are, obviously, the vinyl group at C-13 and the hydroxyl at C-14. Therefore, working backwards, the C-13–C-15 bond and the hydroxyl at C-14 in 14-hydroxycodeinone could be placed *cis* to each other on the middle ring.

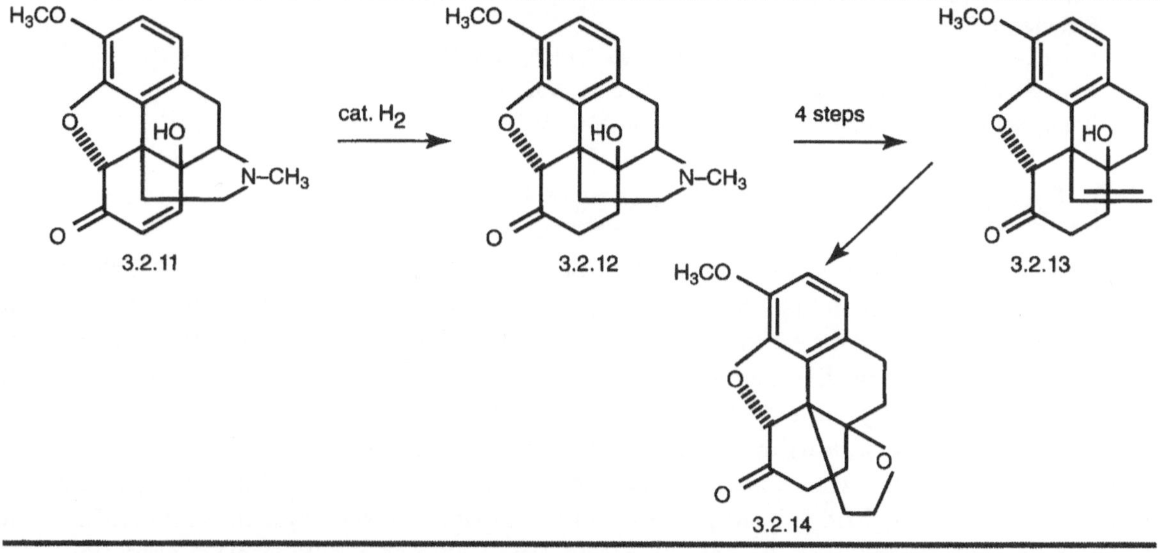

3.2.11   3.2.12   3.2.13

3.2.14

So far so good, even if one has some reservations regarding the logic used in the above interpretation. A more serious objection cropped up when the logic was further extended, as explained below, to find out the configuration at C-14 in codeine. On catalytic hydrogenation, thebaine gave, exclusively and stereospecifically, dihydrocodeine methyl ether (3.2.15). This reaction was considered to proceed stereochemically in the same manner as the hydrogen peroxide reaction and therefore the newly introduced hydrogen at position 14 in the product was placed in the same configurational position as the hydroxyl in 14-hydroxycodeinone (3.2.11). Such a conjecture is, indeed, somewhat far-fetched though it is not altogether devoid of logic.

The isolation of the cyclohexane dicarboxylic acid derivative (3.2.16) confirmed that the conclusion was correct. This compound was obtained from thebaine by a series of reactions involving initial reductive transformations of ring C followed by modification of the nitrogen-containing part and oxidative degradation of the aromatic ring. It was found to possess the absolute configuration shown in the figure below. This result was also confirmed by x-ray diffraction analysis. Therefore, the stereochemistry of morphine and codeine could be delineated as shown.

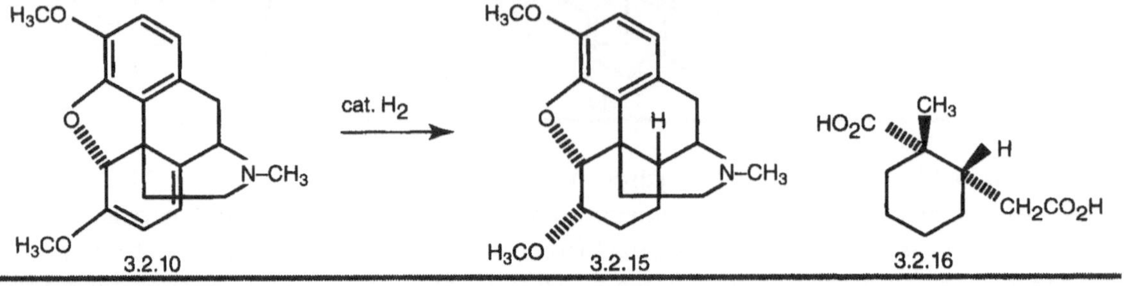

Suggested reading

1.   Bentley KW, In Coffey S, (ed.), Rodd's *Chemistry of Carbon Compounds,* 2nd edn, Vol. 4, Amsterdam: Elsevier, 1978.
2.   Stork G. In Manske RHF, (ed.), *The Alkaloids,* Vol. 6, New York: Academic Press, 1960.
3.   Rapoport H and Payne GB, *J. Amer. Chem. Soc,* 1952, 74: 2630.
4.   Rapoport H and Lavigne JB, *J. Amer. Chem. Soc,* 1953, 75: 5329.

3.3 QUININE AND RELATED ALKALOIDS

The bark of the cinchona tree yields the well known anti-malarial drug, quinine (3.3.1) along with several other related alkaloids. Quinidine (3.3.2) is a diastereoisomer of quinine. Cincho*nine* (3.3.3) is demethoxyqui*nidine* and cincho*nidine* (3.3.4) is demethoxyqui*nine*—the names ought to have been interchanged! Incidentally, every student of organic chemistry should read the book *Organic Chemistry—The Name Game* in which the naming of organic compounds, both systematic and otherwise, is delightfully described. These compounds possess four chiral centres, leaving out the nitrogen, which is also chiral. The chiral nitrogen and C-4 are configurationally related as they occupy bridge-head positions of the bridged bicyclic system, quinuclidine. The other chiral centres are C-3, C-8 and C-9. All the four alkaloids, on oxidative degradation, gave the same optically active meroquinene (3.3.5), representing a part of the quinuclidine ring. This compound retains two chiral centres, namely C-3 and C-4, of the parent compounds. Thus, it could be concluded that all the four alkaloids have identical configurations at these two centres. Further, both cinchonine and cinchonidine gave, on dehydration, the same cinchene (3.3.6), which had lost, along with the elements of water, the chirality at C-8 and C-9. Similarly, both quinine and quinidine yielded the same quinene (3.3.7). Further oxidation of meroquinene yielded cincholoiponic acid (3.3.8) and loiponic acid (3.3.9). The transformation of cinchonine and cinchonidine into cinchotoxine (3.3.10) by the action of hot acetic acid is also shown on the next page.

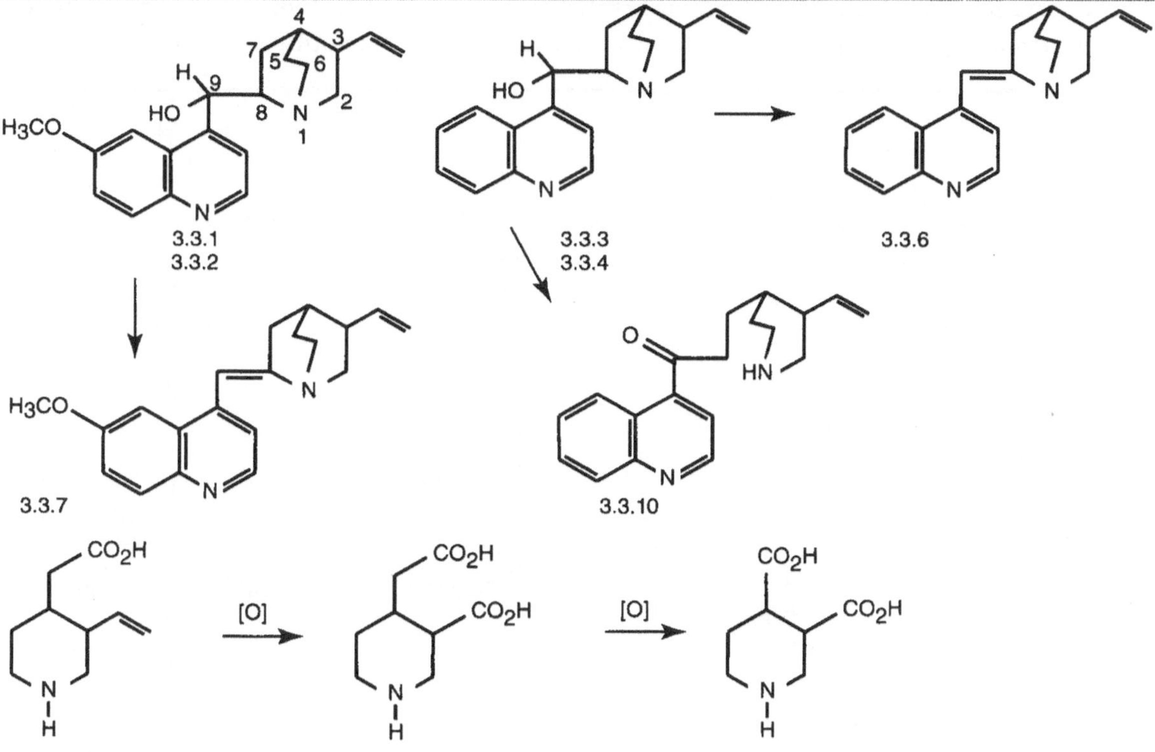

The gross structure of loiponic acid as piperidine-3,4-dicarboxylic acid (3.3.9) was established by Rabe who initiated structural studies on these alkaloids. That loiponic acid was a *cis*-dicarboxylic acid was indicated by the fact that it could be isomerised by treatment with alkali to a thermodynamically more stable geometrical isomer, which should be the *trans*-compound. However, the relative configuration of C-3 and C-4 in quinine was more rigorously proved by Prelog and Zalan who adopted the following experimental strategy. To begin with, dihydrocinchonine (3.3.11) was oxidatively degraded to 3-ethylpiperidine-4-acetic acid (3.3.12), which was then transformed into 3,4-diethylpiperidine (3.3.13) as shown below. von Braun degradation of 3.3.13 gave an optically active dibromoalkane (3.3.14) which was condensed with diethyl malonate and hydrolysed to obtain the optically active 3,4-diethylcyclohexane-1-carboxylic acid (3.3.15). When subjected to the Hunsdiecker reaction, this acid yielded optically active 3,4-diethyl-1-bromocyclohexane (3.3.16). Reductive removal of the bromine, however, yielded the optically inactive 1,2-diethylcyclohexane (3.3.17), which was therefore correctly characterised as the *cis*-compound. Later, the dibromoalkane (3.3.14) was converted into an enantiomorph of 3-ethyl-4-methylhexane which was found to have the absolute configuration 3.3.18. From this observation, therefore, the absolute configurations of C-3 and C-4 in quinine and the related alkaloids could be conclusively determined as 3R, 4S.

This was, indeed, an elegant and eminently satisfying method of determining stereochemistry without resorting to any physical methods, other than recording optical activity.

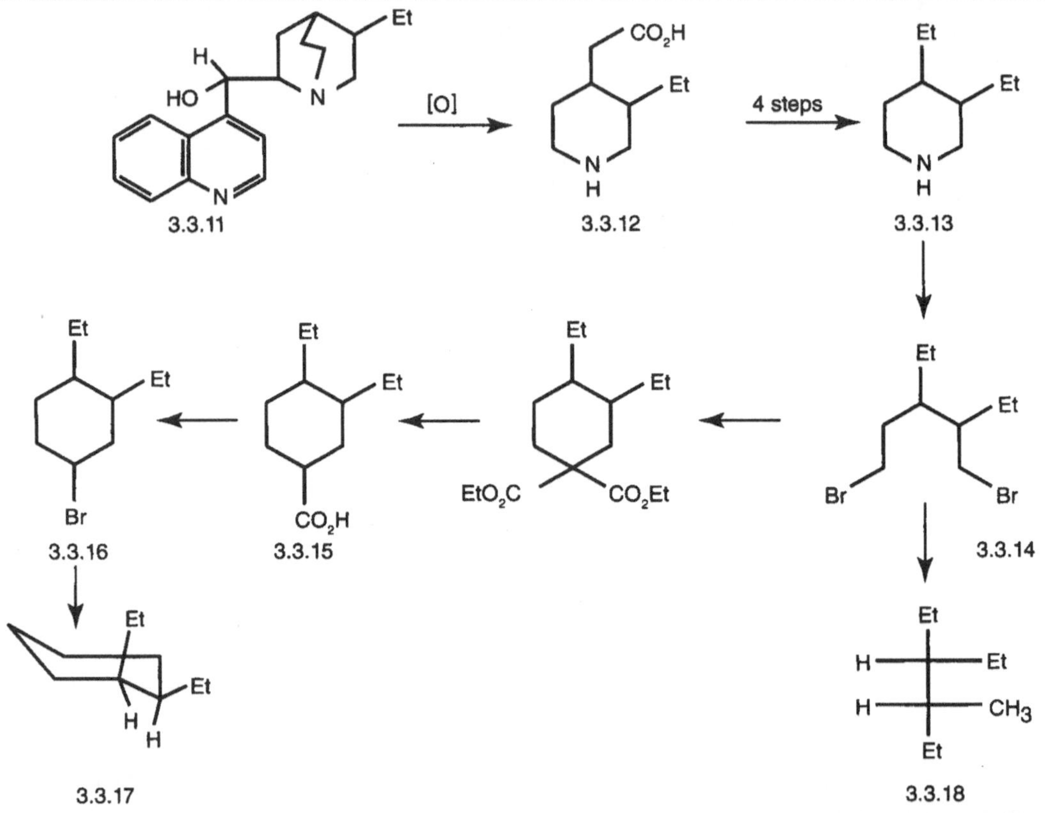

3.3.11

3.3.12

3.3.13

3.3.16

3.3.15

3.3.14

3.3.17

3.3.18

Cinchonine and cinchonidine give, on oxidation, their corresponding ketones (3.3.19 and 3.3.20), which are diastereoisomers. Similarly, the ketones obtained from quinine and quinidine are diastereoisomers. Further, deoxycinchonine (3.3.21) and deoxycinchonidine (3.3.22) are also epimers. These observations clearly show that cinchonine and cinchonidine (and quinine and quinidine) configurationally differ at C-8. On treatment with acids, cinchonine (so also quinidine) undergoes isomerisation to yield the *iso*-base (3.3.23), which does not possess the hydroxyl group or the double bond; the product is a cyclic ether formed by the acid-catalysed, intramolecular addition of the hydroxyl at C-9 to the vinyl double bond at C-3. Such an unusual interaction involving two apparently distant groups is obviously favoured by the stereochemistry which brings these two groups close together—in some conformations—within bonding distance.

This conjecture is supported by the fact that the isomeric cinchonidine, which is epimeric at C-8, is not isomerised in this fashion; in this case, the resulting product (3.3.24) retains the hydroxyl as well as the double bond, which is, however, shifted nearer to the quinuclidine ring. While quinidine behaves, in this reaction, like cinchonine, quinine resembles cinchonidine. Therefore,

3.3.3 3.3.19 3.3.4 3.3.20 3.3.21 3.3.22

in cinchonine and quinidine, the C-8–C-9 bond and the C-3–vinyl bond should be on the same side of the quinuclidine ring system. As a corollary, it follows that these groups are *trans* to each other in cinchonidine and quinine.

3.3.2
3.3.3

3.3.23: R = H or OCH_3

3.3.1
3.3.4

3.3.24: R = H or OCH_3

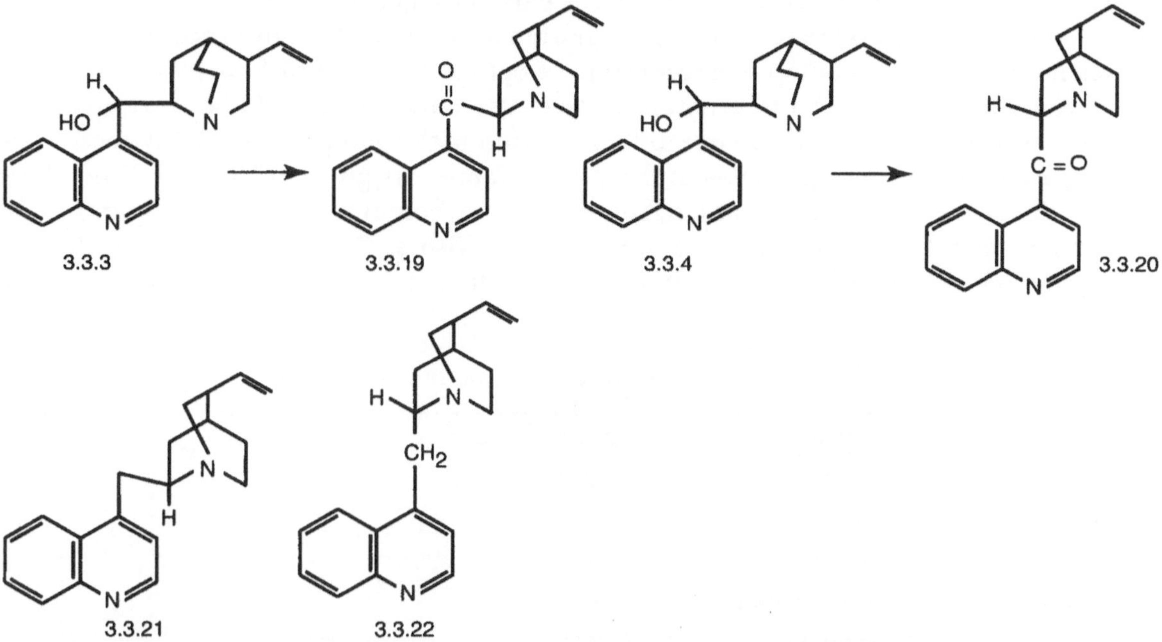

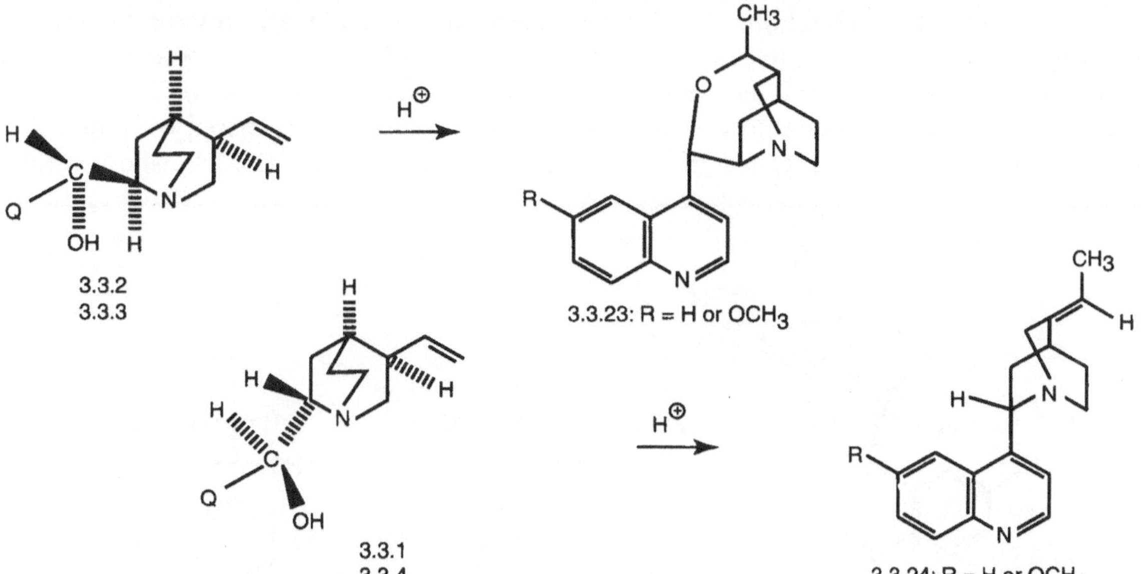

Since the configurational centre C-9 is not a part of a rigid ring system, it was difficult to devise a chemical method to relate its configuration to that of C-8 or any of the other centres. It was necessary to depend on measurements of optical activity and basicity constants for the various diastereoisomers to generate comparative data which could be correlated with configurational differences (of course, by comparison with appropriate model systems). The specific rotations for the eight bases of the quinine family are given in Table 3.1. From these figures, it can be readily seen that cinchonine and cinchonidine (and so also, quinine and quinidine) should differ in configuration at C-9 (as also at C-8). Thus, while cinchonine and quinidine are strongly dextrorotatory, cinchonidine and quinine exhibit large negative rotations.

Table 3.1 Specific rotations for the quinine family

Compound	Specific rotation
cinchonine	(+)224
cinchonidine	(−)111
epicinchonine	(+)120
epicinchonidine	(+)63
quinine	(−)158
quinidine	(+)254
epiquinine	(+)43
epiquinidine	(+)102

Significantly, epicinchonidine (3.3.25) and epiquinine (3.3.26) are dextrorotatory, thus confirming that these are epimeric at C-9. The absolute configuration at C-9 in quinine was deduced as follows. Quinine is less basic (pKa 7.73) than epiquinine (pKa 8.44). The part structure incorporating the basic nitrogen in these compounds could be compared with the pair of diastereoisomeric alkaloids, ephedrine (3.3.27) and pseudo-ephedrine (3.3.28), whose absolute

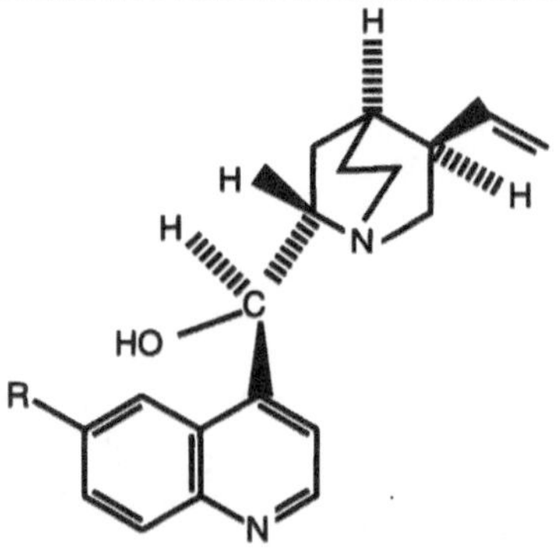

3.3.25: R = H
3.3.26: R = OCH3

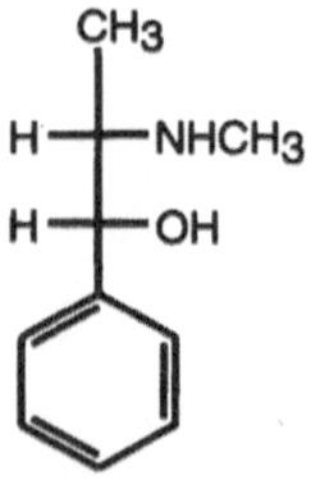

3.3.27

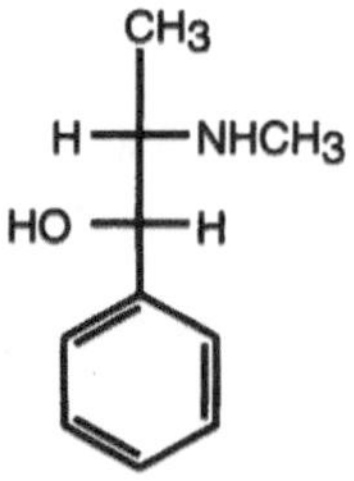

3.3.28

configurations were known. The *erythro*-isomer, ephedrine, is less basic (pKa 9.14) than the *threo*-compound, ψ-ephedrine (pKa 9.22). Therefore, by analogy, quinine was assigned the *erythro*-configuration, while epiquinine was deduced to be the *threo*-isomer. These assignments have been confirmed by x-ray diffraction analysis of quinine sulphate. Thus, the complete stereochemistry of (-)quinine could be defined as 3R, 4S, 8R and 9S.

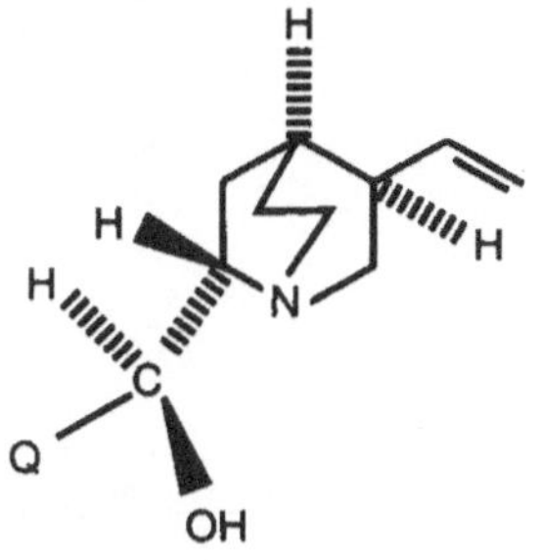

3.3.1	3.3.2
3.3.4	3.3.3

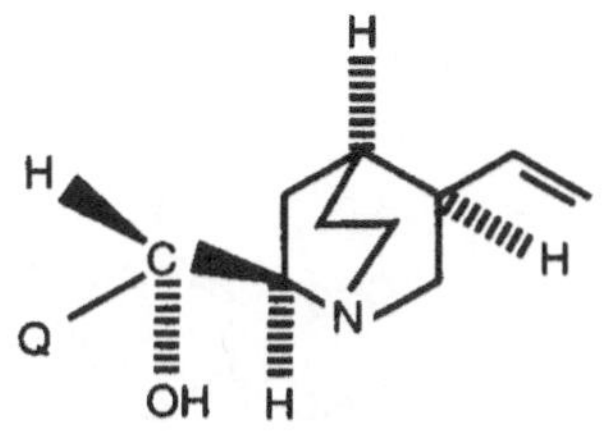

Suggested reading

1. Sainsbury M, In Coffey S, (ed.), Rodd's *Chemistry of Carbon Compounds*, 2nd edn, Vol. 4, Amsterdam: Elsevier, 1978.
2. Turner RB and Woodward RB, In Manske RHF and Holmes HL, (eds), *The Alkaloids*, Vol. 3, New York: Academic Press, 1953.
3. Shamma M, *The Isoquinoline Alkaloids*, New York: Academic Press, 1972.
4. Prelog V, *Tetrahedron Letters*, 1964, 5: 2037.
5. Nickson A and Silver Smith EF, *Organic Chemistry—The Name Game*, London: Pergamon Press, 1987.

3.4 EMETINE

The ipecacuanha root, a widely used folk medicine for the treatment of dysentery, contains the complex alkaloid, emetine (3.4.1). This compound possesses four chiral centres. It was first isolated by Pelletier and Caventou and its structure was investigated initially by Carr and Pyman, and then by Karrer, and later by Spath, Reichstein and their co-workers.

As in the case of morphine and quinine, the stereochemistry of emetine was chemically deduced not only by studying certain relevant reactions of emetine but also those of some related alkaloids like psychotrine (3.4.2), cephaeline (3.4.3) and isocephaeline (3.4.4). Psychotrine is an unsaturated phenolic base and yielded, on hydrogenation, a mixture of cephaeline and isocephaeline which are epimeric compounds. These on *O*-methylation yielded, respectively, emetine and isoemetine (3.4.5).

The *O,N*-dimethyl derivative of psychotrine was converted into its *bis*-benzochloride (3.4.6) which was subjected to oxidation, whereby a betaine (3.4.7) was obtained.

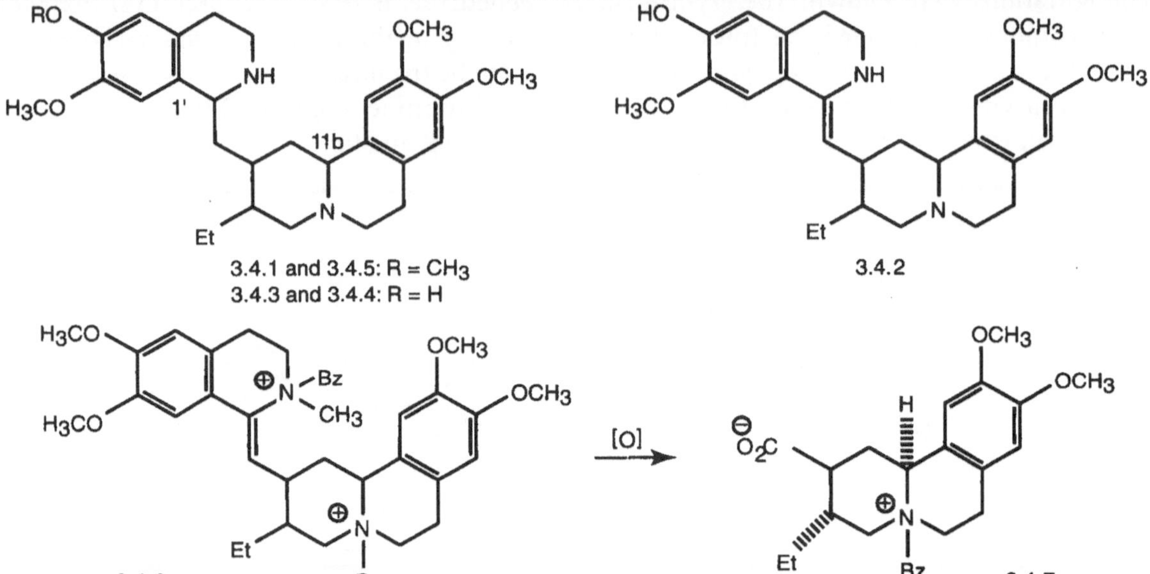

3.4.1 and 3.4.5: R = CH₃
3.4.3 and 3.4.4: R = H

3.4.2

3.4.6

[O]

3.4.7

The betaine was hydrogenolysed to get an acid (3.4.8), the ethyl ester of which was stable even under strong equilibrating conditions. This observation indicated that the carboethoxyl group in the ester must be equatorially oriented. The carboxylic acid (3.4.8) could be reconverted into O-methylpsychotrine by a sequence of reactions beginning with an Arndt–Eistert homologation as shown below. Therefore, it was obvious that the degradation of O,N-dimethylpsychotrine to the betaine had occurred without any inversion at the relevant configurational centres. The homologous acid (3.4.9) mentioned above was also obtained by the mild oxidation (using moist silver oxide) of another ipecacuanha alkaloid, namely, protoemetine (3.4.10). Protoemetine on Wolff–Kishner reduction yielded a tricyclic base (3.4.11), which was also synthesised from trans-3,4-diethylcyclopentanone (3.4.12) as shown below. This synthesis established the relative configurations at three chiral centres which correspond to the positions 2, 3 and 11b of emetine.

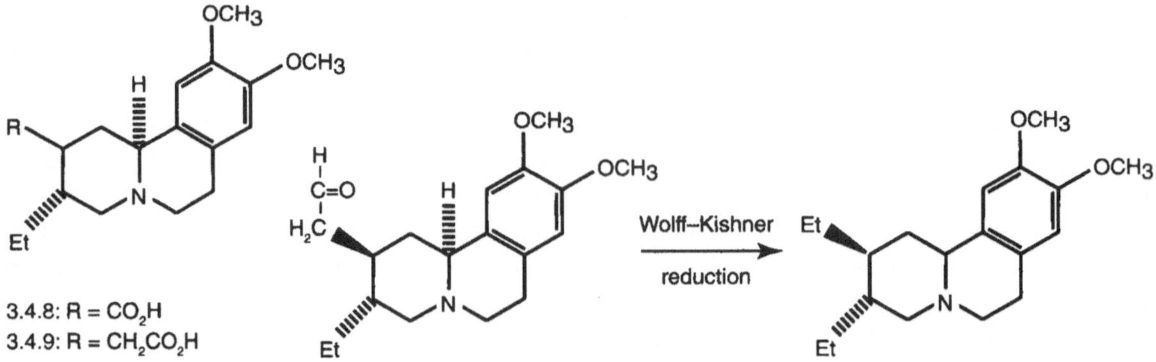

3.4.8: R = CO₂H
3.4.9: R = CH₂CO₂H
3.4.10: R = CH₂CHO

3.4.10

Wolff–Kishner reduction

3.4.11

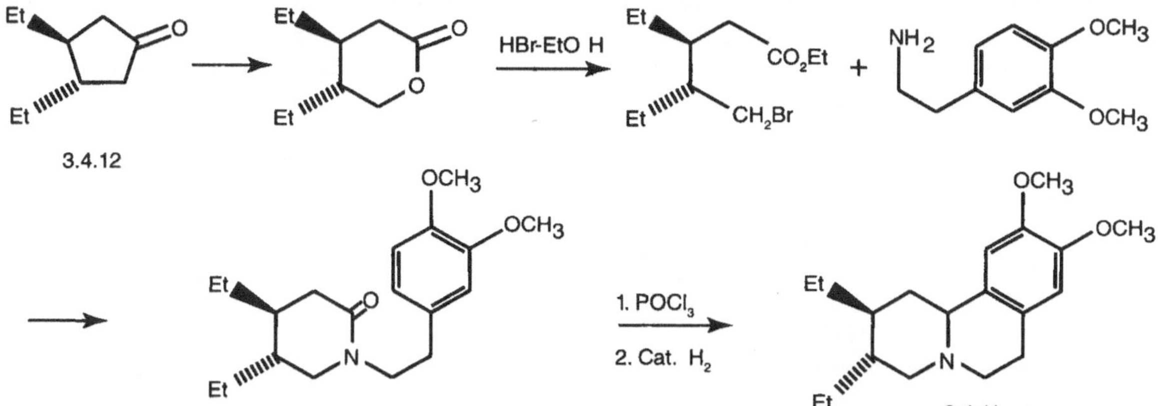

3.4.12

3.4.11

The absolute configuration at CI' position was determined by subjecting N-acetylemetine (3.4.13) to a series of degradation reactions beginning with a Hoffmann exhaustive methylation. The end-product was the dextrorotatory form of N-acetyl-6,7-dimethoxyisoquinoline-l-acetic acid (3.4.14). The absolute stereochemistry of this compound was confirmed by comparing it with (+) calycotamine (3.4.15) whose absolute configuration was established.

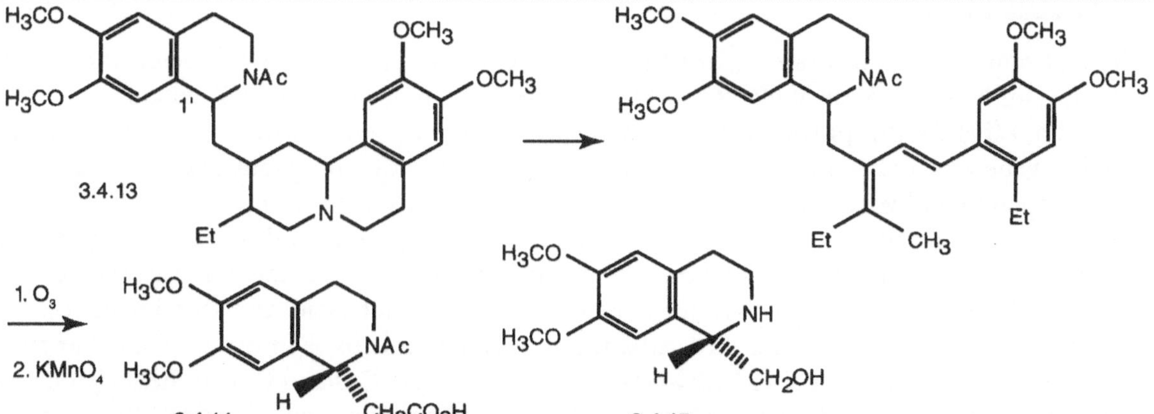

3.4.13

3.4.14

3.4.15

Of the four chiral centres, the two benzylic centres of chirality, namely positions 1' and 11b, can be expected to make significant contributions to the optical rotatory dispersion (ORD) spectra of emetine and its salts, since these are the only centres which are close to the chromophoric parts of the molecule. With this background, van Tamelen observed that the ORD curve of emetine hydrobromide showed negligible change in rotation in the wavelength range 300–700 nm. Therefore, he could conclude that the two benzylic centres must be making contributions of opposite sign to the total optical activity of the compound, thus virtually cancelling out each other's contribution. In other words. these two centres should have opposite configurations, and hence the absolute stereochemistry of emetine could be represented as 3.4.16. This conjecture was supported by the observation that isoemetine hydrobromide, which is epimeric with emetine at

C-1', shows in its ORD spectrum, an increasing rotation down to 300 nm. Obviously, in this case, the two benzylic centres consolidate their contributions, which are in the same direction.

3.4.16   3.4.17

Suggested reading

1.   Dyke SF, In Coffey S, (ed.), Rodd's *Chemistry of Carbon Compounds*, 2nd edn, Vol. 4, Amsterdam: Elsevier, 1978.
2.   Van Tamelen EE, Aldrich PE and Hester JB, *J. Amer. Chem. Soc.* 1957, 79: 4817.
3.   Van Tamelen EE, Aldrich PE and Hester JB, *J. Amer. Chem. Soc.* 1959, 81: 6214.
4.   Terashima M, *Chem. Pharm. Bull.*, 1960; 8:517.

3.5 ENHYDRIN

We shall now take up an example in which the absolute stereochemistry has been worked out predominantly on the basis of NMR spectral data, supported by ORD and CD data. It should be pointed out that for this purpose, suitable derivatives of the compound were required and these could be generated only with the help of well-understood and unambiguous chemical reactions. This compound, enhydrin (3.5.1), is a sesquiterpene lactone isolated from *Enhydra fluctuans* of the family Compositae. For its size, it is a fairly complex compound, and is heavily oxygenated. It possesses as many as six chiral centres on the skeleton and two outside. Further, the skeletal structure, being a cyclodecane system, is conformationally not rigid and this introduces an additional complication. Its gross structure, which was determined as in the case of heliangine, shows that the chiral centres in enhydrin are C-4, C-5, C-6, C-7, C-8 and C-9 on the skeleton and C-2' and C-3' in the acyl side chain. Moreover, the double bond between C-1 and C-10 can either have the *cis*-or the *trans*-configuration. The five-membered unsaturated lactone ring is fused to the ten-membered ring at C-6 and C-7. The primary task was to find out the stereochemistry of the lactone fusion, which was done as explained below.

The Karplus equation has been found to be very useful for finding out whether a lactone is *cis*-fused or *trans*-fused, provided the ring to which the lactone is fused is a small, conformationally not very mobile system like cyclohexane. This equation makes use of the relationship between the dihedral angle and the magnitude of the coupling constant for the hydrogen atoms at ring junctions. Thus, if the *J* value is large (ca. 10 Hz), the dihedral angle could be taken as 180° and, therefore, the lactone should be *trans*-fused. On the other hand, a small *J* value (ca. 3 Hz) would indicate a dihedral angle of 90° and thus a *cis*-fused lactone. However, these inferences lose their credibility when the ring to which the lactone is fused is conformationally flexible, such as a typical medium-sized ring system. In such cases, the average conformation would cover too wide a range of dihedral angles for the Karplus equation to be valid. Such a situation exists in enhydrin and other germacranolides. Therefore, the large coupling constant of 10 Hz between H6 and H7 could not be accepted, per se, as a firm indication of the *trans*-geometry of lactone fusion. Before discussing this matter further, it would be useful to examine the chemical shift values and the coupling constants for hydrogen atoms attached to the various chiral centres in enhydrin. These data are given in Table 3.2

Table 3.2 Chemical shift values of chiral protons in enhydrin and related compounds, in ppm (*J*)

Compound	H5	H6	H7	H8	H9
Enhydrin	2.58, *d*	4.28, *t*(10)	3.03, *m*(10)	6.69, *dd* (8, 1)	5.88, *d*(8)
Pyrazoline (3.5.2)	2.75, *d*(10)	4.56, *t*(10)	3.88, *d*(10)	5.90, *d*(8)	5.65, *d*(8)
Pyrazoline (3.5.3)	2.62, *d*(10)	5.22, *t*(10)	3.10, *d*(10)	5.92, *d*(8)	5.63, *d*(8)
Dihydro enhydrin (3.5.4)	2.58, *d* (10)	4.50, *t* (10)	2.8, *m*	6.30, *dd* (8.5, 1)	5.84, *d* (8.5)
Dihydro enhydrin (3.5.5)	2.58, *d* (10)	4.13,*t* (10)	2.8, *m*	6.30, *dd* (8.5, 1)	5.90, *d* (8.5)

Like other members of this group of unsaturated lactones, enhydrin reacts with diazomethane. In this case the reaction resulted in the formation of two epimeric pyrazolines which could be separated. In the ¹H NMR spectrum of the major product (3.5.2), the signal due to H7 appeared at a lower field compared to its position in enhydrin. On the other hand, the signal due to H6 remained relatively unaffected. The downfield shift of H7 could be attributed to the deshielding influence of the anisotropic diazine group, which, therefore, should be on the same side as H7 in this epimeric pyrazoline. Since H6 was largely unaffected, it should be on the other side, showing that the lactone must be *trans*-fused. This was further corroborated by the observation that in the ¹H NMR spectrum of the other pyrazoline (3.5.3), it was the signal due to H6 which had been shifted downfield, relative to enhydrin, while H7 had remained unchanged (Table 3.2). These observations thus conclusively proved that the lactone was *trans*-fused to the ten-membered ring. The absolute stereochemistry could be deduced by analogy with other naturally occurring related sesquiterpene lactones for which x-ray data were available.

In all these cases, H7 was known to be α-oriented as in 3.5.2; the same conclusion was also indicated by the CD spectra of the two pyrazolines. Thus, the absolute configurations at C-6 and C-7 in enhydrin could be designated as (S) and (S) respectively. Since in this case the *trans*-geometry of the lactone could thus be proved by an alternative strategy, obviously the *J* value for H6–H7 interaction could be accepted as a reliable, independent index. This has an important implication with regard to the configuration at C-5. As can be seen from Table 3.2, the H6 signal appears as a triplet with *J*=10 Hz in the proton NMR spectra of enhydrin and its derivatives. The equal interaction of H6 with both H5 and H7, which flank it on either side, suggests that the dihedral angles in question should also be the same. In other words, H5 can be considered to be *trans* to H6. This conjecture is also supported by the shifts undergone by H5 on conversion of enhydrin into the epimeric pyrazolines (Table 3.2).

From the low *J* value (1.5 Hz) it could be inferred, using the Karplus equation. that the dihedral angle between H7 and H8 should be close to 90°. Since H7 is α-axial with respect to the lactone plane, H8 should be equatorial, and therefore the acyloxy group at C-8 should be roughly perpendicular to this plane. Further, H8 could be located as lying near the exocyclic methylene group as its signal position in the NMR spectra is affected by transformation of the double bond either by hydrogenation into the epimeric dihydro derivatives 3.5.4 and 3.5.5, or by conversion into the pyrazolines (Table 3.2). This implies that the acyloxy group at C-8 can, in some conformations, cover the ten-membered ring and reach the epoxy group bridging C-4 and C-5. This is supported by the observation that this epoxy group is extraordinarily stable and resists opening by hydrolytic or reductive methods.

The large coupling (8 Hz) between H8 and H9 in enhydrin as well as in the glycol (3.5.6) derived from it, suggests a dihedral angle of 180° between these two hydrogen atoms. The NMR spectra of enhydrin and its derivatives as well as their chemical behaviour are in accordance with the stereochemistry shown in 3.5.7. The double bond between C-1 and C-10 should have the *cis*-configuration as indicated by the absence of allylic and homoallylic interactions involving H1. The C-8 acyl unit was shown to be derived from epoxyangelic acid (3.5.8) by the isolation of the methyl ester of this acid by the controlled methanolysis of enhydrin.

The conclusions thus drawn from a detailed analysis of the ^{X}H NMR spectra of enhydrin and its derivatives have also been supported by a 2D NMR study as well as by x-ray diffraction analysis of the bromohydrin (3.5.9).

The ORD and CD spectra of enhydrin, the two epimeric dihydro derivatives and the two epimeric pyrazolines also provided useful information. The octant rule is applicable to enhydrin since the unsaturated lactone chromophore could be considered to be a resonance hybrid of three canonical forms. The symmetry plane is the one that cuts through the lactone carbonyl group. On the other hand, the CD spectra of the dihydro derivatives are analysable by the lactone sector rule.

Suggested reading

1. Kartha C, Goto KT and BS Joshi, *J. Chem. Soc, Chem. Comm.*, 1972, 1327.
2. Krishnaswamy NR, Seshadri TR and Vedantham TNC, *Indian J. Chem.*, 1972, 10:249.
3. Ali E, Ghosh Dastidar PP, Pakrashi SC, Durham LJ and Duffield AM, *Tetrahedron*, 1972, 28:2285.
4. Joshi BS and Kamat VN, *Indian J. Chem.*, 1972, 10:771.
5. Prasanna S, Ph.D. Thesis, Bangalore University, India, 1975.
6. Prasanna S, Krishnaswamy NR and Omana P, in Professor M Shadaksharaswamy Felicitation Volume, Bangalore University, India,1979.

3.6 CONFORMATIONS OF NATURALLY OCCURRING GERMACRANOLIDES

As mentioned in the preceding section, enhydrin is a germacranolide. Naturally occurring germacranolides arise from all-*trans* farnesyl pyrophosphate (3.6.1) by a series of biochemical transformations. A key intermediate is the germacratriene carboxylic acid (3.6.2). It has been suggested that this compound undergoes further enzyme-mediated oxidative transformations to yield the germacranolides. What is relevant in the present context is, however, the configurations of the endocyclic double bonds in the germacranolides thus produced by biosynthesis. For a long time, it was assumed that both the double bonds should have the *trans*-configuration since the primary sesquiterpenoid precursor is the all-*trans* farnesyl pyrophosphate.

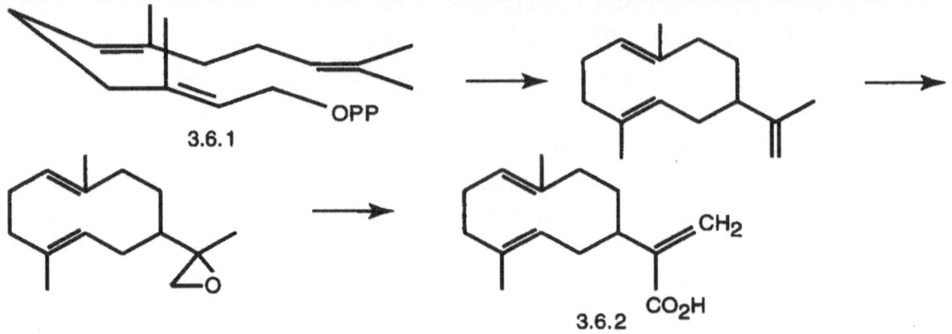

3.6.1

3.6.2

Indeed, the first few germacranolides to be chemically characterised, such as pyrethrosin (3.6.3) (in which one of the endocyclic double bonds has undergone epoxidation, while retaining the *trans*-configuration), costunolide (3.6.4) and related compounds all had the all-*trans* configuration. However, with the discovery of melampodin (3.6.5) and heliangine (3.6.6), in each of which one of the endocyclic double bonds has the *cis*-configuration, it was realised that it was no longer possible to assume *trans*-configurations for the endocyclic double bonds or the corresponding epoxide groups. Obviously, at some stage of the biosynthetic transformation of all-*trans* farnesyl pyrophosphate, a stereomutation can occur.

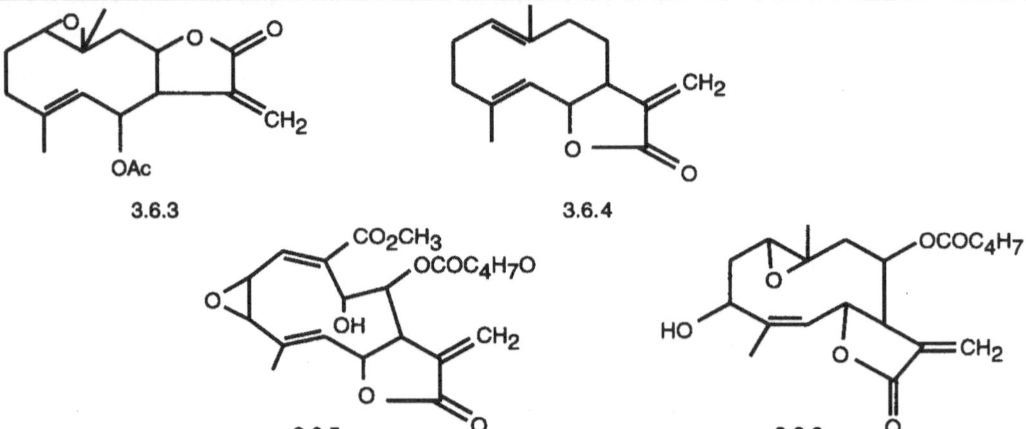

3.6.3   3.6.4

3.6.5   3.6.6

Rogers and co-workers were the first to point out that, granting such a biochemical *trans*-to-*cis* change, a naturally occurring germacranolide could belong to one of four possible stereochemical types. These were designated as germacrolides, in which both the double bonds are *trans*, the heliangolides (Δ^1*trans* and Δ^4*cis*), the melampolides (Δ^1*cis* and Δ^4*trans*) and a fourth type in which both the double bonds have the *cis*-configuration, as in the case of 3.6.7. As a sequel to this classification, they also spelt out clear rules for the nomenclature and for the numbering of the skeletal carbons of a germacranolide. As can be seen from their structural formulae, the four stereochemical skeletal types have distinct conformations, even granting a certain amount of conformational flexibility. Enhydrin, incidentally, is a melampolide.

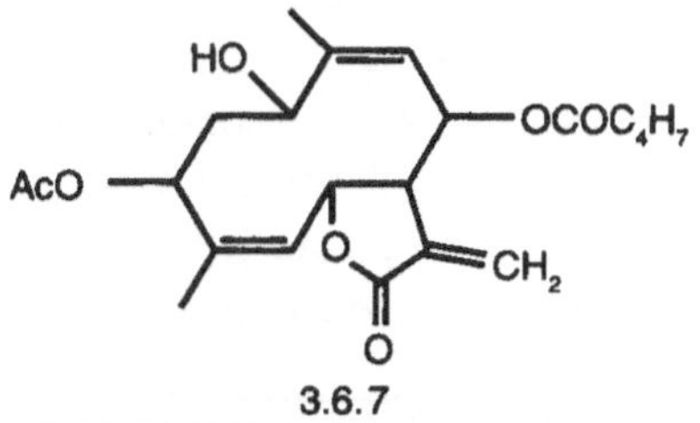

Thus, with the help of these examples, it is possible to recognise the fact that the interannular space in a ten-membered carbocyclic ring system is large enough to accommodate two double bonds in all possible configurations. As a consequence, the skeletal structure can assume four different shapes, each enjoying a fair degree of conformational freedom. We will be discussing in a later section of this chapter the reflections of these stereochemical differences in skeletal structures in their chiro-optical properties.

3.6.7

Suggested reading

1. Geismann TA, In Runeckles VC and Mabry TJ, (eds), *Recent Advances in Phytochemistry*, New York: Academic Press, 1975.
2. Kupchan SM, Kelsey JE and Sim GA, *Tetrahedron Letters*, 1967, 2863.
3. Rogers D, Moss GP and Neidle S, *J. Chem. Soc, Chem. Comm.*, 1972, 142.
4. Neidle S and Rogers D, *J. Chem. Soc, Chem. Comm.*, 1972, 140.
5. Prasanna S, Krishnaswamy NR and Omana P, in Professor M Shadaksharaswamy Felicitation Volume, Bangalore University, India, 1979.
6. Samek Z and Harmatha J, *Coll. Czech. Chem. Comm.*, 1978, 43: 279.

3.7 STEREOCHEMISTRY OF ROTENOIDS

Rotenoids are complex isoflavonoids occurring in several species of *Derris lonchocarpus*, and a few other genera of the family Leguminosae. Rotenone has been long known for its insecticidal activity and as a fish poison. Its gross structure was elucidated independently by Takei and co-workers, La Forge and Haller, Butenandt and Robertson and co-workers, as early as 1932, but its stereochemistry and synthesis remained unsolved problems till recently. Rotenone itself has three chiral centres, positions 6a, 12a and 5', but only the first two are common to all rotenoids. These two carbon atoms determine the stereochemistry of the junction of B and C rings. All naturally occurring rotenoids have these rings *cis*-fused as shown by certain chemical reactions as well as NMR and ORD/CD spectral data.

Naturally occurring (–)rotenone (3.7.1) gets isomerised to (–)isorotenone (3.7.2) on treatment with sulphuric acid. (–)Isorotenone, in which position 5' is no longer chiral, undergoes facile base-catalysed (sodium acetate) racemisation in ethanolic medium. The fact that only racemisation occurs without the formation of diastereoisomers suggests that base-catalysed inversion at position 12a induces inversion at position 6a as well, presumably by the mechanism shown below. As a corollary, one can therefore conclude that of the two possible modes of B–C ring fusion, either *cis* or *trans*, is thermodynamically more stable than the other. Therefore, when inversion occurs at one junction point, it triggers inversion at the other as well. This inference is also supported by the following observation. Similar base-catalysed isomerisation converts (–)rotenone into a mixture of diastereoisomers, known as mutarotenone. This mixture can be separated into (–)rotenone and epirotenone (3.7.4). The latter, on acid treatment, gives (+)isorotenone (3.7.3). Therefore, in epirotenone, positions 6a and 12a have enantiomeric configurations with respect to (–)rotenone, while the configuration at position 5' remains unchanged.

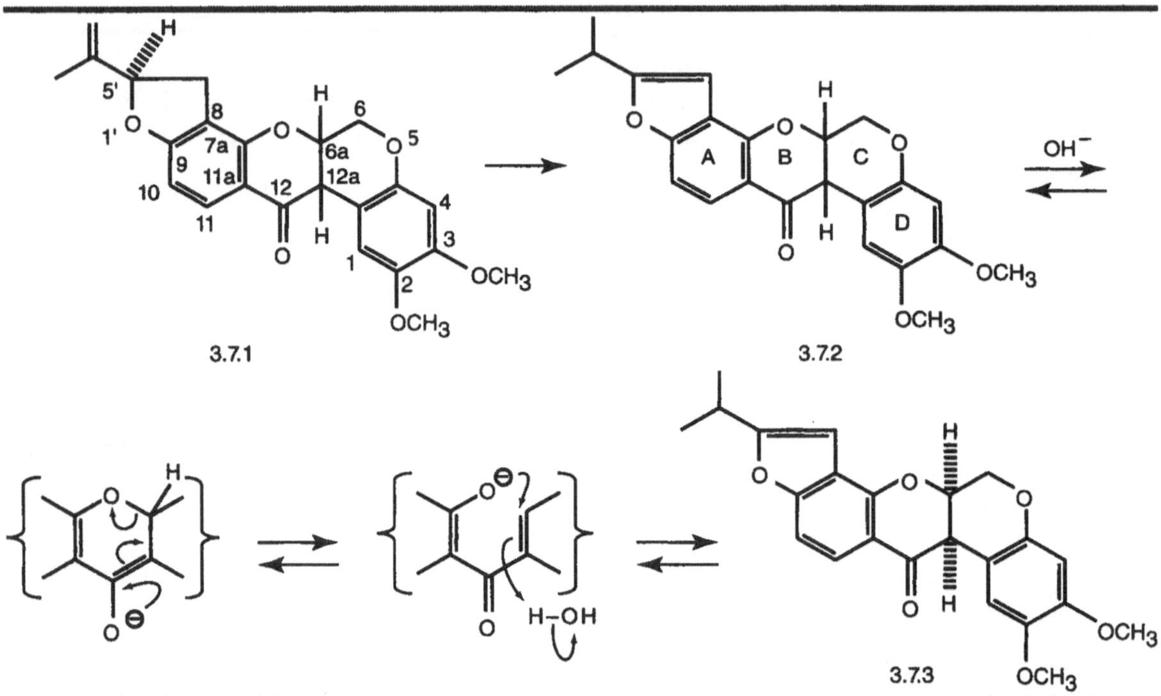

3.7.1   3.7.2

3.7.3

Reduction of racemic isorotenone with potassium borohydride occurs in a stereospecific manner giving rise to a single product (3.7.5) in over 80% yield. The IR spectrum of the compound shows the presence of a strong intramolecularly H-bonded hydroxyl group (absorption at 3566 cm^{-1}). Obviously, this is the hydroxyl formed by the reduction of the carbonyl group and the only other oxygen atom to which this hydroxyl can be linked by a H-bond is at position 5. Models show that this is possible only when the B–C rings are *cis*-fused and when the B ring assumes a quasi-boat conformation. Further, the newly formed hydroxyl at position 12 should be *trans* to the hydrogen at position 12a. This conclusion is also fully substantiated by spin–spin coupling data derived from the ^{1}H NMR spectra of rotenone, isorotenone and dihydrorotenone (3.7.6). Thus, the *J* value for coupling between H6a and H12a was found to be 4 Hz, indicating a dihedral angle of 44° as calculated using the Karplus equation. Models show a dihedral angle of 40° for the B–C *cis*-fused skeletal structure.

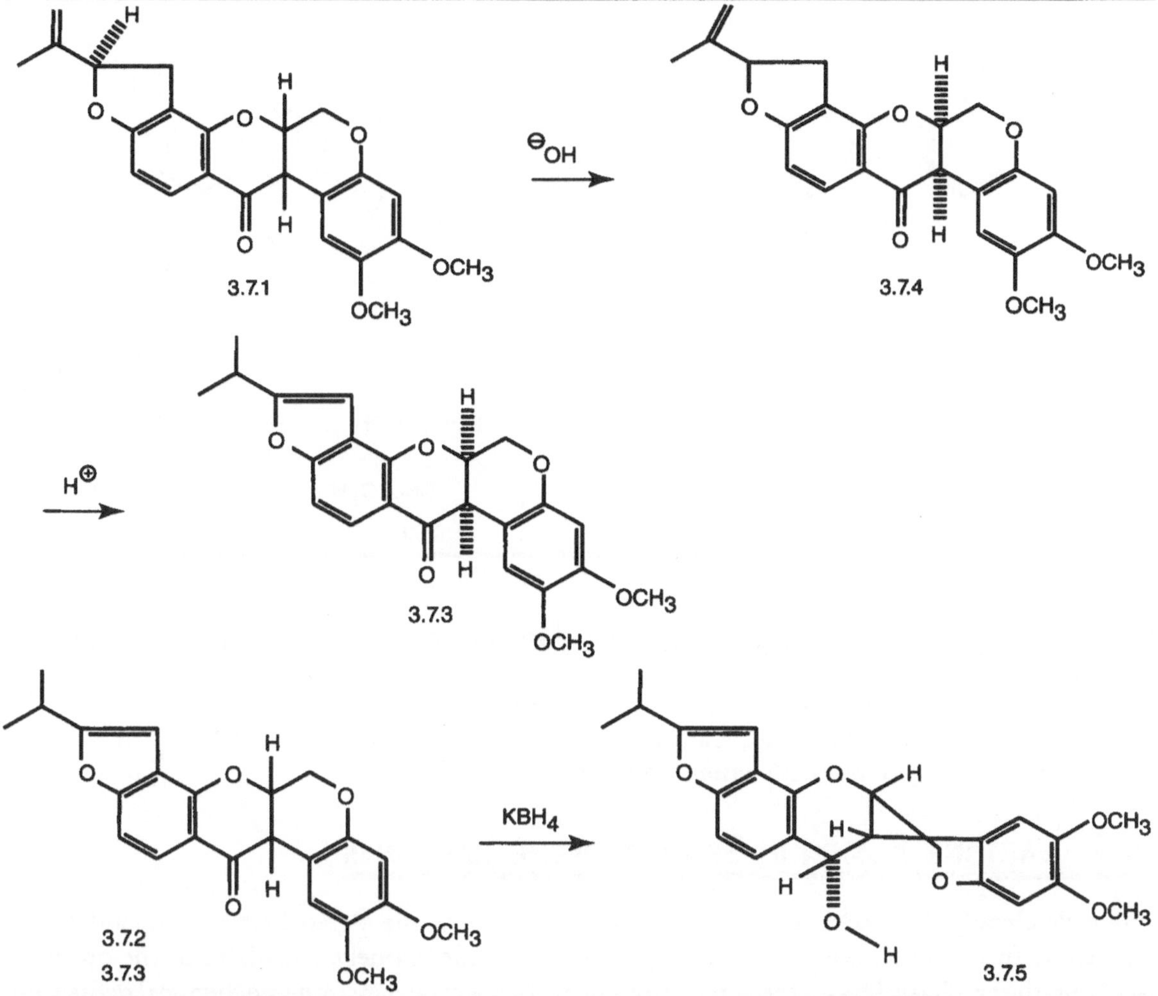

The absolute configuration at C-6a as (S) was unambiguously determined by the isolation of D glyceric acid (3.7.7) by the exhaustive ozonolytic degradation of dihydrorotenone as shown. The chiral carbon atom of D glyceric acid thus obtained is derived from C-6a of rotenone; C-12a of the latter appears as the carboxylic carbon of the glyceric acid.

The absolute configuration at C-5' was established by the isolation of (+)3-hydroxy-4-methylpentanoic acid (3.7.8) as another product of the above-mentioned oxidative degradation of dihydrorotenone. Thus, the absolute stereochemistry of rotenone can be designated as 5' R, 6a S and 12a S and this is supported by the ORD and CD spectra of rotenone, and its derivatives. The ORD and CD spectra of all other naturally occurring rotenoids are similar to those of rotenone, proving that in all of them the B and C rings are *cis*-fused, as in rotenone.

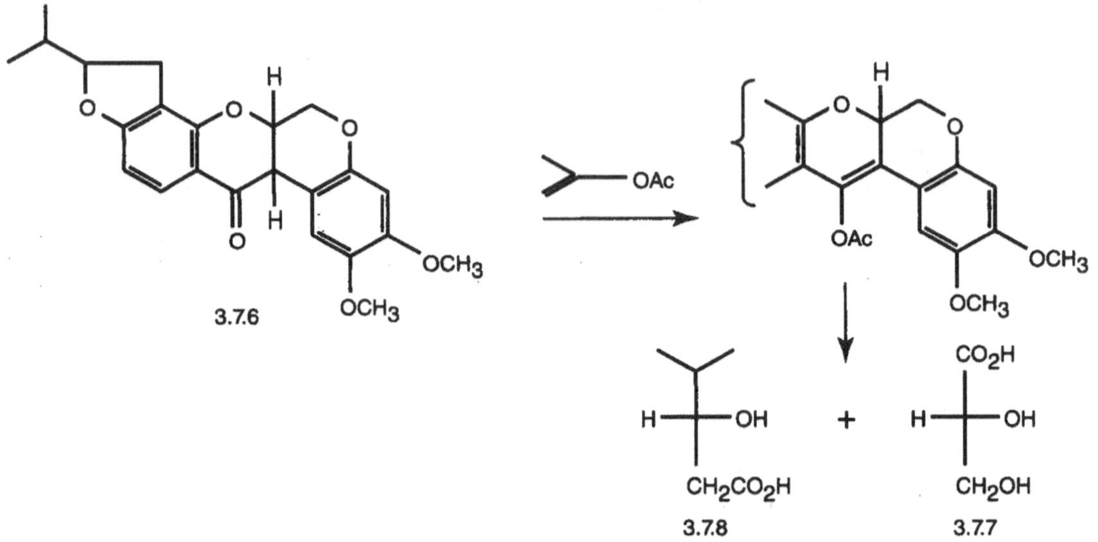

3.7.6

3.7.8   3.7.7

Suggested reading

1.   Livingstone R, In Coffey S, (ed.), Rodd's *Chemistry of Carbon Compounds*, 2nd edn, Vol. 4, Amsterdam: Elsevier, 1978.
2.   Djerassi C, Ollis WD and Russell RC, *J. Chem. Soc.*, 1961, 1448.
3.   Crombie L, Godin PJ, Whiting DA and Siddalingaiah KS, *J. Chem. Soc*, 1961, 2876.
4.   Kostova I and Ognyanov I, *Monatsh.*, 1986, 117:689.

3.8 LACTONE FUSION IN SESQUITERPENE LACTONES

We have already discussed at some length some aspects of the stereochemistry, including that of lactone fusion, of a naturally occurring germacranolide, namely, enhydrin. In the preceding sections, the emphasis was on the applications of NMR spectral data in stereochemical deductions. Among the other physical methods, measurements of optical rotation against wavelength, that is ORD, and dichroic absorptions by means of CD spectra are of direct relevance to stereochemistry. The only limitation of these techniques is the absence of a simple and absolutely sound theoretical framework with the help of which any practising organic chemist could interpret ORD and CD data from first principles. In practice, the data generated by chiro-optical methods are interpreted with the help of rules formulated by a combination of theory and empiricism. However, these techniques have generated a good deal of very useful and unique data which have given valuable insights into the stereochemistry of a wide variety of naturally occurring compounds. The highly successful studies of Djerassi and co-workers on the ORD spectra of steroids deserve mention in this context. In this section, the interpretation of ORD and CD spectra of sesquiterpene lactones will be discussed, with particular reference to germacranolides.

The CD spectra of germacranolides possessing two endocyclic double bonds with *trans*-configurations were studied by Snatzke and co-workers who observed that these spectra had two bands at 200 and 220 nm with opposite signs of the Cotton effect. These could be attributed to the exciton splitting of the band arising from the chiral overlap of the two double bonds as a result of a transannular homoconjugation. Compounds which exhibit such CD spectra include costunolide (3.8.1), juriniolide (3.8.2), tulipinolide (3.8.3) and epitulipinolide (3.8.4). The CD spectra of the last two mentioned compounds are very similar to that of costunolide (with a positive Cotton effect peak at about 220 nm) suggesting that, like costunolide, they also possess a C-6 *trans*-fused lactone ring. It follows that a C-6-closed *cis*-lactone as well as a C-8-closed *trans*-lactone should exhibit negative Cotton effect peaks at about this wavelength.

3.8.1: R = H

3.8.4: R = O Ac

3.8.2   3.8.3   3.8.5

Sesquiterpene lactones lacking either of the two endocyclic double bonds, such as enhydrin (3.8.5), would not be able to exhibit dichroic absorption due to exciton splitting. However, these compounds also gave rise to typical CD spectra in the higher wavelength region, namely 245–265 nm due to the presence of the α-, β-unsaturated lactone group, which underwent $n \rightarrow \pi^*$ excitation in this wavelength region. Systematic analysis of the ORD and CD spectra of a number of sesquiterpene lactones in this wavelength region resulted in the formulation of an empirical rule by Geissman correlating the sign of Cotton effect with the stereochemistry of lactone fusion. According to this rule, C-6-closed *trans*-lactones give rise to a negative Cotton effect around 265 nm. For *trans*-lactones closed at C-8, the Cotton effect is positive; the pseudoenantiomorphic nature of the two alternative points of attachment of the lactone moiety is responsible for the opposite signs. The Cotton effect signs have to be reversed for the corresponding *cis*-lactones.

As mentioned in the section on enhydrin, the rationale for Geissman's rule is quite clear. For the unsaturated lactone chromophore, three mesomeric structures, 3.8.6, 3.8.7 and 3.8.8, can be written. Since each of the four bonds constituting the chromophoric unit, thus, has some double bond character, a vertical plane bisecting the carbonyl group constitutes a symmetry plane. This situation is unlike that in a saturated γ-lactone, where the plane of symmetry is the one that divides the O–CO bond. The space around the unsaturated lactone group can thus be divided as in the case of an enone and the Cotton effect signs predicted by the octant rule.

Thus, in a C-8-closed *trans*-germacranolide, the symmetry plane passing through the lactone carbonyl group and the other two planes perpendicular to it divide the lactone in such a way that the ring residue at C-7 falls in the far upper left octant with the ring residue at C-8 occupying the far lower right octant if one views the compound with the H7 α-oriented, as shown in 3.8.9. Applying the octant rule to this system, one could predict the sign of the Cotton effect to be positive, in accordance with Geissman's rule. The situation in a C-6-closed *trans*-lactone (3.8.10) is the reverse of the above and the sign of the Cotton effect should be negative, as predicted by Geissman's rule.

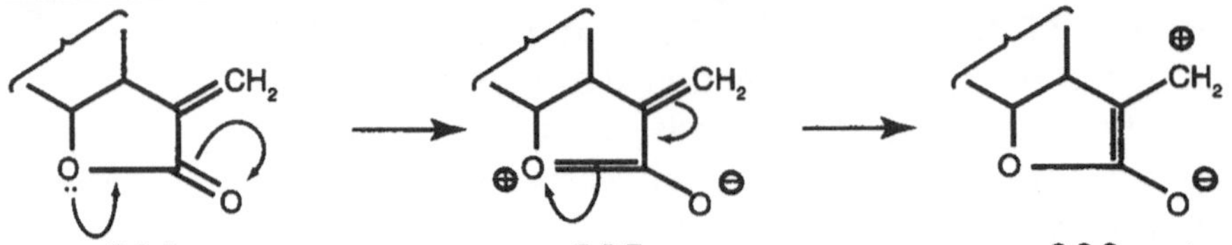

3.8.6   3.8.7   3.8.8

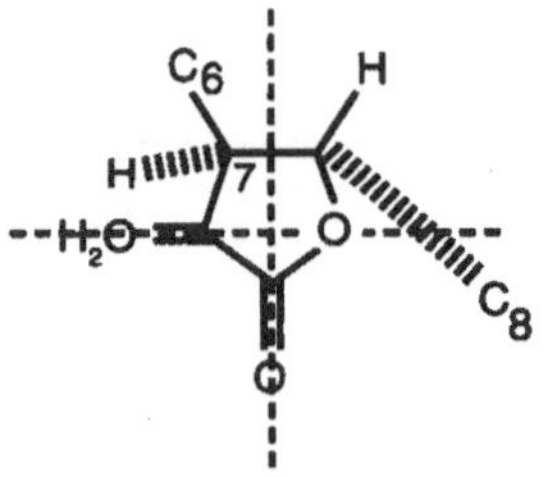

3.8.9 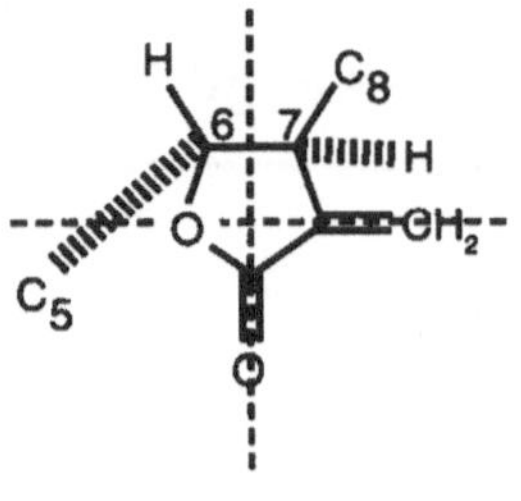   3.8.10

With the discovery of a number of germacranolides and the consequent accumulation of data it was found that there were several exceptions to Geissman's rule. These exceptions include enhydrin (3.8.5), helianginol (3.8.16), pyrethrosin (3.8.11), laurenobiolide (3.8.12), chamissonin diacetate (3.8.13), chrysanolide (3.8.14), woodhousin (3.8.17) and maculatin (3.8.15). All these compounds are *trans*-fused lactones and can be considered under three different types.

C-8-closed lactones with *trans*-4,5-double bond. Compounds 3.8.11 to 3.8.14 belong to this type. All of them have an additional chiral centre at C-8 which carries, in each case, an α-acetoxyl group. Models show that in certain conformations, this acetoxyl group falls in the far lower left octant, thus making a negative contribution to the Cotton effect which could effectively reverse the contribution made by C-7.

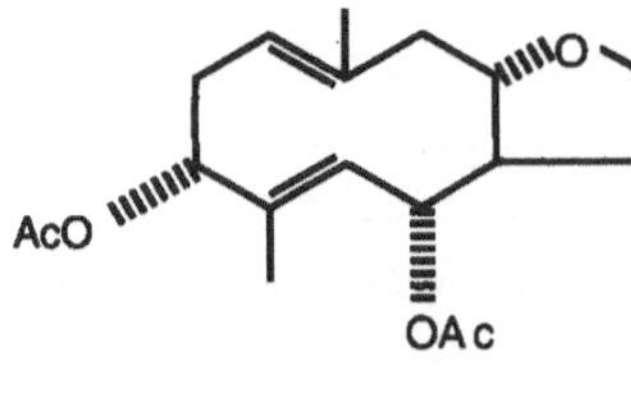

3.8.11 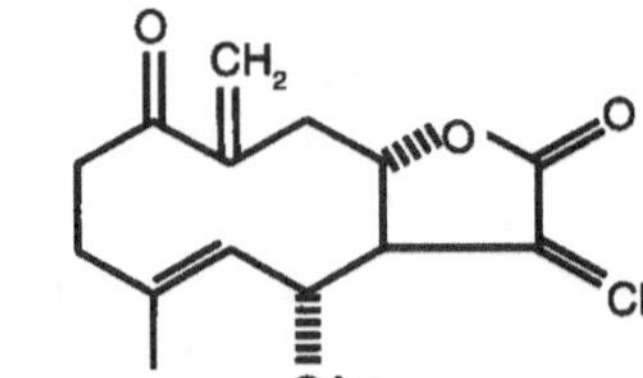   3.8.12

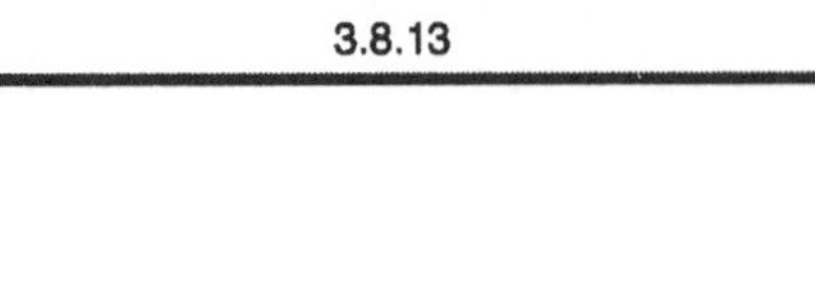

3.8.13 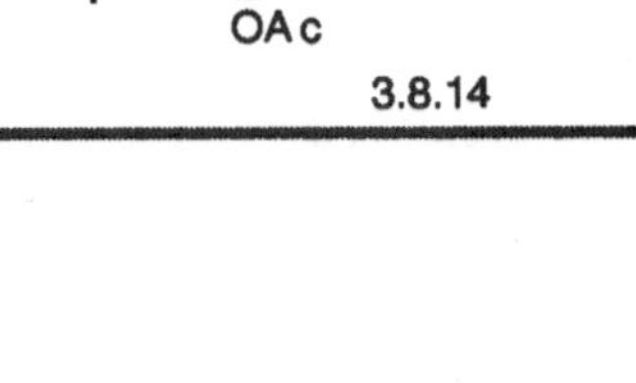   3.8.14

C-6-closed lactones with the 4,5-double bond or its equivalent (epoxide) having the *trans*-configuration. Enhydrin (3.8.5) and maculatin (3.8.15) belong to this type and they also have, unlike costunolide (3.8.1), which obeys Geissman's rule, an additional chiral centre at C-8. While the explanation given above may hold good for these compounds also, it must be pointed out that it may not be so straightforward for the following reason. These compounds possess several other chiral centres and a second, intrinsically symmetric but asymmetrically perturbable chromophore. Without knowing the sign and magnitude of contribution of these to the total Cotton effect of the compound, it

3.8.15: $R_1 = COC_4H_7O$
$R_2 = COCH_3$

will not be possible to give a complete explanation for the observed data. We suggest that readers construct framework models of these compounds and examine the total chiral environment of the lactone chromophore.

C-6-closed lactones with a *cis*-4,5-double bond. Helianginol (3.8.16) and woodhousin (3.8.17) belong to this type. Punctatin (3.8.18), which also belongs to this type, obeys Geissman's rule and exhibits a negative Cotton effect. A model of (3.8.16) shows that the major bulk of the C-7 ring residue, and in particular the epoxide group across C-1 and C-10 lies in the far lower right octant, thus making a positive contribution to the Cotton effect. In this sense, the heliangolide skeleton significantly differs from the germacrolide and melampolide skeletal structures mentioned in Section 3.6. Therefore, the signs predicted by Geissman's rule for the other types of germacranolides have to be reversed for heliangolides. Viewed thus, the behaviour of punctatin is anomalous. In this compound, the acyloxy group at C-8 is β-oriented and perpendicular to the lactone plane. Falling thus in the far upper right octant, this bulky group makes a large negative contribution to the Cotton effect to cancel out the positive effect of the C-7 ring residue.

3.8.16   3.8.17   3.8.18

Suggested reading

1. Suchy M, Dolejs L, Herout V, Sorm F, Snatzke G and Himmeheich J, *Coll. Czech. Chem. Comm.*, 1969, 34: 229.
2. Stocklin W, Waddell TG and Geissman TA, *Tetrahedron*, 1970, 26: 2397.
3. Herz W and Bhat SV, *J. Org. Chem.*, 1972, 37: 906.
4. Doskotch RW and El Feraly FS, *J. Org. Chem.*, 1970, 35: 1928.
5. Prasanna S, Ph.D. Thesis, Bangalore University, India, 1975.

3.9 STEREOCHEMISTRY OF ABIETIC ACID

This much-studied diterpene is the best known among the resin acids which are formed by the action of heat or acids on colophony. Colophony (also known as rosin) is the non-steam volatile residue obtained from the oleo resin formed as a 'wound gum' when incisions are made in the bark of pine trees.

(–)Abietic acid is a tricyclic, di-unsaturated, monocarboxylic acid having the structure 3.9.1. It has four chiral centres: C-4, C-5, C-9 and C-10 (steroid numbering). It is the end-product of a series of heat-induced or acid-catalysed rearrangements of precursory diterpene acids and therefore represents a thermodynamically stable stereo structure. On vigorous oxidative degradation, (–)abietic acid yielded 1,3-dimethylcyclohexane-1,2,3-tricarboxylic acid (3.9.2) which was found to be optically inactive.

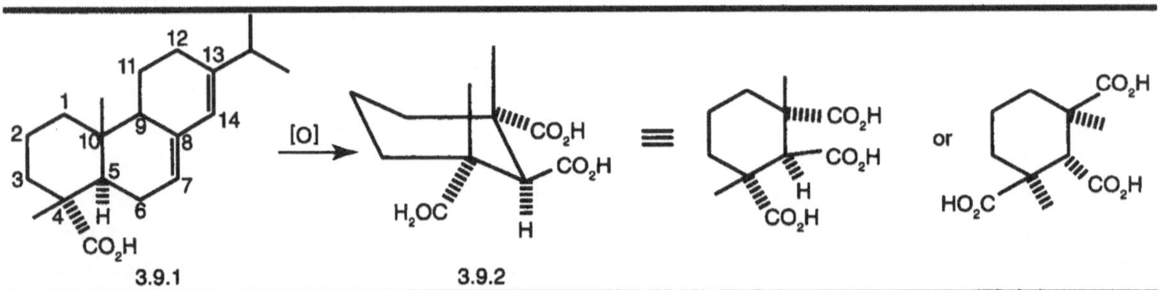

Therefore, 3.9.2 should possess a plane of symmetry, implying thereby that the carboxyls at positions 1 and 3 are *cis* to each other. The thermodynamic stability of (–)abietic acid suggests that the A–B ring system, which forms a decalin unit is, in all probability, *trans*-fused. This would mean that the carboxyl at position 2 in 3.9.2 should be *trans* to either of the other two carboxyls. This inference is supported by a study of the dissociation constants of 3.9.2. Therefore, this tricarboxylic acid should have the stereochemistry shown in 3.9.2 or its enantiomorphic form. The absolute configurations of C-5 and C-10 became evident with the isolation of the dicarboxylic acid (3.9.3) from (–)abietic acid by a series of transformations which left the configurational centres 5 and 10 unaffected. Compound 3.9.3 was also obtained from ergosterol (3.9.4) whose stereochemistry was known.

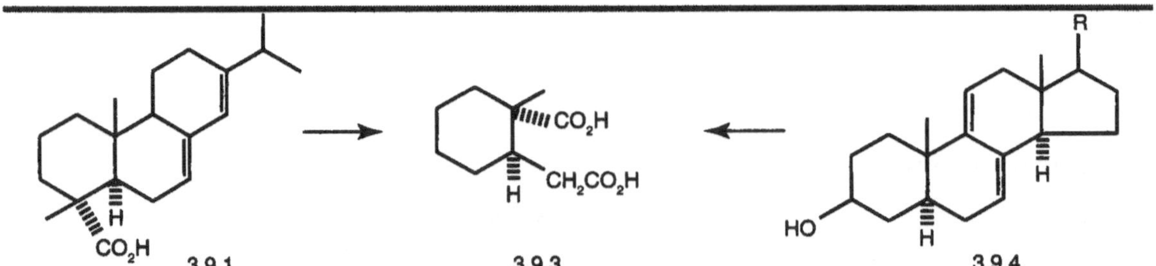

The configuration at C-9 in 3.9.1 was figured out in the following manner. As already mentioned, abietic acid is the most stable among the resin acids and is the end-product of a series of 'shuffling' reactions which, naturally, proceed towards a thermodynamically stable structure. Therefore, it is logical to assume that (–)abietic acid should have the most preferred configurational arrangement, namely, A–B rings trans-fused and the hydrogen at C-9 being anti to the methyl at CIO. Thus, the absolute stereochemistry of (–)abietic acid can be shown as in 3.9.5.

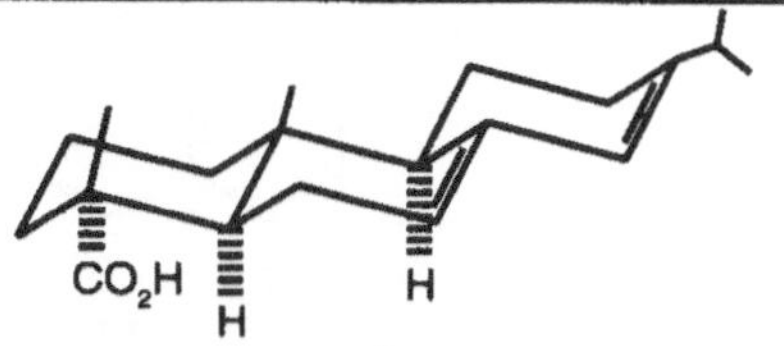

3.9.5

Suggested reading

1. Barton DHR, Quar. Rev. Chem. Soc, 1949, 3: 1.
2. Simonsen J and Barton DHR, The Terpenes, Vol. 3, Cambridge: Cambridge University Press, 1952.
3. Klyne W, J. Chem. Soc., 1953, 3072.
4. Nakanishi K, In Nakanishi K, Goto T, Ito S, Natori S and Nozoe S, (eds), Natural Products Chemistry, Vol.1, Tokyo: Kodansha Ltd, 1974.

3.10 CATECHINS

From the stereochemically complex alkaloids and sesquiterpenoids discussed in some of the preceding sections of this chapter, we may now turn our attention to something simpler but intriguing all the same. Catechins are 3-hydroxyflavans which are the characteristic crystalline components of catechu used by dyers and tanners. The best source for (+)catechin is Gambier catechu obtained from the Malaysian bush, Uncaria gambier. Its structural formula (3.10.1) shows the presence of two chiral centres, namely, positions 2 and 3. When an aqueous solution of (+)catechin was heated to 115°C, the compound underwent epimerisation and yielded (+)epicatechin (3.10.2); the enantiomorph (3.10.3) of (+)epicatechin is a component of Acacia catechu and can be isomerised to (–)catechin (3.10.4) by heating in water. Thus, catechin and epicatechin are diastereomers and should, therefore, differ in configuration either at C-2 or C-3.

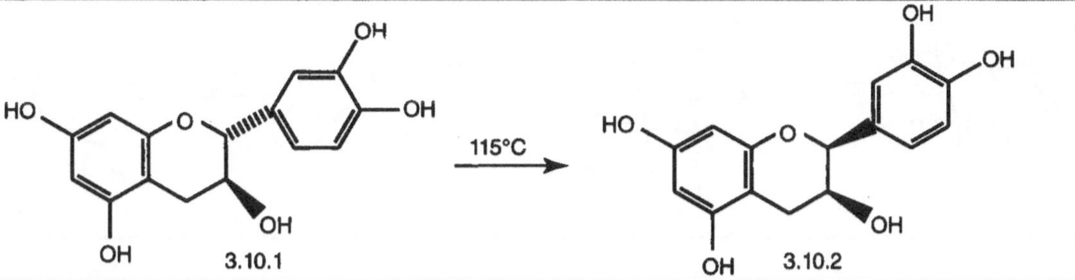

115°C

3.10.1   3.10.2

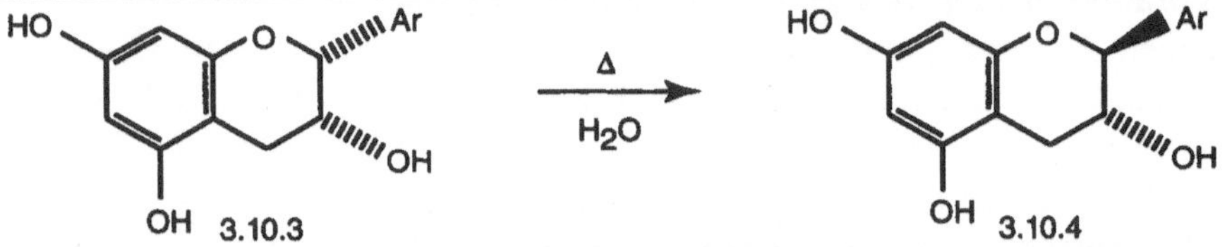

However, till as late as 1977, this apparently simple problem had remained ambiguous; the confusion that existed at that time is clearly revealed by Livingstone's review in Rodd's second edition of *The Chemistry of Carbon Compounds*.The difficulty in arriving at firm configurational assignments in the catechins was due to the conformational flexibility of the pyran ring, which can assume three idealised conformations, if one leaves out the transitionary forms. These are the half-chair (3.10.5), the C-2 sofa (3.10.6) and the C-3 sofa (3.10.7). Therefore, conclusions drawn from an application of the Karplus equation, using NMR *J* values are not fully reliable due to variable dihedral angles as a consequence of conformational mobility. All the same, NMR studies strongly indicated that in the (+) and (–)catechins, the hydrogens at C-2 and C-3 are *trans* to each other whereas in the epicatechins they are *cis*-oriented, as shown. Another complication is the possibility of conformational flipping of each conformational type. For example, the half-chair (3.10.5) in which the C-2 aryl unit is equatorially oriented can switch to the higher energy form, 3.10.8. Similarly, an alternative C(2) sofa (3.10.9) is also possible. Indeed, low-temperature NMR studies have shown that such conformational flippings do occur in solution, in catechin as well as epicatechin, but the preferred conformations are the equatorial forms (for example, 3.10.5). x-ray data on catechin, epicatechin and some of their derivatives have indeed shown that even in the crystalline state, the pyran ring can assume different conformations depending on structural and configurational differences.

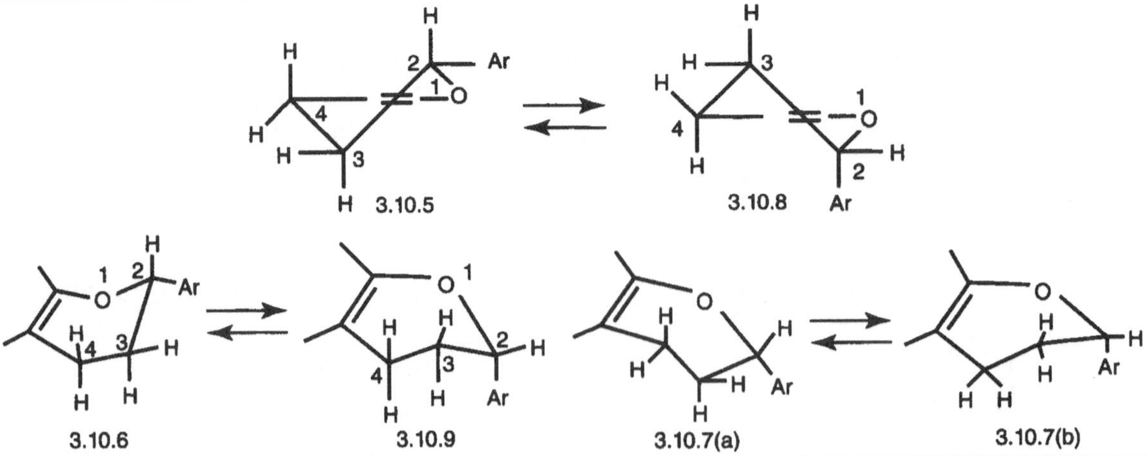

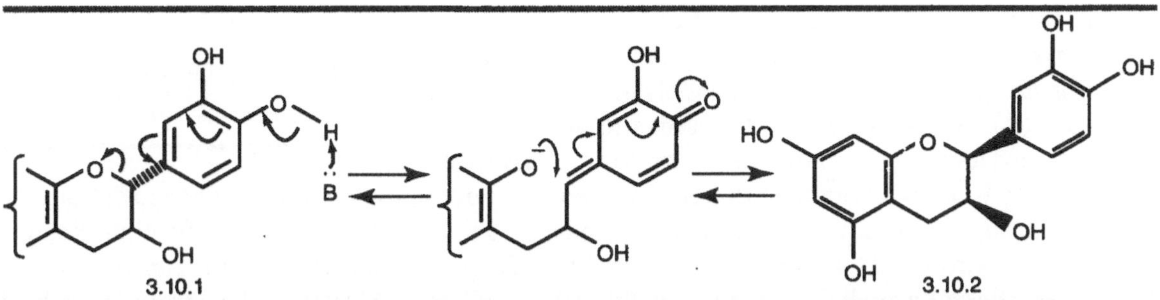

3.10.1   3.10.2

As mentioned earlier, one characteristic feature of catechins is their ready epimerisation even in hot aqueous solutions. Freudenberg, who made a careful study of this phenomenon, observed that only epimerisation occurred and that there was no racemisation. From the structures 3.10.1 to 3.10.4, it is obvious that epimerisation occurs only at C-2 and this has been explained by Whalley as due to a reversible opening of the pyran ring. On the basis of this mechanism it is easy enough to see that the epimerisation can be catalysed either by acids or by alkalis, as shown above.

Suggested reading

1. Livingstone R, In Coffey S, (ed.), In Coffey S, (ed.), Rodd's *Chemistry of Carbon Compounds*, 2nd edn, Vol. 4, Amsterdam: Elsevier, 1978.
2. Haslam E, *Plant Polyphenols—Vegetable Tannins Revisited*, Cambridge: Cambridge University Press, 1989.
3. Engel DW, Hattingh M, Hundt HKL and Roux DG, *J. Chem. Soc, Chem. Comm.*,1975, 695.

3.11 SPHINGOSINE

Our next example is an acyclic compound possessing two chiral centres and a double bond. Though not structurally and stereochemically complex, sphingosine offers ample scope for the application of a fairly wide range of chemical principles in understanding its total structure and is, therefore, a useful example from the instructional point of view. It is also an important natural product, being an integral component of sphingolipids, such as the sphingomyelins, cerebrosides and gangliosides. It was first isolated almost a hundred years ago by Anderson from brain lipids.

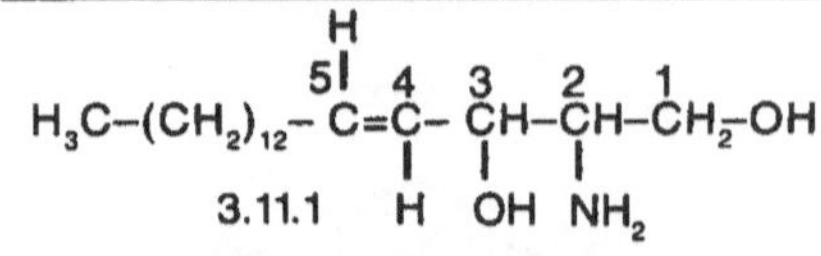

The absolute configuration at position 2 in sphingosine (3.11.1) was determined in the following manner. Triacetyl sphingosine (3.11.2) yielded, on ozonolysis, an aldehyde triacetate (3.11.3) which was hydrolysed and subjected to monobenzoylation to obtain the N-benzoyl derivative,

$$H_3C-(CH_2)_{12}- \overset{5}{C}=\overset{4}{C}- \overset{3}{C}H-\overset{2}{C}H-\overset{1}{C}H_2-OH$$
$$\quad\quad\quad H \quad\quad H \quad OH \quad NH_2$$

3.11.1

3.11.4. It may be pointed out here that under Schotten–Baumann conditions of benzoylation, the hydroxyl groups remain free as the alkaline medium is not conducive for esterification. Being a 2-hydroxy aldehyde, 3.11.4 reacted with periodic acid, losing one carbon atom, and gave 3.11.5. The latter after O-acetylation, oxidation (of the -CHO to -COOH) and hydrolysis, yielded L-serine (3.11.6). Therefore, the configuration at C-2 in 3.11.1 should be the same as that in 3.11.6. This inference was reinforced by another series of reactions, described below.

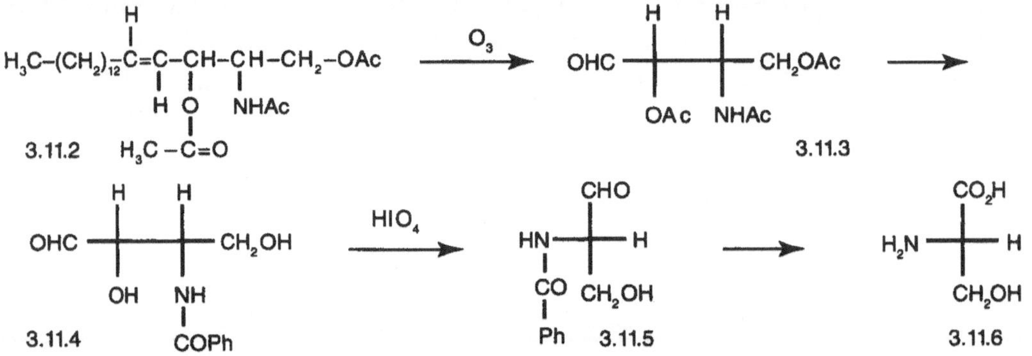

Triacetyl sphingosine (3.11.2) was subjected to hydrogenolysis in the presence of Adam's catalyst (PtO$_2$) when it gave diacetyl sphingine (3.11.7). Allylic and benzylic acetoxyl groups are known to undergo hydrogenolytic cleavage and, therefore, in the above reaction, the selective reductive removal of the acetoxyl group at C-3 presumably precedes the reduction of the double bond. Hydrolysis and monobenzoylation (Schotten–Baumann) of 3.11.7 yielded the N-benzoyl derivative (3.11.8), which could be oxidised to N-benzoyl-α-aminostearic acid (3.11.9). The optical rotation of the latter showed that it had the D configuration. It should be noted that in this case it is the C-1 (which is on the right-hand side of the amino group in 3.11.1) that is oxidised to the carboxyl group, whereas in the formation of 3.11.6, it is C-3 (lying to the left of the amino group) that appears as the carboxyl. That is why the two α-amino acids 3.11.6 and 3.11.9 have opposite configurations.

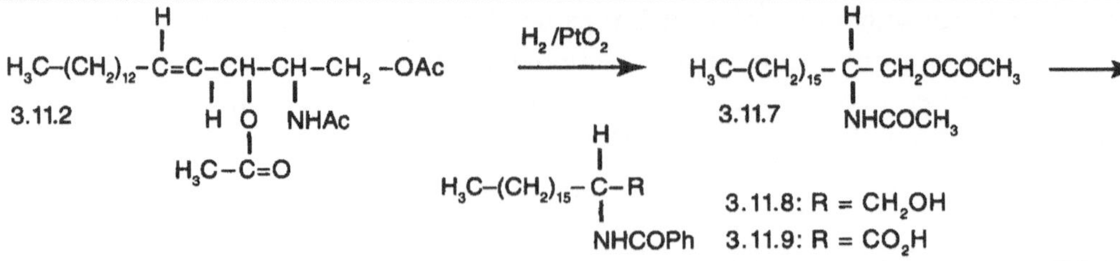

The configuration at C-3 was found to be *erythro* with respect to that at C-2 in the following manner. Synthetic *erythro*-2-amino-3-hydroxystearic acid (3.11.10) gave, on reduction with lithium aluminium hydride (LAH), 3.11.11 which was found to be identical with natural dihydrosphingosine. Since the absolute configuration at C-2 was already known, that at C-3 could therefore be deduced to be 3.11.11.

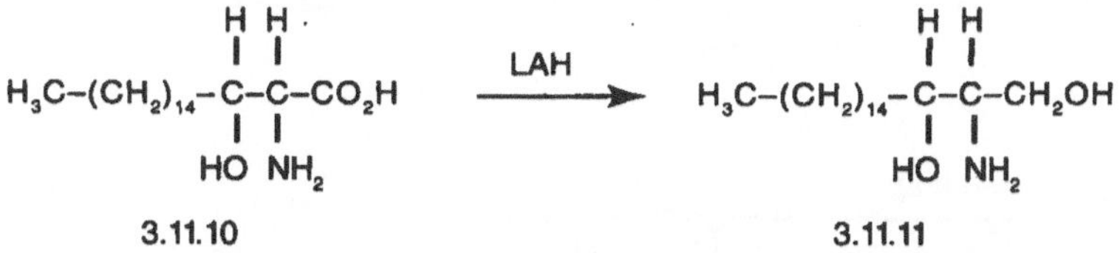

These configurational assignments were confirmed by a stereospecific synthesis of racemic 3.11.11 following the method shown below. Octadec-2-yn-1-ol (3.11.12) gave the *cis*-olefin 3.11.13 on catalytic hydrogenation in the presence of Lindlar's catalyst; the *trans*-isomer could be obtained by the action of lithium aluminium hydride on 3.11.12. Epoxidation of 3.11.13 by perphthalic acid yielded the *cis*-epoxide, 3.11.14, which on ammonolysis gave racemic 3.11.11 as one of the products.

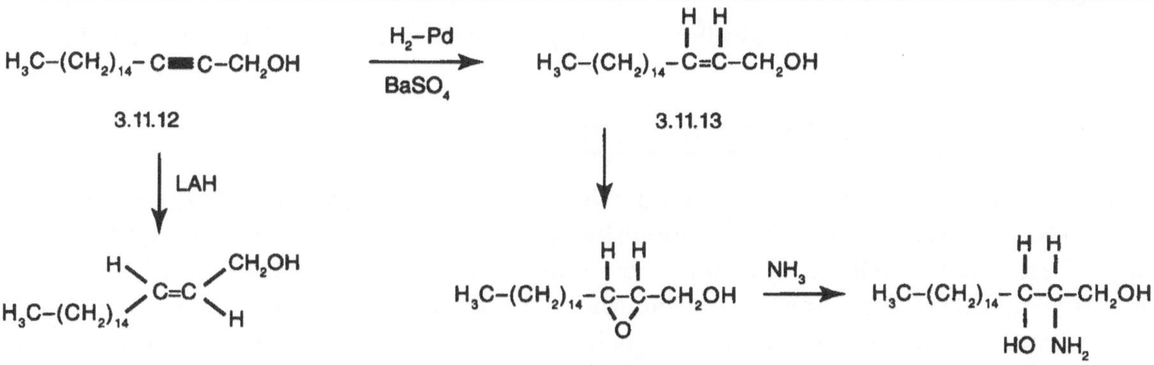

The configuration of the double bond in 3.11.1 as *trans* was indicated by the characteristic absorption band at about 960 cm^{-1} in its IR spectrum as well as by a synthesis by Grob. The condensation product, 3.11.15, of 2-hexadecynal (3.11.16) and 2-nitroethanol, was reduced by zinc and acid to obtain the amine 3.11.17. The latter on reduction with lithium aluminium hydride gave *trans*-sphingosine, identical with the natural sample. It may be pointed out here that LAH can reduce only those triple bonds which are next to a carbinol carbon and that the reaction is stereospecifically *trans* in contrast to catalytic hydrogenation which yields stereospecifically the *cis*-olefin. The LAH reduction presumably involves the participation of the hydroxyl group as shown on the next page.

$$H_3C-(CH_2)_{12}-C\equiv C-CHO \;+\; \underset{\underset{NO_2}{|}}{CH_2}-CH_2OH \quad\xrightarrow[\text{2. Zn/AcOH}]{\text{1. Base}}\quad H_3C-(CH_2)_{12}-C\equiv C-\underset{\underset{OH}{|}}{CH}-\underset{\underset{R}{|}}{CH}-CH_2OH$$

3.11.16   3.11.15: R = NO$_2$

3.11.17: R = NH$_2$

$$\xrightarrow{\text{LAH}}\quad H_3C-(CH_2)_{12}-\underset{\underset{H}{|}}{C}=C-\underset{\underset{OH}{|}}{CH}-\underset{\underset{NH_2}{|}}{CH}-CH_2-OH$$

3.11.1

$$R-C\equiv C-CH- \quad\xrightarrow{\text{work-up}}\quad R-\underset{\underset{H}{|}}{C}=C-\underset{\underset{OH}{|}}{CH}-$$

Suggested reading

1. Gigg RH, In Coffey S, (ed.), Rodd's *Chemistry of Carbon Compounds*, 2nd edn, Vol. 4, Amsterdam: Elsevier, 1976.
2. Shapiro D, *Chemistry of Sphingolipids*, Paris: Hermann, 1969.
3. Gunstone FD, In Haslam E, (ed), *Comprehensive Organic Chemistry*, Vol. 5, Oxford: Pergamon Press, 1979.
4. Gunstone FD, Harwood JL and Padley FB, *The Lipid Handbook*, London: Chapman and Hall, 1986.

3.12 STEREOCHEMISTRY OF MENTHOL

The monocyclic monoterpenoid alcohol, Menthol, is 2-isopropyl -5-methyl cyclohexanol. Since it has three chiral centres, namely carbon atoms 1, 2 and 5, it can exist as eight optically active stereoisomers. All the eight stereoisomers are known, though only (–) menthol occurs in nature. The other stereoisomers are (+) menthol, (+) and (–) neomenthol, (+) and (–) isomenthol and (+) and (–) isoneomenthol. Further, being cyclohexane derivatives, (–) menthol and its stereoisomers exhibit conformational isomerism in addition to configurational isomerism. As mentioned in the introduction, enantiomers can be differentiated from each other only from the signs of their optical rotation. On the other hand, diastereoisomers differ in their physical and chemical properties. Physical constants, such as melting (or boiling) points, refractive index and density can be readily and accurately determined. These properties—and, in particular, density and refractive index—are related to the molecular volume and enthalpy. Based on experimental data von Auwers and Skita formulated a rule which allowed differentiation between epimers among alicyclic compounds. This rule has been modified and is now known as Conformational rule. According to it, between two epimers, which do not differ significantly in their dipole moments, the one having higher enthalpy (thermodynamically less stable) will have a higher boiling point, density and refractive index. Returning to the problem of menthol, we can write the eight possible stereoisomers as (3.12.1) to (3.12.8).

3.12.1   3.12.2   3.12.3   3.12.4

3.12.5   3.12.6   3.12.7   3.12.8

Among these, the enantiomers of menthol, namely (+)and (–)menthol, have the least boiling point, density and refractive index. Therefore, these two compounds can be assigned structures (3.12.1) and (3.12.2), in which all the bulky substituents (hydroxyl, methyl and isopropyl) are equatorially oriented. As the groups in these two structures are spread out, they will have the maximum molecular volume and least enthalpy. One of them is (+)menthol and the other is (–)menthol. We shall revert to this problem later. (–)Menthol and (+)neomenthol yield the same ketone, (–)menthone, on oxidation; (+)menthone is similarly obtained from (+)menthol and (–)neomenthol. These observations show that menthol and neomenthol are epimeric alcohols, differing in the configuration at C-1. Therefore, the two enantiomers of neomenthol should be (3.12.3) and (3.12.4). The two enantiomers of menthone are (3.12.9) and (3.12.10). As noted above, the hydroxyl is equatorially oriented in menthol and therefore in neomenthol, it takes up an axial position.

3.12.9   3.12.10

The above mentioned inferences regarding menthol and neomenthol are further supported by the following observations. The rate of acetylation of racemic menthol is several times greater than that of racemic neomenthol. On the other hand, neomenthol undergoes oxidation at a faster rate than menthol. These facts are in accordance with the known differences between equatorial and axial cyclohexanols. In the acylation reaction, the transition state for the axial alcohol experiences an increase in steric strain due to increase in the size of the axial substituent and hence an increase in energy. In the transition state of the equatorial isomer, on the other hand, this increase in size is inconsequential as the substituent is oriented away from the cyclohexane ring. The position is just the reverse in the oxidation reaction, wherein a tetrahedral carbon atom (the carbinol carbon) gets converted into a trigonal state. The resulting flattening of that part of the ring system in the product relieves any axial–axial interaction present in the starting material. Such an interaction is seen only in neomenthol and not in menthol. Thus, there is an expected and considerable decrease in steric strain, and hence the energy, in the transition state of the neomenthol to menthone oxidation reaction but not in that of the menthol to menthone oxidation. Further, precise estimates have shown that the rates of acylation of the diastereoisomers of menthol follow the order: menthol > isomenthol > neoisomenthol > neomenthol. Therefore, in isomenthol, as in menthol, the hydroxyl group should be equatorially oriented, and in neoisomenthol, it assumes an axial position. The enantiomers of isomenthol are, therefore, represented by the structures (3.12.5) and (3.12.6) and those of neoisomenthol by (3.12.7) and (3.12.8).

Further corroboration of the above mentioned conclusions are provided by the differences in the rates of base catalysed E2 elimination of hydrogen chloride from the menthyl chlorides (3.12.11 and 3.12.12) and neomenthyl chlorides (3.12.13 and 3.12.14). In this reaction, the neomenthyl chlorides react 200 times faster than the menthyl chlorides.

3.12.11   3.12.12   3.12.13   3.12.14

In a E2 reaction in a cyclohexane system, the groups that are eliminated (in this case a hydrogen atom and a chlorine atom on the adjacent carbon atoms), should be anti to each other for a facile elimination. In other words, if one is alpha axial, the other should be beta axial. In neomenthyl chorides (for example, 3.12.13), there are two such arrangements and, therefore, two isomeric menthenes, namely, menth-2-ene (3.12.15) and menth-3-ene (3.12.16) are obtained as products. In the case of menthyl chloride (for example, 3.12.11), it should first undergo conformational change to the alternative, higher energy, chair conformation (3.12.17), wherein the chlorine atom now assumes an axial orientation, before it can undergo the E2 reaction. Therefore, the rate of the reaction is much lower than that of the neomenthyl chloride to menthene reaction. Further, since there is only one diaxial arrangement of hydrogen and chlorine on adjacent carbons in 3.12.17, only one product, namely, menth-2-ene (3.12.15) is obtained in this case.

3.12.15   3.12.16   3.12.17

The final stage in this problem is to determine the absolute stereochemistry of the naturally occurring (–)menthol. As mentioned above it should be either (3.12.1) or (3.12.2). It has also been noted already that (–)menthol yields (–)menthone on oxidation. The ORD and CD spectra of (–)menthone show positive Cotton effect with the amplitude being the same as that of (+) 3-methylcyclohexanone. Therefore, using the Octant rule, (–)methone can be unequivocally assigned the structure (3.12.9). It follows, therefore, that (–)menthol should be 3.12.1. According to the Sequence rule, the configurations at the chiral centres, C-1, C-2 and C-5 are, respectively, R, S and S.

Addendum

An alternative mode of elimination of water from menthol and neomenthol is the Chugaev reaction, which is a thermal reaction proceeding through a cyclic transition state. The substrates are menthyl xanthate (3.12.18) and neomenthyl xanthate (3.12.19). The stereochemical requirement for this elimination is that the groups being eliminated should be syn to each other. There are two such allignments in 3.12.18 whereas there is only one favourable orientation of the groups involved in 3.12.19. Therefore, menthyl xanthate undergoes facile Chugaev reaction to yield a mixture of menth-2-ene (3.12.15) and menth-3-ene (3.12.16), whereas neomenthyl xanthate yields only (3.12.15).

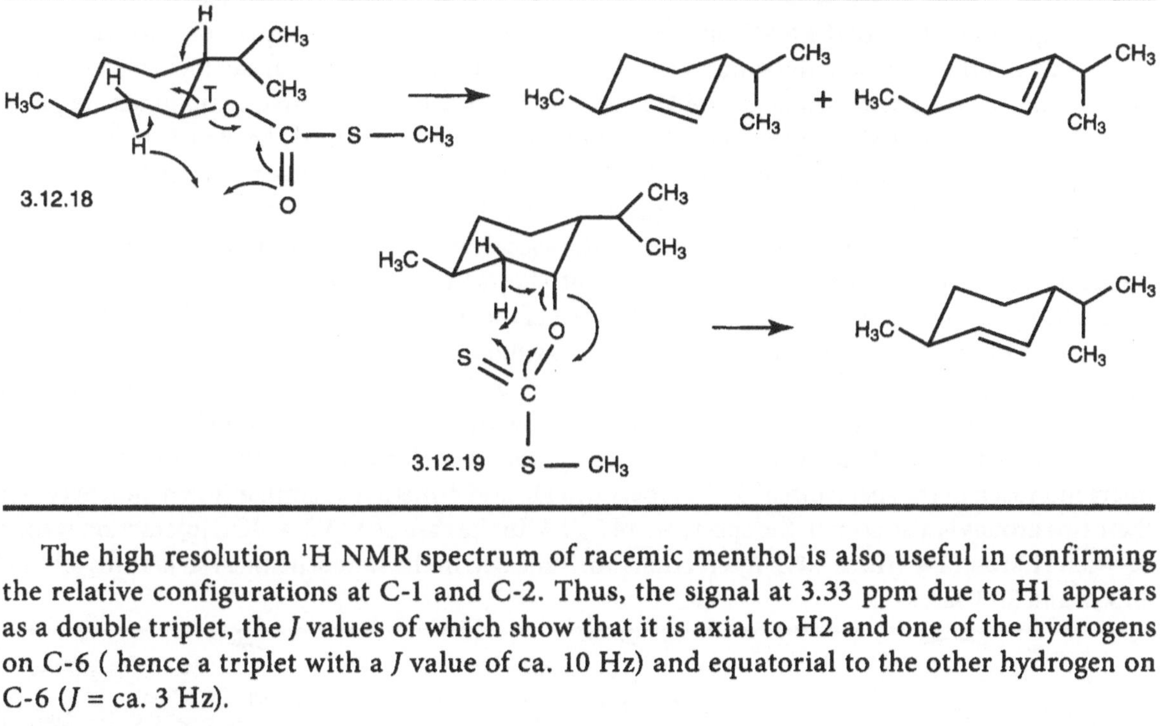

The high resolution ^1H NMR spectrum of racemic menthol is also useful in confirming the relative configurations at C-1 and C-2. Thus, the signal at 3.33 ppm due to H1 appears as a double triplet, the J values of which show that it is axial to H2 and one of the hydrogens on C-6 (hence a triplet with a J value of ca. 10 Hz) and equatorial to the other hydrogen on C-6 (J = ca. 3 Hz).

Suggested reading

1. Finar IL, *Organic Chemistry*, Vol. 2, 6th edn, London: Longmans Green, 1975.
2. Nasipuri D, *Stereochemistry of Organic Compounds: Principles and Applications*, New Delhi: New Age Publishers, 1994.
3. Sell C, *A Fragrant Introduction to Terpenoid Chemistry*, London: RSC Publication, 2003.
4. McCann JL, Rauk A, and Wieser H, *Can. J. Chem.*, 1998, 76: 274–83.

3.13 STEREOCHEMISTRY OF THE KAMAHINES

The diastereoisomeric nor-sesquiterpenoids, kamahines A, B and C, were isolated from the honey of a New Zealand tree, *Weinmania racemosa* (kamahi). The structure of the acetate of kamahine C, which could be obtained in the pure crystalline state, was first established as (3.13.1) by x-ray diffraction and NMR spectroscopy. This was then used as the model for determining the relative and absolute stereochemistry of kamahines A and B as described below.

3.13.1: R_1 = OAc; R_2 = R_3 = H; R_4 = CH$_3$

3.13.2: R_1 = R_3 = H; R_2 = OAc; R_4 = CH$_3$

3.13.3: R_1 = R_4 = H; R_2 = OAc; R_3 = CH$_3$

3.13.4: R_1 = OH; R_2 = R_3 = H; R_4 = CH$_3$

3.13.5: R_1 = R_3 = H; R_2 = OH; R_4 = CH$_3$

3.13.6: R_1 = R_4 = H; R_2 = OH; R_3 = CH$_3$

A comparative study of the NMR spectra of the acetates of the three compounds established that all the three of them possess the same skeletal structure. They differ only in stereochemical details and are, therefore, diastereoisomers. The 1H and 13C NMR data for the three compounds are given in Table 3.3. On the basis of these data, kamahine A acetate is assigned the structure 3.13.2 and kamahine B acetate, the structure 3.13.3. For determining the relative stereochemistries, 1D NOE enhancements and vicinal coupling constants were used. Thus, the enhancement seen in the spectra of all the three compounds in the intensities of the signals due to 2' methyl and H3, when either of them is irradiated in a double resonance experiment, shows that the stereochemical disposition of the spiro tetrahydrofuran moiety with respect to the rest of the structure is the same in all of them. In the spectra of 3.13.1 and 3.13.2, irradiation of H3a brings about an enhancement in the signal intensity of 4-methyl. On the other hand, in the case of 3.13.3, a similar enhancement of 4-methyl ensues on irradiation of H3b. These observations clearly show that while 3.13.1 and 3.13.2 have identical configuration at C-4, 3.13.3 has the opposite configuration at this chiral centre. Another key NOE interaction seen in the spectrum of 3.13.1 is between H5 and 4-methyl. A similar interaction between these two groups is also seen in the spectrum of 3.13.3. In the case of 3.13.2, a NOE interaction is seen between H3b and H4. These NOE interactions are indicated in the stereochemical structures 3.13.1a, 3.13.2a and 3.13.3a.

3.13.1a   3.13.2a   3.13.3a

Since, the stereochemical structure of the acetate of natural kamahine C has been established as 3.13.1, on the basis of x-ray diffraction studies, those of the acetates of kamahine A and kamahine B

Table 3.3 IH and 13C NMR data for kamahines A, B and C

Carbon	3.13.1		3.13.2		3.13.3	
	δ^{13}c	δ' H, J(Hz)	δ^{13}c	δ' H J(Hz)	δ^{13}c	δ' H, J(Hz)
H–CH$_3$	16.1	1.10, d, 6.9	12.0	0.98, d, 6.7	16.8	1.16, d, 7.6
8'–CH$_3$ ax	18.1	1.00, s	18.0	0.99, s	17.9	0.99, s
2'–CH$_3$	20.3	2.06, d, 1.5	20.5	2.08, d, 1.6	20.5	2.08, d, 1.6
C̲H$_3$CO	21.3	2.12, s	21.2	20.9, s	21.3	2.06, s
8'–CH$_3$ eq	21.8	1.27, s	21.7	1.27, s	21.7	1.29, s
4	36.8	2.51, m	35.8	2.59, m	38.2	2.25, m
3	40.5	H3a, 1.66, d 13.0, 9.9 H3b, 2.09 dd, 13.0, 7.3	38.5	H3a, 1.80, dd, 12.6, 9.9 H3b, 1.88 dd, 12.6, 6.7	37.7	H3a, 2.25, dd 12.0, 7.6 H3b, 1.62 d, 12.0
8'	52.0	-	51.9	-	51.7	-
1'	84.9	2.90, s(broad) OH	84.4	2.62, s(br), 1'-OH	84.9	2.66, s(br) OH
5'	89.9	3.98, d, 2.2	89.9	3.94, d, 2.1	90.1	3.96, d, 2.2
5	103.6	5.85, d, 4.8	99.5	6.39, d, 4.8	104.6	6.09, s
7'	114.6	-	115.8	-	117.2	-
3'	125.4	5.91, dq, 2.2, 1.5	125.3	5.89, dq, 2.1, 1.6	125.3	5.89 dq, 2.2, 1.6
2'	164.8	-	165.3	-	165.3	-
CH$_3$C̲O	170.8	-	169.9	-	169.8	-
4'	194.6	-	194.5	-	194.7	-

are deduced as 3.13.2 and 3.13.3. The parent alcohols can, therefore, be given the structures 3.13.4 (kamahine C), 3.13.5 (kamahine A) and 3.13.6 (kamahine B). The absolute configurations at the chiral centres are 1'R and 2R in all the three compounds, 4R, 5R and 5'S in 3.13.4, 4R, 5S and 5'S in 3.13.5 and 4S, 5S and 5'S in 3.13.6.

This example brings out the value of coupling constants and NOE in stereochemical analysis.

Suggested reading

Broom SJ, Wilkins AL, Lu Y, Ede RM, *J. Org. Chem.*, 1994, 59: 6425–30.

3.14 ABSOLUTE CONFIGURATION OF BENZYL ISOQUINOLINE ALKALOIDS

Several benzyltetrahydroisoquinoline alkaloids occur in nature. They possess a chiral centre, namely position 1 in the tetrahydroquinoline part. These include (+)laudonosine (3.14.1) (a minor component of *Papaver somniferum*), (+)romneine (3.14.2), isolated from *Romneya colteri* and (+)reticuline (3.14.3). The *N*-desmethyl derivative of (+)laudanosine (norlaudanosine) (3.14.4), was obtained along with its enantiomer, from papaverine (3.14.5) by electrolytic reduction followed by optical resolution. It is interesting to note that (3.14.4) is laevo rotatory but has the same configuration at C-1 as (3.14.1) as on *N*-methylation it yields the latter.

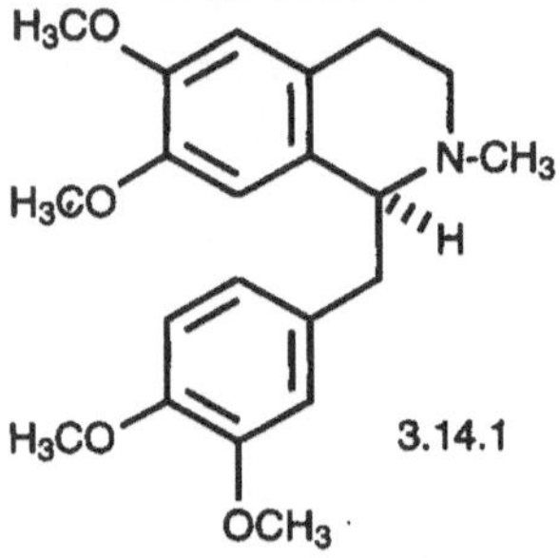

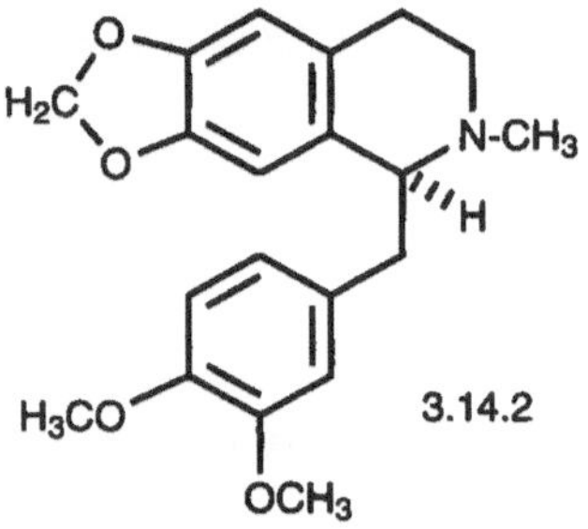

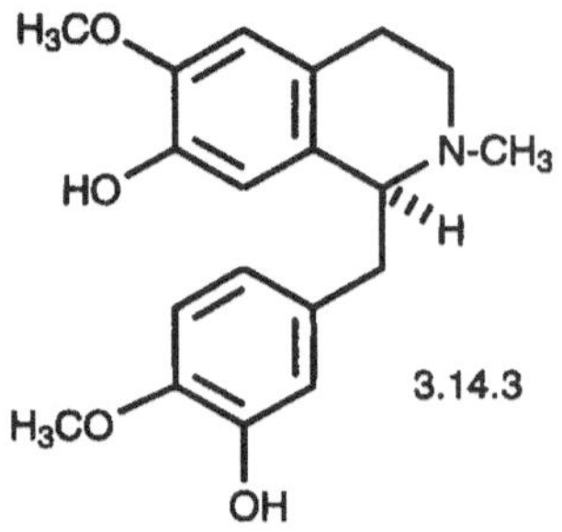

(–)Norlaudanosine was then subjected to *O*-demethylation followed by ozonolysis, when a tricarboxylic acid (3.14.6) was obtained; in this reaction, the two benzene rings are oxidatively opened up. This tricarboxylic acid could be prepared from (+)aspartic acid (3.14.7) by reaction with diazoethane. Therefore, working backwards from (3.14.7), the configuration at C-1 in (3.14.4), and hence, (3.14.1) could be established as S, as shown in the structure. This is an example of the classical, chemical approach for establishing the absolute configuration at a chiral centre by correlation with a reference compound of known stereochemistry (in this example, the reference compound is (+)aspartic acid).

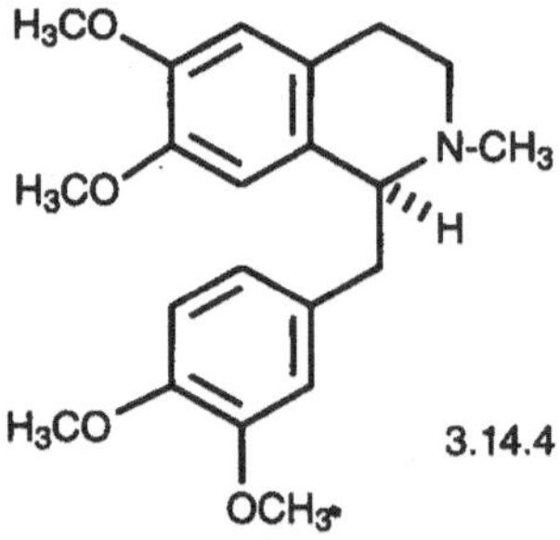

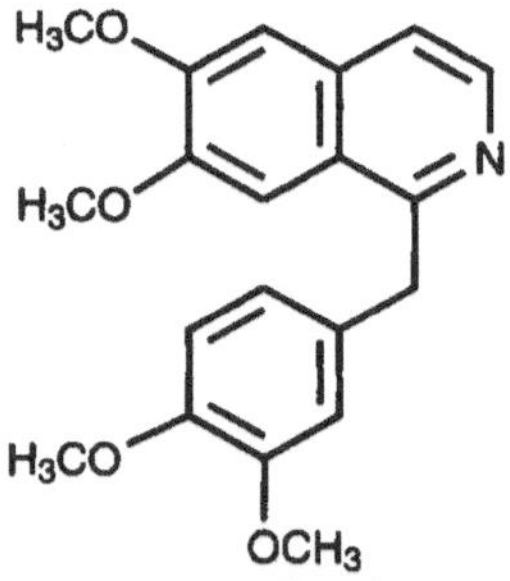

Further, using (+)laudonosine, it is possible to ascertain the stereochemistry of other naturally occurring benzyl isoquinolines by comparing their ORD spectra. Using this method, (+)orientaline is assigned the structure (3.14.8), whereas, (–)armepavine (3.14.9) is shown to have the opposite configuration (R). Benzylisoquinoline alkaloids having the (S) configuration at C-1 exhibit positive Cotton effect in their ORD curves, whereas those belonging to the (R) series show negative Cotton effect. The configuration of (+)coclaurine (3.14.10) deduced as (R) from its ORD spectrum has been confirmed by x-ray analysis.

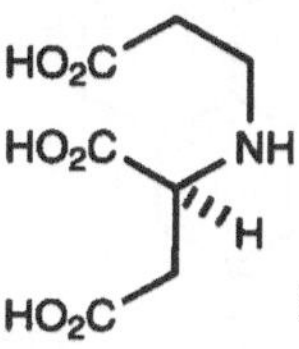

3.14.6

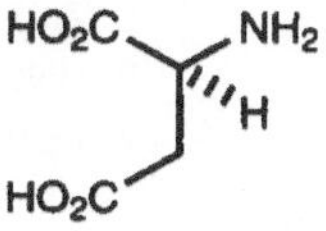

3.14.7

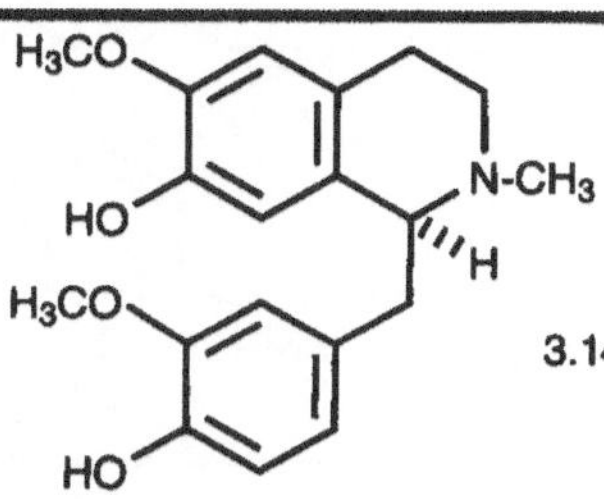

3.14.8

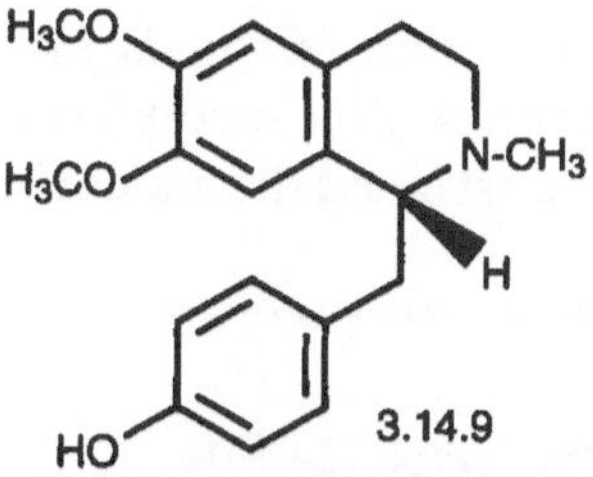

3.14.9

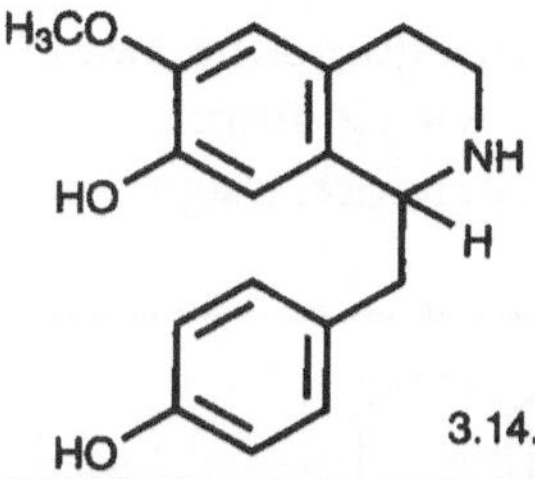

3.14.10

A series of reactions (Scheme 3.2) was used to convert (+)orientaline (3.14.8) into the aporphine alkaloid, (+)isothebaine (3.14.11), thus establishing the absolute stereochemistry of the latter compound. It is interesting to note that a key step in this sequence is an intramolecular oxidative coupling reaction. In this context it should be emphasised that the reactions chosen for such correlations should not involve the chiral centre.

Scheme 3.2

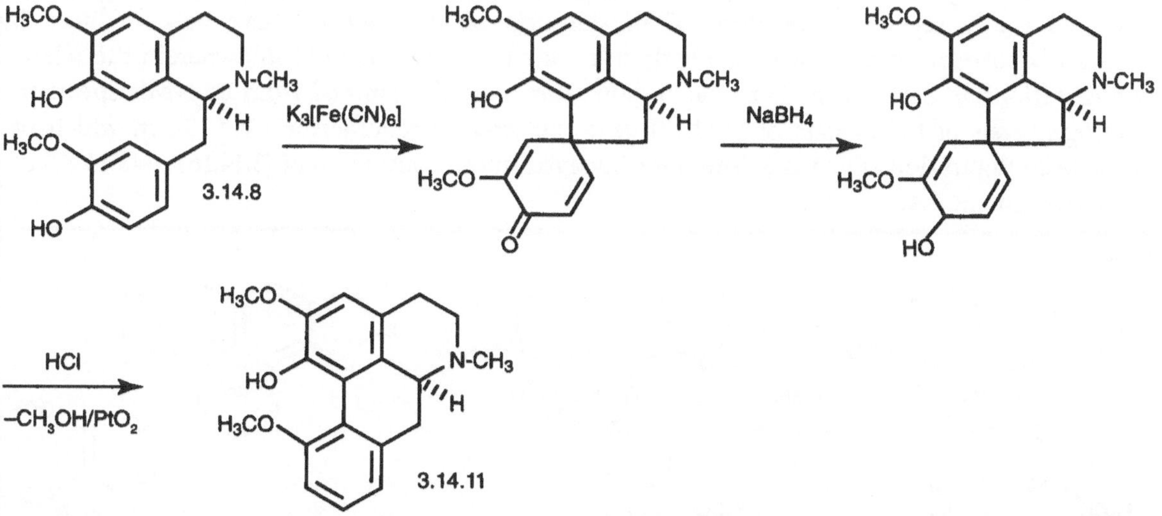

In another series of rections (Scheme 3.3), R (–)laudonosine (3.14.12) was transformed into another aporphine alkaloid, R (–)glaucine (3.14.13).

Scheme 3.3

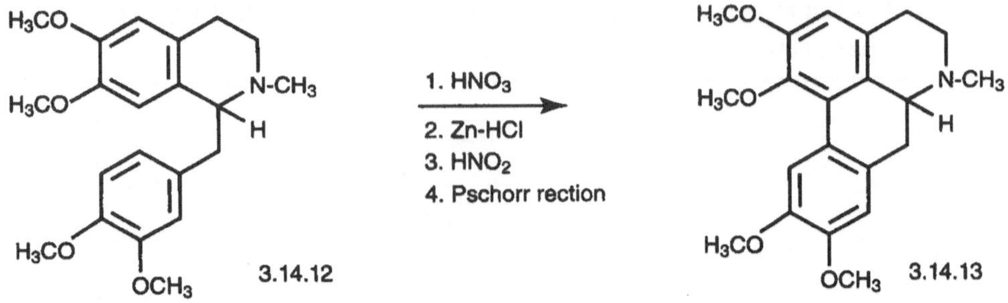

1. HNO₃
2. Zn-HCl
3. HNO₂
4. Pschorr rection

3.14.12   3.14.13

Magnoline (3.14.14) is a *bis*-benzylisoquinoline alkaloid isolated from *Magnolia fuscata*. Its tri-*O*-methyl ether yielded, on cleavage with sodium in liquid ammonia, (+)S *O*-methylarmepavine (3.14.15) and (–)armepavine (3.14.9). Therefore, magnoline could be assigned the stereochemistry shown in (3.14.14).

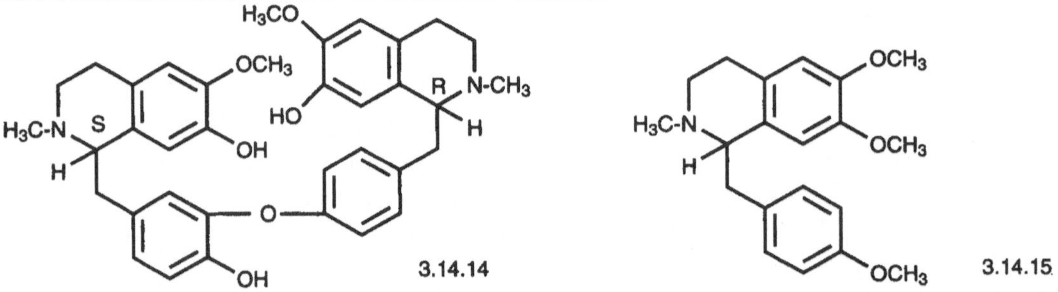

3.14.14   3.14.15

The pthalide tetrahydroisoquinoline alkaloids, such as narcotine (3.14.16), hydrastine (3.14.17) and related compounds possess two chiral centres, namely C-1 of the tetrahydroisoquinoline part and the benzylic carbon atom. Their stereochemical structures have been deduced by chemically correlating them with the hydroxytetrahydroberberine (3.14.18), wherein the relative configurations at C-1 and the benzylic carbon atom were determined from its NMR spectrum. Hydrogenolysis of the acetate of (3.14.18) gave the tetrahydroberberine (3.14.19), in which the absolute configuration at C-1 is S. The steps involved in the conversion of (3.14.16) to (3.14.19) are shown in Scheme 3.4.

Scheme 3.4

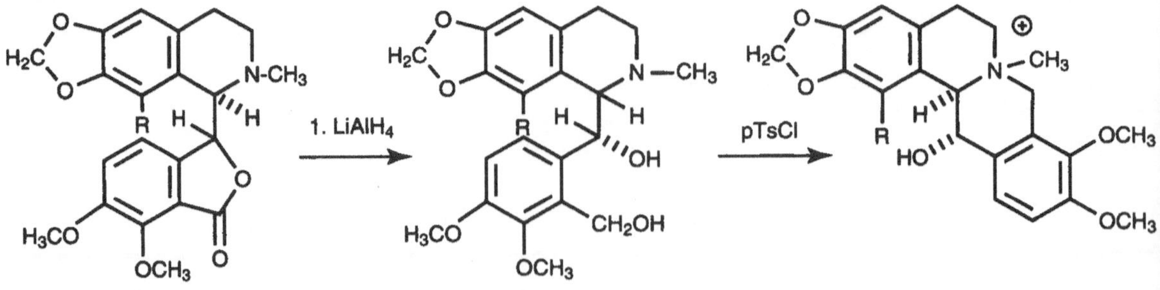

1. LiAlH₄   pTsCl

3.14.16: R = OCH₃
3.14.17: R = H

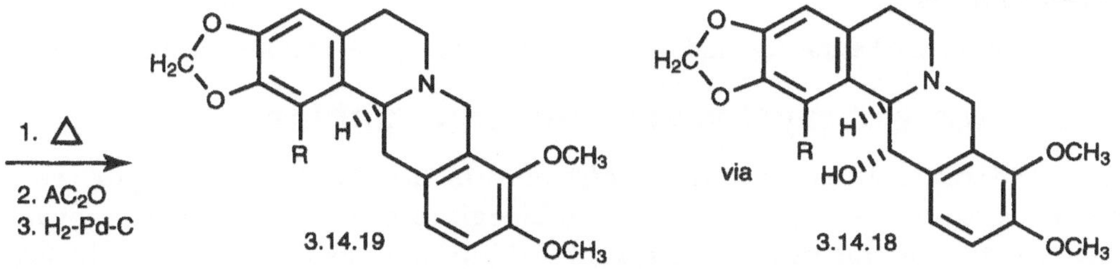

$$1.\ \Delta$$
$$2.\ AC_2O$$
$$3.\ H_2\text{-Pd-C}$$

3.14.19   via   3.14.18

Suggested reading

Nakanishi K, In Nakanishi K, Goto T, Ito S, Natori S and Nozoe S, (eds), *Natural Products Chemistry*, Vol.2, Tokyo: Kodansha Ltd; New York: Academic Press1975.

3.15 STEREOCHEMISTRY OF SOME INDOLE ALKALOIDS

Stereochemistry of the alkaloid peraksine

Peraksine was first isolated from *Rauwolfia perakensis*. Its structure was established as 3.15.1 based on a x-ray crystallographic study of its methiodide. Subsequently , an examination of its 1H nmr spectrum in $CDCl_3$-CD_3OD solution brought to light an interesting fact. The spectrum has two distinct signals, of the same intensity, at 4.58 and 4.98 ppm. The only proton which can give rise to a signal with this chemical shift value is the hydrogen atom attached to C-17 of the cyclic hemiacetal system in 3.15.1. That there are two signals in this region suggests that peraksine is a 1:1 equilibrium mixture (in solution) of two epimers which can be assigned the stereochemical structures 3.15.2 and 3.15.3. The signal at 4.58 ppm is a sharp singlet showing minimal coupling to the neighbouring hydrogen atom on C-16, indicating a dihedral angle of about 90° as in 3.15.2. The signal at 4.98 ppm, on the other hand, shows a small coupling (approximately 2Hz) and appears as a doublet as can be expected from the structure 3.15.3. On acetylation, these signals shift downfield to 5.53 and 5.92 ppm respectively. This example shows how a dynamic stereochemical feature, not apparent even from an x-ray analysis (which is often the ultimate method for structure delineation) is revealed by NMR spectroscopy and emphasise the need for a combination of techniques to unravel nuances of structural features of a compound.

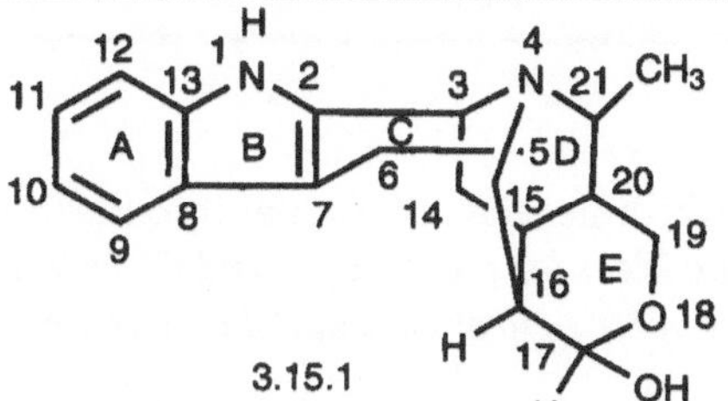

3.15.1 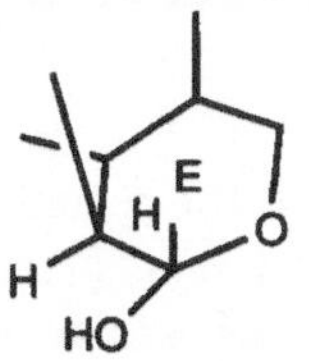   3.15.2 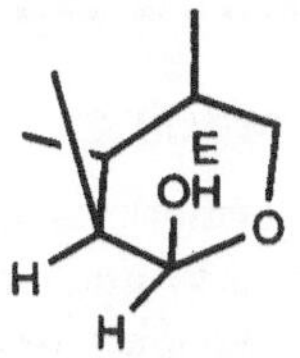   3.15.3

Suggested reading

Arthur HR, Johns R, Lambertonan A and Loot N, *Aust. J. Chem.*, 1968, 21: 1399–1401.

Stereochemistry of α-yohimbine and corynanthine

Yohimbine and corynanthine are epimers which differ in their pharmacological properties. While yohimbine is a selective antagonist of the a2 adrenergic receptor, corynanthine is an antagonist of the a1 receptor. Yohimbine also increases heart rate and blood pressure. Of the four chiral centres in these two compounds, namely C-3, C-15, C-16 and C-20, only the configuration at C-16 is different. In other words, yohimbine and corynanthine are C-16 epimers. From other studies, it is known that the rings D and E are *trans*-fused and H3 is alpha axial. Yohimbine undergoes hydrolysis at a faster rate than corynanthine suggesting that the carbomethoxyl group in yohimbine is equatorially oriented while in corynanthine it is in the axial position. This is supported by the observation that in the presence of methanolic base corynanthine gets converted into yohimbine. Therefore, yohimbine can be assigned the stereochemical structure 3.15.4 and corynanthine 3.15.5. Yohimbic acid (3.15.6) undergoes a base-catalysed diaxial elimination of a molecule of water to yield the olefinic compound (3.15.7), whereas under similar conditions, corynanthic acid (3.15.8) suffers dehydration as well as decarboxylation, as shown in the structure, to give (3.15.9). This example illustrates the application of chemical methods in stereochemical analysis in contrast to the example discussed above.

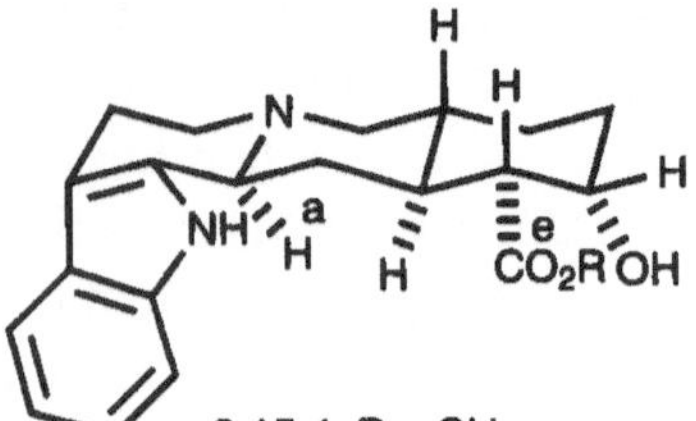

3.15.4: R = CH₃
3.15.6: R = H

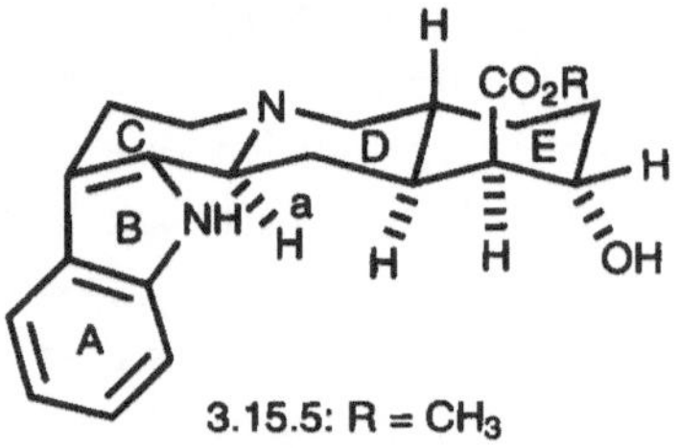

3.15.5: R = CH₃
3.15.8: R = H

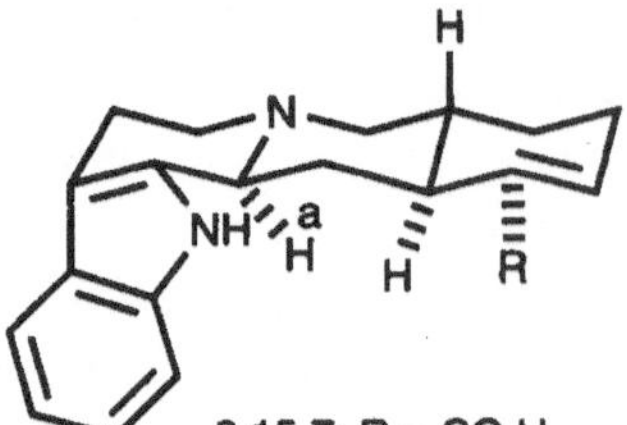

3.15.7: R = CO₂H
3.15.9: R = H

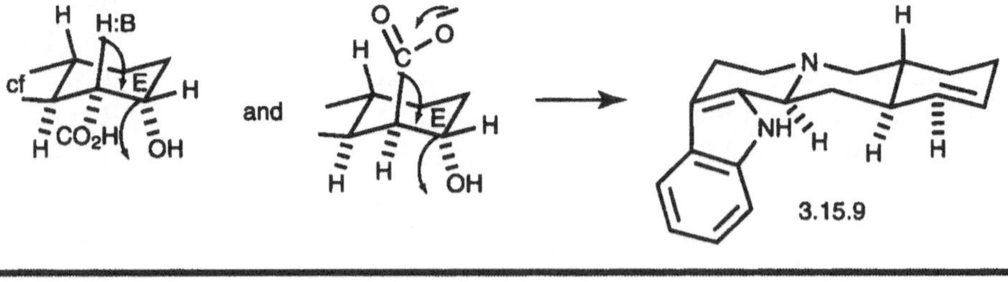

3.15.9

Suggested reading

Eliel EL, Conformational analysis: The elevation of two-dimensional formulas into the third dimension, In Fleischhacker W and Schönfeld T, (eds), *Pioneering Ideas for the Physical and Chemical Sciences: Josef Loschmidt's Contributions and Modern Developments in Structural Organic Chemistry, Atomistics and Statistical Mechanics*, New York: Plenum Press, 1997.

Configuration of C-3 in reserpine

Reserpine (3.15.10) was first isolated from the Indian snake root (*Rauwolfia serpentina*) by Schlittler and co-workers in 1952. Since then, it has been the subject of several studies covering varied aspects of the compound such as its structure, stereochemistry and synthesis. The compound has six chiral centres, namely C-3, C-15, C-16, C-17, C-18 and C-20. While it is closely related to yohimbine in its skeletal structure, it differs from the latter in more than one stereochemical aspect. Thus, while the rings D and E are *trans*-fused in yohimbine, they are *cis*-fused in reserpine. Specific reactions which bring out this stereochemical feature of reserpine are discussed in the next chapter. Another difference between reserpine and yohimbine is the configuration at C-3. As mentioned above, in yohimbine, the hydrogen atom at C-3 is alpha axially oriented. Yohimbine and yohimbic acid are stable and do not undergo any epimerisation. In distinct contrast, reserpic acid (3.15.11) and reserpine diol (3.15.12), which is obtained by the action of lithium aluminium hydride on reserpic acid, undergo epimerisation in acidic as well as basic media to give the corresponding 3-epimers. From these observations, it can be inferred that in reserpic acid, reserpine diol and in reserpine itself, the configuration at C-3 is unstable, with the hydrogen atom at this position taking the beta equatorial orientation as shown in the structures (3.15.10), (3.15.11) and (3.15.12). It is worth mentioning here that isoreserpine lacks the characteristic pharmacolgical properties of reserpine.

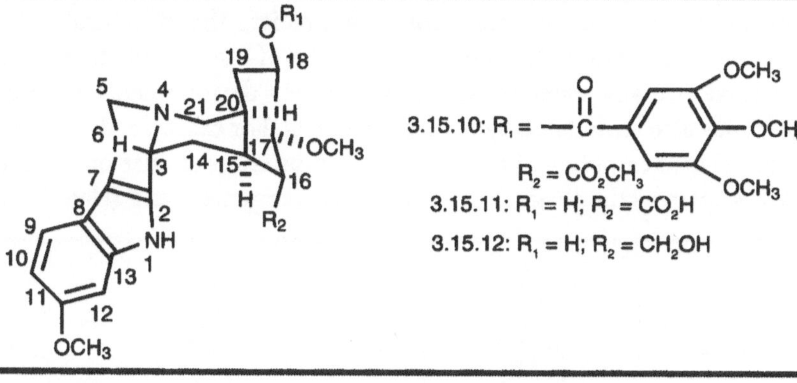

3.15.10: $R_1 = -\overset{O}{\underset{\parallel}{C}}-$ with OCH_3 substituted aromatic ring

$R_2 = CO_2CH_3$

3.15.11: $R_1 = H$; $R_2 = CO_2H$

3.15.12: $R_1 = H$; $R_2 = CH_2OH$

Suggested reading

1. Schlittler E, In Manske RHF, (ed.), *The Alkaloids*, Vol. 8, New York: Academic Press, 1965.
2. Cordell GA, *Introduction to Alkaloids—A Biogenetic Approach*, New York: John Wiley, 1981.

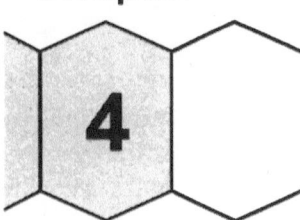

Chapter

4

Reactions and Rearrangements

4.1 Introduction

As the Nobel laureate Sir Robert Robinson once pointed out it is not enough to know the structures of naturally occurring organic compounds. What is more significant is understanding how they behave either in a chemical milieu or in a biological environment. Exposing a compound to a variety of chemical reactions can unravel its chemistry and bring out its chemical potential in full. It is essential to understand the mechanisms of the reactions to appreciate the structural and stereochemical parameters controlling the reactions. The pioneering studies of Ingold and others laid the foundation of modern mechanistic organic chemistry. As with other developments in organic chemistry, research in this area was also stimulated by natural products. For example, the transformation of the terpene borneol (4.1.1) into camphene (4.1.2) triggered numerous studies on the mechanism of the rearrangement now known as the Wagner–Meerwein rearrangement and related reactions.

To rigorously establish the mechanism of an organic reaction, several factors have to be determined. Kinetic experiments, determination of thermodynamic parameters, detection and trapping of reactive intermediates (such as carbanions, carbocations, carbenes and radicals), if any, tracing the reaction pathway using labelled compounds are some of the techniques which provide data needed for this purpose. However, in most cases, the proposed mechanisms are only suggestive pathways based on factors such as the structural and stereochemical features of the substrate undergoing transformation, the reaction conditions, the nature of the reagents and catalysts, if any.

Reactions can be intermolecular or intramolecular. The latter type is particularly influenced by stereochemical dispositions of the groups, within the molecule, which interact. Reactions can be further classified as addition, elimination, substitution and condensation reactions and rearrangements. While all reactions have to proceed through transition states, not all of them proceed through reactive intermediates. Pericyclic reactions, for example, do not go through any intermediates but proceed through cyclic transition states. Reactive functionalities in naturally occurring compounds are the hydroxyl, amino, multiple bonds, the carbonyl group, epoxide group and aromatic ring systems.

Molecular rearrangements are particularly interesting as they often bring out the nuances in structure and stereochemistry of the compounds undergoing such reactions. Apart from the Wagner–Meerwein and the related Nemetkin rearrangements, other rearrangements met with in studies on natural products of different types are the Baeyer–Villiger oxidative rearrangement, the dienone–phenol rearrangement , the Baker–Venkataraman and the Wesley–Moser rearrangements. Examples of the dienone–phenol and Wesley–Moser rearrangements are seen in two of the case studies which follow this introduction. A variation of the Baeyer–Villiger reaction is Dakin oxidation in which an aromatic carbonyl group, such as in salicylaldehyde (4.1.3) is oxidised by alkaline hydrogen peroxide to the formyl ester of catechol (4.1.4). This reaction has been used effectively by Seshadri and his co-workers for the introduction of an additional hydroxyl group in flavonoids. For example, 7-acetoxyflavone (4.1.5) on Fries rearrangement followed by a Dakin oxidation and subsequent hydrolysis yields 7,8-dihydroxyflavone (4.1.6); the intermediate compound is 7-hydroxy-8-acetylflavone (4.1.7). The Baker–Venkataraman reaction constitutes an important method of synthesis of flavones. It involves an intramolecular Claisen reaction and proceeds through a diketone intermediate. For example, when the benzoyl ester of 2-hydroxy-4,6-dimethoxyacetophenone (4.1.8) is treated with a base (potassium carbonate in acetone is a convenient reagent for this purpose), the diketone (4.1.9) is obtained. It then undergoes cyclisation to yield the flavone (4.1.10).

4.1.3   4.1.4   4.1.5   4.1.6

4.1.7   4.1.8   4.1.9

4.1.10

In recent years, light-induced organic reactions of natural products have been extensively studied. One of the earliest examples was the photochemical reaction of ergosterol (4.1.11). As early as 1932, Windaus and his co-workers made a systematic study of this reaction and isolated several products including lumisterol (4.1.12), tachysterol (4.1.13) and ergocalciferol (4.1.14), which possesses antirachitic activity. Subsequent studies have removed some of the anomalies in the earlier schemes proposed by Windaus and others. It is now accepted that the first step is a reversible light-induced conversion of ergosterol to precalciferol (4.1.15). The reaction involved is a electrocyclic ring opening by the conrotatory mode. An alternative electrocyclic ring closure of 4.1.15 results in lumisterol (4.1.12). The conversion of precalciferol to ergocalciferol (4.1.14) is a thermal (1-7) antarafacial sigmatropic rearrangement. On the other hand, tachysterol (4.1.13), which is a geometrical isomer of precalciferol, is formed from the latter on irradiation. Other well known photochemcial reactions of natural products include the photochemical reactions of α-santonin, discussed later in this chapter, and the photochemical conversion of the alkaloid colchicine (4.1.16) into α-, β- and γ-lumicolchicines (4.1.17), (4.1.18) and (4.1.19). β- and γ-lumicolchicines are formed from colchicine by electrocyclic reactions involving two alternative disrotative modes. α-lumicolchicine is a product of cycloaddition of β-lumicolchicine.

With this brief introduction we shall now consider a few case studies which illustrate some of the features mentioned above.

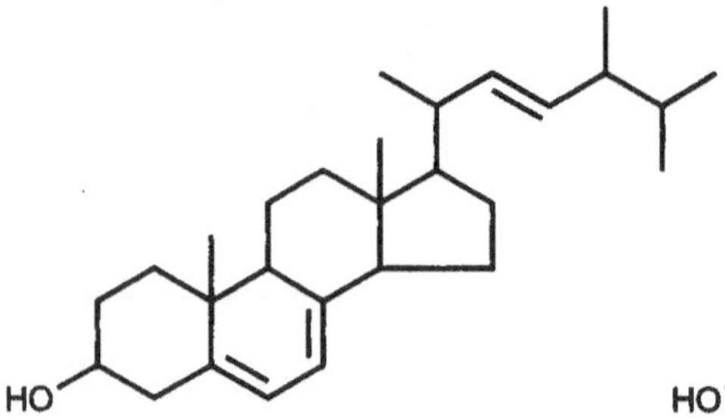

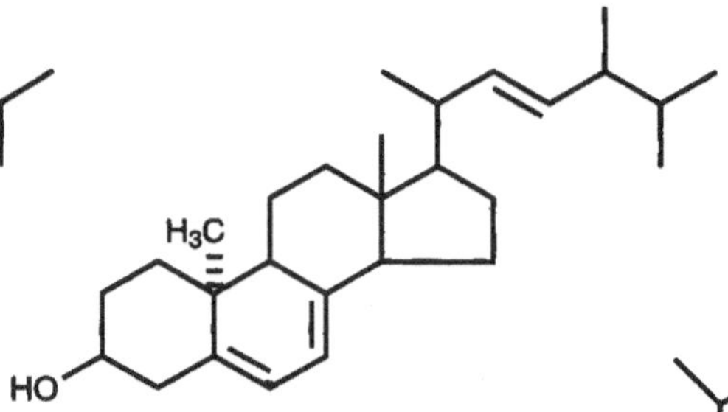

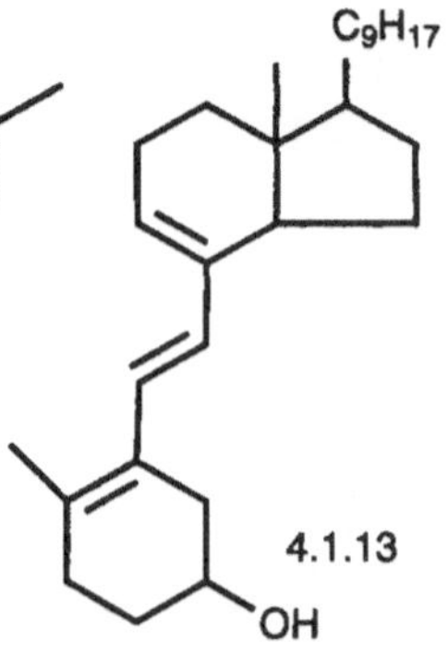

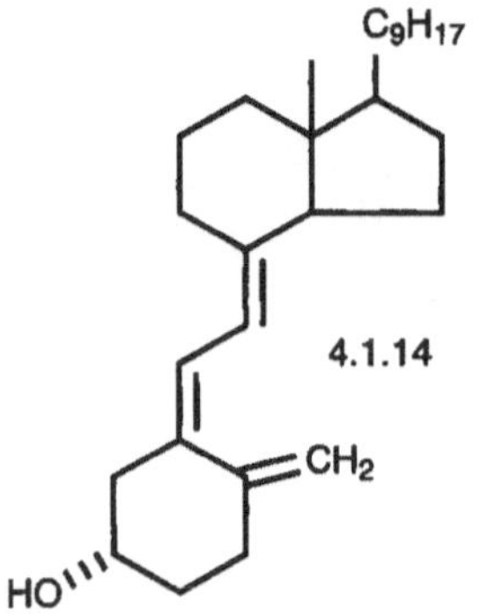

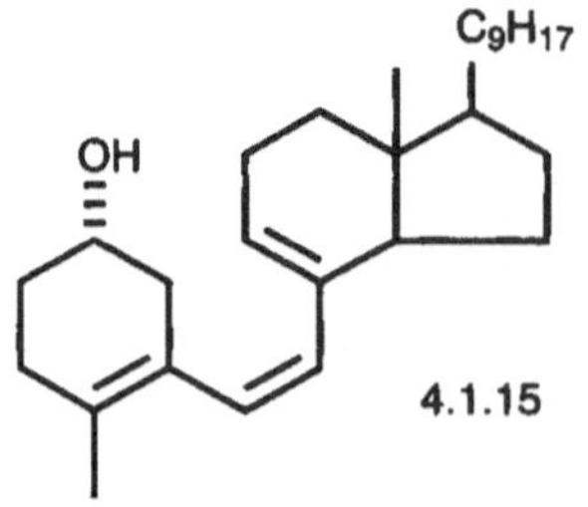

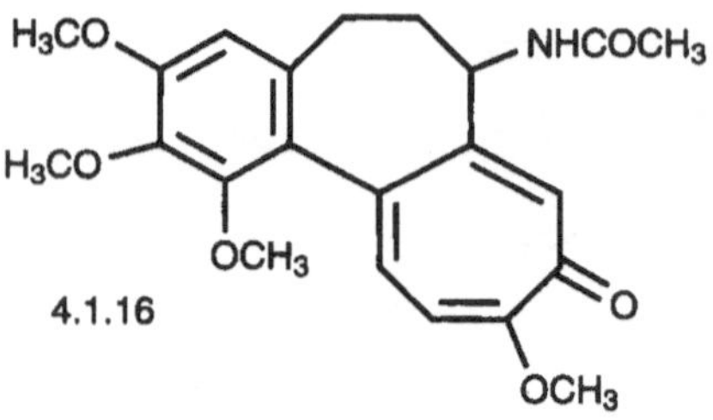

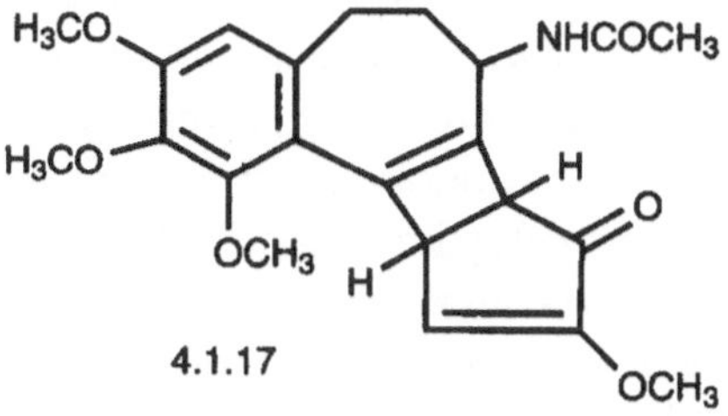

4.1.17

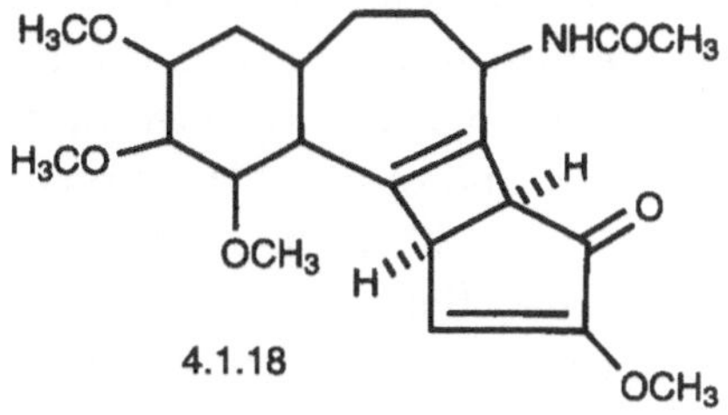

4.1.18

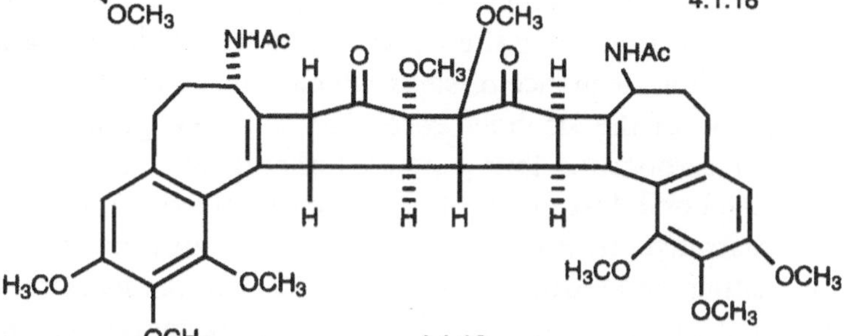

4.1.19

Suggested reading

1. Jacobs A, *Understanding Organic Reaction Mechanisms*, Cambridge University Press, 1997.
2. Sykes P, *A Guidebook to Mechanisms in Organic Chemistry*, Prentice Hall, 1996.
3. March J, *Advanced Organic Chemistry, Reactions, Mechanisms and Structure*, McGraw Hill.
4. Nakanishi K, Goto T, Ito S, Natori S and Nozoe S, (eds), *Natural Products Chemistry*, Vols 1–3, New York: Kodansha, Tokyo and Academic Press, 1975, 1983.

4.2 REARRANGEMENT REACTIONS OF MORPHINE

Morphine (4.2.1) is perhaps the only chemical which has been named after one Greek god and compared to another! It earned its name because it can induce sleep (in therapeutic doses it is an analgesic) and has been compared to Proteus as it shows a remarkable degree of aptitude for changing form and shape. This tendency towards skeletal rearrangement was traced to a 'weak link' in its structure, wherein there is an axially oriented C–C bond. Consequently, when appropriately 'provoked', morphine and its relatives [codeine (4.2.2) and thebaine (4.2.3)] rearrange themselves to assume thermodynamically more stable skeletal structures. We shall now examine a few of these reactions.

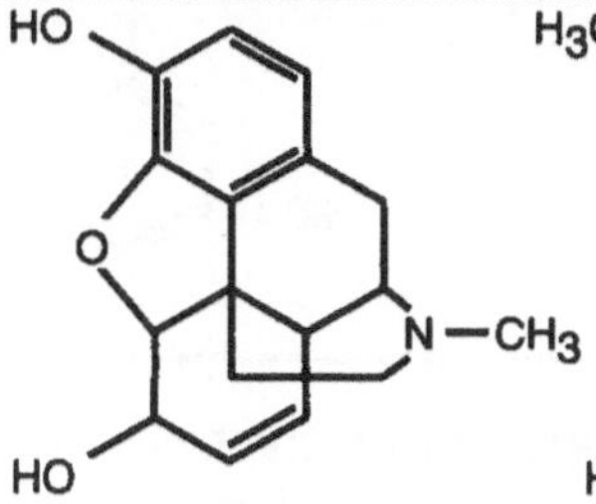

4.2.1

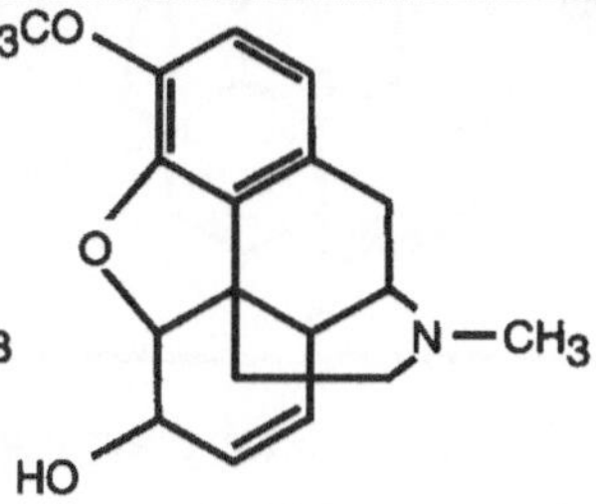

4.2.2

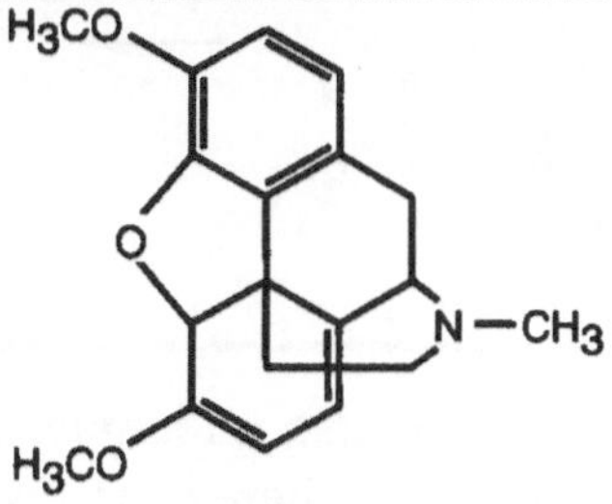

4.2.3

The most well known of the rearrangements of morphine is the acid catalysed reaction resulting in the formation of apomorphine (4.2.4). The process apparently involves the loss of a molecule of water but is accompanied by a skeletal change which was not apparent to earlier investigators (Pschorr and others). The structure of apomorphine was elucidated by routine methods but its correlation with morphine proved elusive. Later, after Robinson had arrived at the correct structure of morphine, the rearrangement to apomorphine could be understood and explained in terms of mechanistic principles. This reaction has instructive value as it is a consequence of the play of certain basic principles underlying the behaviour of allylic alcohols and ethers, and bridged alicyclic systems. The whole sequence of steps which cause a ripple in the original structure is triggered by protonation of the alcoholic secondary hydroxyl group, which is eliminated as a molecule of water. The carbocation thus generated undergoes a series of changes, which include a shifting of the double bond, loss of a proton, proton-catalysed rupture of the ether bridge, and Wagner–Meerwein rearrangements as shown below. It must be noted that this is not a rigorously proved mechanism, either with regard to the precise course or the sequence of events but it reflects the overall chemistry of the molecule worked out into a behavioural pattern.

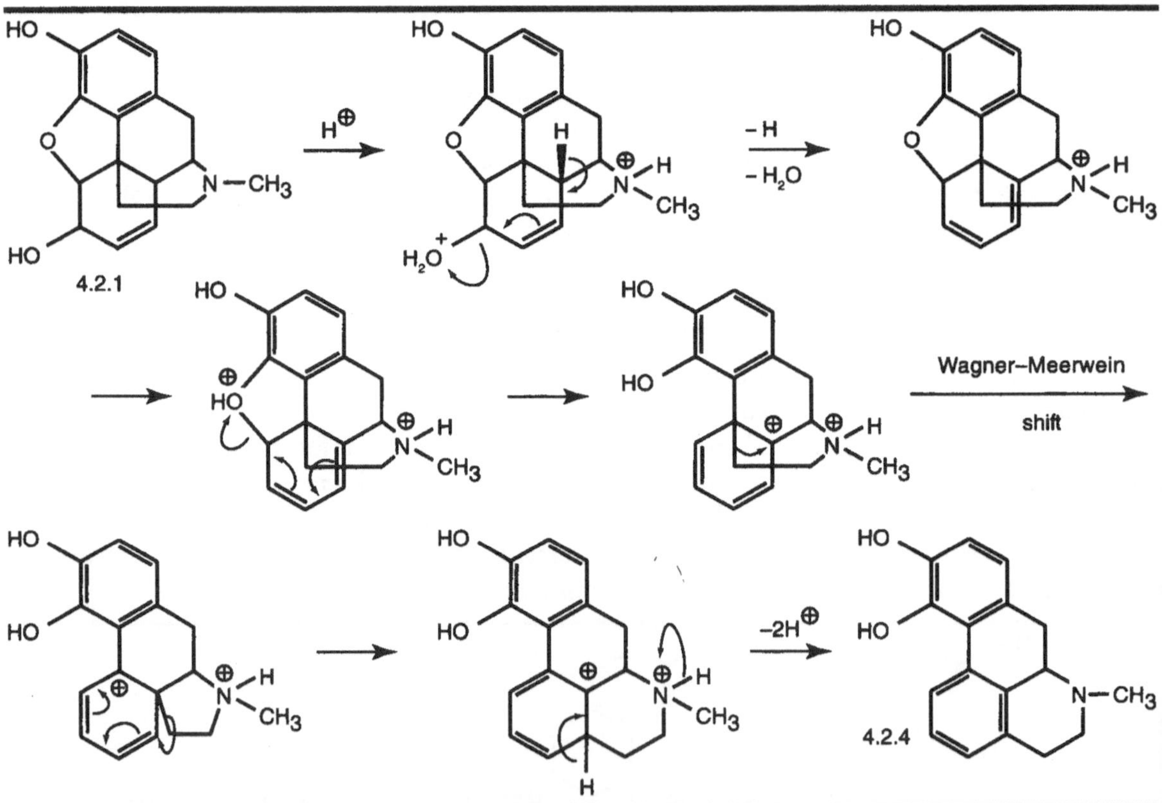

Thebaine (4.2.3) also undergoes a similar skeletal rearrangement on treatment with mineral acids, but in this case there is no loss of any oxygen atom. Apart from skeletal change, the other reactions which occur during this formation of morphothebaine (4.2.5) are the demethylation of the enol methyl ether and the opening of the oxide bridge.

Another interesting rearrangement of thebaine is seen in the formation of thebenine (4.2.6). The initial stages in this reaction sequence are the same as those leading to morphothebaine. In the final stages, instead of a Wagner–Meerwein rearrangement, there is a rupture of the C13–N bond followed by other changes shown below. This reaction is also catalysed by acids.

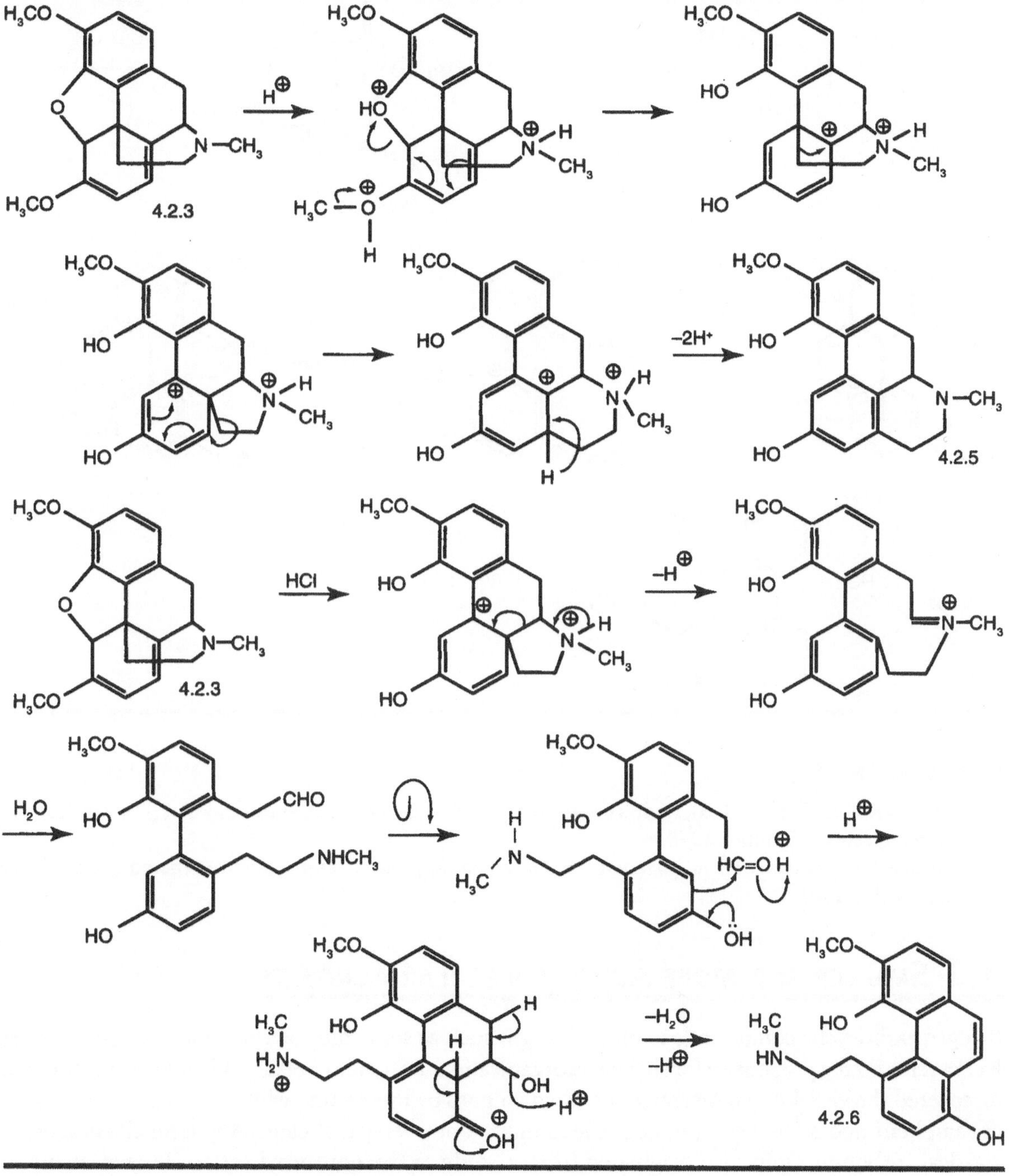

Skeletal transformation of thebaine can also be brought about by the action of phenyl magnesium iodide as shown below. As can be seen from this scheme, this rearrangement initially follows the same course as the one seen in the formation of 4.2.5. The final product in this case is 4.2.7.

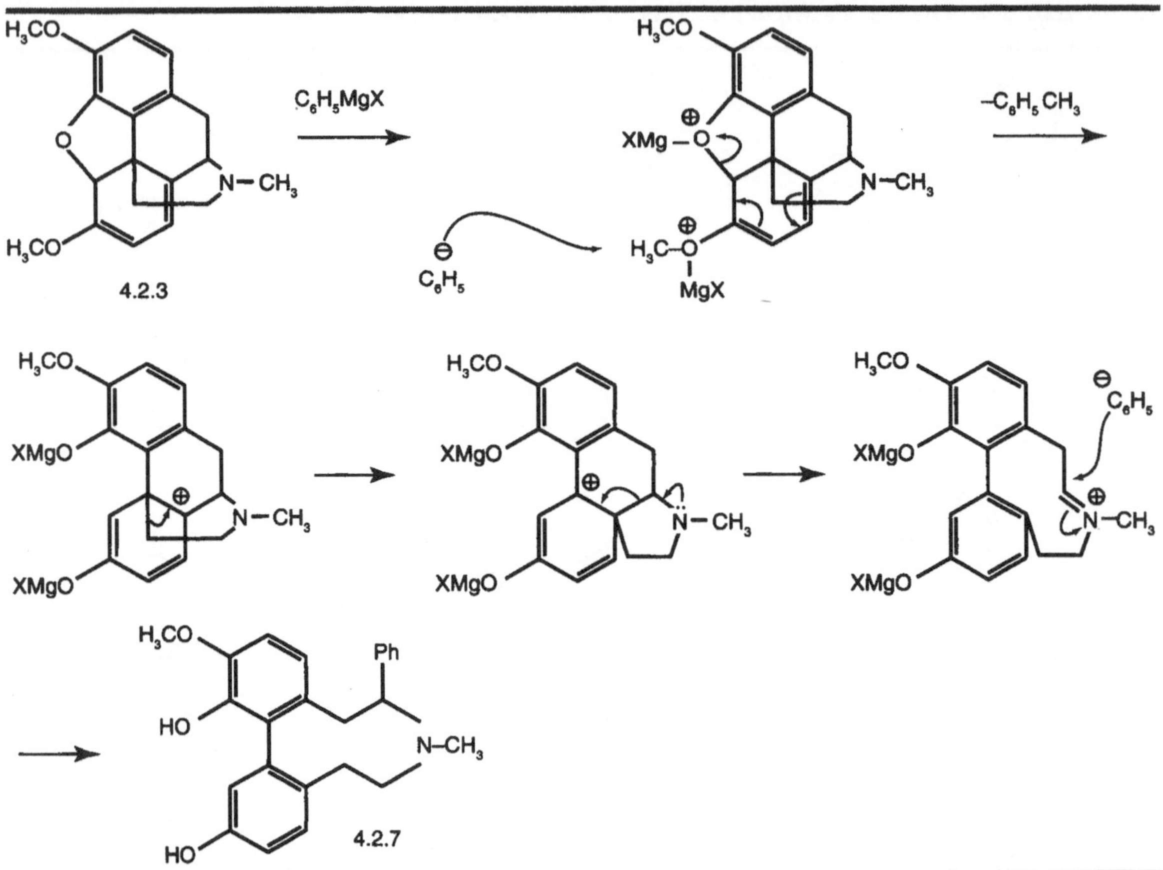

Suggested reading

1.   Bentley KW, In Coffey S, (ed), Rodd's *Chemistry of Carbon Compounds*, 2nd edn, Vol. 4 (G), Amsterdam: Elsevier Science Limited, 1978.
2.   Cordell GA, *Introduction to Alkaloids—A Biogenetic Approach*, (and references therein), New York: John Wiley, 1981.

4.3 SMALLER AND MORE AGILE MOLECULAR ACROBATS

Bridged carbocyclic compounds of natural origin such as some monoterpenes and sesquiterpenes have a remarkable propensity for skeletal reorganisation. As a matter of fact, Wagner and Meerwein discovered the well-known rearrangement which goes by their name while studying the reactions of camphene and related compounds. The famous controversy that centred around classical-non-classical carbonium ions also originated from a study of the norbornyl cation. Indeed, many of the simpler cyclic terpenoids are so agile that the determination of their structures by classical chemical methods proved difficult. Camphor was one such molecule which proved to be a hard nut to crack. In the following paragraphs, we shall examine a few interesting examples chosen from among the monoterpenoids.

Even acyclic monoterpenes have been known to undergo facile rearrangements yielding, for example, cyclic compounds. Such reactions can be brought about either by heat or by irradiation with a suitable light source. The pathways of some of these reactions can be understood in terms of the Woodward–Hoffmann rules of conservation of orbital symmetry. For instance, citral (as a mixture of geranial and neral) (4.3.1) undergoes a [1,5] sigmatropic hydrogen shift when heated to about 200°C. The result is the apparently surprising shift of the conjugated double bond to a position farther away from the carbonyl group. However, 4.3.2 is not the end-product as it suffers a facile *ene*-cyclisation to give a diastereoisomeric mixture of menthadienols (4.3.3). Thus, the two-step rearrangement involves a thermodynamically uphill but kinetically allowed stage followed by a favourable downhill step, the overall reaction being thermodynamically and mechanistically allowed.

Light-induced skeletal rearrangements of acyclic monoterpenes are equally fascinating and instructive. For example, myrcene (4.3.4) on exposure to UV light gives a mixture of three compounds, one of which is the bicyclic monoterpene, β-pinene (4.3.5). The other two are a cyclobutene derivative (4.3.6) and a bicyclo [2.1.1] hexane derivative (4.3.7). This reaction is, evidently, not one single pathway but a composite of three simultaneous and parallel reactions, not all of which are concerted. Thus, two isomeric diradicals, 4.3.8 and 4.3.9, are presumably involved in the formation of 4.3.5 and 4.3.7. On the other hand, the cyclobutene (4.3.6), which is the major product (with a yield of 52%), is probably formed by a symmetry-allowed photochemical electrocyclic reaction. These conjectures are supported by the observation that in the presence of a triplet sensitiser such as acetophenone or benzophenone, myrcene yielded 4.3.7 as the only isolable product (75% yield).

Photochemical geometrical isomerisations are often known to precede further skeletal transformations of appropriately structured olefins. An illustrative example from among natural products is shown. Irradiation of methyl geranate (4.3.10) and methyl nerate (4.3.11) initially brings about a geometrical inversion (E to Z or Z to E as the case may be) and this results in the formation of an equilibrium mixture or photostationary state. Simultaneously and independently, each of the two isomers suffers enolisation and a consequent shift of the conjugated double bond; the ultimate result is the formation of the non-conjugated esters, 4.3.12, 4.3.13 and 4.3.14. Prolonged irradiation leads to the formation of monocyclic and bridged bicyclic products.

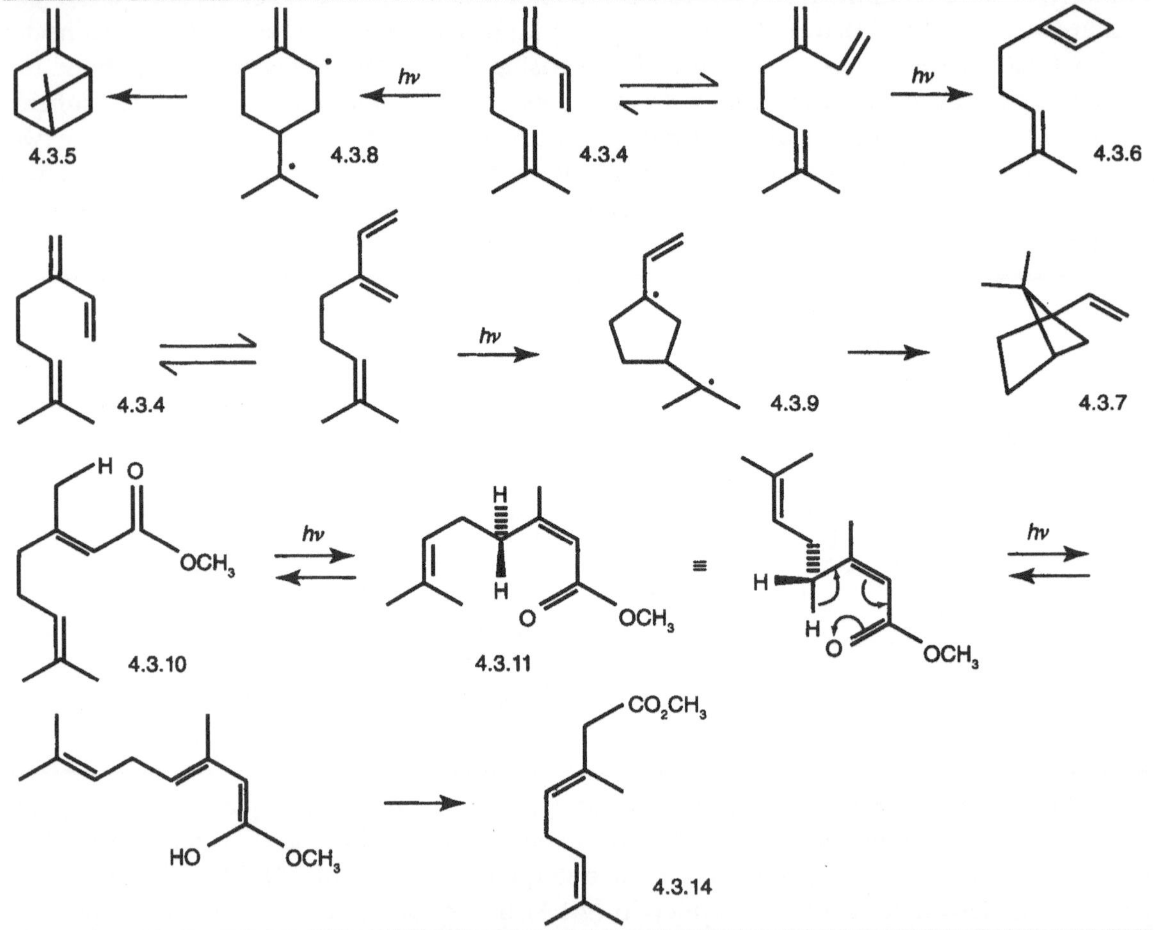

Among monocyclic monoterpene hydrocarbons, one of the most extensively studied compounds is (+)R-limonene (4.3.15), which is a component of orange and lemon oils. One reaction of 4.3.15 which has been very well investigated is the acid-catalysed rearrangement. The products formed depend upon the conditions and the nature of the catalyst used. One major pathway is the proton-catalysed migration of double bonds leading to the formation of as many as seven different dienes, one of which is (−)S-limonene (4.3.16); the others are 4.3.17 to 4.3.22, formed by an initial protonation of the isopropenyl grouping followed by further prototropic shifts as shown on the next page.

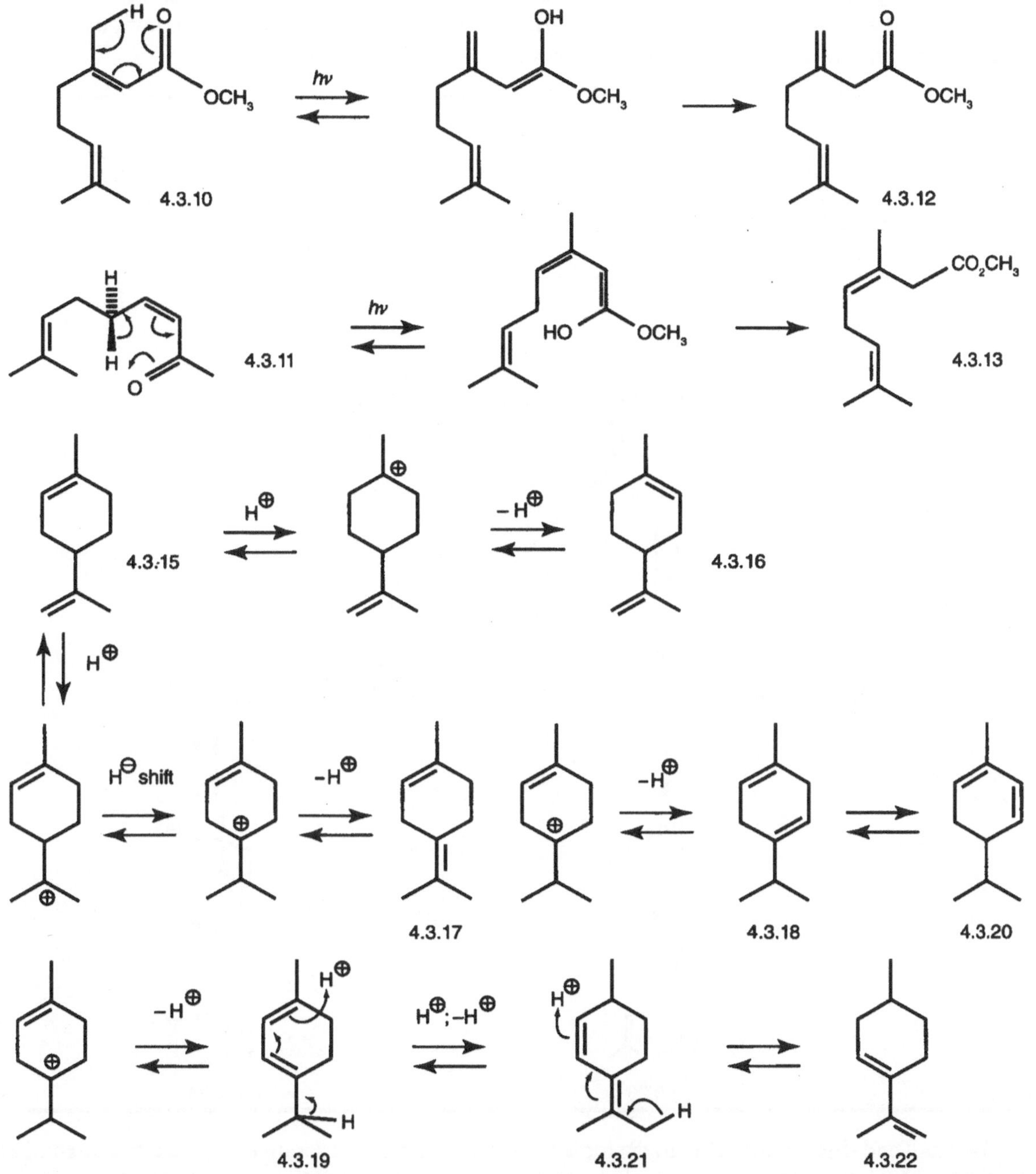

Cyclisation of 4.3.15 to bicyclic compounds under the influence of an acid catalyst has also been observed. Thus, when 4.3.15 is passed over silicophosphoric acid at 200–210°C, four bridged bicyclic compounds, 4.3.23 to 4.3.26 are formed. It has been suggested that the endocyclic double bond in 4.3.15 adds on to the isopropenyl grouping, with the cyclohexene ring assuming a boat conformation, as depicted below. This fact emphasises the importance of conformational mobility in controlling the direction of a reaction, though it is obvious that the catalyst, in some way, has an effect on this conformational control.

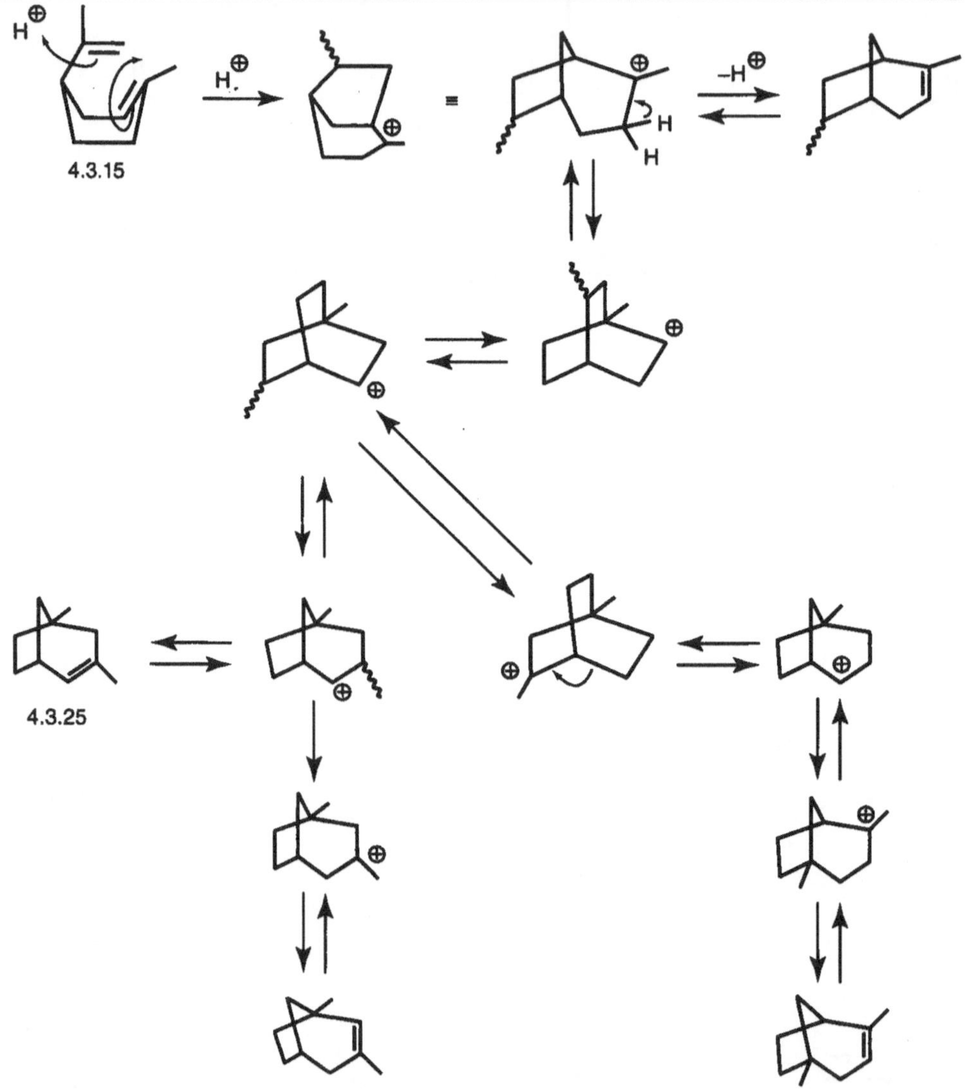

4.3.15

4.3.25

4.3.26   4.3.24

We shall end this section with an example from the numerous rearrangement reactions exhibited by bridged bicyclic monoterpenes such as the bornanes and related compounds. We have already made a passing reference to the much documented original example of the Wagner–Meerwein rearrangement as well as to the classical-non-classical carbonium ion (or the Brown–Weinstein) controversy. We will not discuss these topics any further, but will deal with some reactions of camphor.

The photochemical cleavage of camphor (4.3.27) has instructive value as it illustrates one mode of Norris-I reaction. The experimental strategy involves exposure of a solution of camphor in *n*-heptane to light of wavelength 310–318 nm. Under these conditions, the $n \to \pi^*$ excited state of 4.3.27 is formed first and this undergoes a Norris-type cleavage to yield 4.3.28. This diradical is formed to the exclusion of others as, being a tertiary radical, it is the most stable. Subsequent transformations of 4.3.28 lead to a number of products as shown below. These include the aldehyde, 4.3.29, the bridged bicyclic, saturated hydrocarbon, 4.3.30, the acyclic, doubly unsaturated hydrocarbon, 4.3.31, and the cyclic ether, 4.3.32. While 4.3.29 and 4.3.32 retain the oxygen atom of 4.3.27, the other two products have lost it, presumably as carbon monoxide.

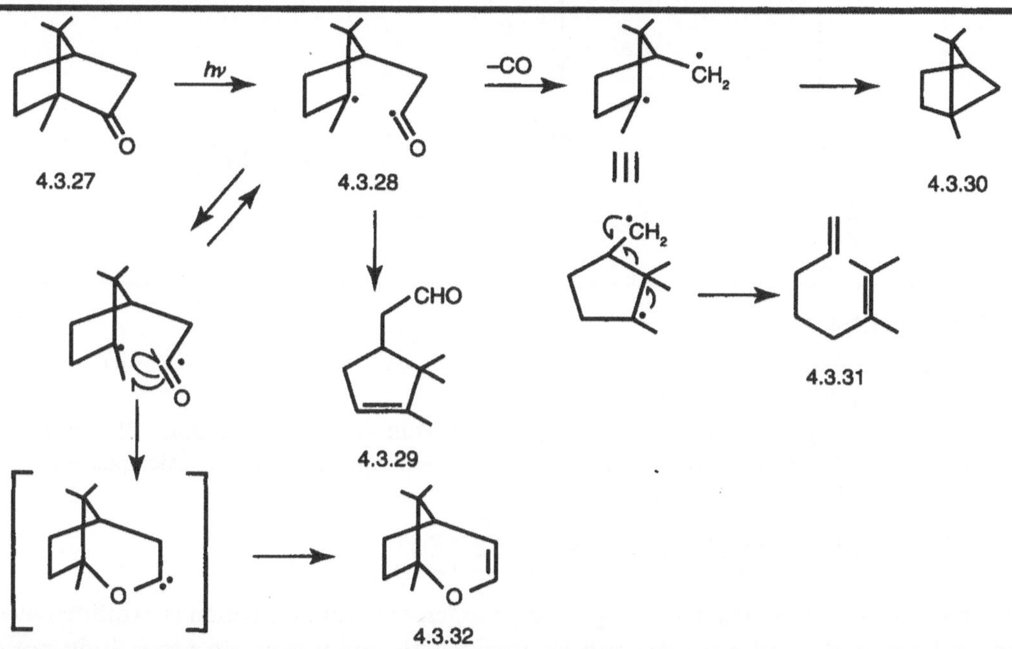

On prolonged irradiation, the aldehyde 4.3.29 undergoes further transformation to yield two diastereoisomeric cyclobutanols, 4.3.33 and 4.3.34, and a compact, tricyclic oxetane, 4.3.35. These changes also go through an excited state of the aldehyde (4.3.29) which initially experiences a Norris-II type reaction (which is analogous to a McLafferty-type migration of a hydrogen atom in a six-membered cyclic transition state).

Thus, this reaction of camphor is a neat package illustrating a variety of photochemical reactions of the carbonyl group exhibited by one single compound. One can also use this example for understanding the behaviour of photochemically generated diradicals.

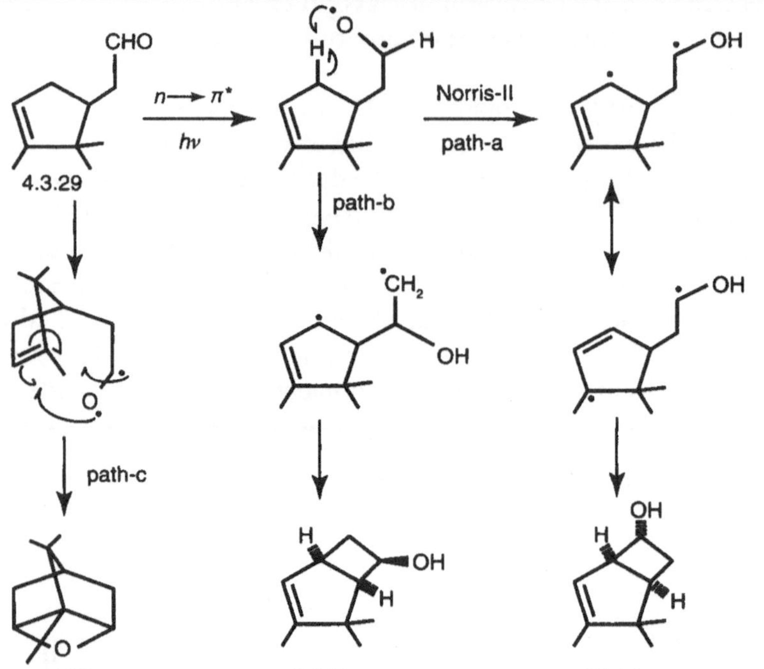

4.3.29

path-a

Norris-II

path-b

path-c

4.3.35   4.3.33   4.3.34

Suggested reading

1. Erman WF, In *Chemistry of the Monoterpenes: An Encyclopedic Handbook (A and B)*, (and references therein), New York: Marcel Dekker, 1985.
2. Gilbert A and Baggott J, *Essentials of Molecular Photochemistry*, London: Blackwell Scientific Publications,1991 (for a good understanding of all types of light-induced organic reactions).

4.4 THE WESLEY–MOSER REARRANGEMENT

For a change, we shall now discuss a simpler rearrangement reaction which is exhibited by certain flavonoids. Flavonoids, and aromatic and heteroaromatic compounds in general, do not exhibit any unexpected behavioural aberrations. The reaction under discussion is an exception and is a mild reminder that under highly provoking conditions even a sedate molecule is apt to perform a somersault!

This reaction involves a skeletal rearrangement of the flavonoid structure accompanying demethylation of 5,8-dimethoxy (4.4.1) and 5,7,8-trimethoxy (4.4.2) flavones and related compounds. The demethylation is brought about by hydriodic acid, and the products are the corresponding 5,6-dihydroxy(4.4.3) and 5,6,7-trihydroxy (4.4.4) flavones. The reaction apparently involves an acid-catalysed opening of the pyrone ring followed by ring closure to yield isomeric products. This conjecture is supported by the observation that HI-induced demethylation of 2′5′-dimethoxyflavone (4.4.5) results in the formation of 6,2′-dihydroxyflavone (4.4.6).

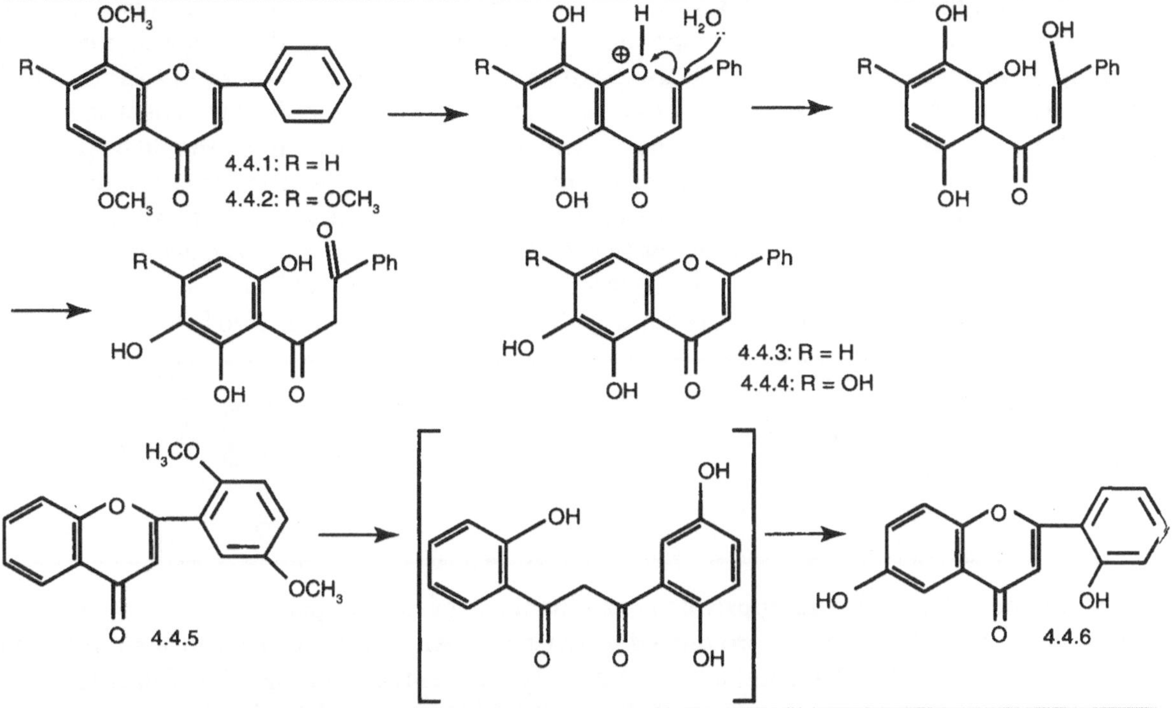

In this context, a brief note on the selective demethylations of poly-O-methylated flavonoids is relevant as these reactions bring out, in a simple demonstration, the subtle differences in the Lewis basicities of the differently located methoxyl groups.

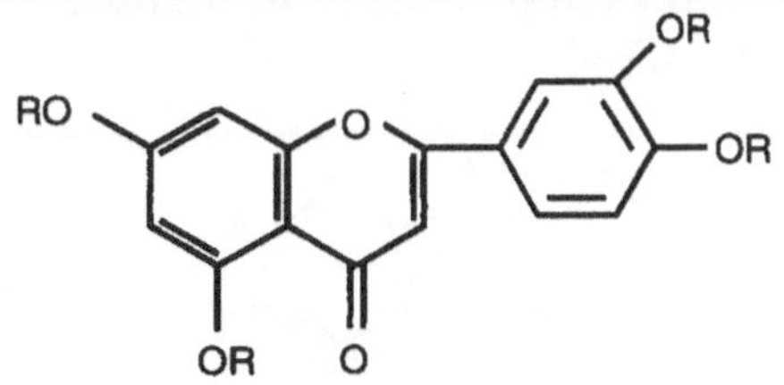

If one carefully examines the structure of the complete methyl ether (4.4.8) of a typical naturally occurring flavone such as luteolin, which is 5,7,3',4'-tetrahydroxyflavone (4.4.7), one would see distinct differences in the basicities of the four methoxyl groups. The least basic is the methoxyl at position 7 since it is in direct conjugation with the pyrone carbonyl group. Though the methoxyl at position 5 is also similarly situated with regard to conjugative interaction, it can be expected to undergo facile acid-

4.4.7 : R = H
4.4.8 : R = CH$_3$

catalysed demethylation since the resulting phenol would be stabilised by the intramolecular H-bonding between the hydroxyl thus liberated and the pyrone carbonyl. Thus, in this case, the thermodynamic driving force is enough to overcome the weak basicity of the 5-methoxyl.

Incidentally, we are assuming that the reader is aware that a prior protonation of the methoxyl group is an essential prerequisite for demethylation to occur and that the extent and ease of protonation would depend on the basicity of the methoxyl group in question. In a planar conformation, the methoxyl at position 4' is also in effective conjugation with the pyrone carbonyl.

However, since the benzene ring carrying this methoxyl is free enough to rotate, this is not the only conformation possible for the compound, particularly at the elevated temperatures used for effecting demethylation. Therefore, it is also possible to predict that the basicity of this methoxyl would be greater than that of the one at position 7. This analysis thus leads to the conclusion that the methoxyl at position 7 should be the most resistant to demethylation and it should be possible to devise a set of experimental conditions conducive to selective demethylation of the other methoxyls. This experimental possibility assumes importance because several naturally occurring flavones do possess a methoxyl group at position 7. That the theoretical approach delineated above is sound is proved by the fact that such selective demethylations have, indeed, been achieved and that this is a widely used strategy in flavonoid synthesis. One example is shown below.

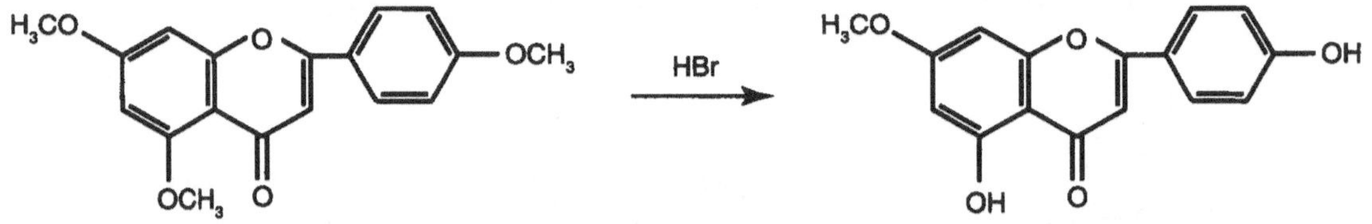

In this context, it is also appropriate to describe the behaviour of papaverine (4.4.9) towards hydriodic acid. In this case, the protonated tertiary nitrogen acts as an electron 'sink' (like the pyrone carbonyl group in flavones) and considerably reduces the Lewis basicity of the methoxyl at position 6 (isoquinoline numbering). Therefore, this methoxyl is the most resistant to demethylation, as observed. The methoxyl at position 7 undergoes easier demethylation as it is not in direct conjugation with the inline group. The two methoxyls on the benzylic moiety are the ones to undergo ready demethylation as they are not affected by the imino function. Thus, in this example, one finds a kinship based on fundamental principles between an alkaloid and a flavonoid!

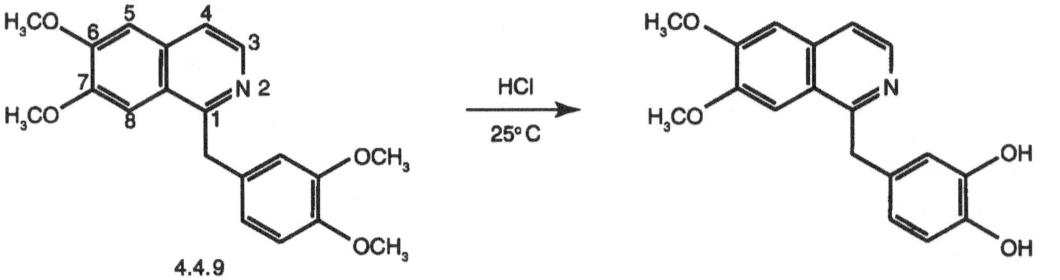

Suggested reading

1. Seshadri TR, In Geissman TA, (ed.), *The Flavonoids*, London: Pergamon Press, 1962.
2. Livingstone R, In Coffey S, (ed.), Rodd's *Chemistry of Carbon Compounds*, 2nd edn, Vol. 4, Amsterdam: Elsevier, 1978.
3. Brossi A and Teitel S, *J. Org. Chem.*, 1970, 35: 1684.

4.5 Some Interesting Reactions of Reserpine

Wc shall now go back to another example from alkaloids—one which has been the subject of several studies—structural, stereochemical and synthetic. The reactions of reserpine and its derivatives which gave a feedback to its stereochemistry are particularly interesting and educative. This alkaloid, which is a component of a well-known medicinal plant, *Rauwolfia serpentina*, has been much investigated because of its pharmacological activities. On hydrolysis, reserpine (4.5.1) yields reserpic acid (4.5.2).

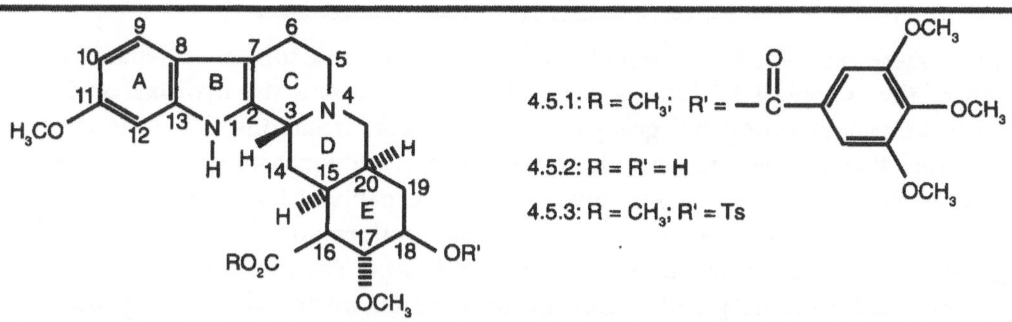

4.5.1: R = CH$_3$; R' = —C(=O)— (3,4,5-trimethoxybenzoyl, OCH$_3$ at positions shown)

4.5.2: R = R' = H

4.5.3: R = CH$_3$; R' = Ts

The tosyl ester (4.5.3) of methyl reserpate undergoes an unusual solvolytic reaction when heated with collidine. The product is a quaternary ammonium compound which retains the tosyl group as the tosylate anion. The structure of the product could be figured out as 4.5.4 as shown below. In this compound, the basic nitrogen of ring C and position 18 in ring E are linked together in spite of their being far away in a two-dimensional projection formula. Apparently, they are close together in space and this is possible only when the rings D and E are *cis*-fused. However, the mechanism of this reaction is perhaps not as straightforward as might appear. That is, it is not a result of a direct attack of the basic nitrogen across the 'depression' formed by the *cis*-fused D–E rings, on the carbon carrying the tosyloxy group. Instead, the reaction has been shown to involve the initial participation of the methoxyl at position 17, which provides anchimeric assistance to the expulsion of the tosyloxy group. The resulting oxonium ion structure (4.5.5) is then opened up by an attack by the basic nitrogen.

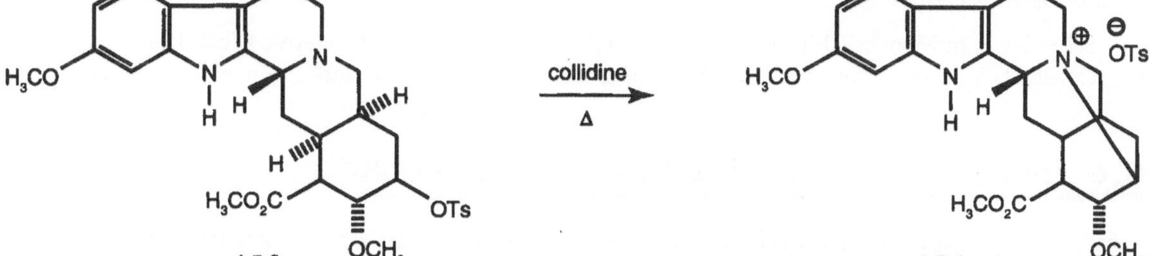

4.5.3   collidine, Δ   4.5.4

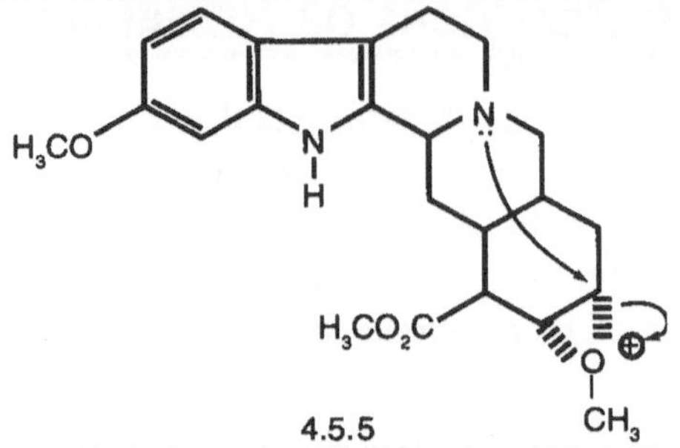

4.5.5

Another interesting reaction of reserpic acid, which can be a subject of debate, is the simple formation of the lactone 4.5.6 linking the carboxyl at C16 and the hydroxyl at C18. Such a lactone is possible only if these two groups assume axial conformations. Other pieces of evidence clearly indicate that in reserpic acid, these two groups are equatorially oriented. This means that ring E which carries these two groups, should be capable of flipping over to an alternative chair conformation, unless this ring assumes a boat conformation in the lactone. This is the point of debate. However, if one accepts that ring E has the chair conformation in the lactone, the implication is far-reaching with regard to the stereochemistry of D–E ring fusion. Anyone familiar with the stereochemistry of decalin would know that ring flipping to an alternative chair–chair conformation is possible only in *cis*-decalin and not in the *trans* diastereoisomer. This result can therefore be taken as further evidence for the stereochemistry of the D–E ring junctions in reserpine, provided one can agree that in the lactone too the preferred conformations for the D and E rings are the chair forms.

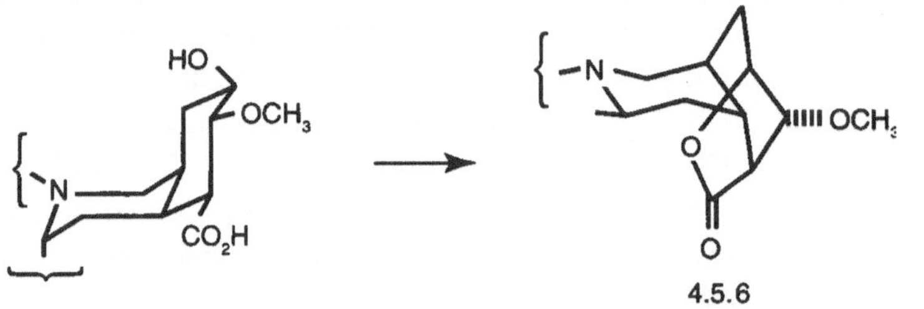

4.5.6

Suggested reading

1. Schlittler E, In Manske RHF, (ed.), *The Alkaloids*, Vol. 8, New York: Academic Press, 1965.
2. Cordell GA, *Introduction to Alkaloids—A Biogenetic Approach*, New York: John Wiley & Sons, 1981.

4.6 MOLECULAR YOGA: REACTIONS OF PAPAVERINE

A fascinating example of a molecule 'drawing in its limbs' in a yogic posture is seen in the following reaction. *N*-methylpapaverine chloride can be regiospecficaly reduced to the dihydro derivative (4.6.2). This compound on treatment with a mineral acid undergoes protonation, not on the nitrogen atom, but at the β-olefinic carbon, as expected in an enamine system. The resulting ammonium ion—carbocation resonance hybrid (4.6.3) then suffers an intramolecular Friedel–Crafts-type reaction to yield the pavine skeleton; the product is *N*-methylpavine (4.6.4) or argemonine, which occurs in *Argemone mexicana*. This, indeed, is an interesting rearrangement which illustrates more than one chemical principle in a neat, concise package!

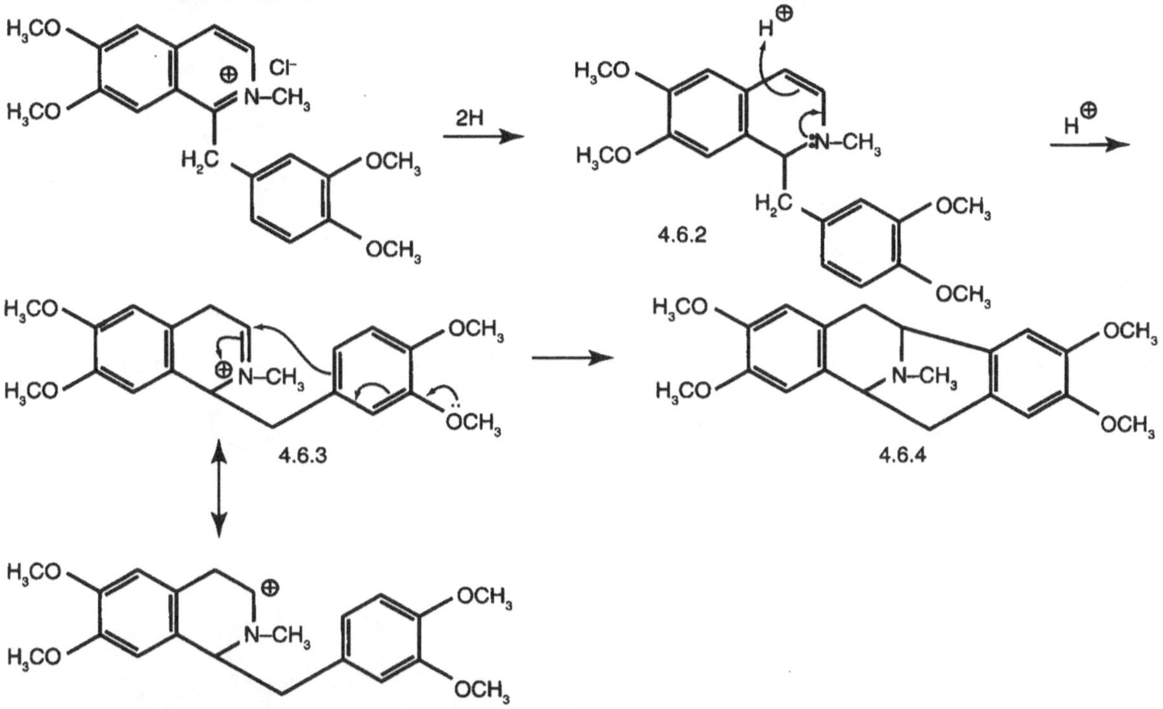

Photolysis of papaverine (4.6.1) or its hydrochloride in methanol and ethanol yields, respectively, 1-methyl-6,7-dimethoxyisoquinoline (4.6.5) and 1-ethyl-6,7-dimethoxyisoquinoline (4.6.6). These compounds are accompanied by the methyl (4.6.7) or ethyl (4.6.8) ether of 3,4-dimethoxybenzyl alcohol. Since the non-bonded electrons on nitrogen are not available in the hydrochloride, initial photochemical excitation must involve a $\pi \to \pi^*$ transition. The excited species then abstracts a hydrogen atom from the alcohol as shown on the next page. The mechanism depicted is supported by the observation that in isopropyl alcohol there is a large increase in the rate of photolysis. On the other hand, the reaction does not occur in benzene medium.

The benzene ring of the benzyl unit in 4.6.1 undergoes electrophilic substitution quite readily. Thus, the 6′-nitro derivative (4.6.9), the corresponding acetyl derivative (4.6.10) and the alcohol (4.6.11) have been prepared. The last-mentioned compound is obtained from 4.6.1 by reaction with formaldehyde and hydrochloric acid in acetic acid; 4.6.11 reacts with papaverine (4.6.1) in the presence of acids to yield the interesting dimer 4.6.12. On the other hand, in the presence of formic acid and formamide, 4.6.11 undergoes cyclisation to give 4.6.13 which can be reduced by sodium borohydride to obtain norcorydaline (4.6.14).

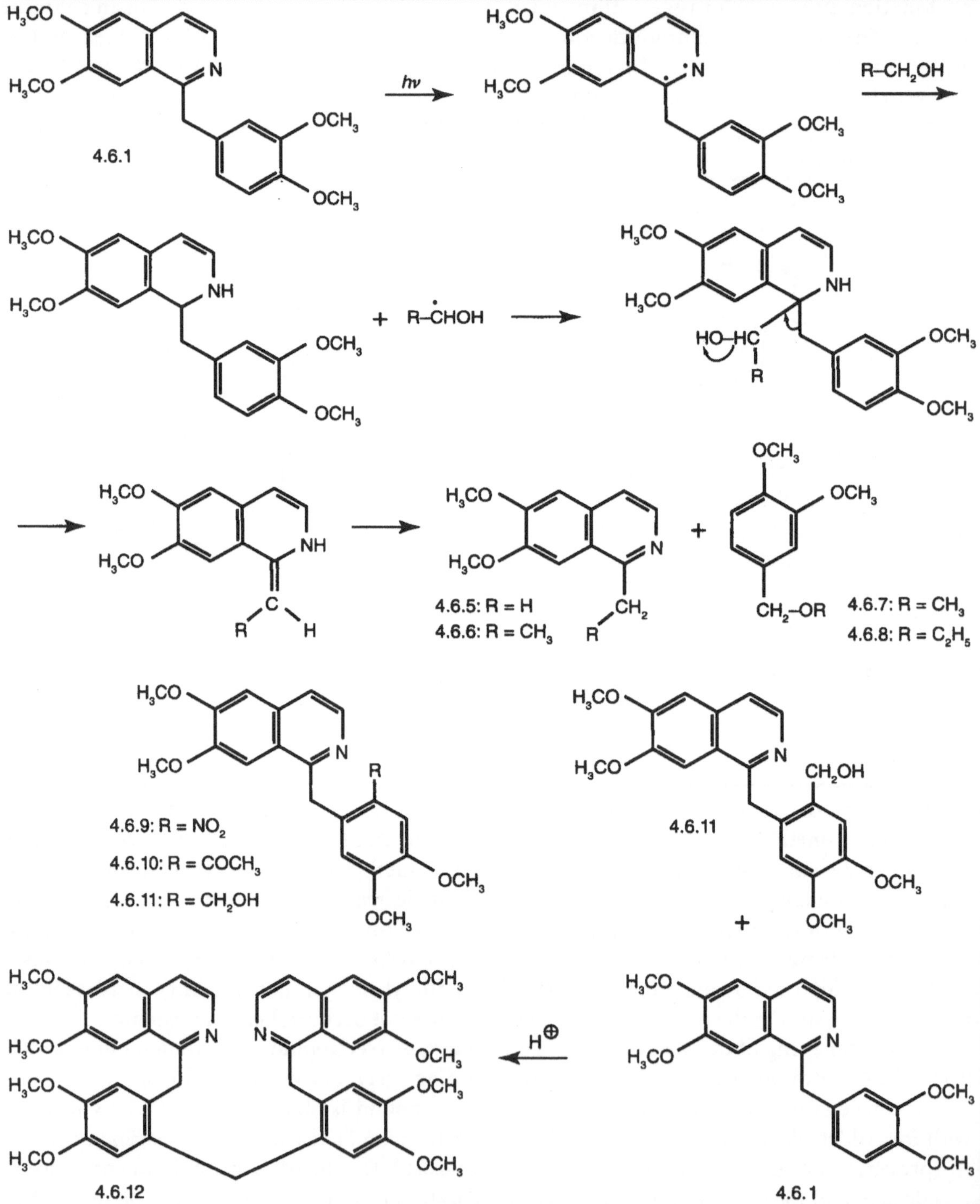
4.6.1
hν
R–CH2OH
+ R–ĊHOH
4.6.5: R = H
4.6.6: R = CH3
4.6.7: R = CH3
4.6.8: R = C2H5
4.6.9: R = NO2
4.6.10: R = COCH3
4.6.11: R = CH2OH
4.6.11
H⊕
4.6.12
4.6.1

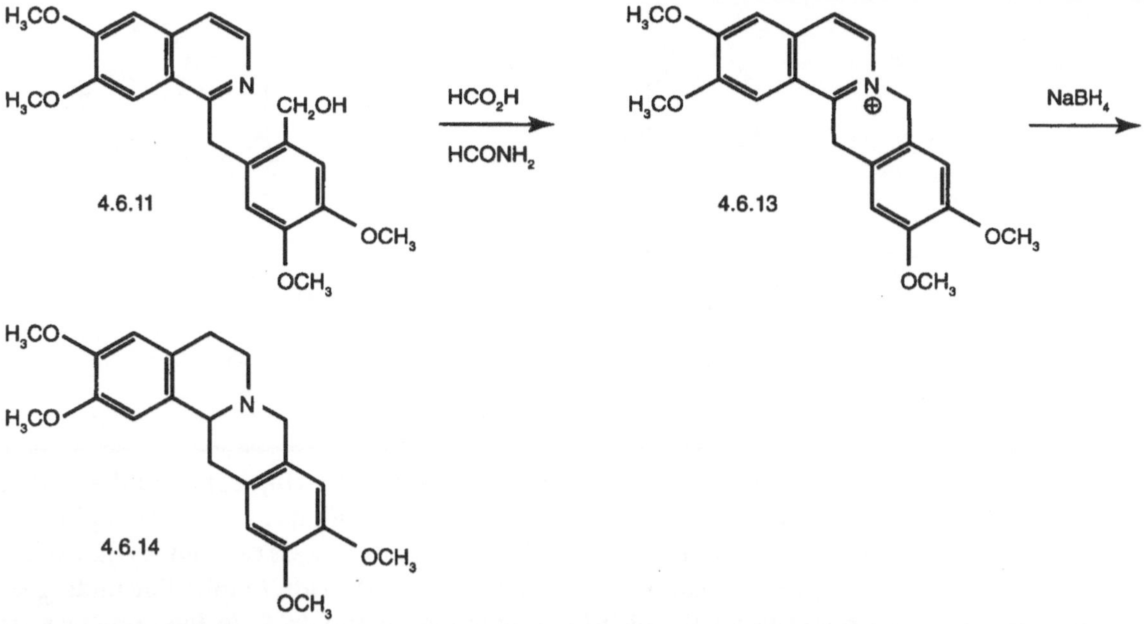

4.6.11 → HCO₂H / HCONH₂ → 4.6.13 → NaBH₄ →

4.6.14

Suggested reading

1. Dyke SF, In Coffey S, (ed), Rodd's *Chemistry of Carbon Compounds*, 2nd edn, Vol. 4 (H), Amsterdam: Elsevier Science Limited, 1978.
2. Shamma M, *The Benzyl Isoquinoline Alkaloids*, New York: Academic Press, 1972.
3. Stermitz FR, Seiber RP and Nicoden DE., *J. Org. Chem.*, 1968, 33: 1136.

4.7 Two Examples of Transannular Reactions

Natural products help us to look deep into several aspects of organic chemistry. There are any number of examples among natural products which can effectively illustrate some dynamic aspect or the other of carbon compounds. The transannular interactions exhibited by medium ring systems are among the most fascinating chemical reactions. They were brought into focus by the pioneering studies of Cope and Prelog. In this section we shall consider two examples of natural products undergoing intramolecular transannular reactions.

Parviflorin (4.7.1) is a phenolic sesquiterpenoid occurring in *Coreopsis parviflora*. Its epoxide (4.7.2) has been shown to undergo a normal, acid-catalysed epoxide ring-opening followed by an intramolecular 1,5-hydride shift (transannular) and the loss of a proton to yield the final product, an ene ol (4.7.3). In this case the ring exhibiting 'articulation' is eight-membered; transannular reactions in cyclooctane systems have been extensively studied by Cope and his co-workers.

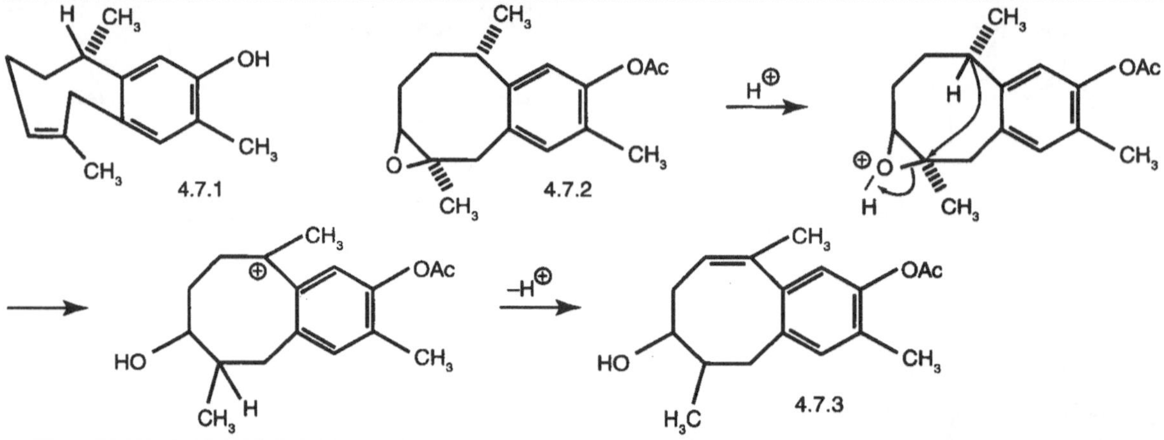

An example of a naturally occurring ten-membered carbocyclic ring compound showing transannular reactivity is tulipinolide (4.7.4) which occurs, along with its epimer, epitulipinolide (4.7.5), in *Lirodendron tulipifera* L. These two cytotoxic compounds possess two endocyclic double bonds besides the α, β- unsaturated lactone unit. On treatment with acids, tulipinolide undergoes intramolecular cyclisation and yields the bicyclic products 4.7.6 and 4.7.7. In the same way, its epimer, epitulipinolide also gives three bicyclic products, namely 4.7.8, 4.7.9 and 4.7.10. It can be readily seen that the reaction should proceed through the intermediate carbocations, 4.7.11 and 4.7.12. These intermediates are presumably formed as a consequence of the transannular interaction (homoconjugative interaction) between the two double bonds in 4.7.4 and 4.7.5, as shown below.

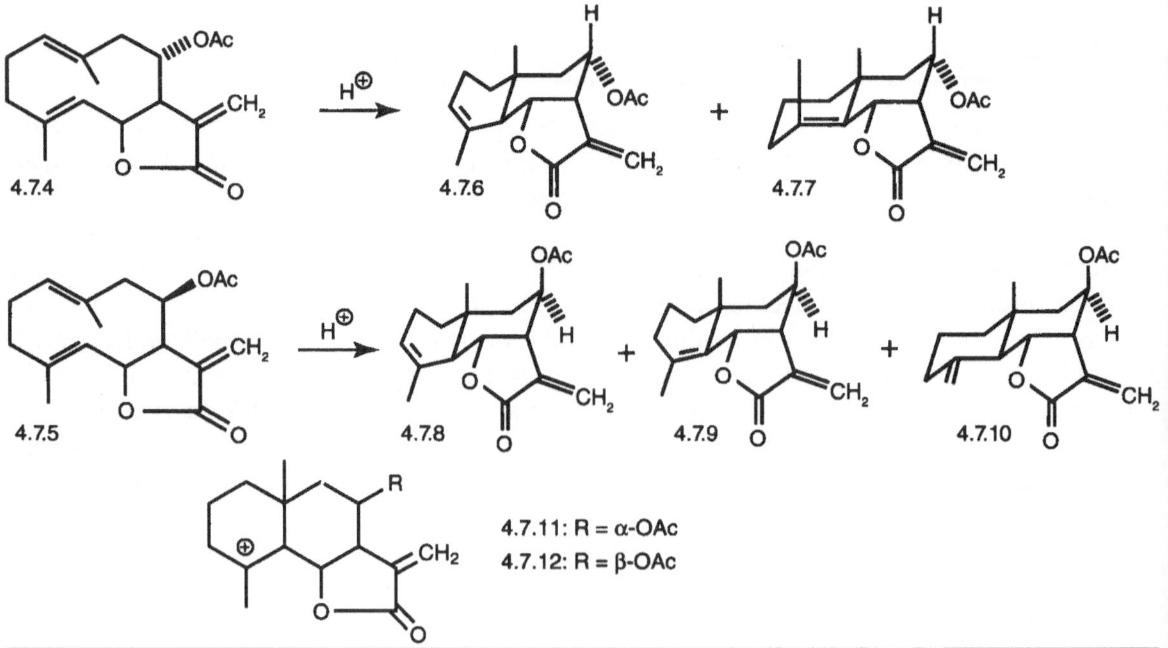

Suggested reading

1.   Graham SH, in Coffey S, (ed.), Rodd's *Chemistry of Carbon Compounds*, 2nd edn, II (B), Supplement, Amsterdam: Elsevier, 1974.
2.   Cope AC, Martin MM and McKerrey MA, *Quart. Rev.*, 1966, 20: 119.
3.   Bohlmann E and Zdero C, *Chem. Ber.*, 1970, 110: 468.
4.   Doskotch RW and El Feraly ES, *Org. Chem.*, 1970, 35: 1928.

4.8 The Epoxide as a Medium of Articulation

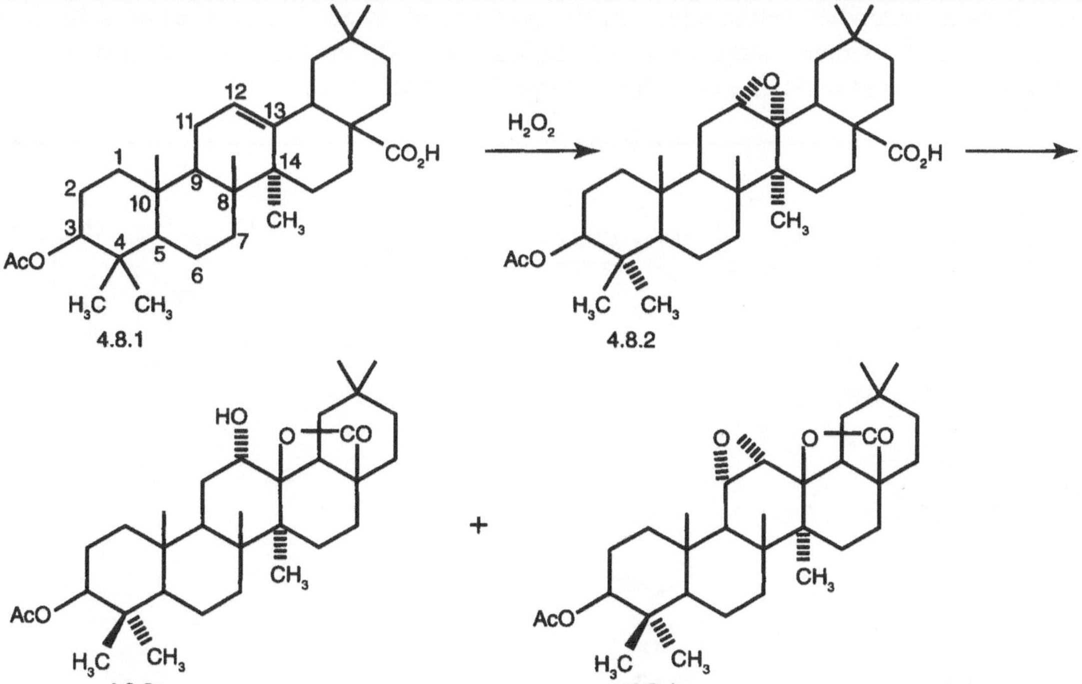

The epoxide group is known to be sensitive towards both electrophiles and nucleophiles and can be opened up under hydrolytic, oxidative and reductive conditions to yield a variety of products. Nature, it appears, is abundantly aware of the potential of the epoxide group as this function is seen in a number of naturally occurring compounds, though we do not know how this chemical reactivity is translated into biological activity. An example of an epoxide group undergoing an intramolecular nucleophilic opening is seen in the following reaction of acetyloleanolic acid (4.8.1). The reaction brought about by hydrogen peroxide proceeds through the initial formation of 12,13-epoxyoleanolic acid (4.8.2). It has been suggested that the epoxide itself is formed by the intramolecular oxidation of the double bond after a prior conversion of the 17-carboxyl group to the peracid. However, this seems to be unlikely, if one accepts as correct the stereochemistry of the final products, 4.8.3 and 4.8.4 shown below. Instead, it appears that the double bond in 4.8.1 has undergone a normal, expected epoxidation by hydrogen peroxide from the less hindered side and subsequently, the epoxide has been opened up by an intramolecular nucleophilic attack by the carboxyl group, as indicated in the reaction. This explanation of the sequence of reactions would account for the observed stereochemistry of the products, as well as for the fact that a free carboxyl group at position 17 is essential for the reaction to occur. (Incidentally, students should learn not to accept tacitly whatever appears in print, but to critically evaluate the inferences drawn from a given set of data. This also applies to the views expressed in this book!)

In this example, there is still one intriguing fact which has not been accounted for, namely the mode of formation of the epoxy lactone 4.8.4. Prima facie, it seems to involve the expulsion of a hydrogen atom from position 11 as a hydride ion. However, the questions that remain are whether such a reaction is possible and if so, what is the hydride ion acceptor. The reader would also, we are sure, like to see if alternative mechanisms are possible for the formation of this compound.

Suggested reading

Mallavarapu GR, In Atta-ur-Rahman, (ed.), *Studies in Natural Products Chemistry*, Vol. 7, (and references therein), Amsterdam: Elsevier, 1990.

4.9 CAPILLARISIN: AN EXAMPLE OF $S_N 2$ DISPLACEMENT

Capillarisin (4.9.1), a choleretic compound isolated from the *Artemisia capillaris* herb is an unusual chromone. It is a phenoxy ether of the rare 2-hydroxychromone structure. The tautomeric 4-hydroxycoumarins are well known as naturally occurring as well as synthetic compounds.

Capillarisin undergoes methanolysis in the presence of an alkali as well as an acid. One of the products under either condition is hydroquinone (4.9.2), but the other product is either a 2-methoxychromone (4.9.3) or a 4-methoxycoumarin (4.9.4).

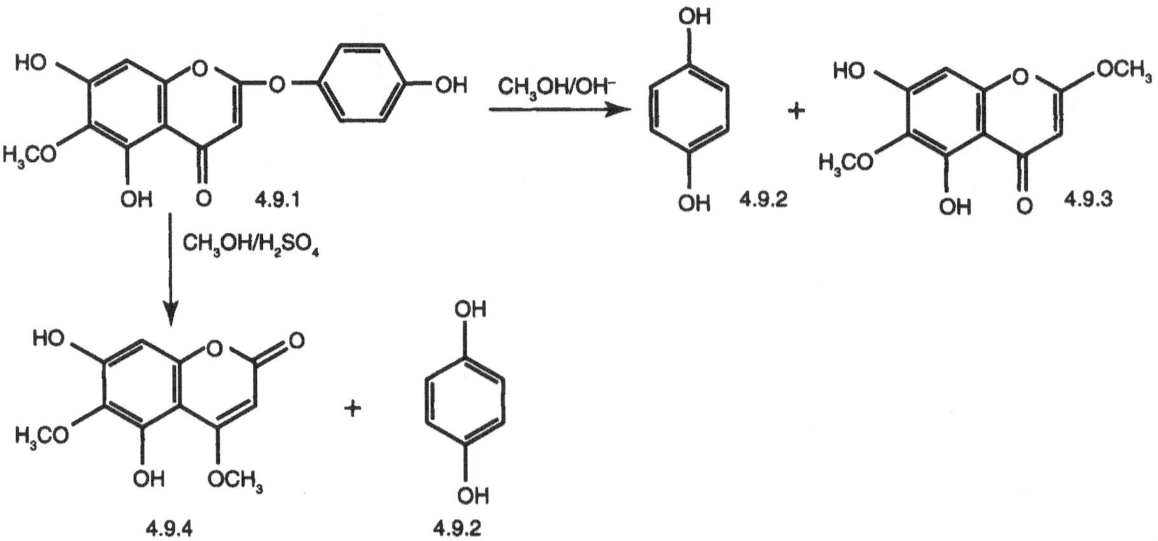

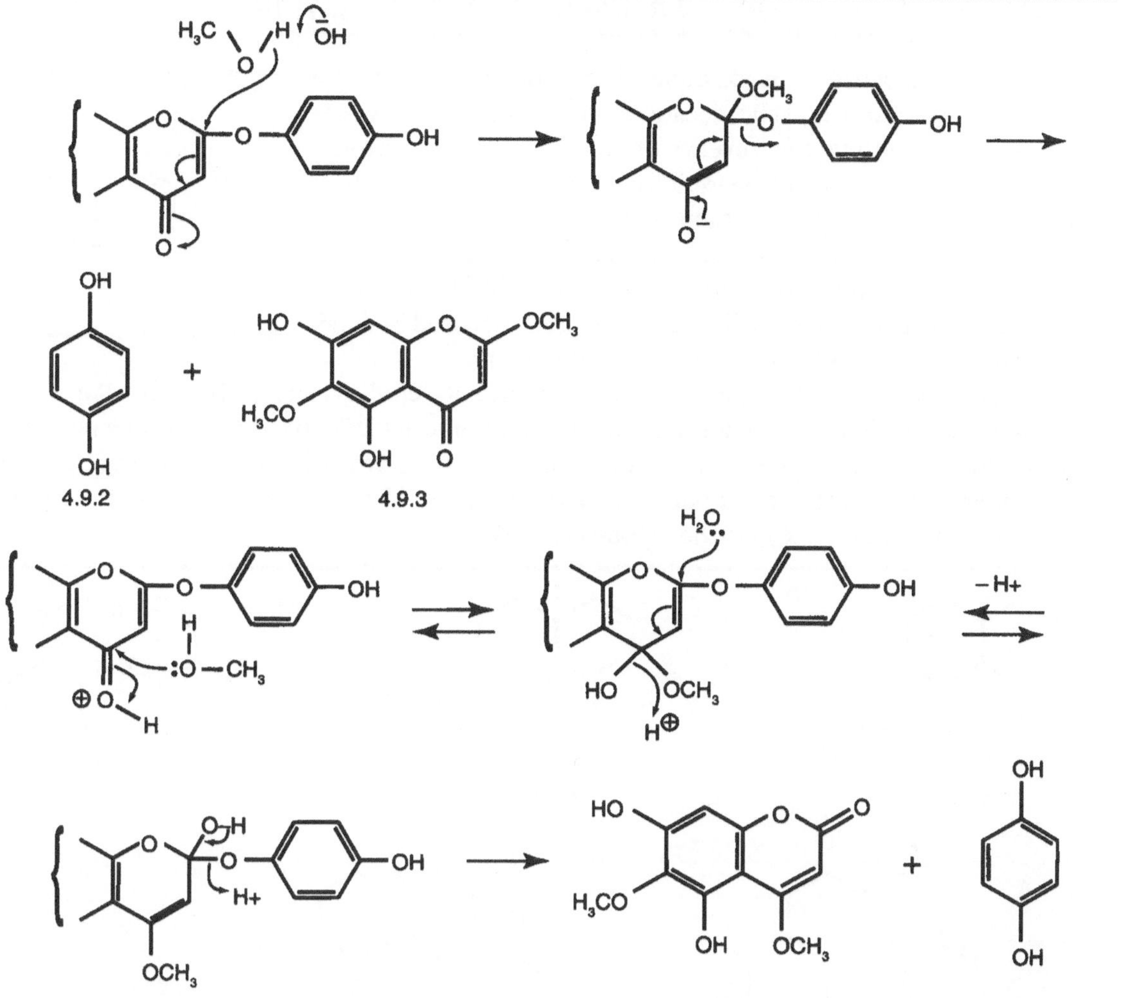

Under alkaline conditions one may presume the initial attack of methanol at position 2, in a Michael-type reaction followed by the elimination of the hydroquinone moiety. Under acidic conditions, it seems, protonation at the carbonyl oxygen triggers the subsequent reactions, with the methanol now attacking position 4 as indicated.

What we have shown is one possible mechanistic pathway and we would like to emphasise that this is not the only possible mechanism. Readers are invited to look at the problem without bias and consider other possibilities. This is a simple and yet sufficiently intriguing example illustrating a nucleophilic attack at an electron-deficient olefinic carbon atom.

Suggested reading

1. Livingstone R. In Rodd's *Chemistry of Carbon Compounds*, 2nd edn, Vol. 4 (E), Supplement, (for recent advances in chromones), Amsterdam: Elsevier, 1990.
2. Kamiya T, Tsukui M and Oshio H, *Chem. Pharm. Bull.*, 1975, 23: 1387.

4.10 THE REVERSE OF A TRANSANNULAR CYCLISATION

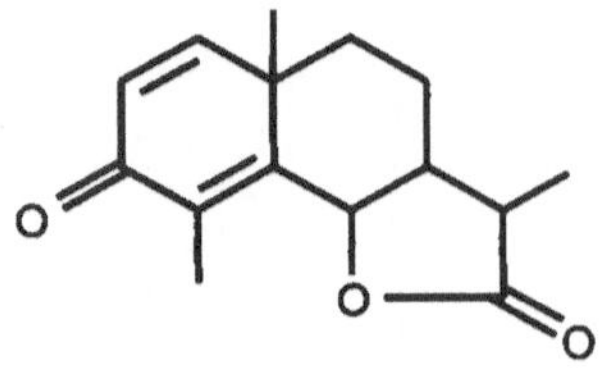

We had earlier discussed an example of the transformation of a germacranolide into a bicyclic decane system. We shall now see an example illustrating the reverse of that process. α-Santonin (4.10.1) is one of the earliest sesquiterpene lactones to have been discovered. This compound, which is a component of the plant *Artemisia maritima*, was at one time considered to be a promising anthelminthic. Its rearrangements brought about by the action of light or acids have been very well studied and are discussed in detail in 4.15. The reaction discussed here is also interesting.

4.10.1

Dihydrosantonin (4.10.2) can be converted into its enol acetate (4.10.3) by the action of isopropenyl acetate in the presence of *p*-toluene sulphonic acid. On irradiation at low temperature in methanolic solution, the enol acetate undergoes electrocyclic ring opening to yield a ten-membered monocyclic compound (4.10.4), which could be converted, by routine reactions, into the dieneacetate (4.10.6) via the ketone (4.10.5).

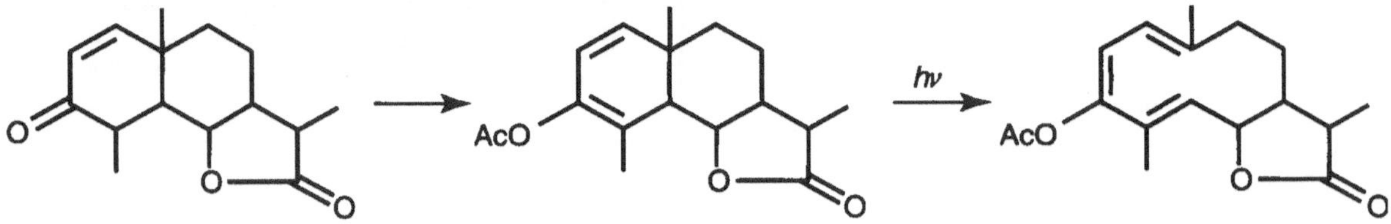

4.10.2   4.10.3   4.10.4

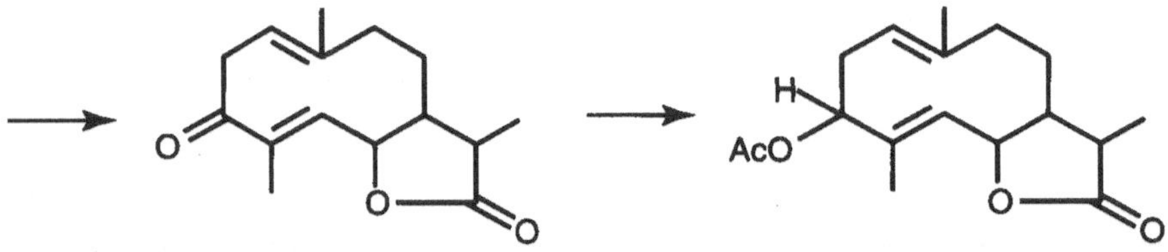

4.10.5   4.10.6

This reaction demonstrates the feasibility of using a light-induced, symmetry-allowed pericyclic reaction for transforming a fairly readily available natural product to a comparatively rare structure. The key step, namely the conversion of 4.10.3 to 4.10.4, is known to involve a photochemically allowed conrotatory ring opening of the cyclohexadiene moiety.

Suggested reading

Bryant R, In *Rodd's Chemistry of Carbon Compounds*, 2nd edn, Vol. 2 (C), and references therein, Elsevier, 1968.

4.11 THE EPOXIDE GROUP AS A 'HANDLE' FOR CONVERSION

Many of the naturally occurring coumarins are structurally interesting because they possess isoprenyl units tagged on to the pyrone skeletal structure. One such compound is auraptenol (4.11.1) which occurs in *Murraya exotica* L. The compound undergoes a facile oxidative rearrangement on treatment wih *m*-chloroperbenzoic acid (MCPBA) in a nitrogen atmosphere. The reaction is, obviously, initiated by epoxidation across the side chain double bond to be followed by an acid-catalysed sequence of steps as shown below. The final product (4.11.2) is the aldehyde, with the epoxide (4.11.3) and the ene-diol (4.11.4) being probable intermediates. As a matter of fact, there is no skeletal rearrangement involved here and the apparent transdisposition of the oxygen function is the result of oxidation followed by elimination of the originally present hydroxyl group.

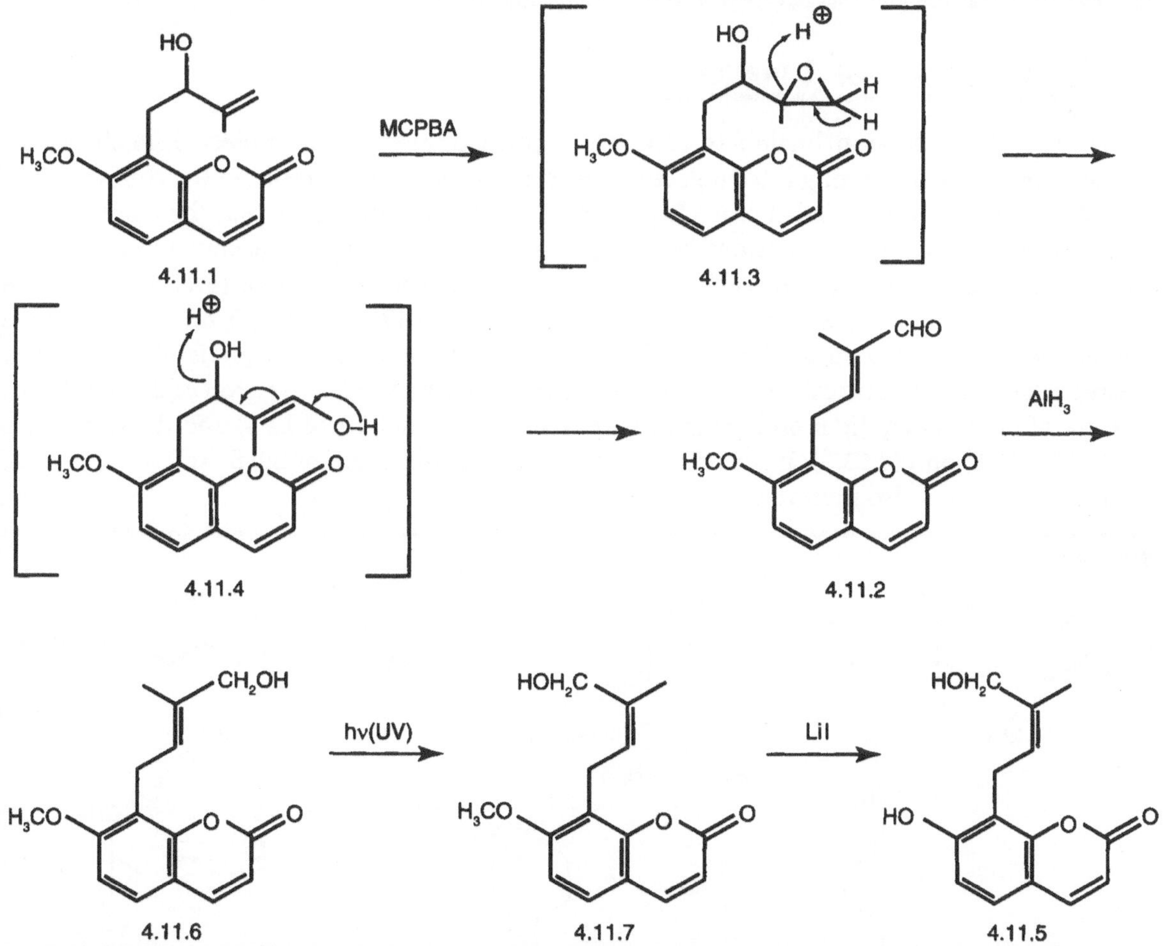

Further transformation of 4.11.2 to arnottinin (4.11.5), which is present in *Xanthoxylum amottianum* Maxim are represented as follows. The first step is a straightforward reduction of the formyl group in 4.11.2 to the corresponding alcohol (4.11.6). The latter undergoes a light-induced geometrical isomerisation to yield, presumably, 4.11.7. This reaction is interesting and intriguing as the driving force for the rearrangement is not clear: there cannot be any significant difference in the molar absorption coefficients of the two isomers and therefore this may not be the reason for the Z-isomer being favoured at the photostationary state. The final demethylation of 4.11.7 to 4.11.5 is also worth noting as lithium iodide is used here as the demethylating agent; this reagent has also been recommended for the selective and mild demethylation of carbomethoxy groups.

Suggested reading

1. Livingstone R. In Rodd's *Chemistry of Carbon Compounds*, 2nd edn, Vol. 4 (E), Supplement, (for recent advances in coumarins), Amsterdam: Elsevier, 1990.
2. Banerji J, Das AK and Das B, *Chem. and Ind.*, 1987, 395.

4.12 REACTIONS OF LINALOOL

Being an acyclic compound, linalool (4.12.1) is a conformationally mobile molecule which enables it to undergo some interesting intramolecular transformations. For example, in the conformation (4.12.2), the two double bonds come close together in space. Threfore, when the compound is heated with acetic anhydride, besides geranyl acetate (4.12.3), α-terpenyl acetate (4.12.4) is also formed. A probable mechanism is that the initially formed linalyl acetate (4.12.5) undergoes rearrangement as shown in Scheme 4.1. For the conversion of 4.12.5 into 4.12.3, one may postulate a homolytic pathway. It must, however, be emphasised that these are not rigorously proved mechanisms. A mechanistically related reaction is the acid catalysed conversion of linalool into a mixture of the monocyclic monoterpene hydrocarbons, terpinolene (4.12.6) and the terpinenes (4.12.7), (4,12,8) and (4.12.9). It is worth noting that in the formation of the β- and γ-terpinenes, a hydride transfer is also involved.

Scheme 4.1

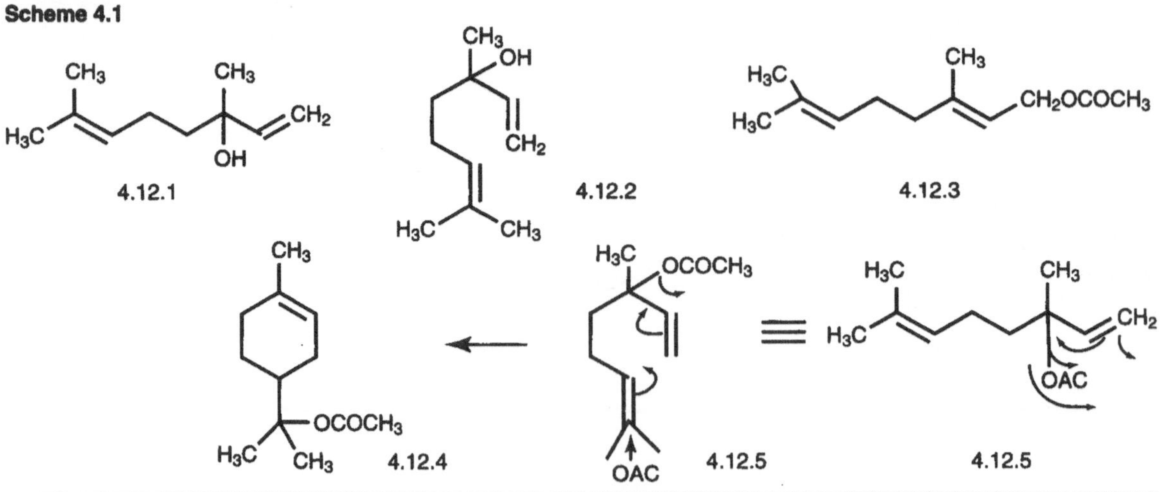

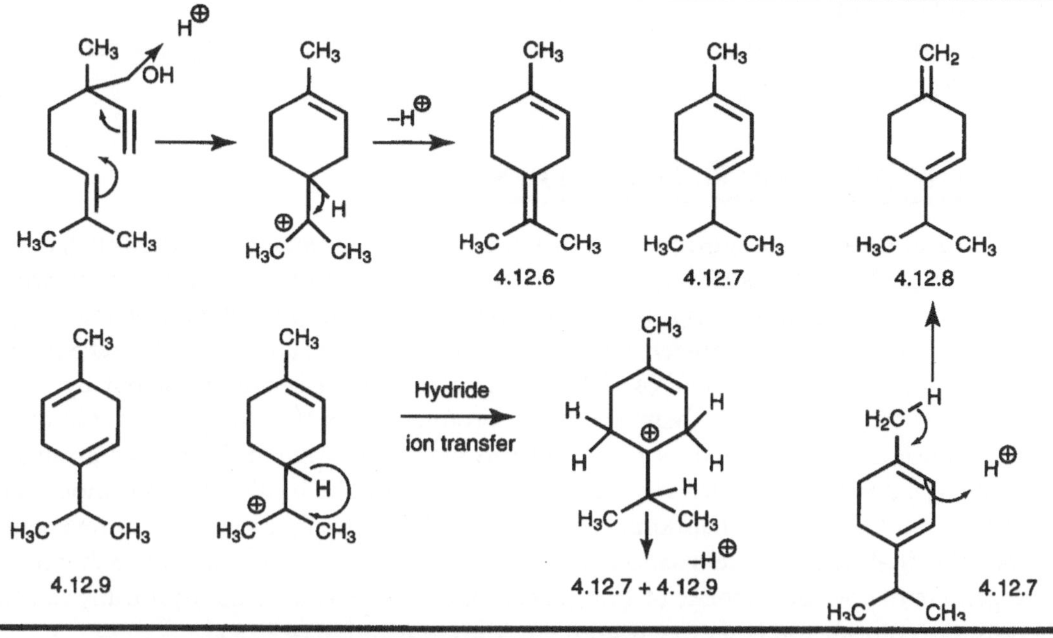

Another interesting chemical transformation of linalool is its conversion into methyl 2,6,6-trimethyl-tetrahydropyranyl-2-carboxylate (4.12.10). The first step in this transformation is the selective epoxidation of the more substituted double bond. The resulting product (4.12.11) is then treated with an acid when the epoxide opens out with attack of the hydroxyl at position 2 on position 6, yielding 4.12.12. The intermediate in this step is probably the tertiary carbocation (4.12.13), though the reaction could have also been a concerted reaction. The subsequent steps in the sequence from 4.12.12 to 4.12.10 are (i) oxidation using chromium trioxide, (ii) Wolff-Kishner reduction, (iii) oxidation using potassium permanganate-periodic acid and (iv) esterification by ethereal diazomethan. The entire process is shown in Scheme 4.2.

Scheme 4.2

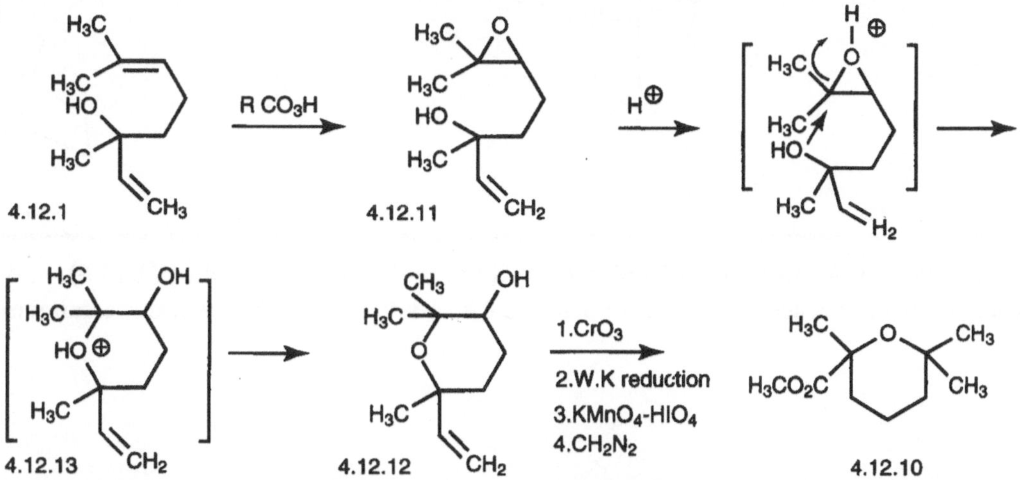

Suggested reading

Erman WF, *Chemistry of the Monoterpenes. An Encyclopedic Handbook (A and B)*, New York: Marcel Dekker, 1985.

4.13 THE NAMETKIN REARRANGEMENT

As briefly mentioned in the introduction, this rearrangement, named after the Russian chemist, SS Nametkin (1876-1950), is closely related to the Wagner–Meerwin rearrangement, which has been the subject of many studies. It is specifically confined to the terpenoids and related compounds and involves methyl group transfer. One of the well known examples is the racemisation of either enantiomorph of camphene (4.13.1). The rearrangement is brought about by the action of an acid and the crucial step, therefore, is the protonation of the exocyclic double bond. The tertiary carbocation intermediate (4.13.2) thus formed attracts one of the methyl groups on the neighbouring carbon atom, which migrates as a nucleophile. The resulting carbocation then loses a proton to yield the other enantiomorph of camphene (4.13.3). Every step is reversible and, therefore, the final result is racemisation of camphene. All the steps are shown in Scheme 4.3. The rearrangement of chlorocamphene (4.13.4) to camphene (4.13.1) does not rquire any catalyst. On solvolysis, the chloride ion is set free resulting in the carbocation (4.13.2).

Scheme 4.3

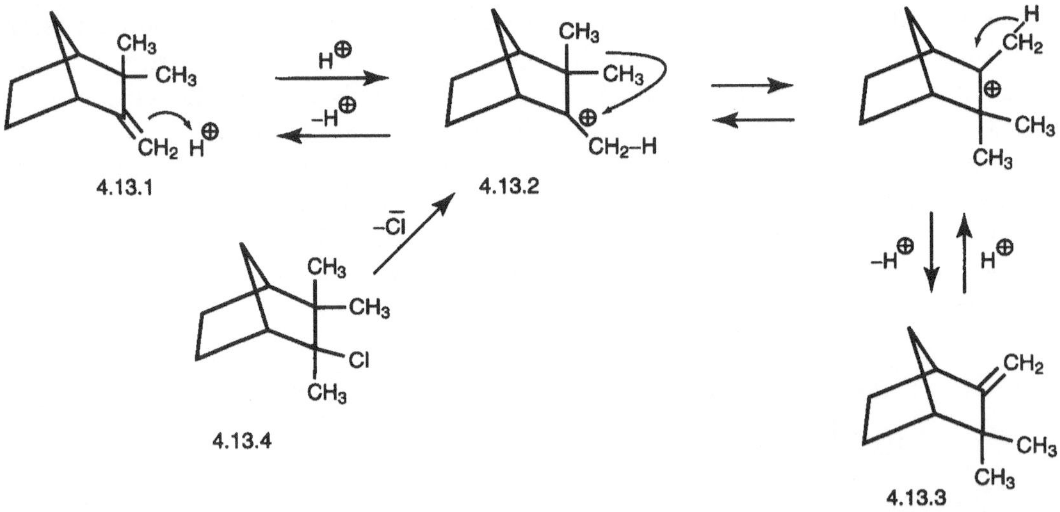

Similarly, (+) camphor (4.13.5) can be converted into (−) camphor (4.13.6), and vice versa, using a series of reactions shown in Scheme 4.4. A key step is a Nametkin rearrangement, which is brought about by the action of chloroformic acid. The subsequent step is a Wagner–Meerwein shift. But, unlike the previous case, all the steps in this reaction sequence are not reversible and, hence, the end result is inversion of configuration and not racemisation.

Scheme 4.4

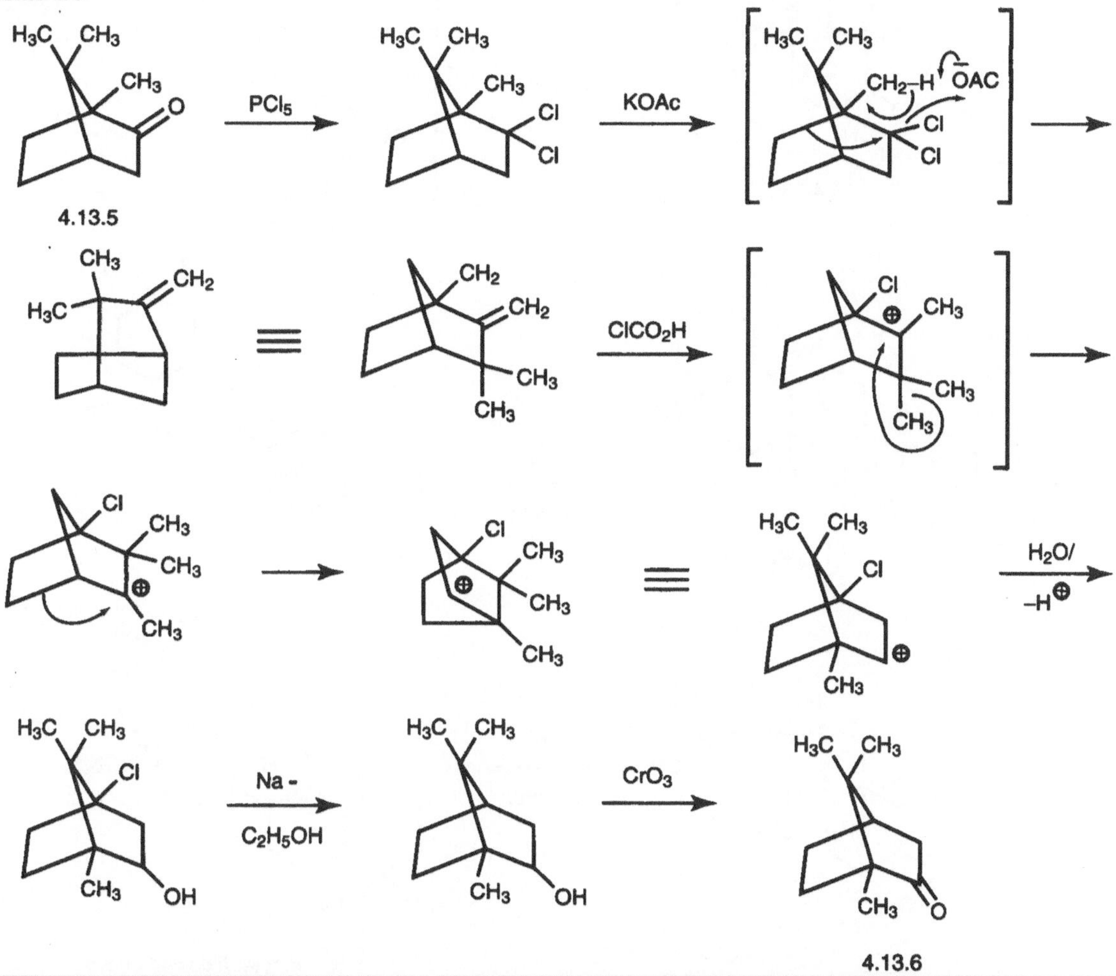

4.13.5

4.13.6

Yet another example is the acid-catalysed rearrangement of beta friedelinol (4.13.7) into (4.13.8), which has the β-Amyrin skeleton. In this case, the initial hydride ion shift triggered by the proton catalysed elimination of the hydroxyl group is followed by a series of alternating methyl and hydride ion migrations all along the skeletal framework. The final step involves the loss of a proton (Scheme 4.5).

Scheme 4.5

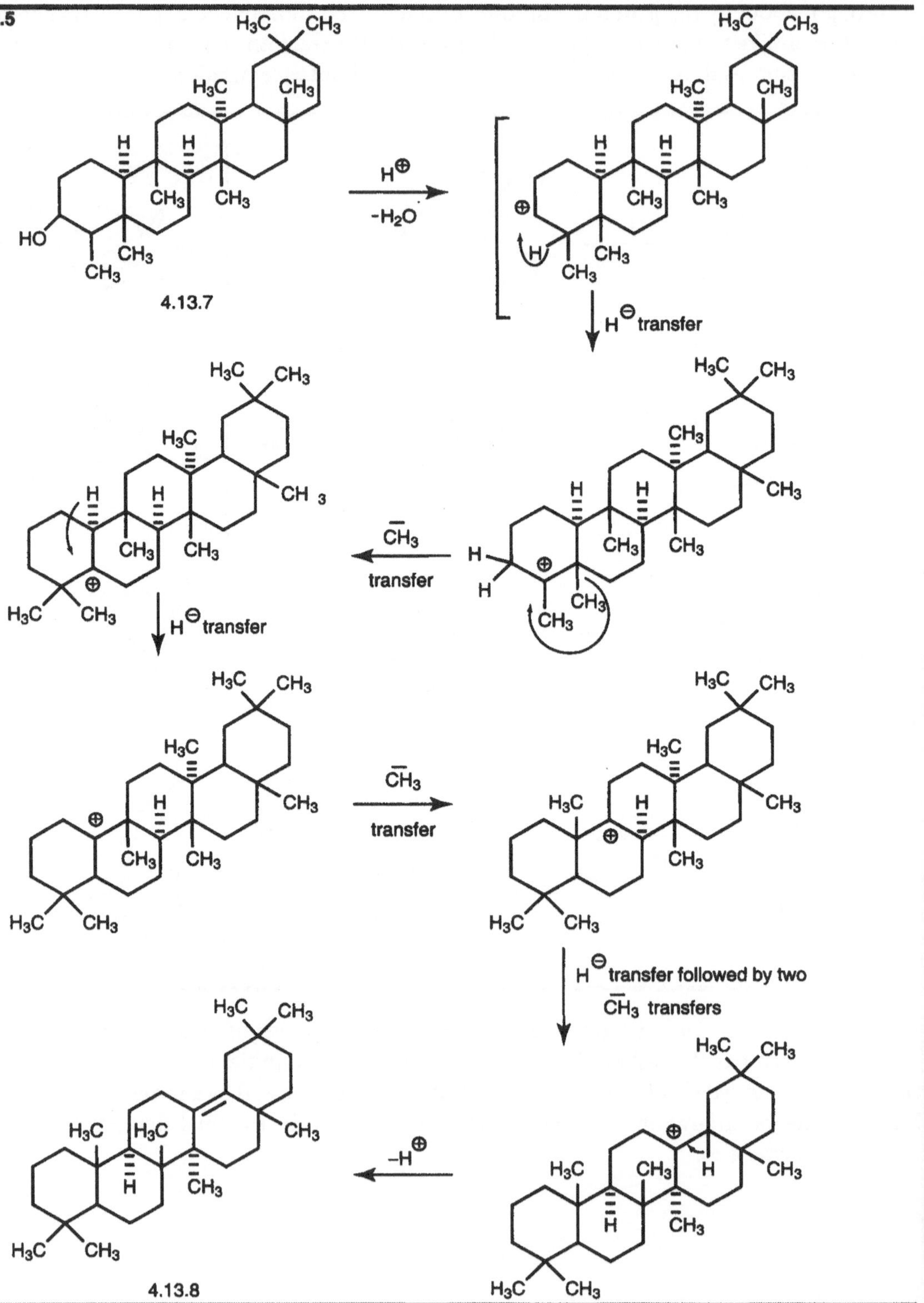
4.13.7
$H^{\oplus}$
$-H_2O$
$H^{\ominus}$ transfer
$\overline{CH_3}$ transfer
$H^{\ominus}$ transfer
$\overline{CH_3}$ transfer
$H^{\ominus}$ transfer followed by two $\overline{CH_3}$ transfers
$-H^{\oplus}$
4.13.8

Suggested reading

1. March J, *Advanced Organic Chemistry, Reactions, Mechanisms and Structure*, 3rd edn, New York: John Wiley, 1985.
2. Li JJ, *Name Reactions. A Collection of Detailed Reaction Mechanisms*, 2nd edn, Berlin: Springer, 2006.

4.14 REACTIONS OF CARYOPHYLLENE

The sesquiterpene hydrocarbon, caryophyllene (4.14.1), which is a component of the oil of lavender and the oil of cloves (*Eugenia caryophyllata*), is an agile molecule. The nine-membered medium-size ring present enables it to exhibit some interesting transannular reactions. In aqueous acid, it gets transformed into a mixture of caryolanol (4.14.2), clovene (4.14.3) and neoclovene (4.1.4.4). If the endocyclic double bond undergoes protonation, the resulting carbocation (4.14.5) can be attacked across the interannular space by the exocyclic double bond leading to the carbocation (4.14.6). Reaction of the latter with water results in caryolanol. On the other hand, if the carbocation (4.14.6) undergoes a Wagner–Meerwein shift involving the cyclobutane ring followed by the loss of a proton, the product is clovene (Scheme 4.6).

Scheme 4.6

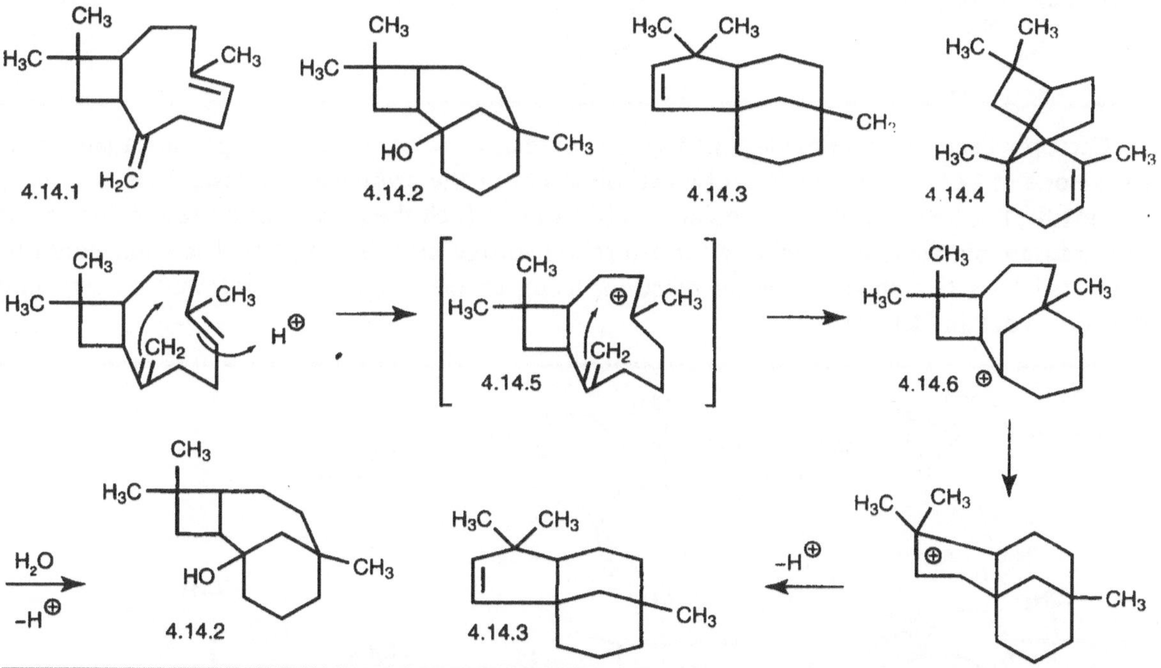

For the formation of neoclovene, the initial step is protonation of the exocyclic double bond, which is followed by the loss of a proton resulting in the diene (4.14.7). The latter then presumably undergoes a proton catalysed transannular interaction between the two double bonds, as depicted in Scheme 4.7. The resulting carbocation (4.14.8) suffers two Wagner–Meerwein shifts before losing a proton to yield neoclovene. It is worth noting that in Schemes 4.6 and 4.7, different conformations of caryophyllene are shown as the reacting species.

Scheme 4.7

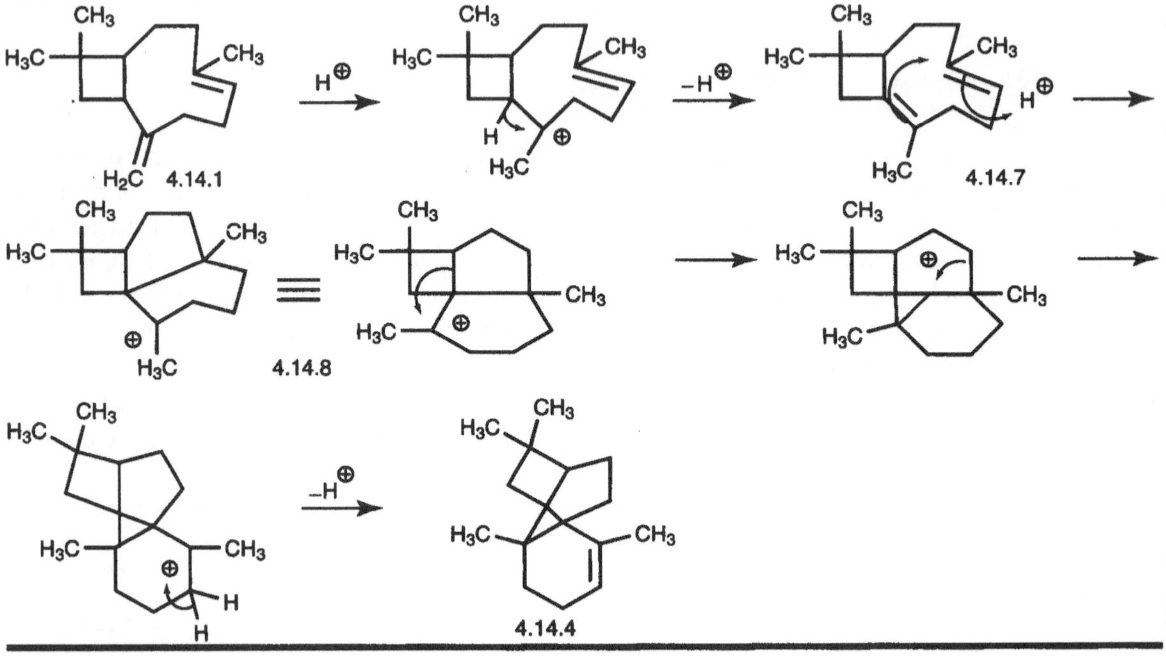

Caryophyllene monoepoxide (4.14.9) on oxidation with potassium pemanganate gives the ketone (4.14.10), with the loss of a carbon atom. In the presence of a base, such as sodium methoxide, this ketone yields the enolate ion (4.14.11), which then reacts with the epoxide group, in a transannular reaction, yielding the tricyclic hydroxy ketone (4.14.12). The latter could be converted by a two-stage oxidation reaction, using chromium trioxide followed by selenium dioxide, to obtain (4.14.13).

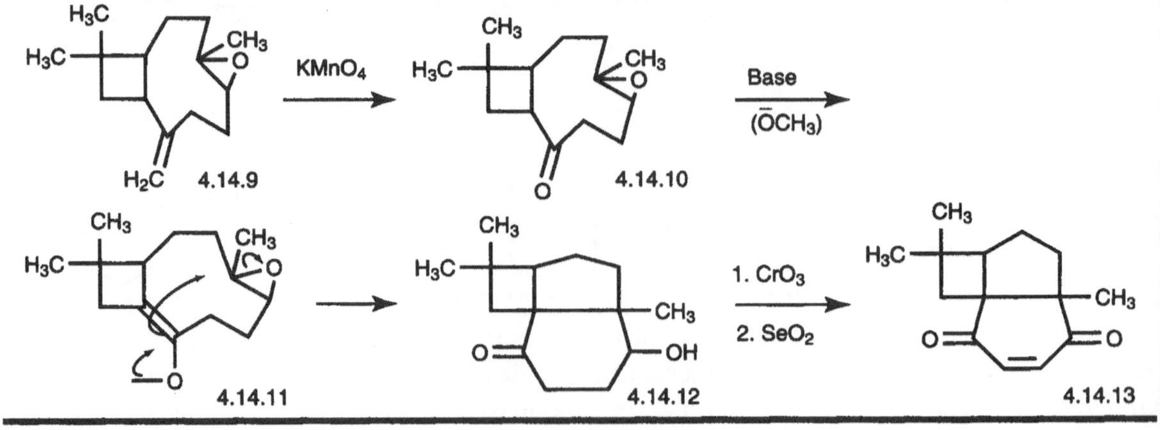

Scheme 4.8

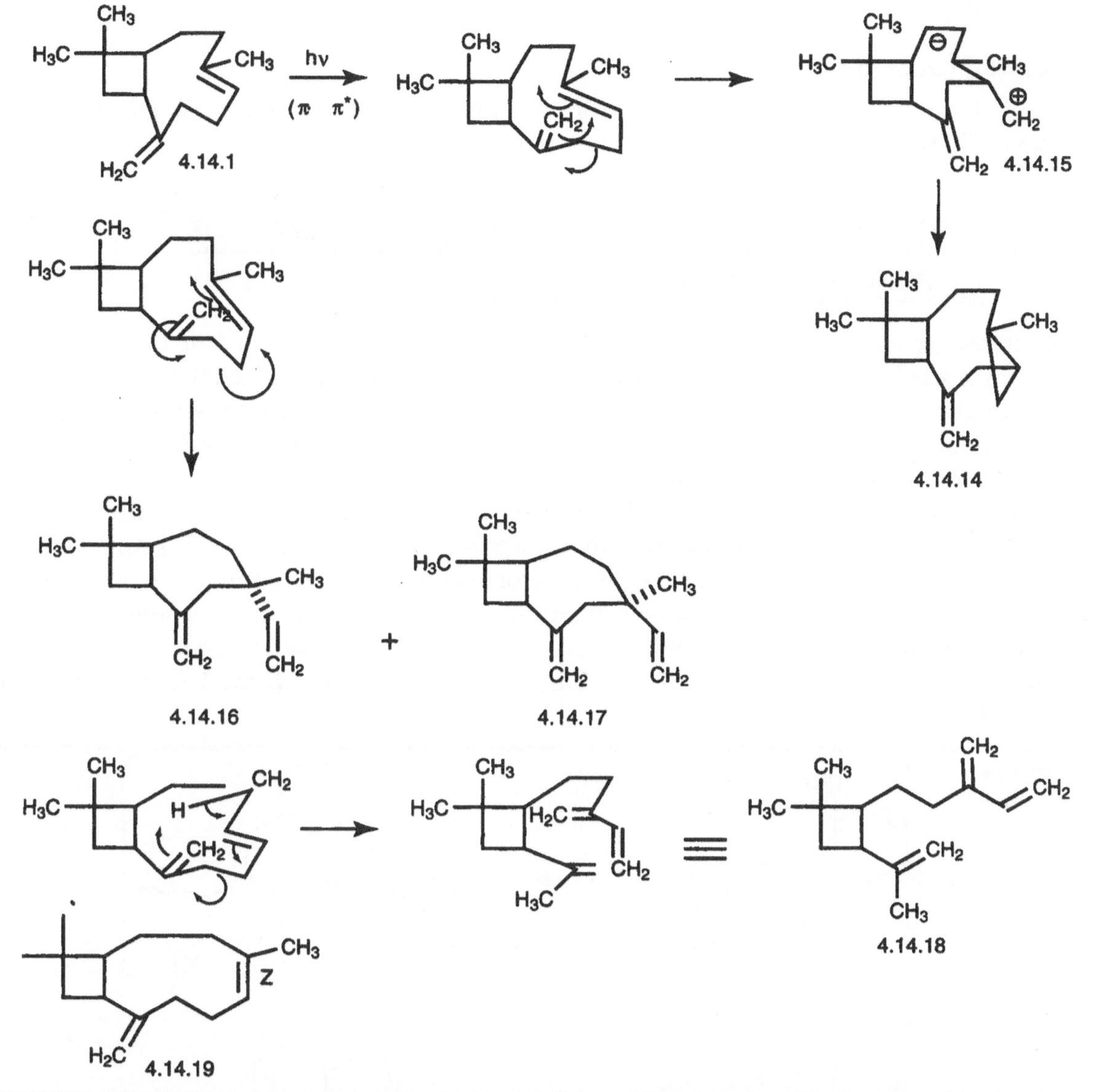

On irradiation with ultra violet light, the homo conjugated diene chromophore in caryophyllene undergoes a $\pi \rightarrow \pi^*$ electronic excitation which triggers a transannular reaction leading to the formation of 4.14.14, which possesses a cyclopropane ring; the zwitter ion (4.14.15) could be an intermediate (Scheme 4.8). Other products are 4.14.16 and its epimer (4.14.17), the cyclobutane derivative (4.14.18), in which the nine-membered ring is opened up and isocaryophyllene (4.14.19). Incidentally, the endocyclic double bond in caryophyllene has the E-configuration whereas in isocaryophyllene it has the Z- configuration. It is well known that light can bring about configurational inversion in geometrical isomers. In the formation of 4.14.16 and 4.14.17 a Cope rearrangement is involved.

Suggested reading

Nakanishi K, Goto T, Ito S, Natori S and Nozoe S, (eds), *Natural Products Chemistry*, Vol 1, New York: Kodansha, Tokyo and Academic Press, 1975 (and references therein).

4.15 SANTONIN TO DESMOTROPOSANTONIN: AN EXAMPLE OF DIENONE–PHENOL REARRANGEMENT

This interesting and much studied sesquiterpene lactone was first isolated in 1830 by Kahler from 'worm seed' (also known as santonica, which are the dried unexpanded flower heads of the Asteracea plant, *Artemisia maritima*). It also occurs in other species of Artemisia. Its chemistry is fascinating and its NMR spectra, discussed in an earlier chapter, are instructive. In this section, its rearrangement to desmotroposantonin is described.

Santonin has the structure (4.15.1). It possesses a dienone moiety which is reflected in the ultraviolet spectrum of the compound with absorption maxima at 227 and 254 nm. Santonin undergoes rearrangement readily in the presence of acids to yield desmotroposantonin (4.15.2). The driving force for this transformation is the aromatisation of the cyclohexadienone A ring. The first step is the protonation of the carbonyl oxygen atom. This is followed by enolisation, shift of a double bond and migration of the angular methyl group as shown in Scheme 4.9. The final step is the loss of a proton.

Scheme 4.9

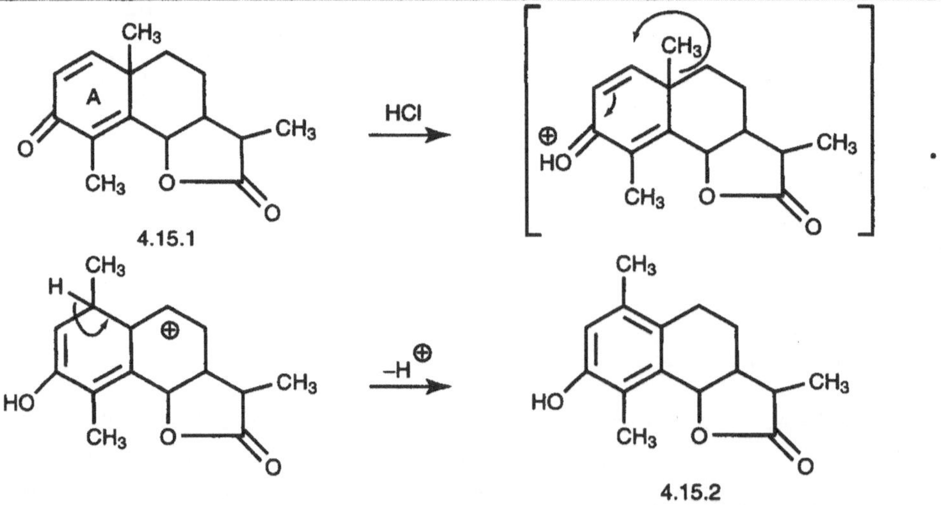

Santonin oxime (4.15.3) (obtained by the action of hydroxylamine on santonin) can be reduced by zinc and acid to get the amino compound (4.15.4). The latter, under acid catalysis, loses a molecule of ammonia followed by shift of the double bond and methyl migration, similar to the sequence in the conversion of 4.15.1. to 4.15.2, to yield hyposantonin (4.15.5) (Scheme 4.10).

Scheme 4.10

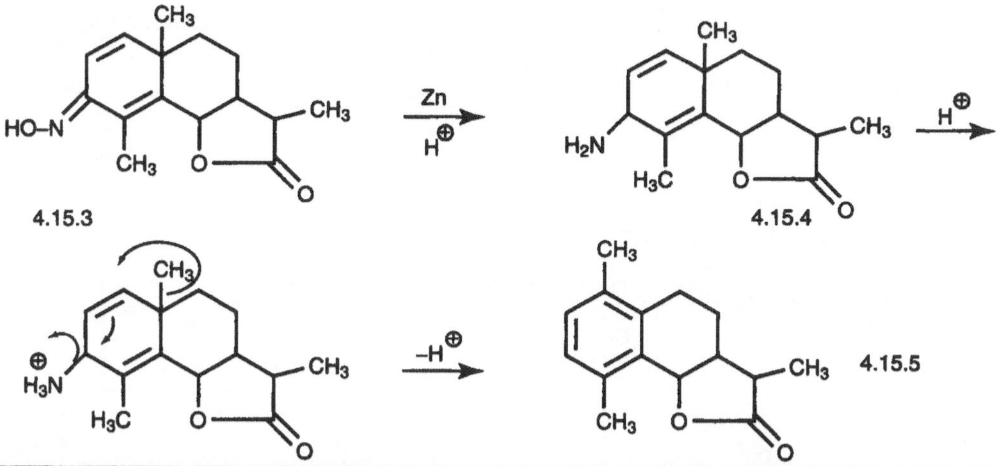

On reduction with hydriodic acid–red phosphorus, santonin yields santonous acid (4.15.6), presumably via desmotroposantonin and the iodo acid (4.15.7). Apparently, in this reaction, hydriodic acid plays more than one role. As an acid catalyst, it initiates the dienone–phenol rearrangement step and thereafter, brings about the opening of the lactone ring, replacement of the hydroxyl by iodine and a final reductive removal of the iodine atom.

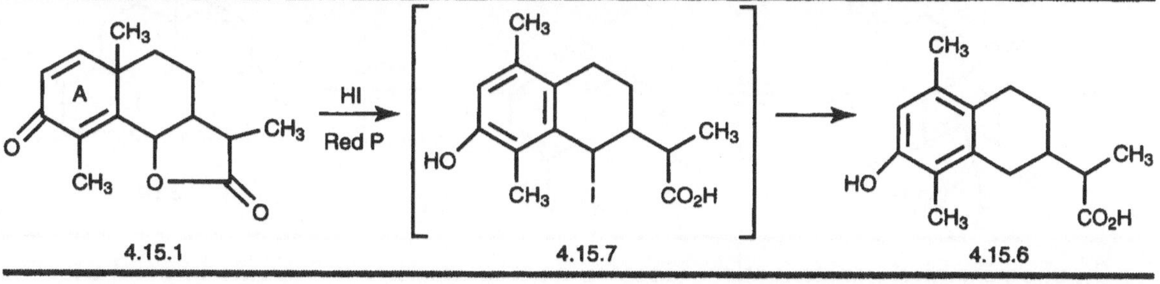

Though not related to the above mentioned rearrangements, the conversion of santonin into santonic acid (4.15.8) on prolonged treatment with alkali also has instructive value. In this reaction, the first step is the hydrolytic opening of the lactone ring. It has been postulated that in the resulting dianion (4.15.9), a hydride ion transfer occurs as shown in Scheme 4.11. One of the subsequent steps is an intramolecular Michael addition of a carbanion to the conjugated double bond. Santonic acid, on treatment with sulphuric acid, yields γ-*m*-santonin (4.15.10). The probable steps involved in this reaction are also shown in Scheme 4.11.

Scheme 4.11

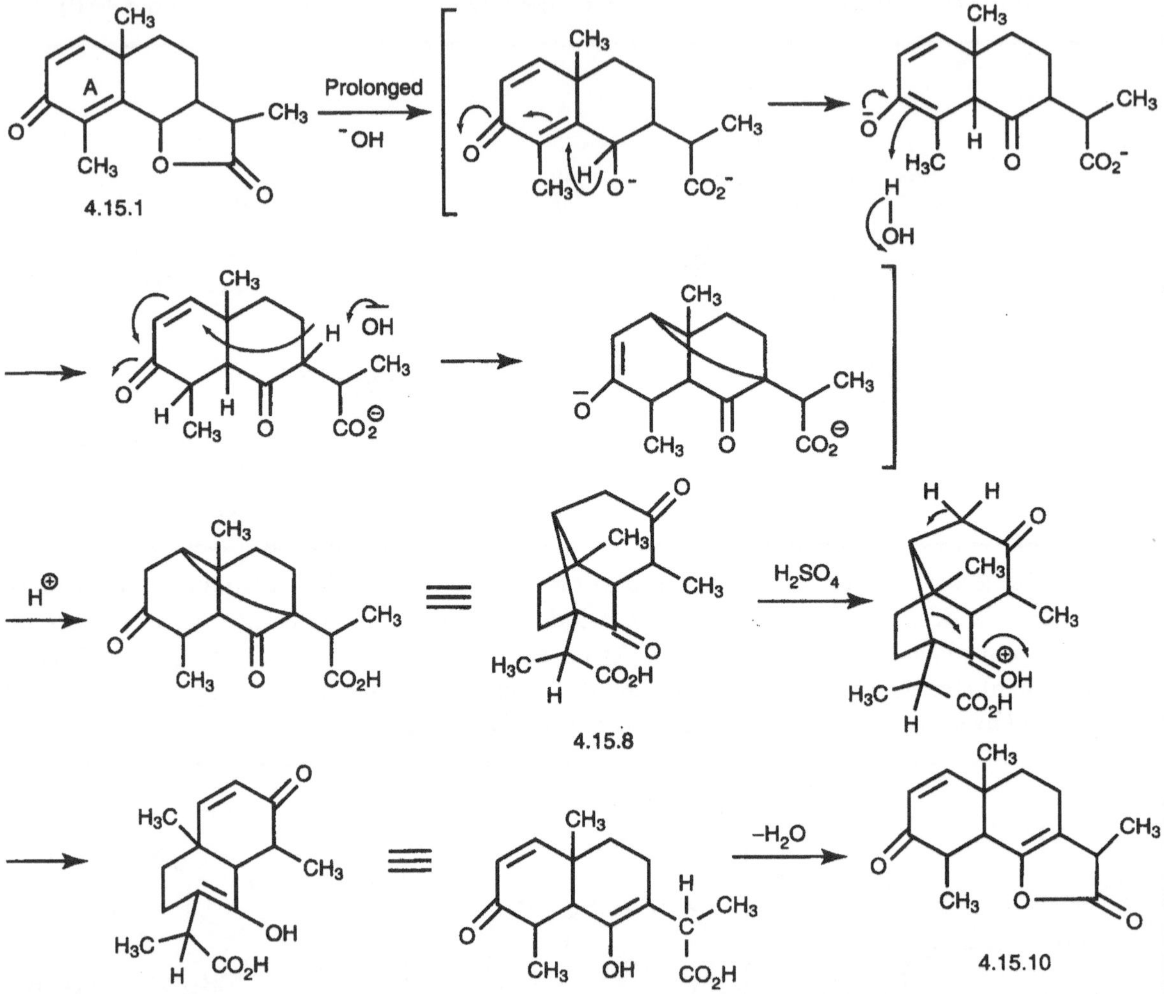

The photochemical transformations of santonin are summarised in Schemes 4.12 and 4.13. The products are Lumisantonin (4.15.11), the dienone (4.15.12), and photosantonin (4.15.13) (when the reaction is caried out in ethanol solution). When santonin is irradiated in acetic acid solution, the product is isosantonic lactone (4.15.14) (Scheme 4.14). A free radical mechanism has been suggested since the excitation involves a n-pi* transition. The mechanisms suggested are based on the extensive studies of DHR Barton, de Mayo and others.

Scheme 4.12

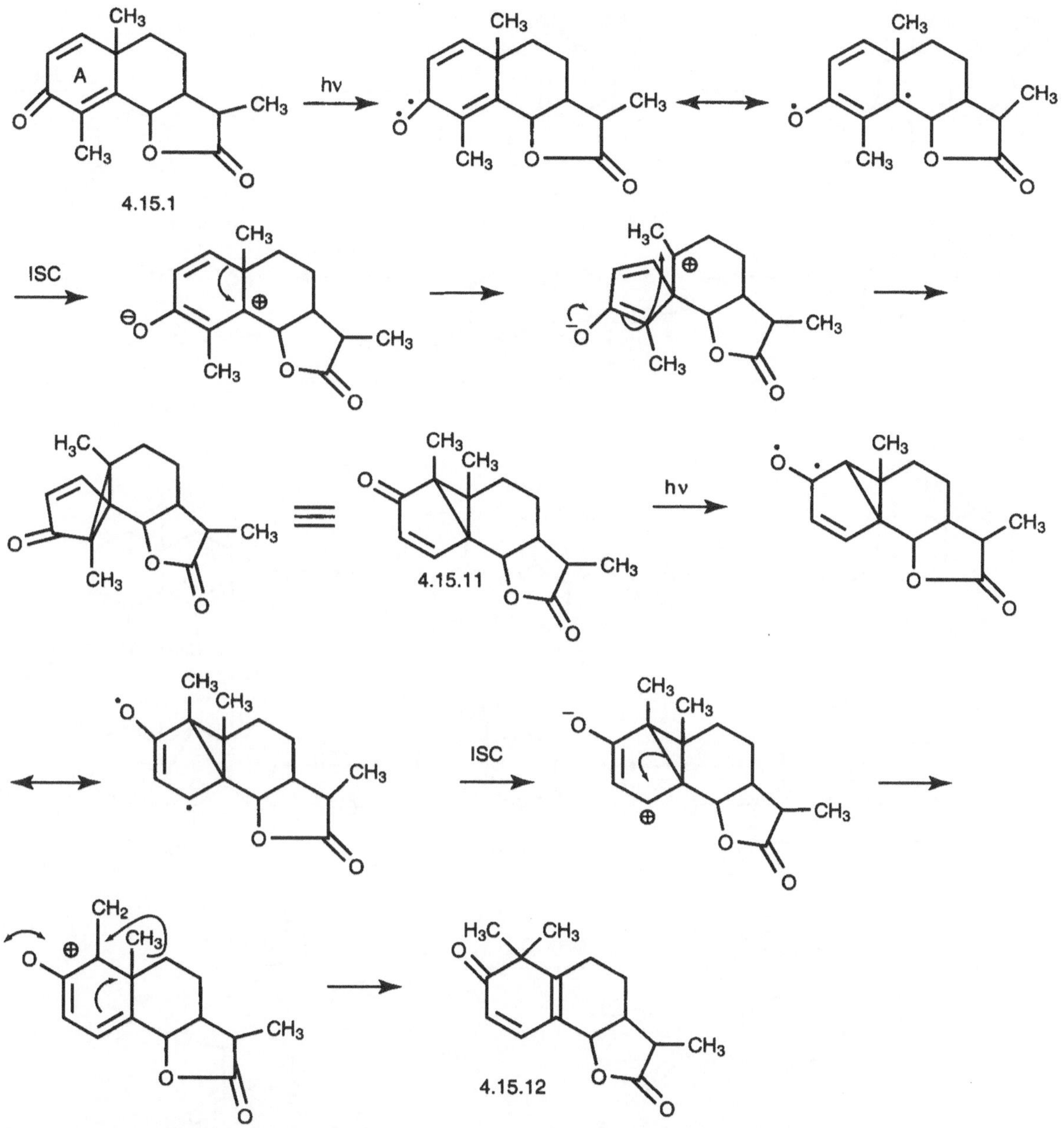

4.15.1

4.15.11

4.15.12

Note: ISC is intersystem crossover

Scheme 4.13

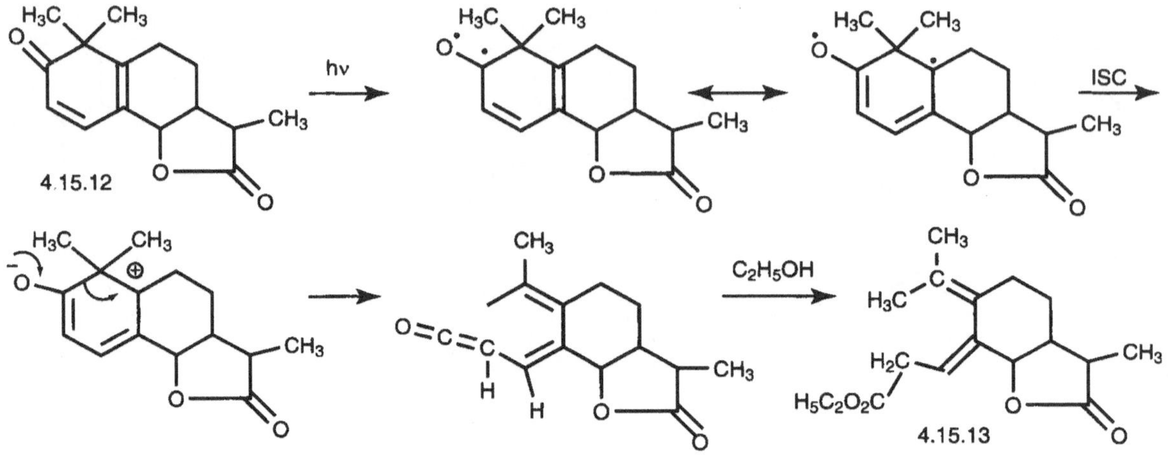

4.15.12

4.15.13

· Scheme 4.14

4.15.1

4.15.14

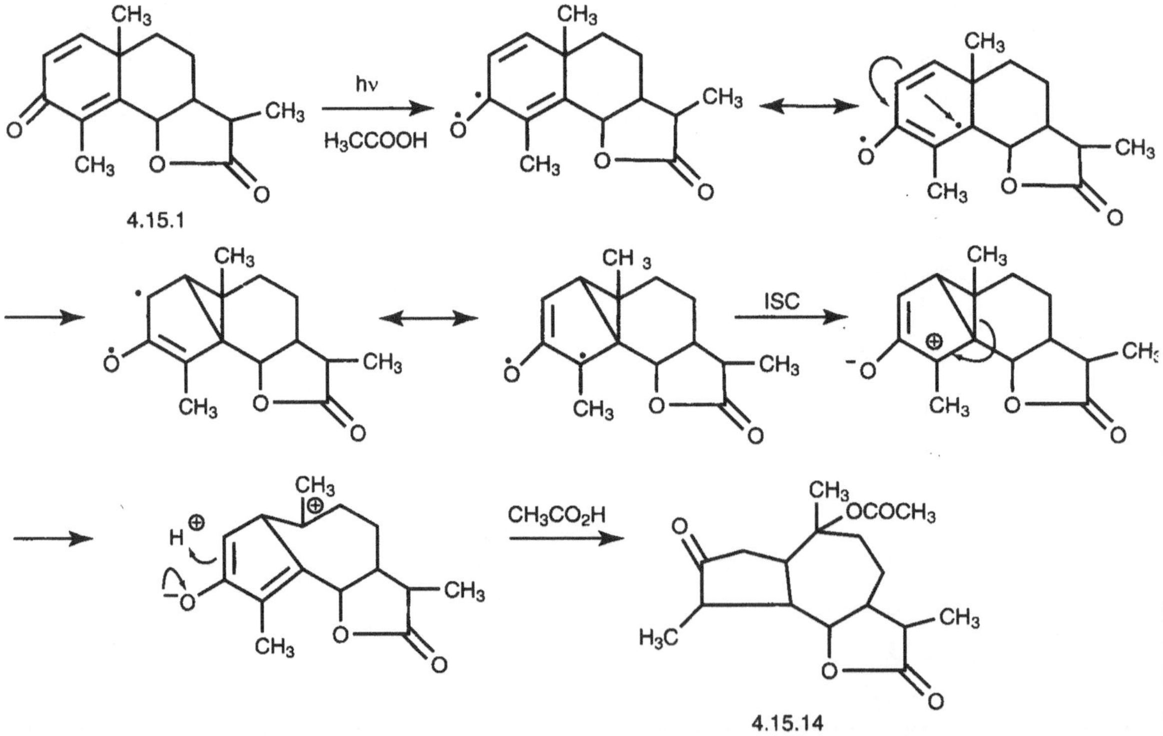

Suggested reading

1.   Sell C, *A Fragrant Introduction to Terpenoid Chemistry*, London: RSC Publication, 2003.
2.   Birladeanu L, *Angewandte Chem.*, 2003, 42: 1201–8.

5

Synthesis

5.1 INTRODUCTION

Analysis and logical deduction are the basis for structure determination, whereas creativity and ingenuity are the essential components of organic synthesis. Over the years this important branch of organic chemistry has transformed from a routine, systematic molecular construction as per a blue print to a work of art involving innovative procedures and new reagents. Very often, the actual practical synthetic plan is preceded by a retrosynthetic analysis of the target molecule. Komppa's synthesis of camphor (5.1.1; Scheme 5.1) and Hans Fischer's synthesis of hemin (5.1.2; Scheme 5.2) are classical examples of a straightforward approach to synthesis, primarily involving already established and unambiguous reactions. On the other hand, Woodward's synthesis of chlorophyll (5.1.3; Scheme 5.3) and Corey's sythesis of the prostaglandins (for example, prostaglandin F2 alpha, 5.1.4; Scheme 5.4) are illustrative of an exploratory and innovative approach.

Scheme 5.1

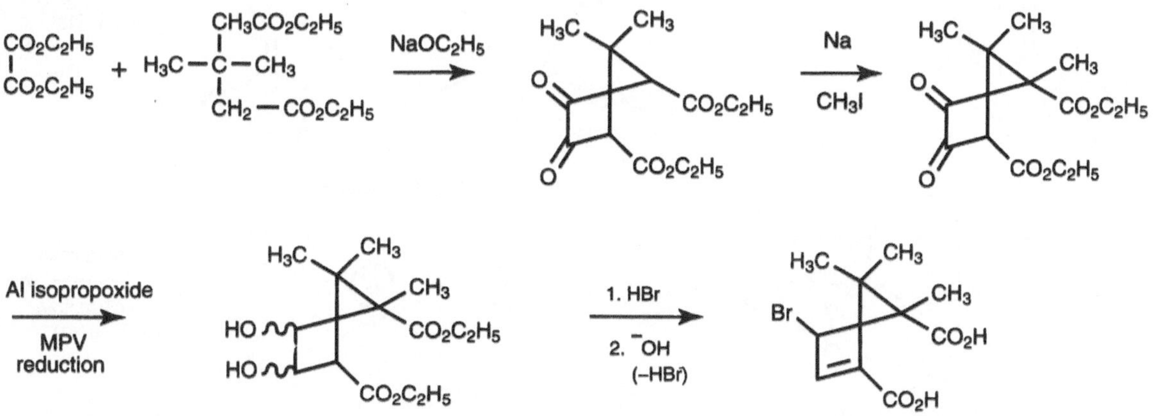

Note: MPV reduction = Meerwein–Pondorff–Verley reduction

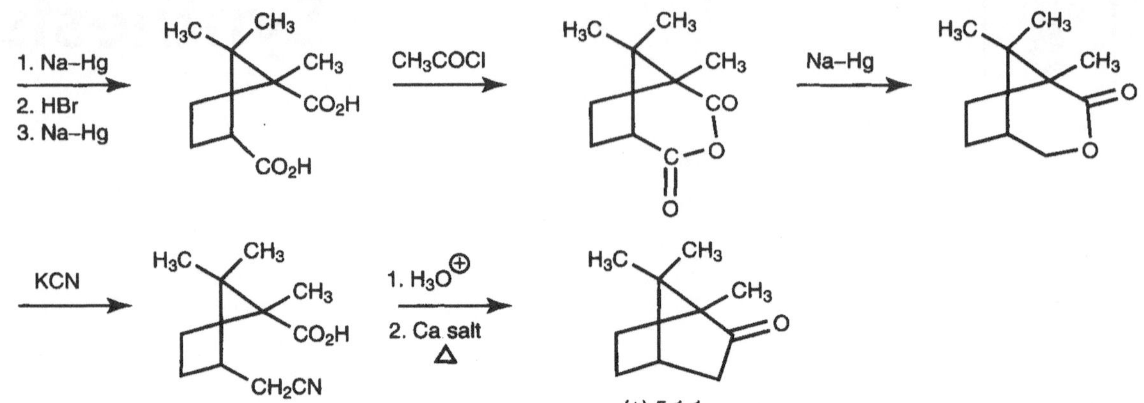

(±) 5.1.1

Scheme 5.2

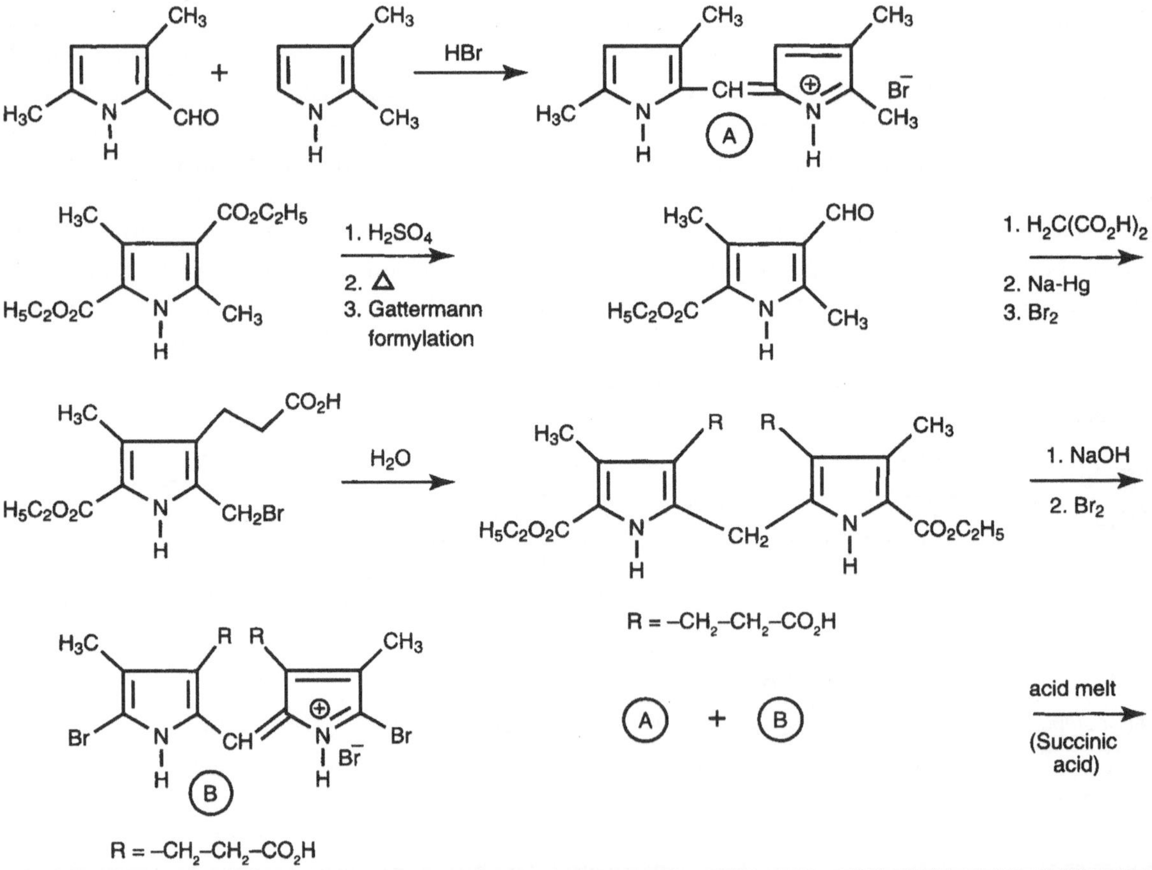

$R = -CH_2-CH_2-CO_2H$

$R = -CH_2-CH_2-CO_2H$

5.1.2

5.1.3

Scheme 5.3

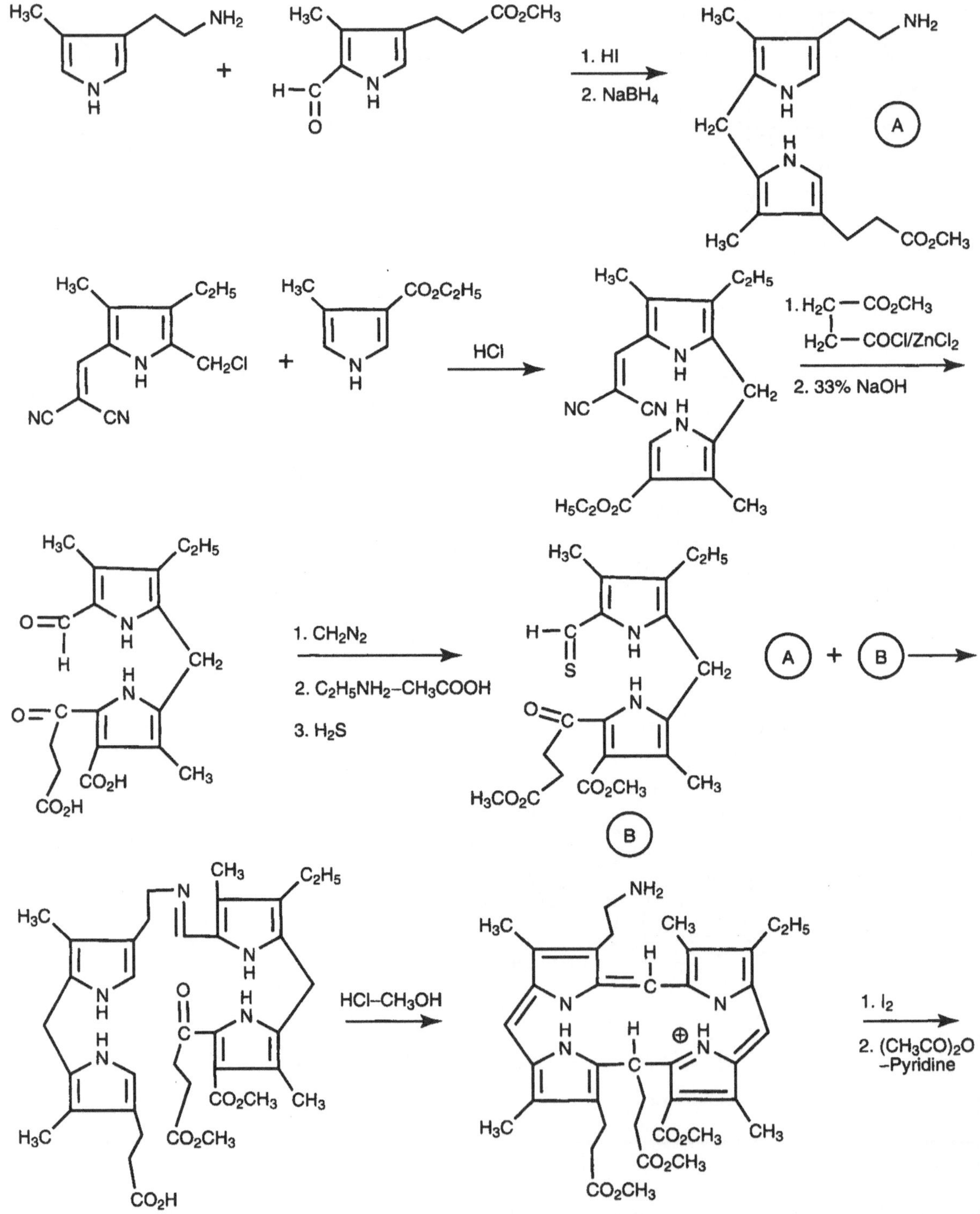

H3C
NH2
N H
+
CO2CH3
H—C
O
1. HI
2. NaBH4
H2C
A
CO2CH3
C2H5
CH2Cl
NC
CN
CO2C2H5
HCl
CH2
H5C2O2C
CH3
1. H2C—CO2CH3
H2C—COCl/ZnCl2
2. 33% NaOH
CO2H
1. CH2N2
2. C2H5NH2–CH3COOH
3. H2S
S
H3CO2C
CO2CH3
B
A + B
HCl–CH3OH
1. I2
2. (CH3CO)2O –Pyridine

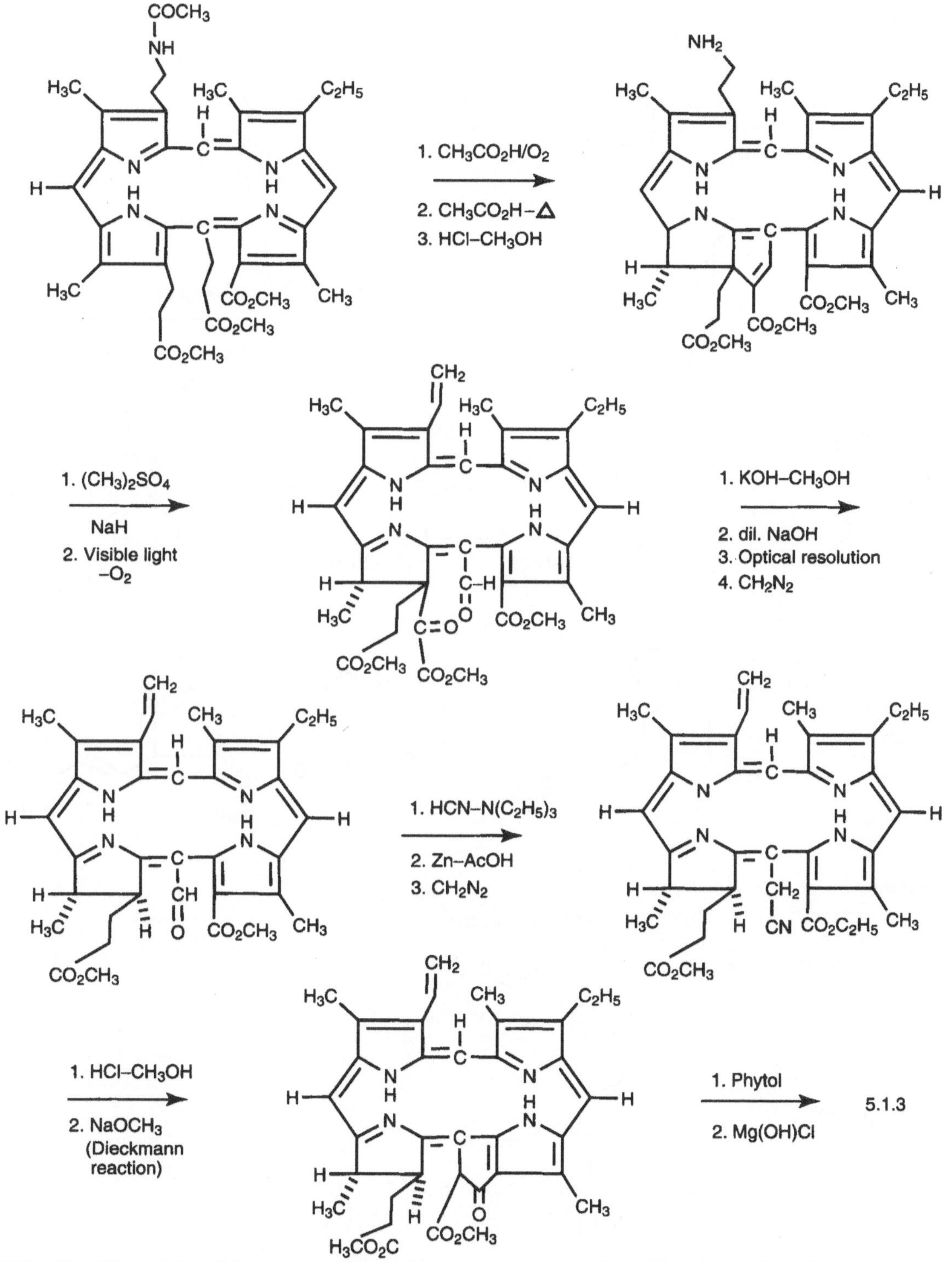
1. CH3CO2H/O2
2. CH3CO2H–Δ
3. HCl–CH3OH
1. (CH3)2SO4
NaH
2. Visible light
–O2
1. KOH–CH3OH
2. dil. NaOH
3. Optical resolution
4. CH2N2
1. HCN–N(C2H5)3
2. Zn–AcOH
3. CH2N2
1. HCl–CH3OH
2. NaOCH3
(Dieckmann
reaction)
1. Phytol
2. Mg(OH)Cl
5.1.3

Scheme 5.4

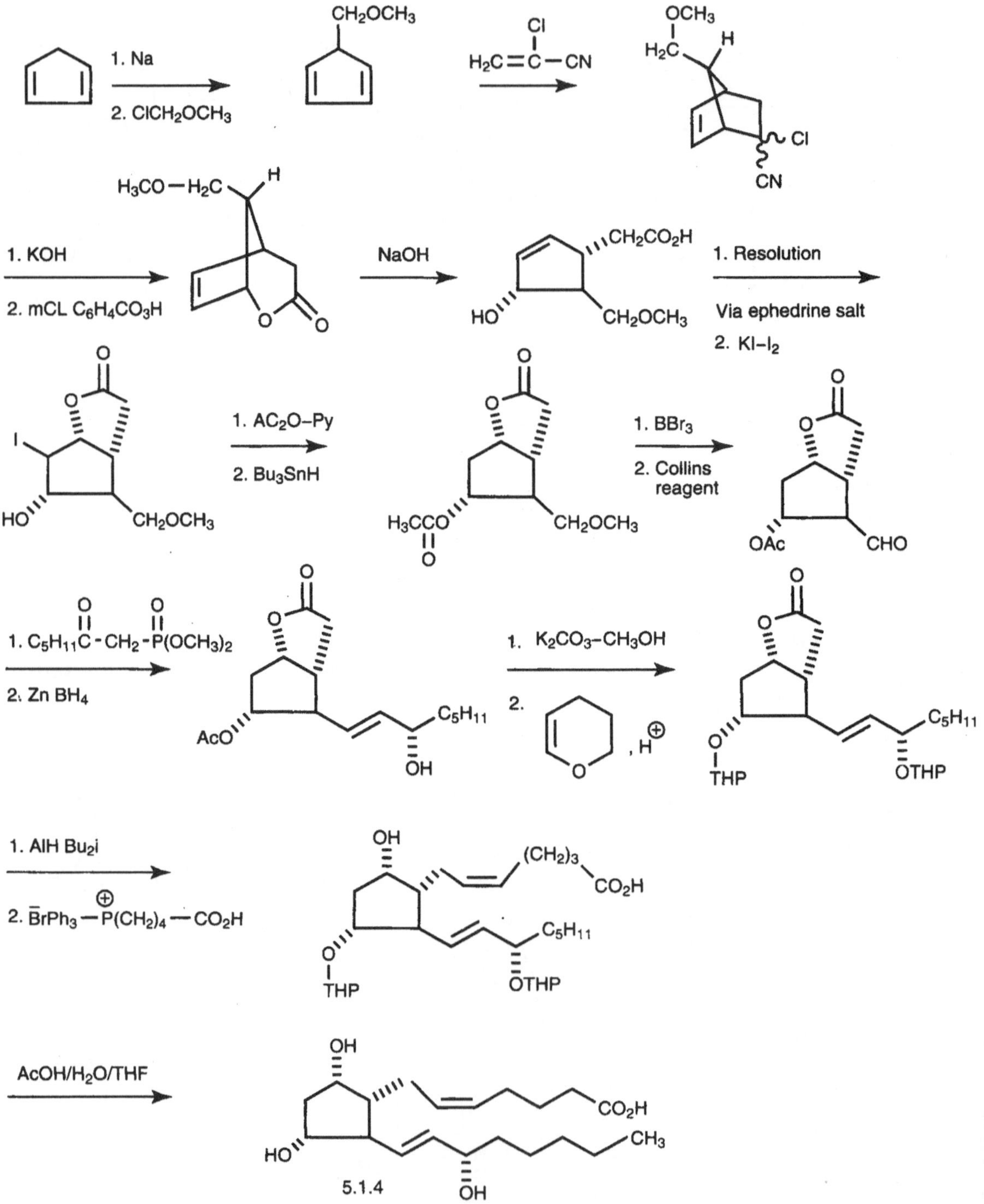

Achieving the objective, often after a large number of steps, and obtaining the end product in poor yield, without stereospecificity, is no longer satisfying and acceptable. Regiospecific and stereospecific reactions are chosen and the reagents selected are those which ensure optimum yields at each stage. Since creativity is the soul of synthetic organic chemistry, the aim is not only to recreate molecules of nature in the laboratory but also to produce new and exotic compounds. Indeed, at the present time, synthetic drugs, dyes, polymers and insecticides far outnumber corresponding compounds of natural origin, though the latter continue to provide the impetus for the creation of such synthetic compounds. However, in this chapter we will be discussing only the total synthesis of some select natural products. Before taking up specific examples as case studies, it will be relevant to give a brief account of the strategies used in a contemporary natural product synthesis.

Retrosynthetic Analysis

The concept of retrosynthesis was propounded by Corey and was considered worthy of a Nobel prize. It is best to summarise it in Corey's own words: " Retrosynthetic (or antithetic) analysis is a problem-solving technique for transforming the structure of a synthetic target (TGT) molecule to a sequence of progressively simpler structures along a pathway which ultimately leads to simple or commercially available starting materials for a chemical synthesis. The transformation of a molecule to a synthetic precursor is accompanied by the application of a transform, the exact reverse of a synthetic reaction, to a target structure. Each structure derived from a TGT then becomes a target for further analysis. Repetition of this process eventually produces... pathways from bottom to top corresponding to possible synthetic routes to the TGT". We shall illustrate this concept with one example. The monoterpene hydrocarbon camphene (5.1.5) can be derived from the alcohol (5.1.6), which, in turn, is obtainable from the carboxylic acid (5.1.7). The latter can be prepared by the action of sodium hypobromite on the methyl ketone (5.1.8). The latter can, by a retro Diels–Alder reaction, yield cyclopentadiene (5.1.9) and mesityl oxide (5.1.10). This retro synthetic analysis and an actual synthesis based on it are given in Scheme 5.5.

Scheme 5.5

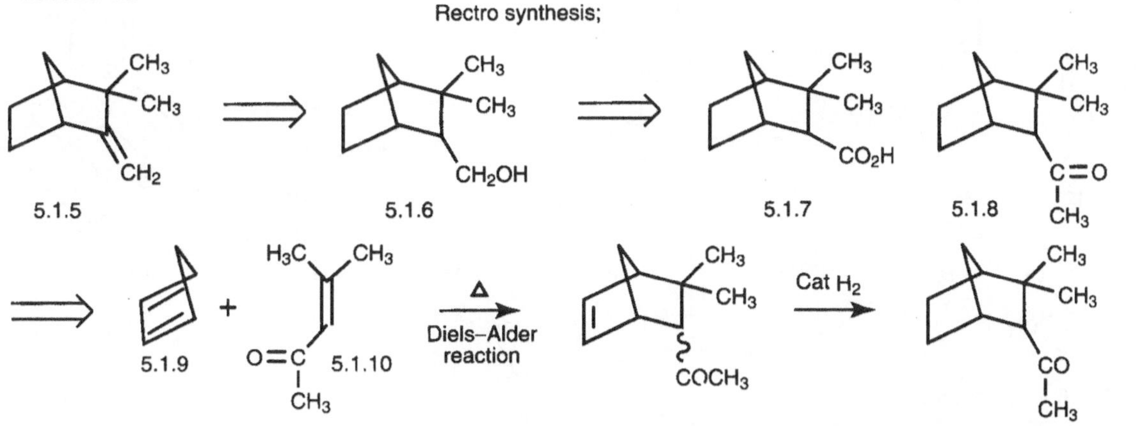

Rectro synthesis;

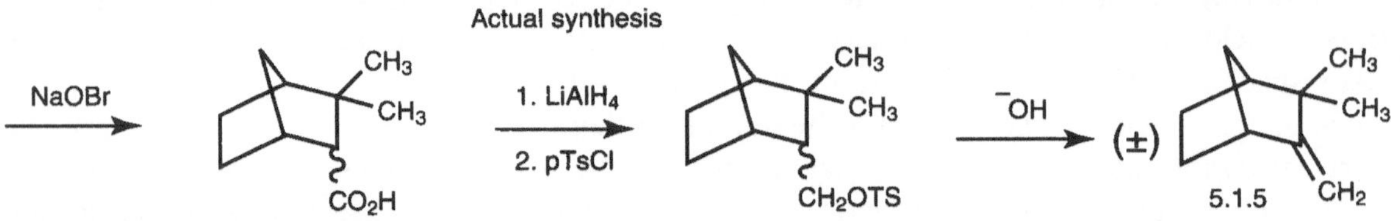

Actual synthesis

Choice of Reactions

Among the large number of reactions available for building molecules of different types, those which are regio- and stereo-specific are preferred in the total synthesis of natural products, the majority of which possess one or more chiral centres. Heading the list of such reactions is the Diels–Alder reaction, which is a pericyclic (cycloaddition) reaction. One example has already been cited in Scheme 5.5 above. Woodward used this reaction as the first step in his synthesis of the alkaloid, reserpine (5.1.11). Thus, the addition of p-benzoquinone (5.1.12) to the methyl ester of butadiene carboxylic acid (5.1.13) stereospecifically yielded the bicyclic compound (5.1.14) in which the two rings are cis-fused. This reaction, which is thermally allowed, was also used by Corey in his synthesis of prostaglandin F2 (Scheme 5.4). Another example of the effective use of the Diels–Alder reaction is seen in the synthesis of the diterpene, taxol (5.1.15), a component of the bark of the yew tree (*Taxus baccata*), which shot into prominence because of its anti-cancer activity. In the construction of the ring A of this compound, Nicolaou and his co-workers added 2-chloroacrylonitrile (5.1.16) to the diene (5.1.17) to obtain the product 5.1.18.

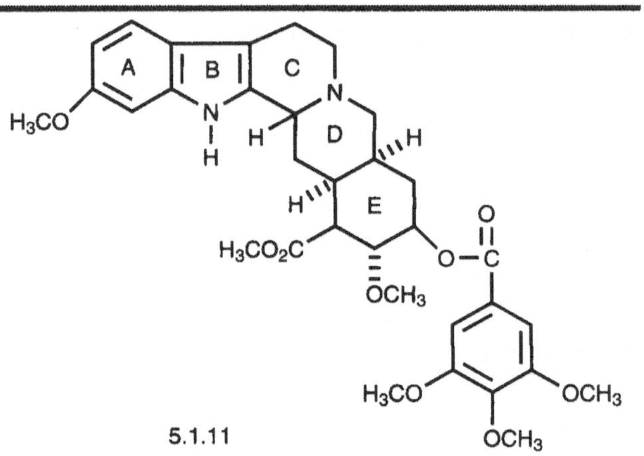

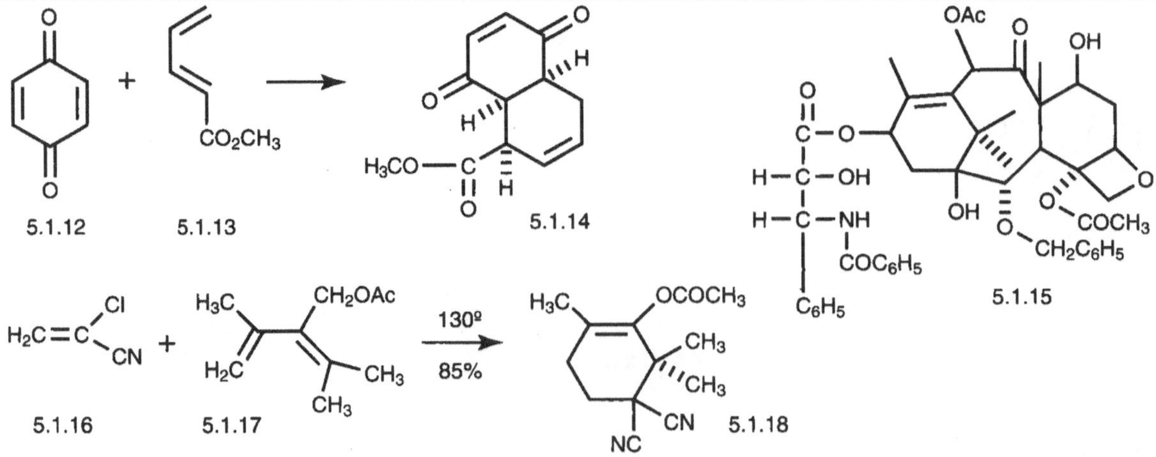

5.1.11

5.1.12   5.1.13   5.1.14   5.1.15

5.1.16   5.1.17   5.1.18

Another reaction which merits a brief description here is the Robinson annulation reaction. This base-catalysed reaction is a tandem Aldol condensation–Michael addition and is very effective for constructiong bicyclic compounds in one step. It was first developed during studies on the synthesis of steroids. Corey used it as the first step in his total synthesis of the sesquiterpene longifolene (5.1.19). Thus, a base-catalysed reaction between methyl vinyl ketone (5.1.20) and 1-methylcyclohexane-2, 6-dione (5.1.21) yielded the bicyclic diketone 5.1.22. The further steps involved in this synthesis also deserve mention here as they bring to light some other important strategies and reactions vital to the success of a multi-step total synthesis of a natural product. Thus, in the next step, the more reactive carbony group was protected by ketalisation and the other carbonyl group made to undergo a Wittig reaction. Selective hydroxylation of the exocyclic double bond in the product (5.1.23) using osmium tetroxide, a versatile reagent for such oxidations, yielded the diol 5.1.24. Selective tosylation of the secondary hydroxyl followed by an intramolecular elimination of the tosyloxy group resulted in 5.1.25, in which ring A is 7-membered. Hydrolysis of the ketal (ketals are stable under basic conditions but are susceptible to hydroysis in aqueous acids) gave the diketone (5.1.26). Treatment of the latter with triethylamine brought about an intramolecular Michael reaction resulting in the tricyclic diketone (5.1.27). The subsequent steps are easy to follow (Scheme 5.6).

Other important reactions met with in natural product syntheses are sigmatropic reactions such as the Cope and Claisen rearrangements, Birch reduction, hydroboration reaction, Sharpless epoxidation, catalytic hydrogentaion, homogeneous catalysis, Dieckmann cyclisation, and Stork's enamine reaction, For example, the Birch reduction was a key step in the synthesis of the diterpene, abietic acid (5.1.28). Thus, 6-isopropyl-2-naphthol (5.1.29) was reduced with sodium in liquid ammonia to obtain the ketone (5.1.30) which was converted—through a number of steps (including a Robinson annulation)—to dehydroabietic acid (5.1.31). The latter was converted into 5.1.32 by another Birch reduction. The final step involved an acid catalysed isomerisation of 5.1.32 to abietic acid.

Scheme 5.6

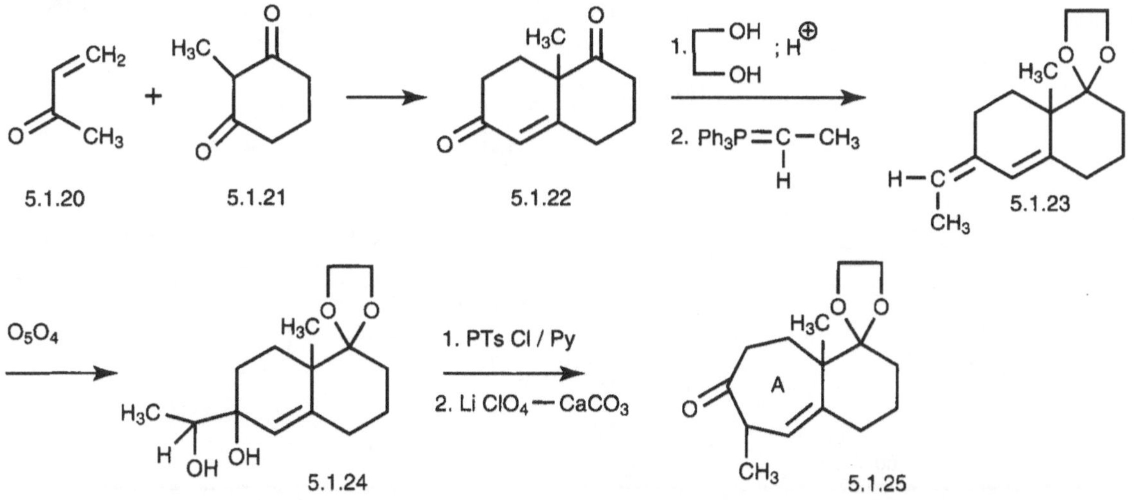

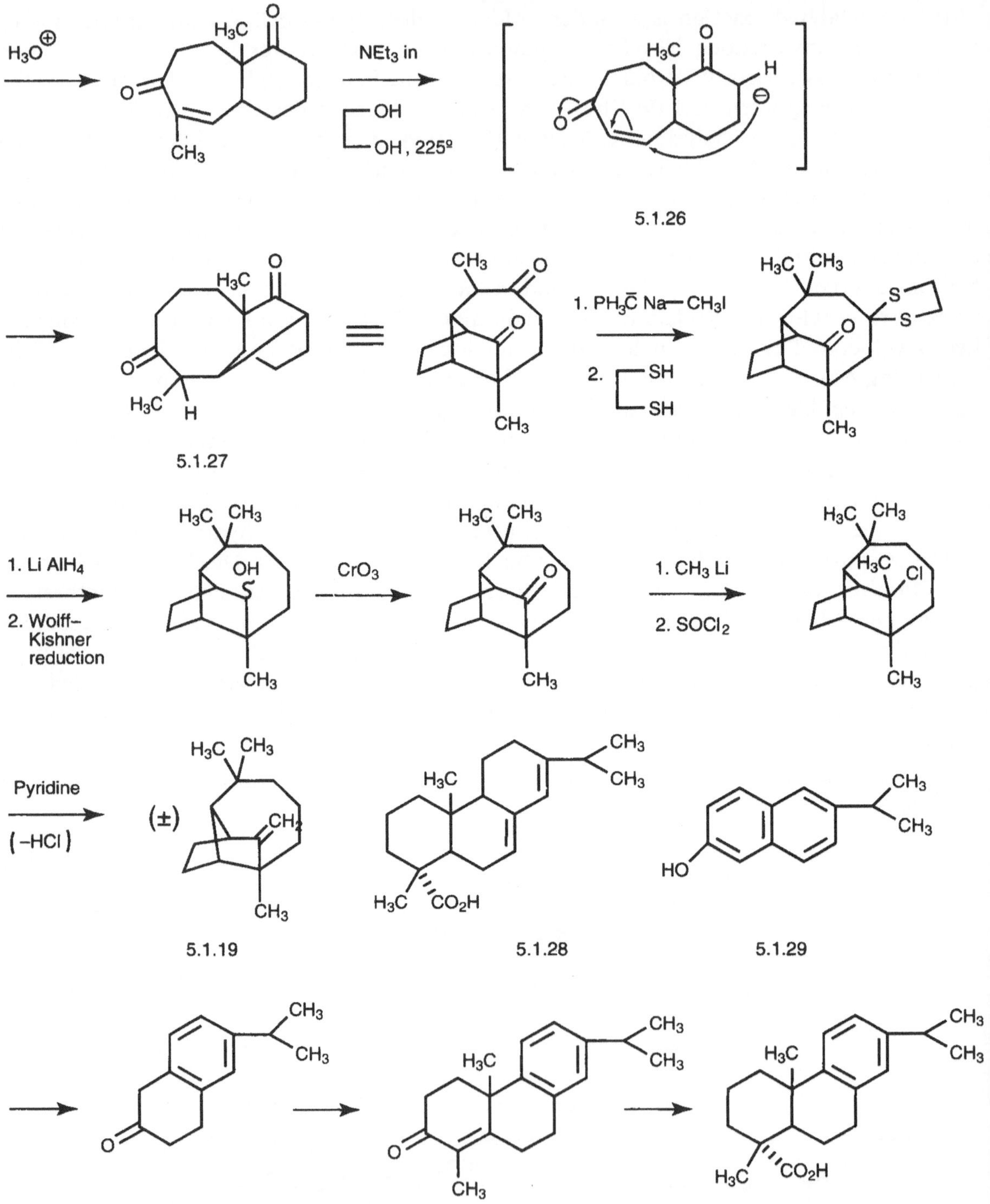

5.1.26

5.1.27

5.1.19

5.1.28

5.1.29

5.1.30

5.1.31

Reagents

Several new reagents have been introduced in recent years for effecting a variety of reactions but they are too numerous to be described in this book. The monumental series *Reagents for Organic Syntheses* by Fieser and Fieser is an invaluable source of information on organic reagents. The Corey catalysts, the boranes developed by Brown and his co-workers, the palladium complexes, such as those used in the Heck reaction (arylation or alkenylation of alkenes), butyl lithium, lithium aluminium hydride, thallium compounds (for the preparation of biaryls, for example) and the oxidising agents, derived from chromium trioxide, developed by Corey and his co-workers are some of the important reagents. These and more are likely to be encountered in the examples which follow this introduction.

Biotransformations

There is an increasing awareness of the adverse effects of chemicals on the environment and hence the need for 'Green Chemistry'. An alternative to the use of chemical reagents is to employ biological agents such as microorganisms and cellular extractives containing enzymes for bringing about a variety of reactions like esterification, hydrolysis, oxidations and reductions. Among these, as far as natural products are concerned, the agents which effect various types of oxidation are perhaps the most important. Oxidative transformations include epoxidation of double bonds, hydroxylation of aromatic rings, phenol oxidation, oxidative coupling, allylic and benzylic hydroxylation. For example, hydroxylation of flavonoids and related compounds has been receiving considerable attention in recent years. For instance, using recombinant *Escherichia coli* strains containing the enzyme biphenyl-2,3-dioxidase, 7,4'-dihydroxyisoflavone (daidzein) (5.1.29) could be converted in 100% yield to 7,2',4'-trihydroxy isoflavone (5.1.30). Similarly, using a strain of *Aspergillus alliaceus*, the chalcone (5.1.31) was converted into a mixture of the chalcones (5.1.32) and (5.1.33) and the flavanones (5.1.34) and (5.1.35), though in poor yields (6%–15%). Biotransformations of other classes of secondary metabolites (for example, the steroids and triterpenes) have also been studied; the books mentioned under suggested reading may be consulted for details regarding these and other biotransformations.

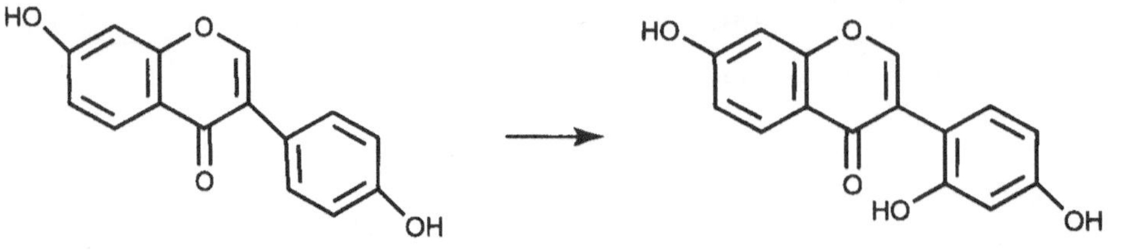

5.1.29

5.1.30

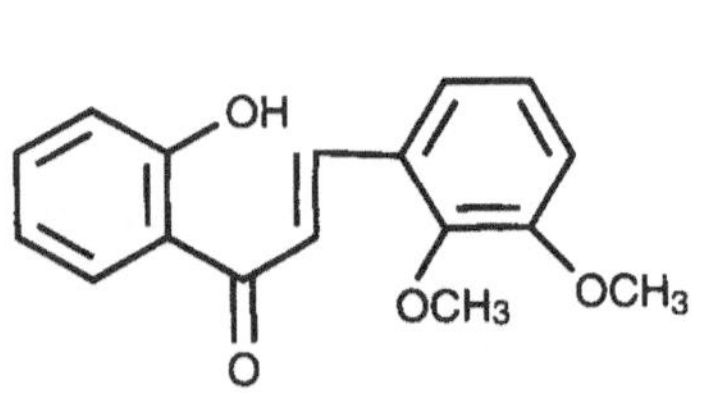

5.1.31

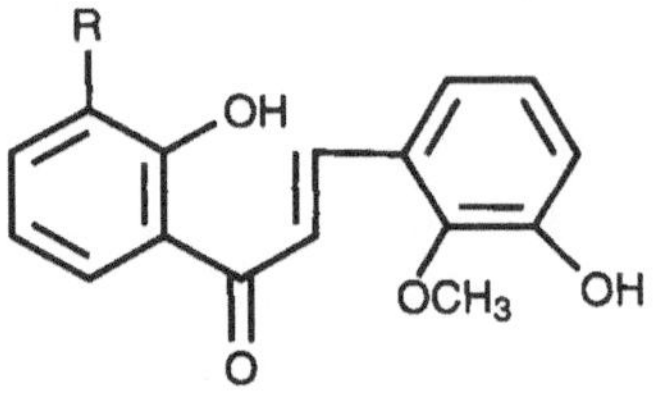

5.1.32: R = H
5.1.33: R = OH

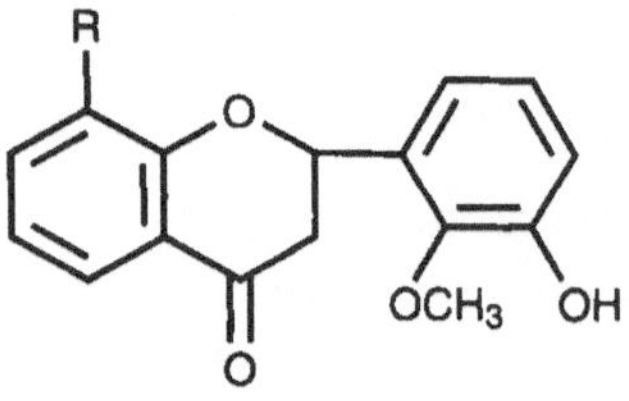

5.1.34: R = H
5.1.35: R = OH

Natural Products and Combinatorial Synthesis

We shall conclude this introduction with a brief account of an emerging area of research in natural product chemistry. Combinatorial synthesis is a methodology which is now well established as a tool for the development of new drugs. Due to the pioneering work of KC Nicolaou and his co-workers, and others, an integrated natural product–combinatorial chemistry has evolved during the past few years. Two strategies have been used for this purpose: the total synthesis of naturally occurring compounds and their use as templates for the construction of combinatorial libraries. We shall cite here only one example. Using a polystyrene based selenyl bromide resin, Nicolau et al prepared the 2,2-dimethylbenzopyran template (5.1.36) for the construction of libraries of pyranochalcones, coumarins, flavones and stibenes. For instance, using 5.1.37, a group of chalcones (5.1.38) possessing important biological activities was prepared by condensation with a variety of benzaldehydes (5.1.39). Using different approaches, natural product-like libraries based on the diterpene taxol (5.1.15), the antibiotic vancomycin (5.1.40) and the animal perfume, muscone (5.1.41) have also been prepared.

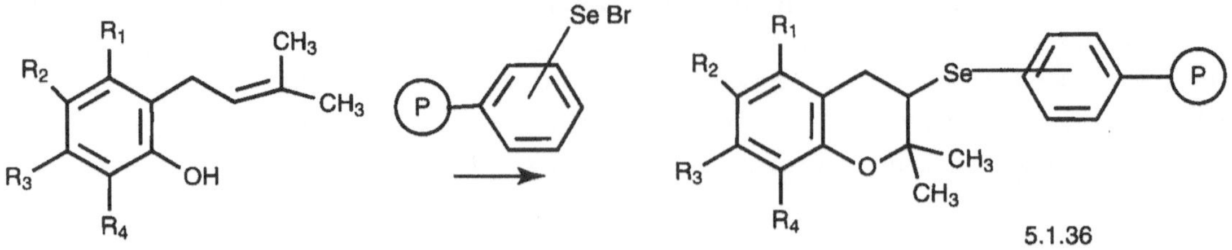

5.1.36

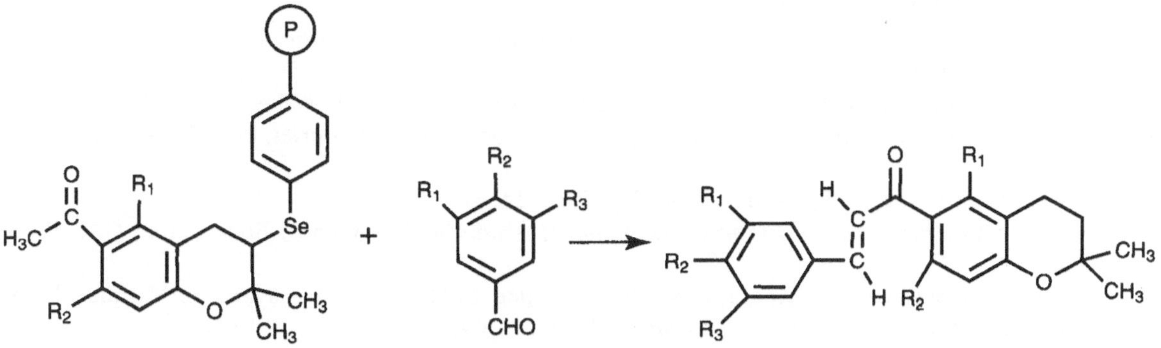

5.1.37

5.1.39

5.1.38

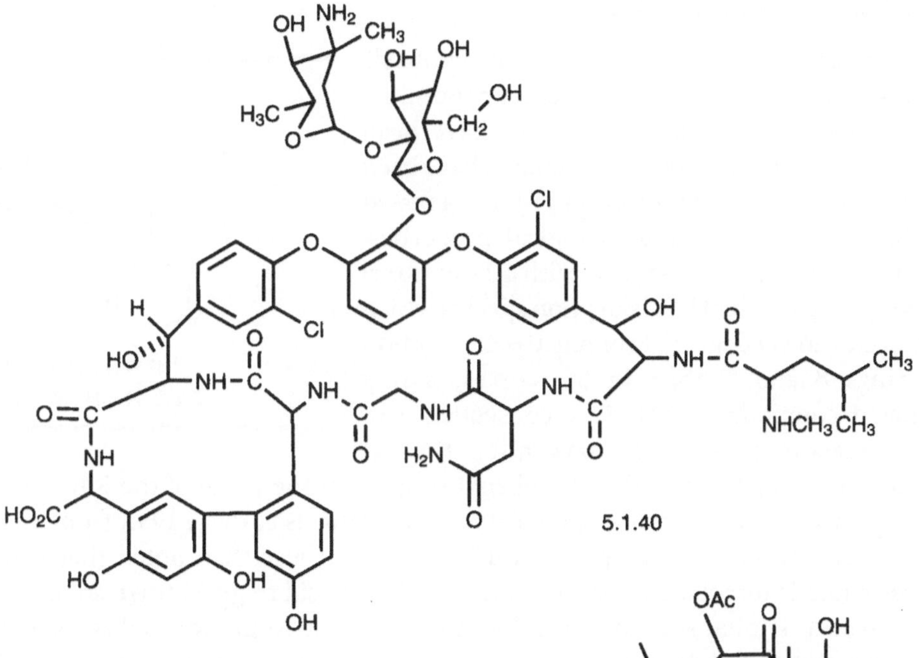

5.1.40

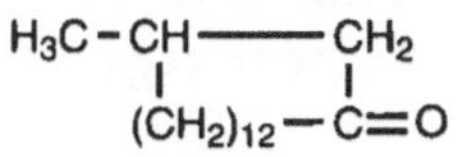

5.1.41

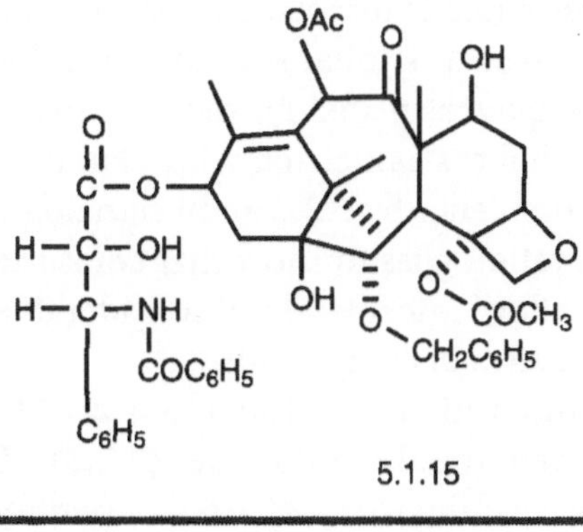

5.1.15

Suggested reading

1. Nicolaou KC and Sorensen EJ, *Classics in Total Synthesis: Targets, Strategies, Methods,* Weinheim: Wiley-VCH, 1996.
2. *Fieser and Fieser's Reagents in Organic Syntheses,* Vols 1–23, New York: Wiley, 1967–2006.
3. Boldi AH, (ed.), *Combinatorial Synthesis of Natural Product based Libraries,* Boca Raton: CRC Press, 2006.
4. Abreu PM and Branco PS, *J. Braz. Chem. Soc.,* 2003, 14: 675–712.
5. Goodwin BL, *Handbook of Biotransformations of Aromatic Compounds,* Boca Raton: CRC Press, 2005.
6. Seeger M, González M, Cámara B, Muñoz L, Ponce E, Mejías L, Mascayano C, Vásquez Y and Sepúlveda-Boza S, *Appl. Environ. Microbiol., 2003,* 69: 5045–50.

5.2 Synthesis of Polyoxygenated Flavones with Uncommon Oxygenation Patterns

Flavones of natural origin with hydroxyls or methoxyls at positions 2' and 5' in the B ring are uncommon. However, in recent years, a few of these have been isolated and characterised. These compounds have been found to differ from the more common 3', 4'-oxygenated flavones in their chemical and biological properties. 3-Hydroxyflavones, that is, flavonols, with a 2'-hydroxyl or methoxyl group exhibit UV absorption spectra which clearly indicate a lower degree of conjugative interaction between rings B and C than in the corresponding 3' oxygenated compounds. This is a consequence of the steric interaction between the oxygen functions at

5.2.1: R = OH, R' = H
5.2.2: R = H, R' = OH

positions 3 and 2' which forces the side phenyl ring out of the plane of the 3-hydroxychromone moiety. An indication of this non-coplanarity was evident, as early as 1957, from an observation made in the Chemistry laboratory of Delhi University. It was then noted that while flavonols like quercetin (5.2.1) dissolved in concentrated sulphuric acid to give brown solutions exhibiting green fluorescence, similar solutions in sulphuric acid of the isomeric 2' substituted compounds, such as isoquercetin (5.2.2), did not exhibit the characteristic fluorescence. Ability to emit fluorescent light is associated with planarity in aromatic systems and therefore, the inability to exhibit fluorescence by 5.2.2 and its analogues can be taken as an indication of non-coplanarity. This observation goes to show that colour tests should not be dismissed as trivial and that very often they afford information of considerable significance with regard to some aspect or the other of the structures of compounds.

Brickelin, which is a flavonoid constituent of the Brickelia species is 2',5-dihydroxy-3,4',5',6,7-pentamethoxyflavone (5.2.3). Earlier, the compound had been assigned the structure 2',5-dihydroxy-4',5',6,6',7-pentamethoxy-flavone (5.2.4), based mainly on spectral properties including nuclear Overhauser effects in the NMR spectra. These two structures were synthesised by unambiguous methods and the identity of brickelin as 5.2.3 was established by comparing it with the synthetic compounds. The synthesis of brickelin is outlined here. The starting material for constructing the A ring of the flavones was 2-hydroxy-4,6-dimethoxyacetophenone (5.2.5) which on persulphate oxidation followed by partial methylation yielded 5.2.6. For constructing the B ring, 2-isopropyloxy-4, 5-dimethoxybenzaldehyde (5.2.7) was chosen. Condensation of 5.2.6 with 5.2.7 in the presence of potassium hydroxide yielded the chalcone 5.2.8. The latter was subjected to the Algar–Flynn–Oyamada reaction with the help of alkaline hydrogen peroxide in ethanol to obtain the flavonol 5.2.9. Methylation of the latter with dimethyl sulphate gave 5.2.10 which on treatment

5.2.3

5.2.4

with boron trichloride underwent selective demethylation of the 5-methoxy group and removal of the isopropyloxy group to yield 5.2.3.

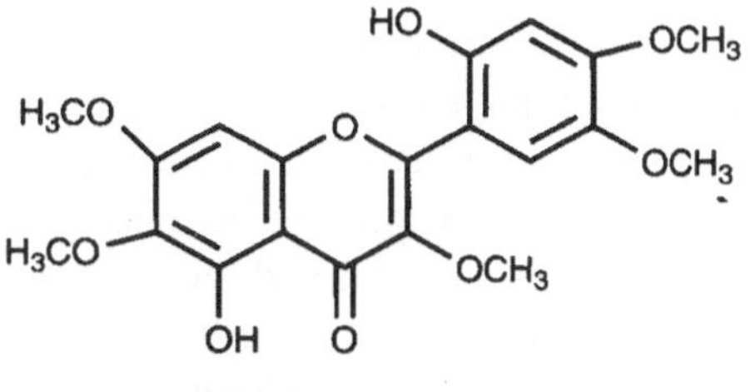

5.2.5

5.2.6

5.2.7

5.2.8

5.2.9

5.2.10

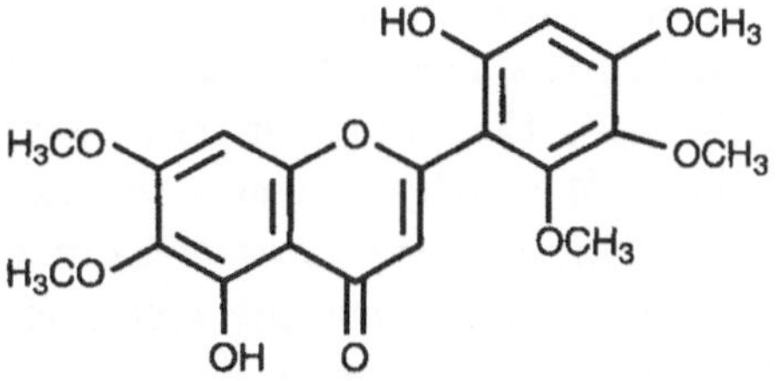

5.2.3

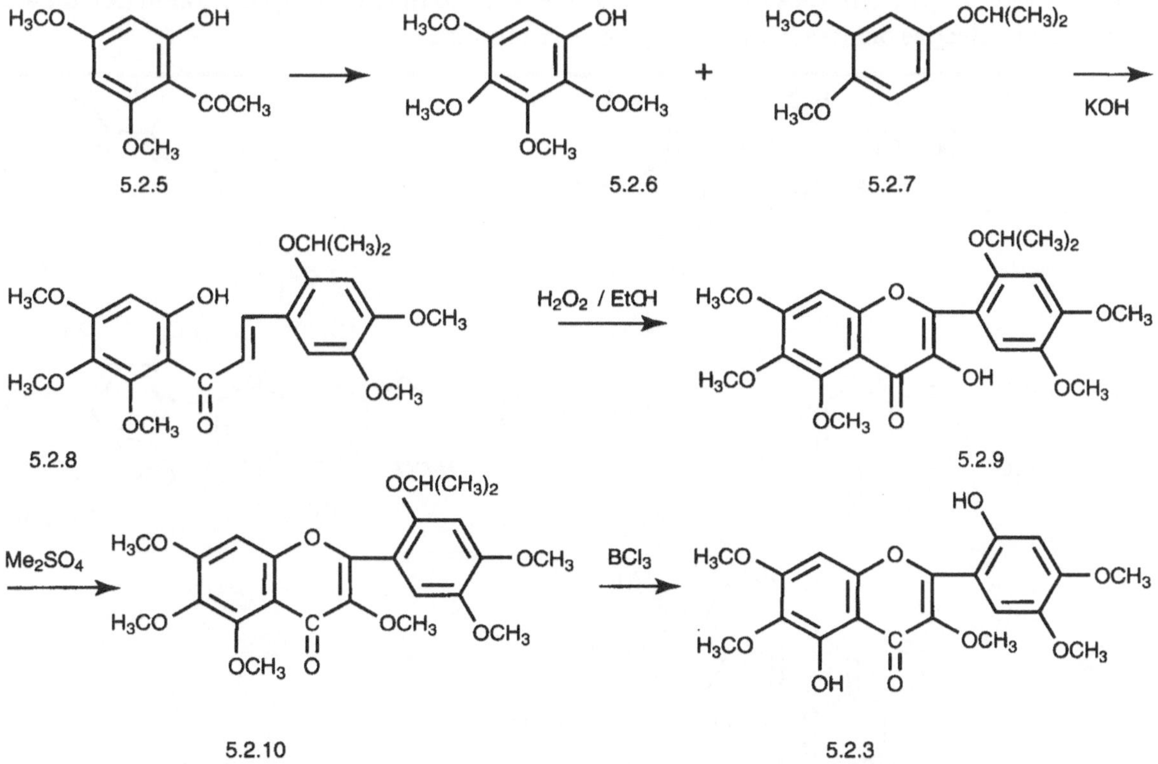

This synthesis is a typical example of an unambiguous experimental strategy designed to provide synthetic support to a structure. The reactions involved are all well known and are based on clearly elucidated mechanistic principles. Perhaps the only novelty is the use of the isopropyl group for O-protection; commonly used protecting groups are the benzyl, methoxymethyl and tetrahydropyranyl groups. Attention may also be drawn to the use of the mild Lewis acid, boron trichloride, as a selective demethylating agent. The fact that the methoxyl at position 3 remains intact is also worth noting; it indicates that the intramolecular hydrogen bonding between the 5-OH and the pyrone carbonyl is stronger than that between the latter and the 3-OH.

In the course of the work described above, which was aimed at confirming the structure of brickelin, a few other partial methyl ethers of polyhydroxyflavones and flavonols were synthesised. One of them was 3',5-dihydroxy-2',4',6,6',7-pentamethoxyflavone (5.2.11), the synthesis of which illustrates the use of appropriate variants of the same basic strategy for the unambiguous synthesis of isomeric structures. In this case too, the starting material for building up ring A was the acetophenone, 5.2.6, which was condensed with 3-benzyloxy-2,4, 6-trimethoxybenzaldehyde (5.2.12) to obtain the chalcone 5.2.13. The aldehyde 5.2.12 was prepared from 2,6-dimethoxyhydroquinone (5.2.14) by a three-step reaction sequence involving a Vilsmeier–Haack formylation to obtain 5.2.15, selective monobenzylation yielding 5.2.16 and a final O-methylation in that order. The chalcone 5.2.13 yielded the flavone 5.2.17 on treatment with 2,3-dichloro-5,6-dicyanobenzoquinone (DDQ) which brought about oxidative cyclisation of 5.2.13. This method is an alternative to the use of selenium dioxide. The benzyl group in 5.2.17 was removed by hydrogenolysis in the presence of Pd–C and the resulting compound (5.2.18) was subjected to selective demethylation using boron tribromide to obtain 5.2.11.

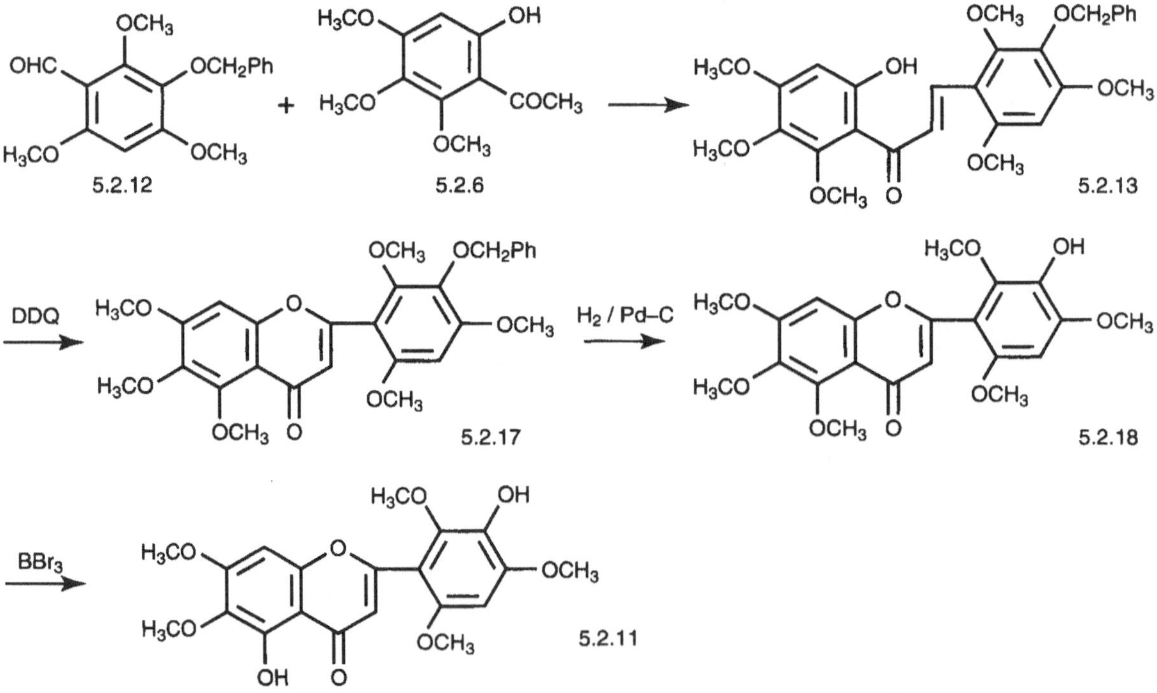

5.2.14   5.2.15   5.2.16   5.2.12

Suggested reading

1. Fang N and Mabry TJ, In Atta-ur-Rahman, (ed.) *Studies in Natural Products Chemistry*, Vol. 5, Amsterdam: Elsevier, 1989.
2. Ferreres E, Tomas-Lorente ER and Tomas-Barberan FA, In Atta-ur-Rahman, (ed.) *Studies in Natural Products Chemistry*, Vol. 5, Amsterdam: Elsevier, 1989.
3. Roberts ME, Timmermann BN and Mabry TJ, *Phytochem.*, 1980, 19: 127.
4. Linuma M, Matoba Y, Tanaka T and Mizuno M, *Chem. Pharm. Bull.*, 1986, 34: 1656.
5. Linuma M, Tanaka T, Ito K and Mizuno M, *Chem. Pharm. Bull.*, 1987, 35: 660.

5.3 TWO CONTRASTING SYNTHESES OF ANTHRACYCLINONES

The discovery of the anti-cancer and antibiotic properties of daunorubicin and related compounds provided powerful motivation for extensive synthetic studies on these compounds. A number of strategies have been developed for the synthesis of the fused tetracyclic skeletal structures present in these compounds and each one of these is elegant and instructive in its own way. In this section, we have chosen two contrasting pathways, each having its own unique features.

5.3.1

The synthesis outlined here has been designed with the Michael reaction playing the key role. The considerable preparative value and regiospecificity of this reaction have been intelligently made use of in this example. The target molecule is 6,11-dihydroxy-4-methoxy-7,8,9, 10-tetrahydro napthacene-5,9,12-trione (5.3.1), which is a late-stage precursor in the synthesis of daunomycinone (5.3.2). What is presented here is the route which finally

5.3.2

proved to be the most efficient in terms of overall yield of the desired end-product. As pointed out earlier, the keynote of the entire procedure is a well-articulated, appropriately 'positioned' Michael reaction which makes a reappearance, in a similar role, further down the synthesis.

The sequence began with a Michael addition of the anion liberated from the methoxy sulphonylphthalide (5.3.3) to the double bond of ethyl hexen-2-oate (5.3.4). Subsequent steps involved exhaustive *O*-methylation to obtain 5.3.5, and benzylic bromination (with the help of *N*-bromosuccinimide) of the latter to get 5.3.6. A base (1,8-diazabicyclo undecene)-catalysed dehydrobromination of 5.3.6 produced an olefin 5.3.7 which on Lemieux–Johnson oxidation, yielded the aldehyde 5.3.8. This compound was converted into the phthalide (5.3.9) by treatment with ethanolic alkali. The ethoxy group in 5.3.9 was then displaced by the thiophenyl group by reaction with benzenethiol in the presence of *p*-toluene sulphonic acid. The resulting product (5.3.10) gave the tricyclic sulphone (5.3.11) on oxidation with *m*-chloroperbenzoic acid. The latter was then condensed with cyclohexenone, in the presence of lithium diisopropylamide (the repeat Michael reaction), to obtain 5.3.12.

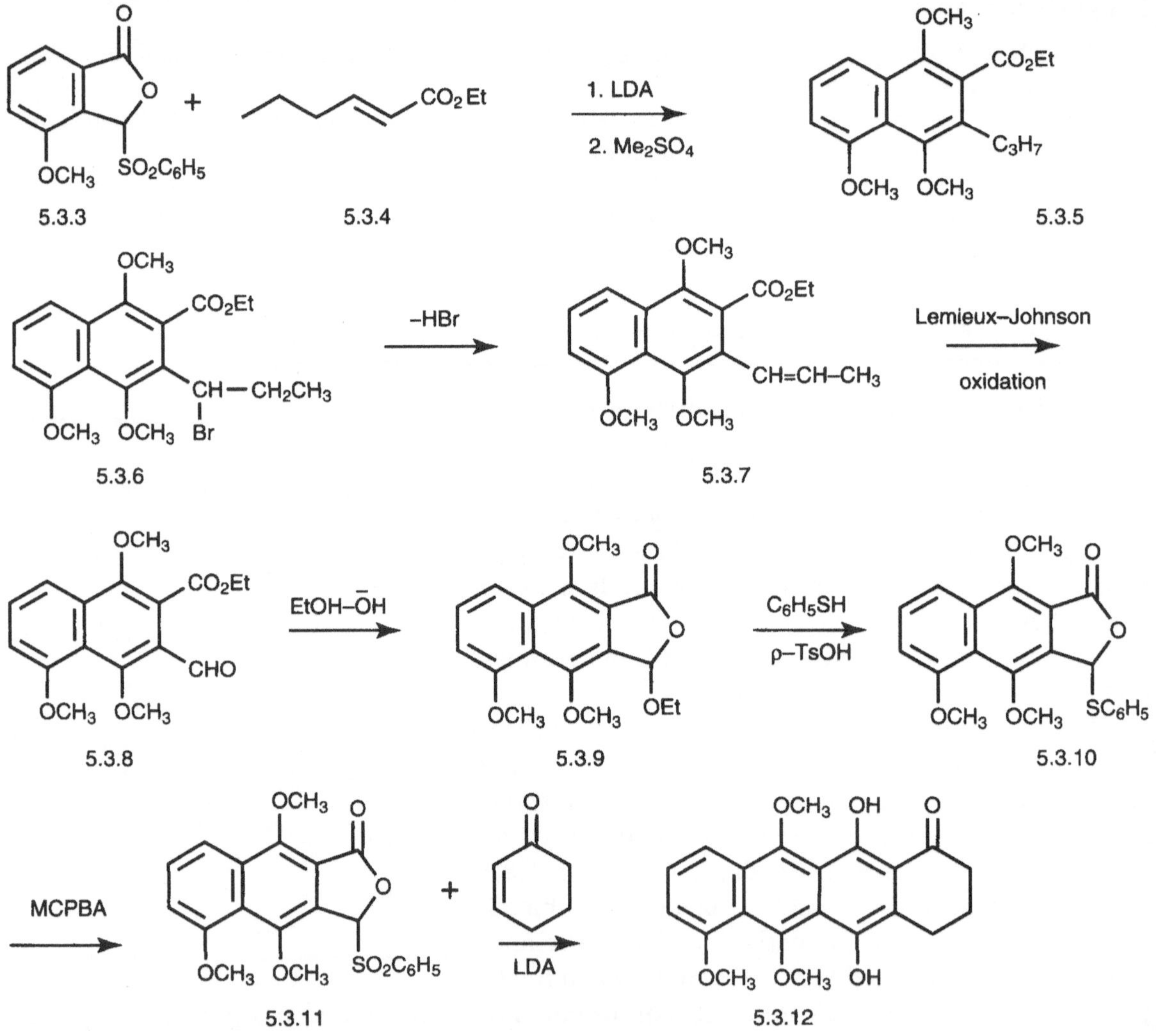

The dimethyl ether of 5.3.12 was converted into its isomer (5.3.13) by a four-step sequence. First, it underwent a sodium borohydride reduction to produce an alcohol (5.3.14) which was then dehydrated using *p*-toluene sulphonic acid and a Dean–Stark water separator to obtain 5.3.15. Conversion of 5.3.15 to the glycol (5.3.16) was achieved using trimethylamine-*N*-oxide and a catalytic amount of osmium tetroxide. The transformation of 5.3.16 to the ketone (5.3.13)

was brought about using *p*-toluene sulphonic acid for the selective dehydrative removal of the benzylic hydroxyl group.

After protecting the carbonyl group in 5.3.13 by ketalisation using ethylene glycol, selective oxidative demethylation was effected with the help of ceric ammonium nitrate in the presence of pyridinedicarboxylic acid *N*-oxide. The product was the quinone 5.3.17. Demethylation of the 6- and 11-methoxyl groups in this compound was brought about using silver oxide in dilute nitric acid followed by treatment with sodium dithionite. This step apparently involved oxidative demethylation and subsequent selective reduction of ring C to obtain 5.3.18. Deketalisation of the latter could be achieved by treatment with trifluoroacetic acid and 10% HCl to reach the desired target, namely, 5.3.1.

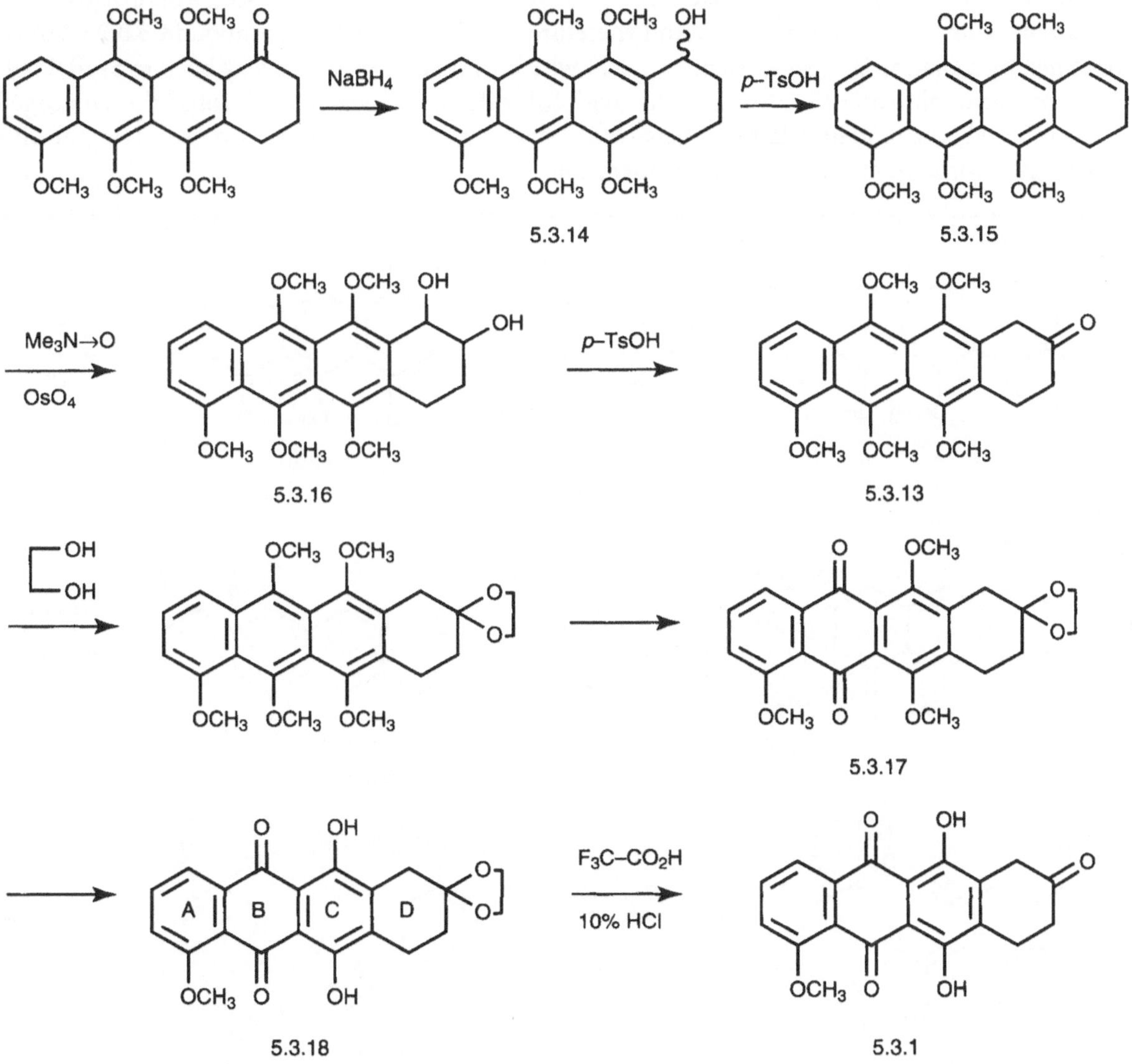

5.3.14

5.3.15

5.3.16

5.3.13

5.3.17

5.3.18

5.3.1

This synthesis is a good example from the instructional point of view as it involves unusual but unambiguous applications of well-known reactions and the use of newer reagents to bring about base-catalysed reactions and oxidative demethylations. The target molecule being largely aromatic in character, no unexpected rearrangements or stereochemical complications were involved.

We now look at an alternative route designed on the basis of a retrosynthetic analysis. This example can be used to illustrate the disconnection approach in organic synthesis which represents a philosophy that has been refined by Corey. If one looks at the target molecule (5.3.19), it is possible to visualise one mode of 'dissection', using a retro-Diels–Alder reaction leading to the tricyclic quinone (5.3.20) and methoxybutadiene (5.3.21). The precursor of 5.3.20 could be the corresponding quinol (5.3.22). Carrying on the disconnection approach further down the retrosynthetic pathway, one can visualise 5.3.23; conversion of 5.3.23 into 5.3.22 would involve an intramolecular Diels–Alder reaction and therefore should be feasible. Compound 5.3.23 can be obtained by a Claisen rearrangement of 5.3.24 which in turn goes all the way back to 5-allyloxy-2-hydroxyacetophenone (5.3.26) and the acid chloride 5.3.27 via 5.3.25. Thus, the envisaged route is based on a tandem Claisen–Diels–Alder strategy which combines two useful preparative reactions to form an effective synthetic operation.

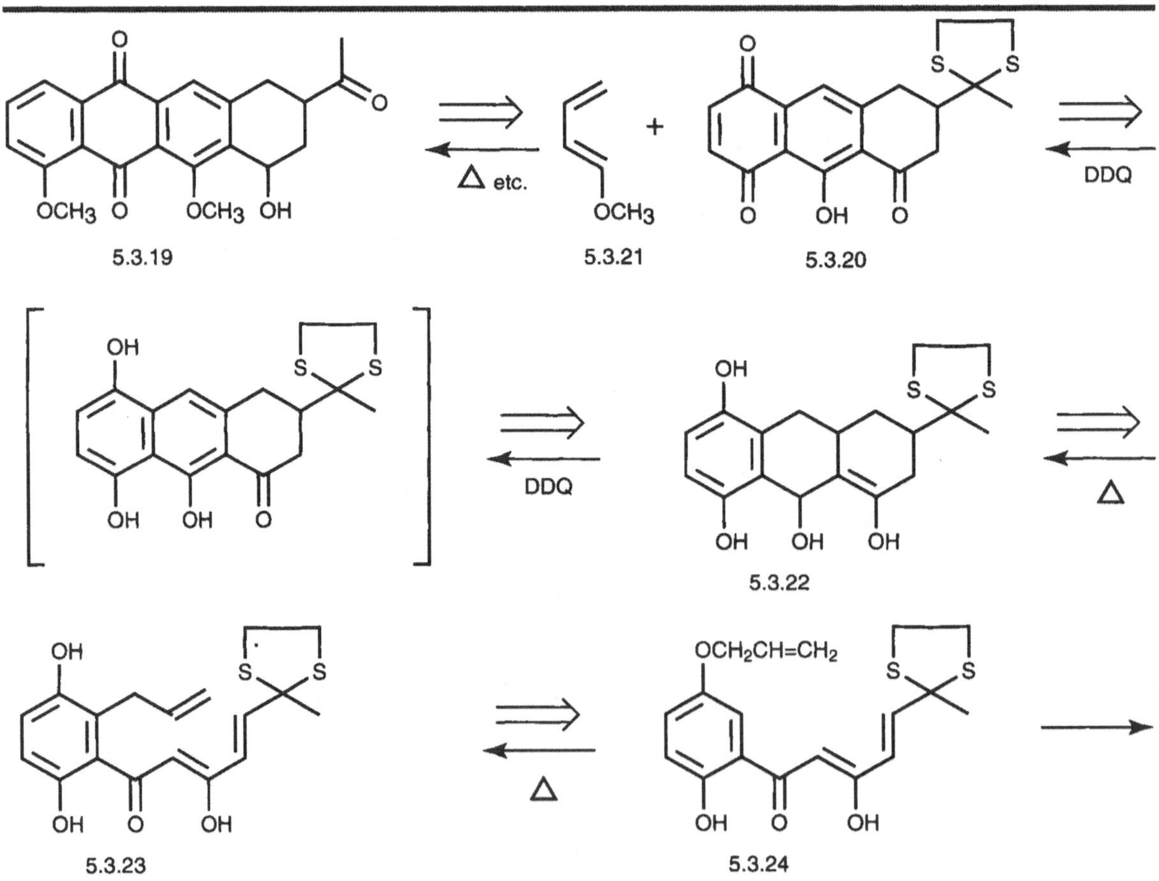

5.3.19   5.3.21   5.3.20

5.3.22

5.3.23   5.3.24

5.3.25   5.3.26   5.3.27

What is more important is the fact that the entire reaction scheme could be translated into practice which proves the feasibilty of the envisaged strategy. We make this point here to emphasise that the real value of any synthetic scheme, however intellectually impressive, can be proved only by actual experimentation which may not always work the way it is expected to! The example in the following section illustrates this point.

Suggested reading

1. Krohn K, In Herz W, Grisebach H, Kirby GW and Ch. Tamm, (eds), *Progress in Chemistry of Organic Natural Products*, Vol. 55, New York: Springer-Verlag, 1989.
2. Nattori S, In Nakanishi K, Goto T, Ito S, Natori S, Nozoe S, (eds), *Natural Products Chemistry*, Vol. 3, Japan: Kodansha Ltd, 1983.
3. Hauser FM and Prasanna S, *J. Amer. Chem. Soc.*, 1981, 103: 6378.
4. Kraus GA and Woo SH, *J Org. Chem.*, 1987, 52: 4841.

5.4 AN APPROACH TO QUASSINOID SYNTHESIS

Quassinoids, such as quassin (5.4.1), are complex and quite heavily oxygenated diterpenoids which occur in plants belonging to the family Simaroubiaceae. Some of them have been found to exhibit anti-leukaemic activity and a few others are known to exhibit anti-viral activity. Polonsky and co-workers have done considerable work on the structures of these compounds, while much elegant work on the synthesis of the quassinoids has come from the laboratories of Greco.

5.4.1

In this section, we will discuss one approach with Hagemann's ester as the starting material. This compound is a useful synthon which has itself been used as an example for illustrating the disconnection approach to synthesis in the highly readable and educative book by Warren. In the attempted synthesis of the quassinoid structure described below, Hagemann's ester (5.4.2) was first protected by ketalisation with ethylene glycol and the ester group was then reduced. The ketalisation step also involved an acid-catalysed shift of the double bond. The resulting alcohol (5.4.3) was then converted into the alkyl chloride (5.4.4) with the help of CCL_4 and $P(NMe_2)_3$. Condensation of 5.4.4

with dimethyl malonate in the presence of sodium hydride yielded 5.4.5. Once again, as in the previous example, we notice familiar transformations brought about by means of new reagents and strategies. For example, the use of hexamethyl phosphorus triamide in CCl_4 for the transformation of 5.4.3 into 5.4.4 is worth noting. Under these circumstances the ketal functionality remains unaffected.

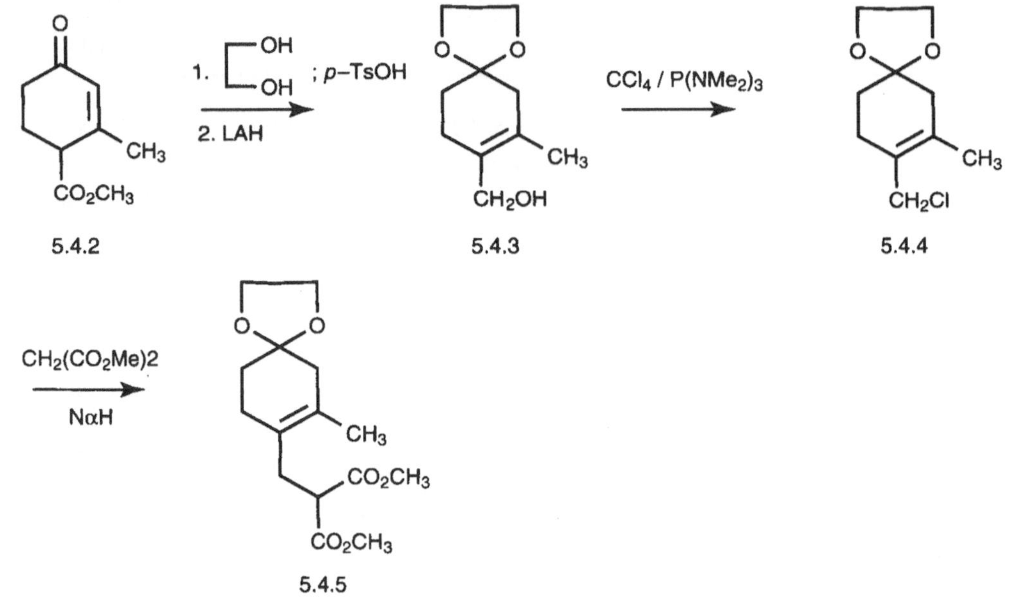

The base-catalysed condensation of 5.4.5 with the iron tricarbonyl salt (5.4.6) to yield 5.4.7 is also an ingenious variation of the familiar reaction between a nucleophile and an electron-deficient reagent. The metal atom in the complex 5.4.7 was removed by treatment with trimethylamine N-oxide and the resulting dienol ether (5.4.8) was selectively hydrolysed with oxalic acid in 'wet' methanol to obtain 5.4.9. These steps emphasise the preparative value of 'control' in experimental conditions for achieving selectivity and serve to remind us that organic chemistry is not entirely intellectual—there is a vital element of skill associated with careful and meticulous experimentation!

Treatment of 5.4.9 with p-TsOH in acetone at room temperature yielded the diketone 5.4.10, which underwent facile cyclisation in acid (HCl on SiO_2) to form 5.4.11. However, the final contemplated acid-catalysed conversion of 5.4.11 into the desired tricyclic target (5.4.12) failed to occur; instead, an intramolecular Michael-type reaction took place resulting in 5.4.13. This example has been chosen here instead of the several successful syntheses of quassinoids to illustrate that schemes may go awry but what is more important than reaching a target molecule is the revelation of variety in molecular behaviour including the unexpected.

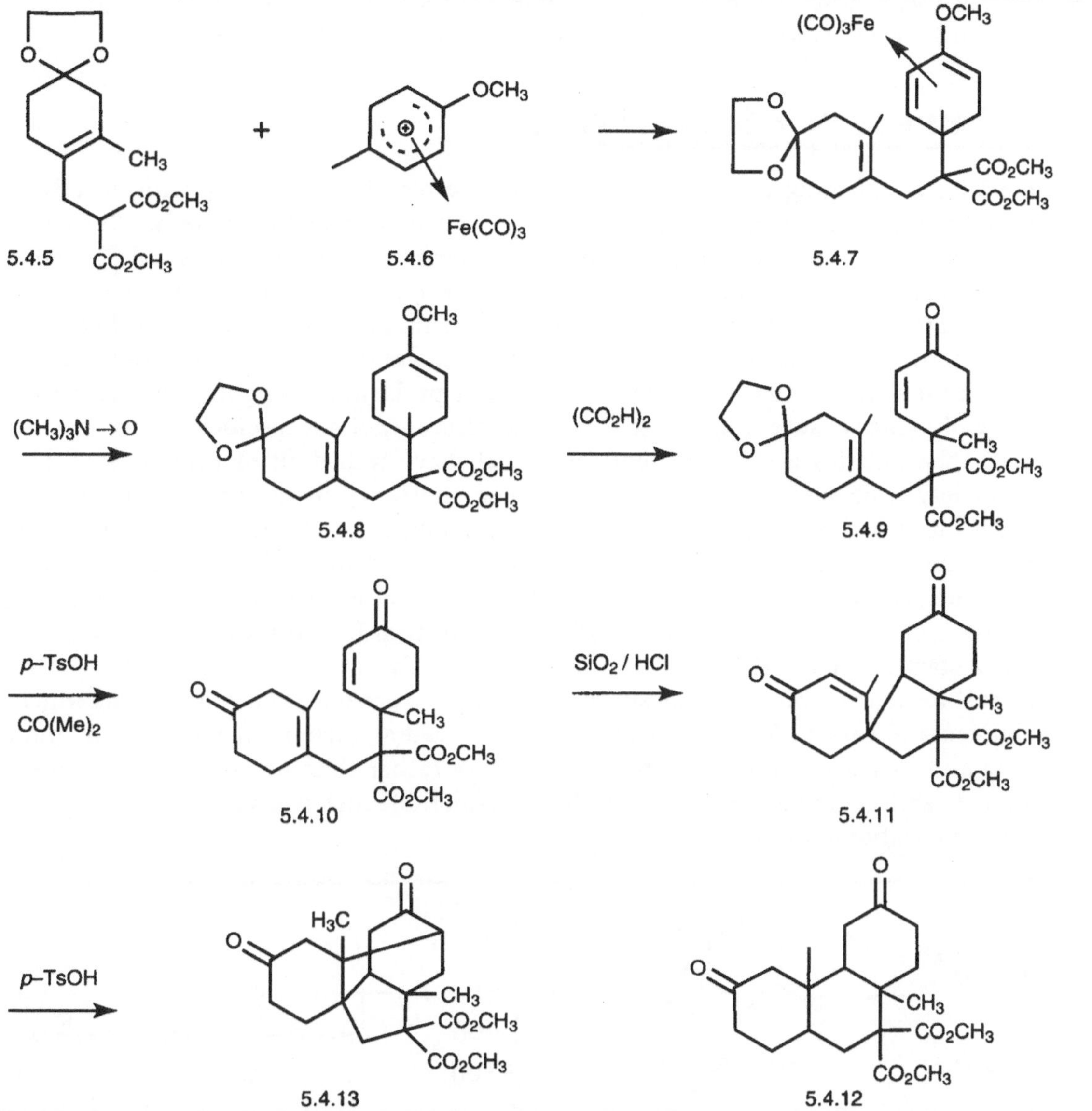

5.4.5 + 5.4.6 → 5.4.7

(CH₃)₃N → O → 5.4.8 → (CO₂H)₂ → 5.4.9

p-TsOH / CO(Me)₂ → 5.4.10 → SiO₂ / HCl → 5.4.11

p-TsOH → 5.4.13   5.4.12

Suggested reading

1. Polonsky J, Quassinoid Bitter Principles II, In Herz W, Grisebach H, Kirby GW and Ch Tamm, (eds), *Progress in the Chemistry of Organic Natural Products*, Vol. 47, New York: Springer-Verlag, 1985.

2. Grieco PA, S Ferrino and Jaw JW, *J. Org. Chem.*, 1982, 47, 601.

3.   Chandler M, Mincione E and Parsons PJ, *J. Chem. Soc. Chem. Comm.*, 1985, 1233.
4.   Warren S, *Organic Synthesis—The Disconnection Approach*, John Wiley & Sons, 1982

5.5 SYNTHESIS OF A SEMIOCHEMICAL

Biosemiotics is emerging as an area of study in the biosciences dealing with effective and articulate communication among life forms through molecular phenomena. We will have an occasion to take a closer look at the topic in Chapter 7. Semiochemicals are the chemical signals through which different species of a genus and various life forms (for example plants and insects) 'communicate' with each other; the communication may be 'friendly' or otherwise. The biological activities of these compounds, many of which are structurally not complex, crucially depend on the correct chirality. Therefore, the major problem in designing the synthesis of a semiochemical is to ensure a high degree of stereoselectivity, if not specificity.

Insect pheromones are semiochemicals which have well-defined functions within an insect society. One such compound is the queen substance of the oriental hornet, *Vespa orientalis*. It has the gross structure 5-hexadecanolide (5.5.1), with one chiral centre. Both the enantiomorphs of this compound have been synthesised and in this section, we describe the synthesis of (S) and (R) 5.5.1 with (R) 2,3-O-isopropylideneglycerol (5.5.2) as the 'chiron'. This compound can be readily obtained from D-mannitol (5.5.3) and has been used as a starting material for the synthesis of optically active glycerophosphoryl choline (5.5.4) and related compounds. The synthesis under discussion involved reactions which are well-known and easy to follow. The reagents employed at various stages are also common reagents and this places focus on the methodology rather than on some exotic reagent or reaction. Therefore, the entire procedure has considerable instructional value and, perhaps, can serve as a laboratory exercise as well.

(S) 5.5.1   (R) 5.5.1   5.5.3   5.5.4

The sequence began with the tosylation of 5.5.2 followed by a nucleo-philic displacement of the tosyloxy group by n-decyl magnesium bromide and hydrolysis of the ketal. The resulting diol (5.5.5) was converted via its monotosylate (5.5.6) into the epoxide 5.5.7. Upto this point, the configuration of the lone chiral centre in 5.5.2 was retained. In the subsequent steps leading to (S) 5.5.1 too, the configurational integrity was maintained.

Reaction of 5.5.7 with the Grignard reagent 5.5.8 yielded the tertiary alcohol 5.5.9 whose acetate 5.5.10 was converted into the carboxylic acid 5.5.11 by oxidative cleavage of the terminal double bond with the help of ruthenium trichloride. The final product, namely (S) 5.5.1, was obtained from 5.5.11 by a two-step sequence as shown below.

For the synthesis of (R) 5.5.1 by an identical sequence of reactions the diol needed was 5.5.12, that is, the enantiomorph of 5.5.5. It could be obtained from the latter by mesylation followed by inversion at the chiral centre accompanying displacement of the mesoxyl group by acetoxyl group. The diacetate (5.5.13) thus obtained was hydrolysed to the diol (5.5.12) which was transformed into (R) 5.5.1 as already described.

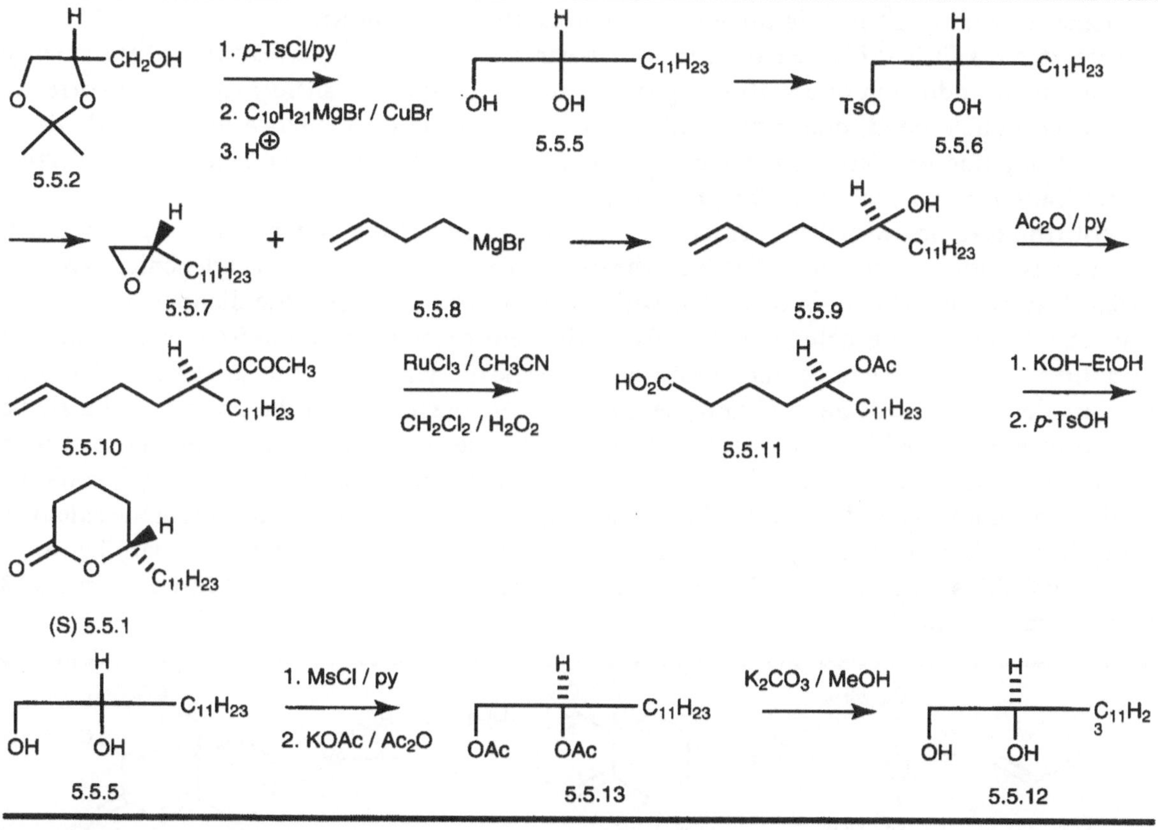

Suggested reading

1. Alves LF, Chemical ecology and the social behaviour of animals, In In Herz W, Grisebach H, Kirby GW and Tamm Ch, (eds), *Progress in the Chemistry of Organic Natural Products*, Vol, 53, New York, Springer-Verlag, 1988.
2. Goldsworthy GI and Wheeler CH, *Endeavour*, 1985, 9: 139.
3. Chattopadhyay S, Mamdapur VR and Chadha MS, *Bull. Soc. Chim. Fr.*, 1990, 127: 108 and references therein.

5.6 A Synthesis of Polygodial

Insect anti-feedant compounds are attracting considerable attention because of their potential value in the control of harmful insects. These are complex sesqui-, di- and tri-terpenoids and possess interesting skeletal and peripheral structural features. Their synthesis has been a fertile ground for the adventurous among synthetic chemists and several of these have been reproduced in the laboratory. Here, we describe the synthesis of a comparatively simple sesquiterpenoid anti-feedant compound, polygodial (5.6.1). The compound has the bicyclic drimane skeleton and has only three chiral centres. In this synthesis,

5.6.1

the first step is a Diels–Alder reaction. The other attractive and ingenious features of the sequence include the introduction of a carbonyl group and its subsequent transformation to a correctly positioned double bond; one may recall here a similar reaction sequence in the synthesis of the anthracyclinones. Perhaps the most innovative step in the entire sequence is the selective derivatisation of an allylic –CH$_2$OH group into the silyl ether.

With this preamble we will now describe the entire synthesis. The first step, as already mentioned, was a Diels–Alder condensation between the diene (5.6.2) and dimethyl acetylenedicarboxylate (5.6.3). The product (5.6.4) on oxidation with chromium trioxide gave the dienone 5.6.5 which was catalytically hydrogenated to obtain 5.6.6. The keto-carbonyl group in 5.6.6 was eliminated by a three-step sequence to get the unsaturated bicyclic diester (5.6.7). The latter was reduced by lithium aluminium hydride to obtain the diol 5.6.8. As mentioned in the preamble, the allylic hydroxyl in 5.6.8 could be selectively converted into the monosilyl ether 5.6.9 whose acetate (5.6.10) was selectively desilylated to obtain 5.6.11. The latter, being an allylic alcohol, could be oxidised by MnO$_2$ to the aldehyde 5.6.12. The subsequent steps involved protection of the aldehyde group by acetalisation, hydrolysis of the acetate and oxidation of the liberated primary hydroxyl group by Collins reagent, to get 5.6.13. Hydrolytic deprotection of the acetal function in 5.6.13 then yielded racemic 5.6.1.

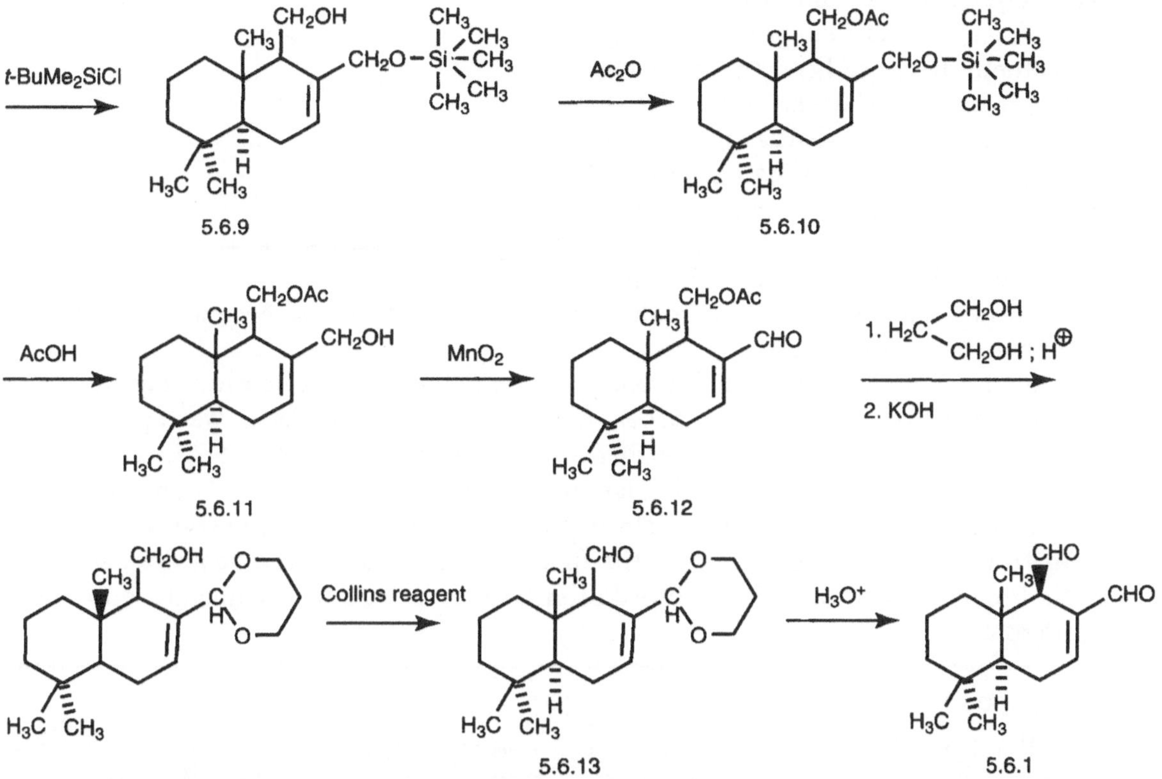

5.6.9

5.6.10

5.6.11

5.6.12

5.6.13

5.6.1

Suggested reading

1. Arnason JT, Philogene BJR and P Morand, (eds), *Insecticides of Plant Origin*, New York: ACS Symposium Series No 387, 1989.
2. de Groot Ae and van Beek TA, Recueil review, *Reel. Trav. Chim. Pays-Bas*, 1987, 106: 1–18 and references therein.

5.7 BALSAMIFERONE AND GRAVELLIFERONE

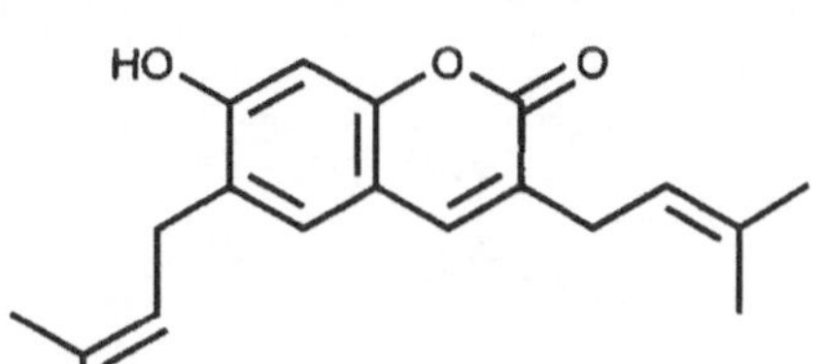

The Claisen rearrangement is one of the oldest reactions known to organic chemists but its mechanism became clear only after Woodward and Hoffmann enunciated their orbital symmetry rules. This reaction, which is a sigmatropic rearrangement, is of considerable value to the chemist dealing with natural products as several isoprenylated flavonoids and coumarins occur in nature and can be synthesised in the laboratory with Claisen rearrangement as a key step. Balsamiferone (5.7.1) and gravelliferone (5.7.2) are two interesting isomeric diisoprenylated

5.7.1

umbelliferone derivatives and have been isolated from
Amyris balsamifera and *Ruta graveolens* respectively.
In the synthesis described here, an interesting feature
is a 'dismantling' of the lactone unit (which is present
in the starting material as well as in the final product)
and re-forming it later after bringing in the additional
isoprenyl unit. The strategy is indeed ingenious and
unambiguous, and makes use of the well-defined
contours and predictability of a pericyclic reaction. A
degradation step is not uncommon in multi-step syntheses and has also been used as a strategy
in one pathway to the rotenoids.

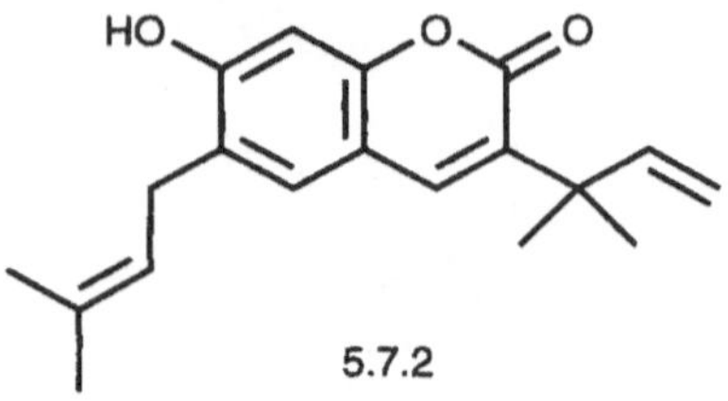

The starting material was umbelliferone (5.7.3) which was converted in three steps with
65% yield into the prenyl coumarin (5.7.4). The pyrone ring in the latter was opened up by
methanolysis and the free hydroxyl thus liberated was prenylated to obtain 5.7.5. When heated
in *N,N*-diethylaniline, 5.7.5 underwent Claisen rearrangement to yield 5.7.6 presumably via
the intermediates 5.7.7 to 5.7.9. The transformation apparently involved a series of sequential
3,3-sigmatropic shifts as shown below. The final debenzylation of the benzyloxy group in
5.7.6 was brought about with the help of boron trichloride at a low temperature to obtain
balsamiferone.

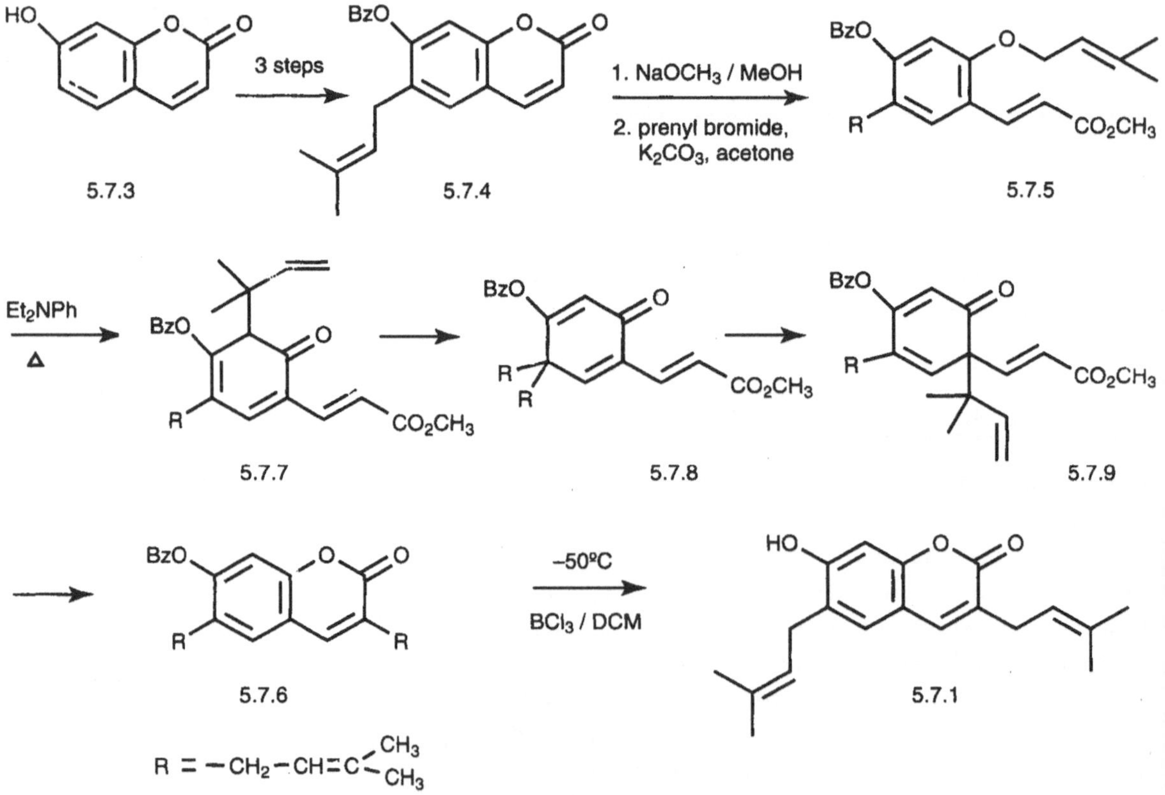

5.7.2

5.7.3   3 steps   5.7.4   1. NaOCH$_3$ / MeOH   2. prenyl bromide, K$_2$CO$_3$, acetone   5.7.5

Et$_2$NPh   Δ   5.7.7   5.7.8   5.7.9

5.7.6   −50°C   BCl$_3$ / DCM   5.7.1

$$R = -CH_2-CH=C\begin{smallmatrix}CH_3\\CH_3\end{smallmatrix}$$

For the synthesis of gravelliferone, the starting material was the *O*-prenyl ether (5.7.10) of 6-prenyl umbelliferone (5.7.4). When heated with *N,N*-diethylaniline, 5.7.10 underwent Claisen rearrangement to yield 5.7.2, presumably via the intermediate 5.7.11. A side product was 5.7.12, obviously formed by an intramolecular cycloaddition of the intermediate 5.7.13.

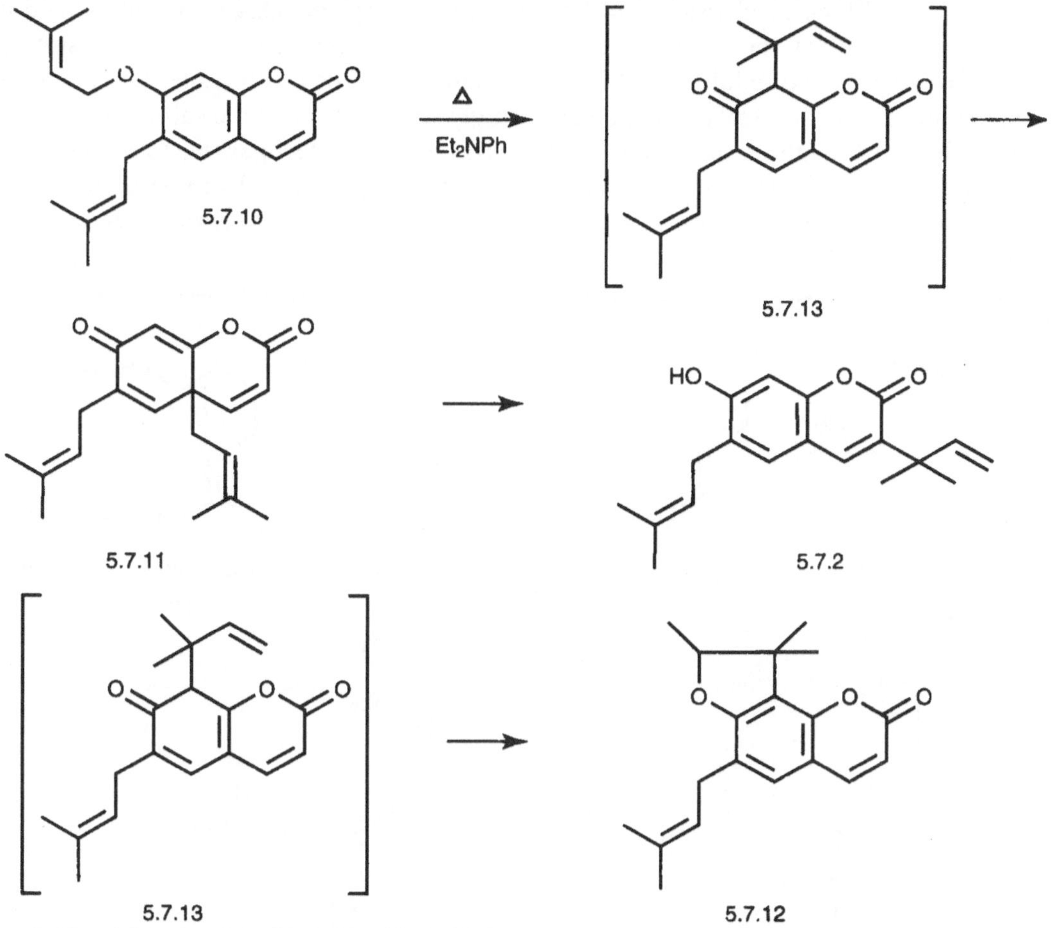

5.7.10

Δ
Et₂NPh

5.7.13

5.7.11

5.7.2

5.7.13

5.7.12

Suggested reading

1.   Burke BA and Parkins H, *Phytochem*, 1979, 18: 1073.
2.   Reisch J, Szendrei K, Minker E and Novak I, *Experientia*, 1968, 24: 992'.
3.   Cairns N, Harwood LM and Astles DP, *J. Chem. Soc. Chem. Comm.*, 1987, 400.

5.8 GYRINAL: DEFENSIVE SECRETION OF AN INSECT

The water beetle, *Gyrinus natator*, secretes a chemical which enables it to glide on water at a speed of upto a metre per second. The main component of the secretion is the norsesquiterpenoid, gyrinal (5.8.1) which possesses three carbonyl groups and three double bonds. It is highly surface-active and toxic to fish. The compound has been synthesised from geranyl acetate (5.8.2). The first step of allylic oxidation was not, obviously, regiospecific and only a 40% yield of the desired isomer of the aldehyde (5.8.3) could be obtained. Incidentally, selenium dioxide is very often the reagent of choice for such oxidations. The subsequent steps followed a textbook pattern and there were no surprises. Thus, condensation of 5.8.3 with the acetylene 5.8.4 gave a 40% yield of the acetylenic diol 5.8.5. The triple bond in the latter could be reduced by LAH and the resulting diol (5.8.6) was hydrolysed and oxidised by MnO$_2$ to obtain 5.8.1. The target molecule, being achiral, did not pose any stereochemical problem. Surprisingly, the yields in all the subsequent steps from 5.8.3 onwards, which ought to have been regiospecinc, were low, perhaps due to the sensitivity of the highly unsaturated skeleton to oxidative polymerisations, particularly under basic conditions. Therefore, there is ample scope for designing improved versions of this synthesis even within the set strategic framework.

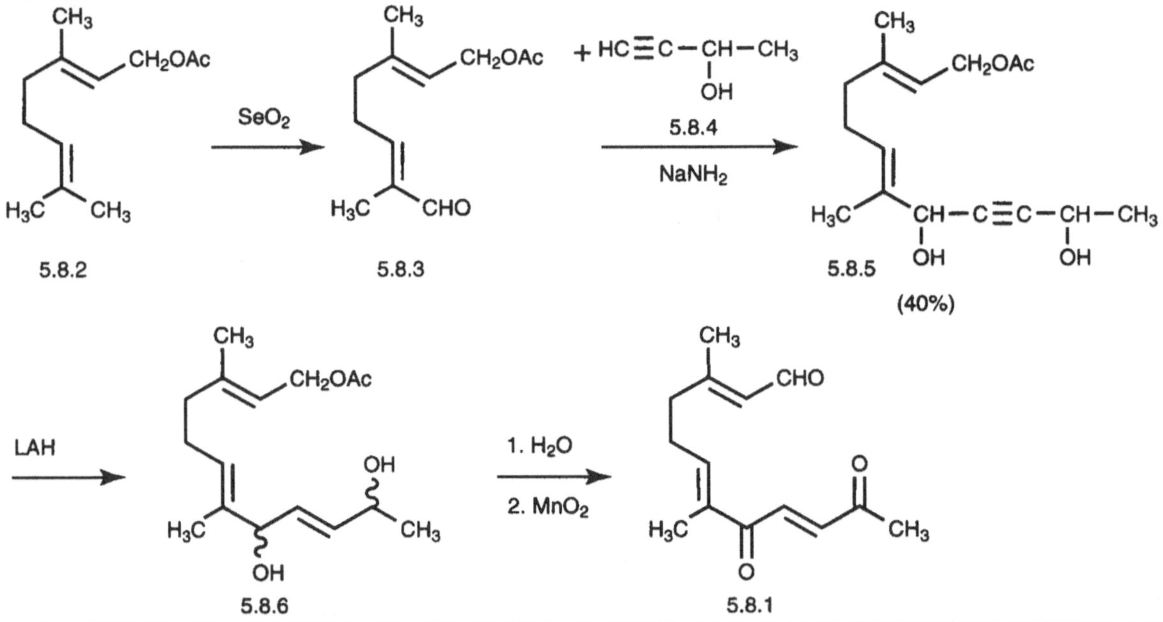

Suggested reading

Schildnicht H, *Angew. Chemie*, International edn, 1976, 15: 214.

5.9 SYNTHESIS OF AN UNUSUAL METHYLTHIOPHENANTHRENE DIOL

Gottlieb and co-workers isolated an unusual sulphur-containing phenanthrene derivative from the trunk-wood of an amazonian Euphorbiaceous tree, *Micandropsis scleroxylon* W. Rodr. The compound, named Micandrol-C, was assigned the structure 5.9.1 which has been confirmed by a synthesis described in this section.

3-Isopropyloxybenzoic acid (5.9.2) was one of the starting materials. Its acid chloride was condensed with 2-amino-2-methylpropanol to obtain the dihydroisooxazole 5.9.3. The latter was converted regiospecifically into the methylthio derivative (5.9.4) by treatment with butyl lithium followed by dimethyl disulphide. The methiodide of 5.9.4 was hydrolytically cleaved to obtain the acid 5.9.5 whose methyl ester (5.9.6) was reduced

5.9.1

by LAH to get the alcohol 5.9.7. The latter was converted into the triphenylphosphonium chloride (5.9.8) which constituted one of the building blocks for the synthesis of 5.9.1.

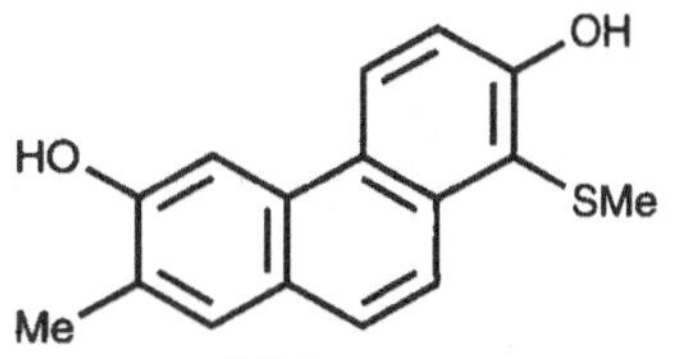

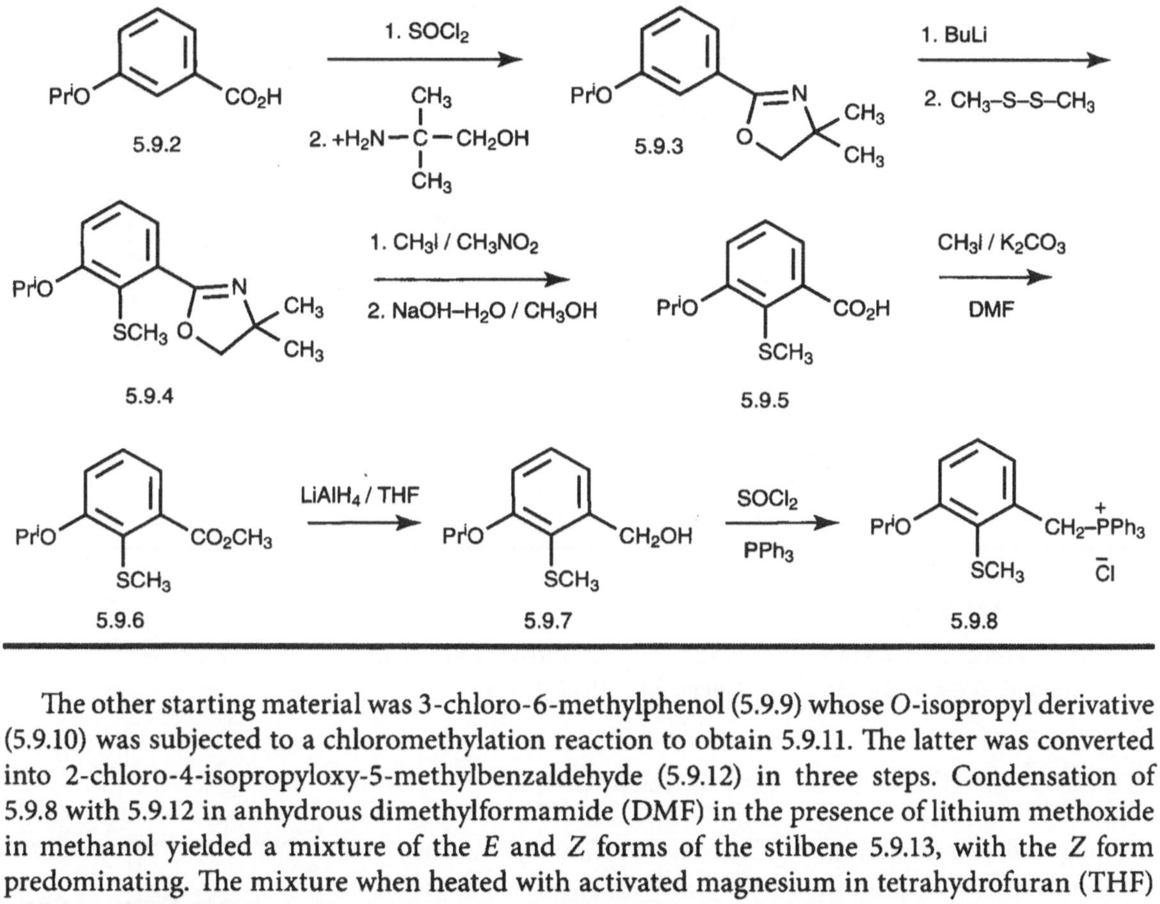

The other starting material was 3-chloro-6-methylphenol (5.9.9) whose O-isopropyl derivative (5.9.10) was subjected to a chloromethylation reaction to obtain 5.9.11. The latter was converted into 2-chloro-4-isopropyloxy-5-methylbenzaldehyde (5.9.12) in three steps. Condensation of 5.9.8 with 5.9.12 in anhydrous dimethylformamide (DMF) in the presence of lithium methoxide in methanol yielded a mixture of the E and Z forms of the stilbene 5.9.13, with the Z form predominating. The mixture when heated with activated magnesium in tetrahydrofuran (THF) gave the desired phenanthrene derivative (5.9.14). Selective O-dealkylation of 5.9.14 could be brought about with the help of BCl$_3$ in dichloromethane (DCM) to obtain 5.9.1.

This synthesis is a classic example of a simple but unambiguous synthesis designed for the purpose of confirming a structure arrived at largely on the basis of limited spectral data. The one novel feature is the use of activated magnesium for the intramolecular coupling of the two phenyl rings of the stilbene 5.9.13. Such cyclisations have been brought about in the past by photochemical methods.

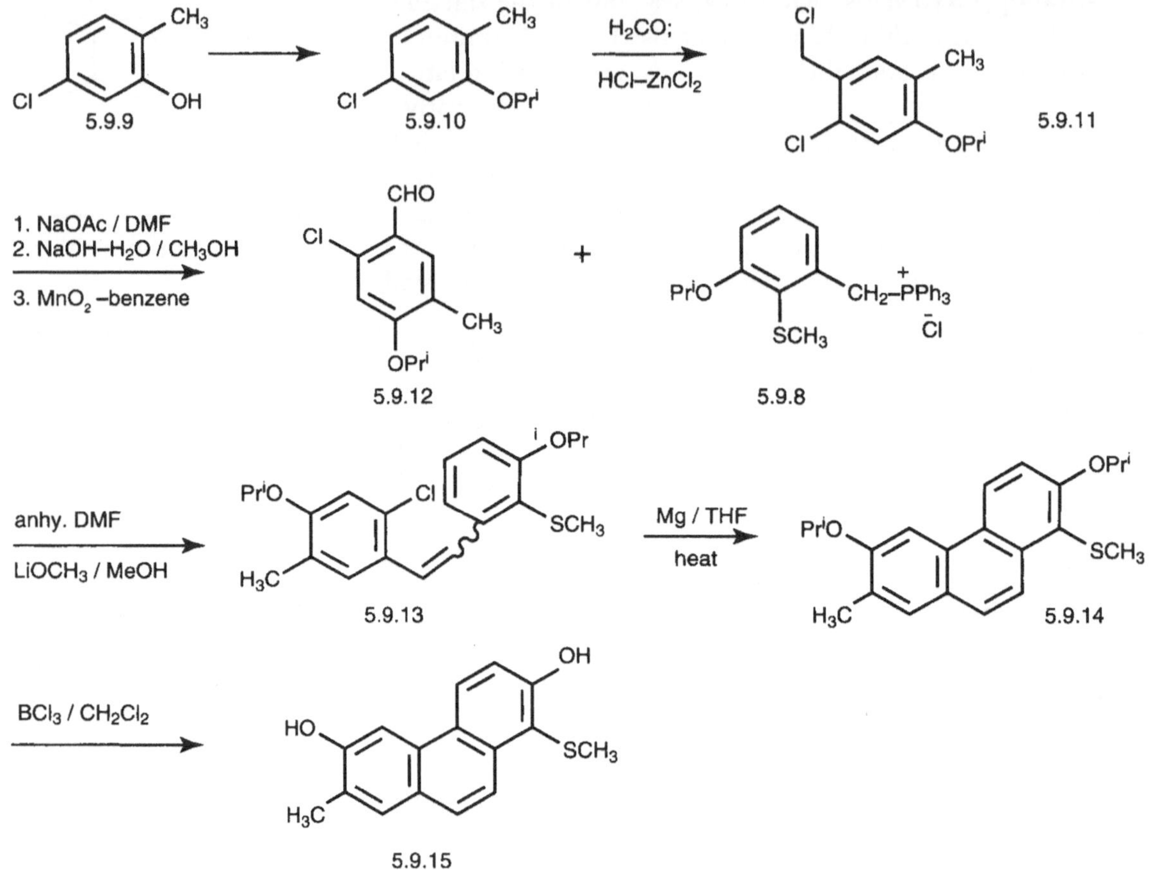
CH3
Cl
OH
5.9.9
CH3
Cl
OPri
5.9.10
H2CO;
HCl–ZnCl2
Cl
CH3
Cl
OPri
5.9.11
1. NaOAc / DMF
2. NaOH–H2O / CH3OH
3. MnO2 –benzene
CHO
Cl
CH3
OPri
5.9.12
+
PriO
SCH3
CH2–PPh3
Cl
5.9.8
anhy. DMF
LiOCH3 / MeOH
PriO
H3C
Cl
OPr
SCH3
5.9.13
Mg / THF
heat
PriO
H3C
OPri
SCH3
5.9.14
BCl3 / CH2Cl2
HO
H3C
OH
SCH3
5.9.15

Suggested reading

1. de Alvarenga MA, da Silva JJ, Gottlieb HE and Gottlieb OR, *Phytochem.*, 1981, 20: 1159.
2. Sargeant MV and Zwicky AB, *J. Chem. Soc.*, 1990, 1: 1713.

5.10 A Synthesis of (–) Khusimone

Among the raw materials used in the perfumery industry one of the most valuable is the essential oil obtained from the roots of *Vetivera ziazanoides*. The commercial vetiver oil contains several sesquiterpenes which are responsible for the strong woody and amber-like fragrance of the oil. One of these compounds is (–)khusimone (5.10.1) which is also a potent insect repellent. The structure of the compound was first confirmed by Buchi and co-workers by a synthesis of racemic khusimone. In this section a recently accomplished synthesis of the (–) form is described.

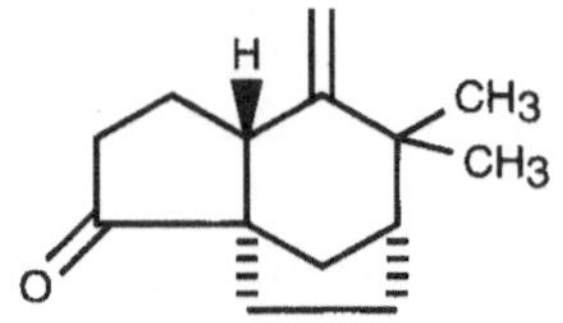
H
CH3
CH3
O

5.10.1

The synthesis began with a highly stereoselective Diels–Alder condensation between isoprene and 6,6-dimethyl-5-methoxycarbonylmethyl-2-cyclohexen-1-one (5.10.2) brought about in the presence of SnCl$_4$ Thus, the stereoisomer 5.10.3 could be obtained in 70% yield. Stereospecific configurational inversion at position 10 was then effected with the help of sodium methoxide in methanol to get 5.10.4. Stereospecific nucleophilic attack on the keto carbonyl group in 5.10.4 by the carbanion of S-methylthiophenol was accompanied by lactone formation. The resulting product (5.10.5) underwent reductive cleavage of the lactone ring and simultaneous reductive desulphurisation to yield 5.10.6 which was subjected to acid-catalysed dehydration and esterification with diazomethane to obtain 5.10.7.

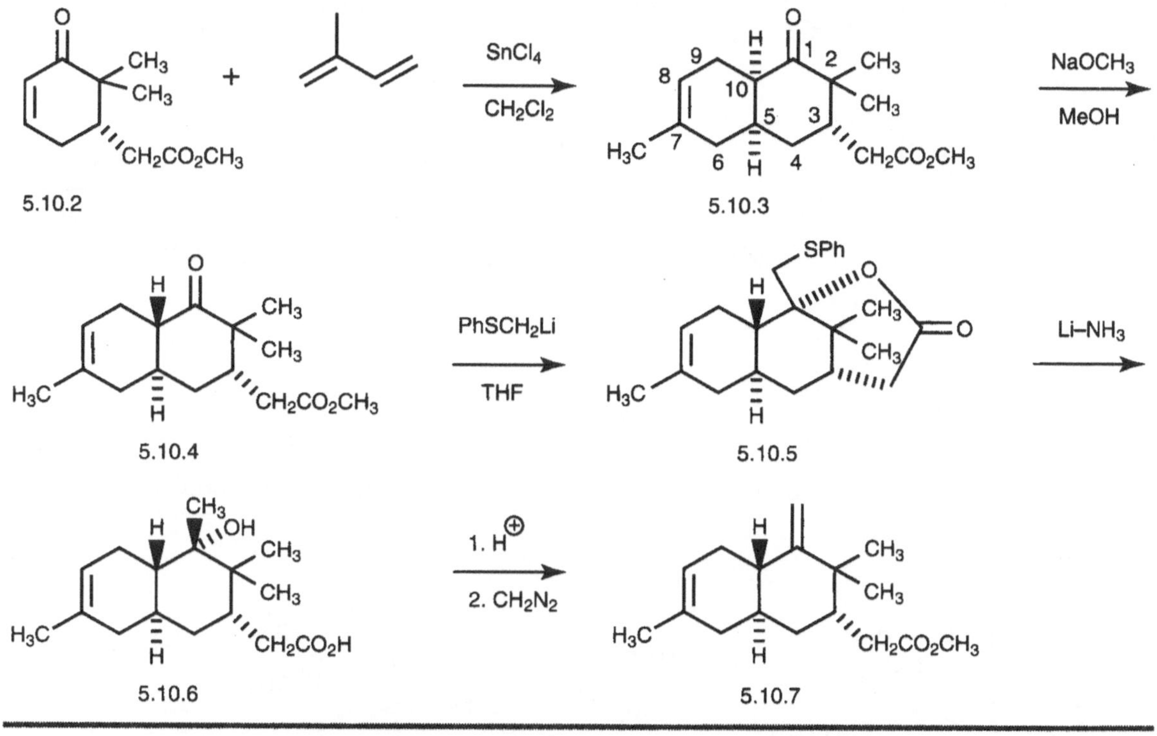

The next two steps involved reduction of the ester group by LAH followed by acetylation. The product (5.10.8) was treated with *m*-chloroperbenzoic acid (MCPBA) and the epoxide (5.10.9) thus obtained made to react with HIO$_4$ and lead tetraacetate in succession to obtain the aldehydo ketone 5.10.10. In the presence of aqueous KOH, 5.10.10 underwent an intramolecular Claisen–Schmidt reaction to yield 5.10.11 which was acetylated to get 5.10.12. It is worth nothing that 5.10.7 undergoes selective monoepoxidation, with the more subsituted double bond reacting, leaving the exocyclic double bond unaffected.

The oxime (5.10.13) of 5.10.12 was then subjected to a Beckmann rearrangement and the product saponified to obtain 5.10.14. The final two steps involved conversion of 5.10.14 into its mesyl derivative and a subsequent base-catalysed intramolecular displacement of the mesyloxy group. (–)Khusimone was thus obtained in 6.9% overall yield through 15 steps.

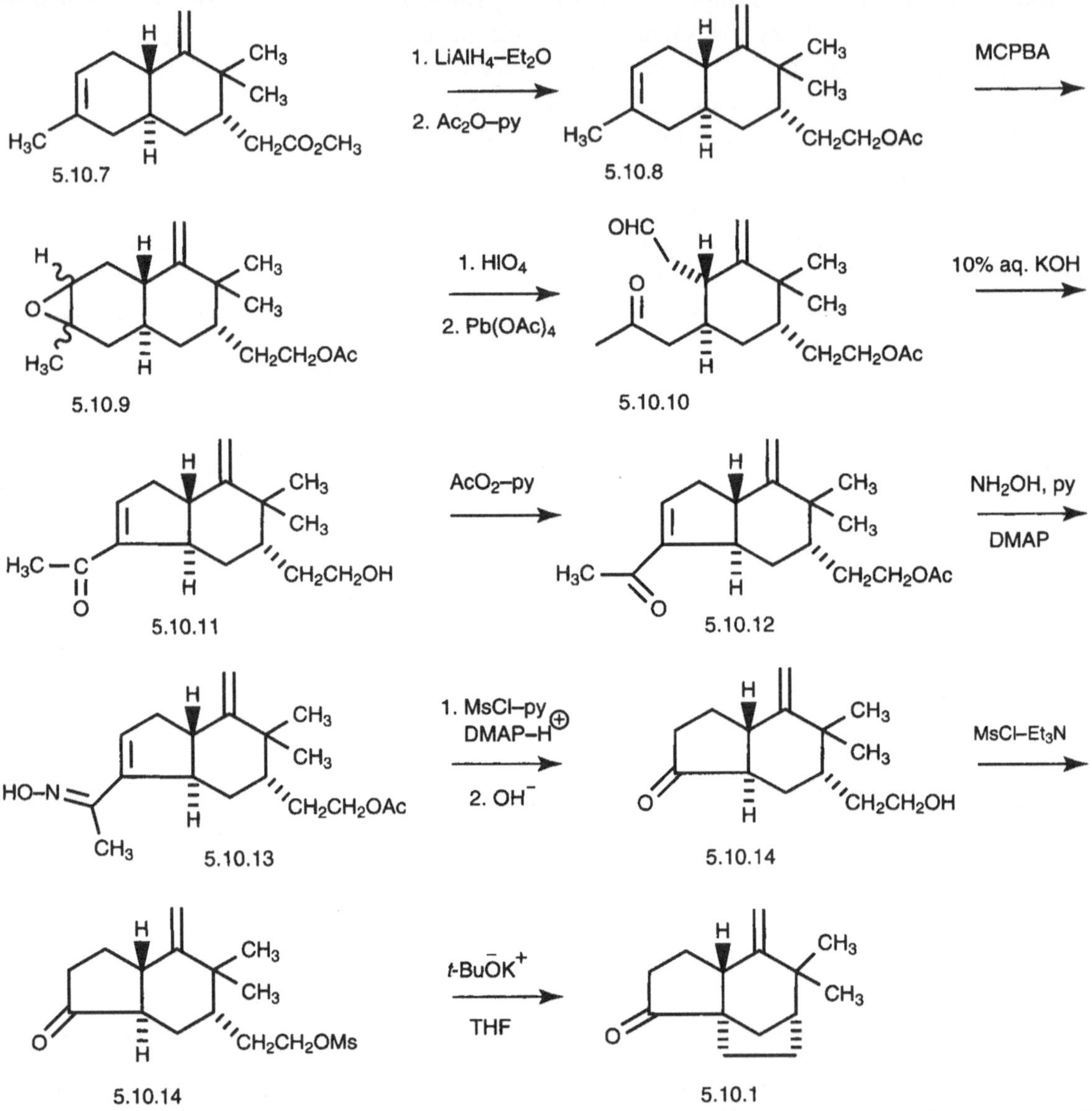

An outstanding feature of this synthesis is the selection of highly stereoselective reactions without resorting to exotic reagents. Thus, while the starting material had only one chiral centre, the final product had three, with the original configurational centre 'guiding' the proper development of the other two under favourable conditions.

Suggested reading

1.   Guenther E, *The Essential Oils*, Vol. 5, NewYork: Robert E Krieger Publishing Co., 1950.
2.   Maurer M, Franchebond M, Grieder M and Ohloff G, *Helv. Chim. Acta.*, 1972, 55: 2371.
3.   Buchi G, Hauser A and Limacher J, *J Org. Chem.*, 1977, 42, 3323.

4. Sakurai K, Kitahara T and Mori K, In Bhattacharya SC, Sen N and Sethi KL, (eds.), *Proceedings of the 11th International Congress of Essential Oils, Fragrances and Flavours*, New Delhi, Vol. **5**, 1989, New Delhi: Oxford and IBH Publishing Co.

5.11 SYNTHESIS OF A CHIRAL MARINE NATURAL PRODUCT

We Shall now discuss a new synthesis of (+)didemnenones A(5.11.1) and B(5.11.2), which were obtained as an inseparable mixture from a Caribbean tunicate, *Trididemnum cf. cyanophorum* by Fenical, Clardy and co-workers. The gross structure as well as the stereochemistry of the compounds were confirmed by an elegant synthesis achieved by Clardy. In this section we will discuss a new synthesis accomplished by Sugahara et al.

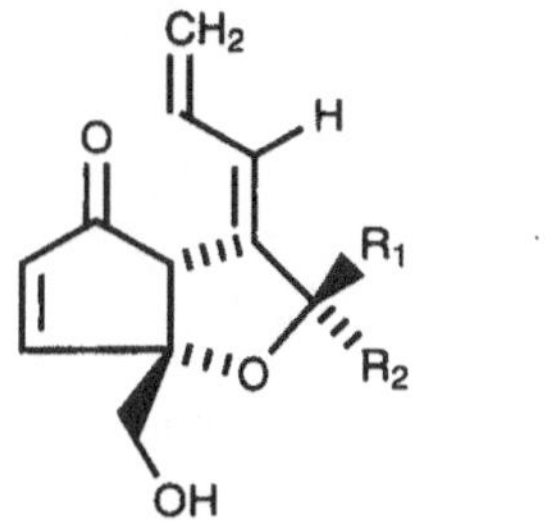

The starting material in this synthesis was the commercially available Corey lactone benzoate, i.e. (–)3α,5α–dihydroxy-2β (hydroxymethyl)cyclo-pentane-1α-acetic acid lactone-3-benzoate (5.11.3) which could be readily transformed into 5.11.4. On treatment with 4-methoxyphenol in the presence of triphenyl phosphine and diethyl aza dicarboxylate (5.11.4) yielded the aryloxy derivative 5.11.5.

5.11.1: R_1 = H; R_2 = OH
5.11.2: R_1 = OH; R_2 = H

Saponification of 5.11.5 followed by reaction with iodine–KI resulted in the formation of the iodolactone 5.11.6. After protecting the hydroxyl group by silylation using *t*-butyldimethylsilyl chloride (*t*BDMS-Cl), the iodolactone was subjected to dehydrohalogenation to obtain 5.11.7. A base-catalysed condensation of the latter with acrolein in the presence of lithium trimethyl-silylamide gave 5.11.8.

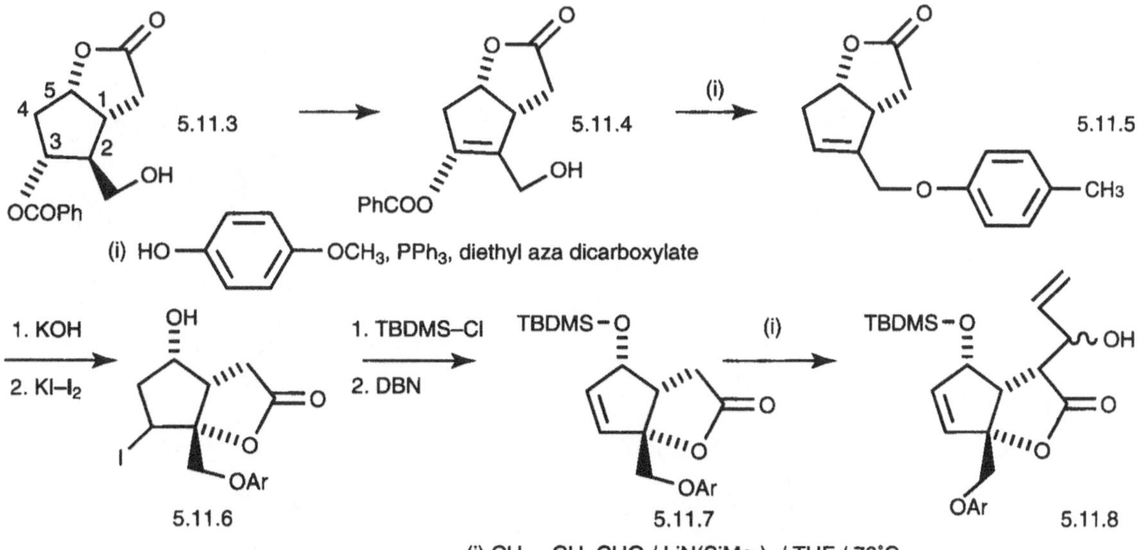

(i) HO—⟨⟩—OCH₃, PPh₃, diethyl aza dicarboxylate

(i) CH_2 = CH–CHO / LiN(SiMe₃)₂ / THF / 78°C

Subsequent steps involved conversion of 5.11.8 to the mesyl ether (5.11.9) and conversion of the latter into a mixture of the two configurationally isomeric dienes, 5.11.10 and 5.11.11. The Z isomer (5.11.10) could be converted into 5.11.11 by treatment with lithium isopropyl sulphide. The E diene (5.11.11) was then reduced with diisobutyl aluminium hydride (DIBAL-H) and the resulting cyclic hemiacetal (5.11.12) converted into its methyl ether (5.11.13). Desilylation of the latter was effected with the help of tetrabutyl ammonium fluoride. The product enol (5.11.14) was oxidised using pyridinium dichromate(PDC) to obtain 5.11.15.

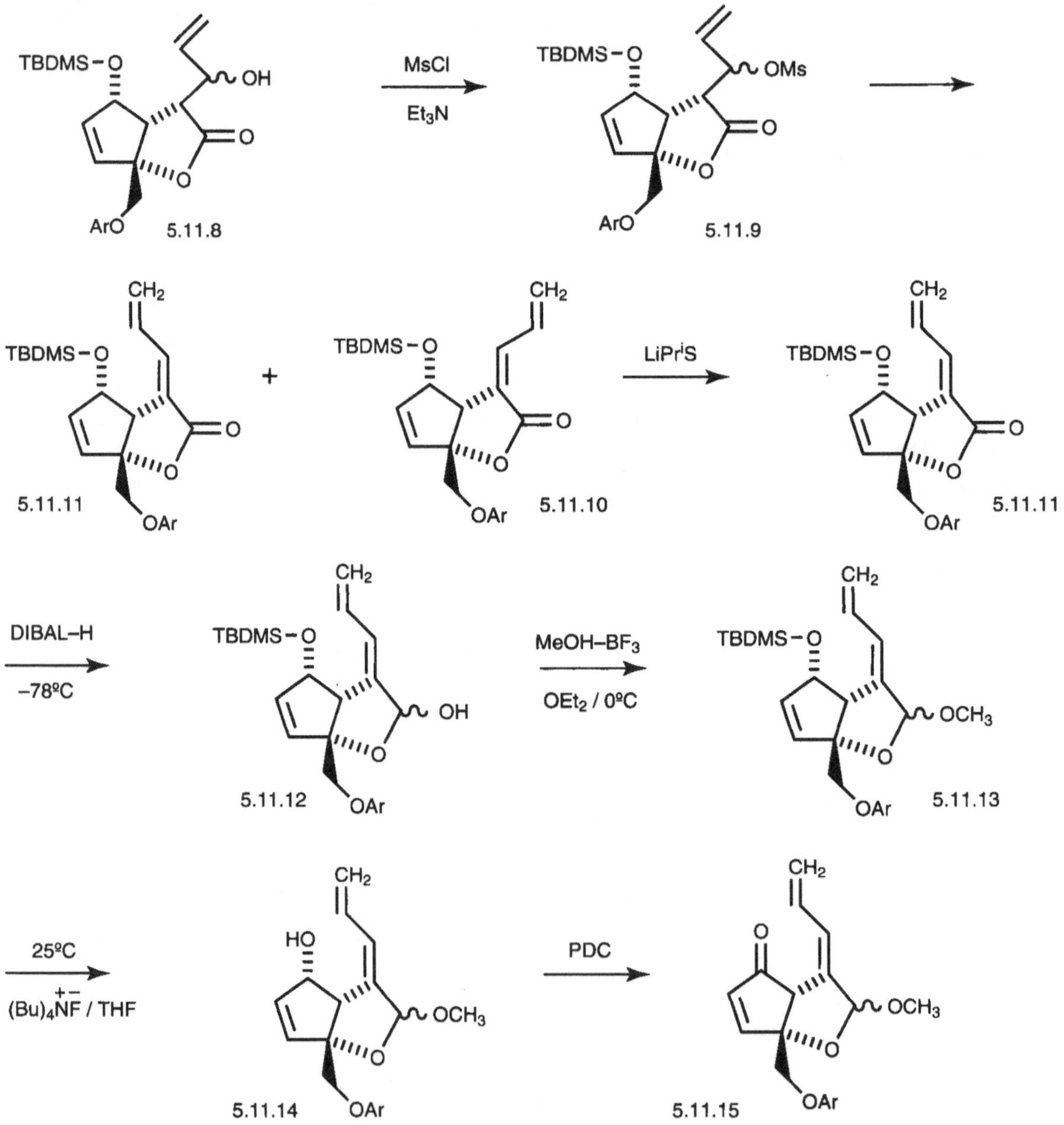

Dearylation of 5.11.15 was brought about by the action of ceric ammonium nitrate to get a 44% yield of an intimate mixture of didemnenones A (5.11.1) and B (5.11.2). The mixture was also accompanied by a separable mixture of the two diastereoisomeric monomethyl ethers 5.11.16 and 5.11.17 which gave 5.11.1 and 5.11.2 on treatment with HCl.

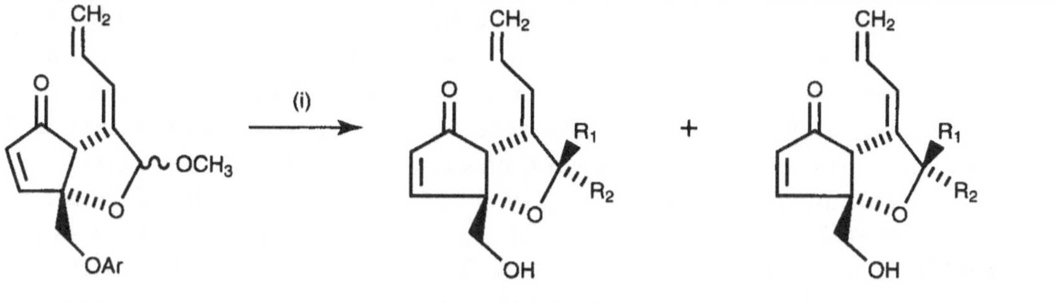

5.11.15

5.11.1: R_1 = H; R_2 = OH
5.11.2: R_1 = OH; R_2 = H

(44%)

5.11.16: R_1 = H; R_2 = OCH$_3$
5.11.17: R_1 = OCH$_3$; R_2 = H

(i) Ceric ammonium nitrate in CH$_3$CN / H$_2$O / 0°C

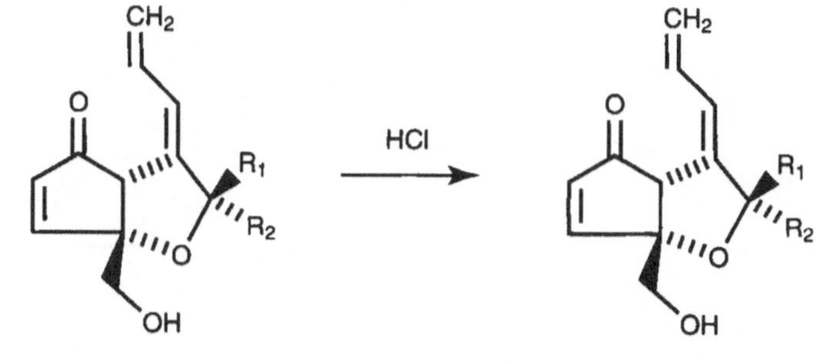

5.11.16: R_1 = H; R_2 = OCH$_3$
5.11.17: R_1 = OCH$_3$; R_2 = H

5.11.1: R_1 = H; R_2 = OH
5.11.2: R_1 = OH; R_2 = H

A striking feature of this synthesis is the use of a wide range of reagents, old and new, for bringing about familiar functional group transformations. This synthesis can constitute material for a classroom discussion as it brings out, under one head, a large number of reactions. This example would require the students to refer to Fieser and Fieser's *Reagents for Organic Synthesis,* enabling them to learn more about the uses of each of the different reagents mentioned above.

Suggested reading

1. Linquist N, Fenical W, Sesin DF, Ireland CM, van Duyne GD, Forsyth CJ and Clardy J, *J. Amer.Chem. Soc.,* 1988, 110: 1308.
2. Forsyth CJ and Clardy J, *J. Amer.Chem. Soc.,* 1988, 110: 5911.
3. Sugahara T, Ohoke T, Soejima M and Takano S, *J. Chem. Soc.,* 1990, 1: 1824.
4. Fieser LF and Fiester M, *Reagents for Organic Synthesis,* Vols 1–16, New York: Wiley Interscience, 1967–92.

5.12 A Biomimetic Synthesis of Morphine

The first total synthesis of racemic morphine (5.12.8) was reported in 1952 by Gates and Tschudi. The following synthesis effected by Morrison et al in 1967 is a biomimetic synthesis which closely follows the biosynthetic pathway. The starting material was 3-methoxyphenylethylamine (5.12.1) which was condensed with 3-hydroxy-4-methoxyphenylacetyl chloride (5.12.2) in the presence of phosphorous oxychloride (Bischler–Napieralski synthesis) to obtain the benzyldihydroisoquinoline (5.12.3). The methiodide of this compound was reduced with sodium borohydride to get the tetrahydroisoquinoline (5.12.4) which was then subjected to a Birch reduction using sodium in liquid ammonia and t-butanol.The product, 5.12.5, underwent a series of proton-catalysed reactions, when refluxed with 10% hydrochlroic acid, to yield 5.12.6, which was also a key intermediate in the Gates–Tschudi synthesis. The subsequent steps were the same as in that synthesis. All the steps are shown in Scheme 5.7. As in the biosynthesis of morphine, codeine (5.12.7) is an intermediate in this laboratory synthesis.

Scheme 5.7

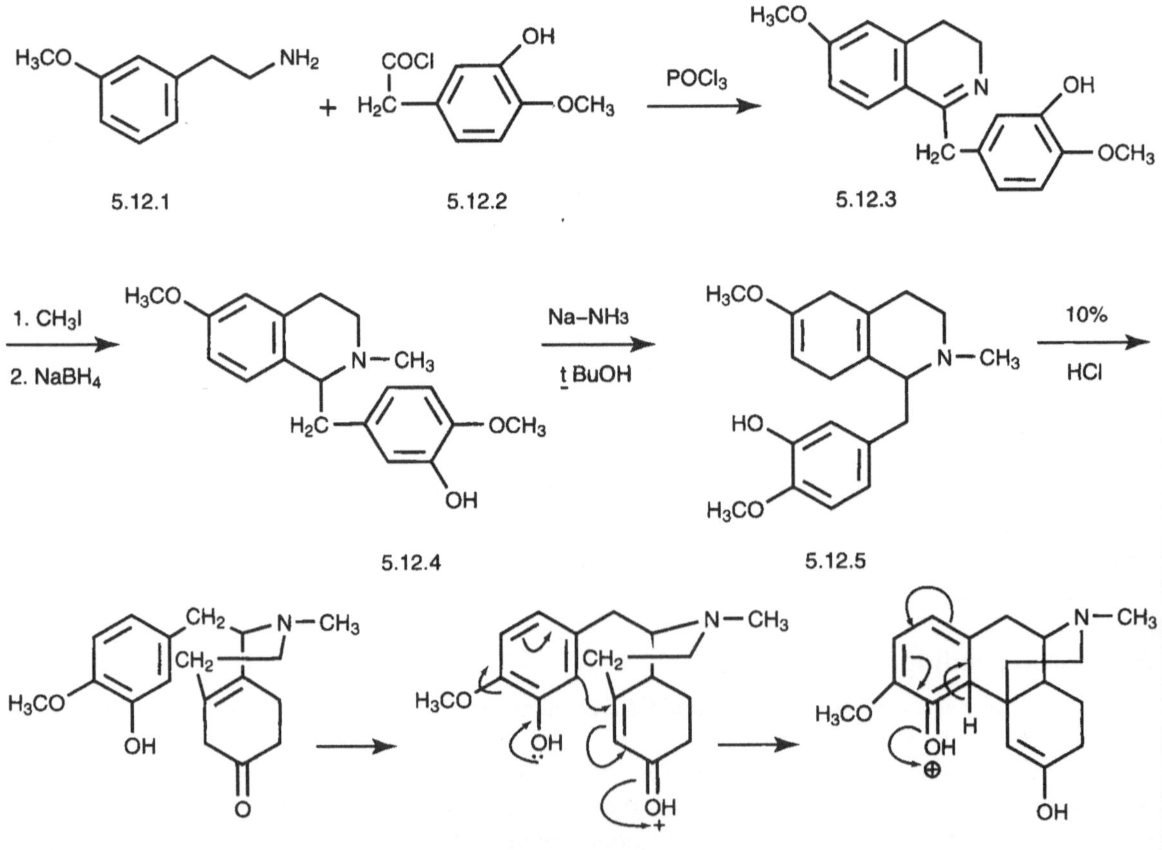

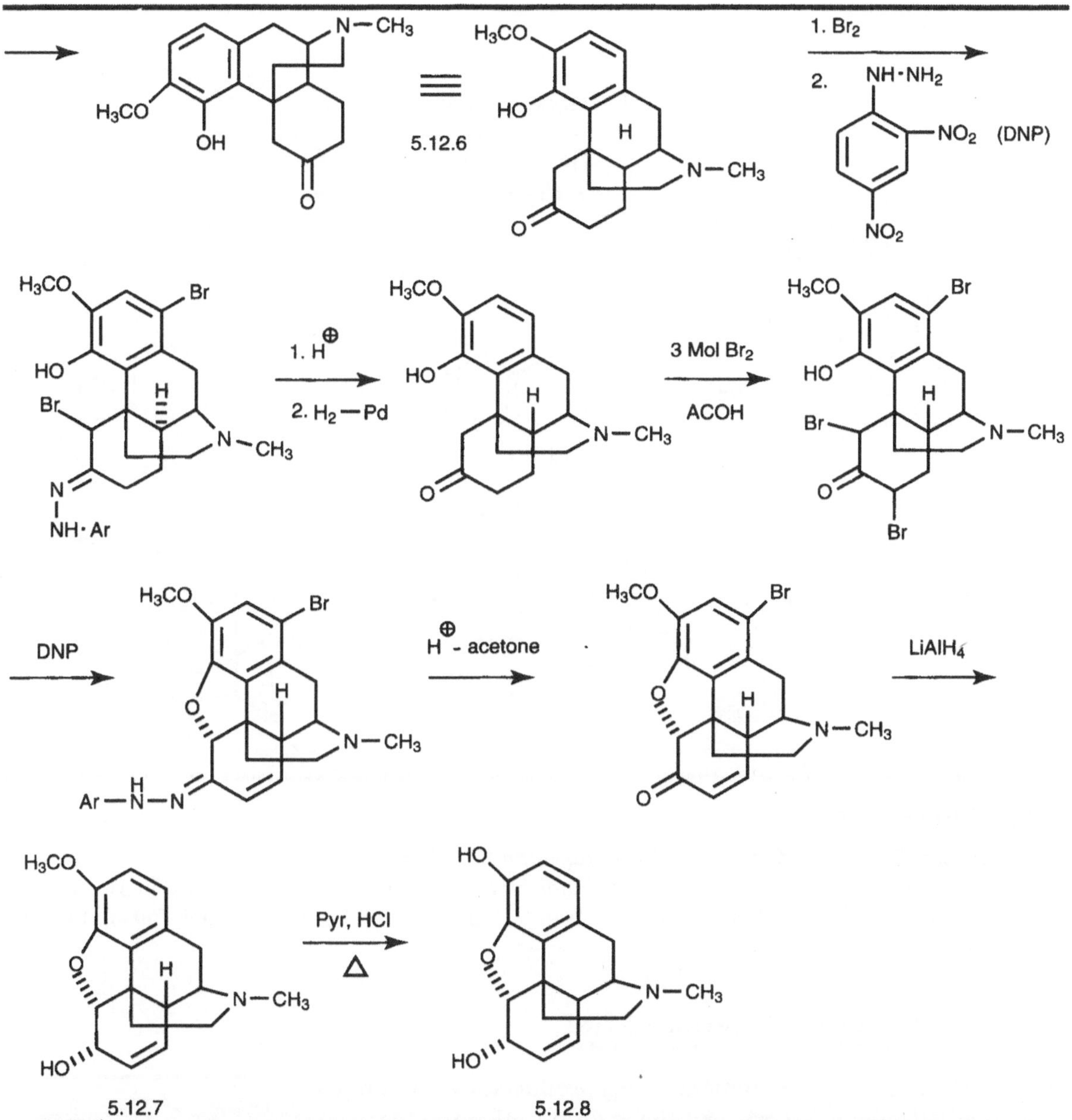

5.12.6

5.12.7   5.12.8

Overman and his co-workers used the intramolecular Heck reaction as a key step in their elegant stereospecific synthesis of (–) morphine, an outline of which is given in Scheme 5.8. Thus, treatment of the compound (5.12.9) with the catalyst derived from Pd(OCOCF₃)2(PPh₃)2 and 1,2,2,6,6-pentamethyl piperidine in refluxing toluene yielded the morphinan derivative (5.12.10). The subsequent steps which converted 5.12.10 into (–) morphine (5.12.11) via (5.12.12) are shown in Scheme 5.8. These involved deprotection of the O-benzyl ether, epoxidation of the double bond, and a *trans* diaxial cleavage of the oxirane ring to yield (5.12.12).

Scheme 5.8

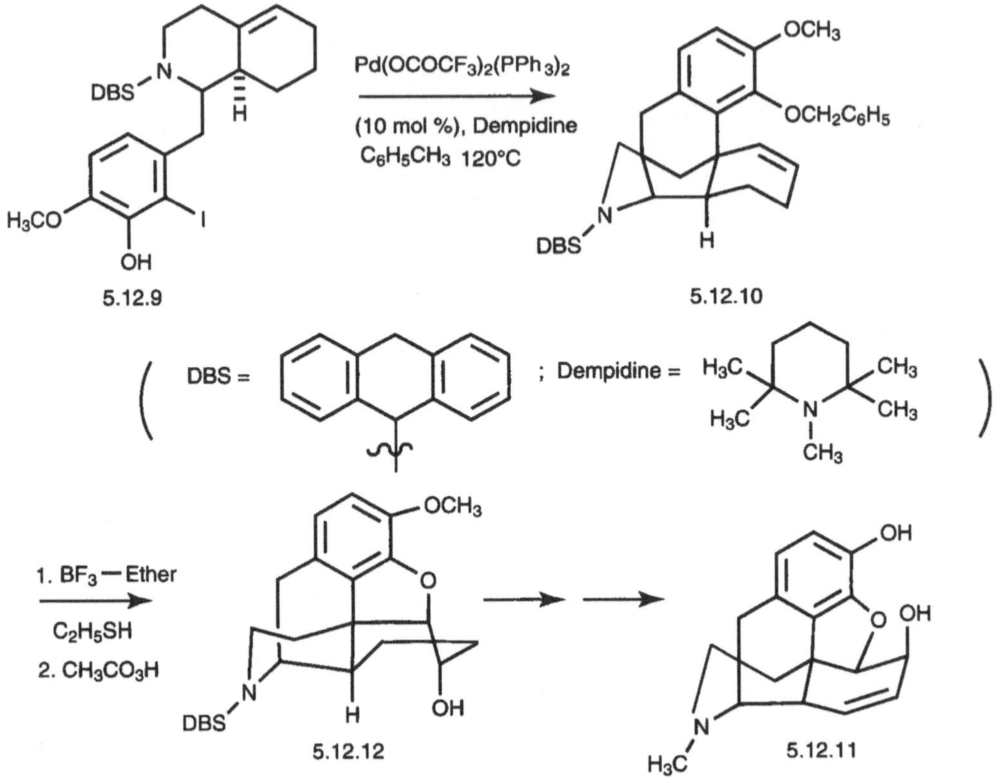
DBS
N
H
H3CO
OH
I
5.12.9
Pd(OCOCF3)2(PPh3)2
(10 mol %), Dempidine
C6H5CH3 120°C
OCH3
OCH2C6H5
N
DBS
H
5.12.10
DBS =
; Dempidine =
H3C
CH3
N
CH3
1. BF3—Ether
C2H5SH
2. CH3CO3H
OCH3
O
OH
5.12.12
OH
O
OH
N
H3C
5.12.11

Suggested reading

1. Morrison GC, Waite RO and Shavel J, *Tetrahedron Lett.*, 1967, 4055.
2. Nicolauo KC and Sorensen EJ, *Classics in Total Synthesis,* Weinheim and New York: VCH, 1996.
3. Overman LE, Abelman MM, Kucera DJ, Tran VD and Riccet DJ, *J. Pure & Appl. Chem.*, 1992, 64: 1813.

5.13 Synthesis of Ergocristine

Among the large number of metabolites produced by the fungus *Claviceps purpurea* (ergot), are the peptide alkaloids derived from lysergic acid and isolysergic acid, which are epimeric at C-8 of the main skeleton (5.13.1). The peptide alkaloids, commonly known as ergot alkaloids, act on the uterus. The simplest among them is ergometrine (ergonovine) (5.13.2), which brings about uterine contraction. It can be prepared from isolysergic acid azide by condensation with 2-aminopropanol. The other ergot alkaloids are cyclic peptides and they include ergotamine (5.13.3) (used in the treatment of migraines), ergosine (5.13.4)

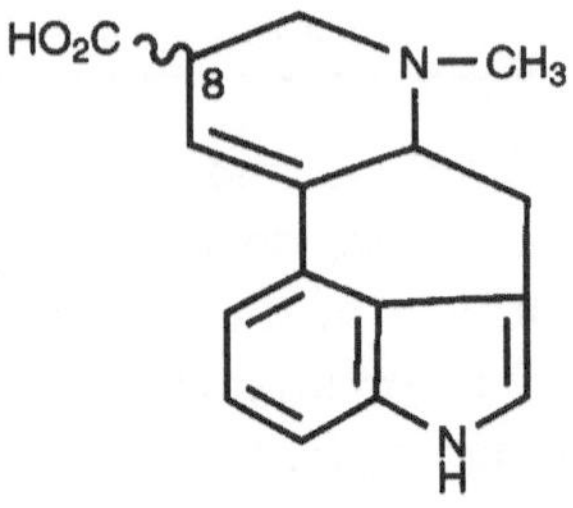
HO2C
8
N—CH3
N
H

5.13.1

and ergocristine (5.13.5). In this section, we will describe a synthesis of ergocrystine from d-lysergic acid, which itself was synthesised by Woodward and co-workers.

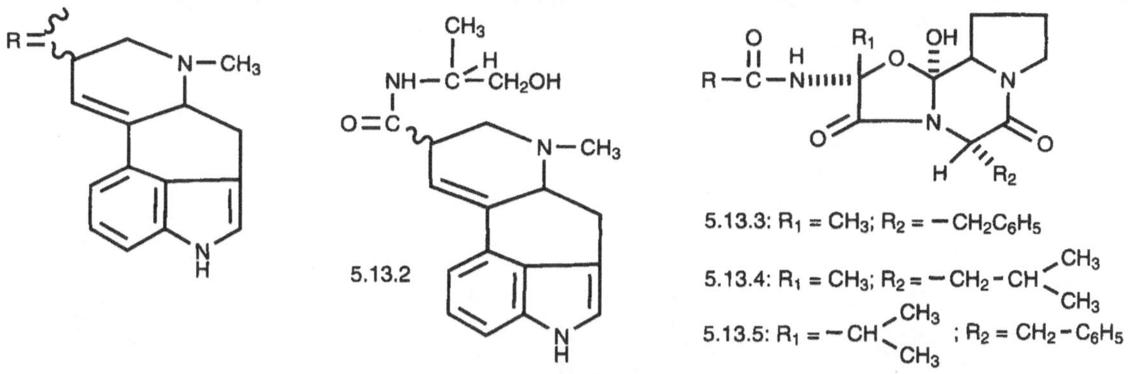

5.13.2

5.13.3: $R_1 = CH_3$; $R_2 = -CH_2C_6H_5$

5.13.4: $R_1 = CH_3$; $R_2 = -CH_2-CH\begin{smallmatrix}CH_3\\CH_3\end{smallmatrix}$

5.13.5: $R_1 = -CH\begin{smallmatrix}CH_3\\CH_3\end{smallmatrix}$; $R_2 = CH_2-C_6H_5$

Ergocristine and ergocristinine were discovered in 1964 as components of 'ergotoxin'. They stimulate smooth muscles and are serotonin antagonists. They were the first examples of peptides containing as a typical amino acid L-alpha-hydroxy-alpha-aminobutyric acid. The main problem in their synthesis was the preparation of S (+) isopropylbenzyloxymalonic acid mono ethyl ester chloride (5.13.6). This compound could be prepared from O-benzyltartronic acid diethyl ester (5.13.7) by reaction with isopropyl iodide in the presence of a base. The latter was hydrolysed and converted to the half ester (5.13.8) which was optically resolved and reacted with thionyl chloride in dimethylformamide to obtain 5.13.6. Reaction of 5.13.6 with the diketopiperazine obtained from L-proline and L-phenylalanine (5.13.9) in the presence of a base yielded, in a stereospecific manner, the cyclic dipeptide (5.13.10). Hydrogenolysis using palladium as catalyst brought about the removal of the benzyl group and simultaneous cyclisation to yield (5.13.11). The latter was subjected to a mild Curtius reaction when 5.13.12 was obtained. Reaction of this peptide with d-lysergic acid chloride yielded a mixture of ergocristine and ergocristinine. The synthesis of lysergic acid itself is not described here as it is well documented.

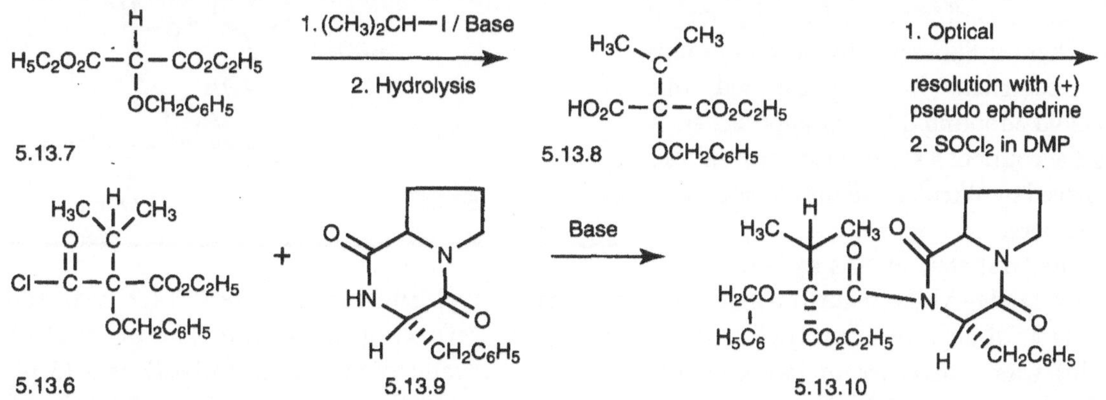

5.13.7

1. $(CH_3)_2CH-I$ / Base

2. Hydrolysis

5.13.8

1. Optical resolution with (+) pseudo ephedrine

2. $SOCl_2$ in DMP

5.13.6

+

5.13.9

Base

5.13.10

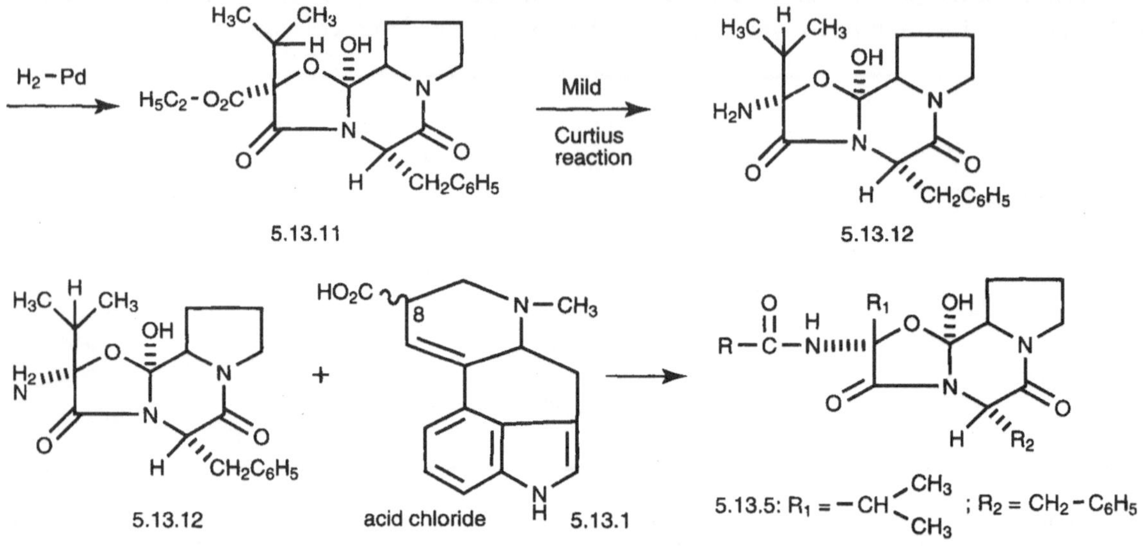

5.13.11   5.13.12

5.13.12   acid chloride   5.13.1   5.13.5: $R_1 = -CH\begin{smallmatrix} CH_3 \\ CH_3 \end{smallmatrix}$; $R_2 = CH_2-C_6H_5$

Suggested reading

1. Komarova EL and Tolkachev ON, The chemistry of peptide ergot alkaloids, *J. Pharm. Chem.*, 2001, 35: 504–13.
2. P Stulz, Brunner R, and Stadler PA, *Experientia*, 1973, 29: 936 and references therein.

5.14 A STEREOSELECTIVE SYNTHESIS OF RESERPINE

Reserpine (5.14.1), an anti-hypertensive alkaloid isolated from *Rauwolfia serpentina*, has been the subject of many studies. Its total synthesis was first effected by Woodward and co-workers; a brief mention of the first step in this synthesis was made earlier in the introduction of this chapter. Subsequently, several groups of researchers have synthesised this alkaloid adopting different approaches. A brief account of a stereoselective synthesis achieved by Wender and his co-workers is given here.

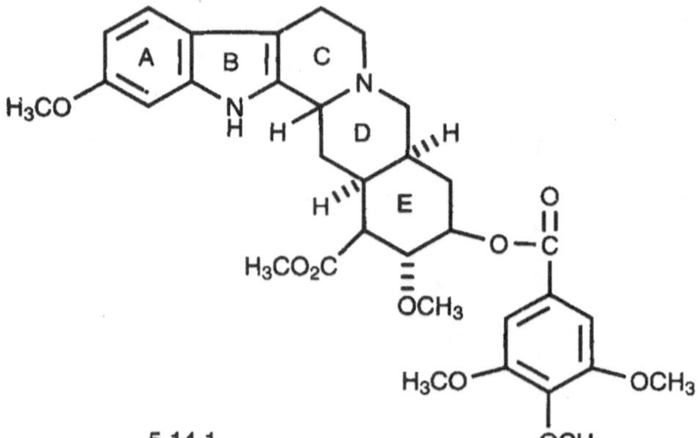

5.14.1

The first step in this synthesis was also a Diels–Alder reaction in which the dihydropyridine carboxylate (5.14.2) was added to the enol acetate of methyl pyruvate (5.14.3) by refluxing in toluene. The product (76% yield) was a mixture of two diastereoisomeric bicyclic compounds (5.14.4) and (5.14.5). After separation, 5.14.4 was reacted with lithium *t*-butyl acetate to obtain (5.14.6). After acetylation of the hydroxyl group, the *t*-butyl ester was hydrolysed by treatment with trifluoroacetic acid and the resulting compound esterified with diazomethane when 5.14.7 was obtained in 64% yield. It is worth noting that the β-keto carboxylic acid unit in the compound exists in the enol form and hence the product is a enol ether ester. When refluxed in xylene, 5.14.7 underwent a oxy-Cope rearrangement to give stereospecifically, in high yield (78%), 5.14.8, which has the correct stereoschemistry needed in the D and E rings of reserpine. The subsequent steps involved catalytic hydrogenation and reduction of the ester group with lithium aluminium hydride. In the process the enol acetate underwent hydrolysis. The product (5.14..9) was then subjected to Jones oxidation and esterified with diazomethane to get 5.14.10 in 79% yield. After reduction with sodium borohydride–ceric chloride combination, the thus generated hydroxyl group was esterified with trimethoxybenzoic acid anhydride in the presence of 4-dimethylaminopyridine (DMAP) to obtain 5.14.11 in 92% yield. Reaction of 5.14.11 with the versatile reagent, iodotrimethylsilane (TMSI) followed by treatment with 3-tosyloxyethyl-6-methoxyindole (5.14.12) resulted in 5.14.13, which was converted into reserpine (5.14.1) in 45% yield by reaction with mercuric acetate followed by sodium borohydride.

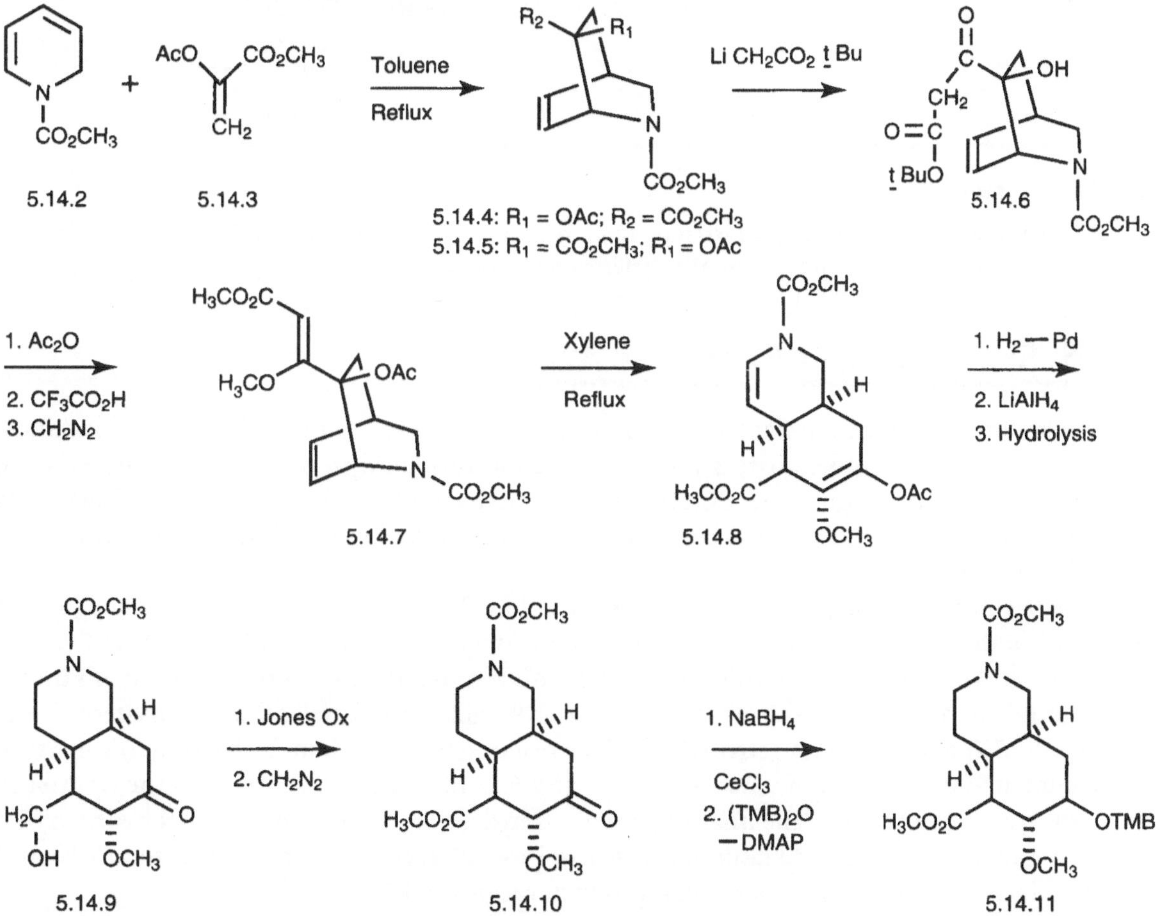

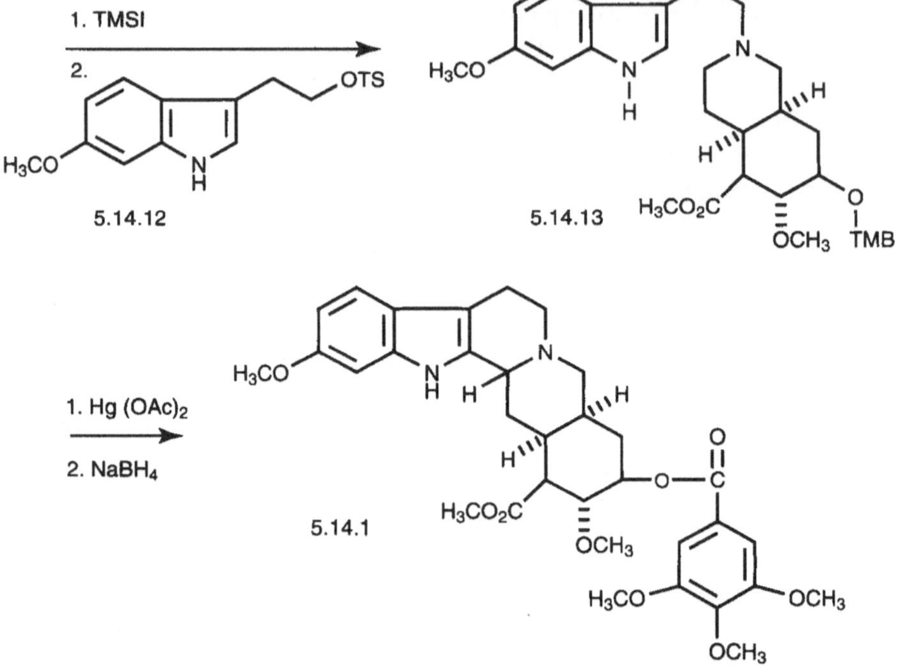

5.14.12   5.14.13

5.14.1

Suggested reading

1.   Nicolauo KC and Sorensen EJ, *Classics in Total Synthesis,* Weinheim and New York: VCH, 1996.
2.   Wender PA, Schaus JM and White AW, *Heterocycles,* 1987, 25: 263.

5.15 SYNTHESIS OF A PARACONIC ACID

Among lichen metabolites, paraconic acids form an important group of bioactive compounds. These gamma butyrolactones possess antineoplastic and antibiotic properties. Two representative examples are (+) roccellinic acid (5.15.1) and (+) methylenolactocin(5.15.2). Several attempts have been made to synthesise these compounds. In this section, a short but elegant synthesis of methylenolactocin is described.

Once again, the first step was a Diels–Alder reaction! The addition of cyclopentadiene (5.15.3) to itaconic anhydride (5.15.4) followed by hydrolysis and esterification yielded 5.15.5, exclusively as the exo-adduct. Treatment of this compound with hexanal (5.15.6) in the presence of lithium diethylamide in tetrahydrofuran at –78°C gave in 79% yield a mixture of the two diastereomeric lactones (5.15.7) and (5.15.8). A retro-Diels–Alder reaction brought about by heating this mixture in vacuum at 500°C yielded in 92% yield 5.15.9 and 5.15.10 as a 2:1 mixture, with the *cis*-isomer being the major component. The final step involved hydrolysis of the ester with 6N hydrochloric acid in butanone (refluxed for 3 hours), which also brought about epimerisation of the *cis*-isomer resulting in a 71% yield of the desired end product, racemic 5.15.2.

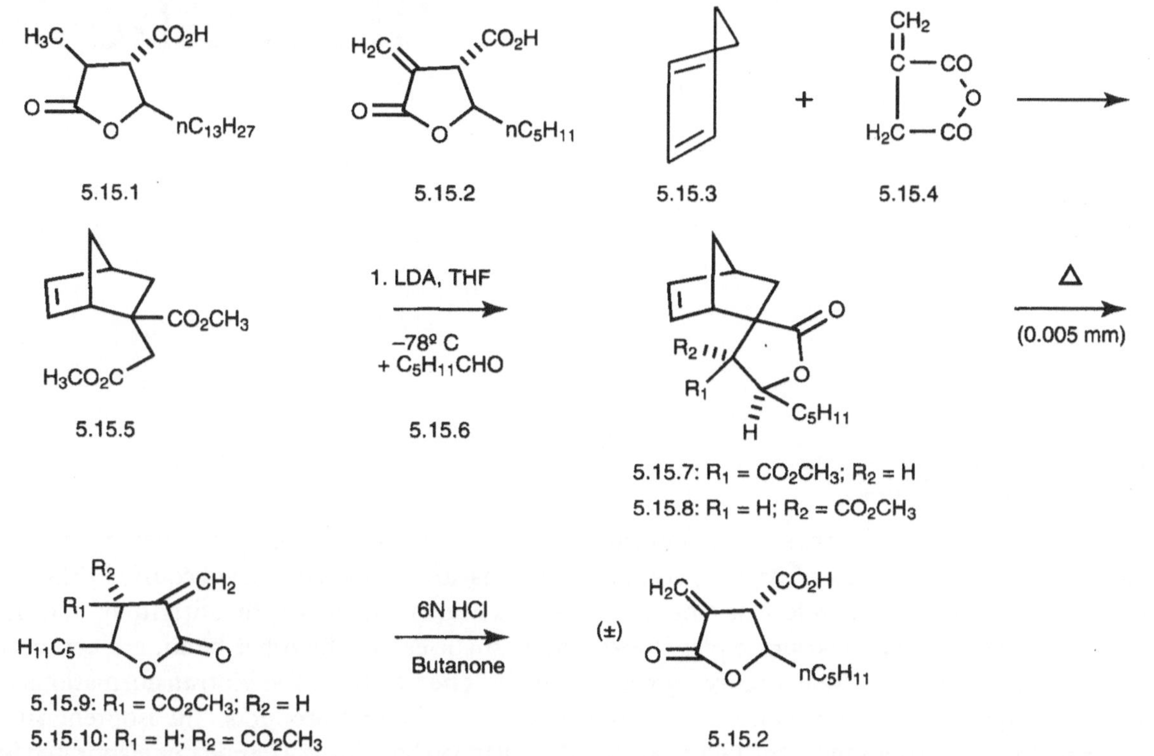

5.15.1

5.15.2

5.15.3 + 5.15.4 →

5.15.5

1. LDA, THF
-78° C
+ $C_5H_{11}CHO$

5.15.6

5.15.7: $R_1 = CO_2CH_3$; $R_2 = H$
5.15.8: $R_1 = H$; $R_2 = CO_2CH_3$

Δ
(0.005 mm)

5.15.9: $R_1 = CO_2CH_3$; $R_2 = H$
5.15.10: $R_1 = H$; $R_2 = CO_2CH_3$

6N HCl
Butanone

($\pm$)

5.15.2

Suggested reading

Bandichhor R, Nosse B and Reiser O, In Mulzer J, Spiezel A, Bandichor R, Basler B and Brando S, (eds), *Natural Product Synthesis 1. Targets, Methods, Concepts,* Berlin: Springer, 2005.

Chapter

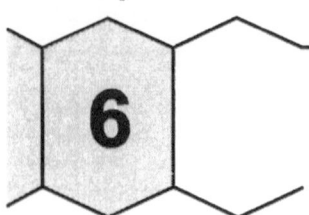

Biosynthesis

6.1 INTRODUCTION

Earlier, the terms biosynthesis and biogenesis were used rather loosely for describing the pathways for the synthesis of organic compounds in nature. They have now acquired different connotations. Biogenesis refers to theoretical speculations regarding the origin of natural products based on chemical principles. Biosynthetic theories, on the other hand, are based on actual experimentation involving living organisms designed to trace the biotransformations of various primary metabolites and probable precursors into the end products. The isoprene rule, for example, first proposed by the German chemist, Otto Wallach in 1897 and later elaborated by Ingold, Ruzicka and others is a biogenetic theory. The actual biosynthetic evolution of terpenoids follows different pathways. This is discussed in detail in this chapter. Similarly, the several schemes put forward by Sir Robert Robinson for the formation of alkaloids and flavonoids fall in the realm of biogenesis. Their actual biosynthetic pathways, however, differ in finer details from the biogenetic theories, as we shall see later. In this chapter only biosynthetic theories which have firm experimental support will be described.

Experimental Methods used in Biosynthesis

Tracer studies involve the use of isotopically labelled compounds which contain isotopes of carbon [carbon 14(C14) and carbon 13 (C13)] and or hydrogen (deuterium or tritium). The preparation of such compounds is a primary requisite and is often quite tedious. Several such compounds, like labelled amino acids, mevalonic acid, acetate and propionate are now commercially available. Prior to 1971, for carbon labelling, the radioactive isotope, C14, was used. In this case the location of the label in the target molecule has to be determined by breaking into smaller molecules by using appropriate degradation methods. However, this is not always feasible. On the other hand, the use of C13 as the label, makes it possible to analyse the location of the label in the product by C13 NMR spectroscopy. This technique has the advantage of being non-destructive and less time-consuming.

The labelled compounds are then fed to carefully selected plants or microorganisms, which are quick-growing, can be cultivated under greenhouse conditions in the laboratory and known to produce the secondary metabolites whose biosyntheis is being studied. We shall illustrate with two examples. The biosynthesis of ubiquinones (6.1.1) in the microorganism, *Escherichia coli*, as well as in the higher plants, *Zea mays* (seeds) and *Phaseolus vulgaris*, was studied by feeding them with C14 shikimic acid or C14 hydroxybenzoic acid. The compounds were isolated by the usual methods and purified by chromatography and then subjected to ozonolysis. Levulinic aldehyde (6.1.2) having C14 was isolated as its 2,4-dinitrophenylhydrazone derivative and assayed for radioactivity using a proportional counter (for other compounds, a scintillating counter could be used). It can be readily understood that this technique involves much experimentation and is time-consuming. The alternative approach using C13 is illustrated by the following example. In an investigation of the biosynthesis of stress metabolites of Solanaceae, 90% enriched, doubly labelled acetate (1,2 C13) was fed to potatoes inoculated with a fungus. A high yield of the sesquiterpene, phytuberin (6.1.3) was isolated and its C13 NMR spectrum critically analysed. The signals of the carbon atoms belonging to those carbon-carbon bonds which arise from the same acetate unit show satellite peaks due to homonuclear (C13–C13) spin–spin coupling–whereas the other carbon atoms show up as singlet signals (after proton decoupling), and are therefore, easily identified. The bonds arising from intact acetate units are shown as thick lines in (6.1.3).

All the reactions involved in biosynthetic processes in nature do not have laboratory analogues or vice versa. Reactions which can proceed under physiological conditions of pH and temperature and which can be catalysed by enzymes are employed by nature to produce a large number of a wide variety of secondary metabolites. The major types of reactions encountered in the biosynthesis of secondary metabolites can be classified under the following heads:

1. Alkylation reactions such as *O*-, *N*- and *C*-methylations;
2. Reactions related to the aldol, Claisen and similar condensation reactions;
3. Oxidative coupling of phenols;
4. Reactions following the Umpolung principle;
5. Transamination;
6. Decarboxylation;
7. Redox reactions.

In nature, *O*-, *N*- and *C*-methylations which are common occurrences in the biosynthesis of several polyphenolic compounds (flavonoids, xanthones and anthraquinones) and benzylisoquinoline alkaloids are brought about through the agency of (S) adenosylmethionine (SAM) (6.1.4), which is the co-factor (prosthetic group) for the enzyme, methyl transferase. It is formed from the amino acid methionine (6.1.5) and adenosine triphosphate (6.1.6) with the elimination of triphosphate. In association with methyl transferase, SAM donates its methyl group to a nucleophile, for example, a carbanion or a phenolate ion, to yield the methylated compound. The mechanism of the reaction is comparable to that brought about in the laboratory with the use of methyl iodide or dimethyl sulphate in the presence of a base.

We presume that the readers are familiar with the mechanisms of the aldol, Claisen and related reactions.

For instance, in the self condensation of two acetone molecules, the primary step is the formation of a carbanion which can tautomerise to the corresponding enolate ion, by the action of a base. In nature, the most important analogue of the aldol reaction is seen in polyketide biosynthesis. In this pathway, the carbanion liberated from a molecule of acetyl coenzyme A (6.1.7) reacts with another molecule of 6.1.7, in aldol fashion, setting free a coenzyme molecule. The resulting acetoacetyl coenzyme A (6.1.8) can then undergo either hydrolysis to yield acetoacetic acid (6.1.9) or condense with another molecule of 6.1.7 resulting in further chain elongation which proceeds further in the biosynthesis of fatty acids.

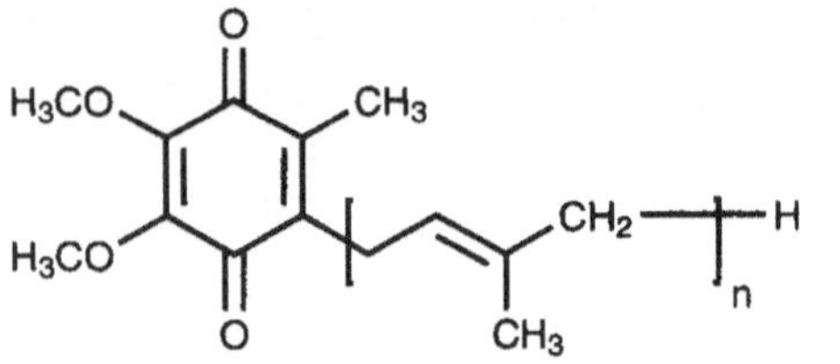

6.1.1 $n = 1$ to 12

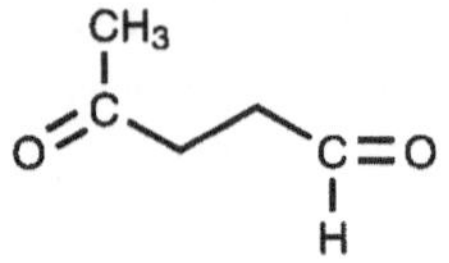

6.1.2

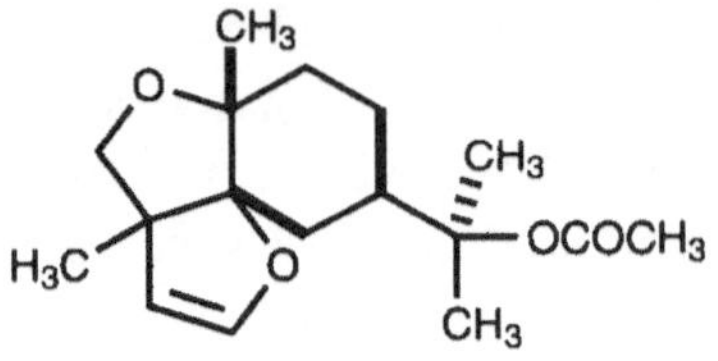

6.1.3

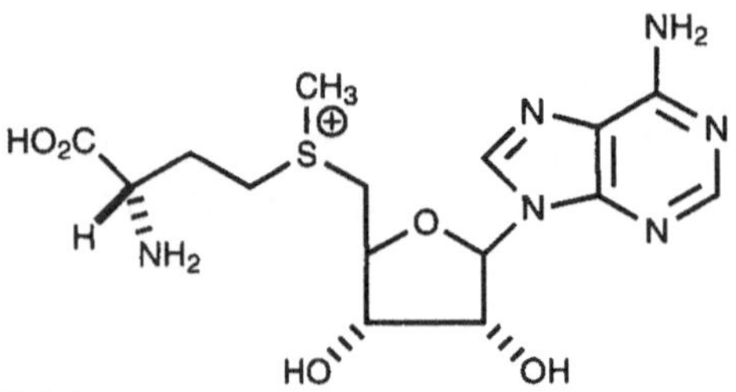

6.1.4

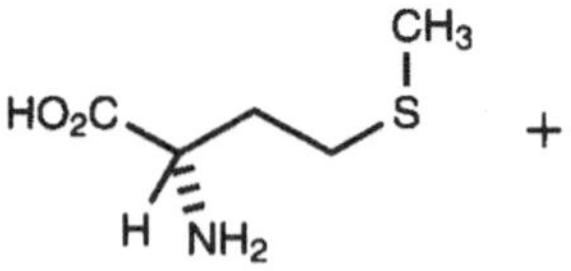

6.1.5

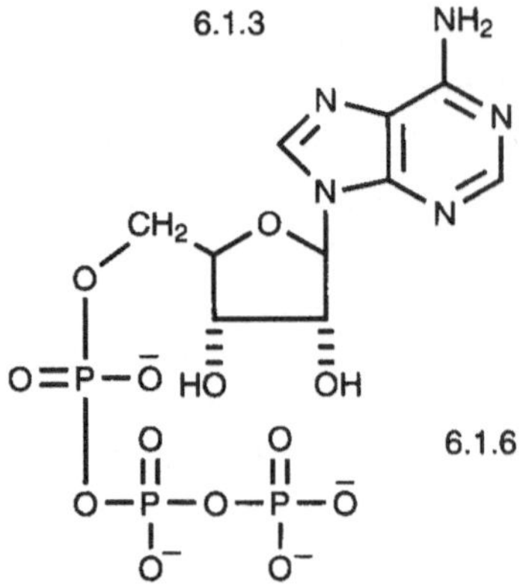

6.1.6

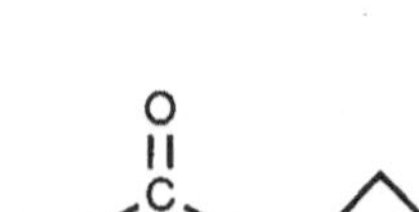

6.1.7

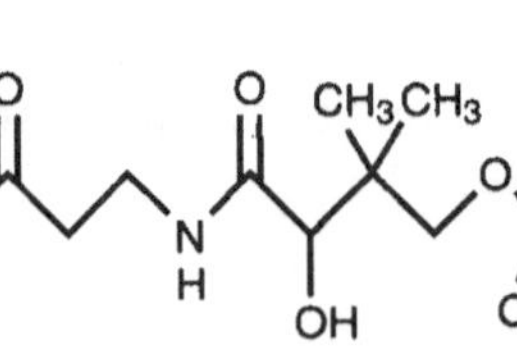

6.1.8

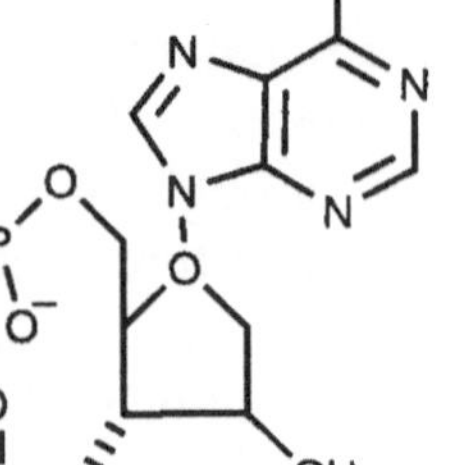

6.1.9

Expanded structure of 6.1.7

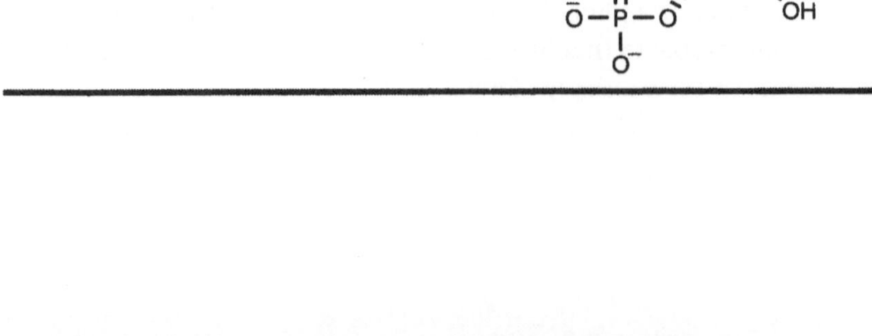

Gilbert Stork developed the enamine reaction to facilitate regioselective aldol type condensations for which there is a natural equivalent. The side chain amino group of the basic amino acid, lysine (6.1.10), in combination with an enzyme through peptide bonds, as shown in (6.1.11), is free to react with a carbonyl group of a substrate (acetaldehyde, for example) to form the corresponding enamine (6.1.12). The latter can, in turn, serve as a substrate for alkylations or aldol type reactions.

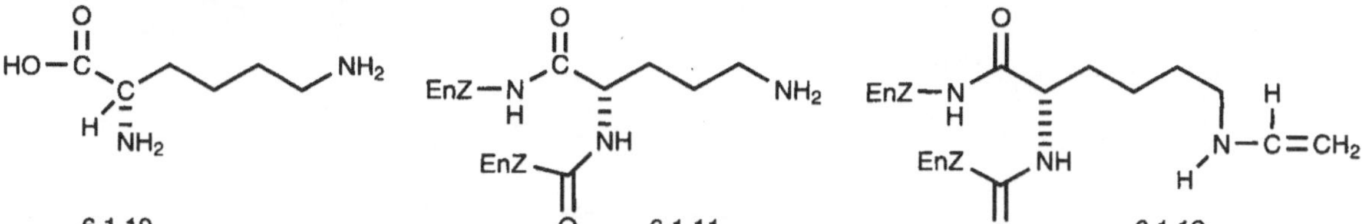

6.1.10   6.1.11   6.1.12

Oxidative coupling of phenols is an important biosynthetic reaction and is responsible for the formation of biflavonoids, bisanthraquinones and similar bimeric compounds as well as some lignans. For example, the bisanthraquinone, skyrin (6.1.13) is presumably formed by the coupling of two chrysaphanol (6.1.14) molecules as shown below. The reaction requires molecular oxygen and is catalysed by ferric ions. Tracer experiments using C14 labelled 5-methyl-2,4,6-trihydroxyacetophenone (6.1.15) have shown that the lichen metabolite, usnic acid (6.1.16) is biosynthesised from this ketone, apparently through a phenolic oxidative coupling reaction. The ketone (6.1.15) itself is biosynthesised from acetyl coenzyme A and malonyl coenzyme A (6.1.17).

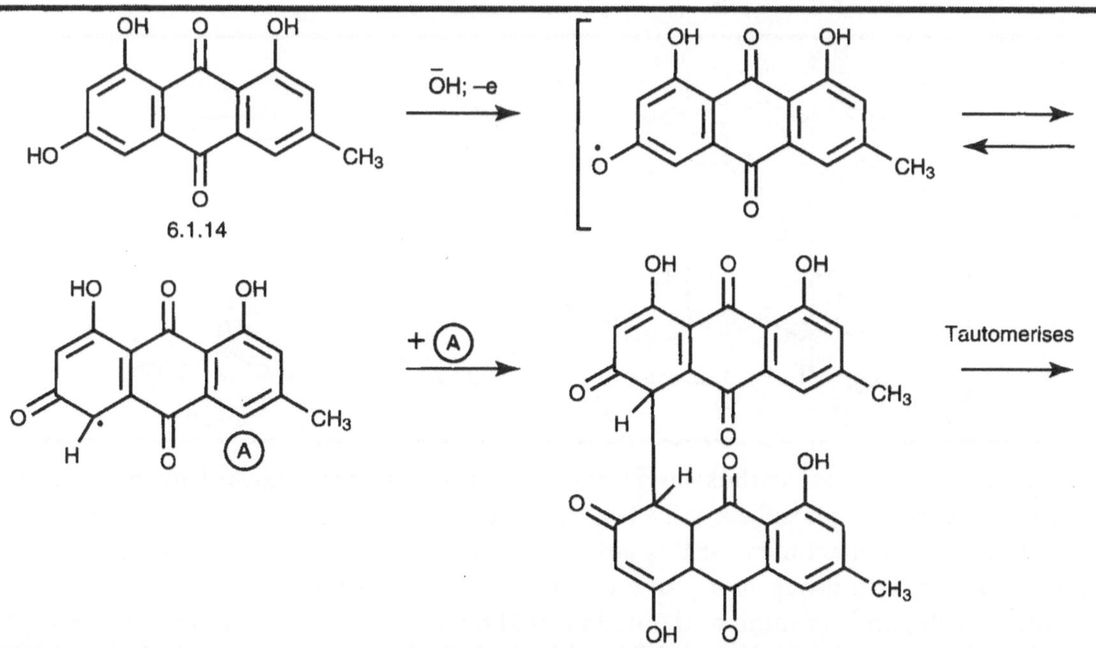

6.1.14

Tautomerises

6.1.13

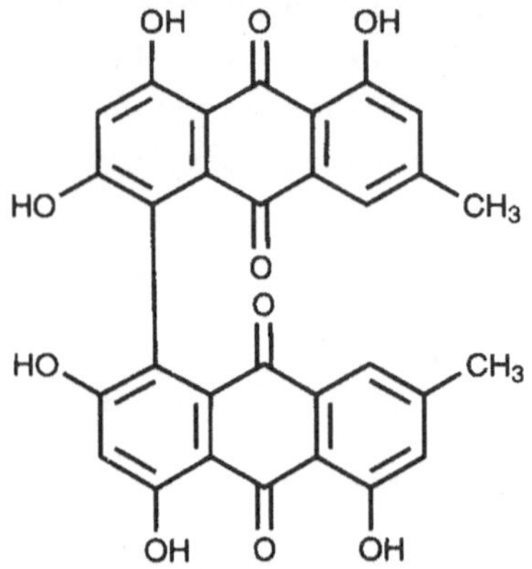

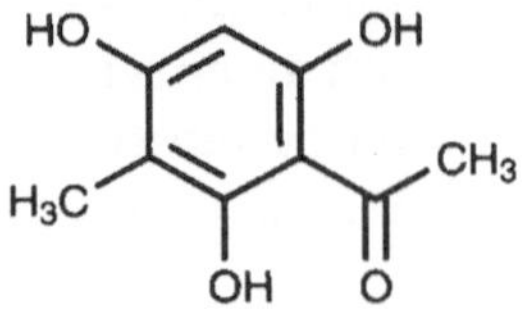

6.1.15

6.1.16

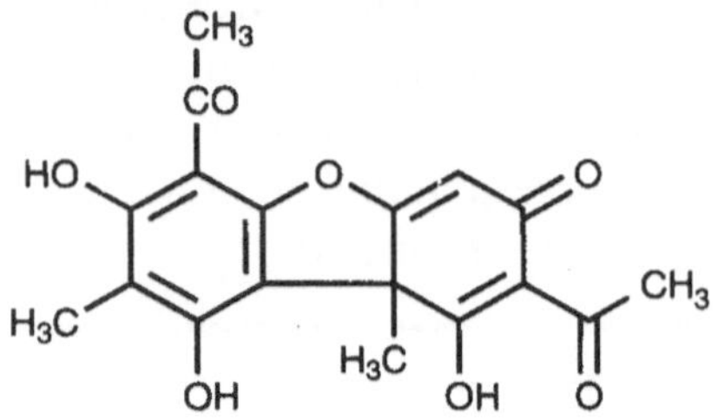

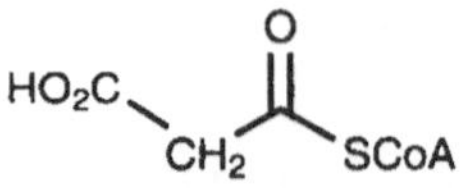

HO₂C–CH₂–CO–SCoA   6.1.17

The phenomenon of Umpolung refers to the reversal of normal polarity at a carbon atom. For example, the natural tendency for the carbon atom of a carbonyl group is to carry a residual positive charge. If it becomes negative the situation is 'Umpolung'. Though it is not common among laboratory reactions, it is used by a number of important enzymes which use thiamine diphosphate (6.1.18) as a cofactor. Under the influence of a basic centre on a enzyme,the hydrogen atom on the thiazolidine ring is lost, generating the zwitter ion (6.1.19). The latter can react, for instance, with the carbonyl group of pyruvic acid (6.1.20) resuting in 6.1.21 via the intermediate 6.1.22. Reaction of 6.1.21 with another molecule of pyruvic acid followed by a base-catalysed elimination of a proton, as shown, results in the formation of the β-keto acid 6.1.23, with the liberation of the zwitter ion 6.1.19.

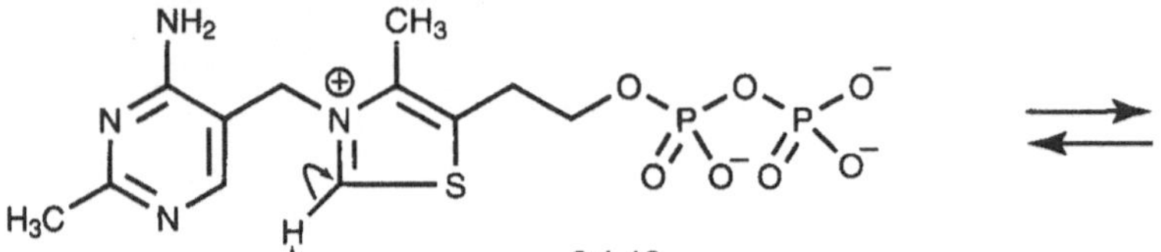

6.1.18

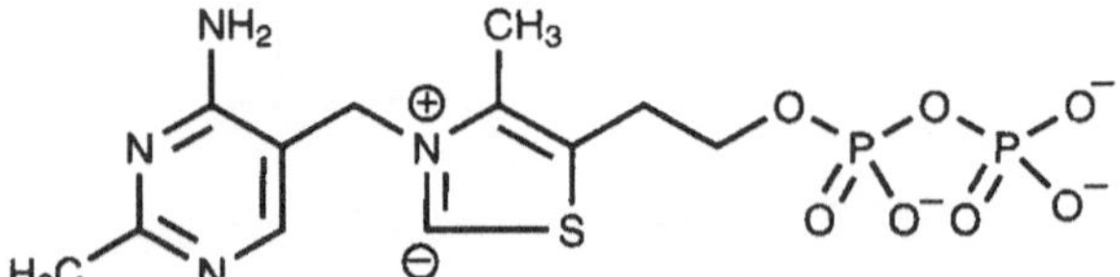

6.1.19

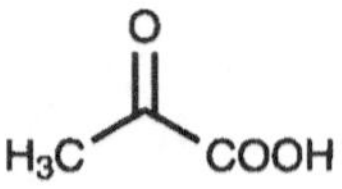

6.1.20

Transamination and decarboxylation reactions in nature are catalysed by enzymes which require pyridoxal pyrophosphate (6.1.24) as the cofactor. In a typical transamination reaction, the first step is the attachment of this co-enzyme to the enzyme through a imino bond with the side chain amino group of a lysine residue in the enzyme as shown in 6.1.25. The latter then reacts with another amino acid 6.1.26 to yield 6.1.27, free from the enzyme; these steps are reversible. Tautomerisation of 6.1.27 followed by hydrolysis results in the alpha keto acid 6.1.28 and pyridoxamine pyrophosphate 6.1.29.

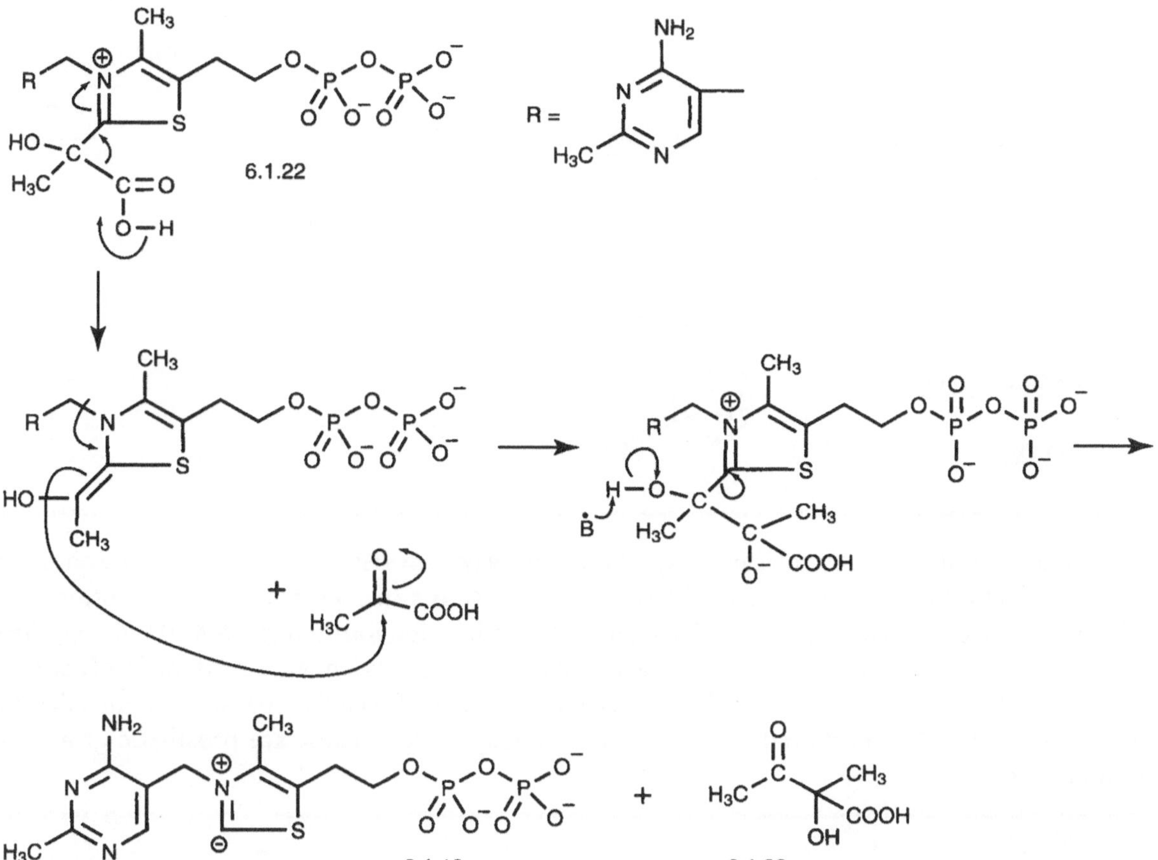

6.1.22

6.1.19

6.1.23

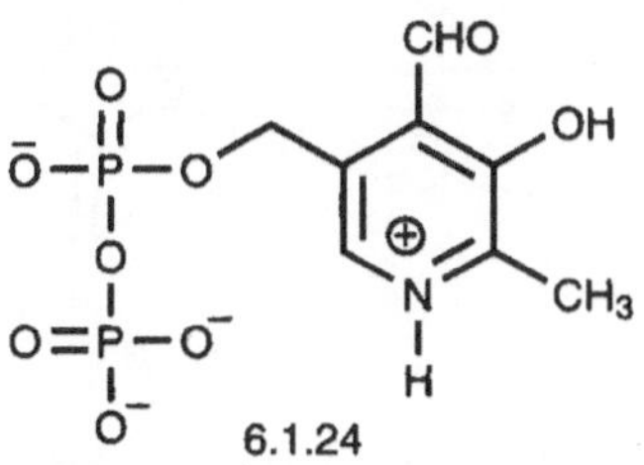

6.1.24

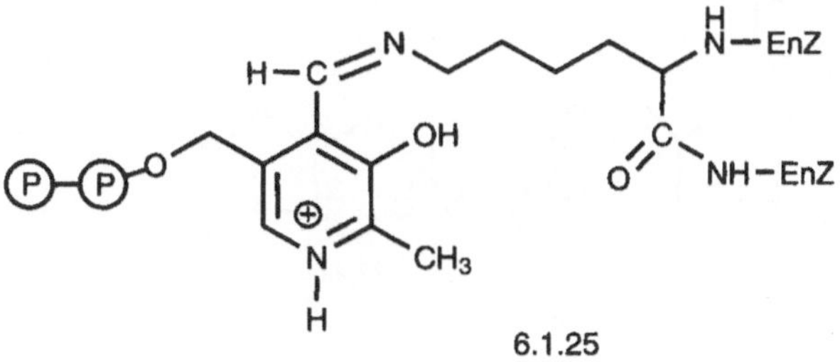

6.1.25

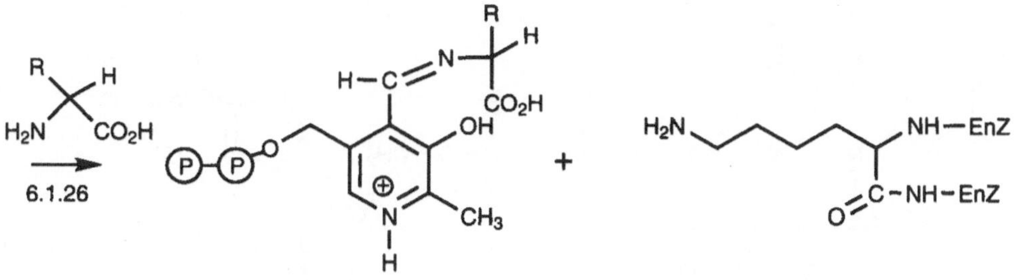

6.1.26

6.1.27

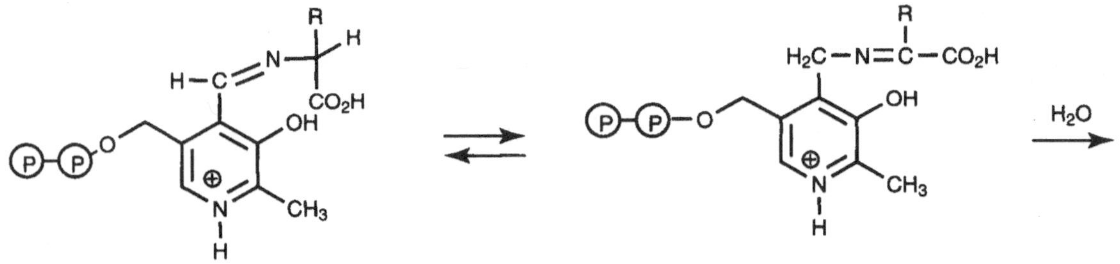

6.1.27

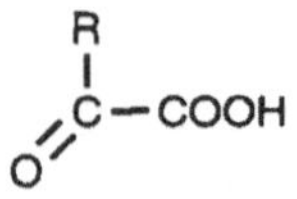

6.1.28

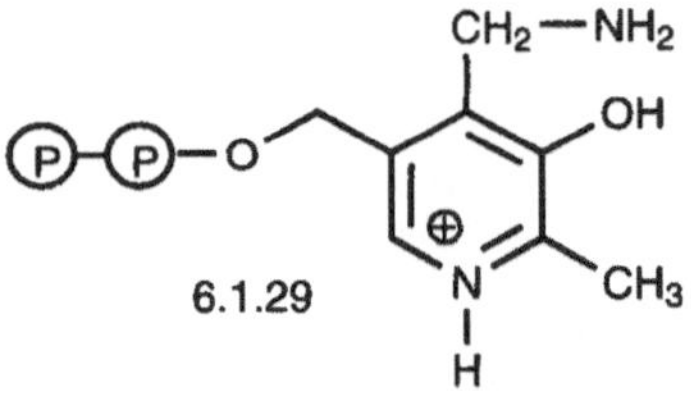

6.1.29

Redox reactions play key roles in biosynthetic pathways. Invariably, these involve the addition or removal of a hydride ion. One of the important cofactors in such reactions is nicotinamide adenine dinucleotide (NAD, 6.1.30). This can accept a hydride ion to form NADH (6.1.31). The latter is a hydride ion donor and is a cofactor involved in many enzyme mediated reductions. Another pair of coenzymes involved in redox reactions is FAD–FADH, namely flavine adenine dinucleotide (6.1.32) and its reduced form (6.1.33), respectively. These are prosthetic groups of flavoproteins.

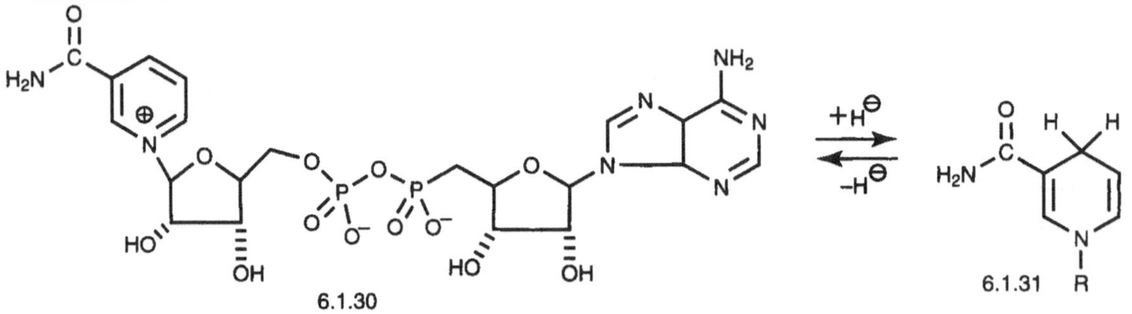

6.1.30

6.1.31   R

R = Adenine dinucleotide residue

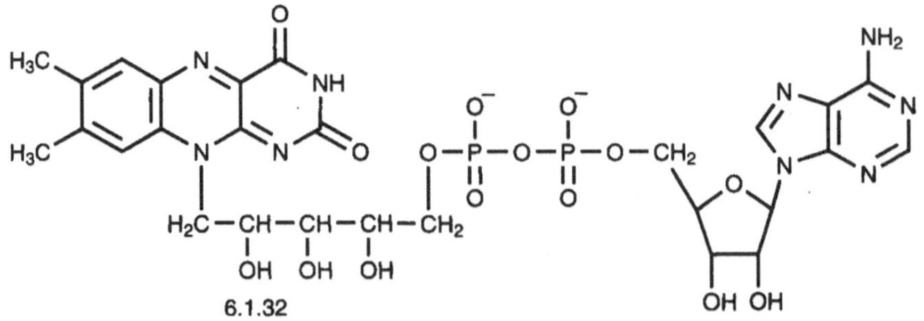

6.1.32

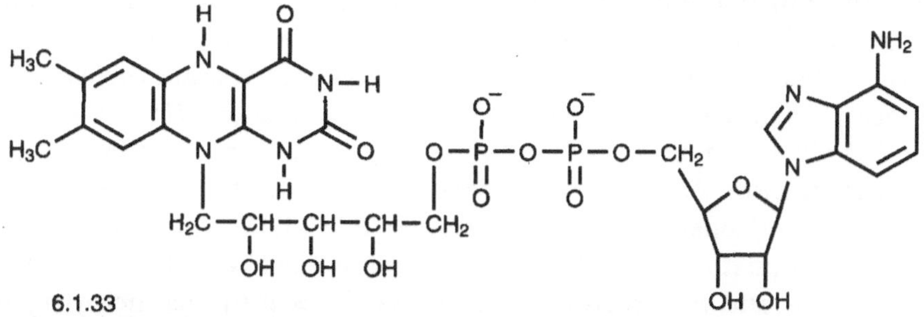

6.1.33

An important biosynthetic oxidation reaction is hydroxylation of aromatic substrates. This is a key step, for example, in the conversion of the amino acid L-phenylalanine (6.1.34) into tyrosine (6.1.35) and then onto dihydroxyphenylalanine (dopa; 6.1.36), which are precursors for the biosynthesis of the benzylisoquinoline alkaloids. The probable mechanism of this hydroxylation reaction has been studied using tritium labelled phenylalanine (6.1.37). The first step is the formation of the arene oxide (6.1.38) Opening of the oxide ring is accompanied by migration of the tritium to yield 6.1.39, which is the tritium labelled tyrosine. The entry of the second hydroxyl group also proceeds, presumably, through the arene oxide 6.1.40, which loses the tritium along with the oxide ring opening resulting in the formation of dopa (6.1.36).

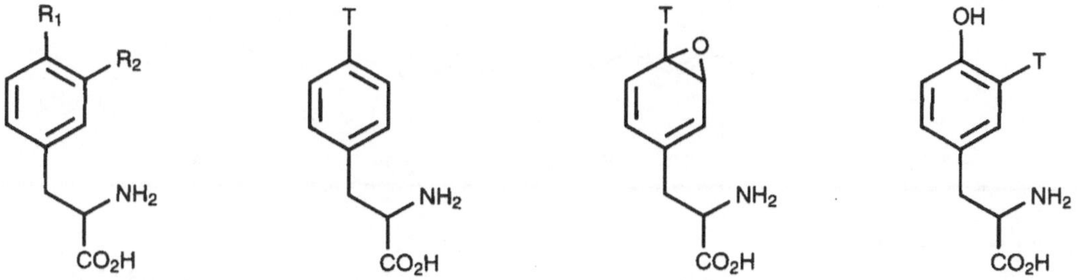

6.1.34: $R_1 = R_2 = H$
6.1.35: $R_1 = OH; R_2 = H$
6.1.36: $R_1 = R_2 = OH$

6.1.37

6.1.38

6.1.39

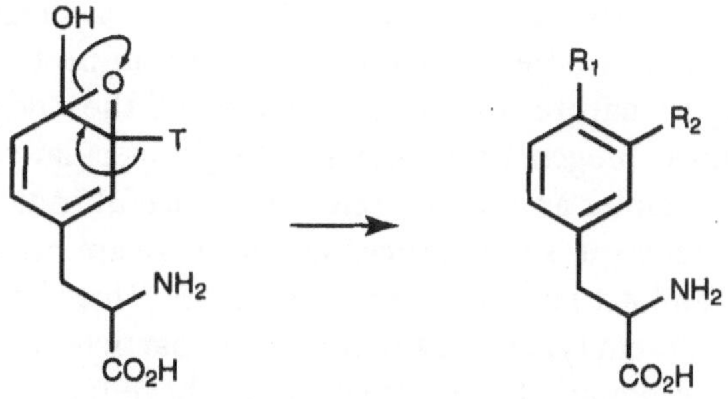

6.1.40

6.1.36: $R_1 = R_2 = OH$

It is, indeed, amazing that only a few primary metabolites are involved in the biosynthesis of the large number of different kinds of secondary metabolites. The major building blocks used by nature for the elaboration of the secondary metabolites are the amino acids (used in aklaloid biosynthesis), acetyl coenzyme A (polyketides), mevalonic acid (6.1.41) (some terpenoids), and shikimic acid (6.1.42) (phenolic compounds). Some examples are described in the sections following this introduction. As discussed in detail in a later section of this chapter, mevalonic acid gets converted into isopentenyl pyrophosphate (IPP; 6.1.43) which, along with its isomer, dimethylallyl pyrophosphate (DMAPP) (6.1.44), serves as the five carbon unit template for the biosynthesis of sesqui- and tri-terpenoids. IPP is also formed by the methylerythritol (MEP) pathway (also known as 1-deoxy-D-xylulose (DOX) pathway) (6.1.45). It has been shown that the latter pathway occurs in the chloroplast and is responsible for the formation of the mono- and di-terpenes.

Suggested reading

Herbert RB, *The Biosynthesis of Secondary Metabolites*, 2nd edn, London: Chapman and Hall, 1989.

6.2 BIOSYNTHESIS OF SOME BENZYLISOQUINOLINE ALKALOIDS

As briefly mentioned in the introduction to this chapter, biogenetic ideas preceded and gave direction to biosynthetic studies which are still continuing to unravel the actual biosynthetic pathways operating in nature for the production of the wide variety and large number of secondary metabolites. Biogenetic theories, though speculative in nature, are based on firm organic chemical principles and are therefore very useful indicators to the synthetic strategies adopted by nature. Hence, it is not surprising that there are marked overall similarities between biogenetic schemes and actual biosynthetic processes. This is particularly so with regard to the biosynthesis of the different types of alkaloids, and in particular, the benzylisoquinoline alkaloids. At the same time, the remarkable advances which have taken place in the past few years with regard to biochemical aspects such as the isolation and characterisation of the enzymes involved, their cloning and over expression and related studies in addition to experiments using labelled precursor compounds have enabled investigators to delineate biosynthetic pathways in greater detail than what was possible ten or fifteen years ago when the first edition of this book was under preparation. Therefore, it has become necessary to revise this and the succeeding sections of this chapter.

In accordance with the biogenetic theory, L-phenylalanine (6.2.1) is, indeed, the primary building block for this group of alkaloids. Phenylalanine itself is derived from shikimic acid (6.2.2) via chorismic acid (6.2.3) and prephenic acid (6.2.4). Regiospecific hydroxylation of phenylalanine results in L-tyrosine (6.2.5), two molecules of which are used for producing (S) norcoclaurine (6.2.6), which is the first member of the benzylisoquinoline alkaloid family. For this purpose, one tyrosine molecule is converted into 3,4-dihydroxylphenylethylamine (dopamine; 6.2.7) and the other into 4-hydroxyphenylacetaldehyde (6.2.8); it is not difficult to figure out the chemical reactions (namely, aromatic hydroxylation, decarboxylation and oxidative deamination) involved in these transformations. The enzymes involved in the biotransformation of tyrosine into norcoclaurine have been investigated and it is now firmly established that this compound is the parent of all benzylisoquinoline alkaloids. The enzyme, norcoclaurine synthase which catalyses the condensation of dopamine with 6.2.8 has been purified after overexpressing the gene encoding for the *Thalictrum flavum* norcoclaurine synthase in *Escheirchia coli*. Using this enzyme and analogues of dopamine, it has been found that the hydroxyl *meta* to the ethylamine side chain in dopamine is essential for the reaction. This suggests that after the formation of the Schiff base, the reaction proceeds through a two-step asymmetric Pictet–Spengler type of condensation reaction. Kinetic studies using deuterated dopamine further indicate that the proton loss resulting in aromatisation of the condensed benzene ring has significant control over the rate of the overall reaction. The probable mechanism of this key step in the biosynthesis of norcoclaurine is given in Scheme 6.1.

Scheme 6.1

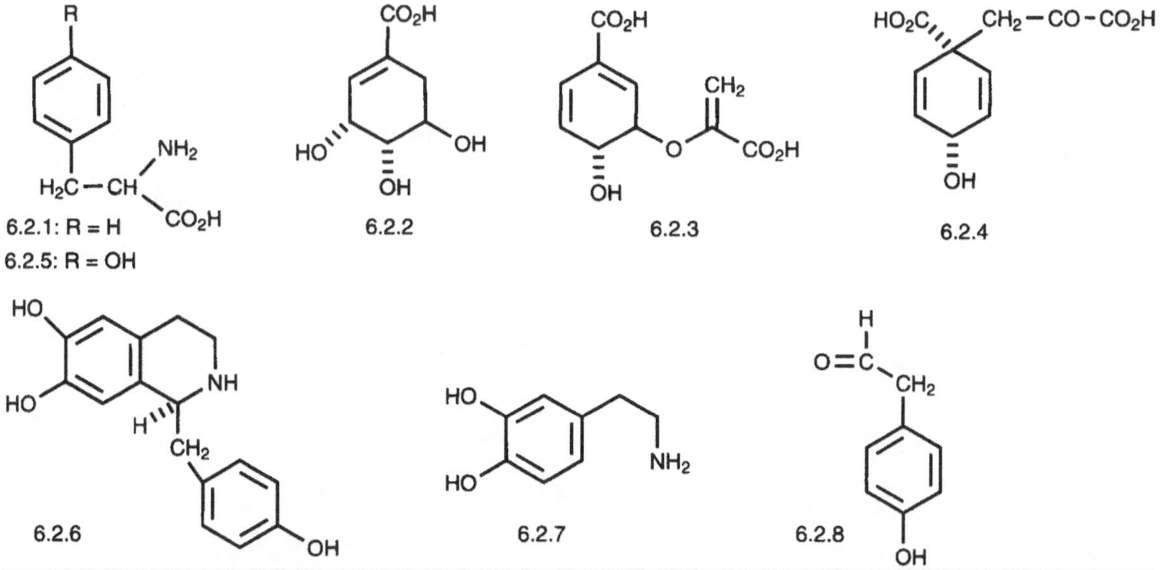

6.2.1: R = H
6.2.5: R = OH

6.2.2

6.2.3

6.2.4

6.2.6

6.2.7

6.2.8

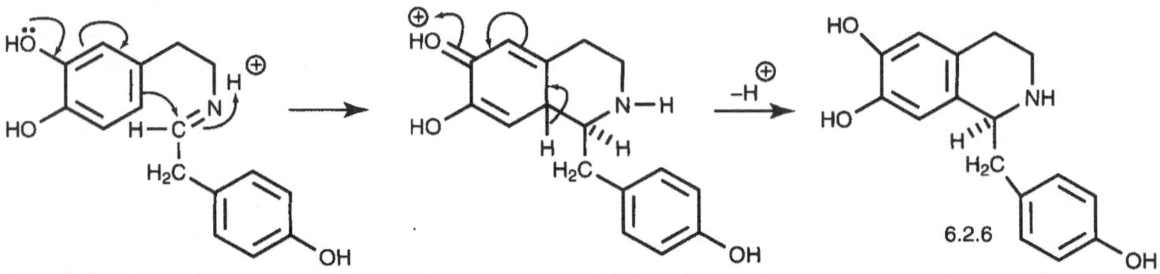

6.2.6

Selective *O*-methylation by an enzyme which uses (S) adenosylmethionine as its cofactor, converts 6.2.6 into coclaurine (6.2.9). *N*-methylation, hydroxylation and further *O*-methylation, in a similar manner, transform coclaurine into (S) reticuline (6.2.10); the intermediate is 6-*O*-methyllaudanosoline (6.2.11). Using cell cultures from plants known to produce benzylisoquin-oline alkaloids, such as, for example, *Annona reticulata*, it has been shown that C14-labelled tyrosine is effectively incorporated into (6.2.9) and (6.2.10). These studies have disproved the earlier belief that (S) norlaudanosoline (6.2.12) was the precursor of reticuline. Further, the sequence, norcoclaurine–coclaurine–*N*-methylcoclaurine–6-*O*-methyllaudanosoline–reticuline is supported by the the distribution of radioactivity in these alkaloids determined by cell culture experiments involving *Berberis stolonifera* fed with C14-labelled tyrosine. The entire process is regio- and stereo-specifically controlled. (S) Norcoclaurine is also the precursor of papaverine (6.2.13), which transformation involves hydroxylation, *O*-methylation and dehydrogenation; the intermediate is norlaudanosine (6.2.14). (S) Norcoclaurine is also the precursor for argemonine (6.2.15).

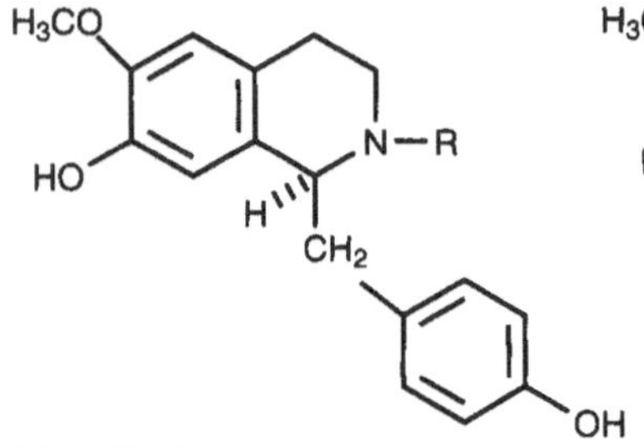

6.2.9: R = H
N-methylcoclaurine: R = CH$_3$

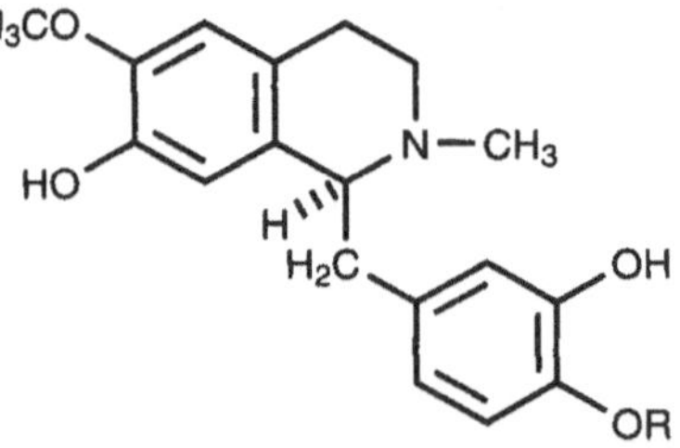

6.2.10: R = CH$_3$
6.2.11: R = H

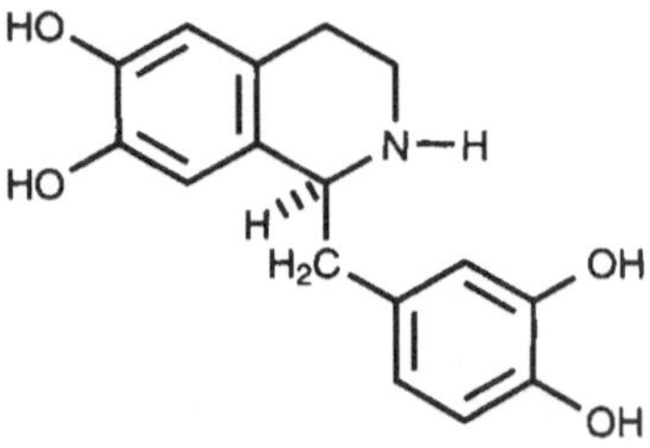

6.2.12

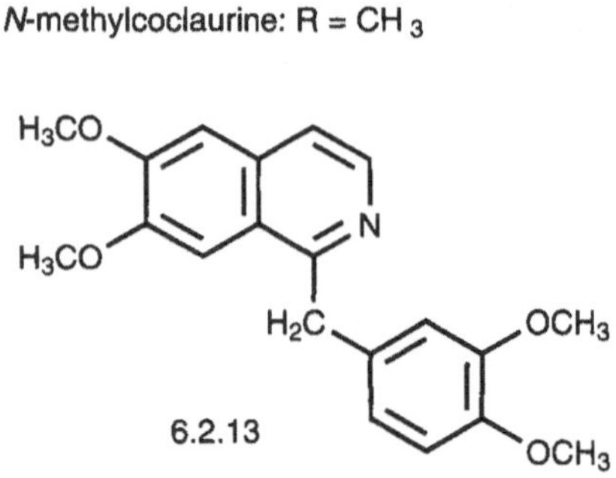

6.2.13

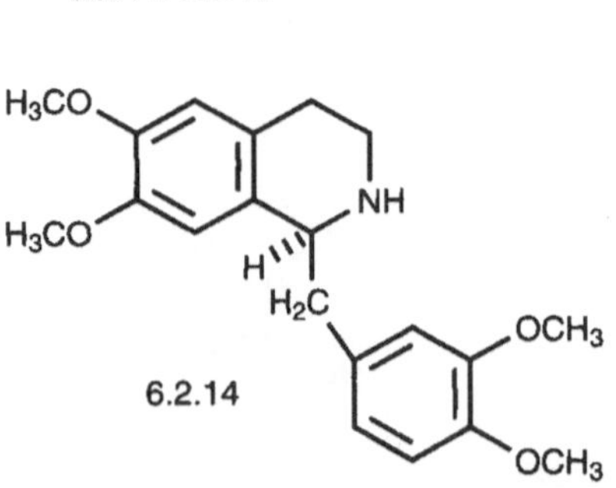

6.2.14

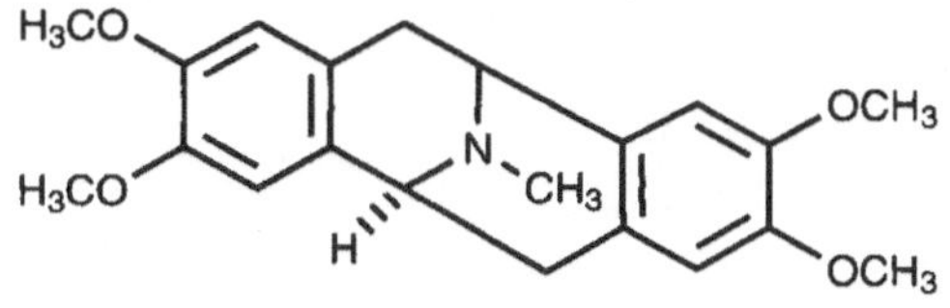

6.2.15

The transformation of (S) norcoclaurine into more complex benzylisoquinoline alkaloids requires enzymes with high regio- and stereo-specifities, particularly after the reticuline stage. The enzymes which bring about, for example, the conversion of (S) reticuline into the protoberberine alkaloid, columbamine (6.2.16) have been isolated and characterised. The intermediates in this sequence are (S) scoularine (6.2.17) and (S) tetrahydrocolumbamine (6.2.18). The additional ring in these compounds in contrast to 6.2.10, is presumably formed by an oxidative coupling type of reaction involving the N-methyl group and the side phenyl ring. The enzyme which brings about this transformation, designated as the berberine bridge enzyme, has been characterised as a P 450 enzyme. A O-methyltransferase catalyses the conversion of (S) scoularine into (6.2.18) which finally undergoes oxidative dehydrogenation to yield columbamine. The enzyme which brings about this change, designated as the tetrahydrobeberine oxidase, has been isolated from Berberis species. (S) Tetrahydrocolumbamine can also get converted into (S) canadine (6.2.19) by the action of the enzyme canadine synthase which is a specific methylenedioxy ring group forming enzyme; this reaction is also an oxidative coupling reaction. Barton had earlier suggested this very mode of formation of the methylene dioxy group in biosynthetic pathways. (S) Canadine is the immediate precursor of berberine (6.2.20). The probable mechanism of the berberine bridge formation is also shown below.

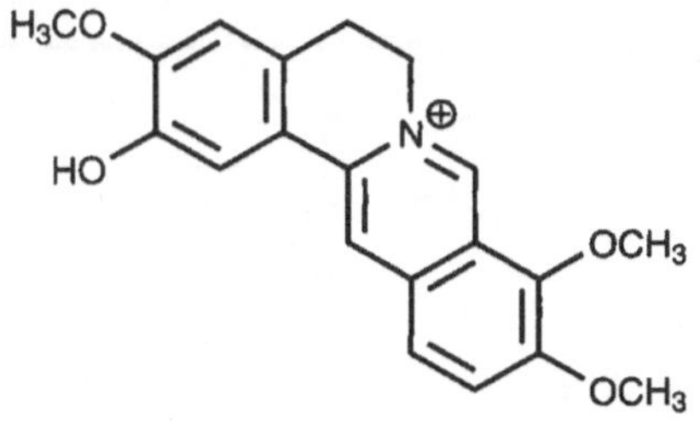

6.2.16

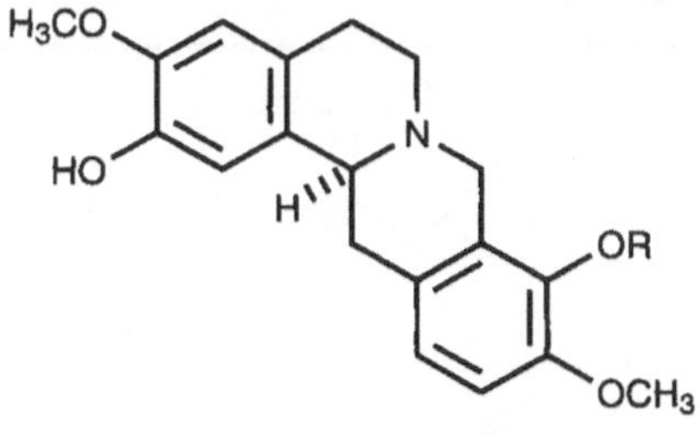

6.2.17: R = H
6.2.18: R = CH$_3$

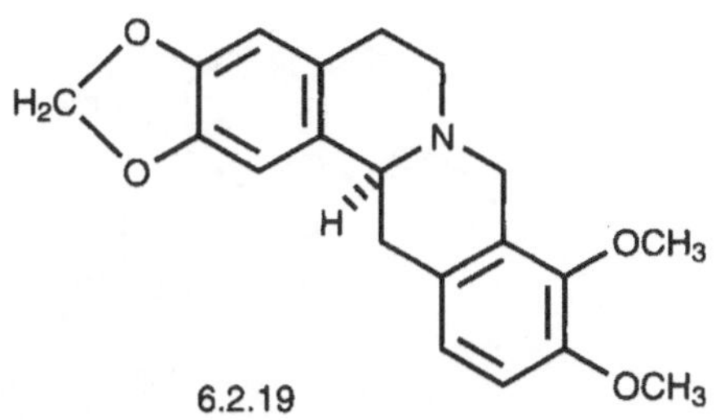

6.2.19

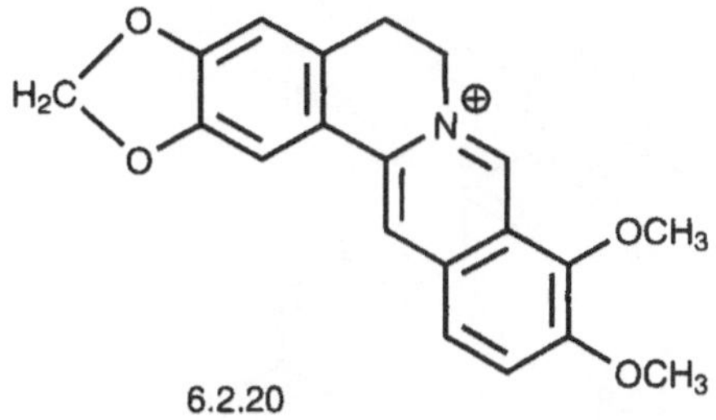

6.2.20

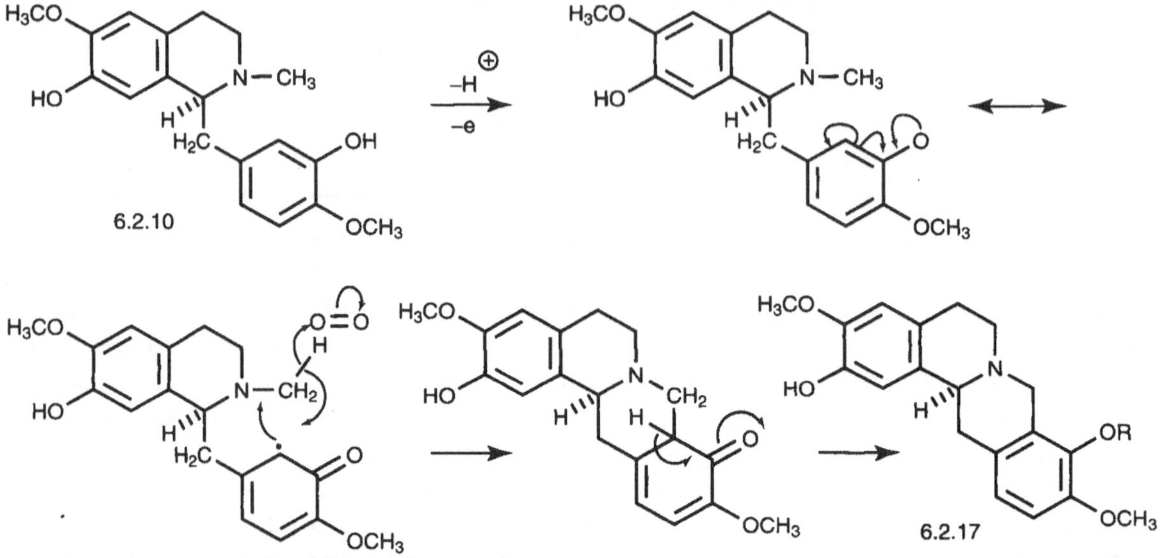

6.2.10

6.2.17

The biosynthesis of the benzophenanthridine alkaloid, sanguinarine (6.2.21) from (S) scoularine goes through some interesting reactions. The first step is the formation of a second methylenedioxy group by a P 450 dependent enzyme and the result is (S) stylopine (6.2.22). The next step involves N-methylation leading to the quaternary ammonium compound, (S) N-methylstylopine (6.2.23). By the action of a hydroxylase, this compound gets converted into protopine for which a ten-membered ring structure (6.2.24) as well as the quaternary ammonium structure (6.2.25) are possible. The latter is supported by the observation that the infrared spectrum of protopine does not show carbonyl group absorption.

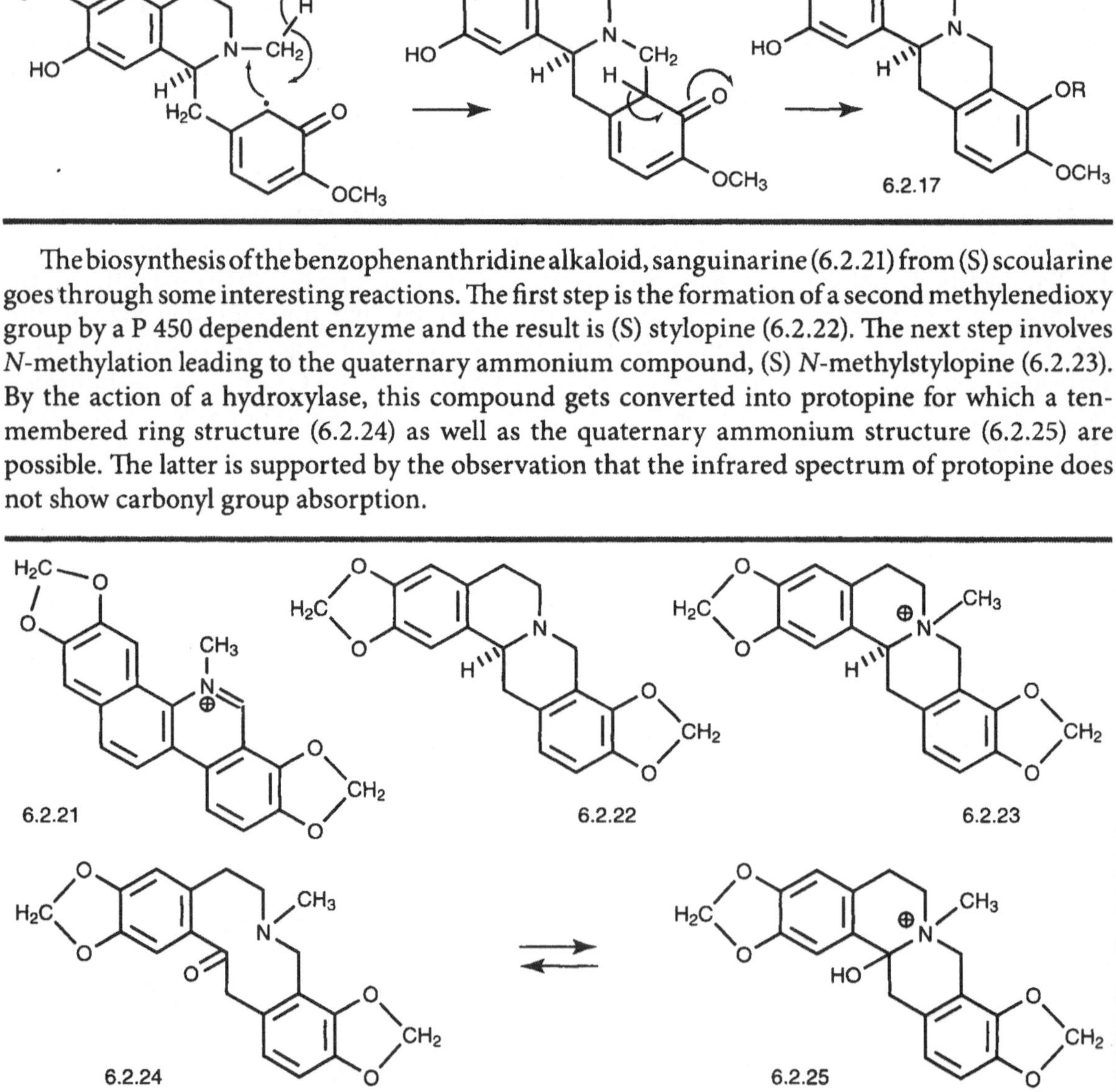

6.2.21   6.2.22   6.2.23

6.2.24   6.2.25

It may be noted here that the structure 6.2.24 can get converted into 6.2.25 by a facile transannular interaction between the nitrogen and the carbonyl group in 6.2.24. The enzyme respnsible for this oxidative conversion of 6.2.23 into 6.2.24/6.2.25 has been identified as a microsomal cytochrome P 450-NADPH dependent hydroxylase. From 6.2.25, it is conceivable to figure out the formation of dihydrosanguinarine (6.2.26) as shown in Scheme 6.2 below.The steps involved are dehydration, oxidative cleavage of the quinoline ring and recyclisation. A final oxidative removal of a hydrogen atom converts 6.2.26 into sanguinarine. The suggested pathway can also account for the formation of the alkaloid, chelidonine (6.2.27) as indicated in Scheme 6.2.

Scheme 6.2

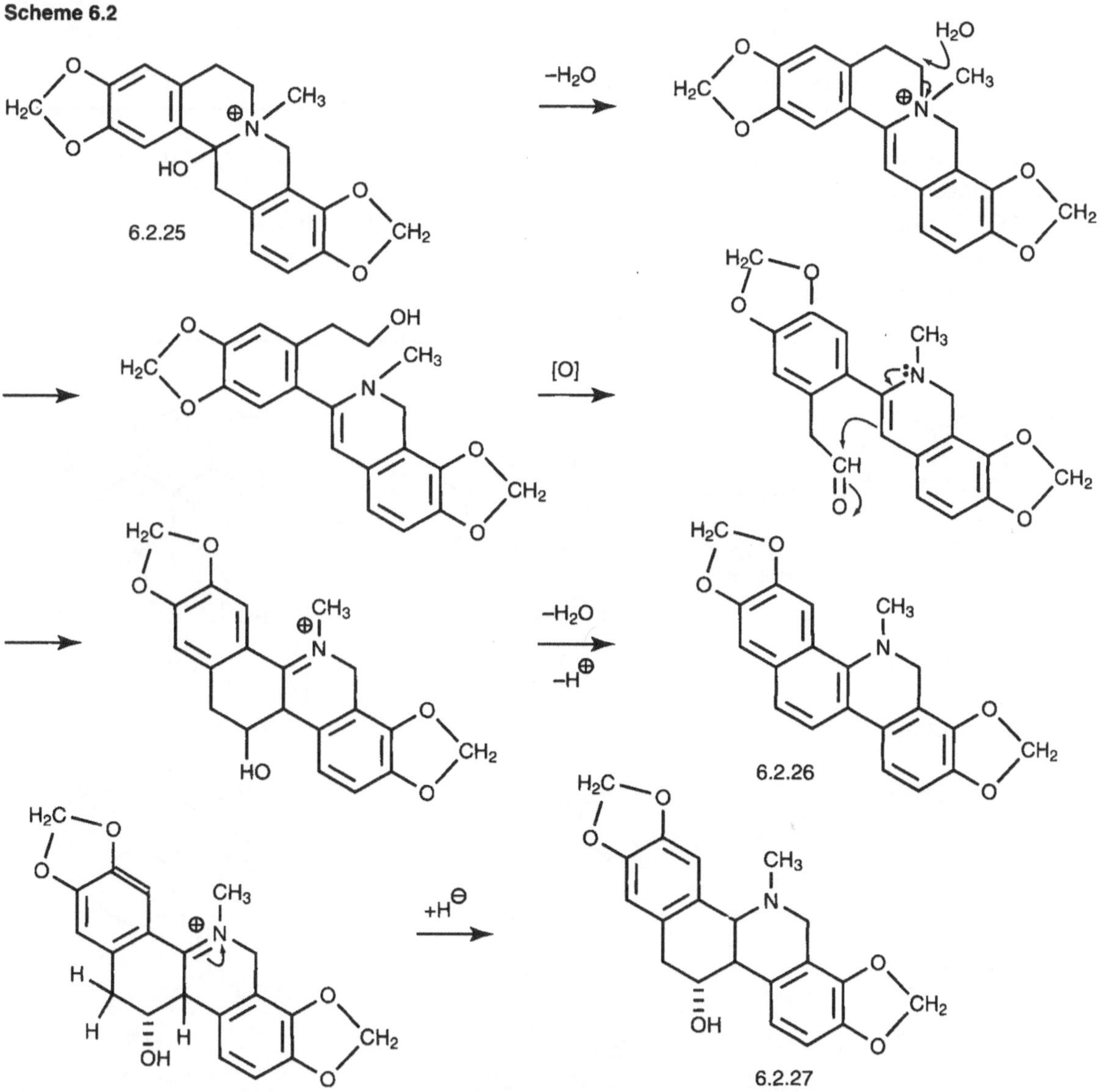

(S) Norcoclaurine is also the primary alkaloidal precursor for the formation of the aporphine alkaloids, boldine (6.2.28), glaucine (6.2.29) and dicentrine (6.2.30).

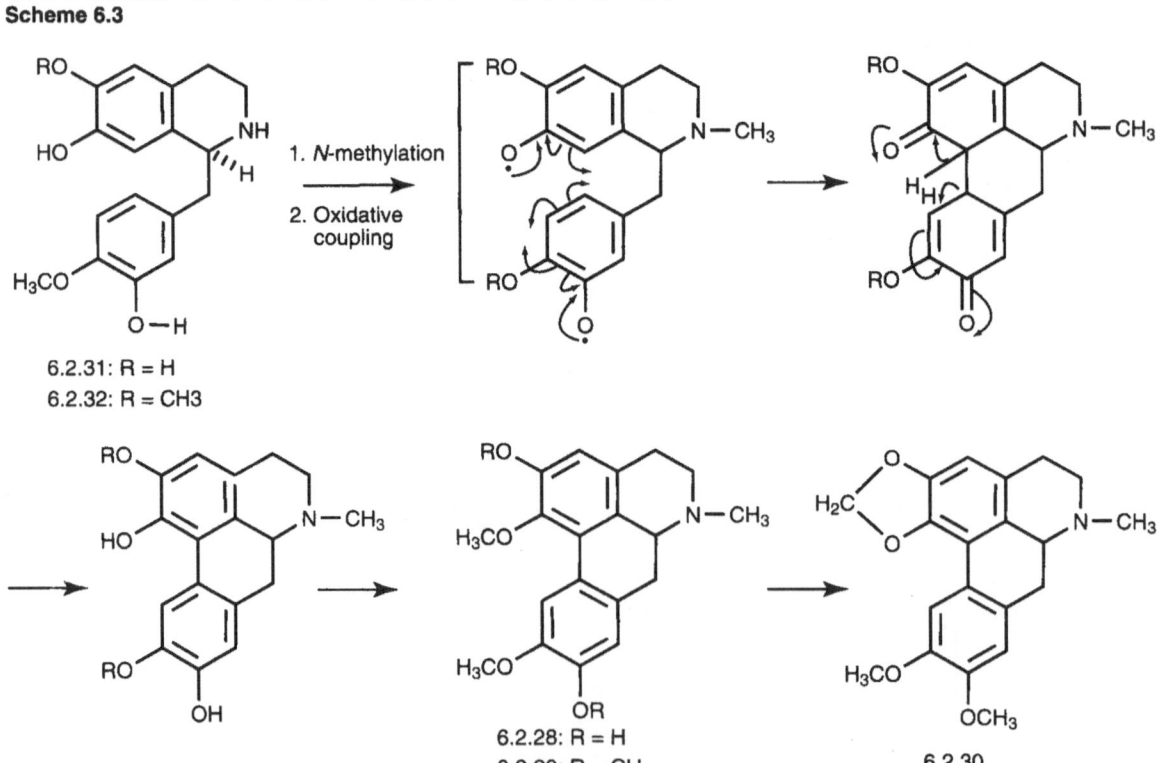

6.2.28: R = H
6.2.29: R = CH₃

6.2.30

Available data indicate that the intermediate is norlaudanosoline (6.2.12) and not reticuline. Thus, radioactive labelled norlaudonosoline as well as its 4'-O-methyl and 7,4'-O-dimethyl ethers (6.2.31 and 6.2.32) get incorporated into the aporphines produced by *Dicentra exima*. In the conversion of the benzylisoquinoline structure into the aporphine structure also oxidative coupling plays a pivotal role as shown in Scheme 6.3.

Scheme 6.3

1. *N*-methylation

2. Oxidative coupling

6.2.31: R = H
6.2.32: R = CH3

6.2.28: R = H
6.2.29: R = CH₃

6.2.30

Suggested reading

1. Luk Louis YP, Bunn Shannon, Liscombe David K, Facchini Peter J and Tanner Martin E, *Biochem.*, 2007, 46: 10153–61.
2. Roberts MF, In Roberts MF and Wink M, (eds), *Alkaloids: Biochemistry, Ecology and Medical Applications*, New York: Plenum Press, 1998.
3. Reuffer M, In Atta-ur-Rahman and Le Quesne PW, (eds), *Biosynthetic Studies of Protoberberine and Related Alkaloidsusing Plant Cell Cultures in Natural Products Chemistry* III, Berlin: Springer-Verlag, 1988.
4. Luckner M, *Secondary Metabolism in Micro-Organisms, Plants and Animals*, 2nd edn, Berlin: Springer-Verlag, 1984.
5. Torsell KBG, *Natural Product Chemistry-A Mechanistic and Biosynthetic Approach to Secondary Metabolites*, New York: John Wiley & Sons, 1983.
6. Cordell GA, *An Introduction to Alkaloids - A Biogenetic Approach*, New York: John Wiley & Sons, 1981.

6.3 RETICULINE TO MORPHINE

The unravelling of the biosynthesis of morphine and related alkaloids is one of the outstanding achievements of modern biosynthetic research. Feeding experiments using labelled compounds have shown that while both the enantiomers of reticuline are incorporated into morphine (6.3.1) in *Papaver somniferum*. the (R) isomer (6.3.2) is more efficently incorporated than the (S) isomer, thus strongly indicating that it is the precursor of the morphine-type of alkaloids. This observation apparently is in contradiction with other evidences (such as those mentioned in the previous section) which suggest that (S) reticuline (6.3.3) is the direct offspring of (S) norcoclaurine, the primary alkaloidal precursor for all types of benzylisoquinoline alkaloids. It is now known that the first step in the biosynthesis of morphine from (S) reticuline, is dehydrogenation of the latter to form the 1,2-dehydroreticulinium ion (6.3.4). This change is catalysed by a flavoprotein which occurs in the leaves as well as roots of *P. somniferum*. Indeed, 6.3.4 is efficiently incorporated into the opium alkaloids thus unequivocally showing that it is an intermediate in the biosynthetic route from (S) coclaurine to morphine. 1,2-Dehydroreticulinium ion is, in turn, readily converted into (R) reticuline (6.3.2). The enzyme responsible for this conversion is a micosomal

Reticuline to Morphine

6.3.1: R = H
6.3.11: R = CH₃

6.3.2

6.3.3

bound P 450 enzyme and needs NADPH as a cofactor. The presence of this enzyme in crude cell preparations from young seedlings of *P. somniferum* has been demonstrated. It has high substrate specificity and there is no evidence for the reverse reaction. Thus, by converting (S) reticuline into (R) reticuline, some species of the genus Papaver are able to channelise the biosynthetic pathway so as to produce morphinandienone alkaloids, such as salutaridine (6.3.5) with defined configuration (R) at the chiral centre. Salutaridine is also a chemical constituent of opium. (R) Salutaridine is then converted stereospeifically into salutaridinol (with 7S configuration; 6.3.6) by a NADPH -7-oxidoreductase. Acetylation of 6.3.6 through the agency of acetyl coenzyme A gives the acetate 6.3.7 which undergoes spontaneous solvolysis at pH 8–9 accompanied by furan ring closure, in a SN2' type reaction to yield thebaine (6.3.8). It is necessary to point out here that 6.3.7 has the correct stereochemistry for an allylic syn displacement of the acetoxy group by the strategically placed phenolic hydroxyl group. A substrate specific acetyl transferase enzyme which converts 6.3.6 into 6.3.7 has been isolated and purified to homogeneity. Thebaine gets converted into morphine through neopinone (6.3.9), codeinone (6.3.10) and codeine (6.3.11). All the enzymes involved in this sequence are yet to be completely characterised. However, the enzyme responsible for the reduction of codeinone, codeine reductase, has been purified and characterised. It requires NADPH as the cofactor. In executing this complex synthesis all the way from phenylalanine to morphine, nature has displayed ingenuity as well as adherence to firm mechanistic principles!

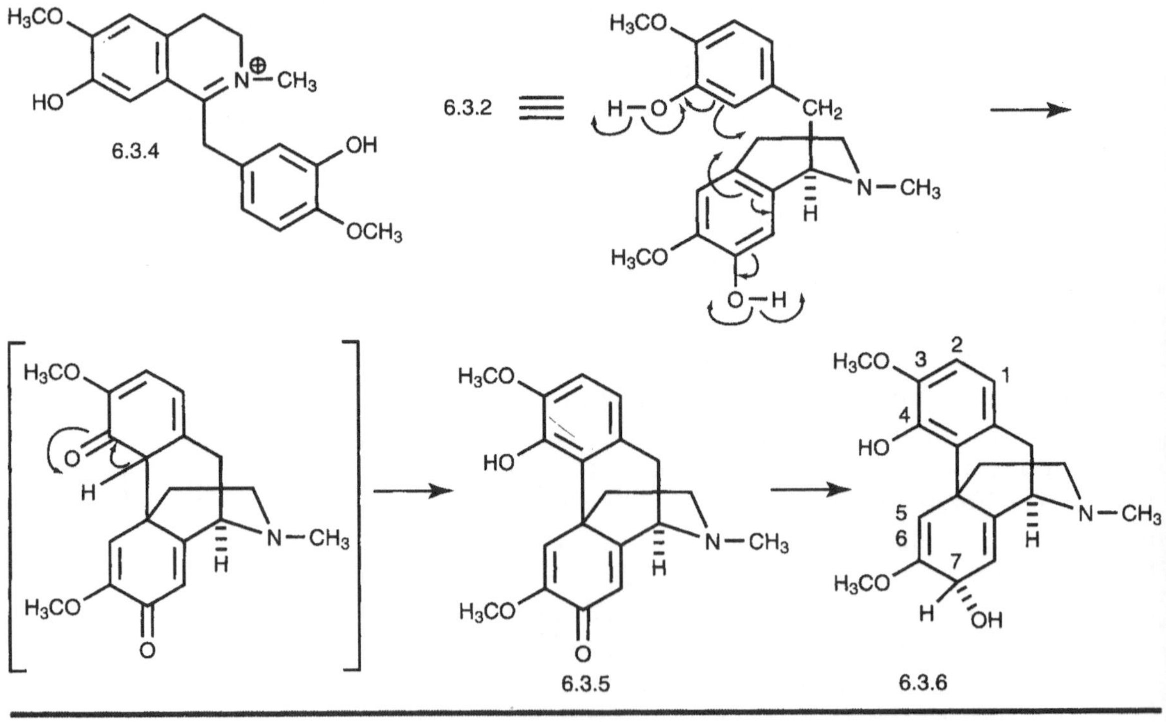

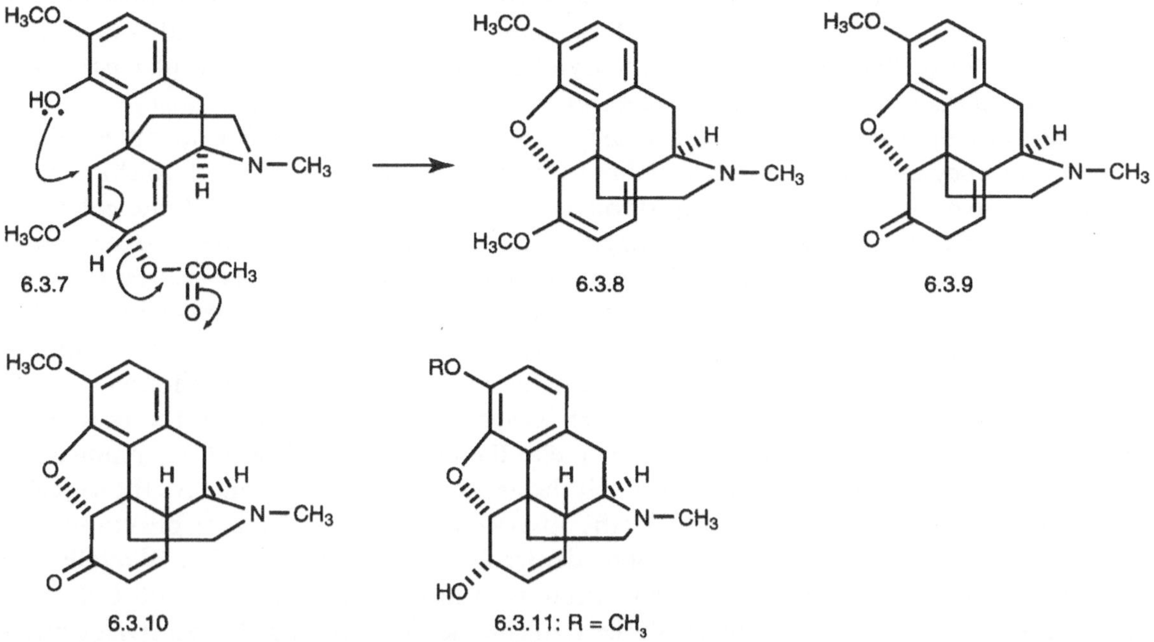

H₃CO

HO

H₃CO

H

O—COCH₃

N—CH₃

H

6.3.7

H₃CO

O

H₃CO

H

N—CH₃

6.3.8

H₃CO

O

H

N—CH₃

O

6.3.9

H₃CO

O

H H

N—CH₃

O

6.3.10

RO

O

H H

N—CH₃

HO

6.3.11: R = CH₃

Suggested reading

In addition to the material cited in the previous section, the folowing sources may also be consulted.

1. McMurry JE and Begley TP, *The Organic Chemistry of Biological Pathways*, Englewood, Colorado: Roberts and Company publishers, 2005.
2. Novak BH, Hudlicky T, Reed JW, Mulzer J and Trauner D, *Curr. Org. Chem.*, 2000, 4: 343–62.

6.4 FROM PHENYLALANINE TO COLCHICINE

The elucidation of the structure of this alkaloid (6.4.1) as a tropolone was discussed in Chapter 2. Its biosynthesis is equally fascinating. Though it is not immediately apparent from its structure, colchicine has biosynthetic kinship with the benzyl isoquinoline alkaloids and like the latter is derived from the amino acid, phenylalanine (6.4.2). Extensive studies by Battersby and co-workers and others have generated a wealth of data on the basis of which the entire biosynthetic route from phenylalanine to colchicine has been delineated. Using C14 and tritium labelled phenylalanine, it has been shown that two units of this amino acid are involved in the biosynthesis of colchicine. One unit gets converted into 3,4-dihydroxyphenylethyl amine (dopamine; 6.4.3) via tyrosine (6.4.4) and the other to 4-hydroxydihydrocinnamaldehyde (6.4.5). A Mannich-type reaction between 6.4.3 and 6.4.5 and selective O-methylation results in the tetrahydroisoquinoline (6.4.6). Further hydroxylation and N- and O-methylation lead to S-autumnaline (6.4.7). That this compound is a key intermediate in the biosyntheis of colchicine has been conclusively established by tracer studies. In this context it is worth mentioning that autumnaline has been isolated from several genera of the family Liliaceae to which *Colchicum autumnale* belongs. The next step in the biosynthesis of 6.4.1 is a intramolecular *para–para* phenolic coupling which transforms S-autumnaline into isoandrocymbine (6.4.8), which is a dienone. O-methylation through the agency of S-adenosylmethionoine converts 6.4.8 into O-methylandrocymbine (6.4.9); androcymbine, which occurs in some species of Colchicum, has the structure 6.4.10. The intriguing transformation of O-methylandrocymbine into colchicine requires the participation of an oxidative enzyme and proceeds through the intermediates, 6.4.11, 6.4.12 and 6.4.13; the last mentioned compound is demecolcine in which the skeletal structure of colchicine is complete. Tracer studies have shown that the N-methyl group of autumnaline and O-methylandrocymbine is fully retained in demecolcine but is lost in colchicine. This fact suggests that demecolcine is a precursor of colchicine and this has been established by carefully designed experiments. The final stages in the biosynthesis of 6.4.1 involve the progressive oxidation of the N-methyl group in demecolcine to the formyl, hydrolysis of the amide group thus generated and a final acetylation through the agency of acetylcoenzyme A. The sequence 6.4.7 to 6.4.13 and 6.4.1 is also supported by the observation that the C-3 carbon of autumnaline is eliminated during the formation of both demecolcine and colchicine but is retained in N-formyldemecolcine (6.4.12), which has been identified as a constituent of *Colchicum corrigerum*. Incidentally, N-formyl-N-deacetylcolchicine (6.4.14) and N-deacetylcolchicine (6.4.15), which are the intermediates between demecolcine and colchicine, are also minor alkaloids of *C. autumnale*. All the steps are shown in Scheme 6.4.

Scheme 6.4 Biosynthesis of Colchicine

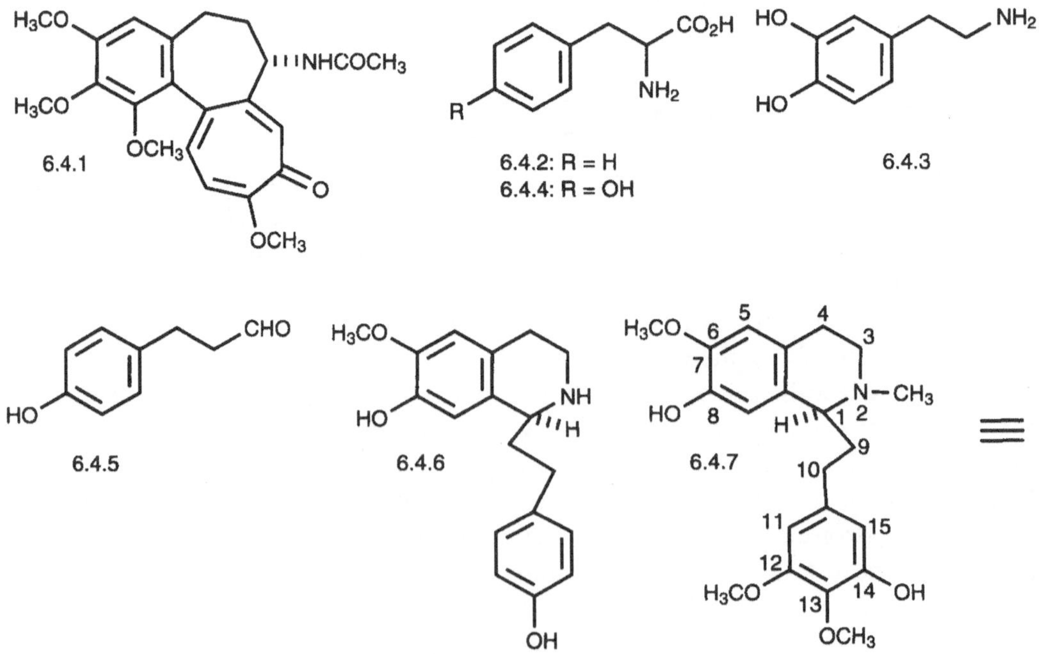

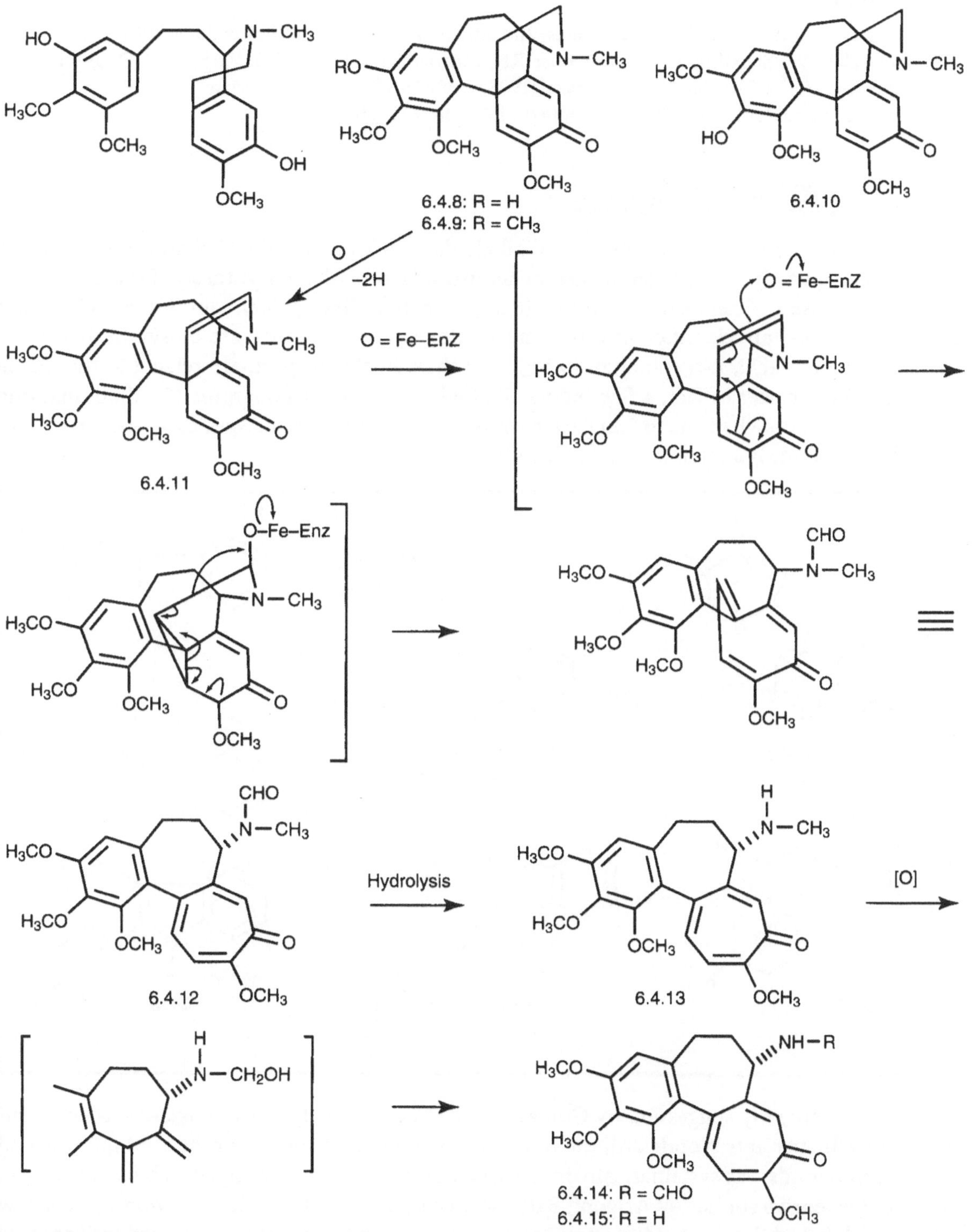
6.4.8: R = H
6.4.9: R = CH3
6.4.10
O
–2H
6.4.11
O = Fe–EnZ
Hydrolysis
6.4.12
6.4.13
[O]
6.4.14: R = CHO
6.4.15: R = H

Suggested reading

1. Dewick PM, *Medicinal Natural Products*, 2nd edn, New York: John Wiley & Sons, 2001.
2. Sheldrake PW., Suckling KE., Woodhouse RN, Murtagh AJ, Herbert RB, Barker AC, Staunton J and Battersby AR, *J. Chem. Soc. Perkins Trans.*, 1998, 1: 3003 and references therein.
3. Maier UH and Zenk MH, *Tetrahedron Lett.*, 1997, 38, 735–60.

6.5 TRYPTOPHAN TO QUININE

A brief introduction to quinine and related alkaloids, also known as the Cinchona alkaloids, has already been given in the chapter on stereochemistry. Herein, the biosynthesis of these alkaloids will be discussed. Chemically, quinine (6.5.1), cinchonidine (6.5.2), quinidine (6.5.3) and cinchonine (6.5.4) are classified as quinoline alkaloids. However, they are biosynthesised from the indole amino acid, L-tryptophan (6.5.5), and thus, in this respect, are related to the indole alkaloids. As a matter of fact, a few indole alkaloids, such as cinchonamine (6.5.6), quinamine (6.5.7) and cinchophyllamine (6.5.8) have been isolated from Cinchona species; co-occurrence of related compounds is a reliable biogenetic indicator.

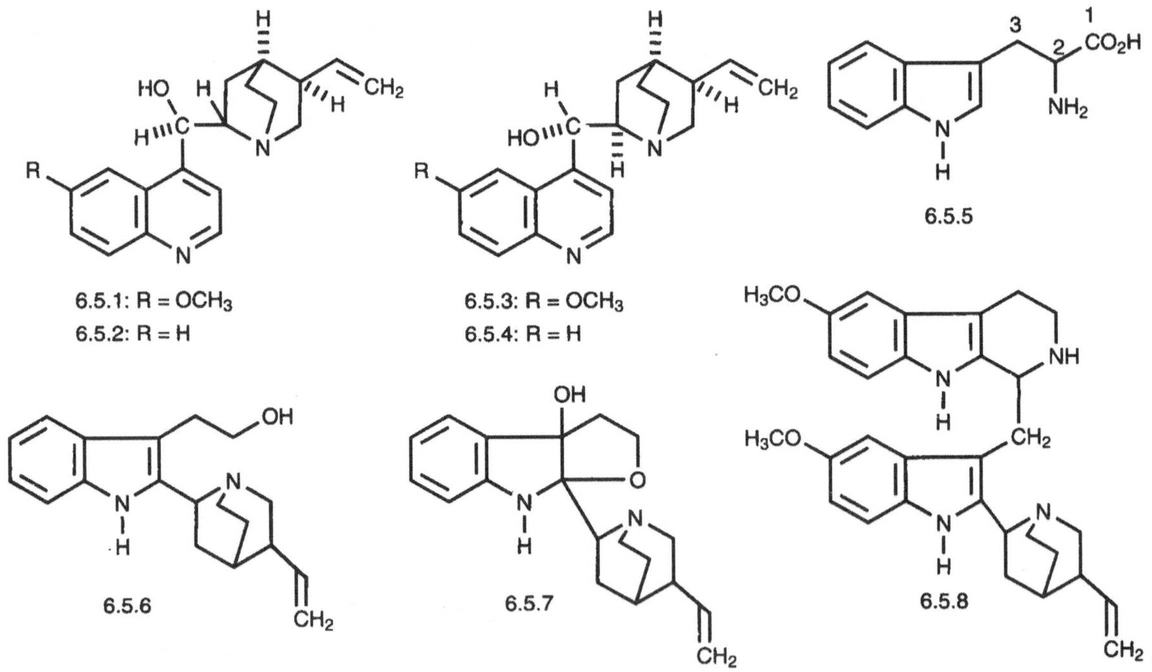

6.5.1: R = OCH$_3$
6.5.2: R = H

6.5.3: R = OCH$_3$
6.5.4: R = H

6.5.5

6.5.6

6.5.7

6.5.8

Following an early suggestion by Goutarel and a later more elaborate biogenetic scheme put forward by Turner and Woodward, the role of L-tryptophan in the biosynthesis of quinine and related compounds was systematically investigated using modern techniques of biosynthesis. As a result of extensive studies by Arigoni, Battersby, Leete and Scott and their co-workers it is now understood that while the quinoline part of quinine and both the nitrogen atoms are derived from tryptophan, part of the molecule also originates from geraniol (6.5.9) via the iridoid glucoside, loganin (6.5.10). Thus, the Cinchona plant utilises two different building blocks for elaborating this antimalarial alkaloid. Incidentally, it has been suggested that quinine is probably a metabolic end product as it is deposited on the outer layer of the bark. Indeed, it accumulates as the tree grows and for commercial extraction of the alkaloid, the plant has to be at least twenty years old. However, even young seedlings produce quinine and therefore, biosynthetic experiments which involve feeding labelled precursors are feasible and have been successfully carried out! For example, racemic tryptophan labelled at position 2 with C14 was fed to *Cinchona succirubra* plants using cotton wicks inserted in the stems. After two months, they were harvested and extracted to obtain radioactive cinchonamine and quinine. The quinine thus obtained was oxidised to quinic acid (6.5.11), which was decarboxylated by heating with copper chromite. Reaction of the resulting 6-methoxyquinoline (6.5.12) with phenyl lithium followed by conversion into methiodide and subsequent oxidation with potassium permanganate yielded benzoic acid which had essentiallly the same specific radioactivity as the quinine. This observation firmly established that the C-2 of the quinoline nucleus of quinine is derived from C-2 of tryptophan. These observations are summarised in Scheme 6.5.

Scheme 6.5

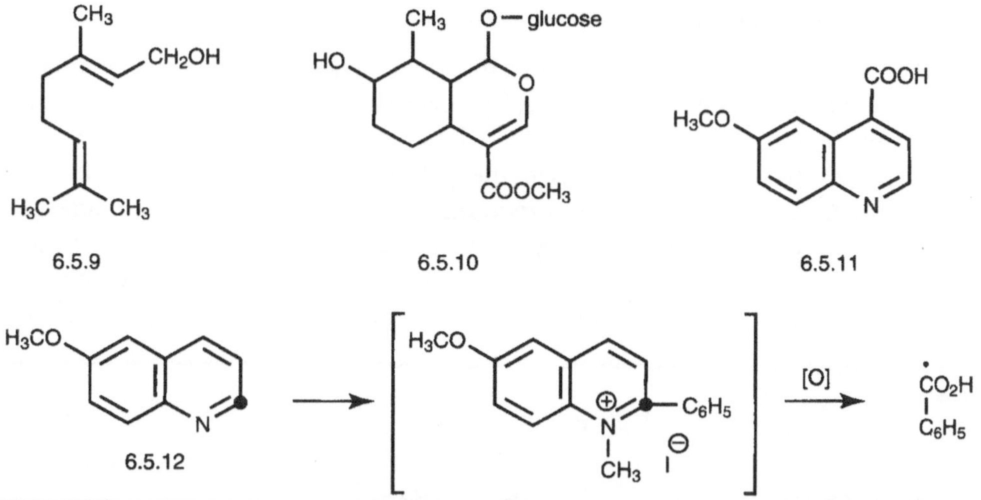

6.5.9   6.5.10   6.5.11

6.5.12

Parallel experiments with geraniol carrying C14 label on C-3, also yielded radioactive quinine with 0.001% incorporation of the label. The quinine thus obtained was hydrogenated and then oxidised by the Kuhn–Roth method (chromic acid), when radioactive propionic acid was obtained. In another set of experiments, doubly labelled tryptophan with N15 and C14 at C-2 of the indole nucleus was prepared as shown in Scheme 6.6 and fed to *C. succirubra* plants. The harvested quinine as well as its degradation products were heated in an evacuated sealed tube with calcium oxide and cupric oxide and the nitrogen gas liberated was analysed using a mass spectrometer. It was found that only the nitrogen of the quinoline moiety got enriched with N15. Further, all the C14 was located at the carbinol carbon.

Scheme 6.6

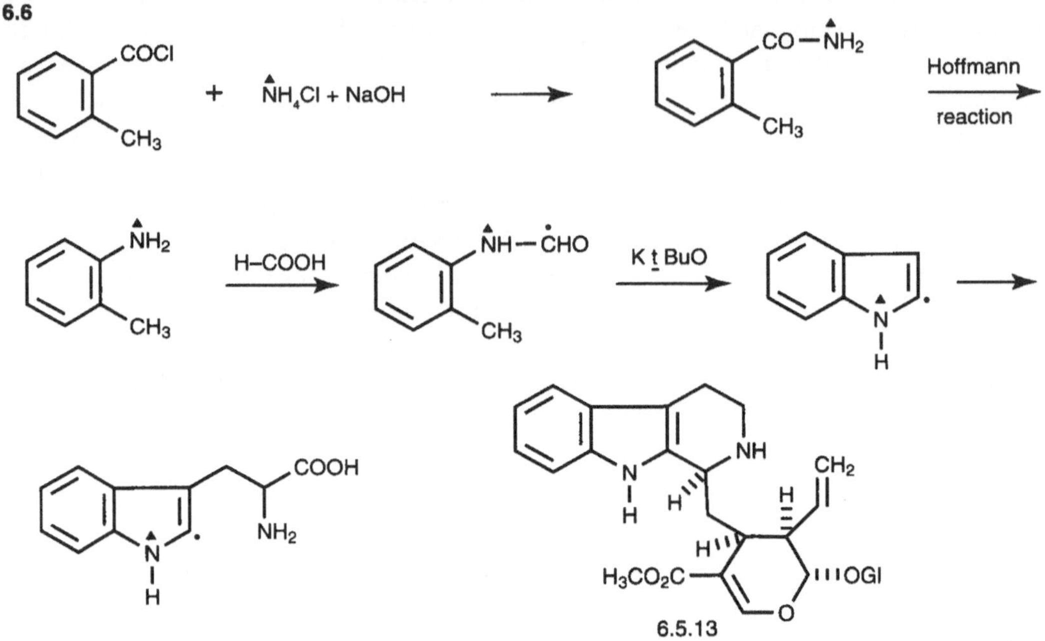

Other experiments established that radioactive loganin and the indole alkaloid strictosidine (6.5.13) were incorporated into quinine. The pathway from geraniol passes through loganin and secologanin (6.5.14). A combination of the latter with tryptophan followed by further changes results in strictosidine. Further down the pathway is another indole alkaloid, corynantheal (6.5.15). The transformation of this compound into quinine involves an oxidative rupture of the indole followed by recyclisation. A key intermediate in these late stages is cinchoninone (6.5.16). The introduction of a phenolic hydroxyl and its methylation are obviously terminal steps. The entire biosynthetic pathway to quinine and the related alkaloids is given in Scheme 6.7.

Scheme 6.7

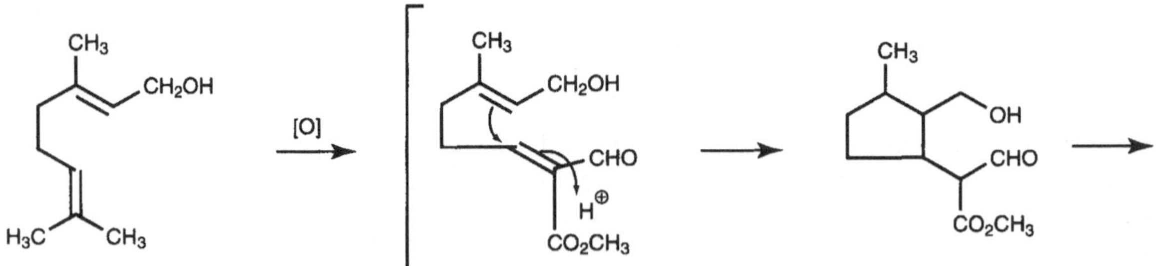

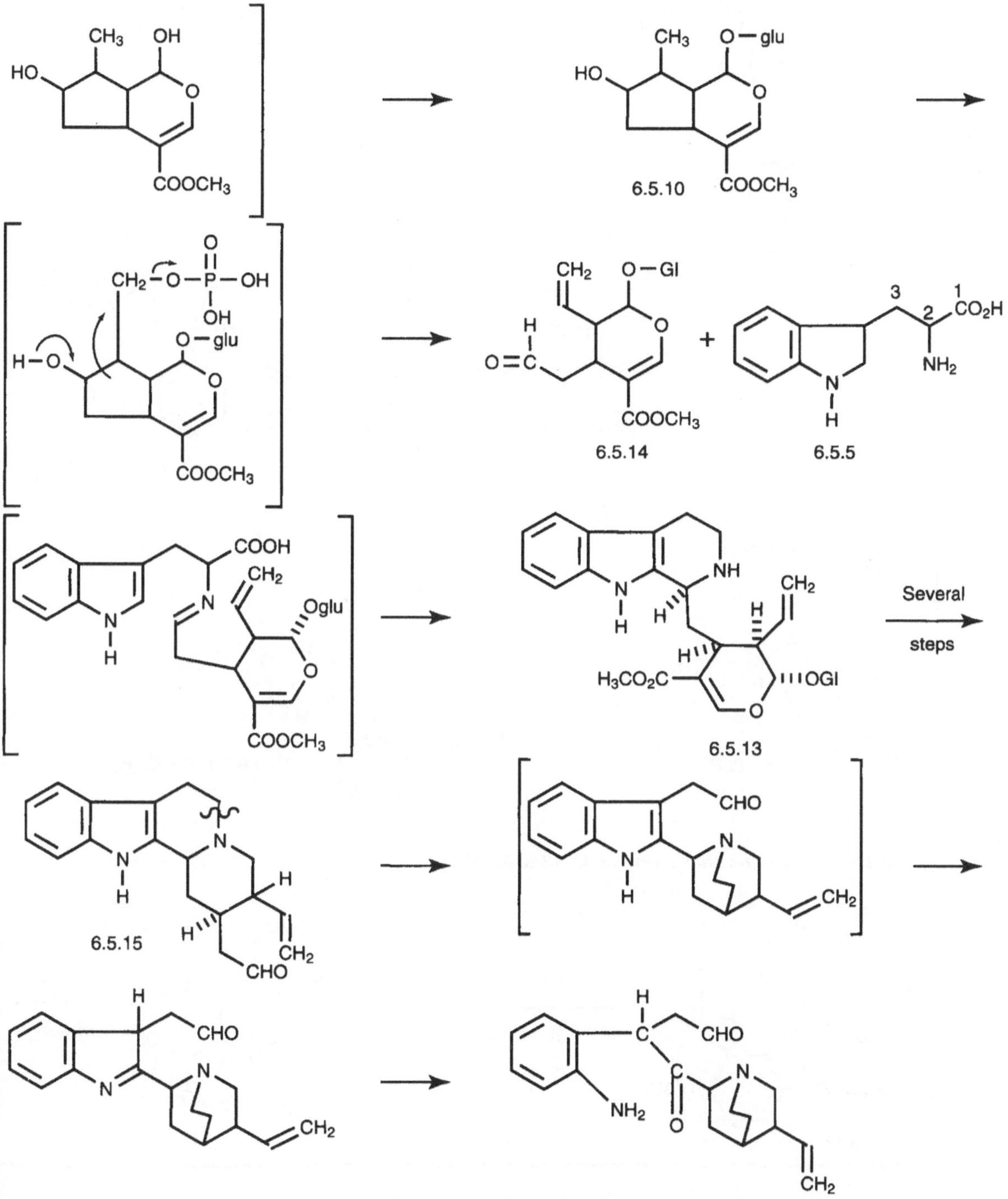

6.5.10

6.5.14   +   6.5.5

Several steps

6.5.13

6.5.15

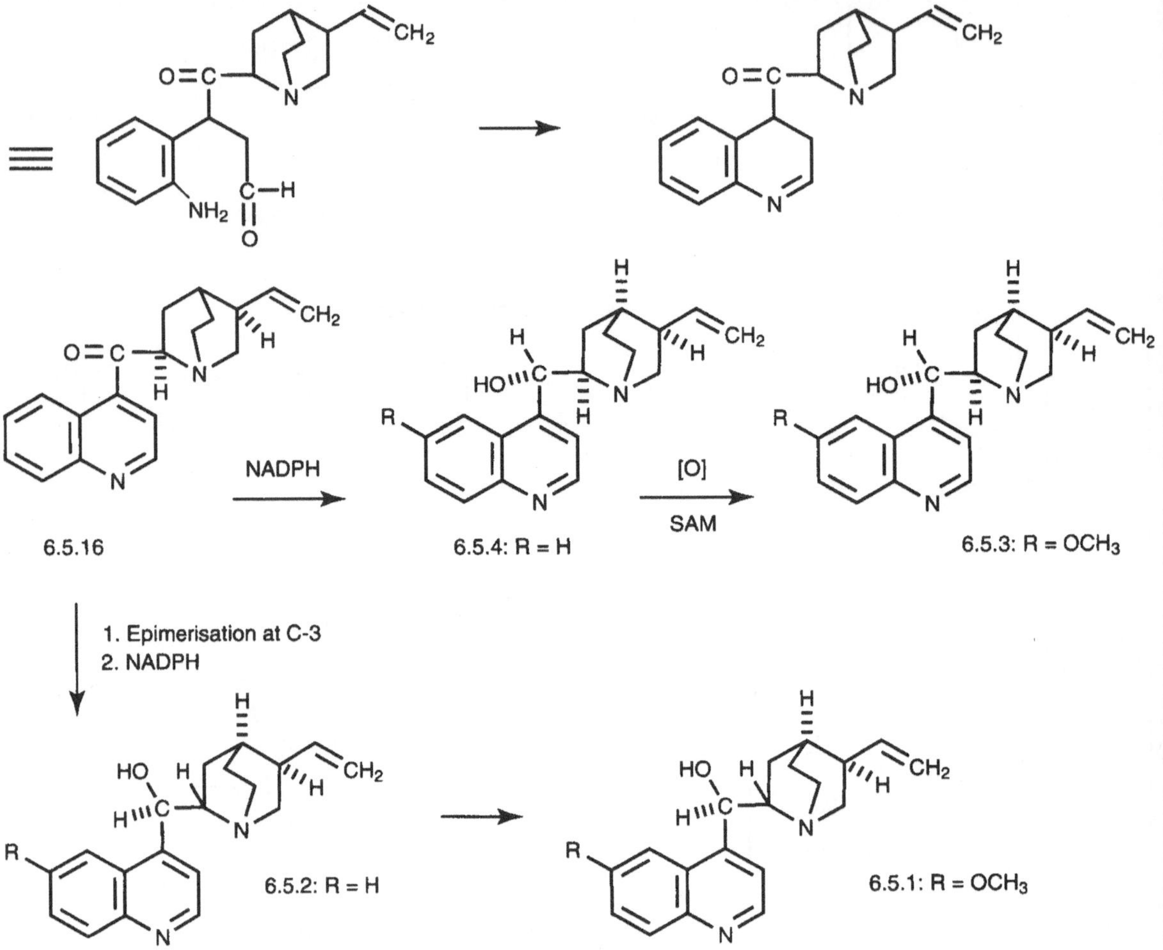

6.5.16   NADPH   6.5.4: R = H   [O]   SAM   6.5.3: R = OCH₃

1. Epimerisation at C-3
2. NADPH

6.5.2: R = H   6.5.1: R = OCH₃

Probable steps involved in the conversion of 6.5.13 to 6.5.15

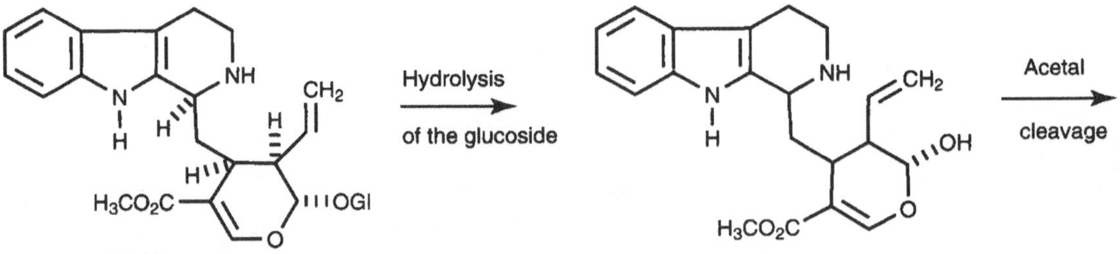

6.5.13   Hydrolysis of the glucoside   Acetal cleavage

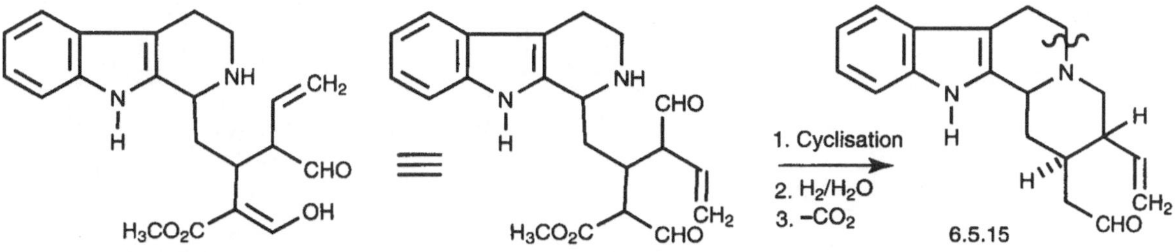

Suggested reading

1. Seigler DS, *Plant Secondary Metabolism*, Berlin: Springer, 1998.
2. Dewick PM, *Medicinal Natural Products*, 2nd edn, New York: John Wiley & Sons, 2001.

6.6 BIOSYNTHESIS OF SOME INDOLE ALKALOIDS

Unlike the alkaloids derived from phenylalanine, those originating from tryptophan are more complex and possess additional structural features which can often be traced to a terpenoid precursor. However, since their basic skeleton is an indole moiety, they are classified as indole alkaloids. Many of these compounds possess medicinal value and occur in plants used in herbal medicines. Some of the important members of this family of alkaloids are reserpine (6.6.1), ajmaline (6.6.2), yohimbine (6.6.3), ajmalicine (6.6.4), vincamine (6.6.5), strychnine (6.6.6), sarpagine (6.6.7) and the dimeric alkaloids of *Catharanthus roseus* such as vinblastine (6.6.8). In this section the salient features of the biosynthesis of a few of these compounds will be briefly discussed.

Biosynthesis of some indole alkaloids

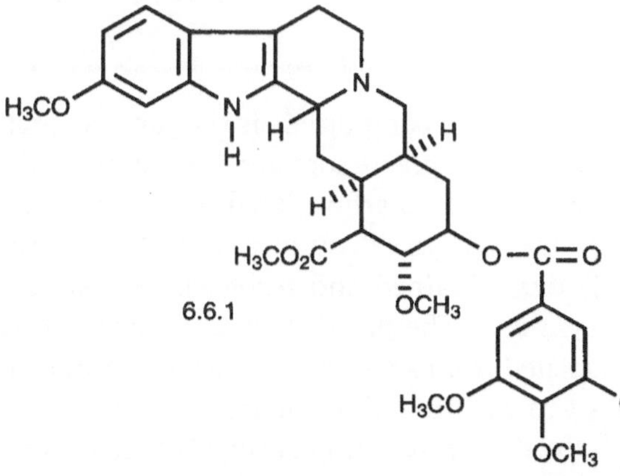

6.6.1

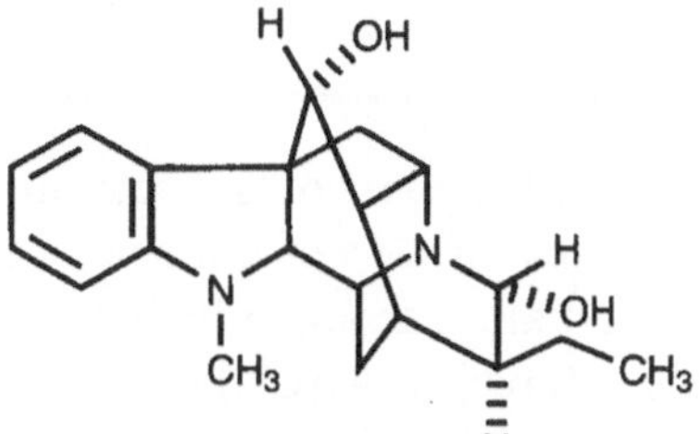

6.6.2

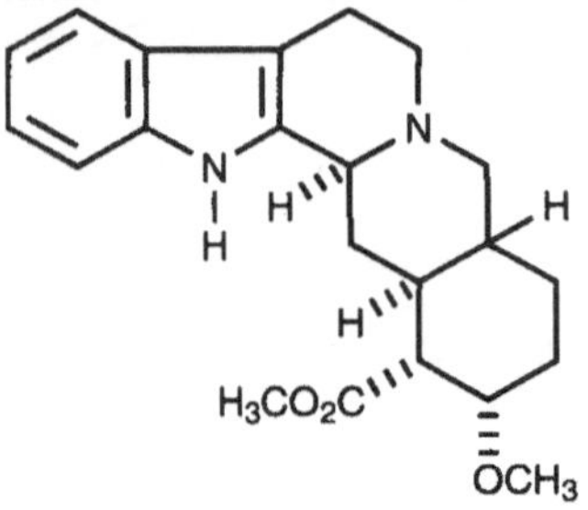

6.6.3

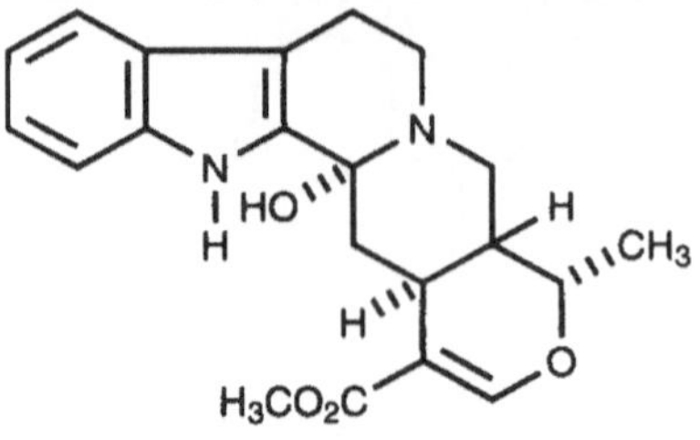

6.6.4

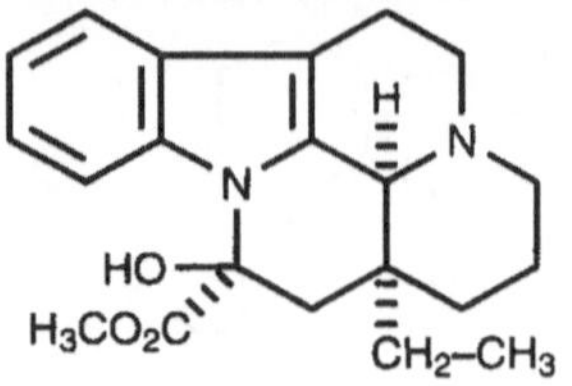

6.6.5

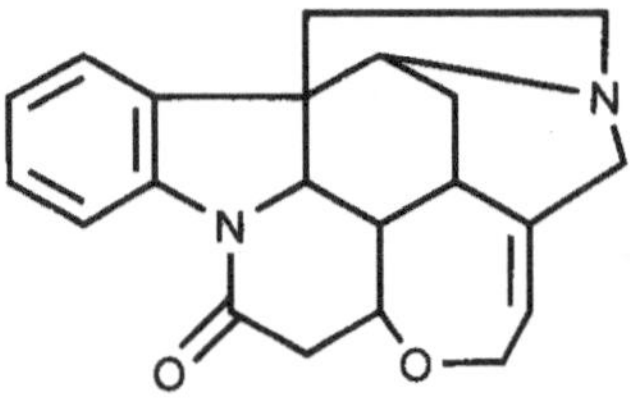

6.6.6

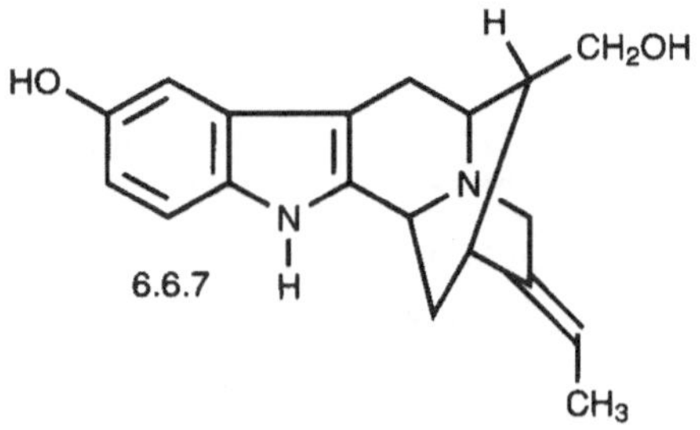

6.6.7

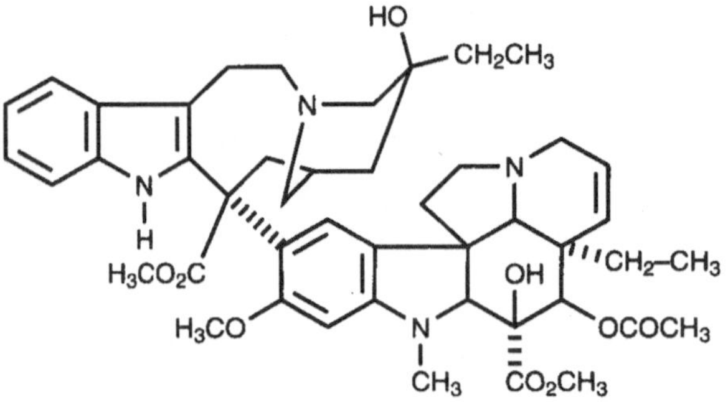

6.6.8

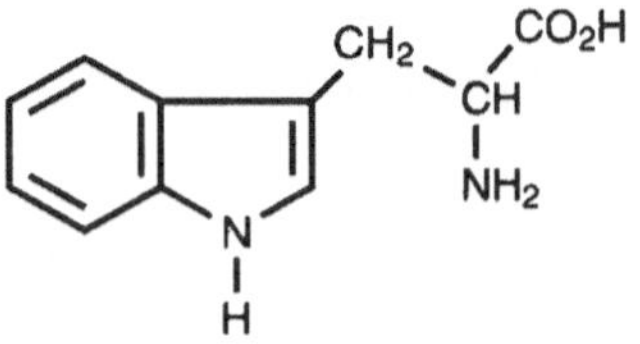

6.6.9

The first part of the bosynthetic pathway to these compounds is the same as that involved in the biosynthesis of quinine. Thus, the two primary precursors are the amino acid, tryptophan (6.6.9) and secologanin (6.6.10). Tryptophan is first decarboxylated by the enzyme tryptophan decarboxylase which has been detected, for example, in cell cultures of *Catharanthus roseus*. The condensation between tryptamine (6.6.11) thus obtained and secologanin is stereospecific and is catalysed by the enzyme strictosidine synthase. The product, strictosidine (6.6.12) occurs in *Rhazya stricta* and the synthase has been isolated from a number of plants belonging to the genera Catharanthus, Rauwolfia and Rhazya. As was mentioned in the last section (6.5), strictosidine can also arise from tryptophan itself, with the decarboxylation takng place at a later stage. Under quinine biosynthesis, the pathway from geraniol to seco loganin was given in pictorial form. The enzyme responsible for this conversion has been characterised as a cytochrome P 450 dependent hydroxylase.

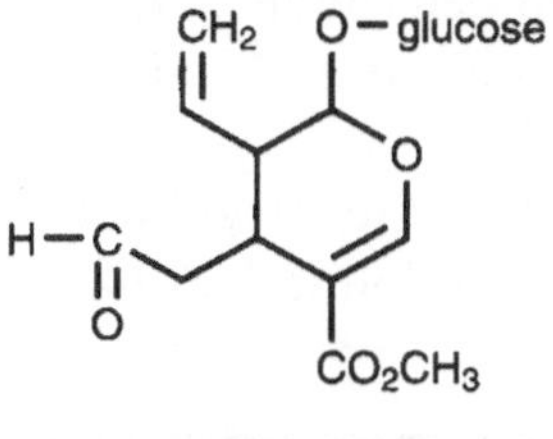

6.6.10 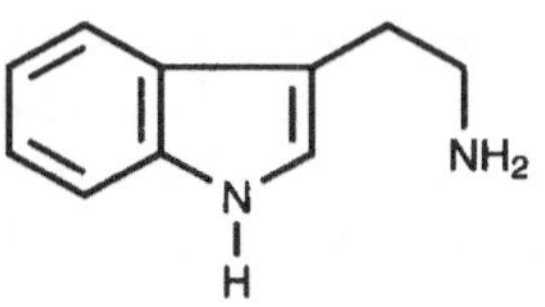   6.6.11 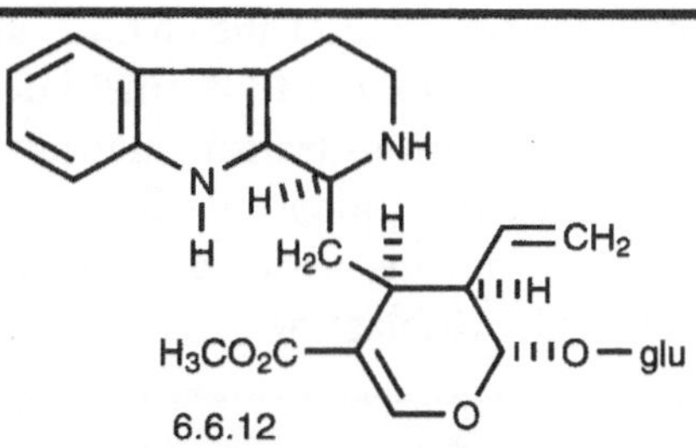   6.6.12

Strictosidine is the central figure in the biosynthetic evolution of the other terpenoid indoles mentioned above. In this process, deglucosylation is a key step and is catalysed by two highly specific gluco-alkaloid beta glucosidases which have been detected in a number of plants of the family Apocynaceae, including *Catharanthus roseus*. The resulting cyclic hemiacetal (6.6.13) can exist as the dialdehyde (mono enol form; 6.6.14). Intramolecular Schiff base formation, reduction and shifting of the double bond convert the dialdehyde to the alkaloid geissoscizine (6.6.15). By the action of a NADP dependent dehydrogenase (6.6.15), gets transformed into dehydrogeissoschzine (6.6.16). This compound then cyclises to 19(R) and 19(S) isomers of cathenamine (6.6.17), which are the precursors for ajmalicine and 19-epiajmalicine.

6.6.13   6.6.14 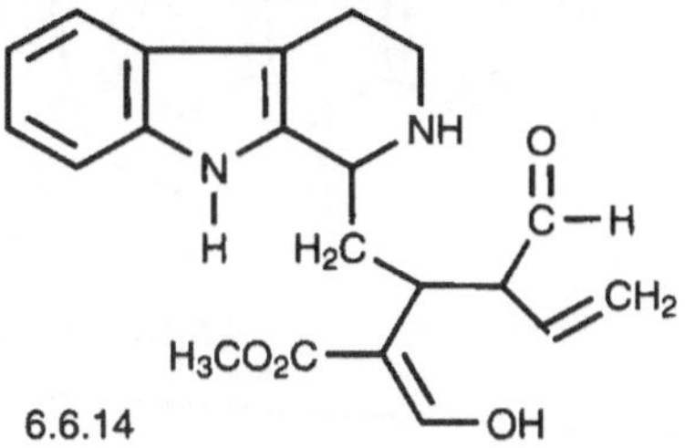   6.6.15

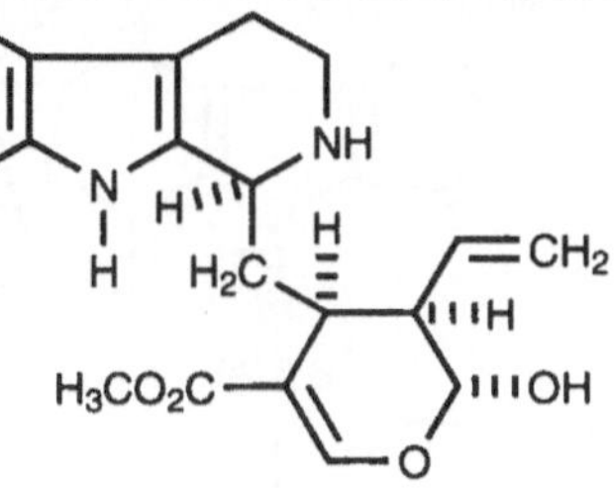

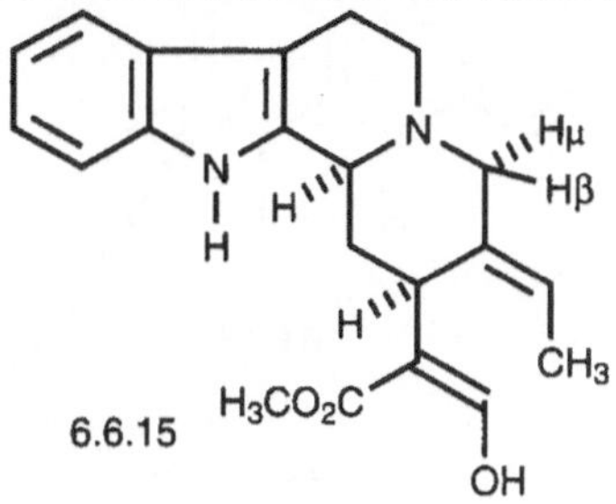

6.6.16 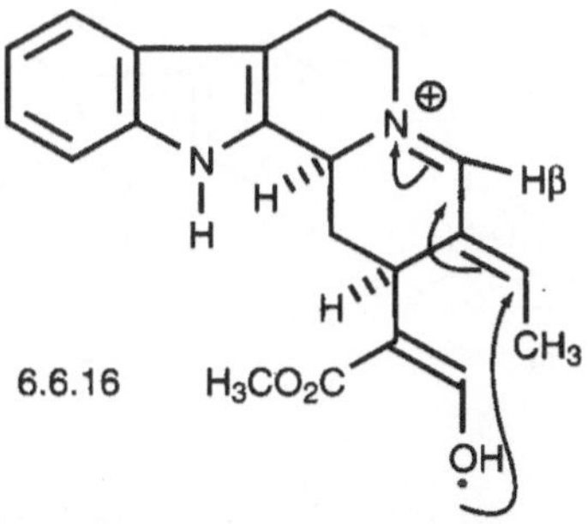   6.6.17

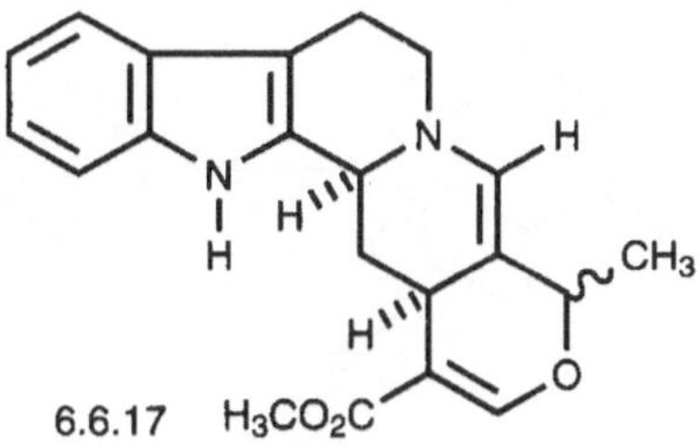

It is easy to visualise the formation of yohimbine (6.6.3) from geissoschizine by an intramolecular aldol-type reaction followed by further changes as indicated below. Strictosidine presumably undergoes transformation to sarpagine (6.6.7) via 4,5-dehydrogeissoschizine (6.6.18), polyneuridine (6.6.19) and epi-vellosimine (6.6.20). The last mentioned compound is also the precursor of deacetylvinorine (6.6.21) and vinorine (6.6.22). The crystal structure of the enzyme which catalyses the conversion of 6.6.21 into 6.6.22, known as vinorine synthase, has been determined. Deacetyl vinorine is an important intermediate in the biosynthesis of the antiarrythmic alkaloid, ajmaline (6.6.2). These various transformations of stricyosidine are shown below.

The dimeric indole alkaloid, viblastine (6.6.8), which is used for the treatment of leukemia, is formed from a combination of catharanthine (6.6.23) and vindoline (6.6.24) via 3',4'- anhydrovinblastine.

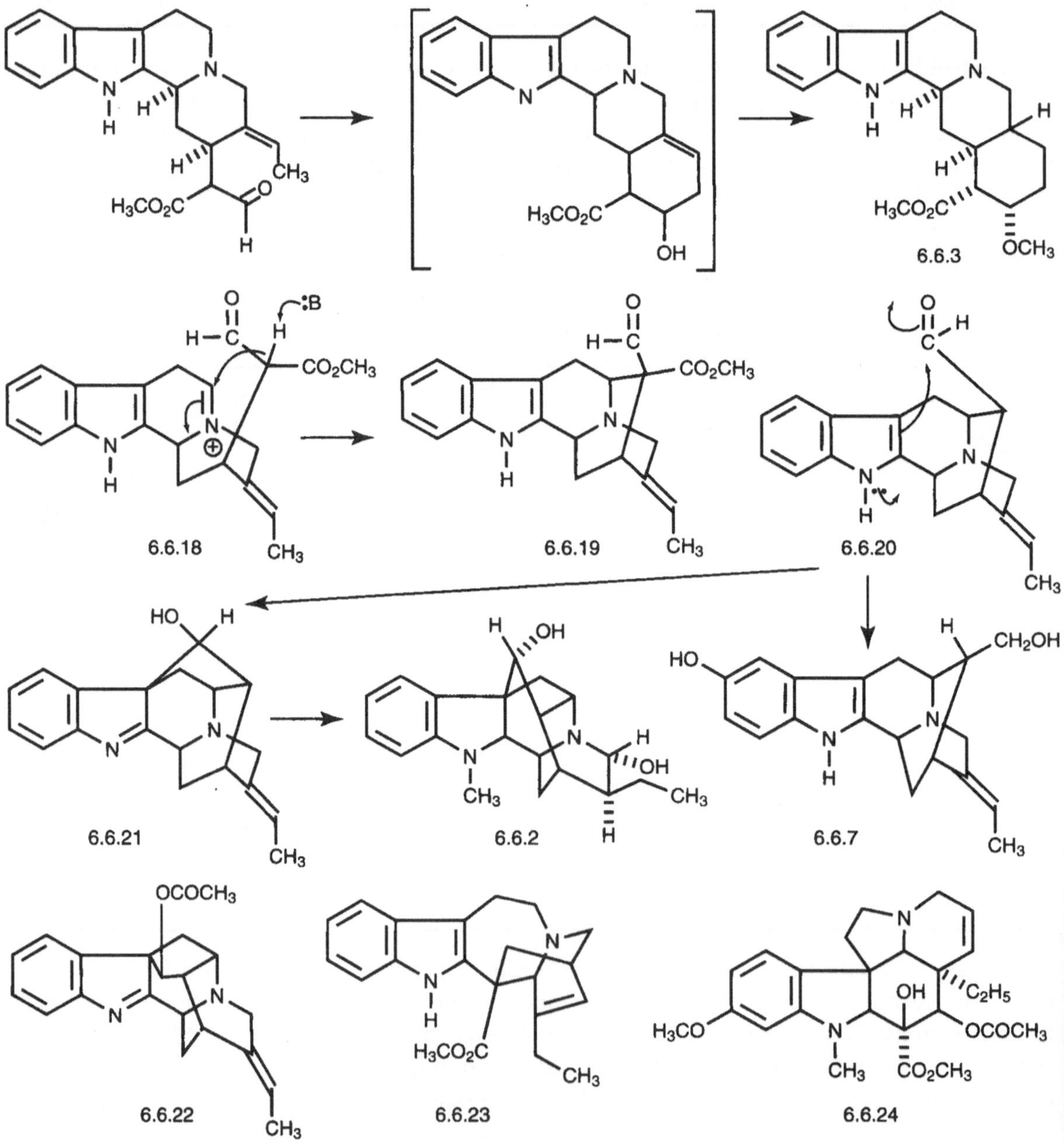

Suggested reading

1. O'Connor SE and Maresh JJ, *Nat. Prod. Rep.*, 2006, 23: 532–47.
2. Ma X, Koepke J, Panjikar S, Fritzsch G and Stockigt J, *J. Biol. Chem.*, 2005, 280(14): 13576–83.
3. Roberts MF, In Roberts MF and Wink M, (eds), *Alkaloids: Biochemistry, Ecology and Medical Applications*, New York: Plenum Press, 1998.

6.7 BIOSYNTHESIS OF ERGOT ALKALOIDS

As mentioned in Section 5.13, the ergot alklaoids are peptide alkaloids derived from lysergic and isolergic acids (6.7.1) which possess the indole skeleton. L-Tryptophan (6.7.2) is the primary metabolite for the biosynthesis of these compounds. In the first step, it condenses with dimethylallylpyrophosphate (DMAPP; 6.7.3). The enzyme which catalyses this reaction has been isolated from *Claviceps purpurea* cultures, purified and characterised. The reaction can be classifed as an aromatic electrophilic substitution reaction with the dimethylallyl group entering position 4 of the indole nucleus. There are ample laboratory analogies for such aromatic prenylation reactions. The product then undergoes *N*-methylation at the amino group of the side chain to yield 6.7.5. Hydroxylation at the benzylic (also, allylic) carbon atom leads to the formation of 6.7.6. The next two steps are dehydration to form 6.7.7 followed by epoxidation to yield 6.7.8. Opening of the epoxide accompanied by a, most likely, spontaneous decarboxylative ring closure, as shown below, results in chanoclavine-I 6.7.9.

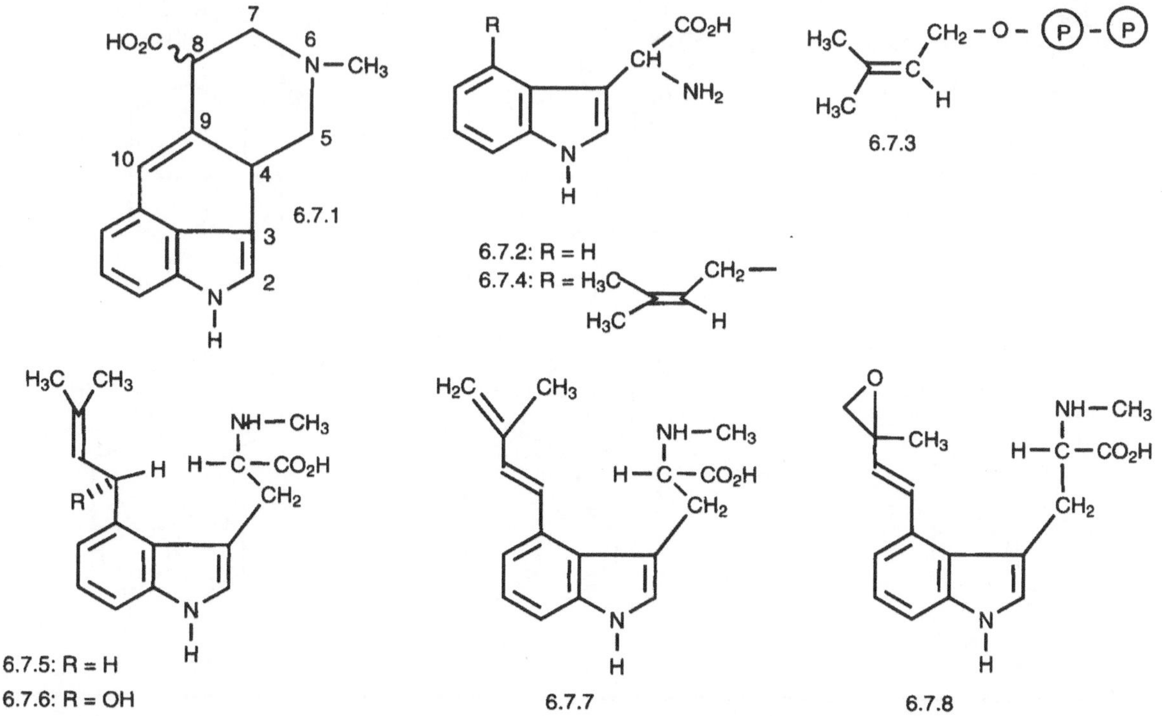

The epoxidation step appears to involve molecular oxygen. Oxidation of chanoclavin-I gives the corresponding aldehyde 6.7.10, which has been detected in a blocked mutant strain of *C. purpurea*. By the action of a NADPH dependent cyclase, chanoclavine-I aldehyde (6.7.10) undergoes transfromation into agroclavine (6.7.11). Oxidation of the allylic methyl to hydroxy methyl converts (6.7.11) into elymoclavine (6.7.12), which is the precursor of lysergic (and isolysergic) acid. During this process, the aldehyde (6.7.13) is presumaly formed with the double bond also shifting from 8,9 to 9,10 positions. This is supported by feeding experiments involving the enol acetate of lysergic aldehyde (6.7.13).

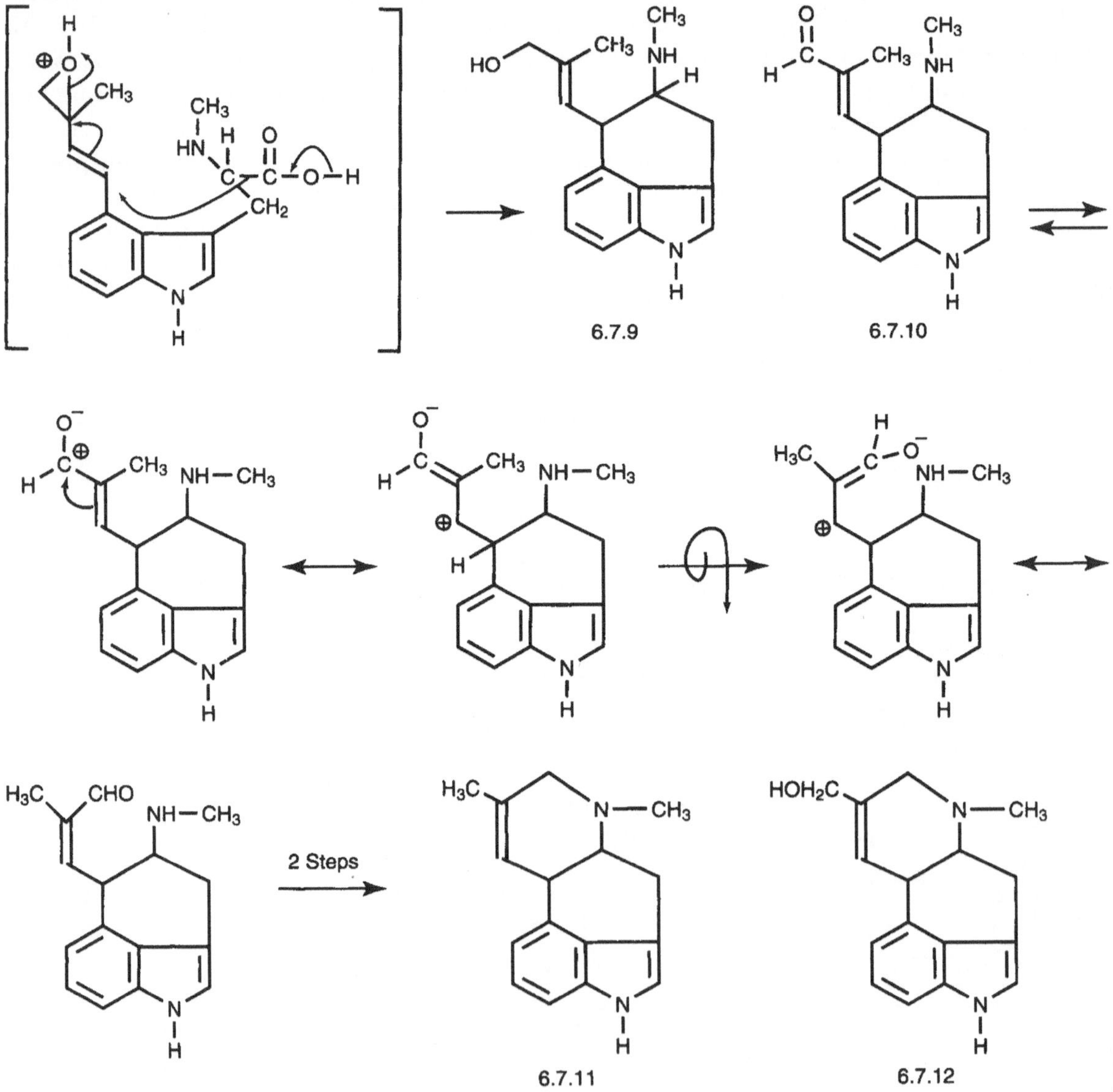

6.7.9

6.7.10

6.7.11

6.7.12

To form the peptide alkaloids, lysergic acid is probably activated by conversion into lysergyl coenzyme A; but this has not yet been experimentally established. In the formation of the peptide part of ergot alkaloids, as for example, ergotamine (6.7.15), the first step is the reaction of activated lysergic acid with L-alanine, followed by peptide bond formation with the amino group of phenylalanine and then with proline. The resulting tripeptide, lysergyl–alanyl–phenylalanyl–proline (6.7.14) then undergoes hydroxylation at the α-carbon atom of the alanyl moeity followed by cyclisation involving the thus introduced hydroxyl group, the carboxyl of proline and the secondary amino group of the phenylalanine unit. The result is the formation of the unique ring system present in ergot alkaloids.

6.7.13   6.7.14   6.7.15

Suggested reading

1. Komarova EL and Tolkachev ON, *J .Pharm. Chem.*, 2001, 35: 504–13.
2. Roberts MF, In Roberts MF and Wink M, (eds), *Alkaloids: Biochemistry, Ecology and Medical Applications,* New York: Plenum Press, 1998.

6.8 EVOLUTION OF THE MONOTERPENES

The biosynthesis of terpenoids is one of the most fascinating stories about the origin of secondary metabolites. The earlier biogenetic ideas, such as the isoprene rule, though speculative in character, did stimulate research studies and gave direction to such investigations. Even biosynthetic theories have been modified as new results accumulated. For example, the earlier view that all terpenoids arise from mevalonic acid (6.8.1) is no longer true. It is now firmly established that in plants there are two different pathways leading to the formation of the main building block of terpenoids, namely, isopentenyl pyrophosphate (IPP; 6.8.2). One of these which takes place in the cytosol does use mevalonic acid for generating IPP. Mevalonic acid itself is formed from acetyl coezyme A via acetoacetyl coenzyme A (6.8.3) and β-hydroxy-β-methylglutaryl coenzyme A (6..8.4). Mevalonic acid gets converted into IPP via its diphosphate (6.8.5) by the simultaneous loss of the carboxyl and the tertiary hydroxyl groups as shown in Scheme 6.8. The IPP thus produced is utilised for building sesqui- and tri-terpenes. On the other hand, in the chloroplast, IPP is obtained by the methylerythritol phosphate (MEP)(6.8.6) pathway (also known as 1-deoxy-D-xylulose (DOX) (6.8.7) pathway). This route leads to the mono-, di- and tetra-terpenes and also operates in certain bacteria.

Scheme 6.8

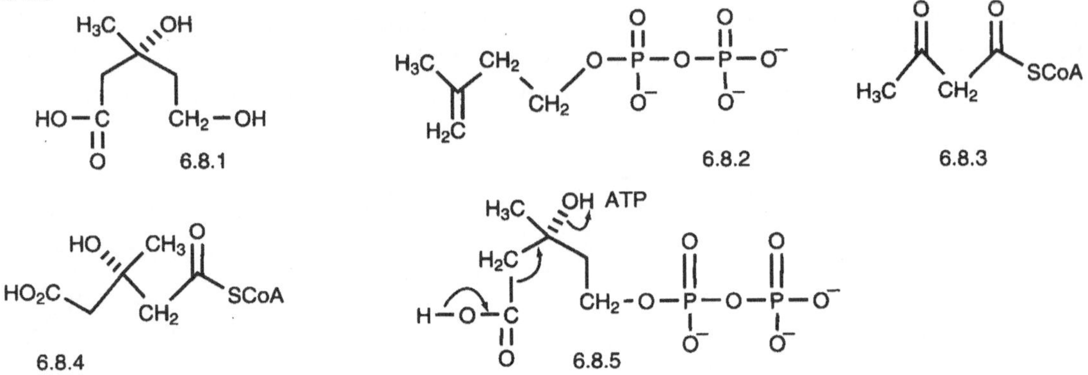

DOX is synthesised from pyruvic acid (6.8.8) and D-glyceraldehyde-3-phosphate (6.8.9) (Scheme 6.9). Thus, it is now clear that terpene biosynthesis in plants is compartmenatlised with the sesqui- and tri-terpenes being produced in the cytosol and the mono-. di- and tetra-terpenes synthesised in the chloroplast. Available evidences at present indicate that sesterterpenes are produced by the mevalonic acid pathway.

Scheme 6.9

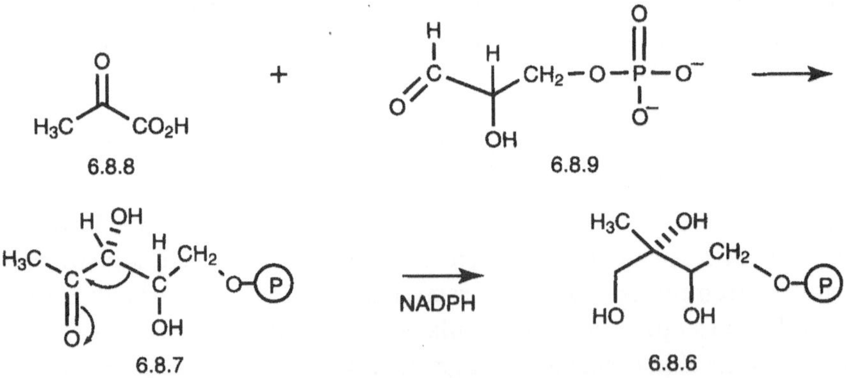

The enzymes involved in the transformation of IPP into the various terpenoids are IPP isomerase, which is responsible for the isomerisation of IPP into dimethylallyl pyrophosphate (DMAPP; 6.8.10), prenyl transferases which catalyse condensation of IPP with DMAPP and further chain lengthening and the cyclases which bring about the formation of cyclic terpenes. In the biosynthesis of monoterpenes, the first step, after the formation of IPP, is the combination between one IPP molecule and one DMAPP molecule, catalysed by a prenyl transferase. The result is geranyl pyrophosphate (GPP; 6.8.11). Hydrolysis of 6.8.11 yields geraniol (6.8.12). The latter can undergo facile oxidation, being an allylic alcohol, to give the aldehyde, geranial (6.8.13)

which has a tendency to isomerise to neral (6.8.14); geranial and neral are E, Z isomers. As a matter of fact they co-exist, and the mixture is known as citral. The reason for the ready *cis–trans* isomerisation was explained in Chapter 1.

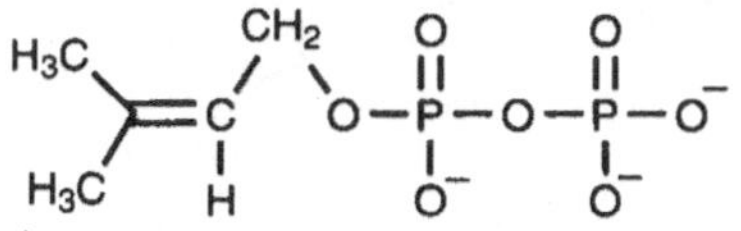

6.8.10

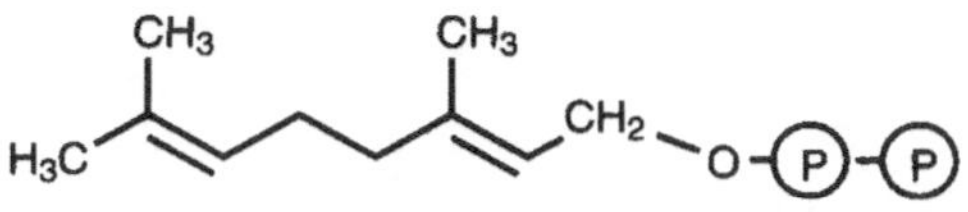

6.8.11

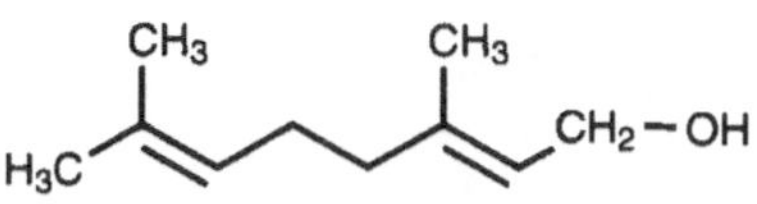

6.8.12

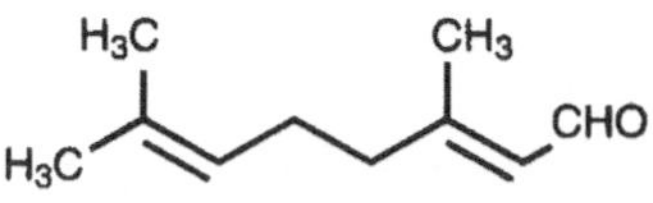

6.8.13

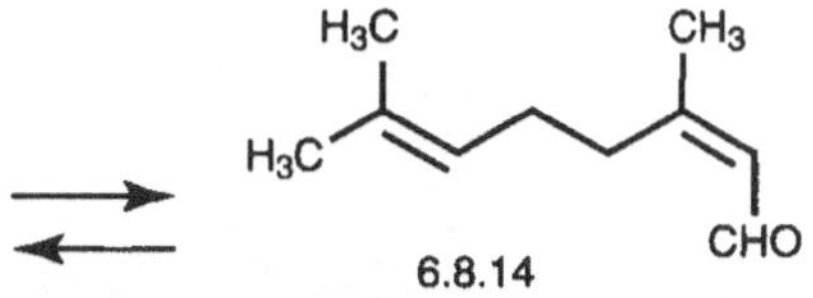

6.8.14

More interesting and intriguing are the cyclisation reactions which convert geranyl pyrophosphate into a variety of mono- and bi-cyclic mono terpenes. Prior to cyclisation, geranyl pyrophosphate undergoes isomerisation to linalyl pyrophosphate (6.8.15). It has been shown that for this change to take place, 6.8.11 first gets stereoselectively bound to the active site of the enzyme catalysing the reaction, in either a right-handed or left-handed helical conformer. Thus bound, geranyl pyrophosphate ionises and the released diphosphate ion migrates in a syn manner to form enzyme-bound (3R) or (3S) linalyl pyrophosphate (6.8.15a and 6.8.15b). Since in these structures, the bond between C-2 and C-3 is a single bond (and not double as in 6.8.11), free rotation is possible about it, thus enabling the two double bonds in the structures to come within bonding distance, in space. The result is cyclisation leading to the monocyclic (4R) or (4S) α-terpinyl carbocations (6.8.16a and 6.16.16b). Which one of these is formed is dictated by the configuration at C-3 of linalyl pyrophosphate which, in turn, is governed by the initial mode of folding of geranyl pyrophosphate at the active site of the cyclase. For the discussion which follows regarding the formation of some represetative members of the monocylic mono terpenes, however, the representation 6.8.16 for the terpinyl cation is convenient. Geranyl pyrophosphate can also lose a molecule of pyrophosphoric acid, as shown below, to form the acyclic monoterpene hydrocarbon, myrcene (6.8.17). Linalyl pyrophosphate (6.8.15) can isomerise to neryl pyrophosphate (6.8.18), which, on hydrolysis, yields nerol (6.8.19).

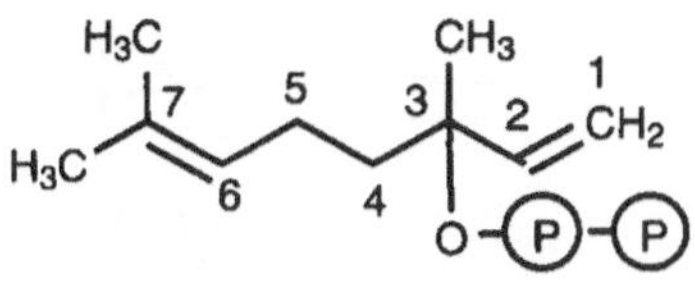

6.8.15

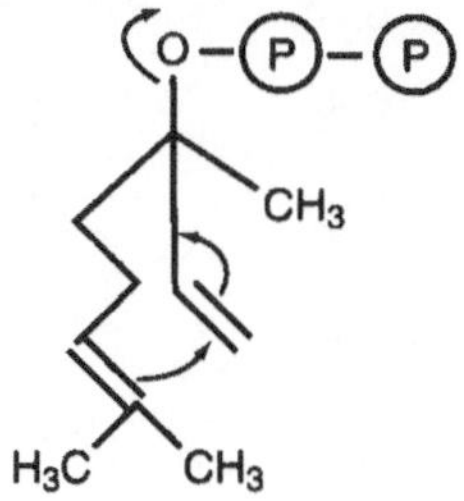

6.8.15a

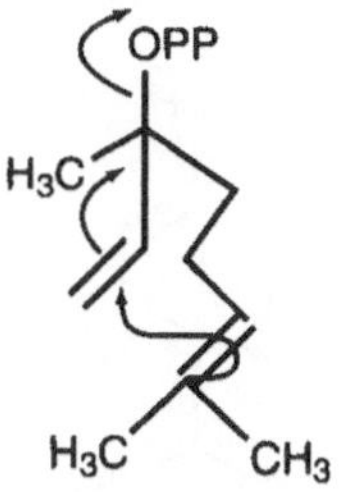

6.8.15b

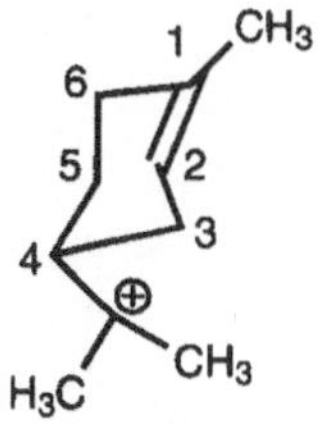

6.8.16a

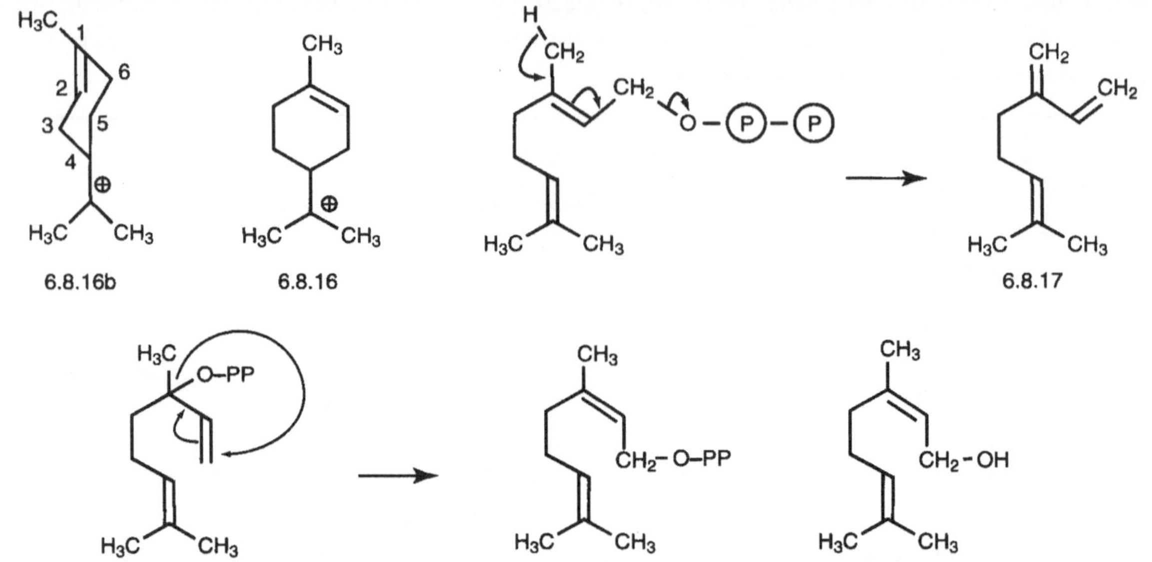

6.8.16b   6.8.16   6.8.17

6.8.18   6.8.19

The terpinyl carbocation is the key precursor for all the monocyclic mono terpenes. Loss of a proton can give either limonene (6.8.20) or terpinolene (6.8.21), whereas attack by water yields alpha terpineol (6.8.22). A 1,2-hydride ion shift converts (6.8.16) into (6.8.23), which can lose a proton to give either terpinolene or α- and γ-terpinenes (6.8.24) and 6.8.25. In a folded conformation, alpha terpineol (6.8.22) can undergo a proton-catalysed reaction to yield the cyclic ether, 1,8-cineole (6.8.26), which is a component of the essential oils of Eucalyptus species. Formation of β-phellandrene (6.8.27) from 6.8.23 involves another 1,2-hydride ion shift, followed by a shift of the endo-cyclic double bond and a final proton loss. as indicated below.

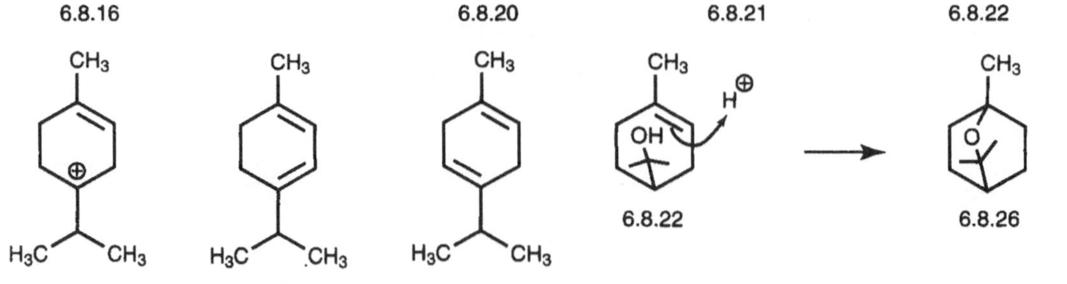

6.8.16   6.8.20   6.8.21   6.8.22

6.8.23   6.8.24   6.8.25   6.8.22   6.8.26

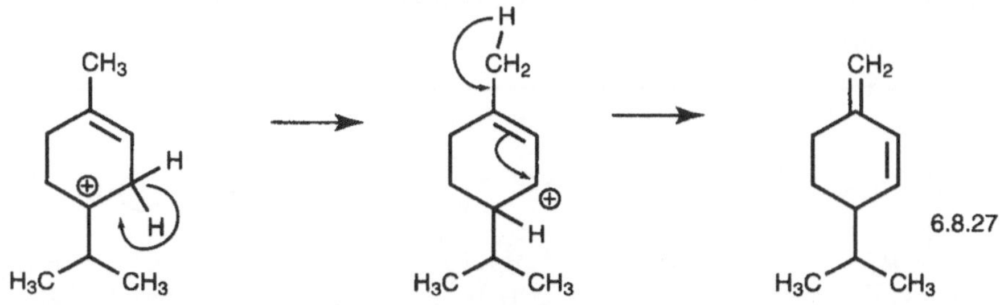

6.8.27

The terpinyl carbocation is also the precursor for the various bicyclic mono terpenes. In the conformation shown below, (6.8.16) can readily undergo cyclisation followed by the loss of a proton to yield α- and β-pinenes (6.8.28 and 6.8.29). In the same conformation, the double bond can react with the cationic centre in a different way to give the bornyl cation (6.8.30) which on reaction with water yields borneol and epi-borneol (6.8.31); further oxidation gives camphor (6.8.32). The formation of 4-carene (6.8.33) and thujene (6.8.34)—which contain cyclopropane rings—from (6.8.16) can also be easily understood in mechanistic terms.

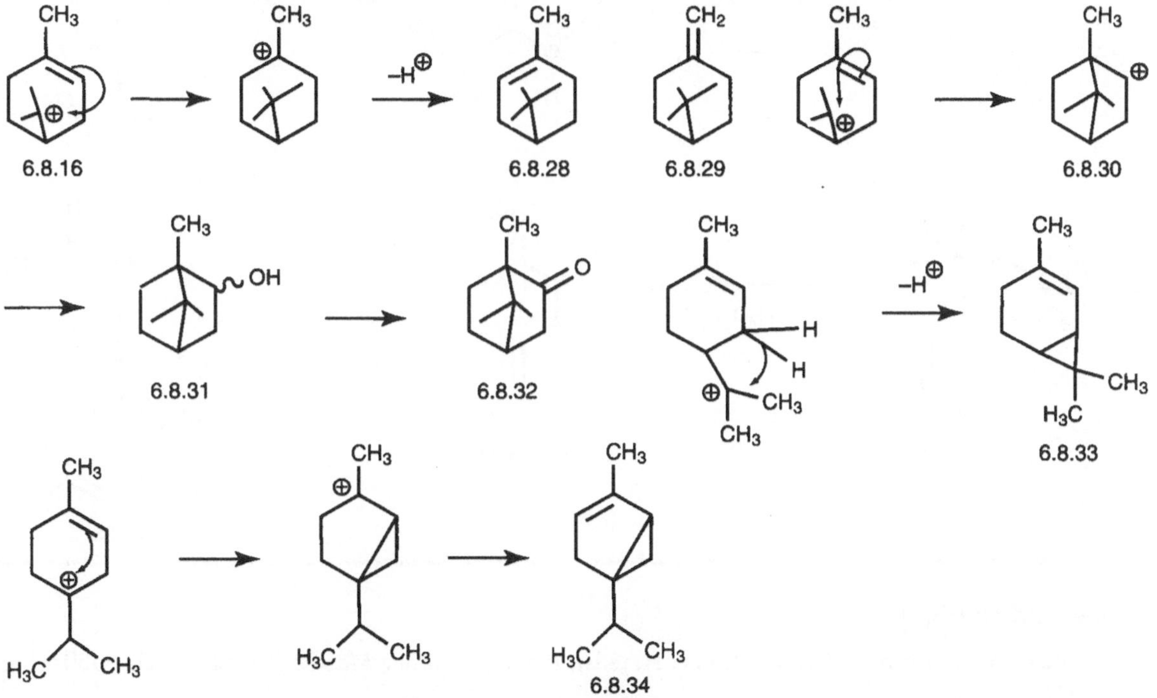

6.8.16   6.8.28   6.8.29   6.8.30

6.8.31   6.8.32   6.8.33

6.8.34

In Chapter 1 a brief reference was made to the so-called irregular monoterpenes which are biosynthesised by a non-head-to-tail combination of the two C-5 units. Such compounds are of rarer occurrence compared to those synthesised by the head-to-tail union, and appear to be confined to some plants of the Asteraceae. Examples are chrysanthemic alcohol (6.8.35), the corresponding acid (6.8.36), and the compounds 6.8.37 and 6.8.38, possessing the lavendulyl and artemisyl skeletons respectivey. The following biogenetic scheme has been proposed for the formation of these compounds.

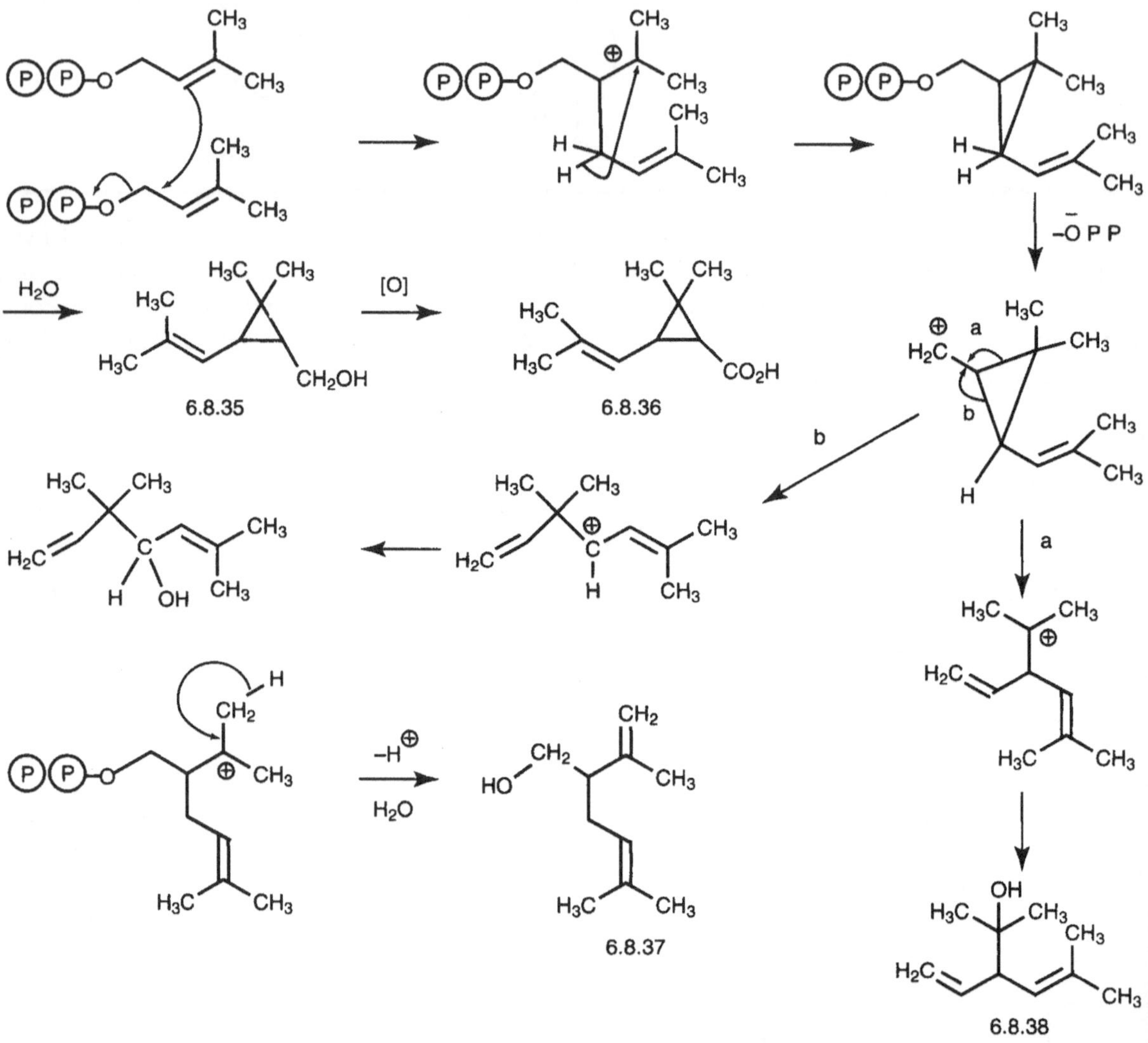

6.8.35

6.8.36

6.8.37

6.8.38

Suggested reading

1. Crozier A, Cliff ord MN and Ashihari H, (eds), *Plant Secondary Metabolites*, Blackwell Publishing, 2006.

2. Davis EM and Croteau R, In Leeper FJ, Vederas JC and Croteau R, (eds), *Biosynthesis of Aromatic Polyketides, Isoprenoids and Alkaloids*, Berlin: Springer, 2000.

3. Erman WF, *Chemistry of the Monoterpenes: An Encyclopaedic Handbook*, Part A, New York: Marcel Dekker, 1985.

4. Luckner M, *Secondary Metabolism in Micro-Organisms, Plants and Animals*, 2nd edn, Berlin: Springer-Verlag, 1984.

5. Porter JW and Spurgeon SL, (eds), *Biosynthesis of Isoprenoid Compounds*, New York: John Wiley and Sons, 1981.
6. Hanson JR, In Barton DHR and Ollis WD, (eds), *Comprehensive Organic Chemistry*, Haslam E, Vol. 5, Oxford: Pergamon Press, 1979.
7. Seigler DS, *Plant Secondary Metabolism*, Norwell: Kluwer Academic Publishers, 1995.

6.9 From Mevalonic Acid to the Sesquiterpenes

As was mentioned in the previous section, sesquiterpenes and triterpenes are biosynthesised in plants in the cytosol from isopentenyl pyrophosphate and dimethylallyl pyrophosphate which are formed from mevalonic acid. The initially formed geranyl pyrophosphate (GPP; 6.9.1) from one unit each of IPP and DMAPP, then reacts with another molecule of IPP to yield the E,E-farnesyl pyrophosphate (FPP; 6.9.2), which is the immediate precursor of all sesquiterpenes. The enzyme involved in this reaction is a prenyl transferase. The loss of a diphosphate unit from FPP, catalysed by sesquiterpene synthase enzymes, results in the formation of a resonance stabilised allylic carbocation 6.9.3, which can be attacked by an extraneous nucleophile, such as water, or by a nucleophilic group, such as a double bond, within the molecule. The result of the former reaction is farnesol (6.9.4), whereas the latter type of reaction leads to cyclic sesquiterpenes. The cation 6.9.3 can also be attacked by the pyrophosphate ion to yield nerolidyl pyrophosphate (6.9.5) in a manner similar to the geranyl pyrophosphate to linalyl pyrophosphate conversion mentioned in the previous section. Loss of the diphosphate ion from 6.9.5 results in the carbocation 6.9.6, which can be attacked by water to yield nerolidol (6.9.7) or undergo cyclisations described later in this section.

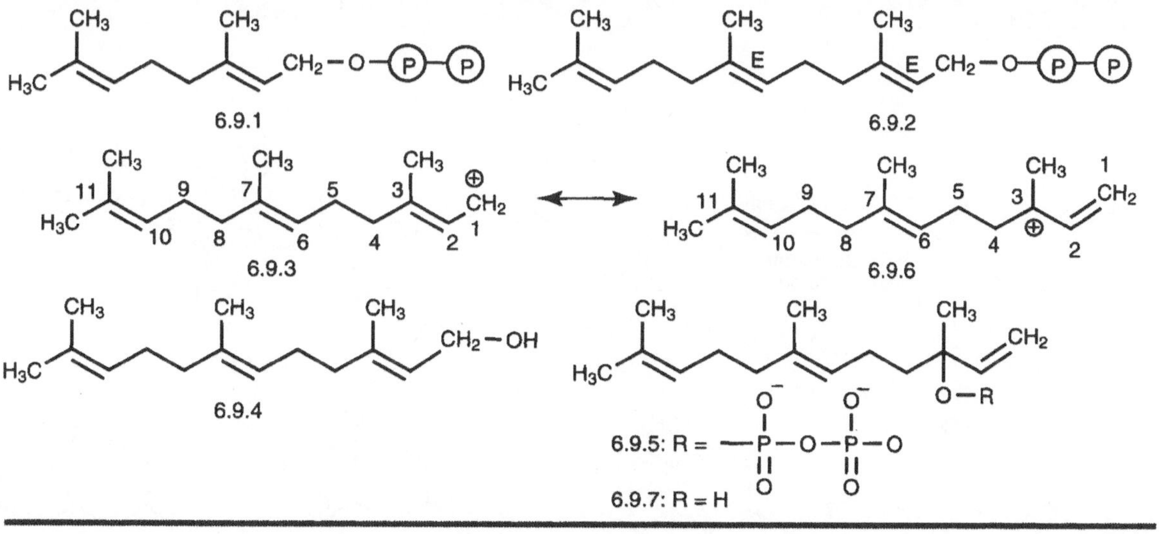

The enzymes which catalyse the intramolecular interaction of either of the two double bonds, namely those between C-6 and C-7 and between C-10 and C-11 with the cationic centres in 6.9.3 and 6.9.6 are known as cyclases. The result is the formation of a fascinating variety of mono-, bi- and tri-cyclic sesquiterpenes. For example, the double bond between C-10 and C-11 can make a nucleophilic attack on the terminal carbocationic centre in 6.9.3 by either path a or path b, as shown below. In the former case, the product is the ten-membered germacryl cation (6.9.8), whereas the alternative mode of attack leads to the eleven-membered humulyl cation (6.9.9). Similar reactions involving the ion (6.9.6) give *cis*-germacryl cation (6.9.10) and *cis*-humulyl cation (6.9.11). Loss of a proton from the germacryl carbocation can give either germacrene A (6.9.12) or germacrene D (6.9.13). Germacrene A and D synthases have been isolated from the Asteraceae plant, *Solidago canadensis* (golden rod), cloned and over expressed in *Escherichia coli*. Germacrene A is the precursor of the sesquiterpene lactone, parthenolide (6.9.14). It is obvious that oxidising enzymes are involved in this transformation. Proton-catalysed transannular interaction of the two endocyclic double bonds in germacrene A can give either the eudesmyl cation (6.9.15) or a cation having the azulene skeleton (6.9.16). The former, then, apparently suffers a 1,2-hydride ion shift, followed by a methyl shift and loss of a proton to form epi-aristolochene (6.9.17), a sesquiterpene hydrocarbon having the naphthalene skeletal structure. On the other hand, the ion 6.9.16 can suffer three synchronised hydride ion shifts followed by the loss of a proton to yield the azulenic sesquiterpene hydrocarbon (6.9.18). The hydrocarbon, γ-humulene (6.9.19) probably arises from the ion 6.9.9 by a transannular hydride ion shift followed by other changes as shown below.

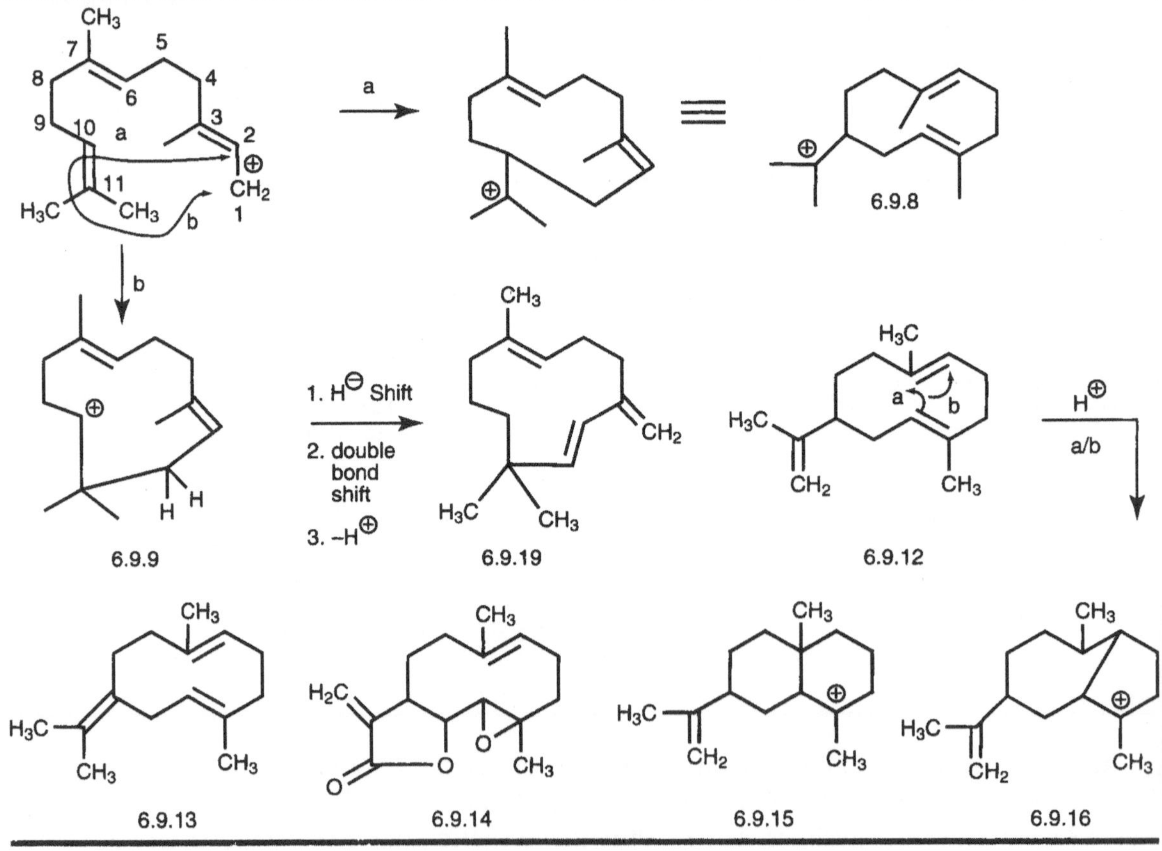

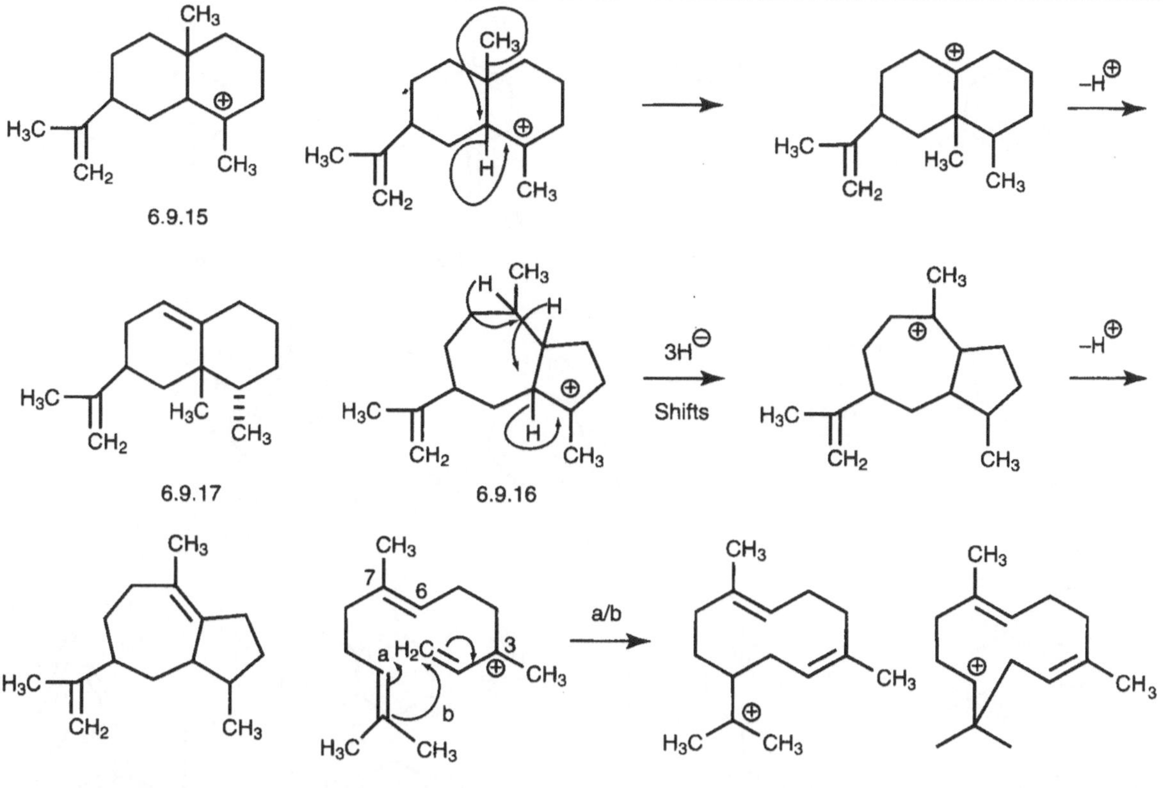

6.9.15

6.9.17   6.9.16   3H⊖ Shifts

6.9.18   6.9.6   a/b   6.9.10   6.9.11

The E,Z-nerolidyl carbocation (6.9.6), in the conformation shown, can readily undergo intramolecular cyclisation involving the double bond between C-6 and C-7 to yield the bisabolyl carbocation (6.9.20), which can then lose a proton to give α-bisabolene (6.9.21) or capture a hydroxyl group to form α-bisabolol (6.9.22). An alternative mode of cyclisation involving the same double bond and the cationic centre, however, results in the ion 6.9.23, which after a second cyclisation, 1,3-hydride shift and reaction with water gives carotol (6.9.24), one of the sesquiterpene components of carrots.

cis-Germacryl cation (6.9.10), mentioned earlier as arising from nerolidyl pyrophosphate by the loss of the pyrophosphate group, can undergo a 1,3-hydride ion shift to yield the ion 6.9.25, which then suffers transannular cyclisation followed by proton loss to yield another naphthalenic sesquiterpene hydrocarbon, namely δ-cadinene (6.9.26). Conformational factors play a key role in the biotransformation of the farnesyl carbocation (6.9.3) into more complex sesquiterpenes such as the tricyclic compound, patchoulol (6.9.27), which is a major perfumery constituent of patchouli oil. The first step is the formation of a ten-membered carbocation (6.9.28) which after two cyclisations and a 1,3-hydride ion shift gets transformed into the ion 6.9.29 as shown below. The latter suffers two successive [1,2] skeletal rearrangements to yield the ion 6.9.30, which finally captures a nucleophile (hydroxyl group) to give patchoulol.

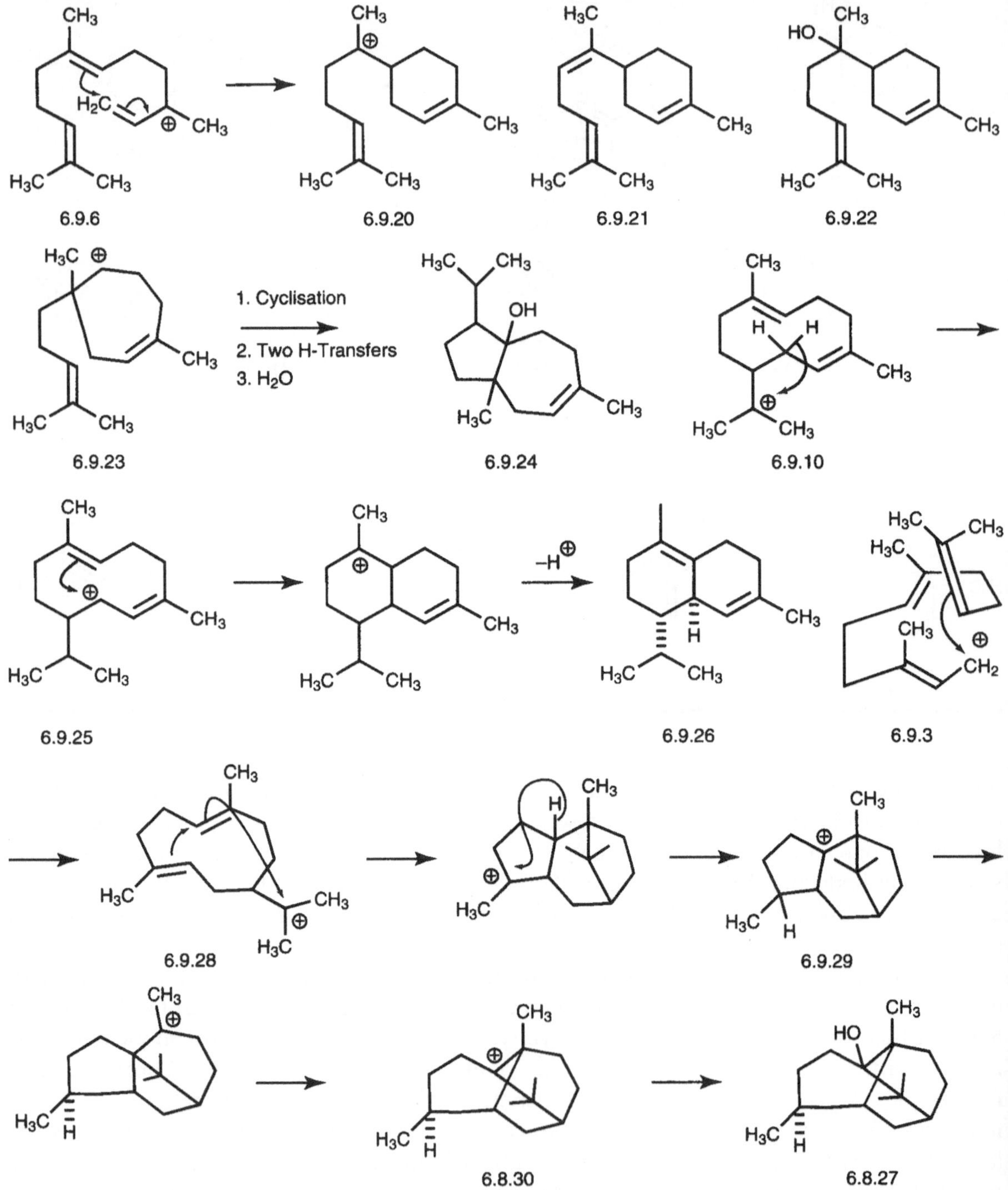

6.9.6   6.9.20   6.9.21   6.9.22

1. Cyclisation
2. Two H-Transfers
3. H₂O

6.9.23   6.9.24   6.9.10

6.9.25   6.9.26   6.9.3

6.9.28   6.9.29

6.8.30   6.8.27

Suggested reading

1. Humphrey AJ and Beale MH, Terpenes, In Crozier A, Cliff ord MN and Ashihara H, (eds), *Plant Secondary Metabolites: Occurrence, Structure and Role in the Human Diet*, Oxford: Blackwell, 2006.
2. Davis EM and Croteau R, Cyclization enzymes in the biosynthesis of monoterpenes, sesquiterpenes and diterpenes, In *Biosynthesis, Topics in Current Chemistry*, Vol. 209, Berlin: Springer, 2000.

6.10 Biosynthesis of some Diterpenes

Like the monoterpenes, the diterpenes are produced in plants, mostly, in the chloroplast and the first member of this group of compounds is geranyl geranlyl pyrophosphate (GGPP; 6.10.1). It is formed from farnesyl pyrophosphate (6.10.2) by combination with one molecule of isopentenyl pyrophosphate. The acyclic diterpene, phytol (6.10.3), is derived from GGPP by hydrolysis and reduction of three of the four double bonds. More interesting are the various cyclic diterpenes in the biosynthesis of which the diterpene cyclases play a vital role. Herein, we shall consider the bioformation of a few of these compounds, starting with the important nor diterpene, gibberellic acid (6.10.4).

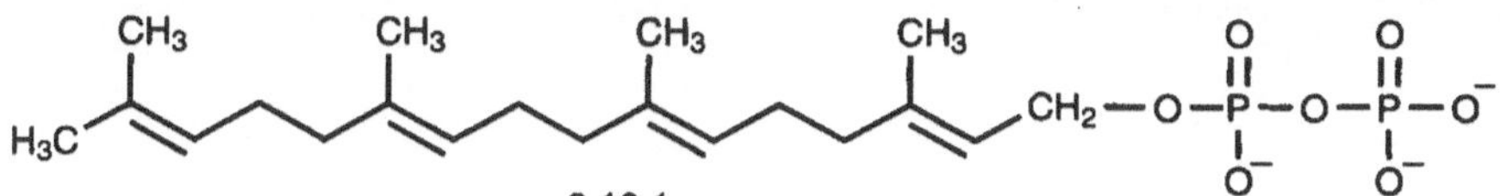

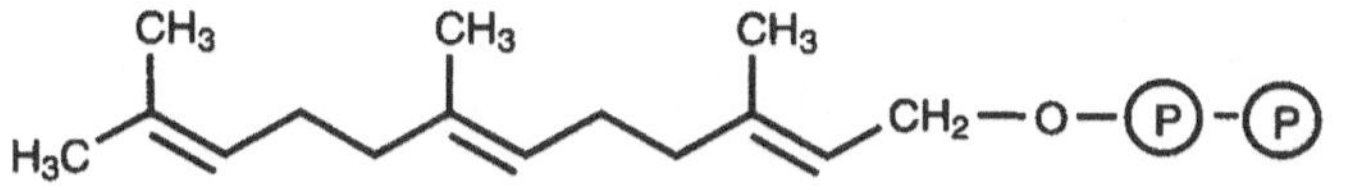

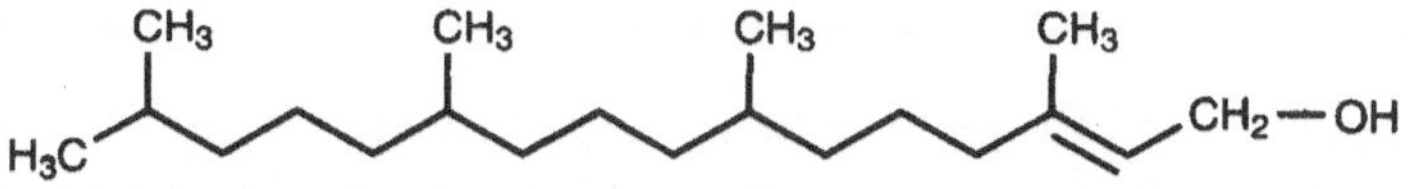

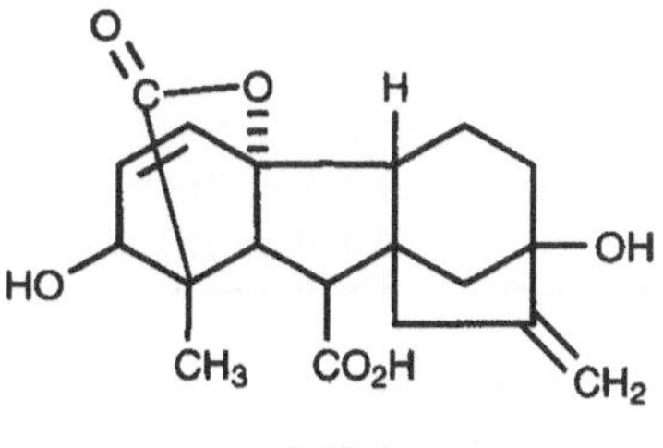

Experiments using C14 and P32 labelled geranyl geranyl pyrophosphate showed that it is a substrate for the biosynthesis of gibberellic acid by the fungus *Gibberella fujikuroi*. In the course of this experiment, doubly labelled copalyl pyrophosphate (6.10.5) was isolated, demonstrating thereby that it is an intermediate in the biosynthesis of gibberellic acid. Using a soluble enzyme isolated from *G. fujikuroi*, 6.10.5 could be converted into (–) kaurene (6.10.6). (–) Kaurene gets oxidised to kaurenoic acid (6.10.7) via kaurenol (6.10.8) and kaurenal (6.10.9). The probable mechanisms for the conversion of GGPP into copalyl pyrophosphate [the diphosphate ester of (–)labda-8(16),13-dien-15-ol] and that of the latter into kaurene are shown below.

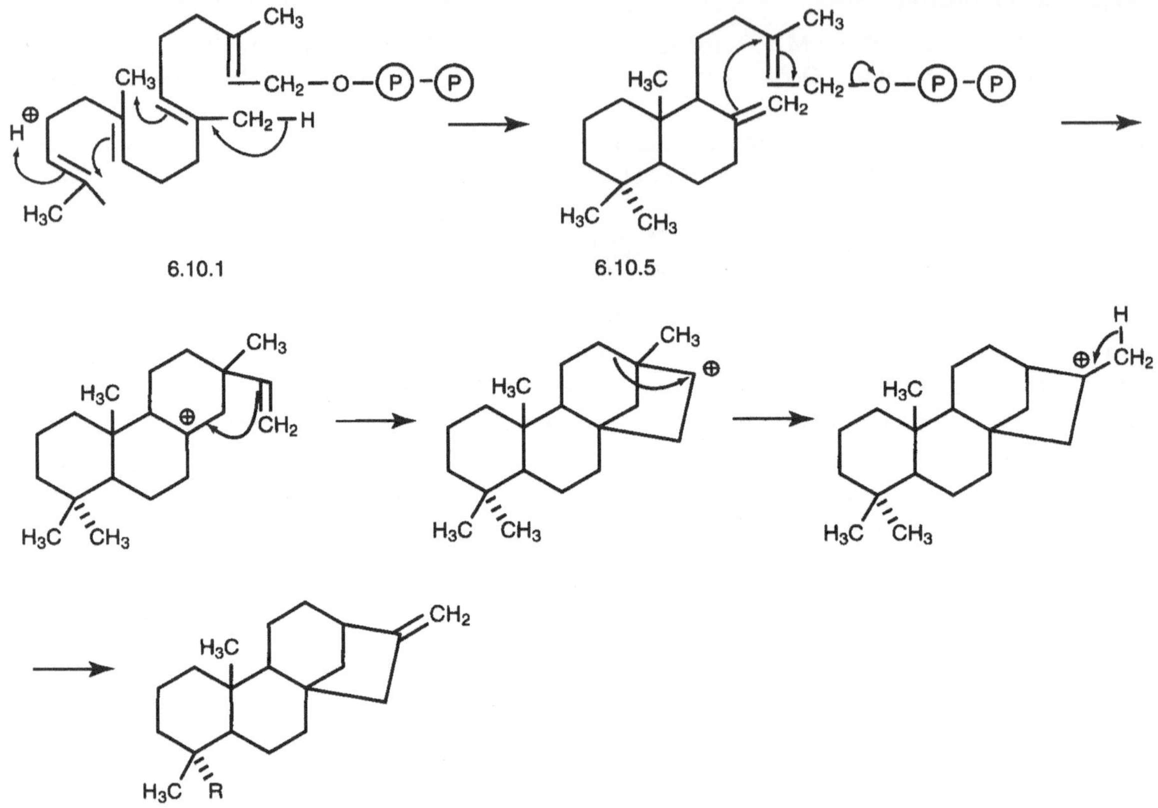

6.10.1   6.10.5

6.10.6: R = CH₃
6.10.7: R = CO₂H
6.10.8: R = CH₂OH
6.10.9: R = CHO

Subsequent conversion of (–) kaurenoic acid into gibberellic acid goes through a number of discrete steps each involving oxidation. In the first stage, dihydroxylation, presumably stepwise, occurs in the B ring to yield 6.10.10 which undergoes a pinacol–pinacolone type rearrangement to give the aldehyde 6.10.11. Oxidation of the formyl to carboxyl, hydroxylation of A ring and oxidation of the angular methyl to a formyl group then take place, in that order, to form 6.10.12. Baeyer–Villiger type oxidation of the angular formyl group gives the formyl ester 6.10.13. The enzymes responsible for these oxidations have been characterised. The angular hydroxyl group released by hydrolysis of 6.10.13 forms a lactone with the carboxyl group on the A ring to yield the intermediate 6.10.14. This compound is designated as GA 4. Removal of hydrogen atoms from C-1 and C-2 creates a double bond between these two carbon atoms to form GA 7 (6.10.15). Gibberellic acid A 3 (6.10.4) arises from 6.10.15 by acquiring an angular hydroxyl group at the junction of C and D rings.

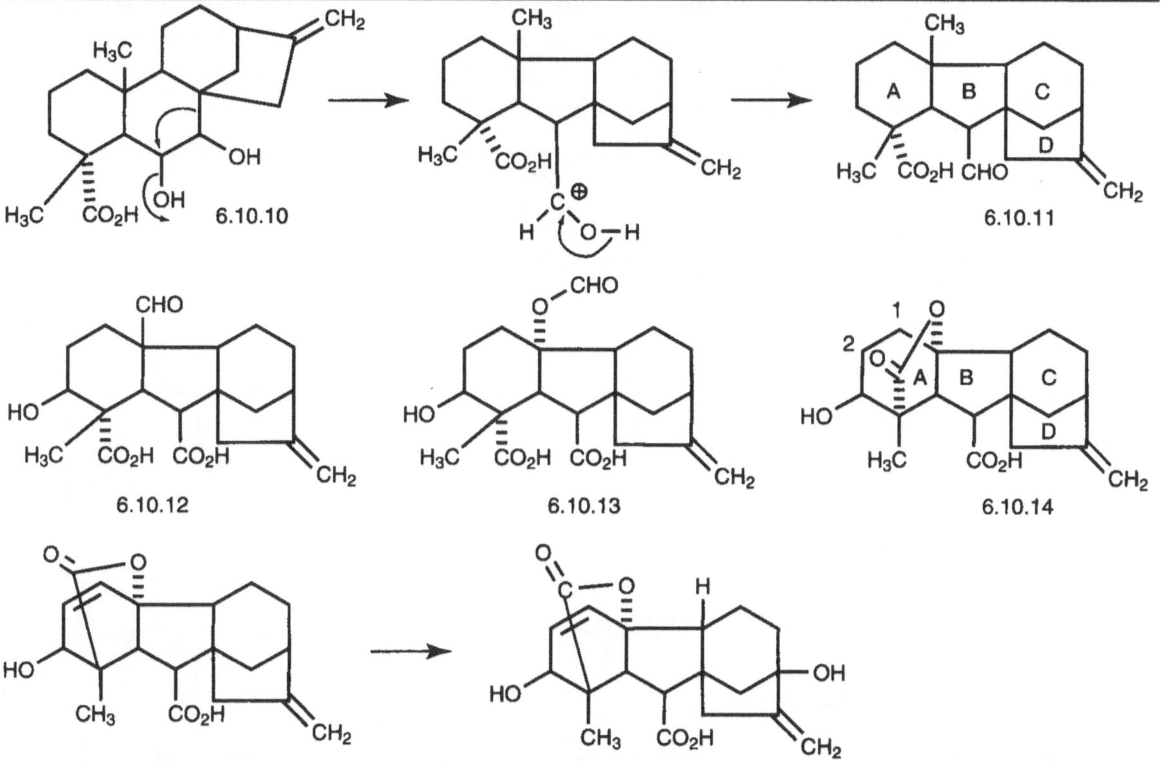

6.10.10 6.10.11

6.10.12 6.10.13 6.10.14

6.10.15 6.10.4

The bicyclic diterpene, scoparic acid (6.10.16) can arise from copalol (6.10.17) by oxidation of the alpha methyl group at C-4 to carboxyl, hydroxylation at C-6 and esterification of the thus created hydroxyl group with benzoic acid. The biosynthesis of the furanoid labdane diterpene, marrubin (6.10.18), which occurs in *Marrubium vulgare* (Lamiaceae) has been studied in considerable detail. It has been conclusively shown that in this case the IPP needed for the first step arises from methylerythritol phosphate and not from mevalonic acid. The labdane skeleton is formed from GGPP as in the case of copalol. The furan ring obviously arises by oxidative transformation of the side chain at C-9. The intermediate is premarrubin (6.10.19).

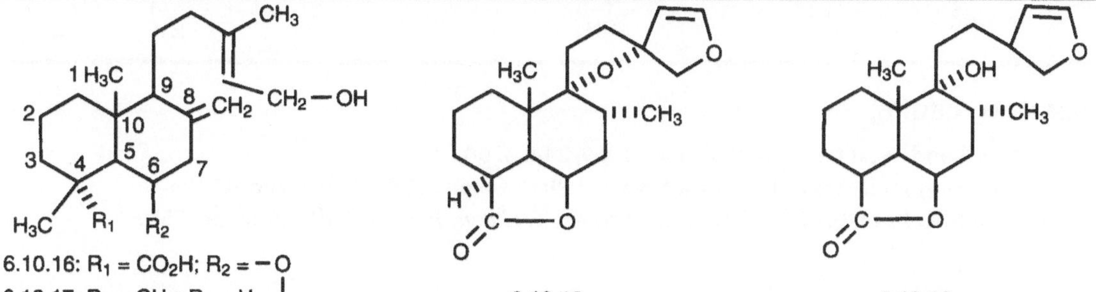

6.10.16: $R_1 = CO_2H$; $R_2 = -O$
6.10.17: $R_1 = CH_3$; $R_2 = H$

$C=O$
C_6H_5

6.10.18 6.10.19

Marine organisms produce diterpenes with novel skeletal structures. For example, gorgonians of the genus Pseudopterogeorgia, otherwise known as sea plumes, contain bicyclic diterpenes of unusual structures. One of these is elisabethatriene (6.10.20), which possesses a serrulatane skeleton. The probable mechanism of its formation from GGPP is shown below.

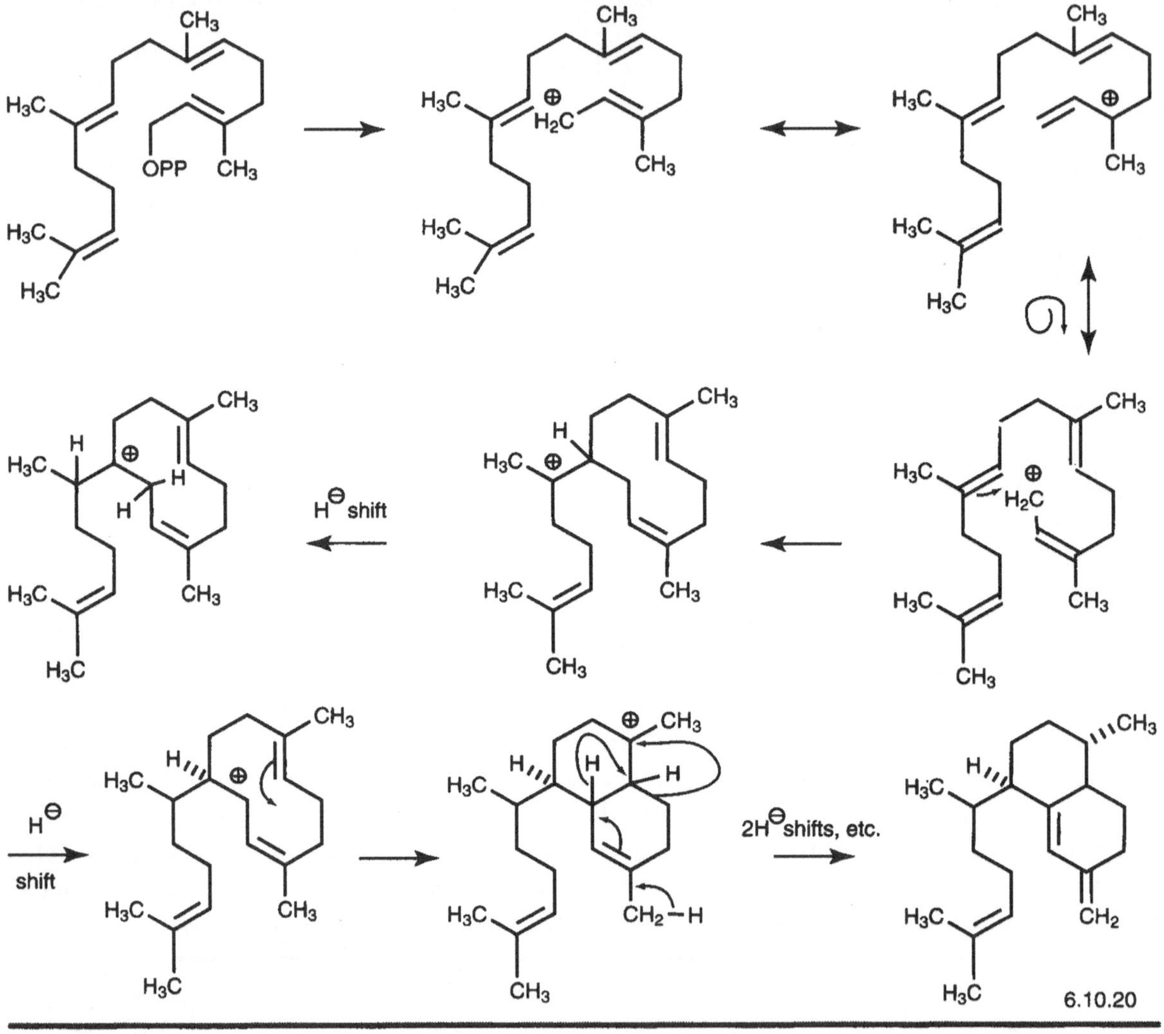

Suggested reading

1. Shechter I and West CA, *J. Biol. Chem.*, 1969, 244: 3200.
2. Kawaide H, Imai R, Sassa T, and Kamiya Y, *J. Biol. Chem.*, 1997, 272: 21706–12.
3. Nkembo MK, Lee J-B, Nakagiri T and Hayashi T, *Chem. Pharm. Bull.*, 2006, 54: 758.

6.11 A BRIEF NOTE ON THE BIOSYNTHESIS OF OPHIOBOLINS

The ophiobolins are sesterterpenes. Their occurrence in nature was first noted in 1965 and since then a number of them have been obtained from diverse sources including fungi, ferns, lichens, marine organisms and insects. Biosynthetic experiments have shown that 3H and 14C labelled mevalonic acid lactone (6.11.1) gets incorporated into ophiobolin A (6.11.2) via geranyl farnesyl pyrophosphate (6.11.3). The last mentioned compound is formed by the union of geranyl geranyl pyrophosphate (6.11.4) and isopentenyl pyrophosphate. As mentioned in the earlier sections of this chapter, appropriate conformations of the prenyl pyrophosphate precursors are crucial for the observed intramolecular cyclisations in nature. Thus, in the conformation of 6.11.3 shown below, the elimination of the pyrophosphate group triggers a chain reaction involving intramolecular nucleophilic attack on the thus generated carbocationic centre by the C-10–C-11 double bond which, in turn, is attacked by the double bond between C-14 and C-15 resulting in the ion 6.11.5. A [1.6] hydride ion shift followed by shuffling of the double bonds and final attack by a hydroxyl nucleophile results in 6.11.6. Oxidative modifications convert it into ophiobolin A (6.11.2).

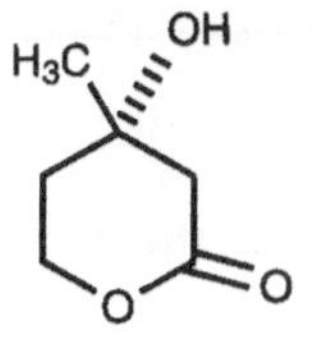

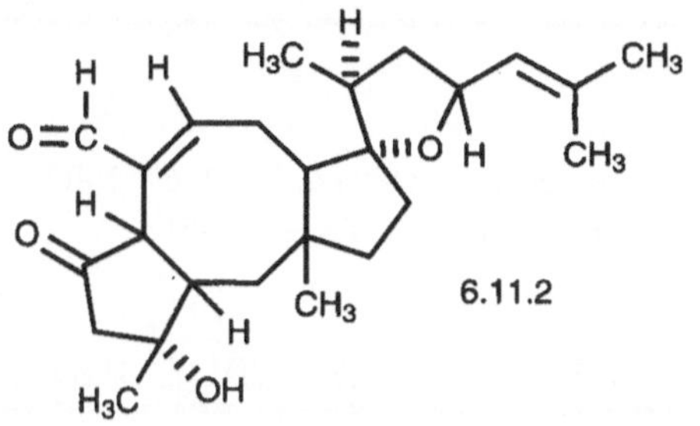

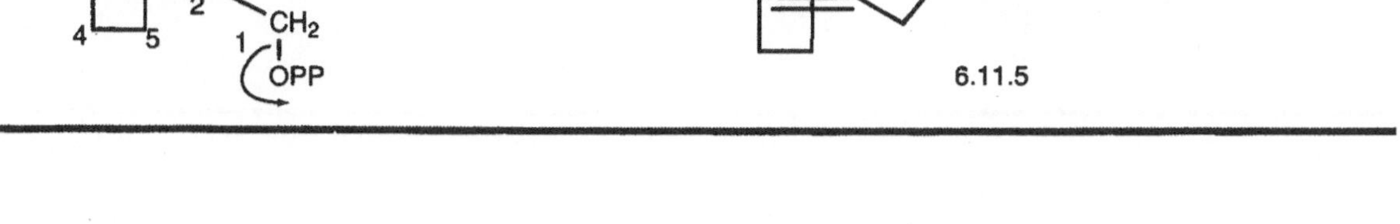

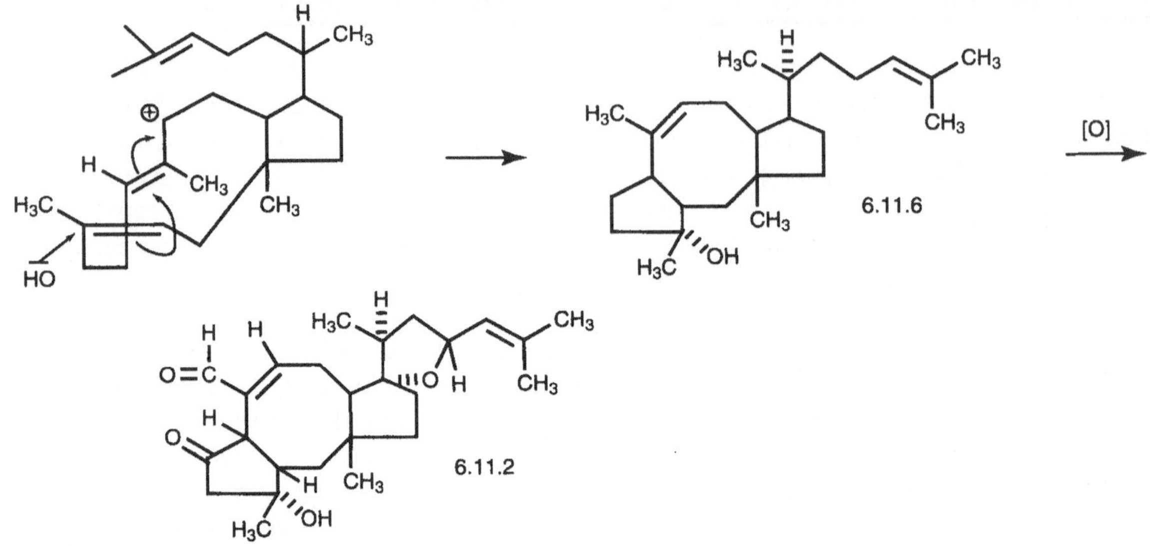

6.11.6

6.11.2

[O]

Suggested reading

1. Ponomarenko LP, Kalinovsky AI, Stonik VA, *J. Nat/Prod.*, 2004, 67: 1507 and references therein.
2. Leung PC, Graves LM, Tipton CL, *Intl. J. Biochem.*, 1988, 20: 1351.

6.12 Geranyl Pyrophosphate to Lanosterol and the Triterpenes

The head-to-tail combination of geranyl pyrophosphate with isopentenyl pyrophosphate yields farnesyl pyrophosphate (6.12.1), which is the primary precursor of all sesquiterpenes. Union of 6.12.1 with one more unit of isopentenyl pyrophosphate results in the acyclic diterpene, geranyl geranyl pyrophosphate (6.12.2). The head-to-head combination of two farnesyl pyrophosphate moieties leads to squalene (6.12.3). The steric course of the entire process has been unravelled in the Nobel prize winning work of Cornforth. In this section we will discuss only the salient features of the biosynthesis of the triterpenes and sterols from 6.12.3, without going into the details of their stereochemistry.

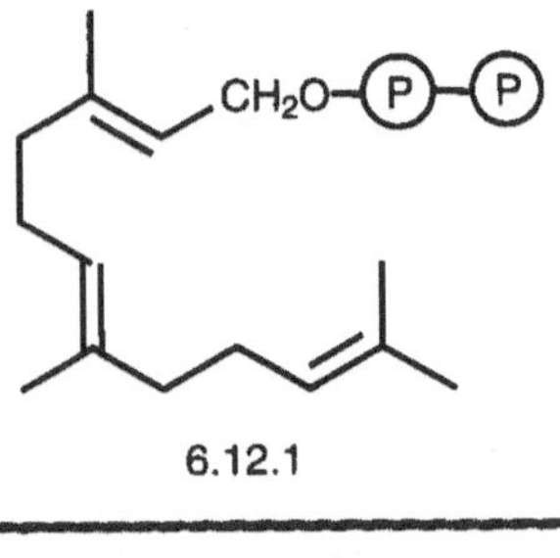

6.12.1

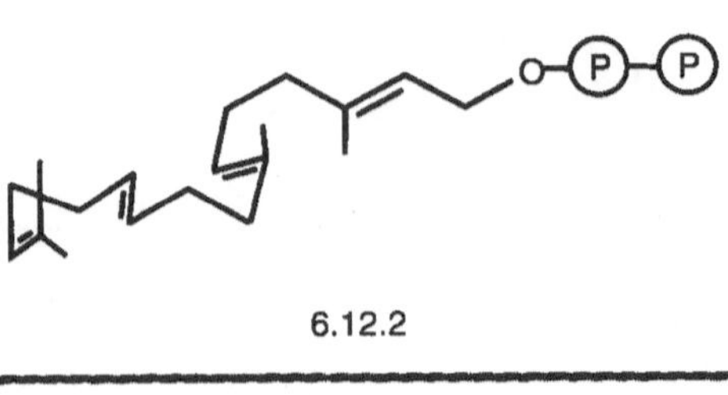

6.12.2

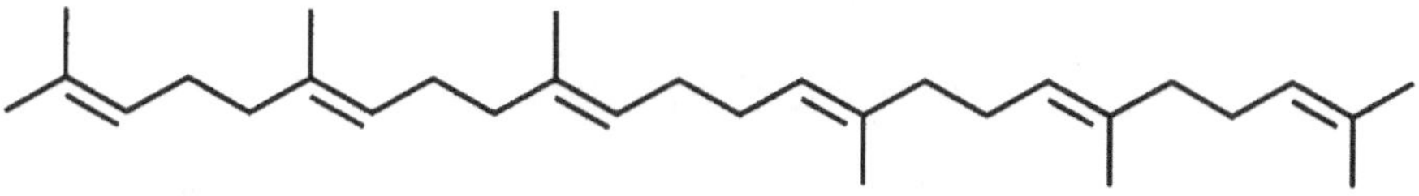

6.12.3

If one looks at the 2,3-epoxide of squalene in a chair–boat–chair–boat conformation (6.12.4), it is possible to visualise a proton-catalysed ring opening of the epoxide which would trigger a sequence of successive homoconjugative interactions among the various double bonds. The result is the formation of a tetracyclic intermediate carbocation (6.12.5). A series of 1,2-hydride and methyl shifts transform 6.12.5 into 6.12.6. The latter loses a proton to yield lanosterol (6.12.7). If 6.12.6 suffers a further [1,2]-hydride shift followed by the loss of a proton from the angular methyl at position 10, the result is the formation of cycloartenol (6.12.8); the intermediate is 6.12.9. Lanosterol is the precursor of cholesterol (6.12.10) and this involves the loss of three methyl groups (at positions 4 and 14), shift of the double bond and modification of the side chain.

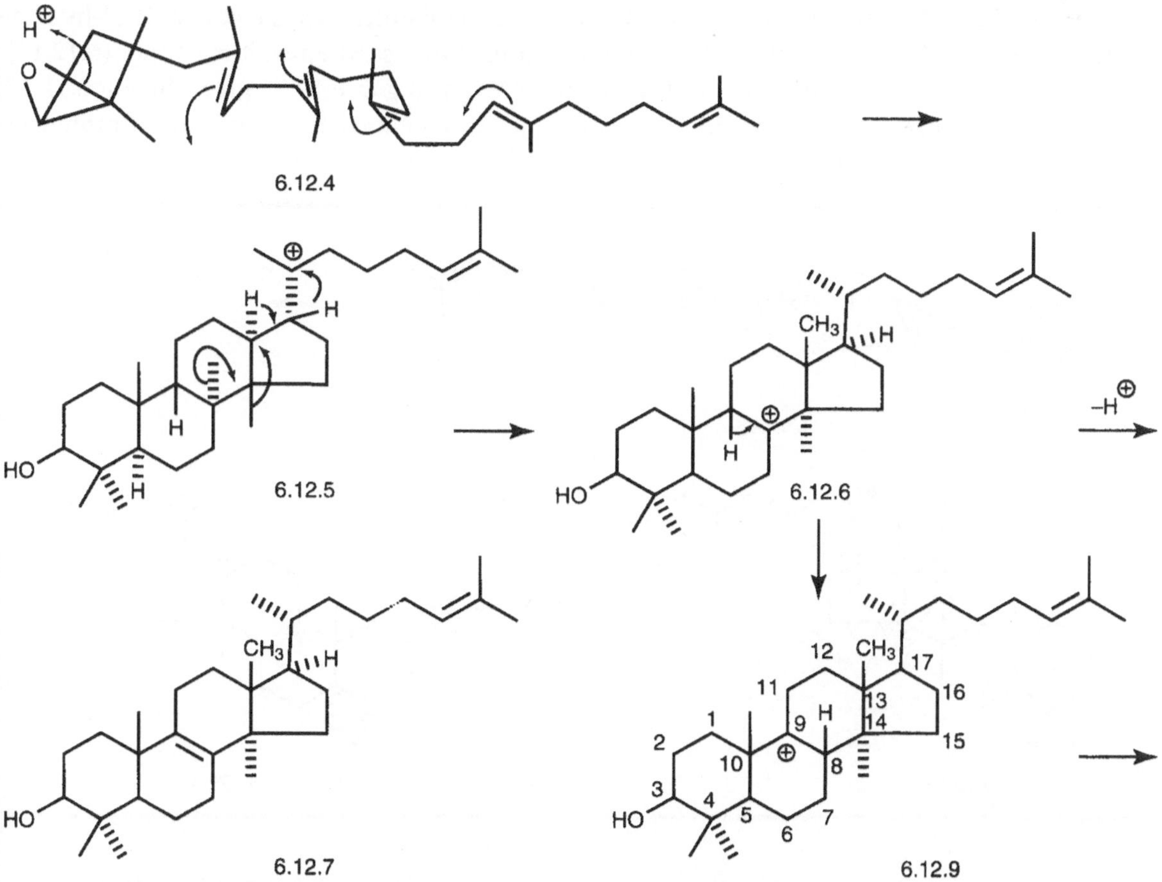

6.12.4

6.12.5

6.12.6

6.12.7

6.12.9

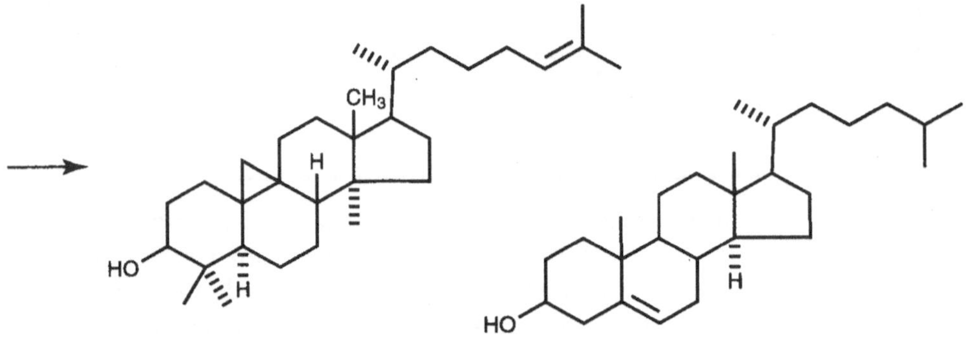

6.12.8   6.12.10

Let us now consider the biosynthesis of euphol (6.12.11), lupeol (6.12.12), friedelin (6.12.13) and β-amyrin (6.12.14), four well-known pentacyclic triterpenes. All of them originate from squalene epoxide in the chair–chair–chair–boat conformation (6.12.15). An acid-catalysed opening of the epoxide group and subsequent electron shifts as in the case of 6.12.4, convert 6.12.15 into the carbocation intermediate 6.12.16. If the latter suffers two successive [1,2]-hydride shifts followed by two [1,2]-methyl shifts and a final proton loss, the result is euphol (6.12.11). On the other hand, a ring expansion of the five-membered ring in 6.12.16 gives the ion 6.12.17, which then rearranges itself into 6.12.18. Lupeol (6.12.12) is obtained by the loss of a proton from 6.12.18. A rearrangement of the E ring in 6.12.18 gives 6.12.19.

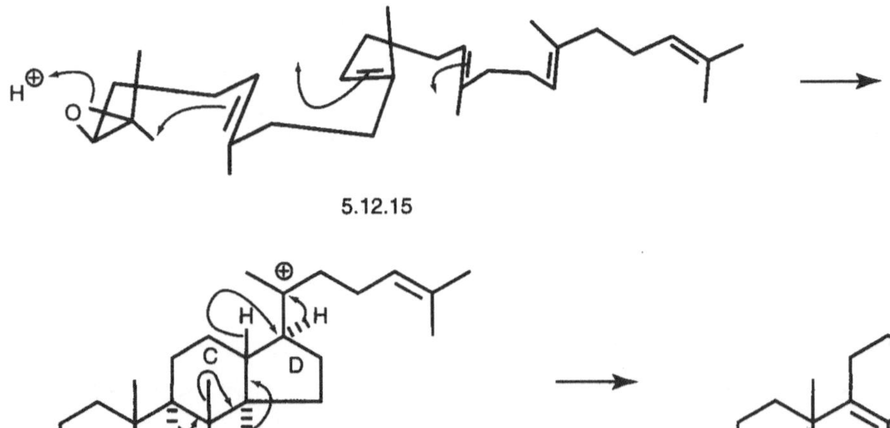

5.12.15

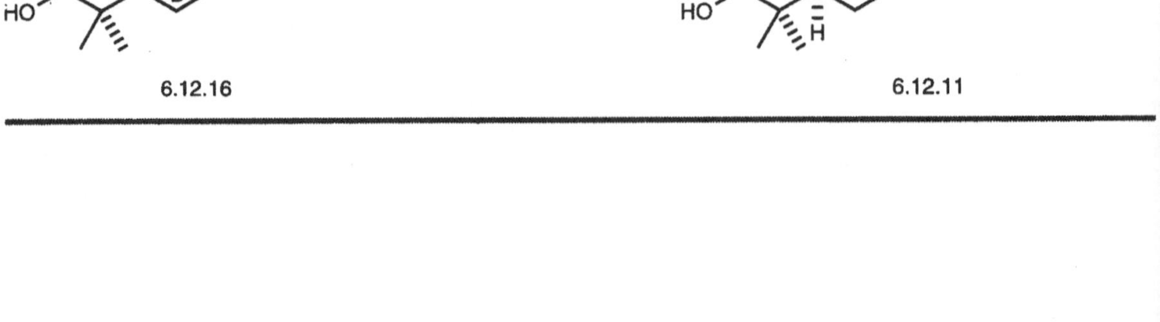

6.12.16   6.12.11

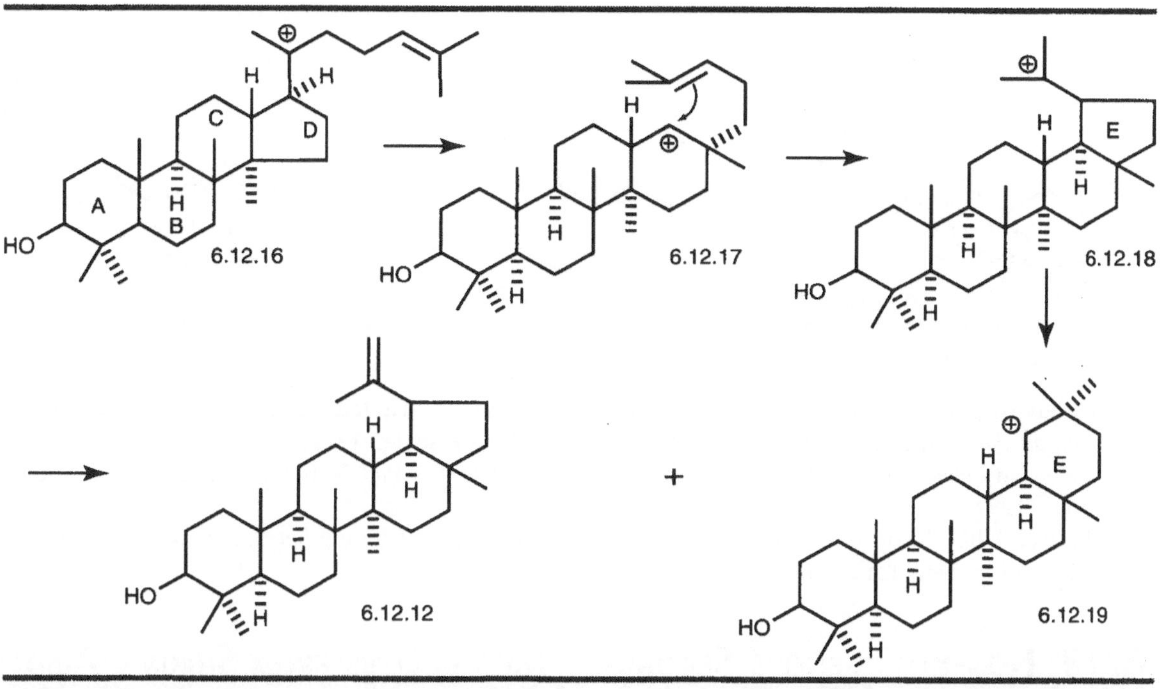

6.12.16   6.12.17   6.12.18   6.12.12   6.12.19

The carbocation 6.12.19 yields β-amyrin (6.12.14) after two successive [1,2]-hydride shifts followed by the loss of a proton. Friedelin (6.12.13) results when the ion 6.12.19 suffers a series of [1,2]-hydride and methyl shifts which shake up the entire skeleton all the way down to ring A. This is indeed a fascinating example of a synchronised chain of successive Wagner–Meerwein shifts appearing as a package! Further peripheral transformations, such as oxidation of one or more angular methyls, and introduction of hydroxyls and other changes convert each skeletal type into a variety of compounds. The books under suggested reading provide further details on the discussion above.

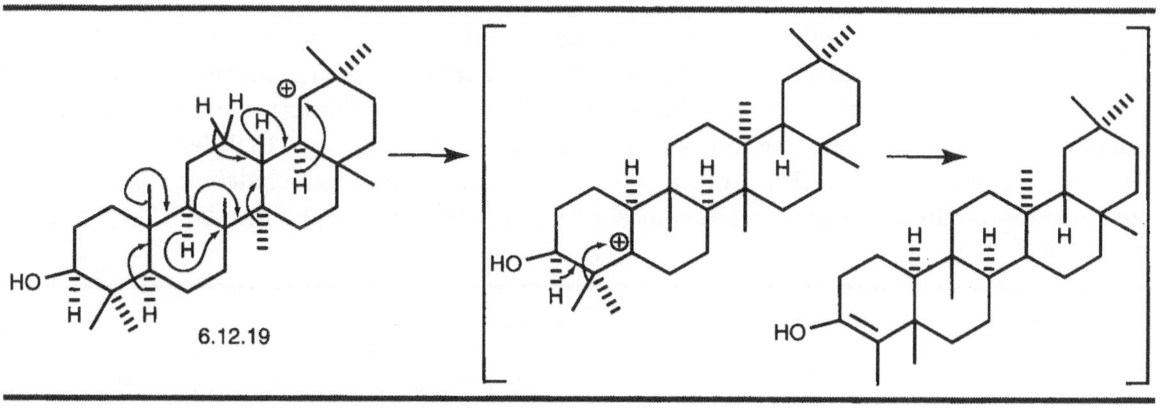

6.12.19

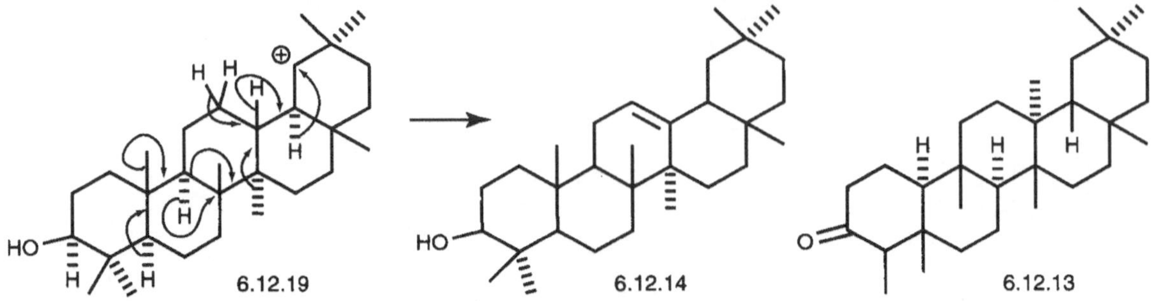

6.12.19   6.12.14   6.12.13

Suggested reading

1. Goodwin TW, Biogenesis of Terpenes and Sterols, In Ansell ME, (ed.), Rodd's *Chemistry of Carbon Compounds,* Supplement to 2nd edn, Vol.11, Amsterdam: Elsevier, 1974.
2. Dev S and Nagasampagi BA, (eds), CRC *Handbook of Terpenoids,* Vols 1 and 2, Boca Raton: CRC Press, 1989.
3. Torssell KBG, *Natural Product Chemistry—A Mechanistic and Biosynthetic Approach to Secondary metabolism,* New York: John Wiley & Sons, 1983.

6.13 NON-NITROGENOUS SECONDARY METABOLITES FROM SHIKIMIC ACID: FLAVONOIDS AND RELATED POLYPHENOLS

As mentioned in Section 6.2, shikimic acid (6.13.1) is a biosynthetic precursor of L-phenylalanine (6.13.2), which by one mode of oxidative deamination gives rise to cinnamic acid (6.13.3), the primary progenitor of coumarins, such as umbelliferone (6.13.4), lignans like pinoresinol (6.13.5) and coniferyl alcohol (6.13.6); the latter is a constituent part of the lignins. Cinnamic acid is also one of the building blocks used for the biosynthesis of flavonoids and related compounds, which form a large and biologically important group of secondary metabolites. The biosynthesis of these compounds has been studied in great detail and a wealth of information is now available not only on the organic chemical aspects but also about the enzymatic processes involved. Cinnamic acid is first converted into p-coumaric acid (6.13.7), a reaction catalysed by the enzyme cinnamate-4-hydroxylase. Combination of p-coumaroyl coenzyme A (6.13.8) with three molecules of malonyl coenzyme A (6.13.9) leads to 4,2',4',6'-tetrahydroxychalcone (6.13.10), under the influence of the enzyme chalcone synthase. The involvement of three malonyl coenzyme A units ensures a phoroglucinol-type oxygenation of ring A.

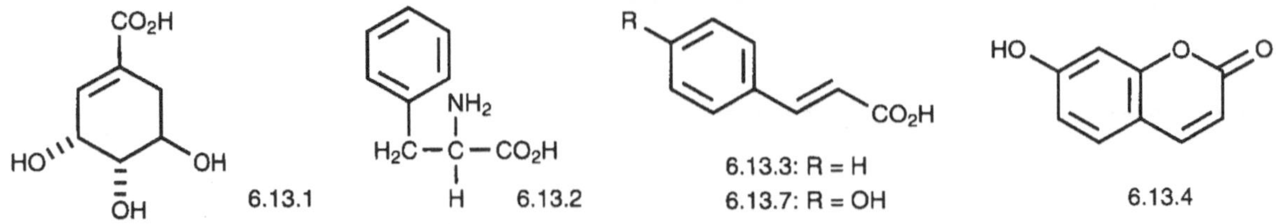

6.13.1   6.13.2   6.13.3: R = H   6.13.4
                        6.13.7: R = OH

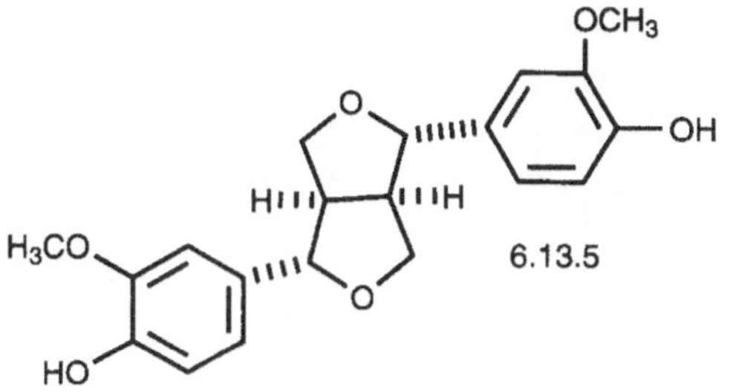

6.13.5 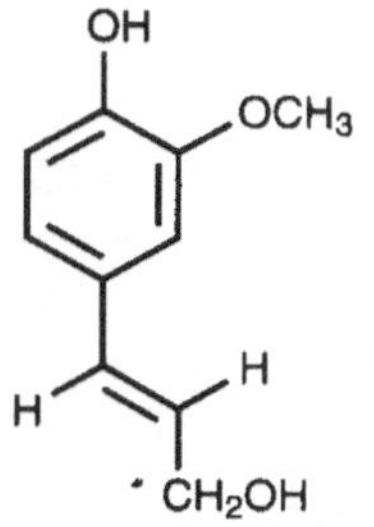 6.13.6 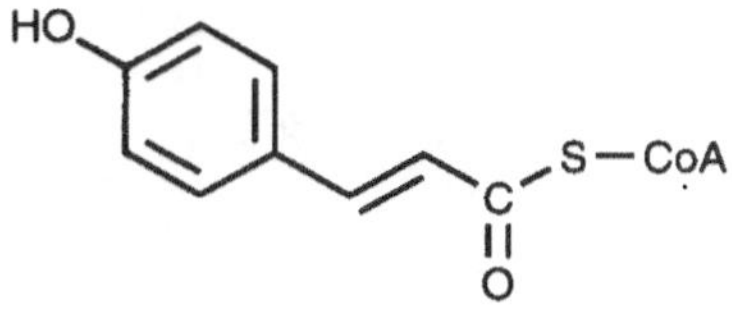 6.13.8

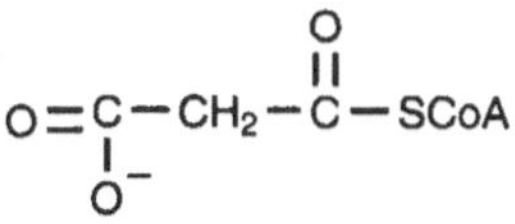

6.13.9 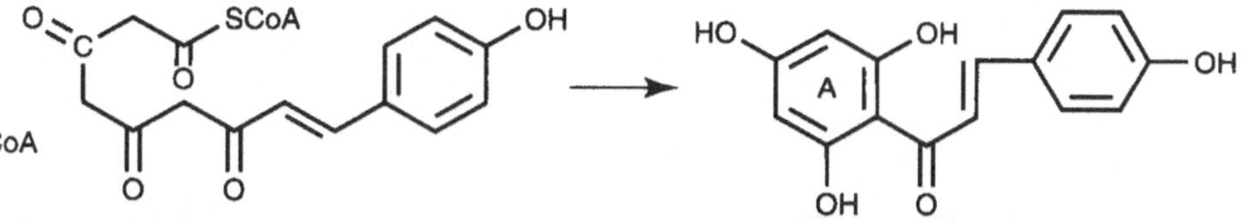 6.13.10

However, chalcones and other flavonoids with resorcinol-type oxygenation of ring A are also known and for their biosynthesis, the condesnation of 6.13.8 with 6.13.9 is catalysed by chalcone synthase acting in tandem with a NADPH dependent polyketide reductase. The chalcone thus formed is 4,2′,4′-trihydroxychalcone (6.13.11). The enzyme chalcone isomerase then converts the chalcones, 6.3.10 and 6.3.11, into the corresponding flavanones, naringenin (6.3.12) and liquiritigenin (6.3.13). The flavanones are subsequently dehydrogenated to the flavones, 6.3.14 and 6.3.15 (apigenin), by a reaction catalysed by flavone synthase, which acts in conjunction with molecular oxygen and 2-oxoglutarate; the other products in this reaction are succinate, carbon dioxide and water. It is presumed that the first step is hydroxylation of the flavanone at position 2 followed by elimination of a molecule of water. The probable mechanistic course is shown below. Hydroxylation in the side phenyl ring of naringenin brought about by the action of the enzyme flavanone 3′-hydroxylase results in eriodictyol (6.13.16). It is worth noting here that while naringenin, eriodictyol and aromadendrin (mentioned in the next paragraph) are chiral compounds, apigenin and kaempferol are achiral.

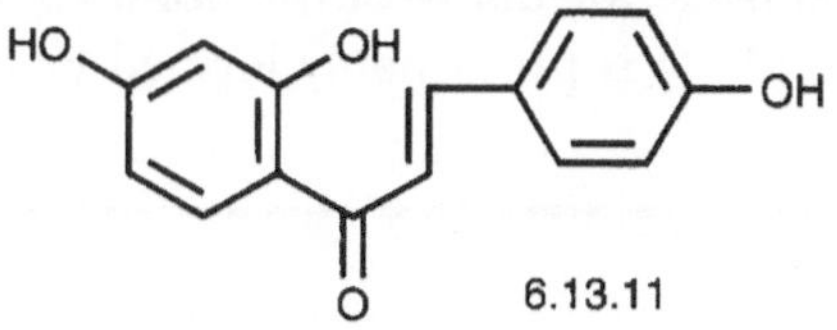

6.13.11 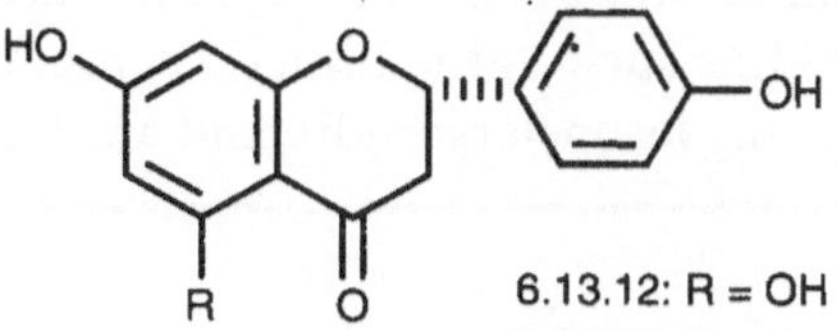 6.13.12: R = OH 6.13.13: R = H

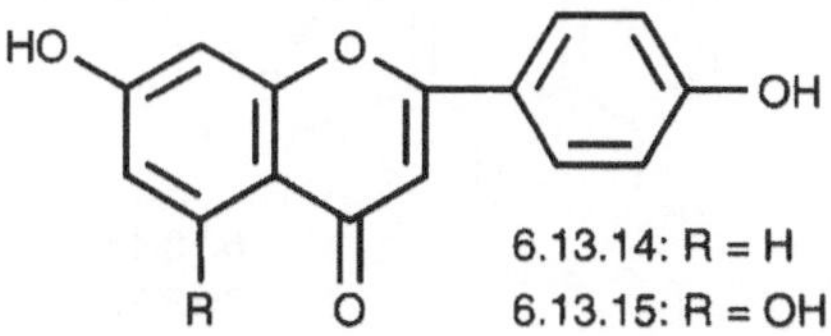

6.13.14: R = H 6.13.15: R = OH via

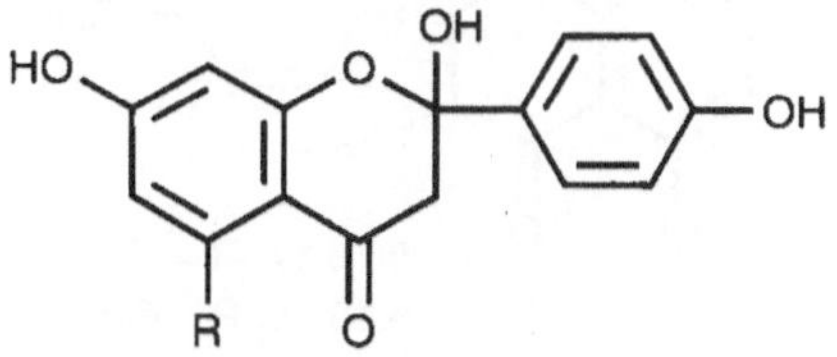

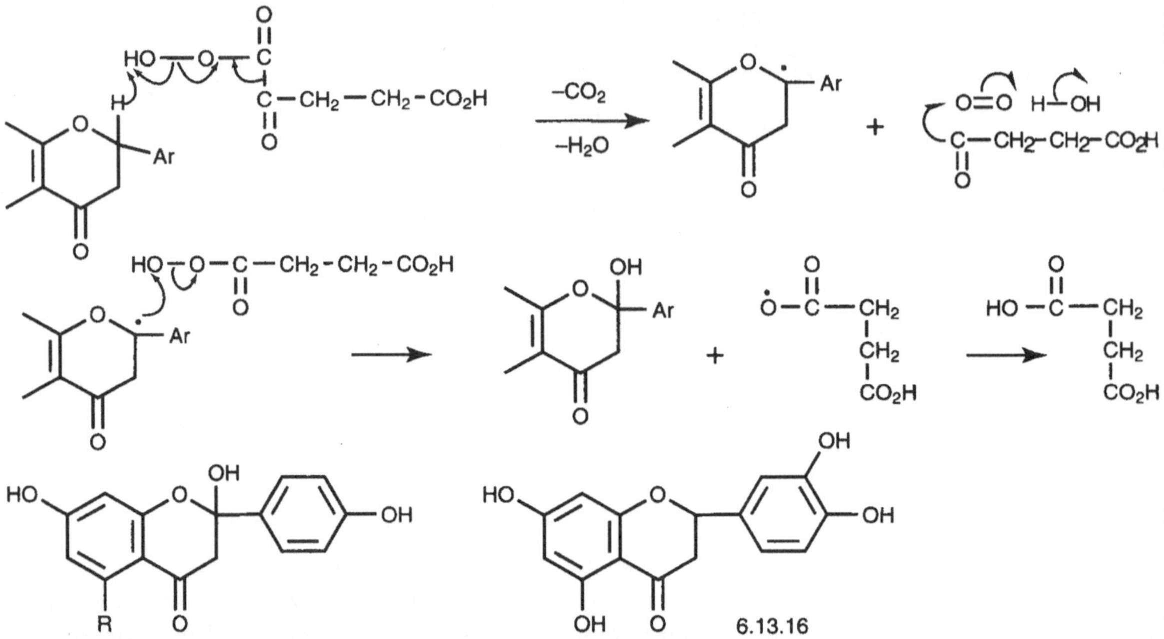

6.13.16

Another enzyme, flavanone 3-hydroxylase catalyses the conversion of naringenin into aromadendrin (6.13.17). A flavonol synthase converts the latter into the flavonol, kaempferol (6.13.18). Aromadendrin acquires an additional hydroxyl group at position 3', by the action of a flavanone 3' monooxygenase to form taxifolin (6.13.19). The widely occurring flavaonol, quercetin (6.13.20) is obtained from taxifolin. It undergoes further transformations, such as O-methylation of one or more hydroxyl groups, glycosidation and further nuclear hydroxylation to yield derivatives such as rhamnetin (6.13.21), quercetrin (6.13.22), gossypetin (6.13.23) and myricetin (6.13.24). Aromadendrin and taxifolin are the precursors, respectively, of leucopalargonidin (6.13.25) and leucocyanidin (6.13.26), which, in turn, are the precursors of the corresponding anthocyanidins, namely, pelargonidin (6.13.27) and cyanidin (6.13.28), as well as of the catechins, afzelechin (6.13.29) and epi-catechin (6.13.30). As mentioned elsewhere in this book, the glycosides of anthocyanidins, known as anthocyanins, are responsible for many of the floral colours. The proanthocyanidins such as the dimeric compound (6.13.31) is presumably formed by oxidative coupling involving lecopelargonidin and afzelechin.

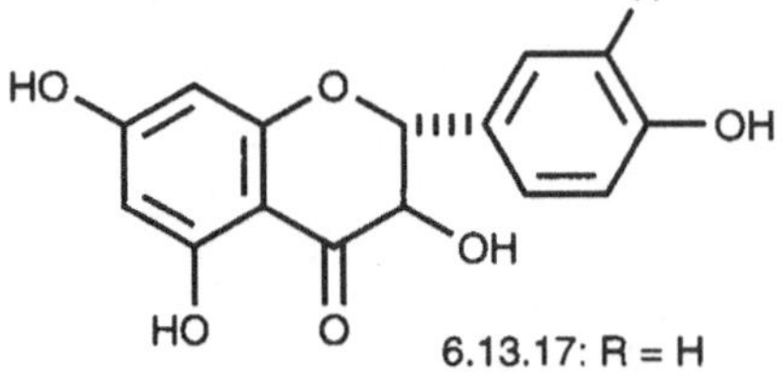

6.13.17: R = H
6.13.19: R = OH

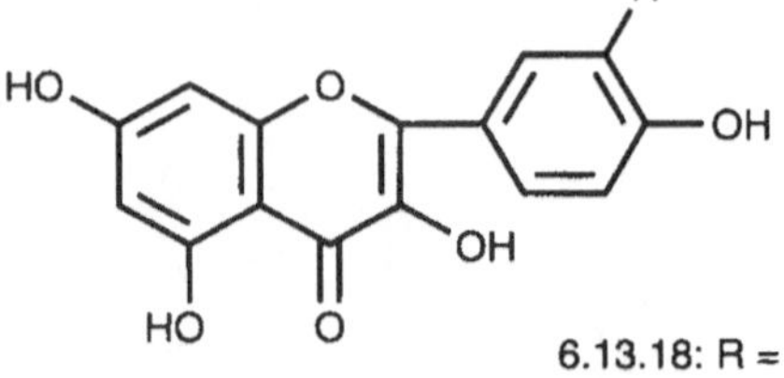

6.13.18: R = H
6.13.20: R = OH

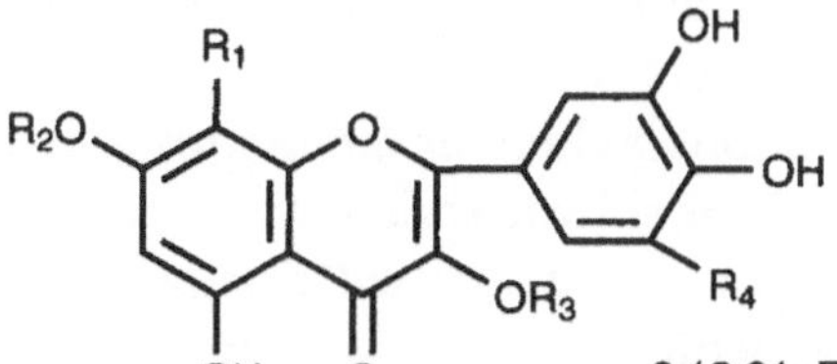

6.13.21: $R_1 = R_3 = R_4 = H$; $R_2 = CH_3$
6.13.22: $R_1 = R_2 = R_4 = H$; $R_3 = $ Rhamnose
6.13.23: $R_1 = OH$; $R_2 = R_3 = R_4 = H^-$
6.13.24: $R_1 = R_2 = R_3 = H$; $R_4 = OH$

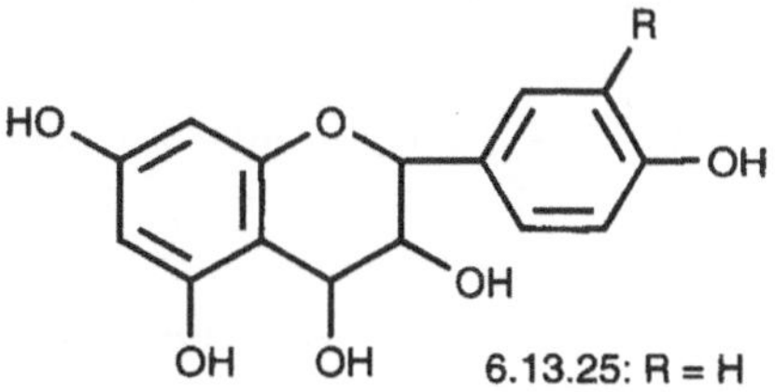

6.13.25: $R = H$
6.13.26: $R = OH$

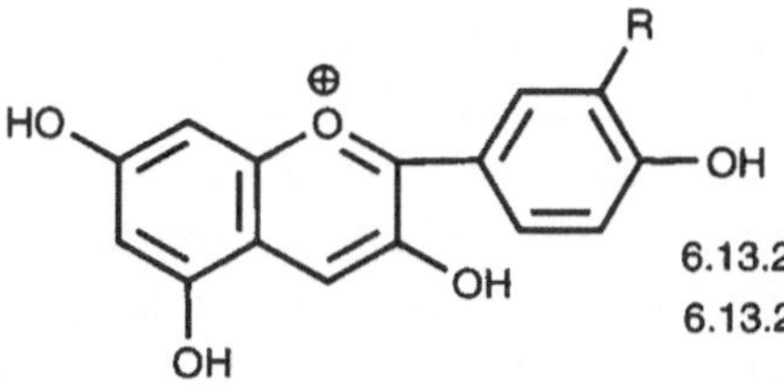

6.13.27: $R = H$
6.13.28: $R = OH$

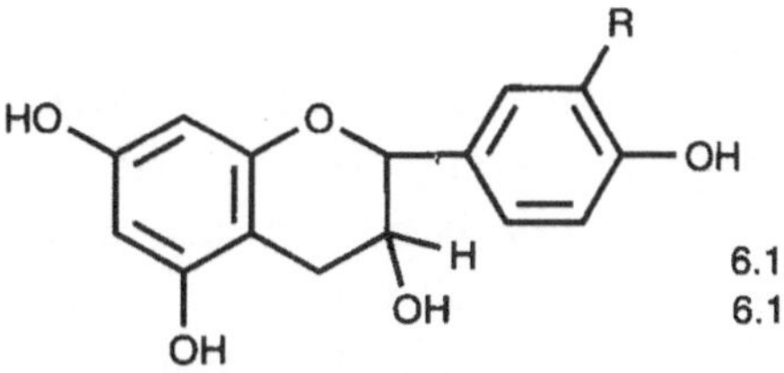

6.13.29: $R = H$
6.13.30: $R = OH$

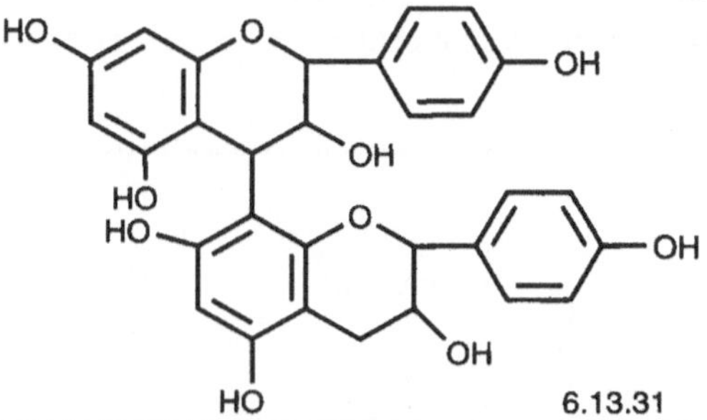

6.13.31

Chalcones are also precursors of the aurones which are not as widely occurring in nature as the other flavonoids. However, they do contribute to the colours of flowers of some plants of the Asteraceae family such as Coreopsis and Bidens. As mentioned in Chapter 1, they arise presumably via chalcone epoxides. For example, the aurone 6.13.32 can be formed from the chalcone 6.13.10 as shown below.

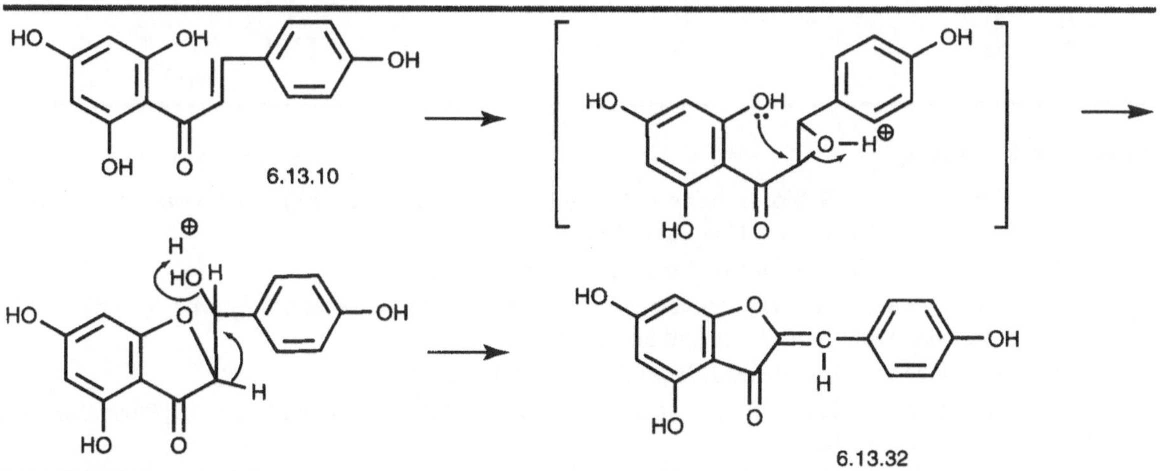

6.13.10

6.13.32

Within each flavonoid type, such as chalcone, aurone, flavanone, flavone and flavonol, peripheral modifications occur, presumably after the skeletal structures are built, to form the wide variety of flavonoids of natural occurrence. These modifications include, as mentioned under quercetin, *O*- and *C*-methylation, *O*- and *C*-glycosylation, additional hydroxylation of either of the two benzene rings and introduction of prenyl groups. In the last mentioned case, another building block, namely, isopentenyl pyrophosphate is also used, which would require catalysis by a prenyl transferase. In many of these terminal modifications, P 450 hydroxylases play a crucial role. *O*- and *C*-Methylations are catalysed by methyl transferases which require as a cofactor (S) adenosylmethionine.

As mentioned earlier, cinnamic acid is also a precursor for coumarins. For example, in the biosynthesis of umbelliferone (6.13.4), a key step is the *ortho*-hydroxylation of *p*-coumaric acid (6.13.7). Though, it is thought that this step is brought about by a P 450 oxidase, the biochemical process of this reaction is not yet fully understood. The formation of the lactone ring has to be preceded by a geometrical inversion of the double bond in the 2,4-dihydroxycinnamic acid (6.13.33). Stilbenes also arise from *p*-coumaroyl coenzyme A and three units of malonyl coenzyme A as shown below. The final product is resveratrol (6.13.34). The enzyme involved has been designated as resveratrol synthase.

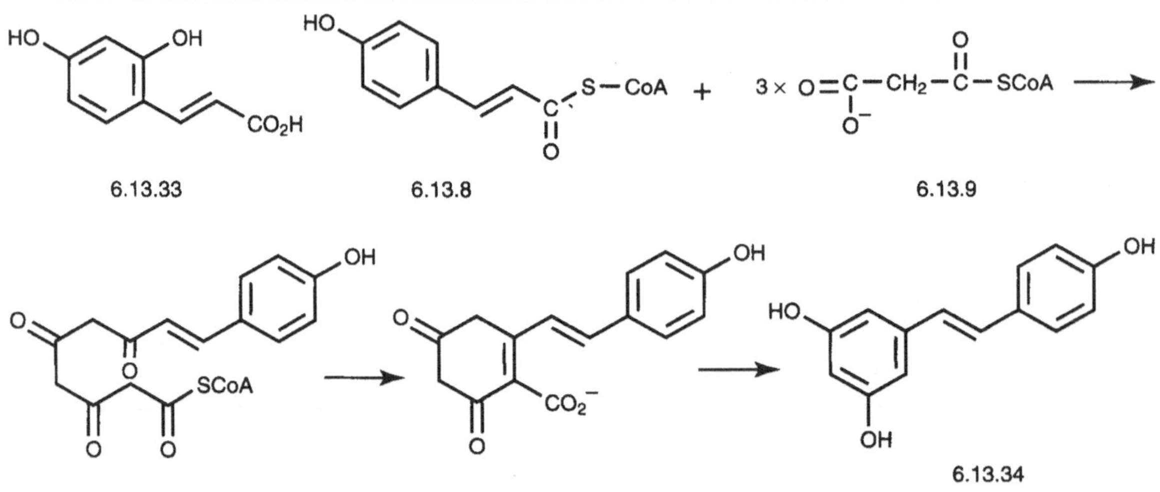

6.13.33   6.13.8   6.13.9

6.13.34

Suggested reading

1.   Heller W and Forkman G, Biosynthesis of flavonoids, In Harborne JB, (ed.), *The Flavanoids: Advances in Research since 1980*, London: Chapman & Hall, 1988.
2.   Stafford HA, *Flavonoid Metabolism*, Boca Raton: CRC Press, 1990.
3.   Torsell KBG, *Natural Product Chemistry: A Mechanistic and Biosynthetic Approach to Secondary Metabolism*, New York: John Wiley and Sons, 1983.
4.   Shirley BW, Flavonoid Biosynthesis, *Plant Physiol.*, 2001, 126: 485–93.
5.   Bourgaud F, Hehn A, Larbat R, Doerper S, Gontier E, Kellner S and Matern U, *Phytochemistry Reviews*, 2006, 5: 293–308.

6.14 BIOSYNTHESIS AND TRANSFORMATIONS OF ISOFLAVONES

Flavone and isoflavone biosynthetic pathways branch off at the flavanone stage thus providing for the formation of two parallel streams of flavonoids possessing isomeric skeletal structures. The enzyme responsible for the transformation of a flavanone to an isoflavone has been identified and characterised. This enzyme, designated as isoflavone synthase, is is a cytochrome P 450 monooxygenase and requires NADPH and molecular oxygen for bringing about the oxidative rearrangement. Thus, (2S) naringenin (6.14.1) gets converted into the isoflavone, genistein (6.14.2) via 2-hydroxydihydrogenistein (6.14.3). This compound, though not very stable, has been isolated and its structure determined by spectroscopic methods. For the conversion of 6.14.3 into 6.14.2, another enzyme, named isflavanone dehydratase, is required; this enzyme has also been obtained in the pure state. Considering that the isoflavone synthase (it would be more appropriate to call it 2-hydroxyisoflavanone synthase) requires molecular oxygen and NADPH, the following mechanism for the rearrangement seems to be probable; it slightly differs from that suggested in the literature. The first step is the formation of a radical by the removal of a hydrogen atom from position 3 by molecular oxygen, which is bound to the enzyme through an iron atom.. The driving force is probably the stabilisation of the radical by resonance as shown below. The aryl group then migrates as a radical thus creating a new radical at position 2. Attack at this point by the peroxide radical bound to the enzyme followed by hydride ion transfer from NADPH leads to 6.14.3. Biosynthetic experiments involving isoflavone synthase, NADPH and 18O$_2$ have shown that the oxygen label is entirely on the 2-hydroxyl group of 6.14.3.

For the formation of daidzein (6.14.4), via the corresponding 2-hydroxyisoflavanone (6.14.5), the flavanone precursor is (2S) liquiritigenin (6.14.6) which is formed from the corresponding chalcone, isoliquiritigenin (6.14.7), by the action of the enzyme chalcone isomerase. For the formation of isoliquiritigenin itself from p-coumaroyl coenzyme A and malonyl coenzyme A, the concerted action of two enzymes, namely, chalcone synthase and chalcone reductase is required. Incidentally, daidzein is the precursor of the more complex isoflavonoids which occur in leguminous plants and the enzyme chalcone reductase is legume specific. Under the catalytic influence of the enzyme, isoflavone 7-O-methyl transferase, the 2-hydroxyisoflavanones, 6.14.3 and 6.14.5 get converted into 6.14.8 and 6.14.9 respectively. These compounds, on dehydration, yield prunetin (6.14.10) and isoformononetin (6.14.11). respectively, On the other hand, the enzyme isoflavone 4'-O-methyl transferase, in conjunction with a dehydratase, is responsible for the production of biochanin A (6.14.12) and formononetin (6.14.13), from 6.14.3 and 6.14.5, respectively. These methyl transferases require (S) adenosylmethionine as the cofactor.

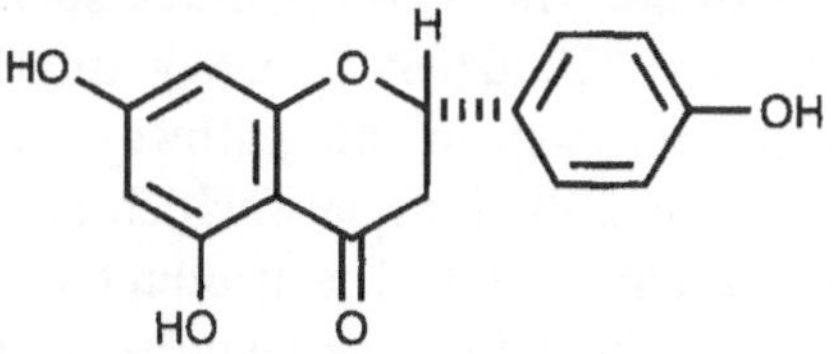

6.14.1 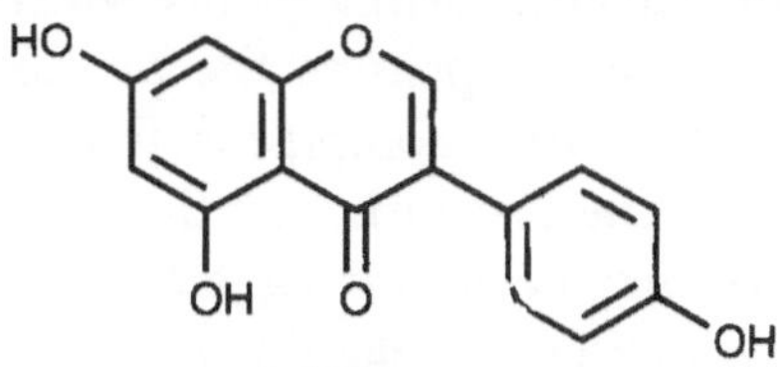   6.14.2

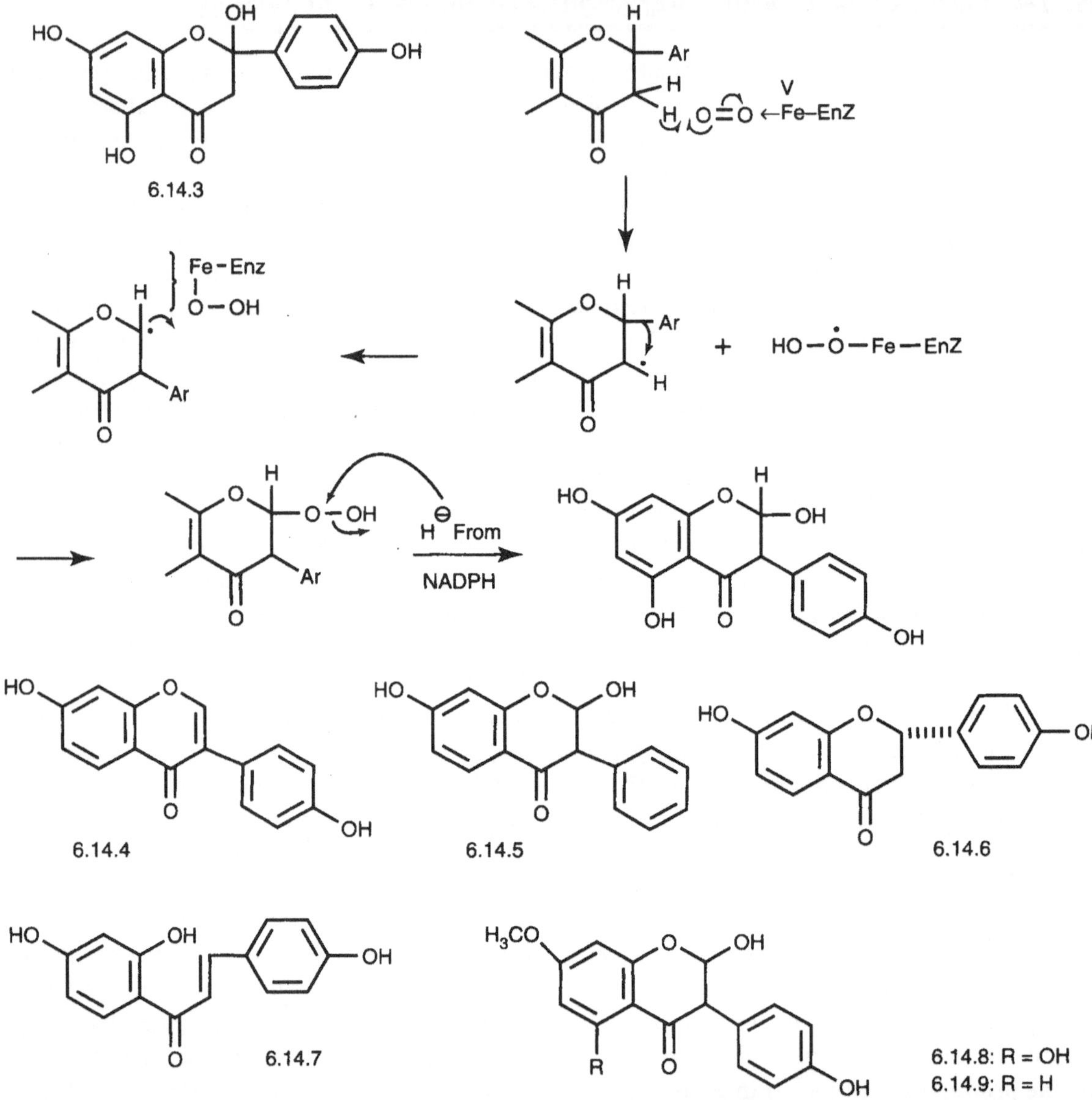
6.14.3
Fe-Enz
O-OH
Ar
HO-O-Fe-EnZ
Fe-EnZ
H From
NADPH
6.14.4
6.14.5
6.14.6
6.14.7
H3CO
6.14.8: R = OH
6.14.9: R = H

Isoflavonoid phytoalexins (see next chapter for a more detailed account of phytoalexins) are crucial to the survival of the plants which produce them. These compounds, such as pisatin (6.14.14), for example, possess the pterocarpan skeleton. Daidzein is a key intermediate in the production of these complex isoflavonoids. In the first step in this pathway, daidzein gets hydroxylated at the 2' position by the action of isoflavone 2'-hydroxylase which is also a P 450 monooxidase enzyme requiring molecular oxygen and NADPH. The product (6.14.15) then undergoes reduction to yield 2'-hydroxydihydrodaidzein (6.14.16). The enzyme responsible for this reduction, isoflavone reductase, brings about a stereospecific reduction of the double bond.

It may be noted here that in the laboratory it is not easy to reduce a conjugated double bond as in isoflavones. In the next step, the carbonyl group is reduced to form the isoflavonol (6.14.17); the reductase catalysing this reaction is NADPH dependent.The final step in the construction of the pterocarpan structure is the dehydrative cyclisation involving the 4- and 2'- hydroxyl groups. The product is 6a, R,11a R(–) 3,9-dihydroxypterocarpan (6.14.18). This pterocarpan is enzymatically hydroxylated at position 6a to form glycinol (6.14.19); the enzyme responsible for this reaction is also a microsomal monooxygenase, which has been, for example, isolated from soy beans. In a manner similar to that described above, the isoflavone, formononetin (6.14.3) gets converted into medicarpin (6.14.20).

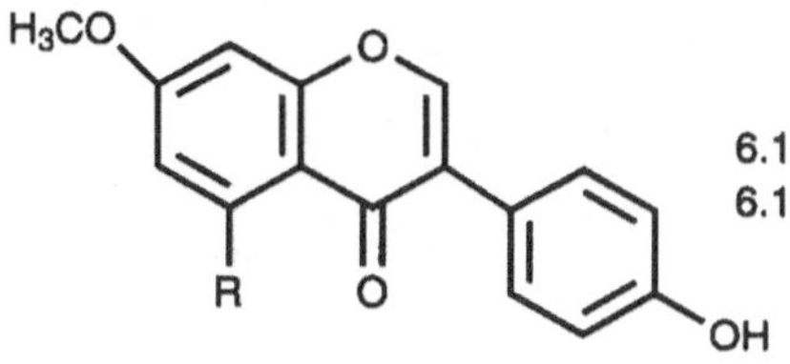

6.14.10: R = OH
6.14.11: R = H

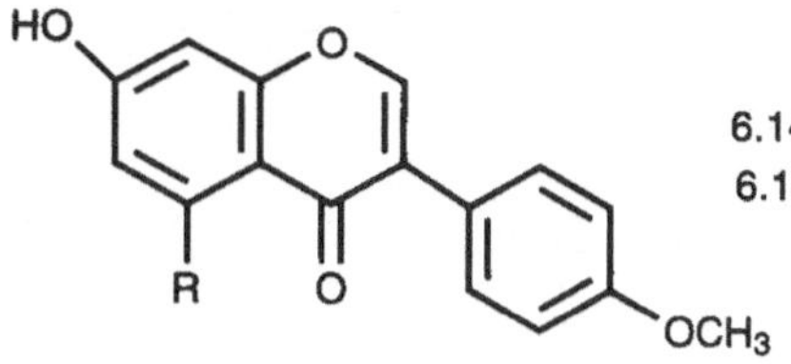

6.14.12: R = OH
6.14.13: R = H

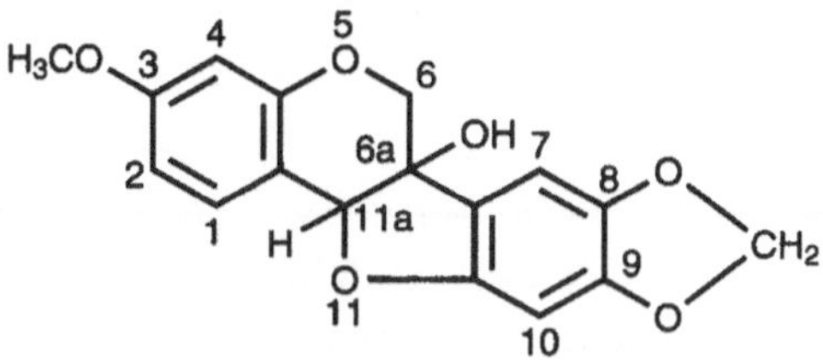

6.14.14

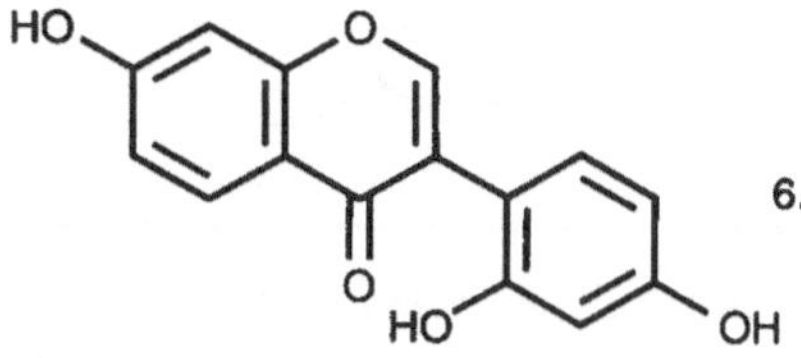

6.14.15

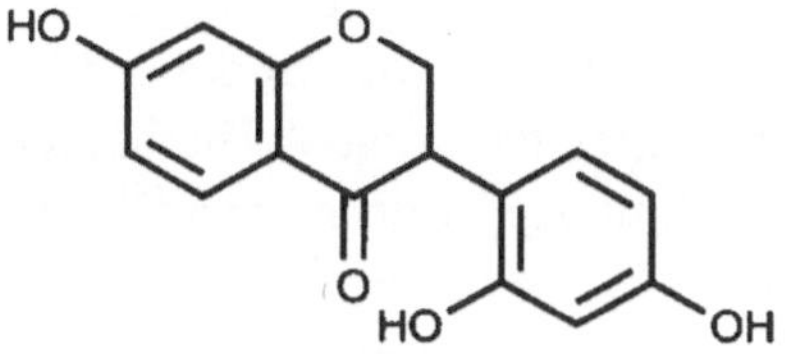

6.14.16

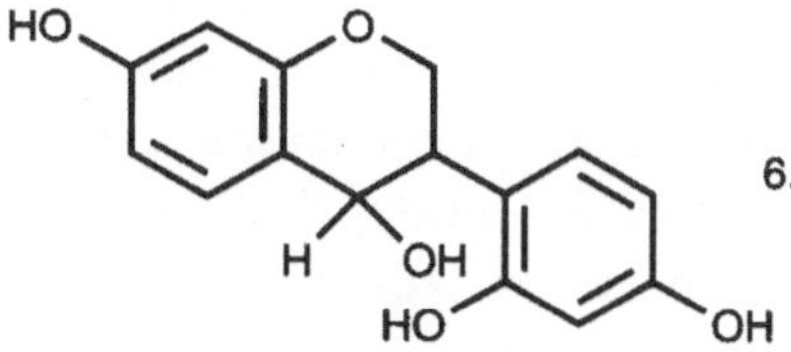

6.14.17

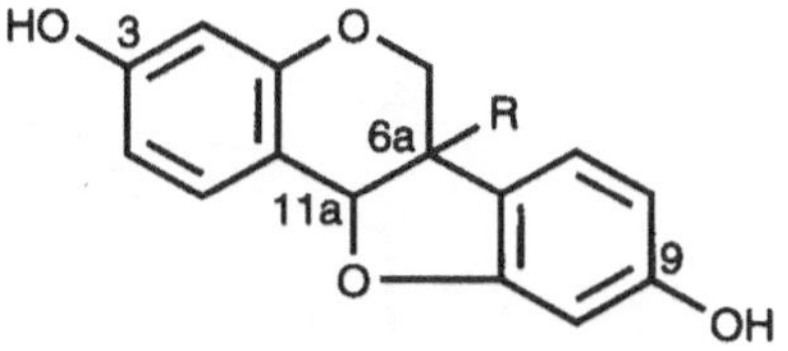

6.14.18: R = H
6.14.19: R =OH

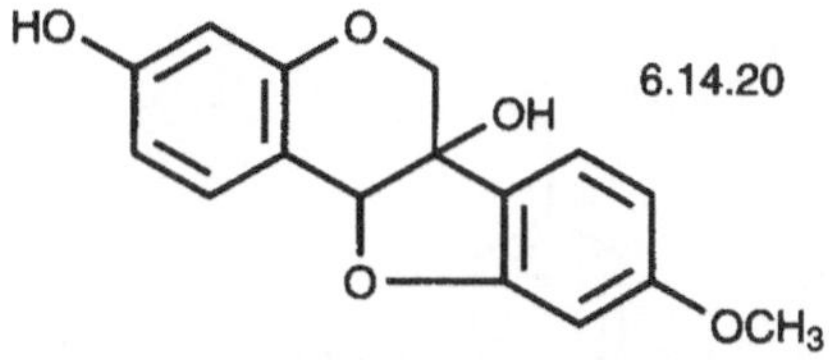

6.14.20

Other biochemical transformations such as additional hydroxylation, partial O-methylation, introduction of prenyl groups, and cyclisation involving a prenyl group and a phenolic hydroxyl *ortho* to it are subsequent reactions which are responsible for the formation of compounds such as (–)maackiain (6.14.21), pisatin (6.14.14), glycollidin (6.14.22), tuberosin (6.14.23), and glyceolin-1 (6.14.24). Coumestrol (6.14.25), which is not a phytoalexin, is presumably formed from daidzein via the pterocarpen (6.14.26). For details regarding the biosynthesis of other isoflavones and the rotenoids, the books mentioned under suggested reading may be referred to.

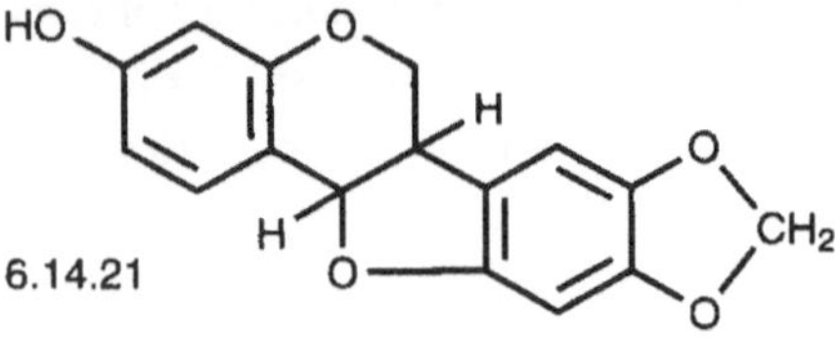

6.14.21

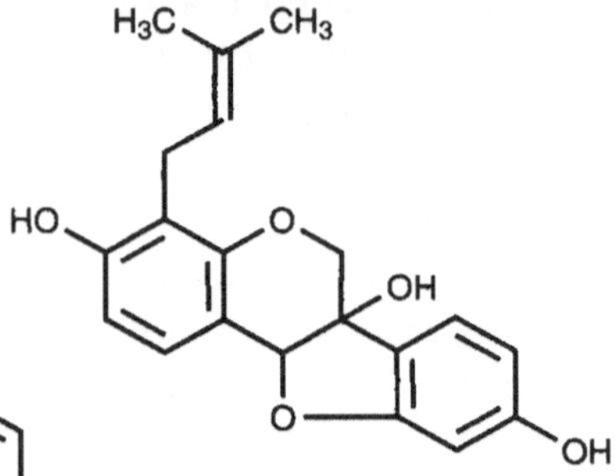

6.14.22

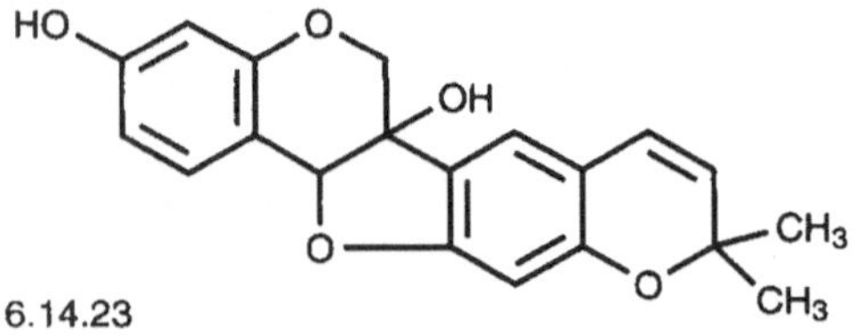

6.14.23

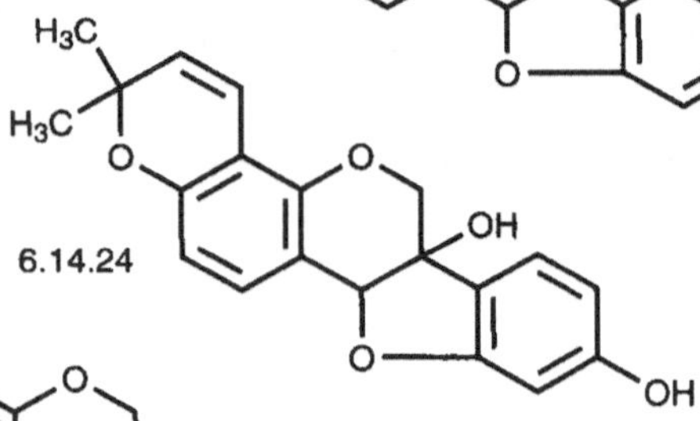

6.14.24

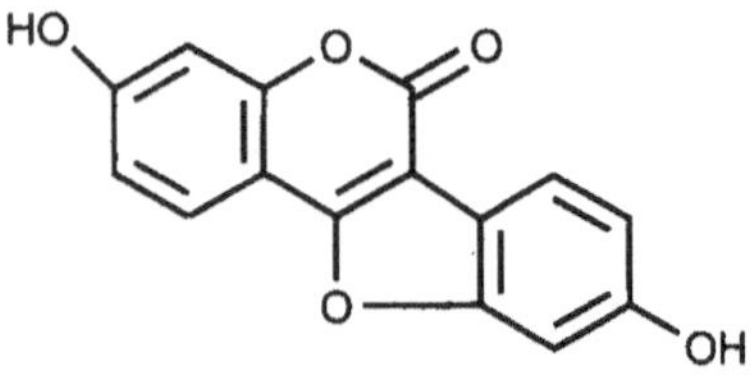

6.14.25

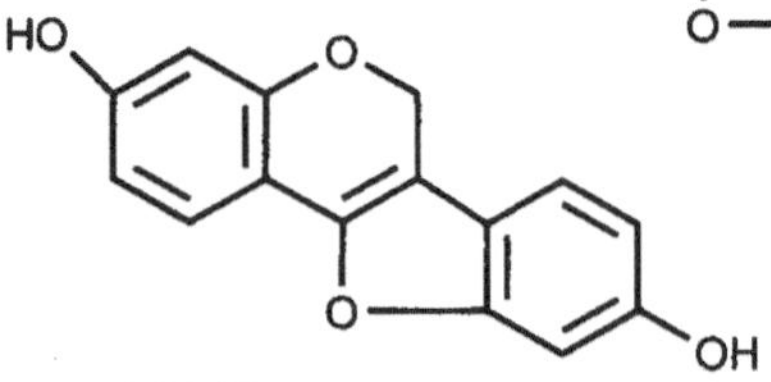

6.14.26

The information available on the biosynthesis of the neoflavanoids, such as dalbergin (6.14.27) and related compounds, is scanty in contrast to the wealth of data on flavone and isoflavone biosynthesis. Some experiments have been done on the biosynthesis of calophyllolide (6.14.28) using labelled phenylalanine, which provides the phenyl group at position 4. The condensed benzene ring arises from malonyl coenzyme A, as in the case of the other flavonoids. The biosynthetic pathway suggested for callophyllolide formation is shown below. Appropriate modifications in this pathway can lead to dalbergin and other neoflavonoids.

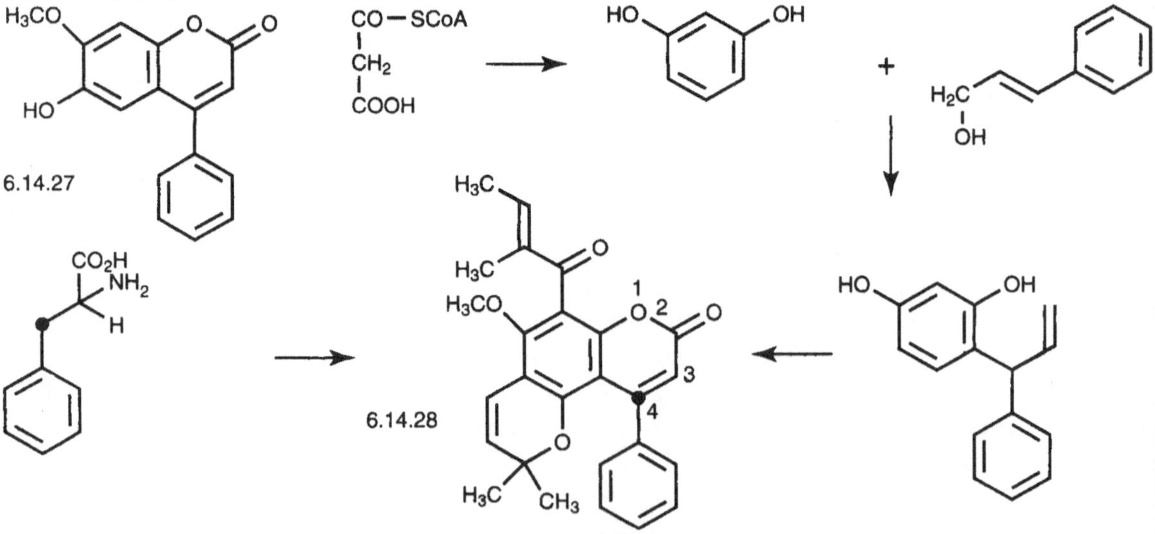

6.14.27

6.14.28

6.14.27

Suggested reading

1. Stafford HA, *Flavonoid Metabolism*, Boca Raton: CRC Press, 1990.
2. Seiger DS, *Plant Secondary Metabolism*, Berlin: Springer, 1998.
3. Hakamatsuka T and Ebizuka Y, Biosynthesis and natural functons of Pueraria isofl avonoids, In Keung WM, (ed.), *Pueraria-the genus Peuraria*, London: Taylor and Francis, 2002.
4. Aoki T. Akashi T and Ayabe Si, *J. Plant Res.*, 2000, 113: 475–88.
5. deVries GE, Isoflavone biosynthesis, *Trends in Plant Science*, 2000, 5: 190.
6. Yu O, Shi J, Hession AO, Maxwell CA, McGonigle B, Odell JT, *Phytochem.*, 2003, 63: 753.

6.15 BIOSYNTHESIS OF ANTHRAQUINONES

Anthraquinones, which exhibit yellow to red colours, occur in fungi, lichens and in some higher plants. Recent research studies have shown that two biosynthetic pathways operate in nature to elaborate these compounds. The polyketide pathway in which acetyl coenzyme A is the basic building block is used for synthesising anthraquinones in which both the terminal benzene rings (A and C) are substituted by hydroxyl and, in some cases, methyl and other groups. This pathway operates in fungi, lichens and in some higher plants belonging to the families Leguminosae, Rhamnaceae and Polygonaceae. On the other hand, the shikimic acid pathway is utilised for constructing anthraquinones having substituents on only one ring and this is the favoured pathway in plants of the Rubiaceae family. In the polyketide pathway, seven malonyl coenzyme molecules and one acetyl coenzyme molecule are condensed together in a linear fashion to form the octaketide (6.15.1). The subsequent conversion of this key intermediate depends on several factors and one of them is the mode of folding of the polyketide chain. For example, the arrangement shown in 6.15.2 is conducive for the formation of the anthraquinone, chrysophanol (6.15.3), whereas the alternative arrangement (6.15.4) leads to aloesaponarin (6.15.5). Chrsophanol occurs in species of Rumex (Polygonaceae) and Rhamnus (Rhamnaceae) and it also co-occurs with aloesaponarin in the roots of *Aloe saponaria* (Liliaceae). The enzymes involved are reductases, aromatases and cyclases.

$$H_3C-CO-SCoA \; + \; 7 \times \overset{-}{O_2}C-CH_2-CO-S-CoA \; \longrightarrow$$

6.15.1

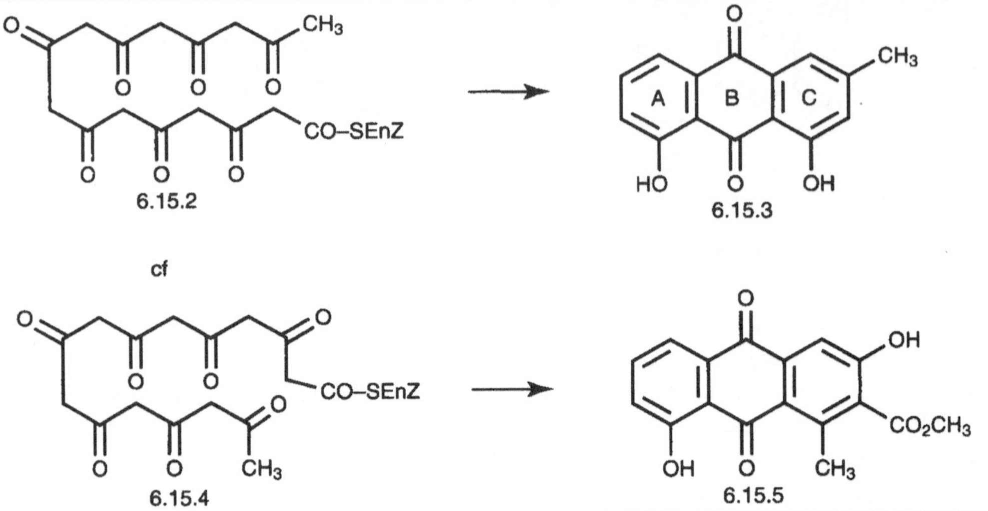

6.15.2

6.15.3

cf

6.15.4

6.15.5

The sequence of steps involved in the shikimic acid pathway has been studied in depth using incorporation experiments with labelled precursors in plants of the Rubiaceae family known to produce anthraquinone pigments. As a result of these studies it is now known that rings A and B of a typical Rubiaceae anthraquinone such as alizarin (6.15.6) or 1-hydroxy-2-methylanthraquinone (6.15.7) are derived from shikimic acid via isochorismic acid (6.15.8) and alpha ketoglutaric acid (6.15.9), whereas ring C is formed by the involvement of dimethylallyl pyrophosphate (6.15.10). An outline of this biosynthetic pathway is given below. Key intermediates are o-succinoylbenzoic acid (6.15.11), 1,4-dihydroxy naphthalene-2-carboxylic acid (6.15.12) and its 3-prenyl derivative (6.15.13). A probable mechanism for the formation of the anthraquinone (6.15.7) from (6.15.13) is shown below.

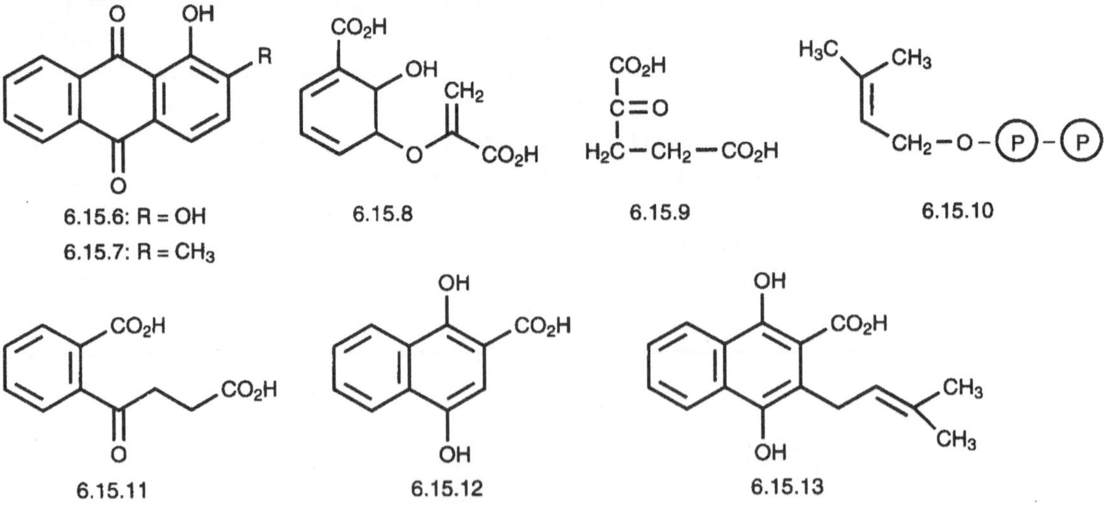

6.15.6: R = OH
6.15.7: R = CH₃

6.15.8

6.15.9

6.15.10

6.15.11

6.15.12

6.15.13

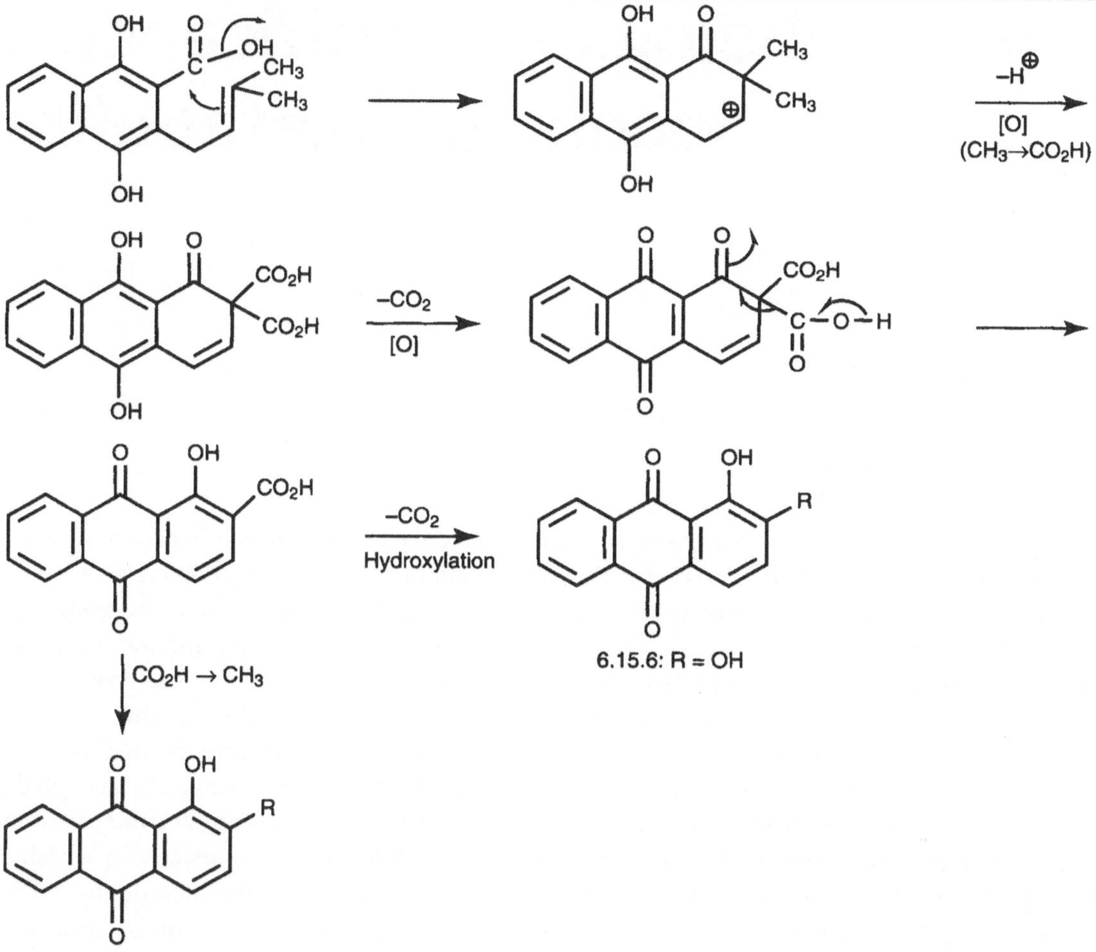

6.15.6: R = OH

6.15.7: R = CH$_3$

Suggested reading

1. Seigler DS, *Plant Secondary Metabolism*, Norwell: Kluwer Academic Publishers, 1995.
2. Han YS, Heijden R, Lefeber AW, Erkelens C and Verpoorte R, *Phytochem.*, 2002, 59: 45–55.

Chapter

7

Biological Significance of Secondary Metabolites

7.1 INTRODUCTION

Not long ago, the large number of organic compounds occurring in nature, other than the primary metabolites and the semantides, were dismissed as waste products of metabolism. However, this description was never fully accepted and there have been, from time to time, suggestions that these compounds do have well-defined biological functions. However, convincing experimental evidences in support of such suggestions have become available only during the past few years. As a result of such studies, the secondary metabolites, particularly of plant origin, have ceased to be merely chemical curiosities; biologists are increasingly taking an interest in them as they help us understand several nuances in physiological functions. It is now generally accepted that while they may not be needed for the primary, primitive and fundamental molecular phenomena inseparable from any form of life, the secondary metabolites endow the species in which they occur with some of their unique and species-specific characteristics. These include the ability to interact in a specific manner with other forms of life in any given environment; in other words, the capacity to 'articulate' in a unique and distinctive manner.

This realisation, backed up by substantial experimental evidence, has thus given a new meaning to the scientific study of secondary metabolites. It is no longer satisfying to know the structure and stereochemistry, however complex and intriguing, of a newly isolated naturally occurring compound and to design and execute ingenious methods of synthesis of these compounds in the laboratory; one would like to know how, when and where they are synthesised in the plant or animal in which they occur, what their functions are and how they perform such functions. The picture available at present is by no means clear but the very nebulous nature of it makes the subject interesting and intriguing. In the following pages we will be discussing some of these fascinating aspects which connect organic chemistry with the life sciences.

During the past decade chemical ecology has emerged as a unique and distinct branch of life sciences at the interphase of chemistry and biology. It is amazing that nature uses a variety of chemically related signals not only for communication within a species but also for interactions between different kinds of organisms in an ecological system. Two such phenomena will be briefly highlighted in this section. One is bioluminescence. Everyone who has seen a firefly at night is fascinated by the yellow-green light emitted by this tiny creature. Fireflies as well as other living organisms which exhibit bioluminescence, such as certain marine animals, annelid worms, mollusks and insects, do so for different reasons. Bioluminescent fish, which are common in deep seas, do so to enable members of a species to recognise themselves in the darkness. One cannot help exclaiming that 'this is cool'!. The light signal may also be used as a sexual attractant, to warn predators and to attract preys. Different organisms use different substances for producing light; these substances are collectively known as luciferins. For example, the firefly luciferin is (4R) 2-(6-hydroxybenzothiazol-2-yl)-1,3-thiazoline 4-carboxylic acid (7.1.1). It is interesting to note that the (S) enantiomer of the compound is ineffective as a light emitter. For light emission the luciferin needs an enzyme, luciferase, adenosine triphosphate, magnesium ion and oxygen. At the end of a two-step reaction, luciferin gets oxidised to oxylucferin (7.1.2), with the emission of carbon dioxide and, more importantly, light.

7.1.1

7.1.2

Another interesting phenomenon is the one exhibited by the touch-me-not plant, *Mimosa pudica*, and several other plants of the leguminoseae family. While the Mimosa plant is sensitive to touch, other plants of this family behave as if they possess a biological clock to regulate their waking and sleeping periods! The chemical mediators of this phenomenon called nyctinasty were first studied by Schildnecht and his co-workers who isolated compounds which they named turgorins. Since then, several important discoveries have been made concerning this phenomenon, particularly as a result of the continuous studies of Ueda, Yamamura and their co-workers. In passing it may be mentioned that this phenomenon was known to the ancients. The observation that the leaves of the tamarind tree, *Tamarindus indica*, go to sleep at night was recorded by Androsthenes in 325 BC! The results of recent studies can be summarised as follows. Leaf movements in plants exhibiting nyctinasty (such plants include species of the genera, Mimosa, Lespedeza, Leucaena, Phyllanthus, Sesbania, Cassia and Albizzia), contain two types of compounds, one, a leaf-closing substance and the other leaf-opening compound. The balance between these two substances is crucial for the leaf remaining either closed or open. This balance is controlled by a biological clock which s operated by the release of an enzyme, a beta-glucosidase. Each plant has its own pair of leaf-closing and leaf-opening compounds which are effective only in that plant. They are effective at very low concentrations—as low as 10^{-6}–10^{-7} M. The majority of these plants close their leaves at sunset and open again early in the morning. Some of the leaf-closing substances which have been isolated and characterised are potassium chelidonate (7.1.3) (from *Cassia occidentalis* and *C.mimosoidae*), potassium D-idorate (7.1.4) (from *Lespedeza cuneata*), phyllanthurinolactone (7.1.5) (from *Phyllanthus urinaria*), potassium 6-*O*-β-glucopyranosylgentisate (7.1.6) (from *Mimosa pudica*) potassium β-D-glucopyranosyl-12-hydroxyjasmonate (7.1.7) (from Albizzia sp.) and potassium 2,3,4-trihydroxy-2-methylbutanoate (7.1.8) (from Leucaena sp). Leaf-opening substances include calcium 4-*O*-β glucopyranosyl *cis*-*p*-coumarate (7.1.9) (from *Cassia mimosoidae*), potassium lespedezate (7.1.10) (from *Lespedeza cuneata*), phyllurine (7.1.11) (from *Phyllanthus urinaria*), mimopudine (7.1.12) (from *Mimosa pudica*) and the *cis*-*p*-coumaroyl derivative (7.1.13) (from Albizzia sp).

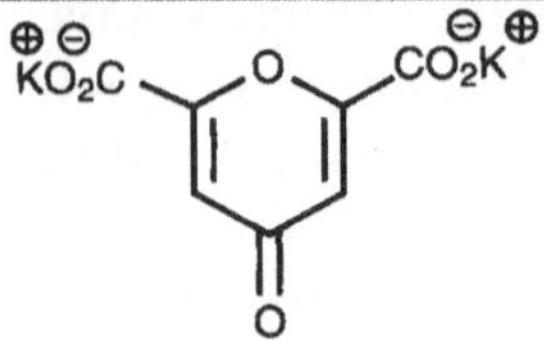

7.1.3 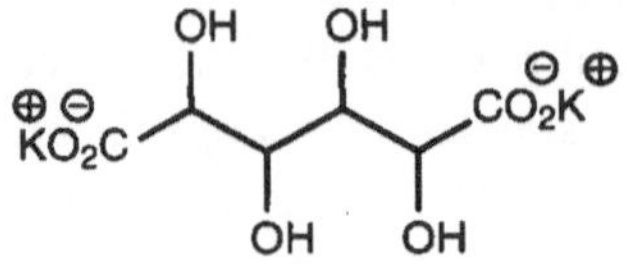   7.1.4 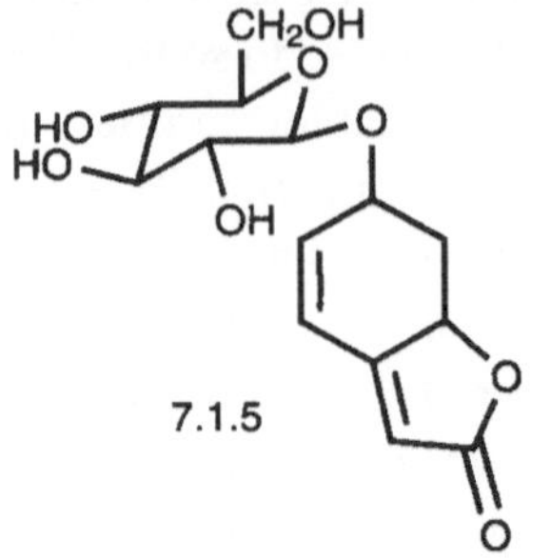   7.1.5 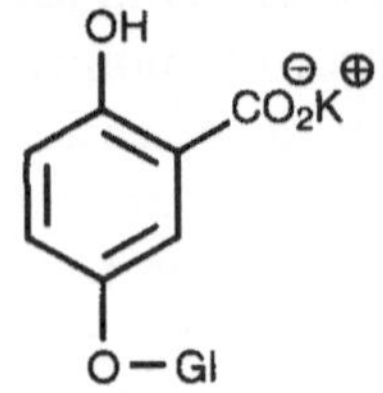   7.1.6

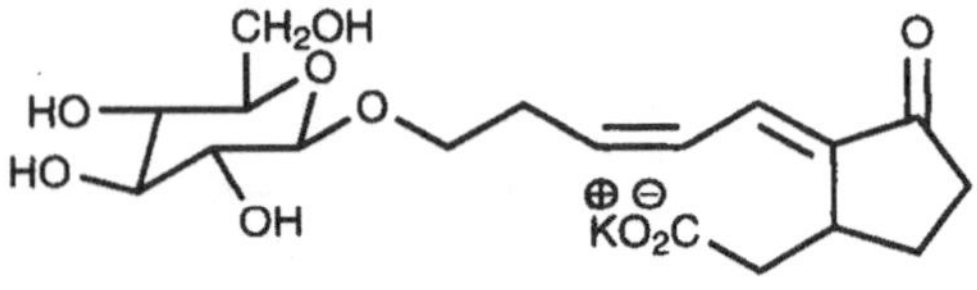

7.1.7 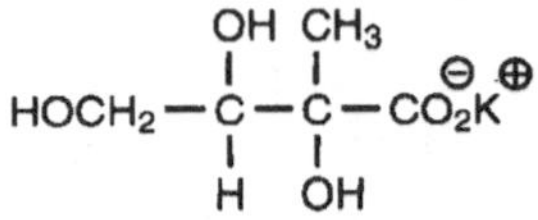   7.1.8

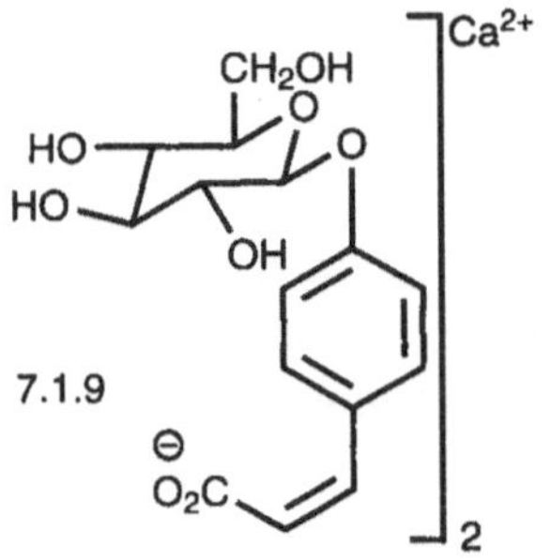

7.1.9 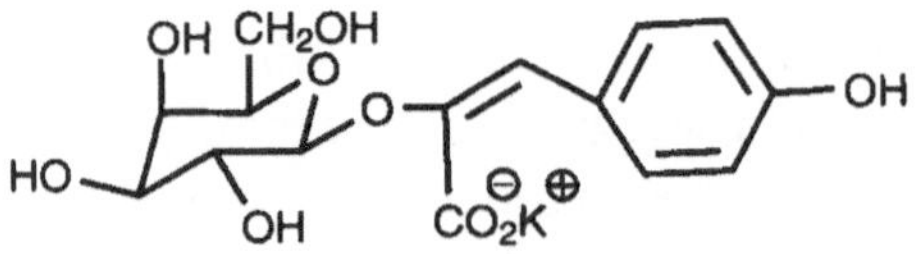   7.1.10 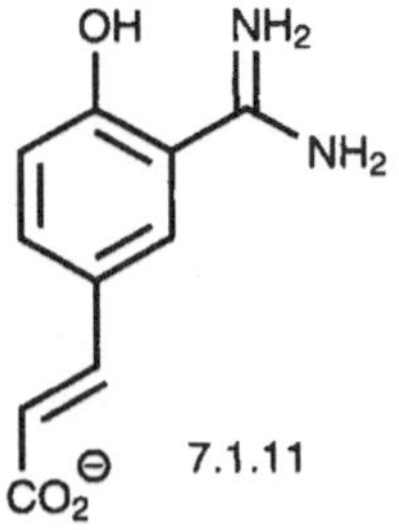   7.1.11

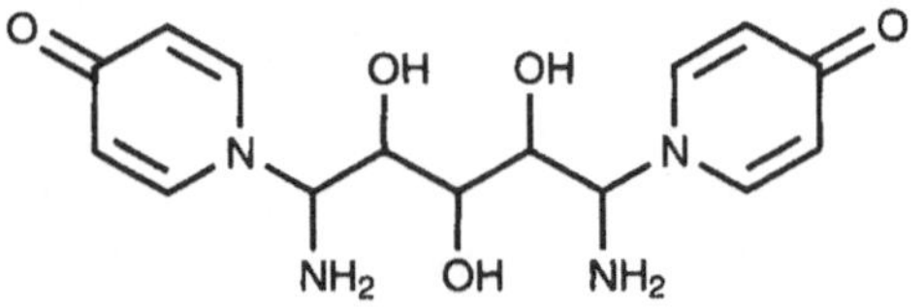

7.1.12 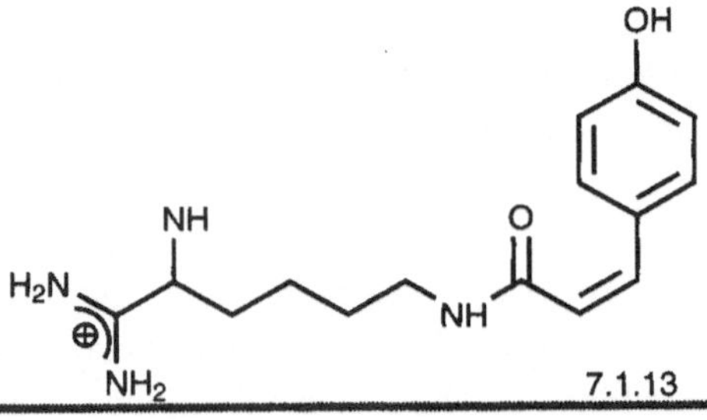   7.1.13

Suggested reading

1.   Shimomura O, *Bioluminescence: Chemical Principles and Methods*, Singapore: World Scientific, 2006.
2.   Ueda M, Takada N and Yamamura S, *Int. J. Mol. Sci.*, 2001, 4: 156–64.
3.   Takada N, Kato E, Ueda K, Yamamura S and Ueda M, *Tetrahedron Lett.*, 2002, 43: 7655–58.

7.2 SEMIOCHEMICALS: AN OVERVIEW

According to Rembold, chemical interactions between plants and insects, which constitute a type of signal exchange, are only a part of the larger phenomenon of biosemiotics. The chemicals which are components of this network of signal communication in any ecosystem are known as semiochemicals. Such chemicals are also responsible for communication between members of the same species. Among insect communities, the secondary metabolites which constitute the complex molecular signal system are collectively known as pheromones. These are further classified as sex pheromones, alarm pheromones, contact pheromones, and so on, depending on their specific functions. In recent years, this area of the science of natural products has been enriched by the research of Schildnicht, Rembold, Towers, Meinwald and others.

In insects, chemical secretions which protect them against predators and an adverse environment, in general, are important for the very survival of the species. In that sense, in an ecosystem, these compounds are as essential as the primary metabolites for the species concerned. One such compound is gyrinal (7.2.1) whose synthesis has been described in Chapter 4. This is one of the chemical constituents of the anal secretion of the water beetle, *Gyrinus natator*. Gyrinal has been found to be toxic to microorganisms living on the surface of water. The compound is also surface-active and enables the small beetle to dash across the surface of water at considerable speed. Thus gyrinal not only protects *G. natator* from waterborne microorganisms, but also prevents it from drowning when the beetle gets carried away by flowing water. The compound is also toxic to fish and its LD_{50} has been found to be just 45 mg/kg in mice.

Another small beetle, the blue-green *Stenus comma* is equally interesting to the natural product chemist. This insect which measures 5 mm and weighs a grand 2.5 mg is able to swim on water at an incredible speed of 40 to 75 cm per second! What enables this tiny creature to perform such an amazing feat is its abdominal secretion which resembles, in odour, the oil of Eucalyptus. The oil obtained from 1000 beetles has been found to contain 0.8 mg of a mixture of 1,8-cineole (7.2.2), isopipertinol (7.2.3), 6-methyl-5-heptenone (7.2.4) and a tertiary amine, stenusin (7.2.5). The first three compounds are stored in a small defensive bladder whereas the amine is secreted in a larger bladder. As can be seen from its structure, stenusin is only sparingly soluble in water but is highly surface-active and has a high spreading pressure. 1,8-Cineole (7.2.2) is an effective, though mild, antiseptic compound. Thus, this tiny beetle knows enough chemistry to enable it to defend itself against the elements (water and wind) and the microflora found in its natural habitat.

Yet another fascinating example cited by Schildnicht in his enchanting story-like review of chemical ecology is the functional chemistry of leaf-cutting ants. These ants have obviously mastered the science and art of growing a fungal garden on a heap of fallen leaves, petals and even refuse. They do this by producing the right kind of chemicals at the right time and in the correct proportion. Entomologists have identified two types among these ants. The smaller kind belong to the worker class and the larger constitute the 'army'. The heads of the worker ants secrete a mixture of about twenty terpenoids which include geraniol (7.2.6), neral (7.2.7), geranial (7.2.8), β-pinene (7.2.9), geranic acid (7.2.10) and farnesol (7.2.11). The secretion also contains phenylethyl alcohol (7.2.12) and *n*-dodecanol (7.2.13). Geranial has been identified as an alarm pheromone. Apparently, the individual members of a group of worker ants communicate with each other through these semiochemicals.

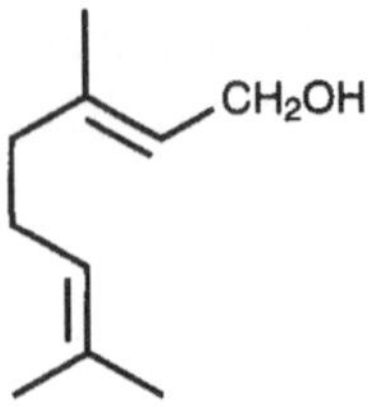

7.2.6 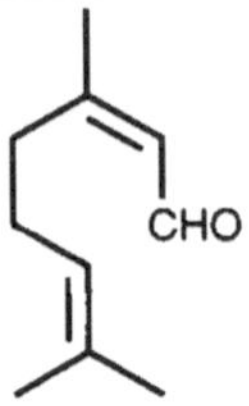   7.2.7 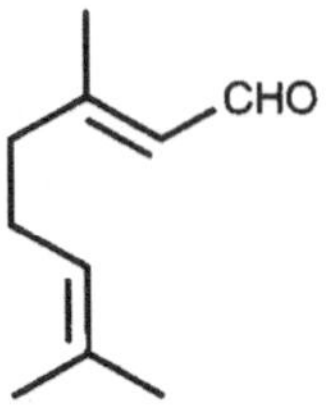   7.2.8 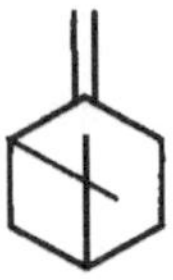   7.2.9 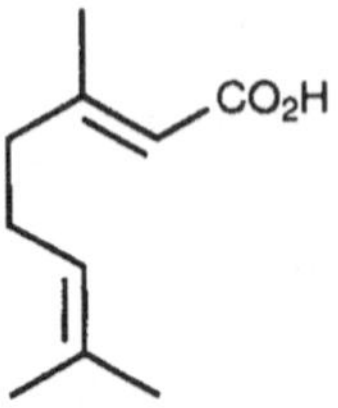   7.2.10

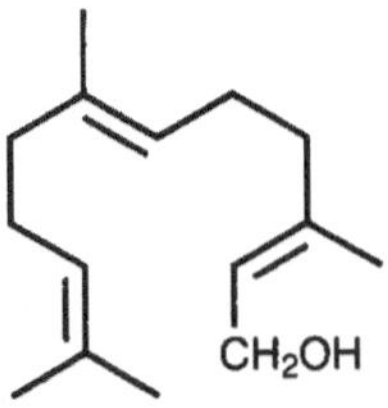

7.2.11 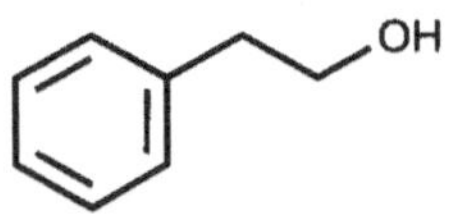   7.2.12 $H_3C-(CH_2)_{10}-CH_2OH$
                                                          7.2.13

Worker ants also possess well-developed glands which produce copious amounts of plant growth promoters and regulators. For example, heteroauxin (7.2.14) and 3-hydroxydecanoic acid (7.2.15) act as plant growth promoters, whereas phenylacetic acid (7.2.16) regulates plant growth. The secretion also contains plant growth inhibitors which help keep the garden trim and prevent it from running wild! The chief inhibitors are 3-hydroxyhexanoic acid (7.2.17) and 3-hydroxyoctanoic acid (7.2.18). The larger soldier ants, constituting the army, produce powerful insect repellents such as *p*-benzoquinone (7.2.19) and *p*-toluquinone (7.2.20). Thus, the worker and soldier ants together make very effective use of chemical knowledge for the welfare of the ant community. One can see the 'seed' of human civilisation in this example!

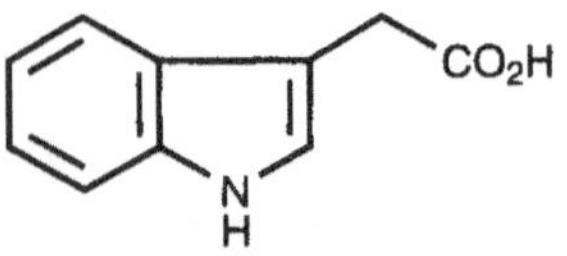

7.2.14

$H_3C-(CH_2)_n-CH-CH_2-CO_2H$
$\qquad\qquad\quad |$
$\qquad\qquad\ OH$

7.2.15: n = 6
7.2.17: n = 2
7.2.18: n = 4

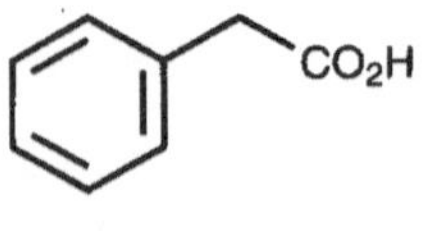

7.2.16

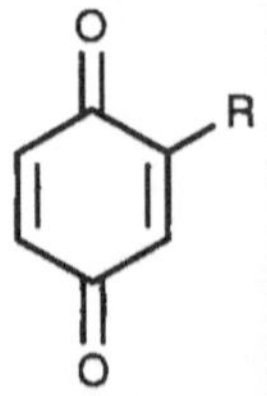

7.2.19: R = H
7.2.20: R = CH_3

Semiochemicals also play a significant role in a plant's response to damage caused by mechanical or other means. For example, the cotton plant, *Gossypium hirsutium*, produces volatile semiochemicals when a part of the plant is damaged by a knife-cut or when a caterpillar eats away a part of a leaf. Interestingly, these compounds are formed not only at the damaged site but in the undamaged parts as well, obviously as a protective measure (they can also be considered as volatile phytoalexins). The mixture of compounds thus produced include the monoterpenes, (E)-β-ocimene (7.2.21) and linalool (7.2.22) as well as a homomonoterpene hydrocarbon (7.2.23); the last mentioned compound is (E) 4,8-dimethyl-1,3,7-nonatriene. Studies with other plants have shown that (E)-β-ocimene, is frequently found in the chemicals produced in response to mechanical wounding or attack by herbivorous insects. It probably serves as an attractant for predators of the plant's natural herbivores and also to attract pollinating insects. Plant-produced semiochemicals are also being increasingly used as safer and species-specific pesticides in crop protection. One example is another monoterpene, *nepeta lactone* (7.2.24), which occurs in *Nepeta catarta* (Lamiaceae). This compound is also a sex pheromone produced by female aphids and also an attractant for aphid predators and parasitoids. Another compound of plant origin which can be used in pest control is the anti-juvenile hormone, precocene-1 (7.2.25), which has been isolated from a weed of the Asteraceae family, *Ageratum houstonianum*.

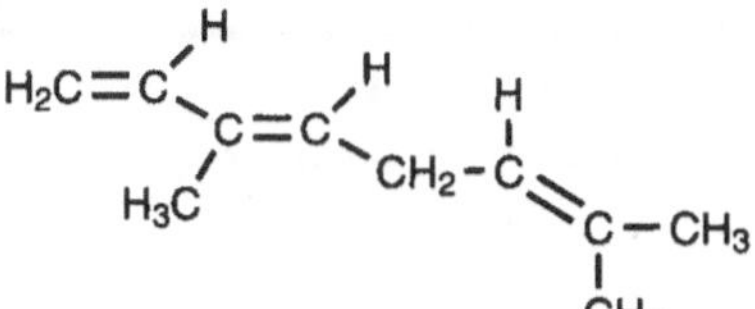

7.2.21

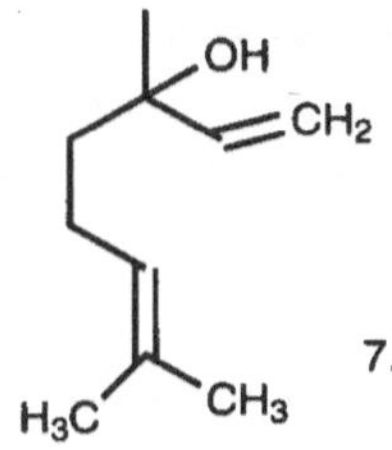

7.2.22

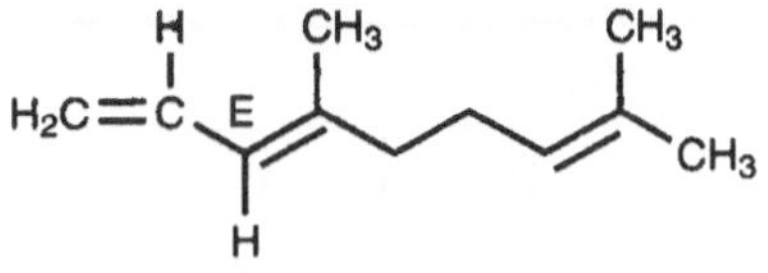

7.2.23

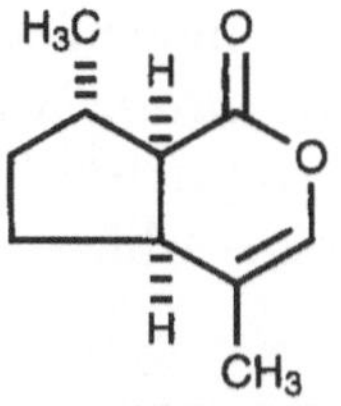

7.2.24

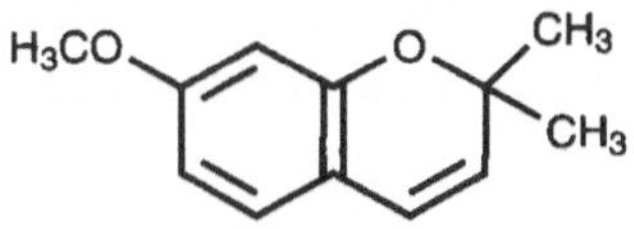

7.2.25

Suggested reading

1. Eder J and Rembold H, *Naturwiss.*, 1992, 79: 60–67.
2. Schildnicht H, *Angew. Chemie, Int. Eng. Ed.*, 1976, 15: 214.
3. Alves LF, Chemical ecology and the social behaviour of animals, In Herz W, Grisebach H, Kirby, GW, and Tamm Ch, (eds), *Progress in the Chemistry of Organic Natural Products*, Vol. 53, Berlin: Springer-Verrlag, 1988.
4. Petroski RJ, Tellez M and Behle RW, *Semiochemicals in Pest and Weed Contro*, ACS Symposium Series, No. 906, Oxford: Oxford University Press, 2005.
5. Rose USR, Manukian A, Heath RR and Tumlinson JH, *Plant Physiol.*, 1996, 111: 487–98.

7.3 Insect Pheromones

Pheromones come under the umbrella of semiochemicals but are associated with specific, well-defined physiological activities in insects. Therefore, it is not surprising that even small changes in the structure or stereochemistry of a pheromone could result in a large change in its physiological profile. Depending on the type of activity, they are classified as sex pheromones or attractants, alarm pheromones, aggregating pheromones and trail-marking pheromones. The majority of them are structurally simple compounds of low molecular weight but possess unique stereochemical features which are essential for their biological activity. Most of them are biosynthetically derived from fatty acids or terpenoids. They occur in nature in minute quantities and are obviously able to exert their action at such low concentrations. For example, it has been found that the male silkworm, *Bombyx mori*, is able to recognise the scent of the female sex pheromone at as low a concentration as just 100 molecules/cm^3 of air. The leaf-cutting ant, *Atta texana*, is even more sensitive to the odour of its trail pheromone, methyl 4-methylpyrrole-2-carboxylate (7.3.1) which is detected at a concentration of 0.08 picogram per cm^1 (equivalent to 3.48 molecules/cm^3!). The trail pheromone of termite worker ants, Z-3, Z-6, E8-dodecatrien-l-ol (7.3.2) is also perceived at about the same level.

7.3.1   7.3.2

It was pointed out earlier that the molecular shape is important for pheromone activity. An interesting example is the major sex attractant secreted by the Sciarid fly, *Lycoriella mali*. This compound is just a hydrocarbon, namely, heptadecane (7.3.3). In an appropriately folded conformation it resembles, in shape, perhydrocyclopentanophenanthrene (7.3.4) which is the skeletal structure of steroids.

7.3.3   7.3.4

Muscalure (7.3.5) is a sex pheromone, with a simple hydrocarbon structure. It has a double bond, more or less in the middle of the molecule, and its *cis* configuration is essential for activity; the *trans* isomer is inactive. This compound has been isolated from the cuticles and faeces of the female housefly, *Musca domestica*. Its structure was determined by a combination of NMR and mass spectral studies and microscale ozonolysis. The latter yielded nonanal (7.3.6) and tetradecanal (7.3.7) which were separated by gas chromatography (GC) and identified by mass spectroscopy (MS).

H₃C—(CH₂)₆—CH₂—C=C—CH₂—(CH₂)₁₁—CH₃ $\longrightarrow$ H₃C—(CH₂)₇—CHO + H₃C—(CH₂)₁₂—CHO

$$H_3C-(CH_2)_6-CH_2-\overset{H}{\underset{}{C}}=\overset{H}{\underset{}{C}}-CH_2-(CH_2)_{11}-CH_3 \longrightarrow H_3C-(CH_2)_7-CHO + H_3C-(CH_2)_{12}-CHO$$

7.3.5   7.3.6   7.3.7

Its structure has been confirmed by synthesis. The starting material for one synthesis was the acetylene derivative (7.3.8) which was converted into 7.3.9. Catalytic hydrogenation of 7.3.9 yielded 7.3.5. Another pathway began with erucic acid (7.3.10) which was converted into the methyl ketone (7.3.11) and then reduced by the Wolff–Kishner–Huang–Minion method to obtain 7.3.5. In the third synthesis an equimolecular mixture of oleic acid (7.3.12) and heptanoic acid (7.3.13) was subjected to Kolbe electrolysis in the presence of sodium methoxide in methanol to get 7.3.5 in a single step.

$$H_3C-(CH_2)_{12}-C\equiv CH \quad \overset{1.\ n\text{-BuLi}}{\underset{2.\ H_3C-(CH_2)_7Br}{\longrightarrow}} \quad H_3C-(CH_2)_{12}-C\equiv C-(CH_2)_7-CH_3 \quad \overset{H_2-Pd\,/\,BaSO_4}{\longrightarrow}$$

7.3.8   7.3.9

$$H_3C-(CH_2)_6-CH_2-\overset{H}{\underset{}{C}}=\overset{H}{\underset{}{C}}-CH_2-(CH_2)_{11}-CH_3$$

7.3.5

$$H_3C-(CH_2)_7-\overset{H}{\underset{}{C}}=\overset{H}{\underset{}{C}}-(CH_2)_{11}-\overset{O}{\underset{}{C}}-R \quad \overset{1.\ CH_3Li}{\underset{\substack{2.\ Huang\text{-}Minion\\ reduction}}{\longrightarrow}} \quad H_3C-(CH_2)_6-CH_2-\overset{H}{\underset{}{C}}=\overset{H}{\underset{}{C}}-CH_2-(CH_2)_{11}-CH_3$$

7.3.10: R = OH   7.3.5
7.3.11: R = CH₃

$$H_3C-(CH_2)_7-\overset{H}{\underset{}{C}}=\overset{H}{\underset{}{C}}-(CH_2)_7-CO_2H + H_3C-(CH_2)_5-CO_2H \quad \overset{\text{Electrolysis}}{\underset{NaOCH_3\,/\,CH_3OH}{\longrightarrow}}$$

7.3.12   7.3.13

$$H_3C-(CH_2)_6-CH_2-\overset{H}{\underset{}{C}}=\overset{H}{\underset{}{C}}-CH_2-(CH_2)_{11}-CH_3$$

7.3.5

The sex pheromone of the Japanese beetle, *Popillia japonica* is more complex. It is a 14-carbon chiral lactone (7.3.14) possessing a *cis*-double bond. The racemic pheromone is inactive whereas the enantiomer is an inhibitor.

7.3.14

The contact pheromones which are responsible for establishing communication between individual members of a group of flies are also aliphatic compounds, usually unsaturated and containing functional groups such as aldehyde, hydroxy or ester groups. They have chain lengths of 12, 14, 16 or 18 carbons.

A long chain diepoxy aliphatic hydrocarbon, (3 Z), *cis, cis*-6,7,9,10-diepoxyheneicosene (7.3.15) has been identified as the sex pheromone in the satin moth. This compound is released by the female of the species to attract the male. The sand fly, on the other hand, uses a homo sesquiterpene hydrocarbon, 9-methylgermacrene B (7.3.16) as a male sex pheromone. Nitrogenous compounds, such as *N*-isopentyl-2-phenylethylamine (7.3.17) and 2',3'-dipyridyl (7.3.18) are used by some ants as trail pheromones.

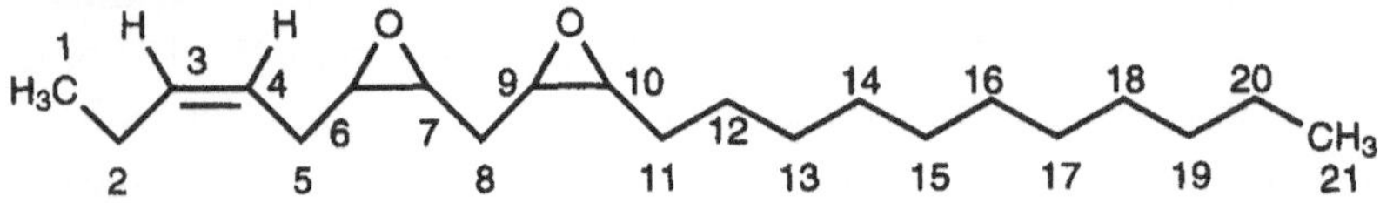

7.3.15

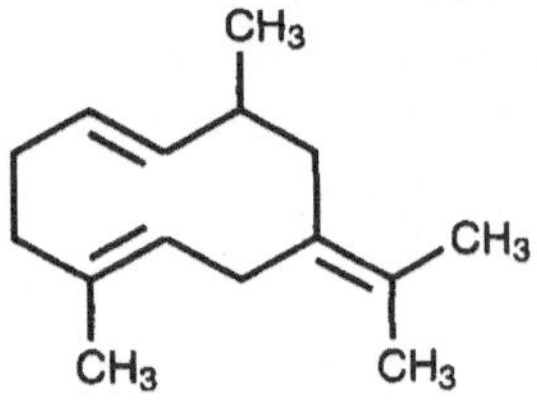

7.3.17

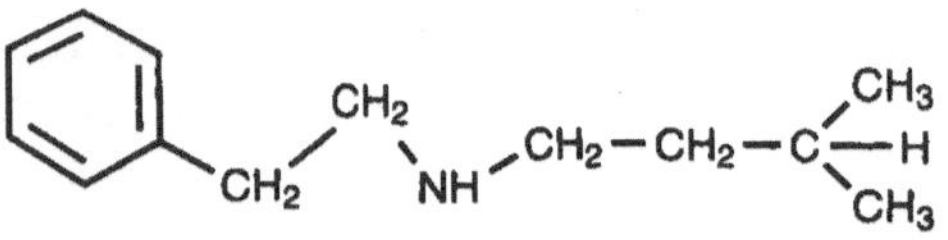

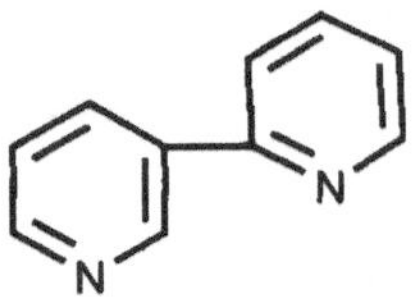

7.3.18

7.3.16

Summing up, the majority of alarm, trail and sex pheromones are chemical compounds which can be air borne and easily detected even at low concentrations. This is because in insects the sense of smell plays a major role in regulating their behaviour. Therefore, it is not surprising that many of these compounds are simple monoterpenes such as nerol (7.3.19), its formyl ester (7.3.20), neral (7.3.21) and geranial (7.3.22). It is worth noting that except geranial, all the other compounds possess a *cis* (Z) double bond. In the mold mite, neral formate has been found to be a very effective alarm pheromone. In some species of mite, nerol functions both as an alarm pheromone as well as a sex attractant. Apart from isolating and characterising the different types of pheromones, studies with regard to their mode of action have also been carried out.

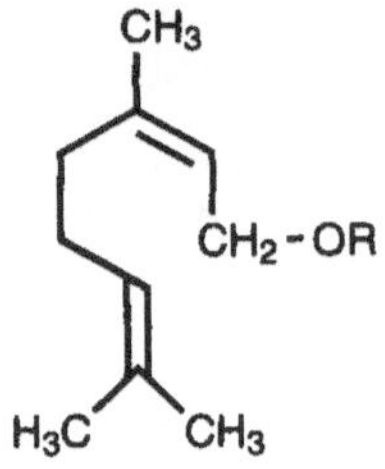

7.3.19: R = H
7.3.20: R = —C—H
                ‖
                O
7.3.21
7.3.22 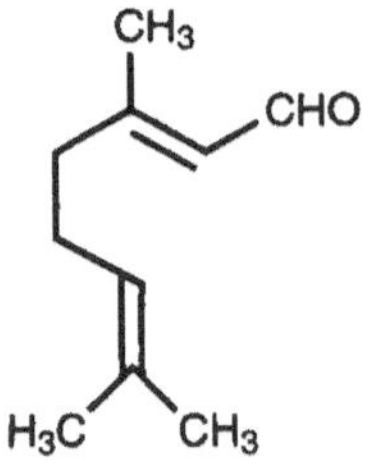

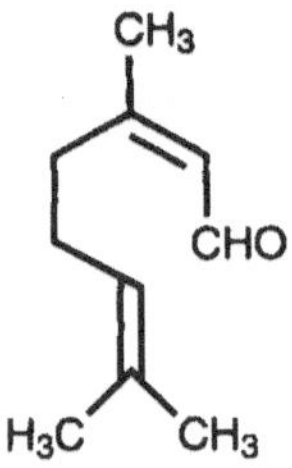

This is only a very brief description of some insect pheromones meant to stimulate interest in this fascinating area of natural products.

Suggested reading

1. Alves LF, Chemical ecology and the social behaviour of animals, In Herz W et al, (eds), *Progress in the Chemistry of Organic Natural Products*, Vol. 53, Berlin: Springer-Verrlag, 1988.
2. Brand JM, Young JC and Silverstein RM, Insect pheromones—A critical review of recent advances in their chemistry, biology and application, In *Progress in the Chemistry of OrganicNatural Products*, Vol. 36, 1979.
3. Wilson EO, *Scientific American*, 1963, 208: 100.

4. Nordlung DA and Lewis WJ, *J. Chem. Ecol,* 1976, 2: 211.
5. Luckner M, *Secondary Metabolism in Microorganisms, Plants, and Animals,* 2nd edn, Berlin: Springer-Verlag, 1984.
6. Carde RT and Millar JG, (eds), *Advances in Insect Chemical Ecology,* Cambridge: Cambridge University Press, 2004.
7. Harborne JB, *Natural Product Reports,* 1999, 16: 509–23.

7.4 PLANT–INSECT INTERACTIONS

Chemical communication between plants and insects in any ecosystem can be both mutually benign and antagonistic. Plants require some insects for the dispersal of their seeds and for protection against certain other pests. They therefore possess insect attractants. They also have insect repellents to ward off the unfriendly among the insect community in the neighbourhood. There are interesting examples of plant–insect interactions, where a plant is friendly to one species of insects while repelling most other insects. For instance, plants of the cabbage family (Cruciferae) produce glycosides of mustard oils which are highly toxic to a large majority of insects. At the same time these compounds seem to stimulate the adult female cabbage-butterfly, *Pieris brassicae,* to lay her eggs on the plant. These compounds serve as feeding stimulants for the larvae of the butterfly. Thus, the cabbage seems to have given exclusive but restricted feeding rights to the cabbage-butterfly while repelling other insects through the agency of isothiocyanate glycosides. Another example of this phenomenon of specific plant—insect cooperation is the following. Plants of the Senecioidae subfamily

of Asteraceae and of the genus Crotolaria of the Leguminoseae family have the unique capacity of producing pyrrolizidine alkaloids which cause severe liver damage in ruminants which feed on them. However, the presence of these toxic compounds does not prevent the cinnabar moth, *Tyria jacobaeae,* from feeding on the leaves of *Senecio jacobaea.* The larvae of these black and yellow caterpillars seem to

have the capacity to assimilate the pyrrolizidine alkaloids of their plant food and to use them later—after they become adult moths—for chemical defence against predators. The structure (7.4.1) of one of these alkaloids is shown here.

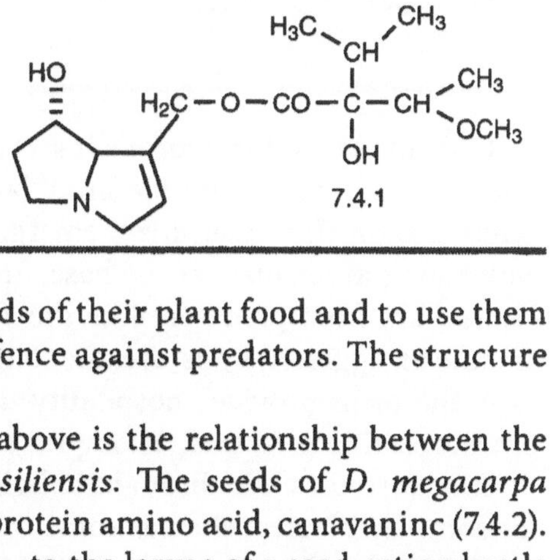

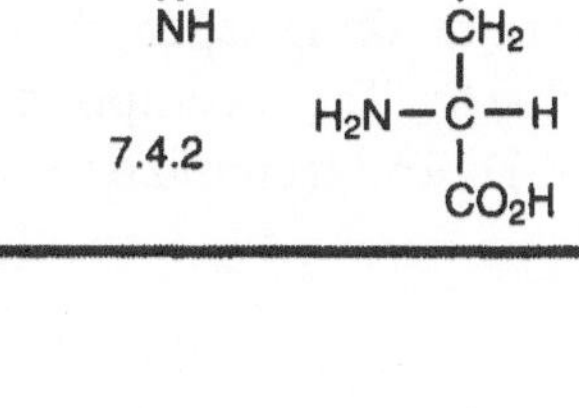

Yet another example of the phenomenon described above is the relationship between the plant *Dioclea megacarpa* and the beetle *Caryedes brasiliensis.* The seeds of *D. megacarpa* contain a fairly large amount (8% and more) of the non-protein amino acid, canavaninc (7.4.2). This serine derivative is toxic—at a concentration of 1%—to the larvae of a seed-eating beetle known as the cowpea weevil, *Callosobruchus maculatis.* At a concentration of 5%, the compound is lethal to the larvae of the above beetle. Interestingly, however, the same seeds serve as the only source of food for *C. brasiliensis,* which is also a seed-eating beetle. Since the structure of canavanine is very similar to that of arginine (7.4.3), it is mistakenly incorporated (in place of arginine) into the proteins of the cowpea weevil larvae. On the other hand, apparently, the RNA synthetase of *C. brasiliensis* has the capacity to differentiate between arginine and canavanine and to avoid the incorporation of the latter in its proteins. The question which remains to be answered is how these RNAs differ.

$$H_2N-\underset{\underset{NH}{\|}}{C}-NH-CH_2-CH_2$$

$$H_2N-\underset{|}{\overset{|}{C}}-H$$

$$CH_2$$

$$CO_2H$$

7.4.3

Several plants are known to synthesise compounds which function as moulting or juvenile hormones in insects. The former group of compounds, the ecdysones (for example, 7.4.4 and 7.4.5) occur only in very minute quantities in insects. For example, from a ton of silkworms, just 0.33 mg of β-ecdysone (7.4.5) could be isolated. The same compound, on the other hand, occurs in much larger quantities in certain plants. Thus, from 2.5 g of air-dried rhizomes of the fern *Polypodium vulgare*, as much as 25 mg of β-ecdysone could be isolated. A plant producing an insect juvenile hormone is the balsam fir, *Abies balsamea*, which accumulates juvabione (7.4.6) in its wood. This tree attracts only one species of an insect family, Pyrrhocosidae.

7.4.4: R = H
7.4.5: R = OH

7.4.6

Certain plants, herbivorous insects and caterpillars form an interesting triangle of symbiotic interactions. The female riodinid butterfly, *Thisbe irenea*, deposits eggs on saplings of the genus Croton. The caterpillars emerging from these eggs start feeding on the tree, the leaves of which have an exudate at the base. This nectary exudate attracts ants which protect the Thisbe caterpillars from predators like the social wasps. The latter are warded off by the ants which assume an aggressive posture due to the release of alarm pheromones by the caterpillars. In this case, the plant provides hospitality to the caterpillars and ants and protects itself from other predators. This is, indeed, a complex triangle of chemical-biological interactions!

Recent studies have brought to light the fact that some flavonoid glycosides act as attractants for some herbivores whereas some other phenolic glycosides have adverse effects on the insect guests. Quercetin-7-O-glucoside (7.4.7) and quercetin-3-glucoside (7.4.8), which are stimulants for the boll weevil, belong to the former category. Advantages offered by such compounds, if any, to the host plant are not known. On the other hand, the phenolic glucosides present in the leaves of the quaking aspen, *Populus tremuloides*, have an adverse effect on the larvae of the swallowtail butterfly. These compounds have been identified as salicortin (7.4.9), tremulacin (7.4.10), salicin (7.4.11) and tremuloiden (7.4.12).

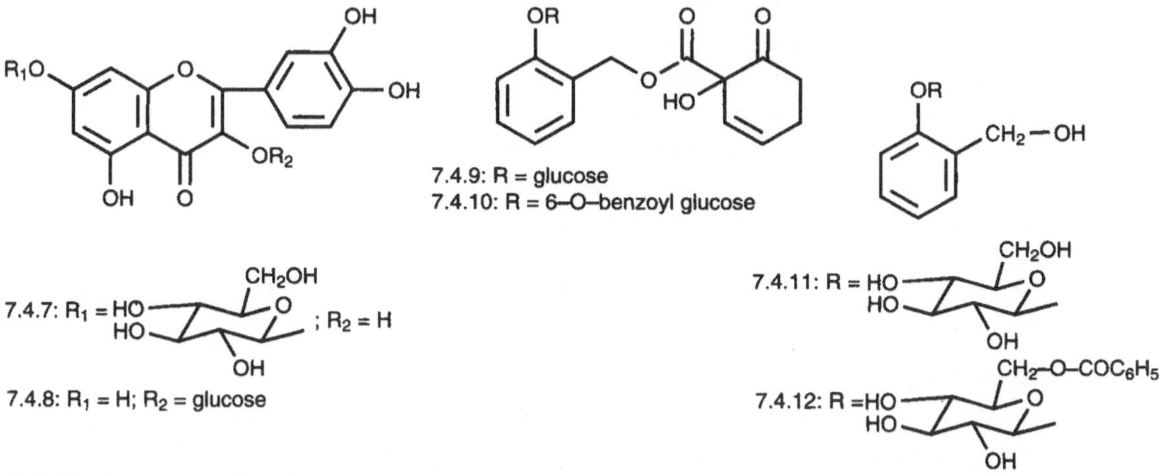

7.4.9: R = glucose
7.4.10: R = 6–O–benzoyl glucose

7.4.7: R$_1$ = HO ; R$_2$ = H

7.4.8: R$_1$ = H; R$_2$ = glucose

7.4.11: R = HO

7.4.12: R = HO

When the leaves are crushed, due, for example, to the chewing action of the insect herbivores, salicortin and tremulacin break down into salicin and tremuloiden respectively, along with 6-hydroxy-2-cyclohexenone (7.4.13) and carbon dioxide. Obviously, crushing releases an enzyme which brings about this hydrolytic reaction and subsequent decarboxylation. 6-Hydroxy-2-cyclohexenone gives catechol (7.4.14) under basic conditions in the presence of air. Salicortin, tremulacin, 6-hydroxy-2-cyclohexenone and catechol are all toxic to the insects. The willow tree,

7.4.13   7.4.14   7.4.15

Salix helix, whose leaves and bark also contain salicin, are the preferred hosts for several species of beetles. In this case, the objective of the herbivores is not dietary but to acquire from the host plant a chemical defence in the form of salicin. The beetles convert salicin into salicyaldehyde (7.4.15) which is used to ward off predators!

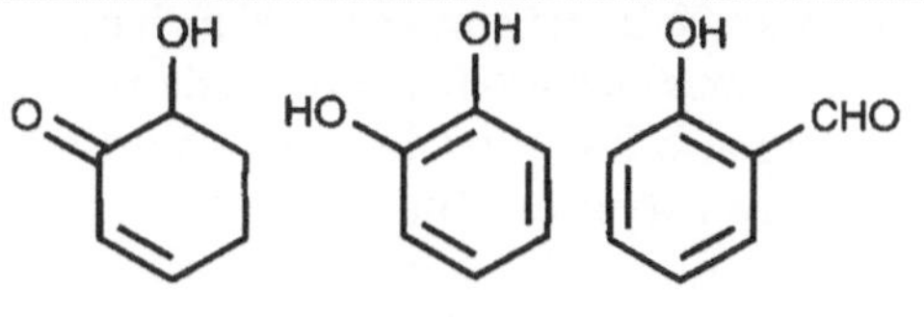

Plants protect themselves against insect attack by using, besides toxins such as the cyanogenetic glycosides decribed earlier in this section, antifeedant and antinutrient compounds. Antifeedants act as feeding deterrents whereas antinutrients are not easily digestible. Some phenolic glycosides are bitter and are thus not palatable. On the other hand, condensed tannins, which occur widespread in several plants, possess antinutrient properties. Being oligomeric in nature, they are not easily digested. They also form complexes with proteins. One example is the proanthocyanidin (7.4.16) which possesses

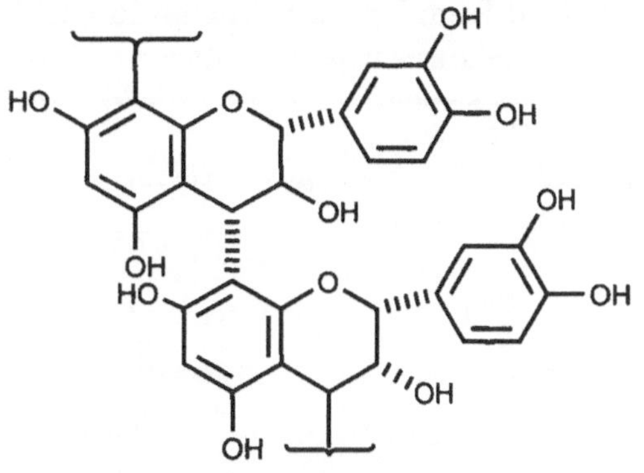

7.4.16

both catechin and epi-catechin as monomeric units. This compound is present in black brush, *Coleogyne ramosissima*, a desert plant, known to prevent other plants growing in its vicinity.

The exploitation of volatile compounds by plants for their survival and propagation is amazing. Mention was made earlier of (E) 4,8-dimethyl-1,3,7-nonatriene as a semiochemical. This compound is used by the flowers of *Yucca filamentosa* as an attractant for some mite and parasitic wasps which are thus recruited by the plant as a second line of defence against herbivores. An interesting and amusing fact is that cactus flowers use the compound, geosmin (7.4.17), which has been termed 'wet earth compound', as it smells of wet earth, to attract earth-nesting bees to help in pollination.

7.4.17

Suggested reading

1. Bell EA, The possible significance of secondary products in plants, In Bell EA and Charlwood BV, (eds), *Encyclopedia of Plant Physiology*, New series, Vol. 8, Berlin: Springer-Verlag, 1980.
2. Swain T, *Ann. Rev. Plant Physiol*, 1977, 28: 479–501.
3. Harborne JB, *Introduction to Ecological Biochemistry*, New York: Academic Press, 1982.
4. de Vries PJ, *Scientific American*, 1992, 267: 56.
5. Inderjit KM, Dakshini M and Foy CL, (eds), *Principles and Practices in Plant Ecology. Allelochemical Interactions*, Boca Raton: CRC Press, 1999.
6. Wada K, Insect antifeedants in plants, In Natori S, Ikekawa N and Suzuki M, (eds), *Advances in Natural Products Chemistry*, Tokyo: Kodansha, 1981.
7. Raguuso RA, Why do flowers smell?, In *Advances in Insect Chemical Ecology*, Cambridge: Cambridge University Press, 2004.

7.5 Plant—Vertebrate Interactions

Plants also need defensive mechanisms for protection from animals, particularly herbivores. Several plants are indeed known to produce secondary metabolites which are toxic to animals. Among such compounds, the most toxic are the cyanogenetic glycosides which occur widely in plants. These are derived from L amino acids by oxidative decarboxylation followed by further modifications of the aldoximes (7.5.1) thus formed. Thus, the glucoside linamarin (7.5.2) is obtained from L-valine. The amino acid precursor of prunasin (7.5.3) and amygdalin (7.5.4) is L-phenylalanine whereas L-tyrosine is the precursor of dhurrin (7.5.5).

One of the well-known sources of these compounds is the bitter variety of cassava, *Manihot esculenta*. However, certain varieties of this plant are palatable, as they contain much lower concentrations of the cyanogenetic glycosides. Several other plants are also similarly polymorphic with regard to these compounds. For example, two forms of bracken, *Pteridium aquilinum*, are known. One, which does not contain cyanogenetic glycosides, is grazed by deer and sheep which avoid the other form which is cyanogenetic. The cyanogenetic glycosides which are intact in the plant, are enzymatically hydrolysed in the gastrointestinal tract of the animals and it is the hydrogen cyanide thus liberated which is toxic. However, the intact glycoside itself is unpalatable, being bitter, and thus the animals can avoid getting poisoned.

Some plants also synthesise organic cyanides, such as, for example, β-minopropionitrile (7.5.6) which was first isolated from the seeds of the sweet pea, *Lathyrus odoratus*, as the γ-glutamyl derivative. This compound causes skeletal deformation in young rats by inhibiting the enzymes which bring about the cross-linking of peptide chains in collagen and elastin. A related compound, β-cyanoalanine (7.5.7) occurs in the free state and also as its γ-glutamyl derivative (7.5.8) in the seeds of the Vicia species. This compound produces convulsions in chicks and rats and is a vitamin B6 antagonist. Yet another toxic nitrogenous compound of plant origin is cycasin (7.5.9) which is an *N*-oxide. It occurs in the leaves and seeds of some species of Cycad and interferes with normal neural function in humans. In experimental animals it has been found to be a potent carcinogen when administered orally. If injected intravenously, however, it is non-toxic. These observations suggest that the glycoside itself is not toxic and that the toxic symptoms become manifest only after the aglycone is liberated by the action of the microflora in the gastrointestinal tract of the animals.

Non-protein amino acids, which occur fairly widely in plants, are considered to be secondary metabolites as they are not needed for building up the structural and functional proteins, without which no form of life is possible. Many of these compounds are indeed highly toxic to animals and humans. One such compound is 2,4-diaminobutyric acid (7.5.10) which occurs in fairly large amounts in the seeds of *Lathyrus latifolius*. When it enters the blood stream of an animal, this amino acid causes convulsions by inhibiting the enzyme ornithine transcarboxylase. Another toxic non-protein amino acid is mimosine (7.5.11) which is produced by certain plants of the family Mimosoideae, a notable example being *Leucaena leucocephala*. Mimosine itself is non-toxic but becomes toxic, after the microorganisms present in the rumen convert it into 3-hydroxy-4(lH)pyridone (7.5.12) and alanine, presumably by a reductive cleavage of the C–N bond. The compound brings about liver damage, loss of hair, and enlargement of the thyroid gland in affected animals.

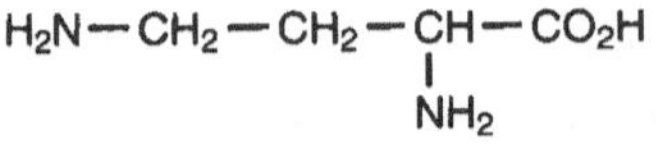

7.5.10 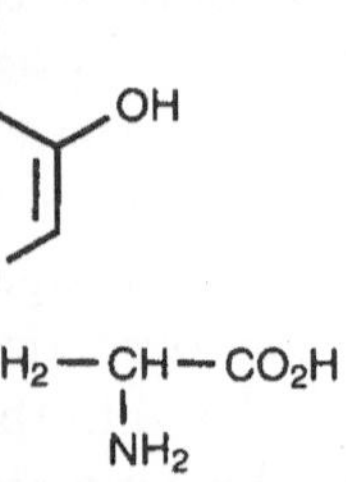   7.5.11 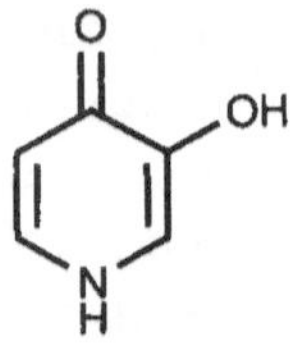   7.5.12

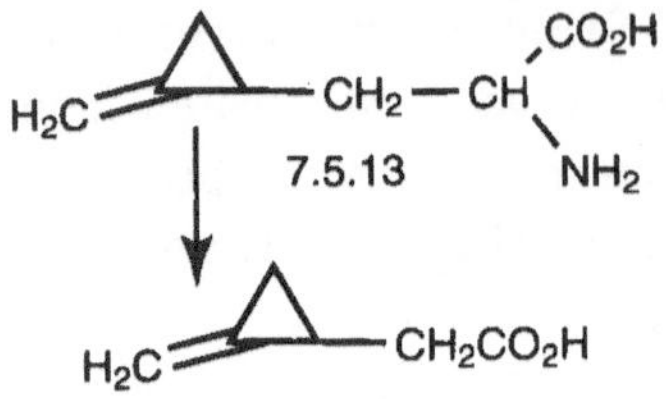

Another example of an unusual toxic amino acid is hypoglycin-A (7.5.13) which is a constituent of the unripe fruit of *Blighia sapida*. In this case too, the toxic compound is not the original amino acid but its metabolite, methylenecyclopropylacetic acid (7.5.14), which presumably arises by a reductive deamination of hypoglycin-A. This compound interferes with carbohydrate metabolism in human beings and can cause death by producing acute hypoglycaemia.

7.5.13

7.5.14

Animals can also bring about the detoxification of toxic compounds they ingest, by metabolic transformations. The toxic pyrrolizidine alkaloids get detoxified by hydrolytic cleavage to the non-toxic necins. For example, the alkaloid senecionine (7.5.15) gets hydrolysed to retronecine (7.5.16) and senecic acid (7.5.17). It may be mentioned here that certain types of butterflies can further transform retronecine into hydroxydanaidal (7.5.18) and danaidone (7.5.19) which function as sex pheromones.

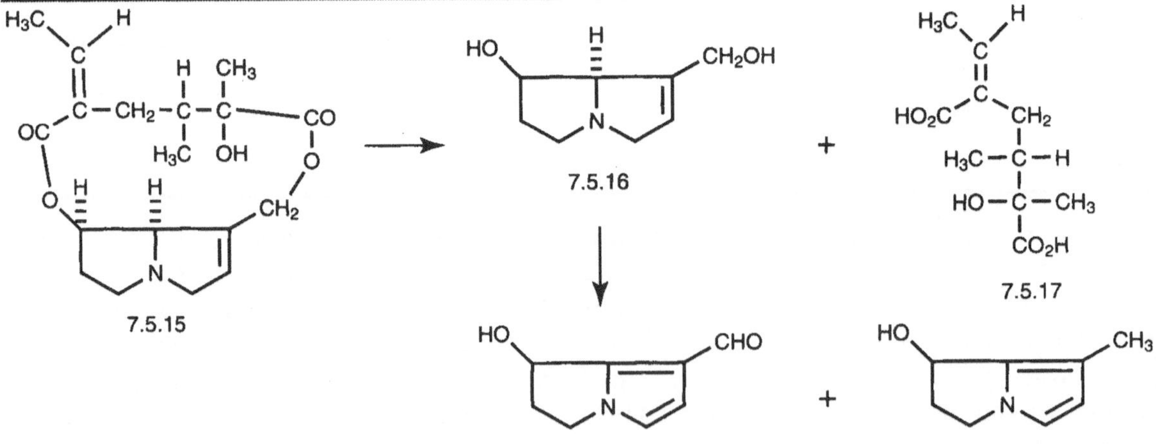

7.5.15   7.5.16   7.5.17

7.5.18   7.5.19

Plant communities growing in particularly harsh conditions such as those prevailing in deserts or in very cold climates require special chemicals to protect themselves against animal herbivores. The bark of the pine tree, growing at higher altitudes. for example, contains the stilbenes, pinosylvin (7.5.20) and its monomethyl ether (7.5.21) which act as deterrents to the snowshoe hare. As mentioned in the previous section, the desert plant, blackbrush, contains proanthocyanidins which serve as feeding deterrents not only for insects but also for animals. In species of the Eucalyptus, some complex phloroglucinol derivatives such as 7.5.22, for example, are present. They prevent Koala bears and possum from feeding on the plant. Even marine flora possess such antifeedant compounds which give them protection against sea animals such as the abalones (Heliotis sp.). One such compound is the bromine containing diarylmethane derivative (7.5.23) which is present in red alga.

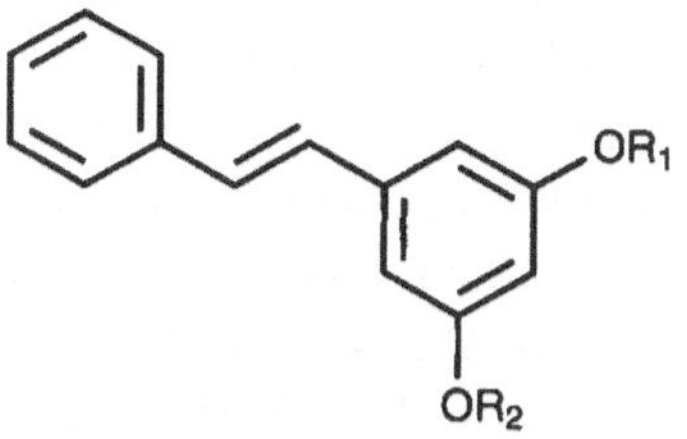

7.5.20: R₁ = R₂ = H
7.5.21: R₁ = H; R₂ = CH₃
7.5.22
7.5.23 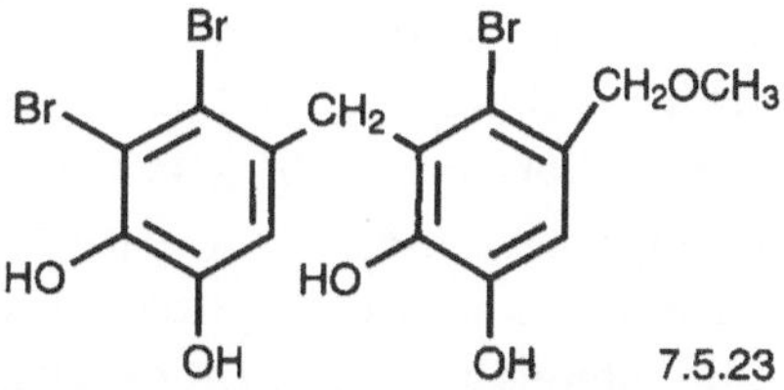

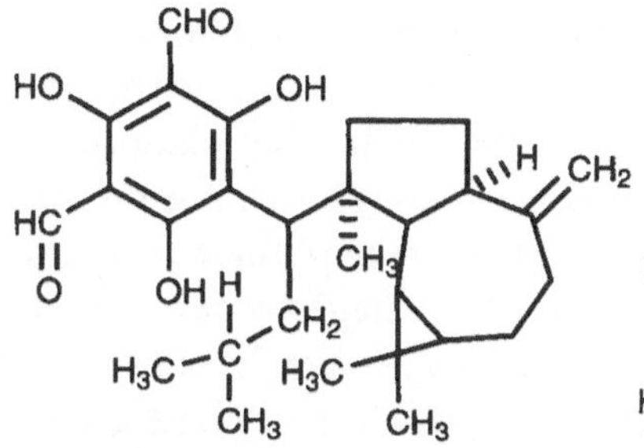

In a similar manner, the tropical brown alga, *Dictyota acutiloba*, uses the diterpenes (7.5.24) and (7.5.25) as feeding deterents against temperate fish and sea urchins. Even the omnivorous slugs and snails keep off the Asteraceae plant, *Petasites hybridus*, which producess the sesquiterpenes, petasin (7.5.26) and furanopetasin (7.5.27).

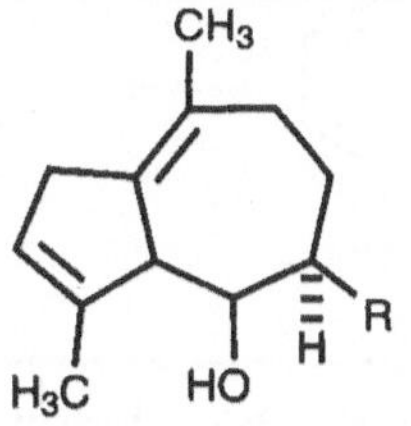

7.5.24: R =
7.5.25: R = 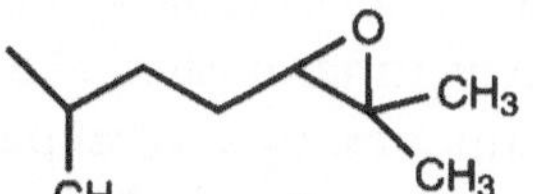
7.5.26 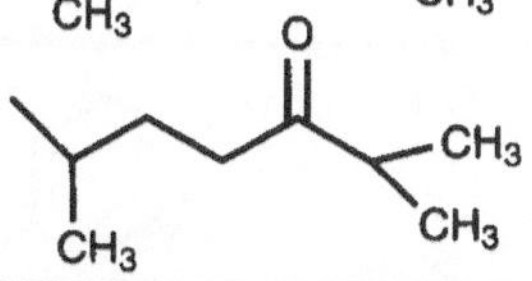 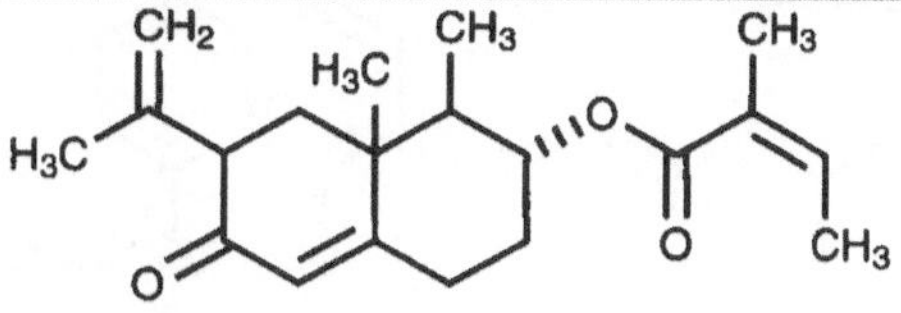
7.5.27 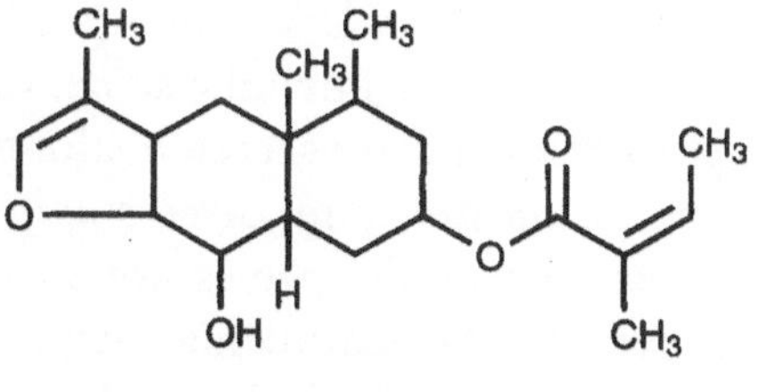

Suggested reading

1. Lindley M, *Nature*, 1977, 266: 776.
2. Luckner M, *Secondary Metabolism in Microorganisms, Plants and Animals*, 2nd edn, Berlin: Springer-Verlag, 1984.
3. Harborne JB, *Natural Product Reports*, 1999,16: 509–23.
4. Reichardt PB, The chemistry of plant-animal interactions, USDA National Wildlife Research Centre Symposium, 1995.

7.6 PLANT—PLANT INTERACTIONS

The system of communication through chemicals among members of a plant community is known as allelopathy. It is a common observation that most plants individually maintain healthy growth in a heterogeneous commune. However, there are some plants which effectively inhibit the growth of other species in their neighbourhood. One such notorious coloniser is *Parthenium hysteroporus* of the family Asteraceae. The plant owes its virulence to the presence of sesquiterpene lactones, such as parthenin (7.6.1) which are plant growth inhibitors. Another plant which has the ability to protect its territorial rights with considerable vigour is the walnut tree, *Juglans nigra*. The leaf canopy of this tree provides a cover under which other plants do not grow. This allelopathic effect is due to juglone, or 5-hydroxynaphthoquinone (7.6.2). It is present in the leaves of the tree as 7.6.3 which is the 4-O-glucoside of 1,4,5-trihydroxynaphthalene. Being a glucoside it has considerable affinity for water and gets washed off the foliage by rain water and dew. When it comes into contact with the microorganisms in the soil it breaks down to the aglucone (7.6.4) which gets readily oxidised to juglone, which is an inhibitor of plant growth. This is indeed an intelligent and interesting use of simple chemical principles by a plant in self-interest! It should, however, be pointed out that juglone does not indiscriminately inhibit the growth of all plants. There are plant species, such as the Kentucky blue grass, *Poa pratensis*, which are able to grow normally under a walnut tree. Such plants apparently possess a biochemical mechanism for neutralising the adverse effect of juglone.

The oak tree, *Quercus falcata*, also produces a simple, water-soluble allelopathic agent, namely salicylic acid (7.6.5). Phenolic acids, such as *p*-hydroxycinnamic acid (7.6.6), do possess allelopathic properties and are secreted, along with some quinones, by several plants including *Adenostema fasciculatum* and *Arctostaphylos glandulosa*. These two plant species are shrubs which are natives of the California chaparral where they inhibit the growth of other herbaceous plants. However, once in about twenty years, the aerial parts of these shrubs get burnt in the California firecycle, thus enabling the dormant

seeds of herbaceous species to germinate. Thus, soon after the breakout of the fire, these herbs flourish, grow to maturity and shed seeds which get embedded in the soil. By this time, the aggressive Adenostema and Arctostaphylos shrubs would have re-established themselves as the dominant plants of the chaparral and their rule would continue for another twenty years till the next fire! This is interesting chemistry of a unique biological phenomenon.

Unlike the walnut and the oak trees which secrete water-soluble allelopathic agents, the leaves of *Salvia leucophylla* produce a volatile oil comprising cineole (7.6.7), camphor (7.6.8) and related compounds. The oil gets volatilised into the atmosphere (like the oil of the Eucalyptus species) and then gets absorbed by the dry soil. Even low concentrations of these compounds which thus get incorporated in the soil are enough to inhibit the germination and growth of grass and herbs which would have otherwise formed a thick undergrowth soon after the onset of the winter rains.

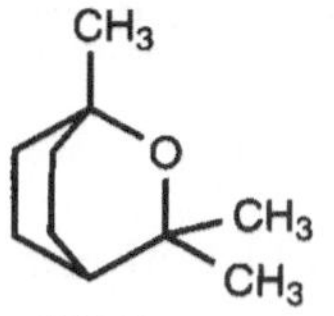

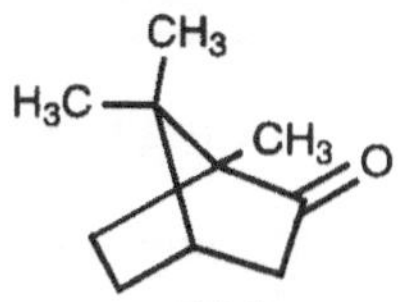

Water-borne plants are particularly susceptible to microorganisms and other water-borne flora. Hence, they produce allelochemicals to give them protection against other living organisms in the aqueous surroundings. One example is the yellow water lily, *Nuphar lutea*, which releases a simple dihydric phenol, resorcinol (7.6.9) into the water. This compound has been found to inhibit the growth of Salvinia, a highly invasive floating fern. Root exudates are particularly effective in ensuring that a plant species gets adequate nutrients from the soil in competition with other plants. For example, the roots of the allelopathic grass *Festuca rubra* produce *m*-tyrosine (7.6.10) which acts as a potent allelopathic agent. Another non-protein amino acid, which is also an effective allelopathic chemical is L-DOPA (7.6.11) which can inhibit the germination of seeds of several plants. In the rice plant, *Oriza sativa*, two chemically unrelated allelochemicals have been identified. These are 5,7,4'-trihydroxy-3', 5'-dimethoxyflavone (7.6.12) and 2-isopropyl-5-acetoxy-cyclohex-2-en-1-one (7.6.13).

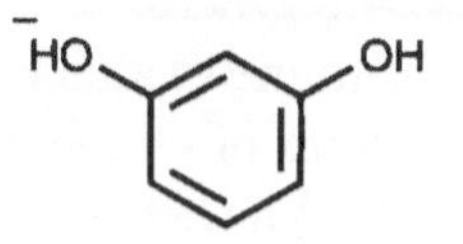

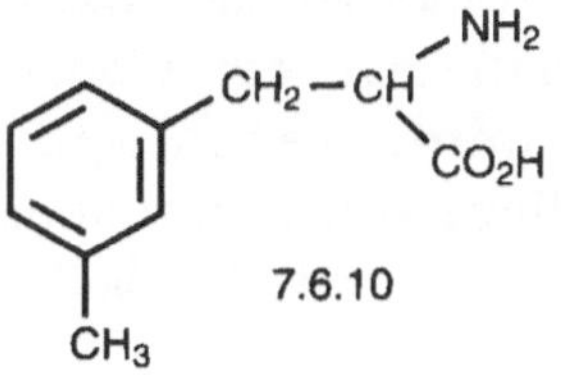

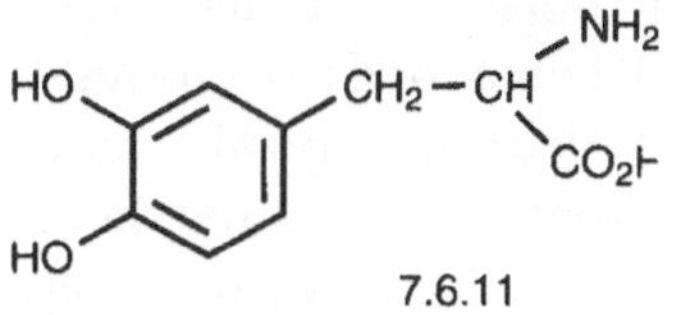

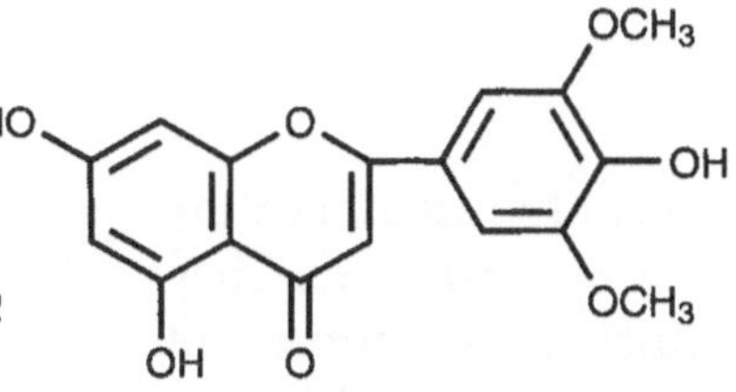

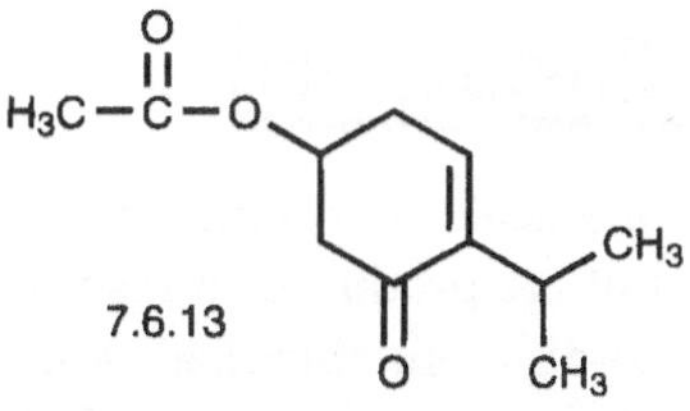

Suggested reading

1.   Rice EL, *Allelopathy*, 2nd edn, New York Academic Press, 1984.
2.   Williams DH, Stone MJ, Hauch PR and Rahman SK, *Nat. Prod.*, 1989, 52: 1189.
3.   Harborne JB, *Introduction to Ecological Biochemistry*, New York: Academic Press, 1982.
4.   Harborne JB, *Natural Product Reports*, 1999,16: 509–23.
5.   Duke SO, *Proc. Nat. Acad. Sci.*, USA, 2007, 104:16729–30.

7.7 PLANT—MICROBE INTERACTIONS

When a fungus or bacterium attacks a plant, three types of chemicals come into the picture.

1. The phytotoxins which are primarily responsible for the pathological symptoms shown by the affected plant;
2. The secondary metabolites produced and stored by the healthy plant to provide defence against the attacking pathogens;
3. The phytoalexins, which are produced by the plant in response to fungal or microbial attack:

We will be discussing phytoalexins in the next section.

In this section, a few examples of the second type will be given. Study of such compounds is important as it provides information regarding antifungal and antibiotic compounds of plant origin which have the potential for use for pest control in agriculture and also for the development of new drugs to counter bacterial diseases. Generally speaking, certain categories of secondary metabolites such as the different types of quinones are known to possess antifungal and antibacterial properties. Several simple monoterpenes and phenolic compounds also act as antifungal and antiseptic agents. The subject is too vast for adequate coverage in this book. Therefore, only a couple of examples will be given to indicate the types of compounds to look for. Cultivated strains of carnation, which are immune to attack by the fungus *Fusarium oxysporium*, contain 3-hydroxyacetophenone (7.7.1). A group of closely related alkyl resorcinols (7.7.2) and the corresponding resacetophenone derivatives (7.7.3) have been isolated from rice seedlings wherein they, presumably, provide the plants protection from fungal attack.

7.8 PHYTOALEXINS

These chemical compounds are not constituents of normal, healthy plants, but are produced in response to infection by a pathogen. They belong to different chemical types, some of which are air-borne and some others root exudates. Methyl salicylate (7.8.1) is a air-borne chemical released from a fungus-infected part of the tobacco plant, which acts as a warning signal to other parts of the same plant or to other plants of the same species in the neighbourhood. Unique sulphur and nitrogen-containing heterocyclic compounds are produced as phytoalexins by members of the Cruciferae family. One such compound is sinalexin (7.8.2) which appears in the white mustard plant, *Sinapis alba*, when infected by the fungus, *Alternaria brassicae*.

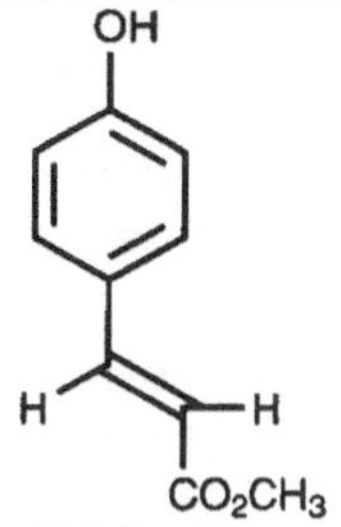

The leaves of the cucumber plant, *Cucumis sativa*, produce, in response to fungal infection, methyl *p*-coumarate (7.8.3), which has also been detected in *Eucalyptus kino*, the gummy exudate from Eucalyptus trees. This compound is toxic to species of the fungus Cladosporium.

In carrots (*Daucus carota*), the major phytoalexin in the root exudate is the isocoumarin, 6-methoxymellein (7.8.4). The flavanone, sakuranetin (7.8.5) is used as a phytoalexin by rice plants (*Oriza sativa*). Perhaps, the most well-known and extensively investigated group of phytoalexins are the pterocarpan type of isoflavonoids. One of the first compounds of this group to be studied was pisatin (7.8.6) produced by the pea plant, *Pisum sativum*, in response to fungal infections. Another examples is maackiain (7.8.7). The leaves of the grape vine, *Vitus vinifera* produce the stilbene, resveratrol (7.8.8), in response to either attack by the fungus *Botrytis cinerea* or exposure to ultraviolet light. Apart from methyl salicylate mentioned above, other air-borne organic compounds which are released from infected or mechanically damaged plant parts include (Z) 3-hexenol(7.8.9.) and its acetate (7.8.10). These compounds also serve as plant–plant communicators.

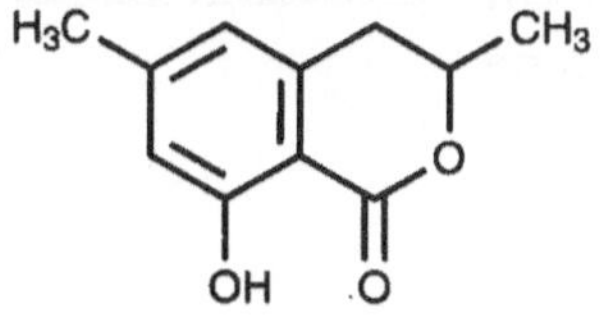

7.8.3

7.8.4

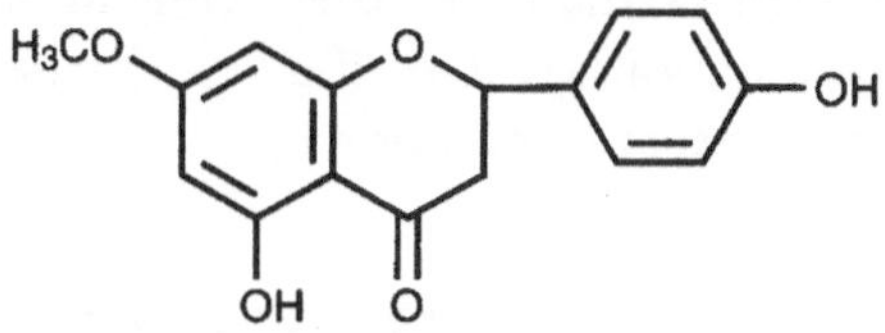

7.8.5

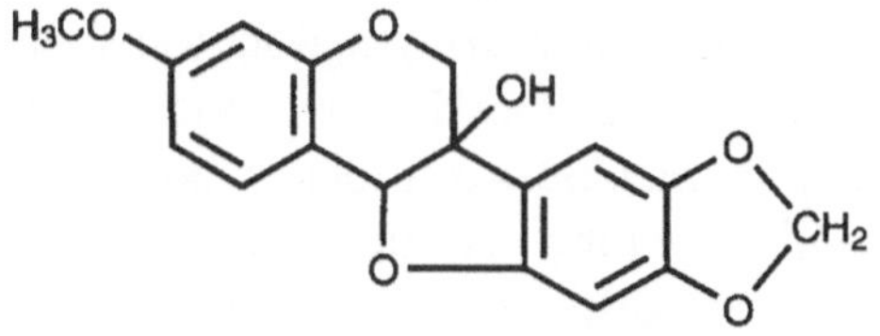

7.8.6

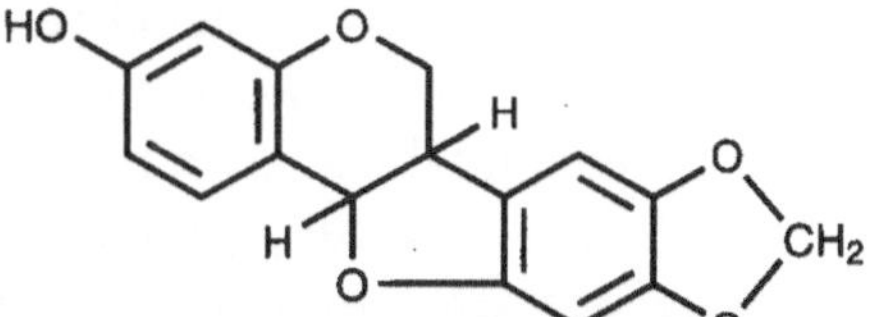

7.8.7

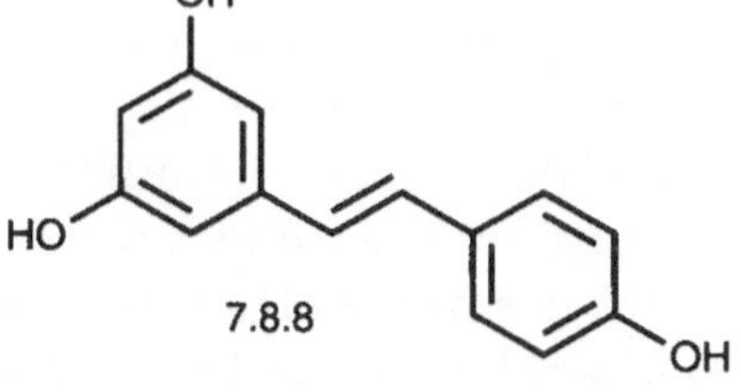

7.8.8

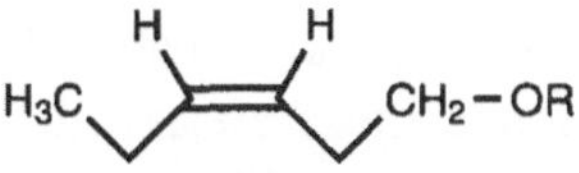

7.8.9: R = H
7.8.10: R = −C−CH₃

Suggested reading

1. Harborne JB, Recent Advances in Chemical Ecology, *Natural Product Reports*, 1999,16: 509–23.
2. Heil M, Lion U and Bolard W, *J. Chem. Ecol.*, 2008, 34, 601–4.
3. Heil M and Silva Beuno JC, *Proc. Natl. Acad. Sci.*, USA, 2007, 104, 5467–72.
4. Heil M, *New Phytologist*, 2008, 178: 41–61.

7.9 Insect—Animal Interactions

Chemical signals are widely used by insects and animals to defend themselves against other insect or animal preadtors. Numerous examples can be gleaned from recent publications in journals such as the *Journal of Chemical Ecology*. Some of these compounds are as simple as naphthalene (7.9.1) and quinoline (7.9.2), whereas some others are much more complex. Termites use naphthalene as an antiseptic to sterilise their nests! This compound also occurs in Magnolia flowers and in the forehead secretion of the white-tailed stag. The Peruvian fire-stick, *Oreophoetes peruana*, uses the more abnoxious quinoline (the nitrogen analogue of naphthalene) to repel a wide range of predators, including cockroaches, the wily spiders and even frogs. The compound is secreted as an aqueous emulsion which is sticky and repusive! Certain frogs store defensive chemicals in their skins, having acquired the compounds from their ant diet. One of these is gephyrotoxin (7.9.3). The compounds described in the next section also come under insect–insect interactions.

7.9.1

7.9.2

C_3H_7 7.9.3 C_4H_9

7.10 Defensive Secretions of Insects

There are several ways in which insects protect themselves against possible predators and adverse climatic conditions in their natural habitats. One mode of defence is physical such as sporting a red colour which serves as a warning signal because of its association with proven vicious arthropods like centipedes. For example, the slow-moving and quite harmless millipedes have this colour so as to confuse a predator which could mistake the millipede for a centipede—initially, at any rate! However, mere colour may not give adequate protection against more aggressive predators and, therefore, several species of millipedes have their own special chemical defence systems to ward off enemies.

Millipedes are found in large numbers in moist places, particularly in areas dominated by deciduous trees. They are larger than centipedes and are much more sluggish in their movement and hence more susceptible to attacks by predators. There are different orders, genera and species among millipedes and the chemicals used for defence vary with the order, if not with the genus and the species. Our knowledge of the chemistry of defensive secretions of millipedes is largely due to the extensive studies of Meinwald, Schildnecht and Towers, and is summarised below.

Millipedes of the order Polydesmus secrete, presumably, glycosides of mandelonitrile (7.10.1). The glycosides (7.10.2) have not yet been unequivocally characterised, but Towers and co-workers have isolated mandelonitrile from these millipedes by conventional methods of extraction and isolation. The secretions of Glomeris millipedes are more interesting from the point of view of chemistry as they contain 1,2-dialkyl-4(3H) quinazolinones (7.10.3 and 7.10.4).

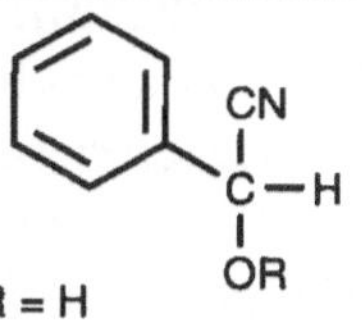

7.10.1: R = H
7.10.2: R = sugar

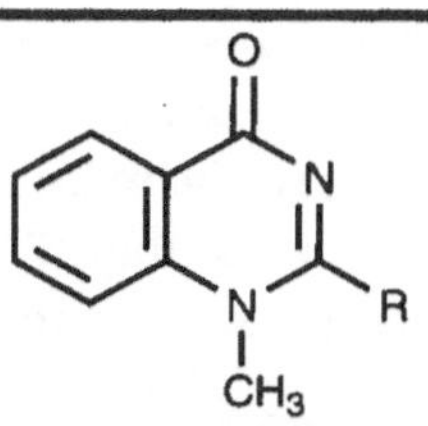

7.10.3: R = CH$_3$
7.10.4: R = C$_2$H$_5$

From one order of millipedes an interesting nitropyrrolizidine (7.10.5) has been isolated and characterised as the compound responsible for its defence against predators. The order Julia which comprises large millipedes, is associated with p-benzoquinone (7.10.6) as the defence secretion. Interestingly, from a single species of a genus, p-cresol (7.10.7) has been isolated and identified as the defence chemical. Another defensive secretion of one order is 2-dodecenal (7.10.8). It may be noted that these special chemicals are ejected from special glands when the millipedes are disturbed.

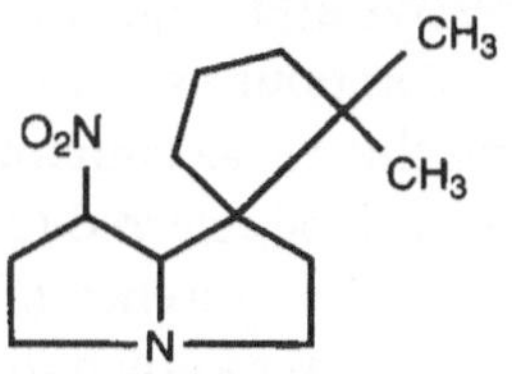

7.10.5

7.10.6

7.10.7

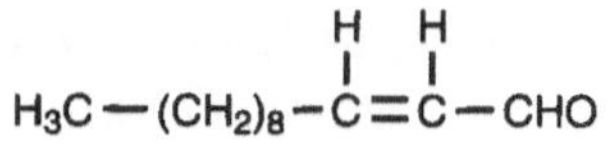

$H_3C-(CH_2)_8-C=C-CHO$

7.10.8

The structures of some of the chemical constituents of various ant venoms are shown below. The piperidine derivative, soleopsin-A (7.10.9), has been isolated from fire-ant venom. Thief-ant venom contains a dialkylpyrrolidine derivative (7.10.10). More complex nitrogenous compounds such as 7.10.11 have been isolated from the venom of the pharaoh ant. A tricyclic nitrogenous compound, precoccinelline (7.10.12) is synthesised by ladybird beetles for defence against predators. In this context, it is interesting to note that compounds related to precoccinelline have been isolated from an Australian shrub belonging to the family Euphorbiaceae. The structures of two of them, namely poranthericine (7.10.13) and poranthenine (7.10.14) are shown below.

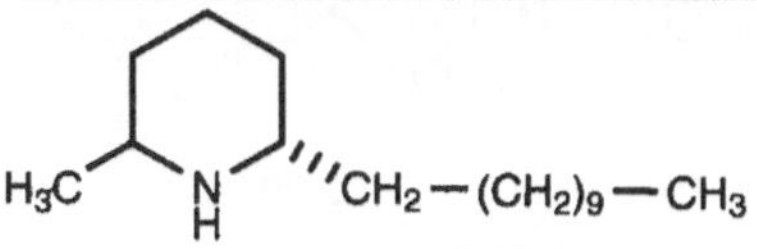

7.10.9

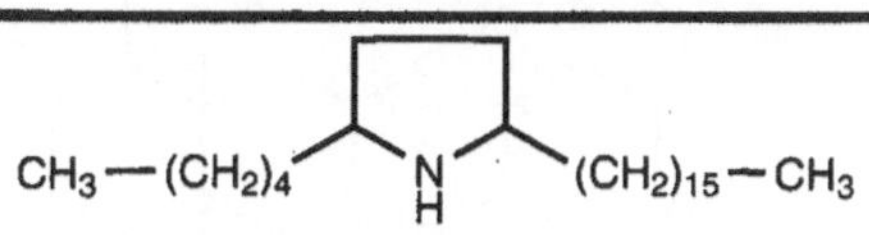

7.10.10

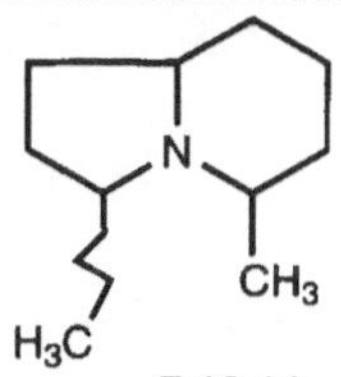

7.10.11

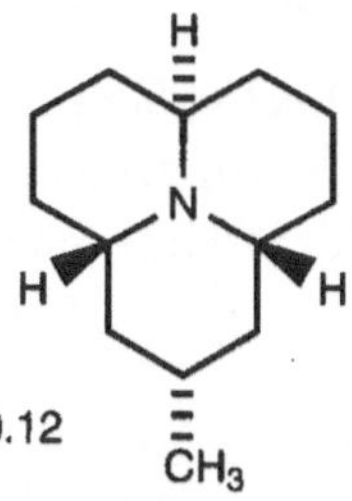

7.10.12

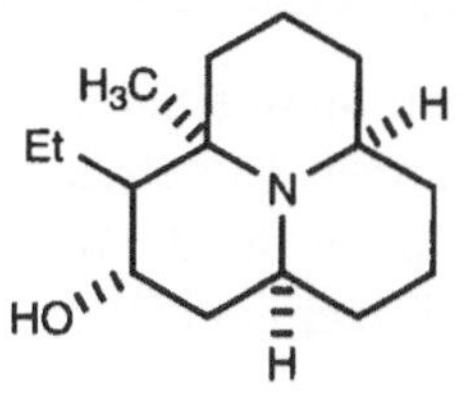

7.10.13

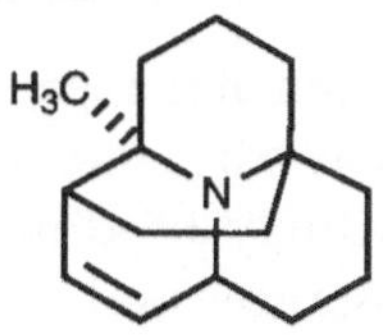

7.10.14

In contrast to the rather aggressive insects mentioned above, the innocuous-looking walking stick insect, *Anisomorpha buprestoides*, defends itself against predators by squirting an odorous spray whose chemical constituents have been analysed and characterised. The main component is the iridoid monoterpene, (+) anisomorphal (7.10.15). Its diastereromers, dolichodial (7.10.16) and peruphasmal (7.10.17) are also present in the spray. The ratios of these isomers vary from individual to individual.

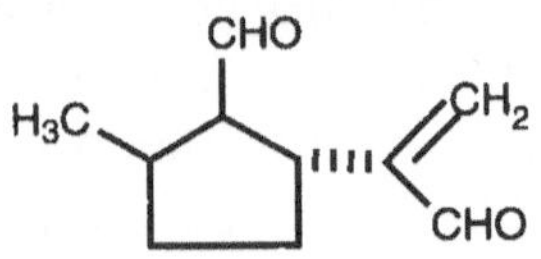

7.10.15 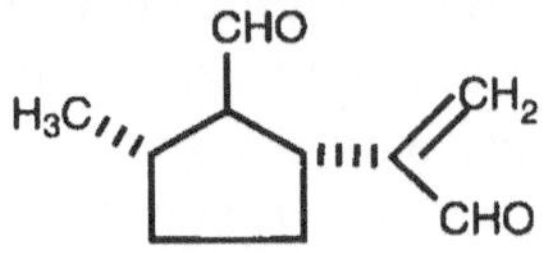   7.10.16 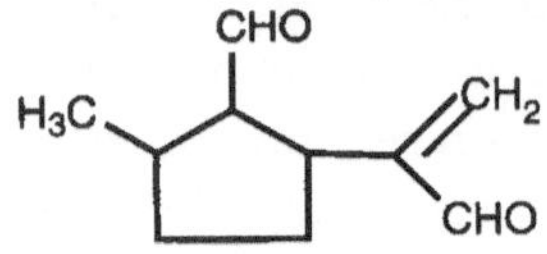   7.10.17

Another harmless insect is the butterly. Its bright colouring seems to act as a visual deterrent to some predators. But to defend themselves against more aggressive and inquisitive predators, butterflies do require more noxious chemicals. They derive such compounds at the larval stage from the plants on which they feed. Therefore, it is not surprising that these compounds belong to different classes of secondary metabolites. Only a few examples will be referred to here. Adult butterflies of the genus Euphydryas are unpalatble to birds due to the presence of the iridoid glycosides, such as catalpol (7.10.18). Some North American butterflies are known to contain the cardiac glycoside, calatropin (7.10.19), which gives them protection against birds. The presence of cyanogenetic glycosides in the defence secretions of millipedes was mentioned earlier. Butterflies belonging to the genus, Heliconius, whose larvae feed on the plant *Passiflora auriculata*, sequester lotaaustralin (7.10.20) and related compounds and use them to defend themselves against predators. The review by JR Trigo may be referred to for more details.

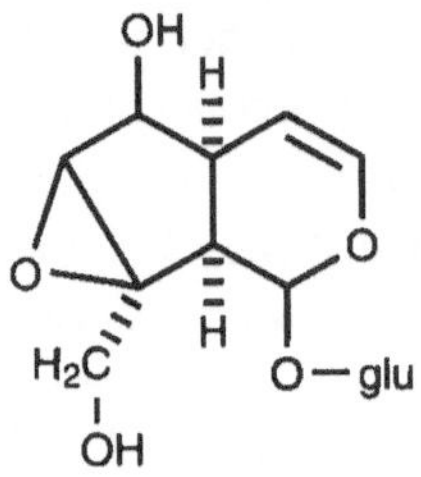

7.10.18 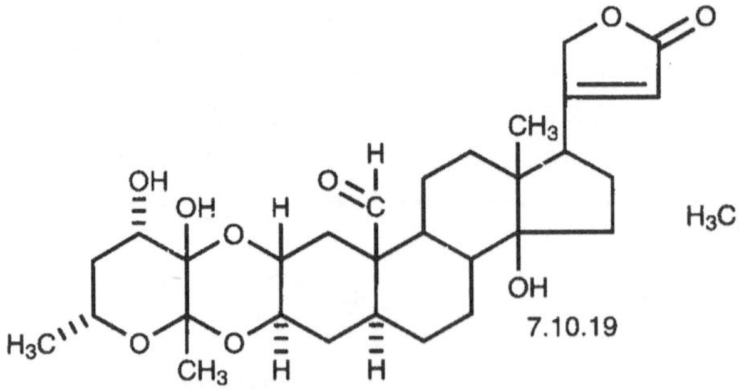   7.10.19 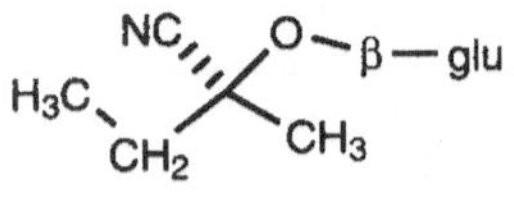   7.10.20

On page 319, the isolation of the semiochemical, stenusin, from the Rove beetle, *Stenus comma* was mentioned. Recently, two benzoic acid derivatives, namely, 3-hydroxy-4,5-dimethoxybenzoic acid (7.10.21) and 4-hydroxy-3,5-dimethoxybenzoic acid (syringic acid; 7.10.22) have been identified as defence substances in the beetle by Bitzer et al (Ref. 8).

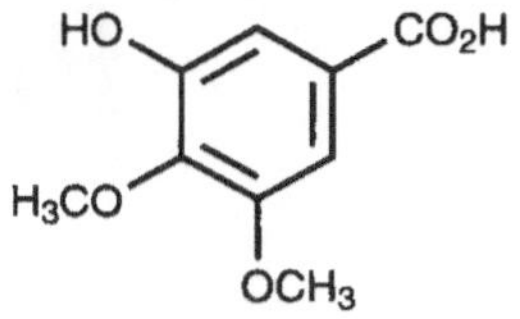

7.10.21 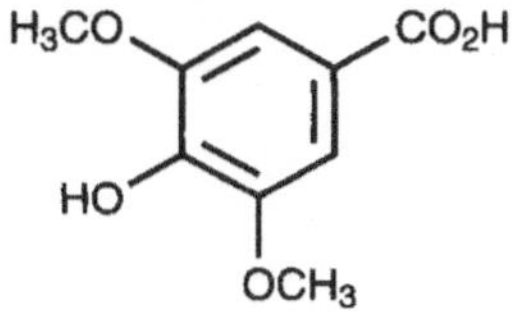   7.10.22

Suggested reading

1. Duffey SS, Underhill EW and Towers GH, Intermediates in the biosynthesis of HCN and benzaldehyde by a polydesmid millipede, *Harpaphe haydeniana* (Wood), *Comp. Biochem. Physiol,* 1974, 47(B), 753 and references therein.
2. Braekman JC, Daloze D, Dupont A, Pasteels JM, Lefeuve P, Bordereau C, Declercq JP and Van Meerssche M, *Tetrahedron,* 1983, 39: 4237.
3. Goldsworthy GJ and Wheeler CH, *Endeavour,* 1985, 9: 139.
4. Jurenko RA, Howard RW and Blomquist GJ, *Naturwiss.,* 1986, 73, 735.
5. Prestwiten GD, Chemical defense and self-defense in termites, In Atta-ur-Rahman, (ed.), *Natural Product Chemistry,* Berlin: Springer-Verlag, 1984.
6. Trigo JR, *J. Braz. Chem. Soc.,* 2000, 11: 1–20.
7. Dossey AT, Walse SS, Rocca JR and Edison AS, *ACS Chem. Biol.,* 2006, 1: 511–14.
8. Bitzer C, Brasse G, Dettner Kand Schulz S, *J. Chem. Ecol.,* 2004, 30: 1591.

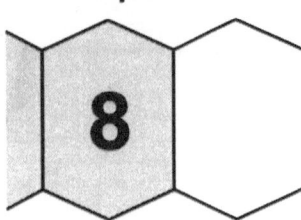

PROBLEM 8.1

A phenolic compound, **A**, isolated from *Helichrysum nannum* of the Compositae family, has the molecular formula $C_{13}H_{18}O_5$. It formed a diacetate and with ethereal diazomethane gave a monomethyl ether, **B**, which was steam volatile. **B** gave a purple colour with alcoholic $FeCl_3$ but failed to react with Gibb's reagent. The UV spectrum of **A** in methanol underwent bathochromic shifts with sodium acetate and with aluminium chloride. Its 1H NMR spectrum had the following signals: δ 0.9 (3H, t), 1.1 (3H, d), 1.3 (2H, dq), 2.5 (1H, m), 3.8 (3H, s), 3.95 (3H, s), and 6.7 (1H, s) besides signals at δ 6.1 (1H, br) and 11.2 (1H, s). Suggest structures for **A** and **B**.

Analysis: The formation of a diacetate shows the presence of two hydroxyl groups in **A**, one of which resists methylation by diazomethane, indicating that it is not sufficiently acidic to react with diazomethane (see Note 8.1.1). The properties of **B** (purple colour with $FeCl_3$ and volatility with steam) indicate the presence of an intramolecularly hydrogen-bonded phenolic OH as in 2-hydroxyacetophenone, salicylaldehyde and related compounds (see Note 8.1.2). The negative response to Gibb's reagent shows that the position *para* to the OH in **B** is blocked (see Note 8.1.3). The effect of shift reagents on the UV spectrum of **A** indicates

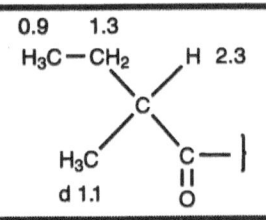

- the presence of an acidic OH as in 4-hydroxyacetophenone which is ionisable by the weakly basic sodium acetate;
- an intramolecularly hydrogen-bonded OH which can form a chelate complex with $AlCl_3$ (see Note 8.1.4).

The 1H NMR spectrum shows the presence of the unit shown above. The assignments are shown beside.

Compound **A** also has two Ar-OCH_3 groups (δ 3.8 and 3.95), an aromatic H (δ 6.7) and two phenolic OH groups (δ 5.1 and 11.2), one of which is hydrogen-bonded (δ 11.2) (*see* Note 8.1.5). Therefore, the possible structures for **A** and **B** are Ia, Ib or IIa, IIb (Ia, Ib are the correct structures). The position of the aromatic proton signal in the NMR spectrum is also in better agreement with Ia.

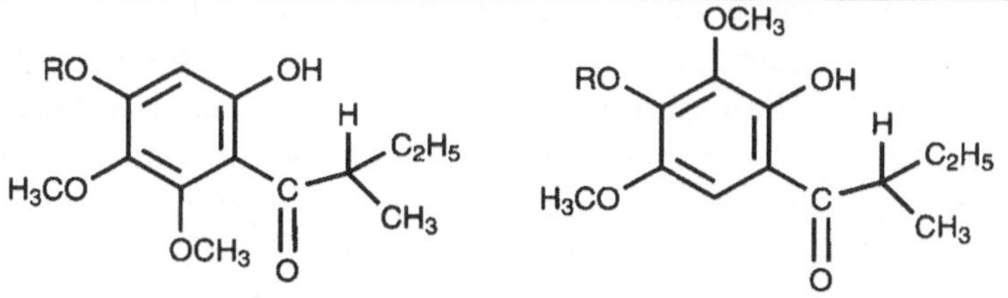

I a: R = H
  b: R = CH₃

II a: R = H
   b: R = CH₃

Notes

8.1.1 Diazomethane, $H_2C=\overset{+}{N}=\overset{-}{N}\leftrightarrow H_2\overset{-}{C}-\overset{+}{N}\equiv N$, can bring about facile *O*-methylation of phenols which are sufficiently acidic for the following reaction to occur first.

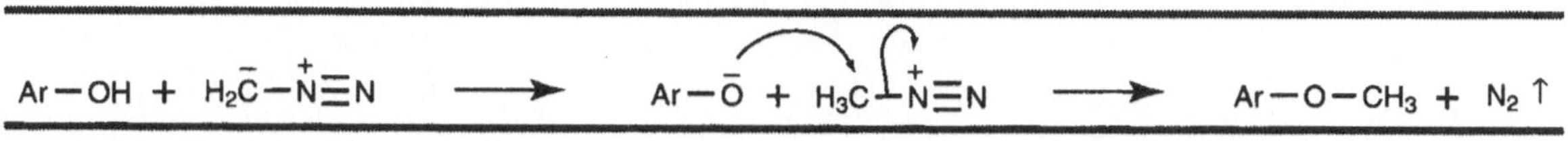

Phenols containing hydrogen-bonded OH groups are not sufficiently acidic for this reaction to take place.

8.1.2 Colour formation on reaction with alcoholic FeCl₃ has been a very useful diagnostic test for phenols, and the colour obtained can give valuable clues with regard to the disposition of the OH group(s). Ferric chloride is an oxidising as well as chelating agent and the colours formed may be due to either of these factors or a combination of the two. Thus, catechols which can undergo ready oxidation to *o*-quinonoid systems, give a green colour with FeCl₃. 2-Hydroxyacetophenone, salicylaldehyde and similar compounds possessing hydrogen-bonded OH groups give purple colour with FeCl₃. Interestingly, highly acidic phenolic OH groups, as in 4-hydroxybenzoic acid, 4-hydroxyacetophenone and related compounds give only a brown colour with FeCl₃. Many students may not be aware of the fact that certain non-phenolic compounds also give dark colours with FeCl₃. For example, *N*, *N*-dimethylaniline gives a blue colour due to oxidation.

8.1.3 Gibb's reagent is 2, 6-dichloroquinonechlorimide. Phenols that have the *para* position free react with this reagent in alkaline medium to give a green condensation product.

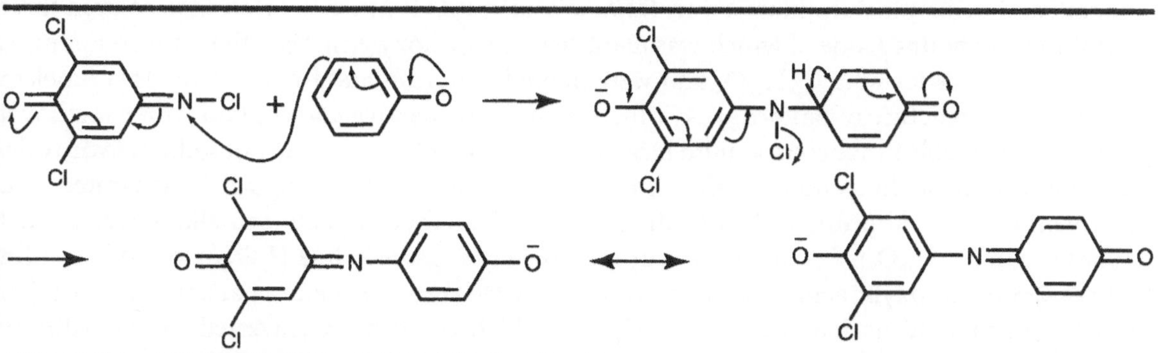

8.1.4 (i) The acetate ion, being a resonance-stabilised species, is a weak base and can bring about ionisation of only those phenols in which the OH group is *para* to an electron withdrawing functionality, most typically the carbonyl group. This ionisation is reflected as a measurable bathochromic shift in the UV spectrum of the compound and is of considerable diagnostic value. (ii) AlCl$_3$ forms 'stable' coloured chelate complexes bridging a phenolic OH and a carbonyl group placed *ortho* to it. The result is a large bathochromic shift in the absorption maximum of the parent compound. This effect has found an important application in the study of flavonoids and related polyphenols.

8.1.5 In 1H NMR spectra, the position of the OH protons is generally variable and is solvent- and concentration-dependent. Also, if the spectra are recorded in CDCl$_3$, the OH proton signals may not show up due to deuterium exchange. Deuterated DMSO is a more suitable medium. The far downfield position of the hydrogen-bonded OH proton signal is noteworthy and of considerable diagnostic value.

Reference

Bohlmann F and Suwita A, *Phytochem.*, 1979, 18: 2046.

PROBLEM 8.2

The bark of the Australian rainforest tree, *Halfordia scleroxyla* F. Mucll, was found to contain three crystalline compounds, one of which was identified as xanthoxyletin; the other two compounds were isomers with formula C$_{14}$H$_{12}$O$_6$ and were named halfordin and isohalfordin. This problem deals with the structure of halfordin, **A**, which is optically inactive and contains three methoxyl groups. It is insoluble in aqueous sodium bicarbonate but dissolves slowly in hot sodium hydroxide. On acidification of the yellow alkaline solution, halfordic acid (C$_{14}$H$_{14}$O$_7$), **B**, separated out. Treatment of both halfordin and halfordic acid with dimethyl sulphate and alkali gave methyl halfordic acid (C$_{15}$H$_{16}$O$_7$), **C**, which has four methoxyl groups. Alkaline H$_2$O$_2$ oxidised halfordin to furan-2,3-dicarboxylic acid. Treatment with boiling HCl brought about a partial demethylation of halfordin to yield norhalfordin (C$_{13}$H$_{10}$O$_6$), **D**, which could be reconverted to halfordin by treatment with ethereal diazomethane. Hydrogenation of halfordin in the presence of Adam's catalyst or Pd–C gave the dihydro derivative, **E**. Prolonged ozonolysis of halfordin followed by methanolysis resulted in a diphenolic aldehyde ($C_{10}H_{10}O_6$), **F**, whose 1H NMR spectrum had signals at δ 4.00 (3H, s), 4.08 (3H, s), 6.41 (1H, s), 10.23 (1H, s), 12.36 (1H, s) and 12.54 (1H, s). **F** could be hydrolysed to obtain an acid, **G**, $C_9H_8O_6$ Decarboxylation of **G** yielded an aldehyde which was identified as 2,4-dihydroxy-6-methoxybenzaldehyde. Suggest structures for halfordin, halfordic acid, their derivatives and degradation products (**A** to **G**).

Analysis: The solubility of **A** in hot aqueous alkali indicates the presence of a lactone unit. The co-occurrence of the compound with xanthoxyletin suggests that **A** is also probably a coumarin derivative; xanthoxyletin is 5-methoxy-6,7-(2', 2'-dimethylpyrano-5', 6') coumarin (I) (see Note 8.2.1). The formation of furan dicarboxylic acid (II) on oxidative degradation in alkaline medium indicates the presence of a furan unit. Therefore, **A** is most probably a furanocoumarin-like psoralen (III) and related compounds (see Note 8.2.2). It is the furan double bond and not the pyrone double bond which undergoes reduction on catalytic hydrogenation. The facile demethylation of one of the methoxyl groups indicates its location at position 4 of the coumarin moiety; 4-methoxycoumarin is known to undergo easy demethylation (see Note 8.2.3). At this stage halfordin (**A**) can be assigned the partial structure IV.

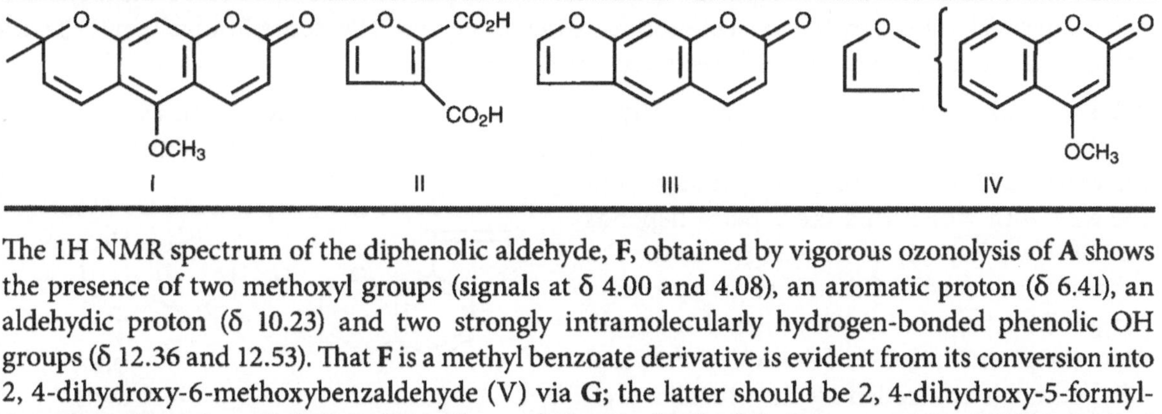

I   II   III   IV

The 1H NMR spectrum of the diphenolic aldehyde, **F**, obtained by vigorous ozonolysis of **A** shows the presence of two methoxyl groups (signals at δ 4.00 and 4.08), an aromatic proton (δ 6.41), an aldehydic proton (δ 10.23) and two strongly intramolecularly hydrogen-bonded phenolic OH groups (δ 12.36 and 12.53). That **F** is a methyl benzoate derivative is evident from its conversion into 2, 4-dihydroxy-6-methoxybenzaldehyde (V) via **G**; the latter should be 2, 4-dihydroxy-5-formyl-6-methoxybenzoic acid (VI) and **F** is the methyl ester (VII) of **G**. Since compound VII is derived from halfordin by the ozonolytic cleavage of the furan as well as the coumarin rings, halfordin can be formulated as VIII. The following diagram sums up the various transformations of halfordin described in this problem. The structure was confirmed by a synthesis achieved by Fukui and co-workers.

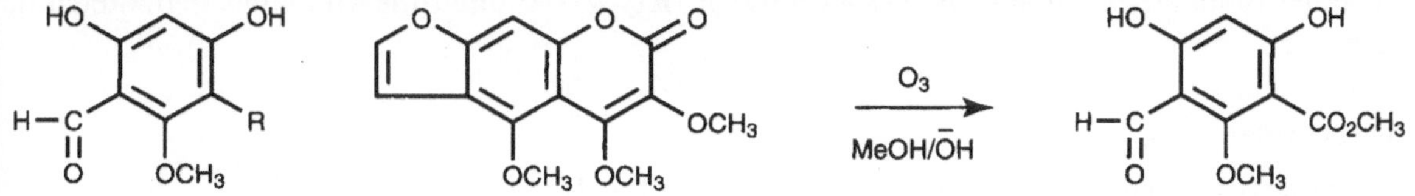

V: R = H
VI: R = CO$_2$H

VIII

VII

VIII

B: R = H
C: R = CH₃

D

Notes

8.2.1 Several furano and pyrano coumarins have been isolated from plants. Xanthoxyletin was first isolated as early as 1829 by Staples from the bark of the Rutaceous plant, *Xanthoxylum americanum*. Incidentally, the genus Halfordia also belongs to the family Rutaceae (subtribe Toddalinae). The only other member of this genus is *H. kendack* F. Muell. Both the trees grow in northern Queensland. *H. kendack* is also known in New Caledonia. The structure of xanthoxyletin was established by A Robertson and co-workers.

8.2.2 Among furano coumarins, psoralen is the simplest and best known. This compound which occurs in the seeds of *Psoralea corylifolia* is fluorescent and photodynamic. Its use in the treatment of leucoderma has been traced to its photosensitising properties. The isomeric angelicin (IX) with an angular furan ring has been isolated from Angelica species.

8.2.3 and 8.2.4 Aryl alkyl ethers generally require a strong mineral acid such as HI or HBr or a Lewis acid like $AlCl_3$ for dealkylation. The first step is protonation (or coordination with the Lewis acid) of the ether oxygen atom and the ease with which this takes place depends on the Lewis basicity of the oxygen atom. The subsequent removal of the alkyl group is also governed by the stability of the phenol liberated. Thus, the extent and ease of dealkylation of aryl alkyl ethers is controlled by structural as well as experimental parameters. In the present context, it is to be noted that of the three

X: R = CH₃
XI: R = H

methoxyls in halfordin one undergoes selective demethylation under relatively mild acidic conditions (with HCl). This is characteristic of enol methyl ethers, such as 4-methoxycoumarin (X), which can be demethylated even under alkaline conditions as it resembles an ester, being the ether of an acidic enol; 4-hydroxycoumarin (XI) was at one time known as benzotetronic acid.

References

1. Hegarty MP and Lahey FN, *Aust. J. Chem.*, 1956, 9: 12–131.
2. Lahey FN and MacLeod JK, *Tetrahedron Lett.*, 1968,447.
3. Fukui K, Nakayama M, Fujimoto S and Fukuda O, *Experientia*, 1969, 15: 354.

PROBLEM 8.3

The genus Diospyros is large, comprising evergreen deciduous trees, and belongs to the family Ebnaceae. The genus includes the well-known ebony wood, *D. ebenum*. Several other species are known in India and Australia. From an Australian species, *D. hebecarpa* A. Cunn (a small or medium tree common in Queensland), several interesting quinonoid compounds have been isolated along with a colourless compound which is the subject of this problem. The colourless compound, **A**, isolated from the fresh leaves and fresh immature fruits of *D. hebecarpa* has the molecular formula $C_{11}H_{10}O_3$ and is steam volatile. Treatment of **A** with hot acetic anhydride and sodium acetate gave a triacetate. When heated with dimethyl sulphate and alkali, **A** gave the trimethyl ether while the compound was recovered unchanged from attempted methylation with ethereal diazomethane. Oxidation of **A** with $FeCl_3$ in acidic medium yielded a red coloured compound, **B**, $C_{11}H_8O_3$, which could be converted into a colourless triacetate, **C**, $C_{17}H_{15}O_6$, by reaction with zinc and acetic anhydride. The mixed melting point of **C** with the triacetate of **A** was undepressed. The UV spectrum of **A** had absorption maxima at 235, 248, 270 and 348 nm whereas **B** absorbed at 253, 350 and 425 nm. Compound **B** exhibited a number of interesting colour reactions:

1. with aqueous alkali, it gave an unstable violet colour,
2. with dilute aqueous nickel acetate, a clear mauve solution was obtained,
3. with conc. H_2SO_4 it developed a brilliant red colour changing to dull purple on heating,
4. with Dimroth's reagent, a red colour, changing to purple-red on warming, was obtained.

The structure of **C** was determined by the following synthesis. Condensation of 4-chloro-*m*-cresol with maleic anhydride in the presence of molten $AlCl_3$ and NaCl at 180–200°C gave an orange-red product, **D**, $C_{11}H_7O_3Cl$. **D** gave a colourless triacetate, $C_{17}H_{15}O_6Cl$, on acetylation with acetic anhydride in the presence of zinc dust and sodium acetate. Reductive removal of the chlorine atom from this triacetate by catalytic hydrogenation in the presence of Pd–C resulted in **C**. Deduce the structures of **A** and its transformation products.

Analysis: The formation of a triacetate and a trimethyl ether indicates the presence of three hydroxyl groups in **A**, but its steam volatility is intriguing. As described under Problems 8.1 and 8.2, this observation strongly suggests the presence of a hydrogen-bonded OH as in 2-hydroxyacetophenone and similar compounds. **A**, which is colourless, gets converted into the deeply coloured **B** by a simple dehydrogenation brought about by the action of $FeCl_3$, which is a rather mild oxidising agent. That **B** is a quinone is evident from its colour reactions (see Note 8.3.1). The colourless triacetate, **C**, is a product of a reductive acetylation of **B**. This observation indicates the presence of a quinonoid moiety in **B** (see Note 8.3.2). The difference in the UV spectra of **A** and **B** is also significant. The removal of just two hydrogen atoms from **A** has brought about a bathochromic shift of more than 75 nm. If **B** is a quinone, **A** could be the corresponding *p*-quinol or something closely related to it (see Note 8.3.3). The structure of **C** becomes evident from its synthesis. **D** is a product of a Friedel–Craft acylation and should be 5-chloro-8-hydroxy-6-methyl-l, 4-naphthoquinone (I). Its formation and further transformation to **C**, which should therefore be l, 4, 5-triacetoxy-7-methylnaphthalene (II), are shown on the following page.

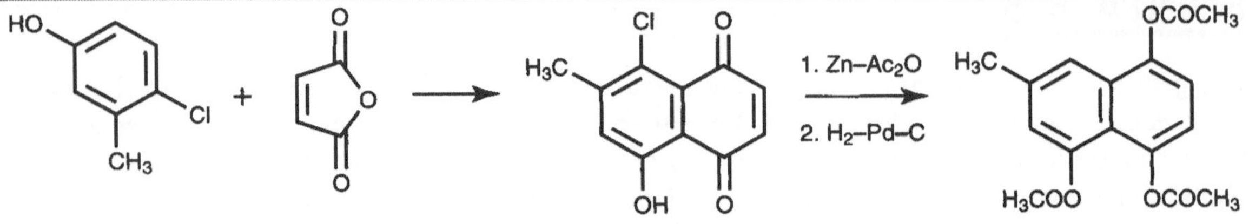

I (D)   II (C)

From the above discussion it logically follows that **B** is 5-hydroxy-7-methyl-l, 4-naphthoquinone or 7-methyljuglone (III) (see Note 8.3.4). One may be tempted to conclude from this that **A** should be l, 4, 5-trihydroxy-7-methylnaphthalene (IV) but this structure is not in accordance with some of the properties of **A**. As already mentioned, this structure does not account for the steam volatility of the compound. A more serious objection is that remains unchanged with diazomethane. As discussed under Problem 8.1, diazomethane fails to bring about the methylation of intramolecularly hydrogen-bonded phenolic OH groups as in 2-hydroxyacetophenone and related compounds but effectively methylates non-chelated phenolic OH groups. Therefore, the structure which can be assigned to **A** and which is compatible with its properties is l, 4-diketo-5-hydroxy-7-methyl-1, 2, 3, 4-tetrahydronaphthalene (V).

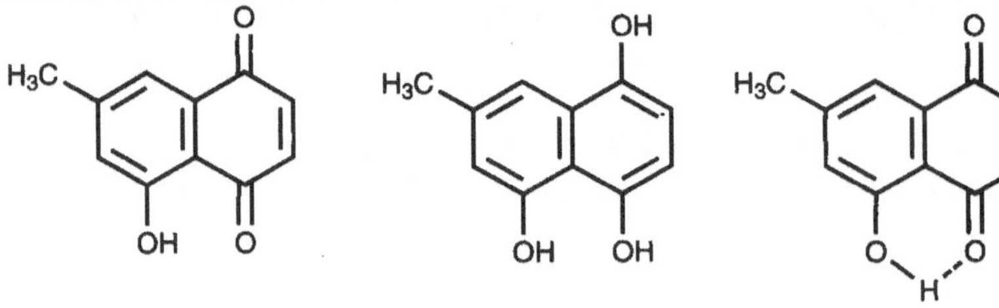

III   IV   V

It is unusual for a phenol to exist in the keto tautomeric form, particularly in a *p*-dihydric phenol (see Note 8.3.5). In the present case, the structure is stabilised by the intramolecular hydrogen bond involving the OH and one of the carbonyl groups.

Notes

8.3.1 The colour reactions mentioned here are characteristic of naphthoquinones of the juglone type; juglone is 5-hydroxy-l, 4-naphthoquinone (VI).

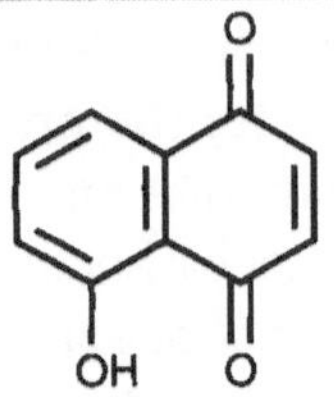

8.3.2 The mechanism of reductive acetylation can be depicted as shown below, with the electrons being supplied by the metal atom.

8.3.3 The observation that B absorbs at 425 nm, which means it is coloured yellow, indicates that it is a *p*-quinone; *o*-quinones absorb at longer wavelengths and are coloured orange or red.

VI

8.3.4 Juglone is an allelopathic agent (see Chapter 7) and occurs as the 4-*O*-glucoside of 1, 4, 5-trihydroxynaphthalene in the leaves of the oak tree, *Juglans nigra*.

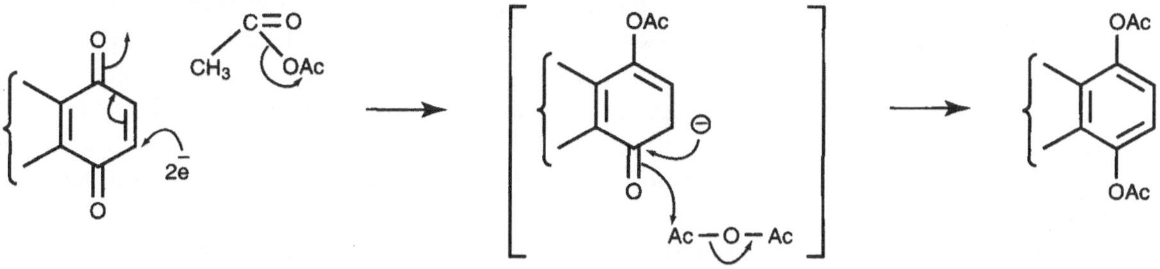

8.3.5 Among dihydric phenols, resorcinol is the only one which shows a tendency to react, under suitable conditions, as the diketo tautomer (VII). However, it should be noted that cyclohex-2-en-l, 4-diones have been prepared by the Diels–Alder reaction using *p*-benzoquinone as the dienophile. The following reaction which constituted the first step in Woodward's synthesis of reserpine is an example.

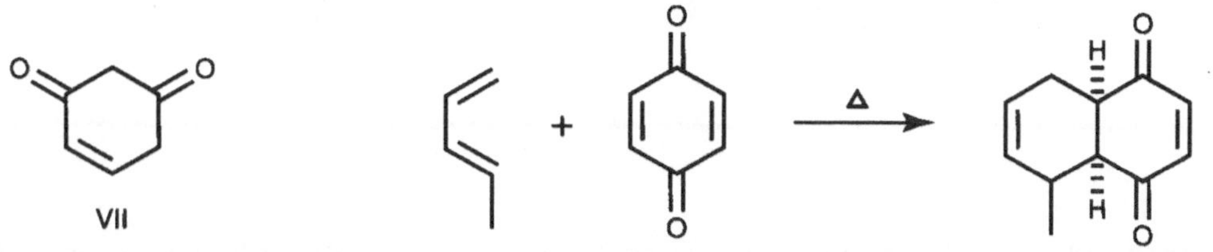

Reference

Cooke RG and Down H, *Aust. J. Chem.*, 1952, 5: 760–67.

PROBLEM 8.4

A blue fluorescent compound, **A**, isolated from the Indian medicinal plant, *Eclipta alba* Hassk., has 34.5% sulphur. It forms a monoacetate and occurs in *E. alba* not only in the free form but also as the angelate. **A** underwent facile oxidation under mild conditions (for example aerial oxidation in the presence of light) to yield **B**, which exhibited a greenish yellow fluorescence and had a higher R_f value than **A** on TLC plates. The 1H NMR spectrum of **B** had a singlet signal at δ 9.66. In the mass spectrum of **A**, the molecular ion signal appeared at m/z 278. Vigorous oxidation of **A** using aqueous $KMnO_4$ yielded thiophene-2, 5-dicarboxylic acid and a monocarboxylic acid, **C**, $C_9H_6S_2O_2$. **C** could be synthesised from 2, 2′-bithienyl by a Vilsmeier–Haack formylation followed by oxidation with Ag_2O. Interpret the above observations and deduce the structures of **A**, **B** and **C**.

Analysis: Since the molecular weight of **A** is 278 and it contains 34.5% of sulphur, the compound should have three sulphur atoms. The compound possesses a primary hydroxyl group which is most probably benzylic in nature as indicated by its facile oxidation to the aldehyde **B**. That **B** is the corresponding aldehyde is shown by its higher R_f value on TLC plates (see Note 8.4.1) and the signal at δ 9.6 in its 1H NMR spectrum. The results of oxidative degradation show that **A** and **B** are terthienyl derivatives. One terminal thiophene ring carries a –CH$_2$OH group in **A** and a –CHO group in **B** and this unit appears as thiophene-2, 5-dicarboxylic acid on oxidative degradation. It can be seen from its synthesis that compound **C** is bithienyl-2-carboxylic acid (see Note 8.4.2). Therefore, **A**, **B** and **C** can be assigned the structures I, II and III respectively (see Note 8.4.3).

Compound I was the first terthienyl derivative to be isolated from a natural source, though the simple, unsubstituted terthienyl had been isolated earlier from African marigold flowers (*Tagetes erecta*). Thiophene compounds are widespread in Compositae plants and biosynthetically arise from appropriately substituted polyacetylenes (see Note 8.4.4). The structure of **A** was confirmed by a synthesis with terthienyl as the starting material. Formylation of the latter by the Vilsmeier method gave **B**, reduction of which by NaBH$_4$ yielded **A**.

I: R = CH$_2$OH
II: R = CHO

III

Notes

8.4.1 Bithienyl and other polythienyl compounds exhibit brilliant fluorescence and can be readily detected by this means, on thin layer chromatograms. The colour of fluorescence (to be more precise, the emission maximum) depends on the size of the oligomer and also on the nature of the functional groups attached to the skeletal structure. While terthienyl methanol exhibits a bright blue fluorescence, the corresponding aldehyde shows a brilliant greenish yellow fluorescence. The light-induced oxidation of the alcohol to the aldehyde is brought about by singlet oxygen for the *in situ* generation of which the alcohol itself acts as a sensitiser. Available evidence shows that a sulphur-oxidation precedes the oxidation of the carbinol group.

8.4.2 The Vilsmeier–Haack formylation is very convenient for the introduction of a formyl group at position 2 of a thiophene ring. This is a consequence of the superior reactivity, compared to benzene, of thiophenes towards electrophiles. The preferential position of attack is α to the sulphur atom; β-Substitution takes place only when the α-position is blocked.

8.4.3 The conformations shown in the structural formulae I, II and III are supported by spectroscopic and diffraction data. These are to be preferred to the structures IV and V often found in most books and even research publications. The dipolar repulsive interactions between the sulphur atoms of the neighbouring rings are probably responsible for the higher energy of conformations IV and V.

IV   V

8.4.4 α-Terthienyl was first isolated by L Zechmeister and JW Sease, *J Amer. Chem. Soc,* 1947; 69: 273. Subsequently, it has been isolated from a number of plants of the Compositae family (for a recent review, see *J* Kagan, Naturally Occurring di- and tri-thiophenes, in W Herz et al (eds.), Zechmeister's *Progress in the Chemistry of Organic Natural Products,* 1991; 56: 87–179, Springer-Verlag, Wien-New York).

References

1. Krishnaswamy NR, Seshadri TR and Sharma BR, *Tetrahedron Letters,* 1966, 4227–30.
2. Krishnaswamy NR, Ch. Siva Sai Ramana Kumar and Prasanna S., *J. Chem. Res.,* 1991, 166 (S); 1991, 1801–30(M).

PROBLEM **8.5**

A yellow-coloured, water-insoluble compound, **A**, having the molecular formula, $C_{18}H_{16}O_7$, was obtained from the epigeal part of common betony *(Betonica officinalis)* by chromatography over polyamide. On chromatograms it appeared as a dark spot which turned yellow on spraying with aqueous alkali. Its UV spectrum had absorption maxima at 204, 271 (sh), 310 (sh) and 331 nm. Bathochromic shifts were observed on the addition of alkali, and of $AlCl_3$ with and without HCl. The addition of sodium acetate–boric acid had no effect on the spectrum. The 1H NMR spectrum of the compound had the following signals: δ 3.68 (6H, s), 3.78 (3H, s), 6.72 (1H, s), 6.72 (1H, d, J = 2 Hz), 6.95 (1H, d, J = 2 Hz), and 8.12 (2H, s). Demethylation of the compound followed by alkaline degradation yielded phloroglucinol and gallic acid. Deduce the structure of the compound.

Analysis: The molecular formula (7 oxygens) and the yellow colour given with alkali on chromatograms indicate that the compound is a polyphenol. The UV spectrum is characteristic of flavonoids (see Note 8.5.1). The bathochromic shift with alkali confirms the presence of phenolic OH groups. An intramolecularly hydrogen-bonded OH is revealed by the bathochromic shift with $AlCl_3$. The stability of the Al complex in HCl shows that this intramolecular hydrogen-bonding is as strong as in 5-hydroxyflavones. The absence of a catechol group is indicated by the absence of any shift on addition of sodium acetate–boric acid (see Note 8.5.2). The singlet signals at δ 3.68 and 3.78 totally integrating for nine protons in the 1H NMR spectrum show that the compound has three methoxyl groups; this is also supported by the observation that the compound could be demethylated. The formation of phloroglucinol (I) and gallic acid (II) on alkaline degradation of the product of demethylation reveals the oxygenation pattern of positions 5, 7, 3', 4', 5' in the compound (the remaining two oxygen atoms are of the middle pyrone ring). Therefore, the product of demethylation (not isolated and characterised) should have been III, that is, 5, 7, 3', 4', 5' pentahydroxyflavone.

The three methoxyl groups in **A** could be located by identifying the hydrogen atoms responsible for the signals between δ 6.72 and 8.12. The singlet at δ 6.72 could be attributed to H-3 and the doublets at δ 6.72 and δ 6.95 to H6 and H8 (*J* value agreeing for *m*-coupling), respectively. The 2H singlet at δ 8.12 is unambiguously assigned to H-2' and H-6'. Since the compound does not have a catechol group but has a free OH at position 5, the three methoxyls could be located at positions 7, 3' and 5'. Thus, the compound **A** is 5, 4'-dihydroxy-7, 3', 5'-rimethoxyflavone (IV) (see Note 8.5.3).

I: R = H; R' = OH
II: R = OH; R' = CO₂H
III
IV

Notes

8.5.1 UV spectra are of great value in structural studies on flavonoids. All flavones and flavonols exhibit two well-defined maxima, with one or more shoulders, in their UV spectra. The shorter wavelength band is due to the chromophoric part in the A ring; whereas the longer wavelength band arises by excitation of the chromophoric part of the B and C rings as indicated below (see JB Harborne, *Phytochemical Methods*, Chapman & Hall, London, 1973 and KR Markham, Techniques of flavonoid identification in Harborne JB and Mabry TJ, (eds), *The Flavonoids*, London: Chapman & Hall, 1982.

8.5.2 The addition of AlCl₃ brings about large bathochromic shifts in the spectra of 3- and 5-hydroxyflavones. These shifts remain unaffected on further addition of HCl. On the other hand, catechol groups register a bathochromic shift with AlCl₃ alone as the metal complex in this case is unstable in the presence of HCl. The magnitude of the shift gives valuable clues as to whether both 3- and 5-OH groups are present or only one. The diagnostic reagent for detecting the presence of a catechol group is sodium acetate–boric acid. In this case, any bathochromic shift observed is due to the formation of a borate complex.

III: R = H
V: R = CH₃

It may be mentioned here that, generally, vicinal 1, 2-diols form such stable borate complexes. Another way this may be detected is to measure the conductivity of boric acid which is increased by the addition of a 1, 2-glycol.

8.5.3 5, 7, 3', 4', 5' -Pentahydroxyflavone (III) is known as tricetin. Tricin (V) is the 3', 5'-di-O-methyl derivative of tricetin. Partial methyl ethers of flavones are fairly widespread in the plant kingdom, particularly in higher plants.

Reference

Kobzar AYa. and Nikonov GK, *Chemistry of Natural Compounds* (English translation of Khimiya Prirodnykh Soedinenii), 1986, 22: 600–601.

PROBLEM 8.6

Ergolide, A, $C_{17}H_{22}O_5$, is a constituent of the aerial parts of *Erigeron khorasannicus* of the Compositae family. The compound is strongly dextrorotatory. On heating with selenium at a high temperature (350°C) it yielded a blue-violet liquid. The IR spectrum of ergolide has absorption bands at 1770, 1740 and 1670 cm^{-1}. Its 1H NMR spectrum has signals at δ 1.08 (3H, s), 1.12 (3H, d, J = 6.6 Hz), 1.86 (1H, m), 1.98 (3H, s), 2.28 (1H, m), 2.50 (1H, octet, J = 13.3, 4.4 and 2.8 Hz), 3.03 (1H, m), 4.49 (1H, octet, J = 11.7, 10.3 and 2.8 Hz), 5.50 (1H, d, J = 7.8 Hz), 5.86 (1H, d, J = 3.1 Hz) and 6.21 (1H, d, J = 3.5 Hz). Sixteen lines are seen in the 13C NMR spectrum at δ 18.4 (q), 19.8 (q), 21.1 (q), 24.5 (t), 30.0 (d), 38.1 (t), 44.7 (t), 46.6 (d), 52.3 (d), 56.2 (s), 75.3 (d), 76.4 (d), 120.6 (f), 139.0 (s), 169.4 (s) and 214.6 (s). Saponification of ergolide yielded a compound, **B**, with the molecular formula $C_{15}H_{20}O_4$ which gave ergolide on treatment with acetic anhydride. Suggest an acceptable structure for ergolide.

Analysis: The molecular formula strongly suggests that ergolide is a sesquiterpenoid. Sesquiterpene lactones are widespread in plants of the Compositae (also called Asteraceae) family (see Note 8.6.1). Thus, on phytochemical grounds, and keeping in view the oxygen content, it could be further inferred that **A** is a sesquiterpene lactone. The spectral data lend strong support to this conjecture. Thus, the absorptions in the IR spectrum at 1640 and 1770 cm^{-1} indicate the presence of an α, β-unsaturated-γ-lactone moiety (see Note 8.6.2). The band at 1740 cm^{-1} shows the presence of an ester function which is also supported by the observation that **A** could be saponified to **B**, which is deacetyl **A**. These

arguments account for four of the five oxygen atoms. Since the compound does not contain any OH group (notice that there is no absorption in the 3500–3600 cm^{-1} region of the IR spectrum), the fifth oxygen could either be an ether function or a carbonyl group whose absorption band overlaps that of the ester group; in such a case it should be part of a five-membered ring structure.

Strong support for the presence of a carbonyl group is provided by the 13C NMR spectrum (see Note 8.6.2) which has a singlet signal at δ 214.6. The bluish violet product of selenium dehydrogenation of ergoline is most definitely an azulene (I) (see Note 8.6.3) and this fact defines the skeletal structure of ergoline. The 1H and 13C NMR spectra provide the following information: the presence of an exocyclic methylene group of the unsaturated lactone moiety [characteristic signals at δ 5.86, 6.21, 120.6 (t) and 139.0 (s)], an angular methyl group at one of the bridge-head positions of the azulene skeleton [δ 1.08 (3H, s) and δ 18.4 (q)], and a secondary methyl group [1.12 (3H, d) and 19.8 (q)]. These data, in conjunction with biosynthetic considerations (see Note 8.6.4), led to the partial structure II for **A**. The carbonyl group, which is present in the five-membered ring, and the acetoxyl group can be located by a careful analysis of the other NMR signals. The assignments are shown in the final structure, III.

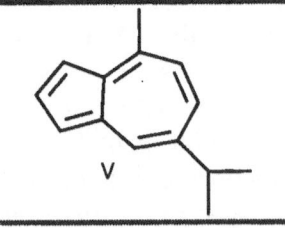

Notes

8.6.1 Plants of the family Compositae have been known to produce a wide range of sesquiterpene lactones of different skeletal types. A large number of these compounds have been isolated and characterised by W Herz, a pioneer in this field, Sorm F, Kupchan SM, Bohlmann F, Geissman TA, Mabry TJ and several others. Some of these compounds, like vernolepin, exhibit anti-cancer activity.

8.6.2 α, β-Unsaturated five-membered lactones exhibit characteristic absorptions at about 1770 cm⁻¹ and 1640 cm⁻¹ due to the lactone carbonyl and the double bond respectively. The absorption at 1740 cm⁻¹ is obviously a composite band due to the carbonyl group in a five-membered ring structure and a saturated ester function. The 13C NMR signal beyond δ 200 is a clear and unambiguous indication of the presence of a carbonyl group.

8.6.3 Azulenes are blue aromatic hydrocarbons which were first discovered by the Swiss chemist L Ruzicka. These compounds are interesting as they are isomeric with, but less aromatic than, the corresponding naphthalenes. Azulenes are also dipolar and react with both electrophiles and nucleophiles with equal ease. The azulene obtained from ergolide has been identified as guaiazulene (2-methyl-5-isopropylazulene, V).

8.6.4 The identification of guaiazulene fixes the skeletal structure as well as the positions of the secondary methyl and the isopropyl groups in ergolide. The position of the angular methyl follows from an application of the isoprene rule of Ruzicka.

Reference

Ovezdurdyev A., Abdullaev ND, Kasymov ShZ and Akyev B, *Chemistry of Natural Compounds*, 1986, 22: 532–35.

PROBLEM 8.7

From an ethyl acetate extract of the red alga, *Laurencia pinnatifida*, a compound, **A**, with the molecular formula $C_{20}H_{40}O$ was obtained as a mobile oil. Its IR spectrum showed strong absorption at 3350 cm^{-1}. In its mass spectrum, besides the molecular ion signal at m/z 296, there were peaks at m/z 278 and 269. Its 1H NMR spectrum had the following signals: δ 0.83 (3H, d, J = 6.56 Hz), 0.84 (3H, d, J = 6.56 Hz), 0.86 (6H, d, J = 6.64 Hz), 1.27 (3H, s), 5.03 (1H, dd, J = 10.76, 1.30 Hz), 5.18 (1H, dd, J = 17.36, 1.30 Hz) and 5.91 (1H, dd, J = 17.36, 10.76 Hz). The 13C NMR spectrum had signals at 19.7 (q), 21.3 (t), 22.6 (q), 22.7 (q), 24.8 (t), 27.7 (q), 28.0 (d), 32.7 (d), 32.8 (d), 37.3 (t), 37.4 (t), 39.4(t), 42.7 (t), 73.3 (s), 111.4 (t) and 145.3 (s). Deduce the structure of **A**.

Analysis: The oily nature of the compound, its solubility in nonpolar solvents, and its molecular formula indicate that A could be a long-chain aliphatic alcohol. The presence of a hydroxyl group is strongly supported by the 3350 cm^{-1} absorption band in the IR spectrum and the M$^+$–18 peak in the mass spectrum (see Note 8.7.1). The 1H NMR spectrum shows the presence of five methyl groups, including an isopropyl group (6H doublet signal at δ 0.86), two

13C	SIGNALS
3	73.3
3a	27.7
7a	19.7
11a	22.6
15a	22.7
16	
1	111.4
2	145.3

secondary and one tertiary methyl groups, and a terminal double bond (as shown by the 1H dd signals at δ 5.03, 5.18 and 5.91). The presence of the vinyl grouping is also supported by the M$^+$–27 peak in the mass spectrum. A mature consideration of the above facts strongly suggests that **A** is an acyclic, mono-unsaturated diterpenoid tertiary alcohol; the 3H singlet signal at δ 1.27 may be assigned to the methyl group on the carbinol carbon atom. In the 13C NMR spectrum, the signal due to the alcoholic carbon atom appears at δ 73.3 and those due to the olefinic carbon atoms at δ 111.4 (terminal carbon) and 145.3. The compound could therefore be identified as isophytol with the structure I shown (see Note 8.7.2). The 13C NMR assignments are also indicated.

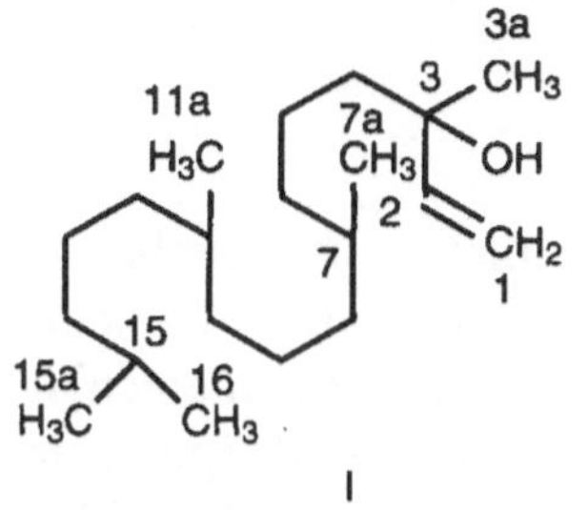

Notes

8.7.1 The loss of a molecule of water can be explained by invoking a McLafferty-type fragmentation process as shown below. The resulting ion radical may further undergo a 1, 2-hydrogen atom shift to yield the more stable species, as indicated below.

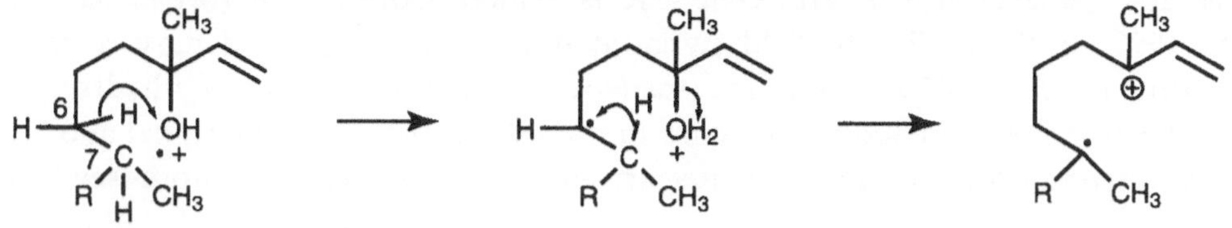

8.7.2 *Trans*-phytol (II), which is a primary allylic alcohol, is a well-known diterpene alcohol with wide occurrence in terrestrial as well as marine plants; chlorophyll is a phytyl ester. However, the isomeric isophytol described in this problem, with a terminal double bond and a tertiary allylic OH, is not so common. It has also been isolated from jasmine (Demole E and Lederer E, *Bull. Soc. Chim. France*, 1958, 1128).

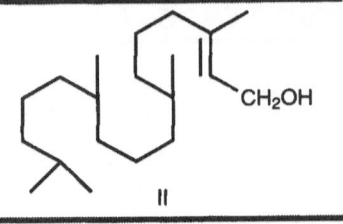

Reference

Ahmad VU and Ali MS, *Phytochem.*, 1991, 12: 4172–74.

Problem 8.8

From the toluene-soluble fraction of a methanolic extract of the sponge *Penares* sp., a colourless, optically active compound, A, named as penasterol, was obtained as needle-shaped crystals from hexane–ethyl acetate. It had the molecular formula $C_{30}H_{43}O_3$. Its 13C NMR spectrum was similar to that of lanosterol (I). The IR spectrum of penasterol had absorption bands at 3600–2500 and 1695 cm^{-1}. Its 1H NMR spectrum had a low-field signal at δ 11.7 and a broad signal at δ 3.39, which disappeared on addition of D$_2$O. In the mass spectrum, there were signals at m/z 456 (due to the molecular ion) and 411. In the 13C NMR spectrum, there was a singlet signal at δ 176.4. A remained unaffected when attempts were made to bring about an intramolecular dehydrative ring-closure reaction. The 13C NMR spectrum of A had signals attributable to seven methyl groups and these appeared at δ 0.69 (3H, s), 0.70 (3H, s), 0.88 (3H, d, *J* = 6.8 Hz), 0.89 (3H, s), 0.95 (3H, s), 1.55 (3H, s) and 1.63 (3H, s). Other signals in the spectrum were at δ 0.96–2.28 (23H, m), δ 2.99 (1H, dd, *J* = 10.3, 5.7 Hz) and 5.06 (1H, t, *J* = 7 Hz). Penasterol could be converted into lanosterol-*O*-methyl ether by the following sequence of reactions: (i) diazomethane in methanol, (ii) NaH, MeI in THF (reflux), (iii) LAH in THF (reflux); (iv) pyridinium chlorochromate in dichloromethane (room temperature), and (v) 80% hydrazine–KOH in ethylene glycol, 180°C. Interpret the data given above and arrive at an acceptable structure for penasterol. The structure of lanosterol (I) is given.

Analysis: That penasterol is a carboxylic acid is evident from the IR (broad band at 3600–2500 and 1695 cm^{-1}), NMR (low-field signal at δ 11.7 and 176.4) and mass spectral (peak corresponding to an M$^+$–45 species) data (see Note 8.8.1). The presence of a hydroxyl group is shown by the broad band at about 3600 cm^{-1} in the IR spectrum and the signal at δ 3.39 in the 1H NMR spectrum, which is due to a deuterium-exchangeable hydrogen atom (see Note 8.8.2).

The similarity to lanosterol (I) (see Note 8.8.3) has been mentioned. It is therefore clear that penasterol differs from I in having a carboxyl group in place of one of the methyls. That the side chain in penasterol is the same as that in lanosterol is shown by the presence of a secondary methyl and two methyls on an olefinic carbon (as indicated by the signals at δ 0.88 (d), 1.55 and 1.63 in the 1H NMR spectrum) in penasterol.

The resistance towards a dehydrative ring-closure, that is, lactonisation, rules out positions 4 and 10 for the carboxyl group (see Note 8.8.4). These observations allow the carboxyl group to be located at position 13 or 14 in I. With the available data it is not possible to make a choice between these two positions but other data show that 14 is the correct position (structure II).

The sequence of reactions which brings about the conversion of penasterol into lanosterol methyl ether involves the step-wise transformation of the carboxyl group to a methyl group in four stages; in the course of this, the hydroxyl is converted to a methoxyl. An outline of the entire sequence is given below.

Notes

8.7.1 The broad absorption band between 3600 and 2500 cm^{-1} in combination with the sharp band at 1695 cm^{-1} is diagnostic of the carboxyl group. This is amply supported by the low-field proton signal beyond δ 10 in the 1H NMR spectrum, and the M$^+$−45 peak (due to loss of the carboxyl group) in the mass spectrum.

8.7.2 The deuterium exchange technique is a convincing way of identifying signals due to OH protons in 1H NMR spectra. In the IR spectrum, the absorption band due to O–H stretching overlaps that due to the OH of the carboxyl group.

8.7.3 Lanosterol is a key intermediate in the biosynthetic transformation of the acyclic triterpene hydrocarbon, squalene, into the tetra- and penta-cyclic triterpenoids and the sterols. It was first isolated from lanstin (wool fat).

8.7.4 A carboxyl at position 10 could be expected to readily interact with the hydroxyl at position 3 to form a six-membered lactone. A carboxyl at position 4 can form only a four-membered lactone. Positions 13 and 14 are too far away from the OH at position 3 for any type of interaction to take place.

Reference

Cheng JF, Kobayashi J, Nakamura H, Ohizumi Y, Hirata Y and Sasaki T, Penasterol, *Penares* sp., *J. Chem. Soc, Perkin*, 1988, 1: 2403–6.

PROBLEM 8.9

From the hexane and ether extracts of freeze-dried limpet, *Collisella limatula,* a colourless, oily compound, limatulonc, A, was isolated. This compound had very potent anti-feedant inhibitory action on fish and crabs. It had the molecular formula $C_{30}H_{45}O_4$ but showed only 15 signals in its 13C NMR spectrum. Its IR spectrum had absorption bands at 3500 (br) and 1700 cm^{-1}. The 13C NMR spectrum had signals at δ 215.4 (s) and 70.0 (d), apart from other signals. The 1H NMR spectrum of limatulone had the following signals: δ 0.88 (3H, s), 1.35 (1H, m), 1.45 (1H, br, d, J = 13 Hz), 1.50 (1H, m), 1.58 (3H, br, s), 1.63 (1H, m), 1.67 (3H, br, s), 1.83 (1H, br, d, J = 13 Hz), 2.06 (1H, m), 2.19 (1H, td, J = 13, 4 Hz), 2.54 (1H, td, J = 13, 4 Hz), 3.18 (2H, br, d, J = 6.5 Hz), 4.70 (1H, br, s), 5.25 (1H, t, J = 7 Hz) and 5.59 (1H, br, t, J = 6.5 Hz). A 2D COSY (correlation spectroscopy) experiment showed that the signal at δ 5.59 was coupled to the 3H signals at 1.58 and 1.67. W-coupling between the signals at δ 0.88 and 2.19 was also noticed. Irradiation of the signal at δ 4.70 caused NOE enhancement of the signals at δ 3.18 (4%), 2.06 and 1.83 (9% combined) and 0.88 (5%). Irradiation of the 5.25 signal brought about NOE enhancement of the signal at δ 1.83. It has been suggested that limatulone is biosynthetically derived from squalene. Limatulone undergoes an ene-reaction with molecular oxygen to give a hydroperoxide which possesses an α, β-unsaturated carbonyl group. Interpret the above data and suggest an acceptable structure for limatulone.

Analysis: The molecular formula and the biosynthetic origin from squalene show that limatulone is a triterpene (see Note 8.9.1). The presence of only 15 signals in its 13C NMR spectrum indicates that the compound is made up of two identical subunits, each of which contains a carbonyl group (absorption in the IR region at 1700 cm^{-1} and 13C NMR signal at δ 215.4) and a secondary hydroxyl group [absorption in the IR region at 3500 cm^{-1} and 13C NMR signal at 70.0 (d)].

Each subunit also has two double bonds, each of which is trisubstituted as shown by the one-proton signals at 5.25 and 5.59. The fact that each of these signals is a triplet further indicates the presence of two units of the type I. This fact as well as the IR stretching frequency of the carbonyl group shows that neither double bond is conjugated with the carbonyl group. However, the presence of the unit II is indicated by the observation that limatulone undergoes an ene-reaction with oxygen which can be interpreted as follows (see Note 8.9.2).

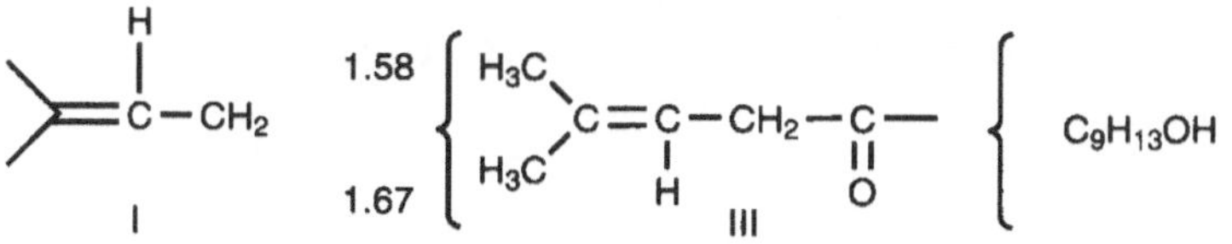

The results of the 2D COSY experiment (see Note 8.9.3) show the presence of a terminal γ, γ-dimethylallyl unit which should be next to the carbonyl group. Therefore, the partial structure of each subunit of limatulone can be further expanded to III.

The formula of the connecting unit, namely, $C_9H_{13}OH$, in the above part structure, shows that this unit has two degrees of unsaturation, one of which is a trisubstituted double bond as mentioned earlier. Therefore, this connecting bit of the subunit of limatulone should possess a monocyclic ring moiety, most probably a six-membered ring which carries the secondary hydroxyl group. This part of the molecule also has a tertiary methyl group as shown by the 3H singlet signal at δ 0.88. The observation that this signal suffers an NOE enhancement when the signal at δ 4.70 is irradiated allows one to place this methyl group on a carbon atom next to the carbinol carbon; the signal at δ 4.70 is assigned to the hydrogen atom on the carbinol carbon (see Note 8.9.4). A mature consideration of the above arguments leads to the structure IV for each subunit of limatulone and structure V for the whole molecule. The other 1H NMR assignments are shown in IV.

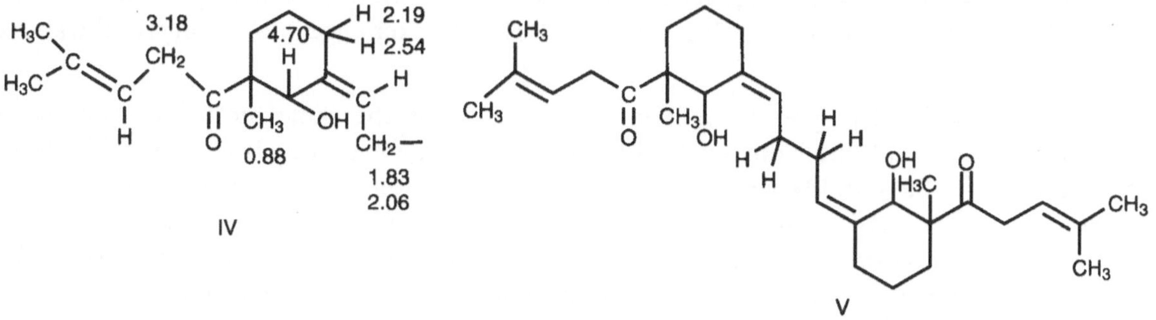

Notes

8.9.1 Limpets are marine molluscs found in intertidal zones. They are endowed with tough shells for protection against the harsh physical environs of their habitat. Among the five dominant limpet species found off the southern Californian coast, *Collisella limatula* is unique. It possesses a chemical defence system to protect itself against predators like fishes and crabs. The molecular formula of limatulone suggests that it is a triterpene and there are indications that it is derived from squalene.

8.9.2 The ene-reaction between an allylic system and molecular oxygen is most typically brought about using singlet oxygen which can be generated in situ using light and a dye-sensitiser such as Rose Bengal.

8.9.3 A 2D COSY experiment involves the use of homonuclear shift correlation through *J* or scalar coupling and has become an invaluable tool in structural studies on complex naturally occurring compounds.

Reference

Albizati KF, Pawlik JR and Faulkner DJ, *J. Org. Chem.*, 1985, 50: 3428–30.

PROBLEM 8.10

Dacus cucurbitae, the melon fly, is one of the most active and destructive among fruit fly pests. From the rectal gland secretions of sexually mature male melon flies, an aromatic compound, **A**, was isolated (~4 mg per insect). The amount was sufficient initially only for a mass spectroscopic study on the basis of which it was characterised as 2-ethoxybenzoic acid. However, a direct comparison (GC retention times and mass spectral fragmentation patterns) of **A** with authentic 2-ethoxybenzoic acid showed significant differences. Thus, while the mass spectra of both the compounds exhibited common peaks at m/z 166 (due to the molecular ion) and m/z 138, the base peak in the spectrum of 2-ethoxybenzoic acid occurred at m/z 120 and this was absent in the spectrum of **A**; in the spectrum of the latter the base peak was at m/z 121. The structure of **A** was finally established by a direct comparison with all possible reference compounds synthetically available. The 1H NMR spectrum of synthetic **A** had signals at δ 1.4 (3H, t, *J* = 6 Hz), 4.33 (2H, q, *J* = 6 Hz), 6.83 (2H, d, *J* = 8 Hz), 7.87 (2H, d, *J* = 8 Hz) and 8.9.5 (1H, s, D-exchangeable). Interpret the above data and identify **A**.

Analysis: From the data presented, it is clear that **A** has the same molecular formula as 2-ethoxybenzoic acid, I (see Note 8.10.1). The presence of an ethyl group is also unambiguously shown by the M⁺-28 (m/z 138) peak in the spectrum of **A** as well as that of I. The chemical shift values of the $-CH_2-CH_3$ protons in **A** indicate that the ethyl group is attached to an oxygen atom as in an ethoxy or a carbethoxy group. The absence of a signal beyond δ 10 clearly rules out the presence of a free carboxyl group. That the phenolic hydroxyl is in conjugation, through the benzene ring, with a carbonyl group is indicated by the low-field position of its signal. The chemical shift values and multiplicity of the aromatic protons further show that the compound is a *para*-disubstituted benzene and that it is ethyl 4-hydroxybenzoate (II) (see Note 8.10.2). The mass spectral fragmentation of this compound and that of I can be explained as shown below (see Note 8.10.3).

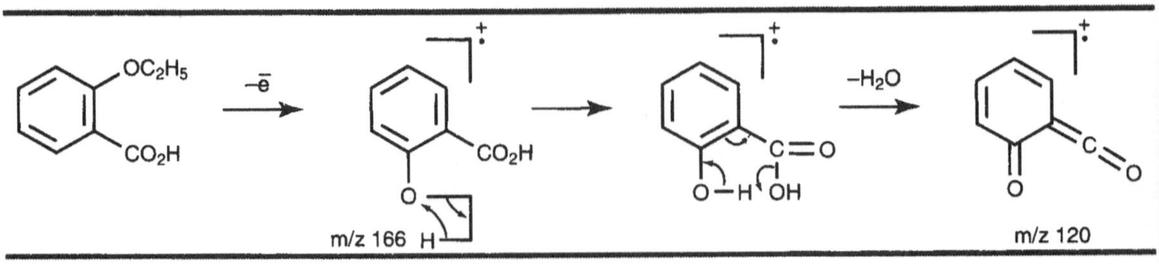

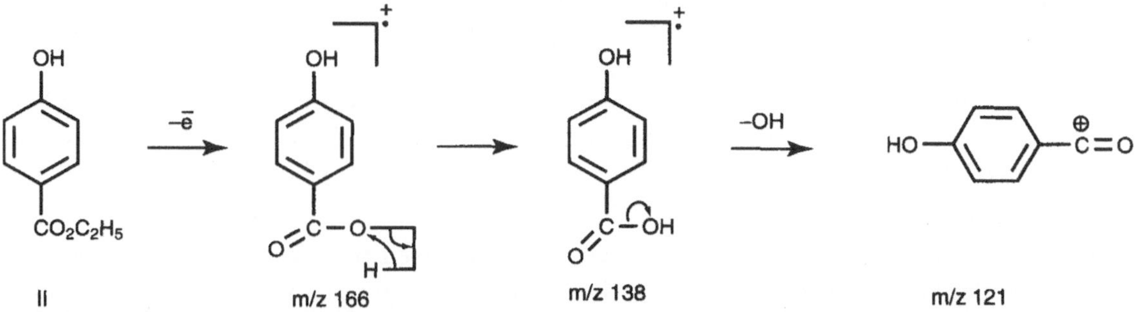

Notes

8.10.1 The melon fly is one of the tropical fruit flies which possess a well-developed system of chemical cues to regulate their biology. It is found in several parts of the world, ranging from East Africa and Southeast Asia to Hawaii and the South Pacific islands. Earlier studies by Baker, Herbert and Lomer (*Experientia*, 1982, 32) showed that sexually mature male melon flies secreted through their rectal glands a complex mixture of compounds which included three alkyl acetamides, three pyrazine derivatives and the ethyl esters of a number of fatty acids. The major component, however, was the compound discussed in this problem.

8.10.2 A minor component of the rectal gland secretions of male melon flies is propyl 4-hydroxybenzoate (III) indicating that simple alkyl esters of 4-hydroxybenzoic acid are used as semiochemicals by these fruit flies.

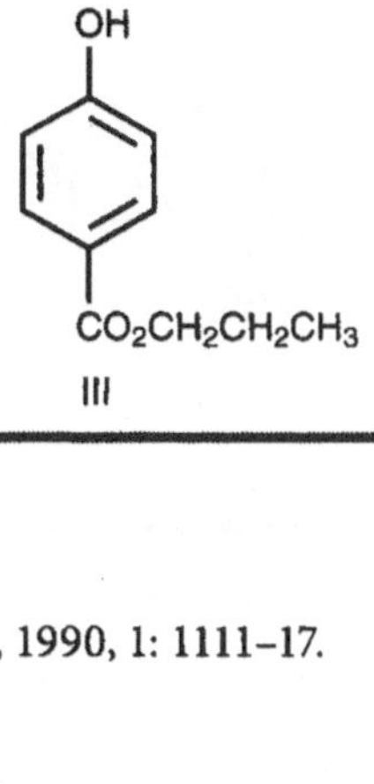

8.10.3 The mass spectral fragmentation pattern of 2-ethoxybenzoic acid is characteristic of *ortho*-substituted systems, with the formation of the ion responsible for the base peak involving a double hydrogen transfer (see S Tajima et al, *Org. Mass Spectrom.*, 1979, 14: 499).

Reference

Perkins MV, Kitching W, Drew RAI, Moore CJ and Konig WA, *J. Chem. Soc. Perkin*, 1990, 1: 1111–17.

PROBLEM 8.11

An ethanolic extract of the orange fruit bodies of an Australian toadstool, a species of the genus Dermocybe, yielded an orange, optically active, crystalline compound, $C_{19}H_{14}O_7$, **A**, as the major pigment; it was named austrocorticin. With acetic anhydride and a trace of concentrated sulphuric acid it yielded a diacetate. The UV absorption spectrum of **A** had maxima at 227, 232 (sh), 282, 350 (sh) and 430 nm. The last band shifted to 510 nm on the addition of NaOH. In the IR spectrum, the pigment absorbed at 3435, 1760, 1673 and 1634 cm^{-1}. In its mass spectrum, the peak due to the molecular ion (m/z 354) was quite prominent (81%) though the base peak occurred at m/z 336. Other peaks were at m/z 308 (40%) and 280 (14%). The 1H NMR spectrum of A had the following signals: δ 1.69 (3H, d, *J* = 6.5 Hz), 4.02 (3H, s), 4.05 (3H, s), 5.55 (1H, d, *J* = 6.5 Hz), 6.84 (1H, d, *J* = 2.2 Hz), 7.47 (1H, d, *J* = 2.2 Hz), 7.73 (1H, s) and 14.21 (1H, D-exchangeable). Irradiation of the signal at δ 6.84 brought about an NOE enhancement of the 3H signals at δ 4.02 (4.3%) and 4.05 (10.4%). Irradiation of the signal at δ 7.47 resulted in an NOE enhancement of the δ 4.02 signal. NOE effects were seen on the δ 1.69 and 5.55 signals when the signal at δ 7.73 was irradiated. Suggest an acceptable structure for **A**.

Analysis: The UV spectrum of **A** is typical of a condensed polycyclic aromatic compound and, in particular, an anthracene derivative. The yellow colour of the compound and the absorption maximum at 430 nm indicate that it is a 1, 4-quinone (see Note 8.11.1). The presence of an intramolecularly hydrogen-bonded OH group is shown by the IR absorption band at 3435 cm^{-1} and the NMR signal at δ 14.21 (see Note 8.11.2). The presence of two methoxyls is indicated by the signals at δ 4.02 and 4.05. Thus, five of the seven oxygen atoms can be accounted for. The IR absorption bands at 1673 and 1634 cm^{-1} can be assigned to the quinone carbonyls; the band at the lower wave number is due to the hydrogen-bonded carbonyl group (that is, the one peri to the hydroxyl) (see Note 8.11.3). The IR maximum at 1760 cm^{-1} can be attributed to the carbonyl group of a five-membered lactone moiety (see Note 8.11.4). The 1H NMR spectrum shows the presence of a secondary methyl group. Further, the chemical shift value of the methine proton, adjacent to this methyl group, indicates that the compound has the structural unit I.

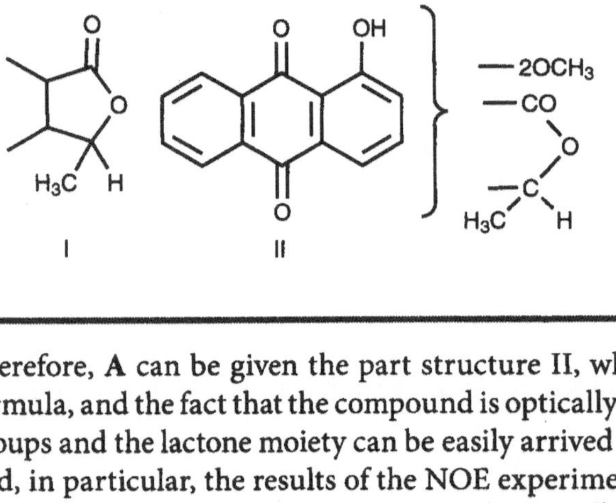

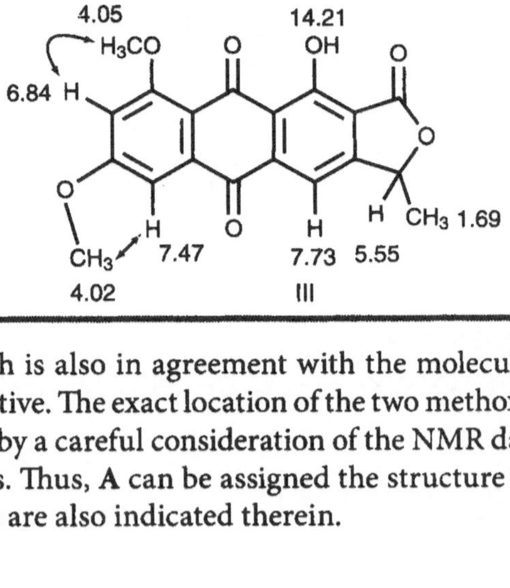

Therefore, **A** can be given the part structure II, which is also in agreement with the molecular formula, and the fact that the compound is optically active. The exact location of the two methoxyl groups and the lactone moiety can be easily arrived at by a careful consideration of the NMR data and, in particular, the results of the NOE experiments. Thus, **A** can be assigned the structure III shown above. The NMR assignments and NOE effects are also indicated therein.

Notes

8.11.1 *p*-Quinones are coloured yellow whereas *o*-quinones are more deeply coloured (orange to red). This difference can readily be attributed to the greater degree of conjugation in *o*-quinones.

8.11.2 The presence of a signal, due to an OH proton, beyond δ 12 affords clear evidence for the presence of an intramolecularly hydrogen-bonded OH as in salicylaldehyde and similar compounds.

8.11.3 Conjugation and hydrogen-bonding are known to shift carbonyl stretching frequencies to lower values and this example brings out these features in a fluent manner.

8.11.4 In contrast to the effects mentioned above, reduction in the ring size of a cycloalkanone brings about a shift of the carbonyl stretching frequency to a higher frequency. For a five-membered saturated lactone, the average value is 1750–1760 cm^{-1}.

Reference

Gill M and Gimenz A, *J. Chem. Soc. Perkin*, 1990, 1: 1159–67.

PROBLEM **8.12**

Cherylline, **A**, is an optically active, alkali-soluble alkaloid having the molecular formula $C_{17}H_{19}NO_3$. It reacts with diazomethane to form an *O,O*-dimethyl derivative, **B**. Its UV spectrum shows maxima at 280 and 285 nm. On addition of alkali the maxima shift to 299 nm. Its 1H NMR spectrum has the following signals: δ 2.24 (3H, s), 3.51 (3H, s), 6.23 (1H, s), 6.41 (2H, d, J = 7 Hz), 6.49 (1H, s) and 6.64 (2H, d, J = 7 Hz) besides other signals between δ 2.5 and 3.5. Compound **B** has been synthesised with *N*-benzyl-4-keto-6,7-dimethoxy-1, 2, 3, 4-tetrahydroisoquinoline (**C**) as a starting material. Treatment of **C** with 4-methoxyphenyl magnesium bromide yielded **D**, which underwent a facile acid-catalysed dehydration. The product **E** thus obtained was reduced with NaBH$_4$ to get **F**. The latter was catalytically hydrogenolysed to obtain **G** which yielded **B** on *N*-methylation. Deduce the structure of cherylline (**A**) and the other compounds mentioned in the problem.

Analysis: The presence of two phenolic hydroxyls in cherylline is revealed by the solubility of **A** in alkali, by the alkali-induced bathochromic shift of the UV absorption maxima, and by the formation of a di-*O*-methyl derivative. That the third oxygen is present as a methoxyl group is shown by the three-proton singlet at δ 3.51 in the 1H NMR spectrum (see Note 8.12.1). The three-proton singlet at δ 2.24 can be attributed to an N-methyl group (see Note 8.12.2). Unspecified signals between δ 2.5 and 3.5 strongly indicate that **A** is a tetrahydroisoquinoline derivative. The presence of a *p*-disubstituted benzene ring is evident from the pair of doublets, each integrating for two hydrogens, at δ 6.41 and 6.64. The very first step in the synthesis of the di-*O*-methyl ether of **A**, that is, **B**, confirms that these compounds are tetrahydroisoquinolines with a 4-methoxyphenyl substituent at position 4; the pattern of oxygenation on the isoquinoline moiety is also evident from the structure of the starting material. The entire synthesis leading to **B** can be delineated as shown below. The strategy adopted for bringing about the *N*-methylation in the final step is worth noting (see Note 8.12.3). On the basis of the above data, cherylline can be assigned any one of the structures I to III; III is the correct structure.

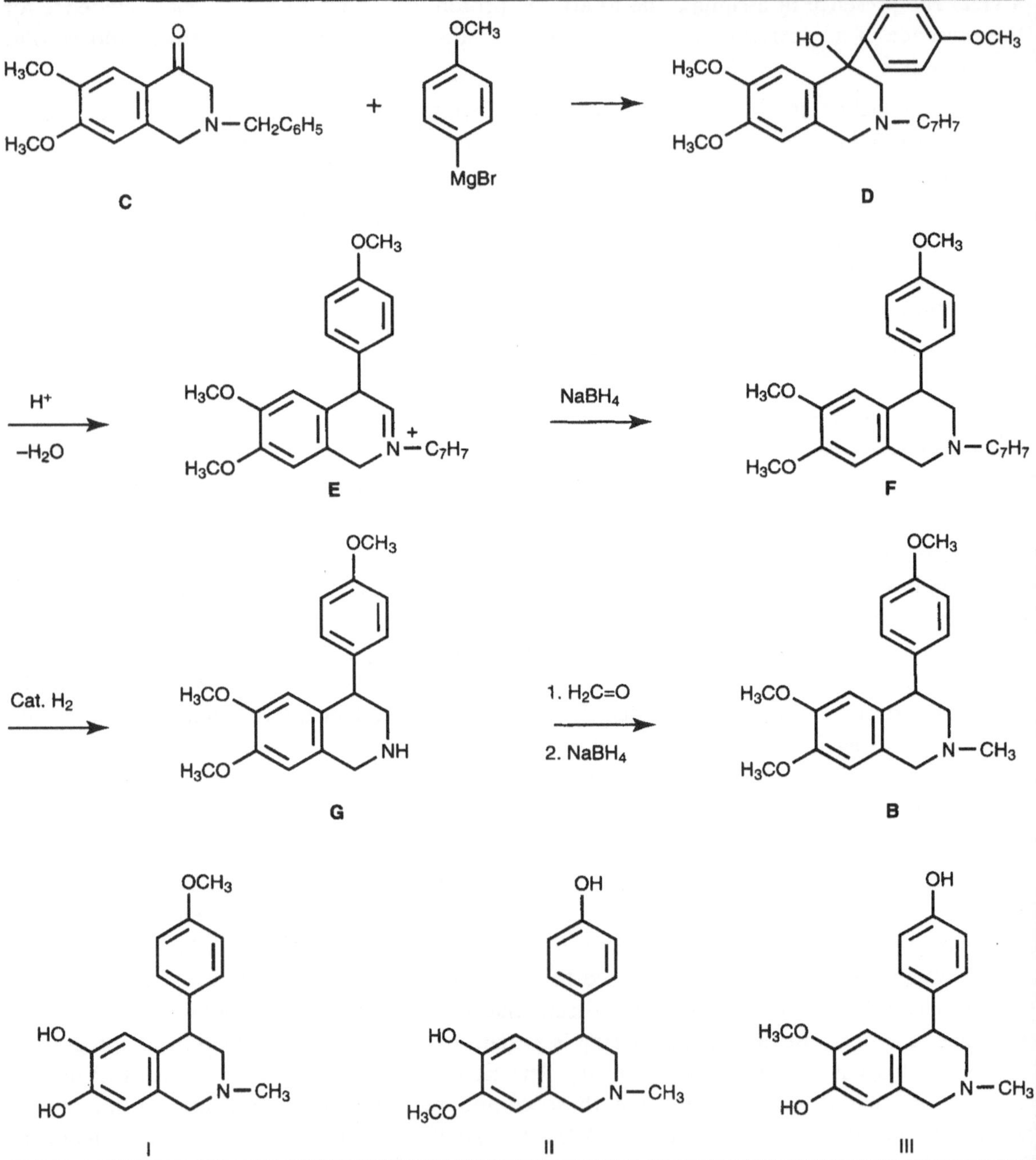

Notes

8.12.1 The position of the methoxyl protons' signal is somewhat upfield. Normally, one would expect this signal to appear nearer to δ 4.0; signals due to aliphatic methoxyls appear below δ 3.5. The fact that there are exceptions shows that the chemical shift values are only approximate.

8.12.2 A three-proton singlet at about δ 2.0–2.5 is an indication of a methyl group on a nitrogen, on a benzene ring, on an olefmic carbon or on a carbonyl group. In this case, the last three possibilities can be ruled out on the basis of other data.

8.12.3 Methylation of **G** with methyl iodide would have proceeded beyond **B** and resulted in the methiodide of the latter. The two-step reaction involved here proceeds as follows:

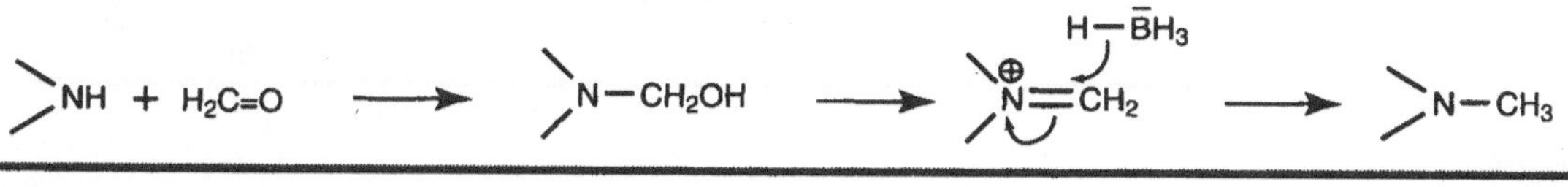

Reference

Wildman WC, Brossi A, Grethe G, Teitel S, Wildman WC and Bailey DT, *J. Org, Chem.*, 1970, 35: 1100.

PROBLEM 8.13

The essential oil of *Hibiscus syriacus* L. (shrub althea) contains as many as 65 identified chemical components; one of these, **A**, is a derivative of 2-phenylethanol. Its IR spectrum has absorption bands at 1080, 1360, 1600 and 1670 cm⁻¹. Its 1H NMR spectrum has signals at δ 1.65 (3H, s), 1.74 (3H, s), 2.69 (2H, t), 3.61 (2H, t), 3.96 (2H, d), 5.34 (1H, m) and 7.26 (5H, m). Deduce the structure of **A** and suggest a method for its synthesis from chlorobenzene.

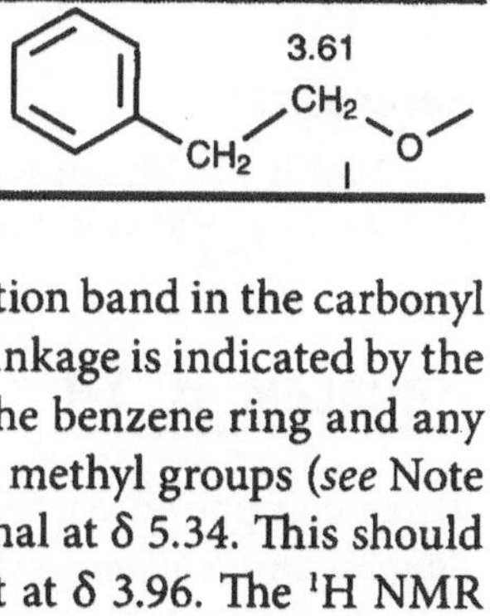

Analysis: It is evident from the NMR data that **A** is a derivative of 2-phenylethanol. The 5H complex signal at δ 7.26 and the two 2H triplets at δ 2.89 and 3.61 clearly show the presence of the part unit I in **A**.

One could infer that **A** is not an ester from the absence of any absorption band in the carbonyl region of the IR spectrum. On the other hand, the presence of an ether linkage is indicated by the band at 1080 cm⁻¹. The absorptions at 1600 and 1670 cm⁻¹ are due to the benzene ring and any additional double bond, and the band at 1360 cm⁻¹ can be attributed to methyl groups (*see* Note 8.13.1). The presence of an olefinic proton is shown by the ¹H NMR signal at δ 5.34. This should be adjacent to a –CH₂–O– group as the latter appears as a 2H doublet at δ 3.96. The ¹H NMR spectrum also reveals the presence of two methyl groups at an olefinic carbon as indicated by the two 3H singlets at δ 1.65 and 1.73. Putting these facts together, the presence of the structural unit II in **A** can be deduced.

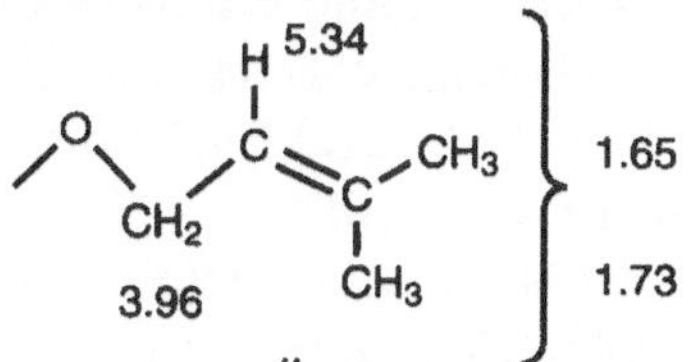

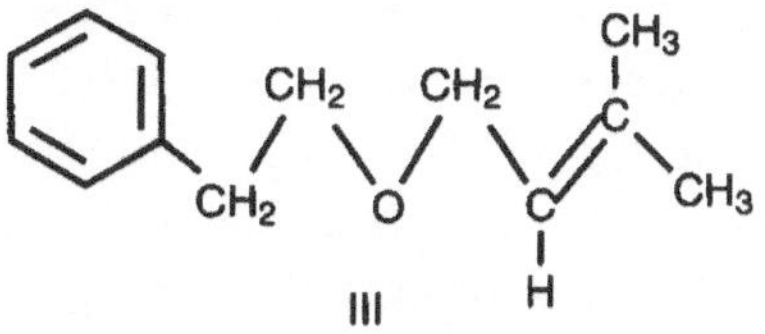

Therefore, **A** should be the prenyl ether (III) of 2-phenylethanol. It can be synthesised from chlorobenzene following the synthetic scheme given below (see Note 8.13.2).

Notes

8.13.1 Absorptions in the IR spectrum in the region 1360–1380 cm^{-1} are due to bending vibrations of methyl groups and have considerable diagnostic value. Geminal dimethyl groups usually give rise to a pair of doublets in this region but in this case only one absorption band has been recorded.

8.13.2 2-Phenylethanol is a valuable perfumery chemical and is a component of several natural essential oils including rose and geranium oils. Its methyl ether is the characteristic odoriferous component of Khewda oil which is the essential oil obtained from the leaves of *Pandanus odoratisimus*. The prenyl ether discussed in this problem has also been found to have a high degree of perfumery value. The method of synthesis described herein has been patented (*Chem. Abs.*, 1979, 90: 127414). It has also been prepared from 2-phenylethanol and prenyl chloride using a phase-transfer catalyst (Andreev and Bibicheva, 1985).

References

1.   Hanny BN, Thompson AC, Gueldner RC and Hedin PA, *J. Agr. Food. Chem.*, 1973, 21: 1001.
2.   Andreev VM and Bibicheva AI, *Chemistry of Natural Compounds*, 1985, 21: 254.

PROBLEM **8.14**

Three compounds, named sphorochnols **A**, **B** and **C**, were isolated from the Caribbean alga, *Sphorochnus bolleanus*. The major metabolite, sphorochnol A (**A**), was obtained as a noncrystalline solid having the molecular formula $C_{16}H_{22}O$. Its UV spectrum in methanol had a maximum at 277 nm which shifted to 285 nm on the addition of a few drops of dilute KOH. Its IR spectrum had absorption bands at 3360, 1608 and 1506 cm^{-1}. The 1H NMR spectrum showed signals at δ 1.36 (3H, s), 1.53 (3H, s), 1.67 (3H, s), 1.72 (1H), 1.78 (1H), 1.79 (1H), 1.87 (1H), 5.03 (1H, dd), 5.08 (1H, dd), 5.10 (1H, br, t), 6.01 (1H, dd), 6.78 (2H, d, J = 7.5 Hz), and 8.18 (211, d, J = 7.5 Hz), 1H homonuclear COSY NMR experiments established the following correlations. The signal at δ 6.01 (dd, J = 17, 11 Hz) correlated with those at δ 5.08 (dd, J = ll, 1 Hz) and δ 5.03 (dd, J = 17, 1 Hz). The signal at δ 5.10 (br, t, J = 7 Hz) correlated with those at δ 1.87 and δ 1.79. The latter two signals also correlated with the signals at δ 1.78 (dd, J = 4.5, 12 Hz) and δ 1.72(dd, J = 5.5, 12 Hz). The 13C NMR spectrum had signals at δ17.5 (q), 23.3 (t), 25.0 (q), 25.6 (q), 41.1 (t), 111.4 (t), 114.8 (d), 124.7 (d), 127.8 (d), 139.6 (s), 148.1 (d) and 153.3 (s). Acetylation of **A** yielded a monoacetate in whose 1H NMR spectrum the signals at δ 6.78 and 7.18 of **A** appeared at 7.0 and 7.3 respectively. Electron-impact mass spectrum of **A** showed a base peak at m/z 147. Analyse the above data systematically and suggest a probable structure for sphorochnol **A**.

Analysis: Having solved some of the earlier problems, it should be clear to the reader that sphorochnol **A** is a phenol, from the following data: IR absorptions at 3360 and 1506 cm⁻¹, KOH-induced bathochromic shift in the UV absorption maximum and the formation of a monoacetate (see Note 8.14.1). The molecular formula indicates six degrees of unsaturation, four of which are due to a benzene ring. From this, the presence of two olefinic double bonds in **A** can be inferred. This is supported by the four 1H NMR signals between δ 5.03 and 6.01. The presence of a *p*-disubstituted benzene

ring is also obvious from the A₂B₂ pattern exhibited by the signals at δ 6.78 and 7.18. That one of the substituents is a hydroxyl is shown by the downfield shift of these signals on acetylation. Therefore, at this stage, one may write the part structure I for **A**.

The $C_{10}H_{17}$ unit has three methyl groups (3H signals at δ 1.36, 1.53 and 1.67 and 13C NMR signals at δ 17.5, 25.0 and 25.6). Two of these are most probably on olefinic carbons as indicated by their chemical shifts (*see* Note 8.14.2). The presence of a terminal vinyl group is shown by the ¹H NMR signals at δ 5.03, 5.08 and 6.01 and more emphatically by the triplet signal at δ 111.4 in the ¹³C NMR spectrum. This signal can be assigned to a terminal olefinic methylene group (*see* Note 8.14.3). Since the signal at δ 5.01 appears only as a double doublet (due to spin—spin interactions with the protons on the adjacent terminal carbon atom), it should be adjacent to a completely substituted carbon atom. This inference is also supported by the correlation between the signal at δ 6.01 with those at 5.03 and 5.08 and with none other in the COSY spectrum. We have already noted the presence in **A** of the moiety =C(CH₃)₂. The other olefinic carbon atom of this unit has a hydrogen atom as shown by the broad triplet signal at δ 5.10. Its triplet nature shows that it is next to a methylene group; the broadening of the signal is a result of allylic interaction with the methyl groups. This inference is also well supported by the correlation between the δ 5.10 signal and those at δ 1.79 and 1.87. The methylene protons appear as separate signals as they are magnetically non-equivalent. This methylene group is adjacent to another methylene group which gives rise to signals at δ 1.72 and 1.78. This inference can be drawn from the COSY correlations between the signal pair at δ 1.79 and 1.87 with that at δ 1.72 and 1.78. Therefore, the $C_{10}H_{17}$ moiety of **A** can be expanded to the expression II.

The structure of sphorochnol A itself can now be deduced as III. The mass spectrum shows the base peak at m/z 147 which corresponds to a loss of 83 mass units from the molecular ion. In structural terms this is equivalent to the unit $-CH_2-CH_2-CH = C(CH_3)_2$, thus lending strong support to the assignment of structure III to sphorochnol A.

Notes

8.14.1 Phenolic hydroxyl groups can be readily distinguished from alcoholic hydroxyl groups by the alkali-induced bathochromic shifts in their UV spectra. In this case, the observed shift of 8 nm is characteristic of a monohydric phenol of the *p*-cresol type. The absorption band at 1506 cm^{-1} in the IR spectrum of **A** is also a characteristic feature of phenols and has diagnostic value.

8.14.2 Hydrogen atoms of methyl groups attached to a double bond generally give rise to signals between δ 1.5 and 2.0. The 13C NMR chemical shifts are also correspondingly shifted downfield from 10–15 ppm to 25 ppm and beyond. The third methyl group in A, though not directly attached to a double bond, is also in the vicinity of unsaturated groups as indicated by its signal positions in the 1H and 13C NMR spectra; these values are in between those expected for a methyl in a saturated environment and that on a double bond.

8.14.3 At one time, off-resonance decoupled 13C NMR spectra were the only source of information regarding the position of a carbon atom (that is, whether it is primary, secondary, tertiary or quaternary). This information can now be obtained conveniently from 2D NMR experiments. In this particular case, DEPT spectra provided the required information.

8.14.4 Sphorochnols **B** and **C** have been assigned structures IV and V respectively. Sphorochnol A is used by the alga *Sphorochnus bolleanus* as a chemical defence against marine herbivores such as the Caribbean surgeon fishes (Acanthurus sp.) which avoid *S. bolleanus* while feeding on the red alga, *Acanthophora specifera*. Interestingly, sphorochnol A has no antifeedant effect on the sea urchin *Diadema antillarum*, or the amphipod *Cymadusa filosa*.

IV   V

Reference

Shen TC, Tsai PI, Fenical W and Hay ME, *Phytochem.*, 1993, 32: 71.

PROBLEM 8.15

From a methanolic extract of the young shoots of the tea plant, a yellow crystalline compound A was isolated. It had the molecular formula $C_{21}H_{20}O_{10}$, was optically active and responded to the Shinoda colour test. Its UV spectrum in methanol had maxima at 270 and 331 nm, which shifted to 279, 303 (sh) and 378 nm on addition of $NaOCH_3$. A did not give any colour with Gibbs reagent and was stable when refluxed with 10% aqueous HCl. However, under the action of Killiani's mixture, it broke down to yield a mole each of glucose and compound B, $C_{15}H_{10}O_5$, which formed a triacetate and reacted with Gibbs reagent. The UV maxima of B suffered bathochromic shifts with sodium acetate and aluminium chloride (separately) but not with sodium acetate–boric acid. On vigorous oxidation B yielded 4-hydroxybenzoic acid. The mass spectrum of A showed a peak at m/z 414 (20%). The base peak occurred at m/z 283. Other prominent peaks were seen at m/z 270 (49%), 165 (36%), and 121 (28%). Deduce the structure of A.

Analysis: The positive Shinoda test and the UV spectrum show that compound A is a flavonoid (see Note 8.15.1). The bathochromic shifts in the UV absorption maxima due to the addition of $NaOCH_3$ confirm the presence of phenolic hydroxyls. The optical activity of the compound and the liberation of glucose under Killiani conditions (see Note 8.15.2) indicate that A is a glucoside. However, it is not an O-glucoside since it is stable under acid-catalysed hydrolytic conditions. The molecular formula of the aglucone shows that it is a trihydroxyflavone. This is also supported by the formation of a triacetate. The three hydroxyls can be located at positions 5 (bathochromic shift with $AlCl_3$), 7 (shift caused by sodium acetate) and 4′ (formation of 4-hydroxybenzoic acid on vigorous oxidation) (see Note 8.15.3). Therefore, the aglucone should be 5, 7, 4′-trihydroxyflavone (I), commonly known as apigenin (see Note 8.15.4). A should be a C-glucoside of I. The location of the glucose unit can be inferred from the observation that A does not react with Gibbs reagent indicating that all *para* positions to the phenolic hydroxyls are blocked. A careful examination of I would reveal that the only available position for the sugar moiety which would account for this observation is position 8. Compound A is, therefore, 8-C-glucosylapigenin (II), also known as vitexin. It occurs fairly widely in plants belonging to phylogenetically primitive families (see Note 8.15.5). The mass spectrum of

A does not show the molecular ion peak expected at m/z 432. The signal with the highest m/z value corresponds to M^+–H_2O. The peak at m/z 270 can be attributed to the molecular ion of the aglucone. The base peak at m/z 283 corresponds to the benzylic ion (III) and supports the direct attachment of the glucose unit to the flavone moiety by a C–C linkage (see Note 8.15.6).

Notes

8.15.1 The Shinoda test is a sensitive, characteristic colour test for flavonoids, in which a solution of the compound under examination, in methanol, is treated with powdered magnesium followed by concentrated HC1. The development of orange to magenta colours indicates that the compound is a flavonoid. The colour is due to the reductive conversion of the flavone into the corresponding anthocyanidin pigment. For UV spectral characterisation of flavonoids, see Problem 8.5 and also chapters 1 and 2.

8.15.2 The stability under usual hydrolytic conditions (7 to 10% aqueous acid under reflux) shows the absence of an acetal linkage as in O-glycosides. Killiani's reagent is used for cleaving C-glycosides.

8.15.3 See Note 7.5.1 under Problem 7.5 for the effects of shift reagents on the UV spectra of flavonoids. A combination of sodium acetate–boric acid brings about a bathochromic shift if the flavone contains a catechol group as in 3′, 4′-dihydroxyflavone.

8.15.4 For the biosynthesis of flavonoids see Chapter 6.

8.15.5 Flavonoids have been used in taxonomy as chemotaxonomic markers and phylogenetic indicators. For instance, flavone C-glycosides and biflavonyls are generally found in primitive plants and are either rare or absent in phylogenetically advanced plant families.

8.15.6 Glucosides rarely give rise to detectable molecular ions on electron-impact. In this case, the base ion obviously arises from the cleavage of the sugar residue accompanied by a hydrogen transfer. The ion III owes its abundance to resonance stabilisation. The readers may work out possible fragmentation pathways to account for the peaks at m/z 165 and 121.

Reference

Chkhikvishvili ID, Kurkin VA and Zaprometov MN, *Chemistry of Natural Products*, 1985, 21: 117.

PROBLEM 8.16

Borrerine (A), an alkaloid, is present in the underground parts of the plant *Borreria venticillata*. It has the molecular formula $C_{16}H_{20}N_2$ and is optically active. It readily yielded a dihydro derivative, **B**. On oxidation with $KMnO_4$ it gave a mole each of acetone and a carboxylic acid, **C**, which underwent decarboxylation to yield D, $C_{12}H_{12}N_2$. **A** and **D** behaved as monoacidic bases and reacted with one mole of methyl iodide to give the corresponding methiodides. The methiodide, **E**, obtained from **D** gave **F**, $C_{13}H_{16}N_2$; on treatment with a base. On $KMnO_4$ oxidation, **F** yielded formic acid and **G**, $C_{12}H_{14}N_2O_2$. The latter behaved as a monobasic acid and gave **H** on decarboxylation. **H** has been identified as 2-(N,N-dimethylaminomethyl) indole. Deduce the structures of **A** and the other derivatives mentioned above.

Analysis: Compound **A** is likely to be an indole derivative as shown by the structure of **H** (I) which is obtained from **A** following a series of reactions. The non-basic nitrogen of borrerine is the indolic nitrogen (see Note 8.16.1). The location of the other basic nitrogen, in relation to the indolic NH group is also indicated by the structure of **H**. The presence of a double bond in **A** is shown by its ready conversion to its dihydro derivative **B**. The formation of acetone on vigorous oxidation indicates the presence of a $=C(CH_3)_2$ moiety in **A**. Working backwards from **H**, **G** and **F** can be assigned structures II and III respectively. F is obtained by a Hoffmann elimination from the methiodide **E** of the tertiary amine, **D**; hence **E** and **D** can be formulated as IV and V respectively. There are three possible structures for **C**, namely, VI, VII and VIII. Of these three, VI is to be preferred since **C** undergoes facile decarboxylation; the carbanion thus formed from VI can be stabilised by resonance as shown below. Since **C** arises from **A** by oxidative cleavage of the double bond whose terminal carbon atom is fully substituted, A can be assigned the structure IX; **B** should be X. Thus, borrerine is a tetrahydrocarbazole derivative. Alkaloids of this skeletal type are fairly common in plants belonging to the family Rubiaceae which comprises the genus Borreria (see Note 8.16.2).

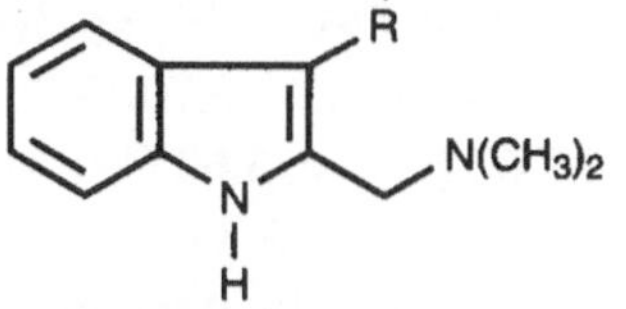

I: R = H
II: R = CO_2H

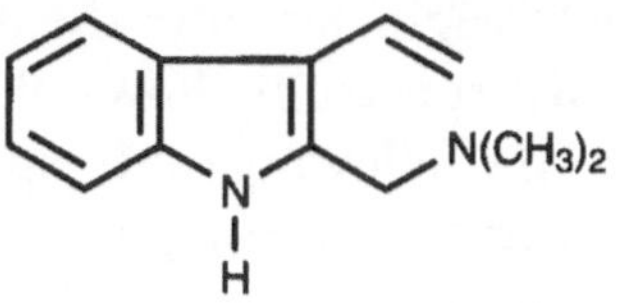

III

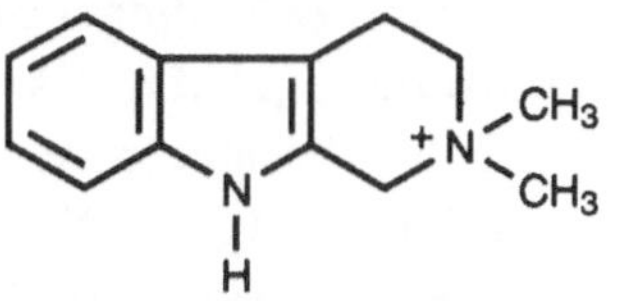

IV

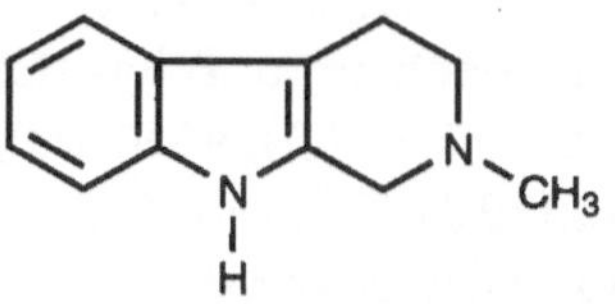

V

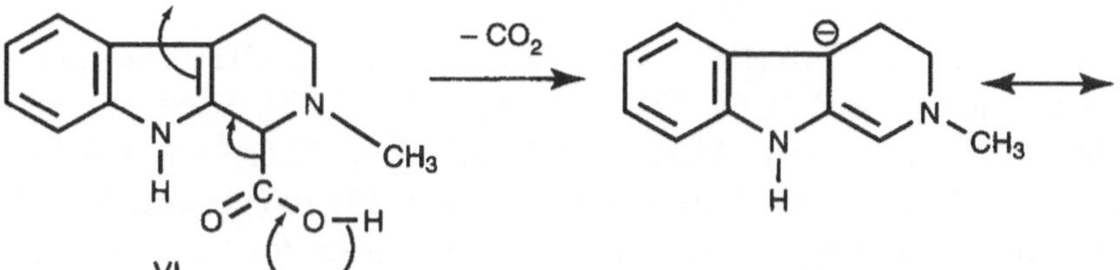

VI

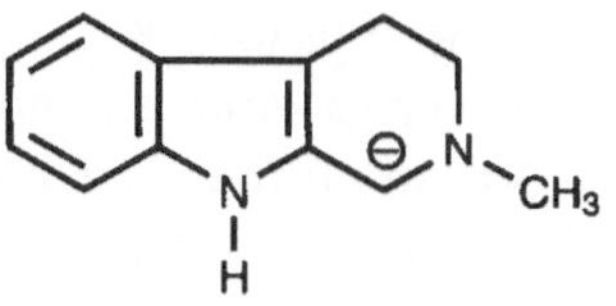

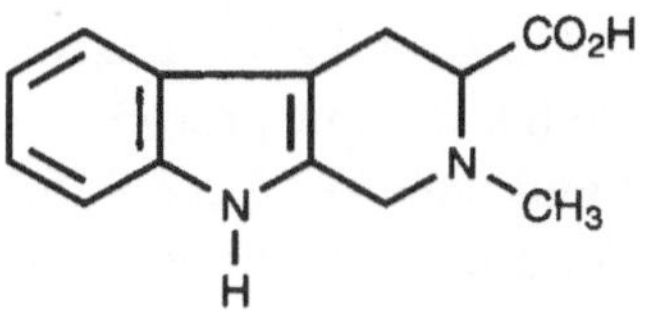

VII

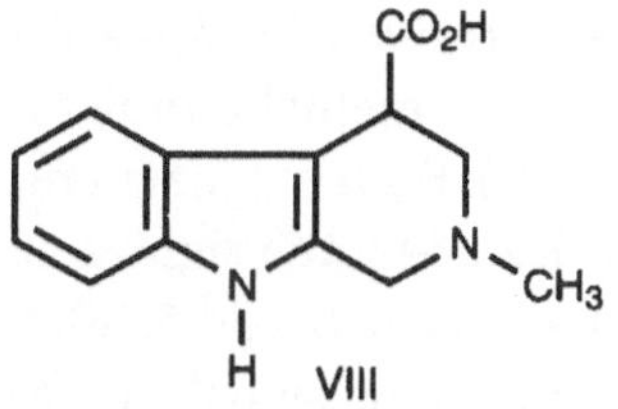

VIII

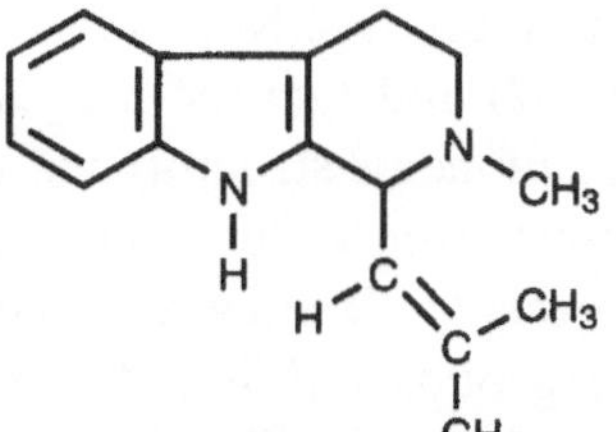

IX

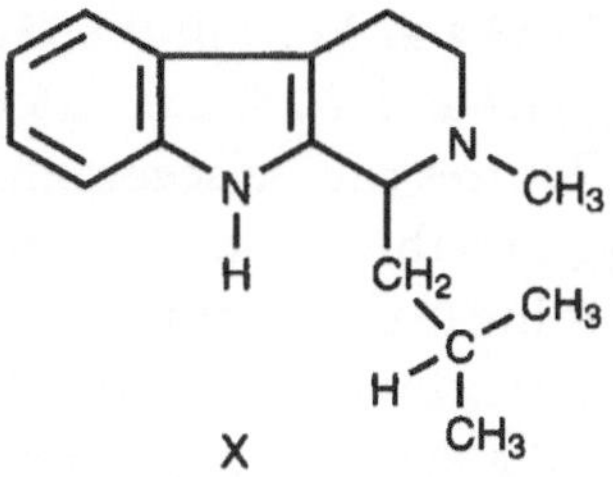

X

Notes

8.16.1 Pyrrole and indole are non-basic due to the participation of the lone pair of electrons of nitrogen in the aromatisation process. Therefore, borrerine, **A**, and compound **D** behave as monoacidic bases and form methiodides with one mole of methyliodide.

8.16.2 Indole alkaloids have considerable value as chemotaxonomic markers. The genus Borreria has recently been renamed as Spermacose.

Reference

Pousset JL, Kerharo J, Maynart G, Moneur X, Cave A and Goutarel R, *Phytochem.*, 1973, 12 : 2308.

PROBLEM 8.17

Desertorin **C** (**A**) is one of the three bicoumarins isolated from the mycelium of *Emericella desertorum* Samson & Mouchacca, which originates from the desert soils of Egypt. It has the molecular formula $C_{24}H_{22}O_8$. Its 1H NMR spectrum shows 3H singlets at δ 2.27, 2.70, 3.70, 3.77, 3.94 and 3.95. The spectrum also has the following other signals: δ 5.62 (1H, s), 5.69 (1H, s), 6.97 (1H, s) and 6.98 (1H, s). When refluxed with 10% aqueous KOH in dioxane, **A** yielded **B**, $C_{20}H_{22}O_6$, whose mass spectrum has the molecular ion signal at m/z 358, with the base peak appearing at m/z 343. Its IR spectrum has absorption bands at 1560 and 1600 cm^{-1}. The 1H NMR spectrum of **B** shows signals at δ 2.25 (3H, s), 2.63 (3H, s), 2.67 (3H, s), 3.71 (3H, s), 3.78 (3H, s), 6.41 (2H, br, s), 13.18 (1H, s, OH) and 13.22 (1H, s, OH). The 13C NMR spectrum of **B** has the following signals: δ 16.64 (Q), 22.57 (Qd), 32.33 (Q), 32.64 (Q), 55.18 (Q), 55.50 (Q), 96.91 (D), 106.19 (Dq), 110.91 (St), 113.49 (Sm), 118.81 (Sm), 121.63 (Sm), 136.35 (Sq), 139.24 (Sq), 156.15 (Sm), 158.3 (Sd-like), 158.97 (Sm), 159.97 (Sd-like), 204.10 (Sq) and 205.54 (Sq). Deduce the structures of **A** and **B** and suggest an acceptable mechanism for the conversion of **A** into **B**.

Analysis: It is given that **A** is a bicoumarin. The 1H NMR spectral data indicates that the compound contains two *C*-methyl groups (signals at δ 2.27 and 2.70) and four methoxyl groups (δ 3.70, 3.77, 3.94 and 3.95). Each coumarin moiety also has a single proton on the pyrone ring (positions 3 and 3′) as indicated by the signals at δ 5.62 and 5.69 (see Note 8.18.1) and a proton on the benzene ring (signals at δ 6.97 and 6.98). **B** is a diketone as shown by the 13C NMR signals at δ 204.10 and 205.54. Since these signals appear as Sq, they can be attributed to the carbonyl carbons of two CH_3CO- units (see Note 8.18.2). Each of these groups should be *ortho* to a phenolic hydroxyl to account for the low-field signals at δ 13.18 and 13.22 in the 1H NMR spectrum (see Note 8.17.3). **B** also has two *C*-methyl groups (signals at δ 2.25, 2.63, 16.64 and 22.57) and two methoxyls (δ 3.71, 3.78, 55.18 and 55.50). These features can be accommodated in a diphenyl structure with each benzene ring carrying a *C*-methyl, a methoxyl, an acetyl and a hydroxyl *ortho* to the carbonyl. The presence of twenty lines corresponding to the twenty carbon atoms present in **B** shows that the compound cannot have a symmetrical structure. The hydroxyl and acetyl groups in **B** should have arisen from the hydrolytic cleavage of the pyrone rings. Such a susceptibility to alkaline cleavage is characteristic of 4-methoxycoumarins (see Note 8.17.4). As pointed out under Problem 8.2, 4-methoxycoumarin can undergo alkali-induced demethylation to yield 4-hydroxycoumarin.

The keto tautomer of the latter undergoes lactone ring opening and subsequent decarboxylation as shown below.

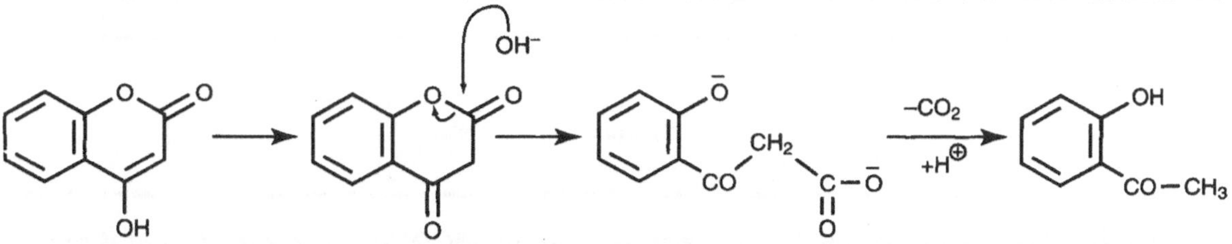

The location of two of the four methoxyls at positions 4 and 4′ rules out these positions as points of linkage in the bicoumarin **A**. As pointed out earlier, positions 3 and 3′are free. Therefore, the two coumarin moieties should be linked together through positions on their benzene rings. On biosynthetic grounds, the other two methoxyls (see Note 8.17.5) can be placed at positions 7 and 7′. Since the compound is not symmetrical, the points of linkage can be either 5, 6′or 5, 8′ or 6, 8′. Therefore, the correct structure can be arrived at if the methyl group positions can be correctly identified. This information is provided by a careful analysis of the NMR data on the diketone **B**. The chemical shifts of the two aromatic hydrogen atoms show that they are not *ortho* to the carbonyl groups (see Note 8.17.6). In the 13C NMR spectrum, only one of the C-methyl groups gives rise to a Qd signal indicating that it is next to an aromatic hydrogen atom. Keeping in view the other points, that **A** is a bis-7-methoxycoumarin and that **B** is unsymmetrically substituted, structure I can be assigned to **B**. Therefore, **A** should have the structure II. The monomeric unit of II is, then, 4,7-dimethoxy-5-methyl coumarin (III); two such units are linked together through position 6 of one unit and position 8 of the other. When one examines all other possible structures for **A** and **B**, it is posssible to convince oneself that only structures I and II are compatible with all the data.

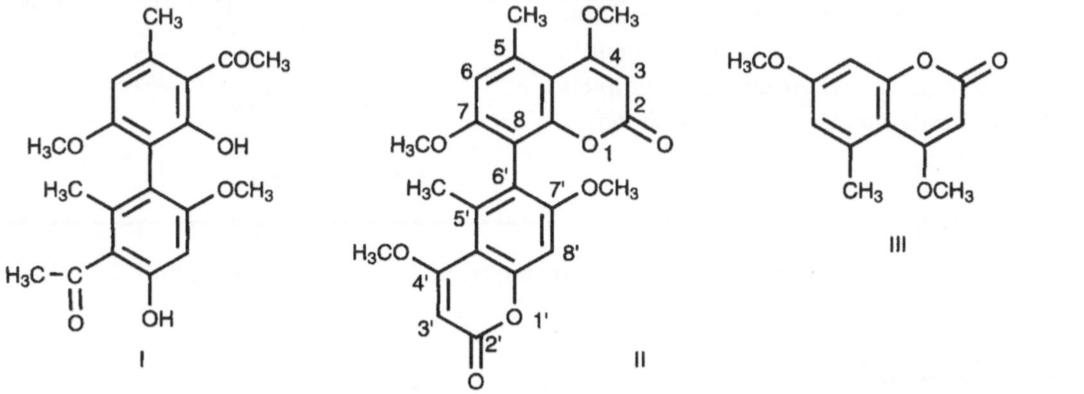

Notes

8.17.1 The pyrone ring consists of an α, β-unsubstituted ester group. In such systems an olefinic hydrogen on the β-carbon atom would be expected to resonate at a lower field than a hydrogen on the α-carbon due to the development of a greater positive charge on the β-carbon as shown below. This is well-supported by experimental data. Generally, β-olefinic hydrogens give rise to signals beyond δ 7.0.

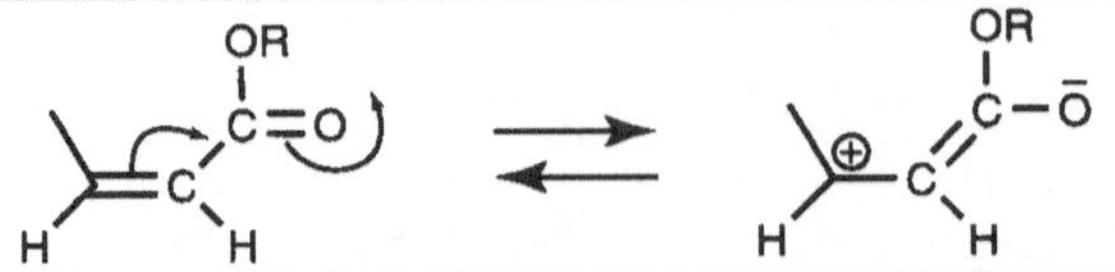

8.17.2 Capital letters in the coupling patterns seen in the 13C NMR spectrum data of **B** refer to the pattern resulting from directly bonded coupling, and small letters to long-range coupling. In this particular case, the latter are useful in locating the different functional groups relative to each other.

8.17.3 The shifting of OH protons to lowfield positions beyond δ 10–11 in 1H NMR spectra due to intramolecular hydrogen-bonding has been referred to under some of the earlier problems. In this case, the shift is considerable.

8.17.4 As mentioned in Problem 8.2, 4-methoxycoumarins behave as esters, being enol ethers of acidic enols. Therefore, they can be hydrolysed not only under mild acidic conditions but also under alkaline conditions.

8.17.5 Oxygenation at position 7 in coumarins is very common, as in flavonoids. Simple 7-methoxycoumarin is known as herniarin and is the methyl ether of umbelliferone, a widely occurring coumarin.

8.17.6 Aromatic hydrogens *ortho* to a carbonyl group generally give rise to signals on the lowfield side of δ 7.0 due to deshielding by the adjacent carbonyl group. The value observed in the present case sugests that these hydrogens are adjacent to hydroxyl or methoxyl groups. The complete 1H NMR assignments of A (II) are shown in the structure.

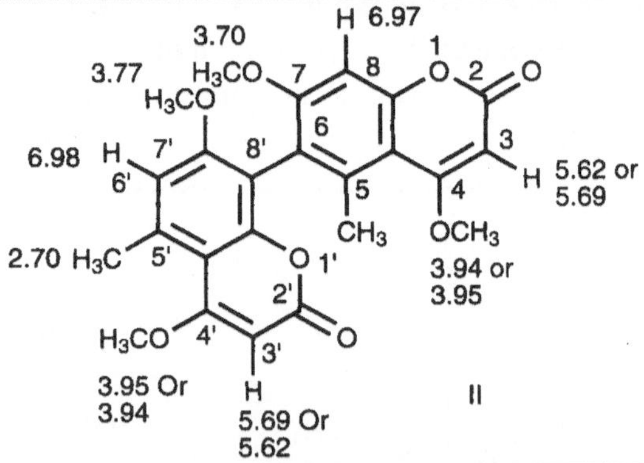

Reference

Nozawa K, Sayea H, Nakajima S, Udagawa SI and Kawai KI, *J. Chem. Soc, Perkin*, 1987, 1: 1735.

PROBLEM 8.18

Rotenone (I) can be converted into rotenonone (II) by a two-step oxidation reaction. On treatment with methanolic KOH, rotenonone (II) undergoes a skeletal rearrangement to yield rotenonic acid, which has the following spectral characteristics. IR maxima at 3440, 1735 and 1640 cm^{-1}; 1H NMR: δ 3.81   (3H, s),   3.94 (3H, s) 5.93 (1H, br, s),   6.29 (1H, d, $J = 8.8$ Hz),   6.77 (1H, s),

7.07 (1H, s), 8.26 (1H, d, J = 8.8 Hz), and 12.38 (1H, s); MS: m/z 424 (M⁺). When heated with sodium acetate and acetic anhydride, rotenonic acid (**A**) gave β-rotenonone, **B**, which had the same molecular formula as rotenonone. The IR spectrum of **B** had absorption bands at 1640 and 1730 cm⁻¹. Its 1H NMR spectrum had signals at δ 1.81 (3H, s), 3.25 (1H, dd, J = 9, 16 Hz), 3.64 (1H, dd, J = 9, 16 Hz), 3.99 (3H, s), 4.02 (3H, s), 4.98 (1H, s), 5.12 (1H, s), 5.41 (1H, t, J = 9.2 Hz), 6.86 (1H, d, J = 8.4 Hz), 8.12 (1H, s), 7.87 (1H, s), 8.19 (1H, d, J = 8.4 Hz). Compound **B** reacted with hydroxylamine to form an oxime (**C**), which on treatment with polyphosphoric acid underwent a rearrangement to yield compound **D**, $C_{22}H_{19}NO_5$. Its 1H NMR spectrum had the following signals: δ 1.39 (6H, d, J = 6.4 Hz), 3.12 (1H, sept, J = 6.4 Hz), 3.95 (3H, s), 4.06 (3H, s), 6.63 (1H, s), 7.43 (1H, d, J = 8.8 Hz), 7.62 (1H, d, J = 8.8 Hz), 7.82 (1H, s), 8.29 (1H, s). Deduce the structures of compounds **A** and **B** and account for the formation of **A** from II and **D** from **C**.

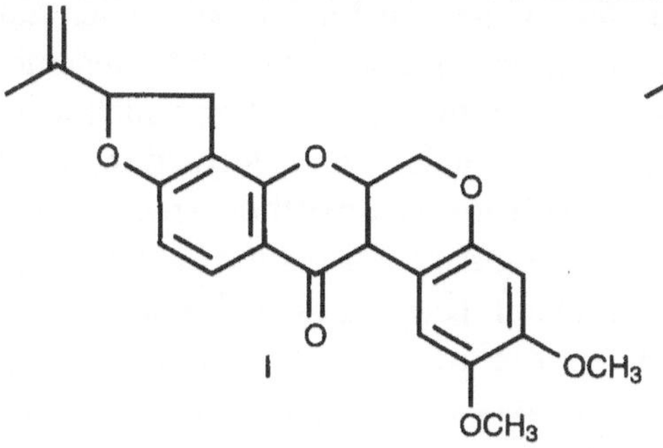

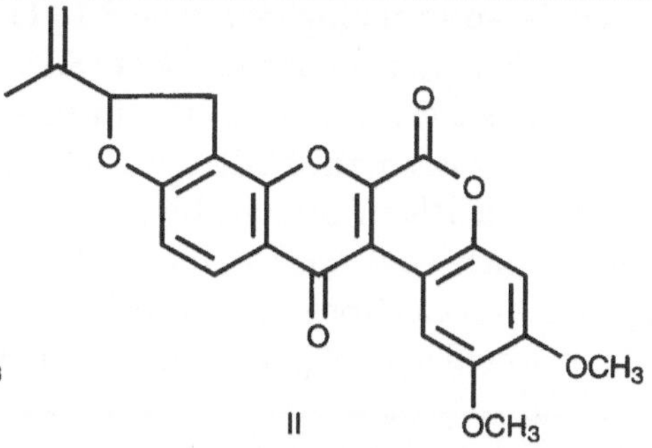

Analysis: The IR spectrum of **A** shows the presence of hydroxyl and carbonyl groups. The signal at δ 12.38 in the 1H NMR spectrum indicates that **A** has an intramolecularly hydrogen-bonded hydroxyl group. This should be in addition to the OH of the carboxyl group, because **A** is an acid. The 1H NMR spectrum of **B** shows that in the latter the dihydrofuranoid moiety, with the isopropenyl side chain, is intact. The signal assignments are given in the structural formula at the end of this analysis. Compound **B** does not possess the hydrogen-bonded hydroxyl group of **A** and is, most probably, the lactone of **A**. The IR absorption at 1640 cm⁻¹ in **A** and **B** can be attributed to an α, β-unsaturated keto carbonyl group. A structure which is compatible with the spectral properties of **A** is III. Its formation from II can be understood in mechanistic terms, as shown below. The first step is the hydrolytic opening of the lactone and this is followed by an intramolecular Michael-type addition of the phenolate ion thus liberated on the conjugated double bond. Compound **B** should, therefore, have the structure IV.

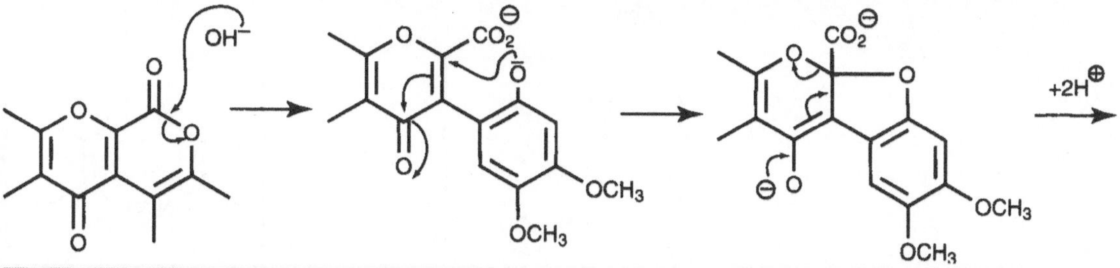

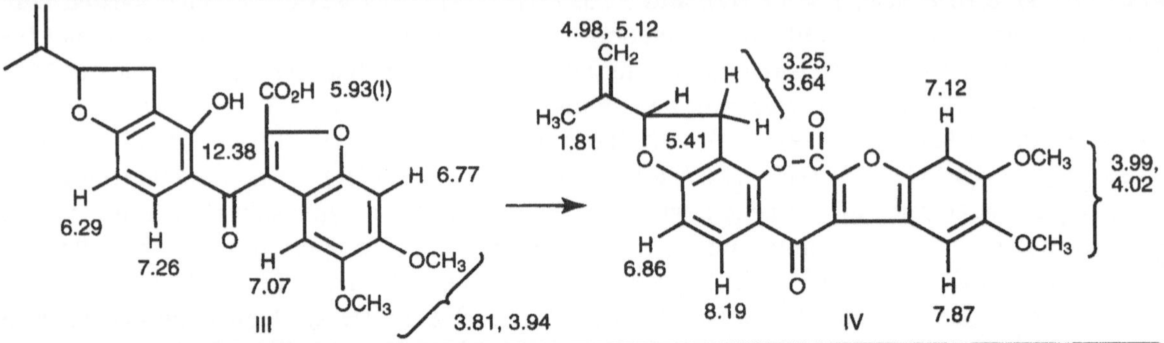

The unusually upfield position (δ 5.93) of the carboxyl proton in the 1H NMR spectrum of rotenonic acid is worth noting (see Note 8.18.1). The oxime of β-rotenonone, which has the same molecular formula as rotenonone (see Note 8.18.2), undergoes a Beckmann rearrangement to yield, initially, an intermediate which undergoes further transformation to yield **D** for which structure V can be assigned on the basis of the NMR data. A probable mechanism for its formation has been shown below. It may be noted that V presumably arises by the decarboxylation of the corresponding carboxylic acid derivative which, however, was not isolated; in the original publication it has been suggested that V arises by the loss of CO from an intermediate acylium ion. This suggestion is not acceptable because such a reaction would have required a hydride ion donor in the reaction medium.

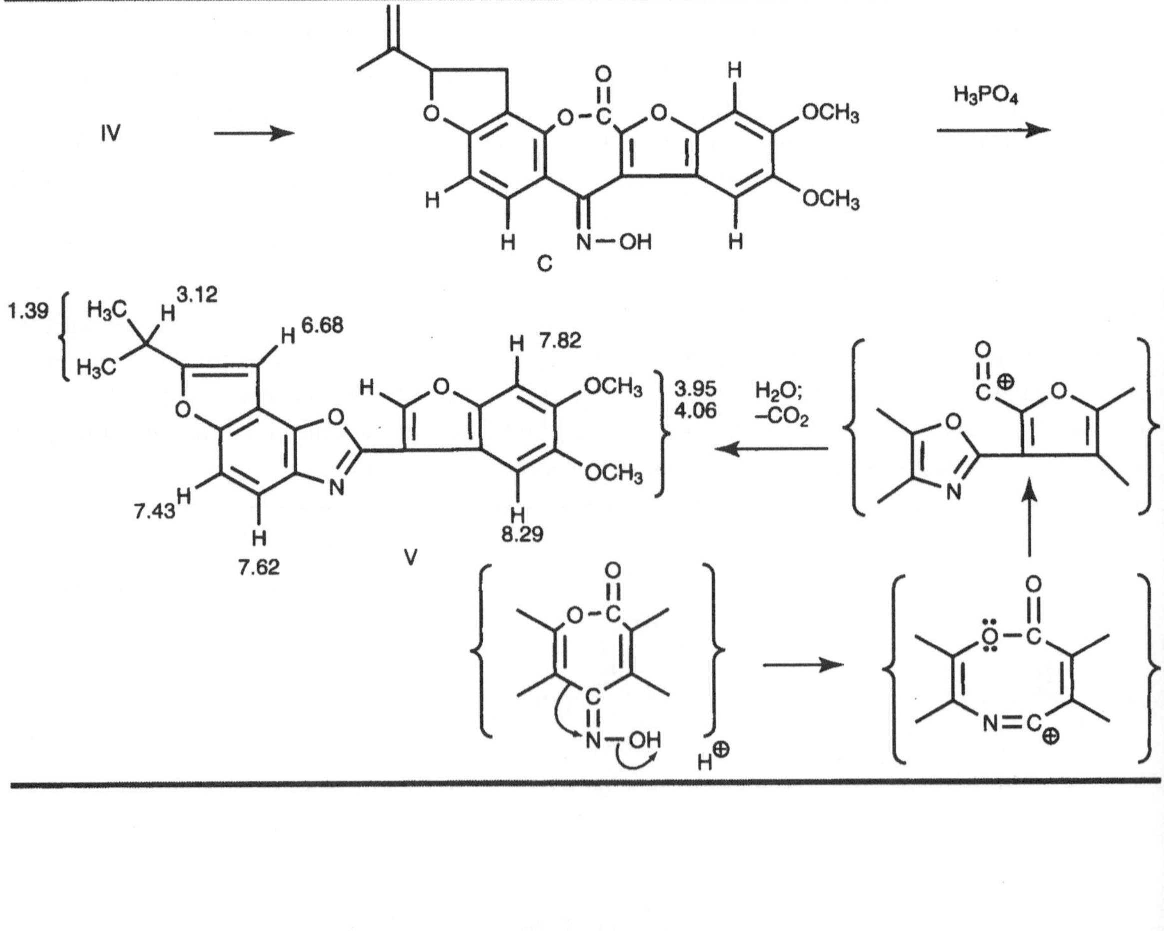

Notes

8.18.1 The proton of a carboxyl group normally gives rise to a signal beyond δ 10. It is difficult to offer a ready explanation for the abnormal highfield position seen in this case; presumably the neighbouring furan oxygen/hydroxyl is (are) responsible for this effect.

8.18.2 Rotenone is a well-known insecticide which occurs in several species of the genus Derris. It is a complex isoflavonoid; its stereochemistry has been discussed in Chapter 3. It has been converted into rotenonone by oxidation using NOCl.

Reference .

Sakakibara J, Nagai SI, Akiyama T, Ueda T, Oda N and Kidouchi K, *Heterocycles*, 1988, 27: 423.

PROBLEM 8.19

The seed oil of neem, *Azadirachta indica*, contains a potent insect anti-feedant compound, azadirachtin **A**, which has the structure I shown below. This compound has attracted considerable attention in recent years because of its growth-inhibiting effect on insects which affect crops. Several reactions of this compound have been studied to find out the structural features responsible for the biological activity. One such reaction forms the subject matter of this problem.

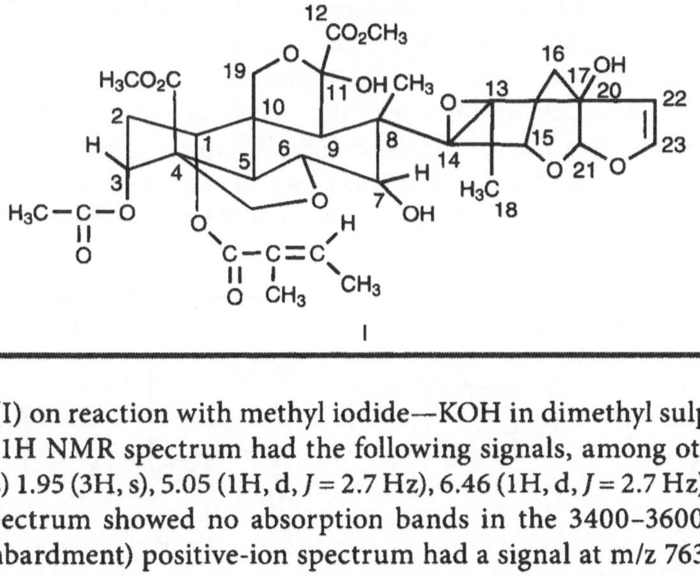

I

Azadirachtin A (I) on reaction with methyl iodide—KOH in dimethyl sulphoxide—yielded a product **A** whose 1H NMR spectrum had the following signals, among others: 1.78 (3H, d, $J = 7$ Hz), 1.86 (3H, s) 1.95 (3H, s), 5.05 (1H, d, $J = 2.7$ Hz), 6.46 (1H, d, $J = 2.7$ Hz), and 6.93 (1H, q, $J = 7$ Hz). Its IR spectrum showed no absorption bands in the 3400–3600 cm^{-1} region. Its FAB (fast atom bombardment) positive-ion spectrum had a signal at m/z 763. Treatment of **A** with KCN in methanol at 60°C gave **B** whose 1H NMR spectrum did not contain the signal at δ 1.95. Its IR spectrum, on the other hand, showed an absorption band at 3500 cm^{-1}. On hydrogenation in the presence of Pd–C, **A** yielded **C** which on reaction with KCN in methanol gave **D**. **D** had a lower R_f value than **C** on TLC (silica gel) plates and was transparent at 215 nm at which wavelength, **A** exhibited strong absorption. In the 1H NMR spectrum of **D**, the signals at δ 6.93, 6.46, 5.05, 1.95, 1.86 and 1.78 seen in the spectrum of **A** were absent. However, the spectrum of **D** had two doublet signals which were not seen in the spectrum of **A**; these two signals disappeared on the addition of D_2O. Interpret the above data and suggest acceptable structures for compounds **A** to **D**.

Analysis: An examination of the complex structure (I) of azadirachtin **A** reveals the presence of three hydroxyls, which are apparently absent in **A** as shown by the absence of any absorption band in the 3400–3600 cm^{-1} region of its IR spectrum. Methyl iodide–KOH is indeed a good O-methylating agent (see Note 8.19.1) and therefore, **A** can be expected to be the tri-O-methyl ether (II) of I. This conjecture is supported by the molecular weight of the compound, which was shown to be 762, by the FAB spectrum (see Note 8.19.2). The 1H NMR signals mentioned in the problem are clearly due to the acetyl (1.95, 3H, s) and tigloyl groups (6.93, 1H, q; 1.86, 3H, s and 1.78, 3H, d) as well as the olefinic hydrogens on carbons 22 and 23 of the dihydrofuran moiety (signals at δ 6.46 and 5.05). Thus, compound **A** retains these structural features of I. KCN in methanol is a mild reagent for bringing about hydrolysis of esters. The IR spectrum of **B** does show the presence of a hydroxyl group (absorption at 3500 cm^{-1}), which has presumably arisen by the hydrolysis of the acetoxyl group (as evident from the absence of the signal at δ 1.95 in the 1H NMR spectrum). Therefore, **B** is desacetyl-tri-O-methylazadirachtin **A**. Thus, **A** and **B** can be assigned structures II and III respectively. Compound **C** is clearly the tetrahydro derivative of **A**, resulting from the hydrogenation of the C-22–C-23 double bond as well as the double bond of the tigloyl moiety. Therefore, **C** can be formulated as IV. Compound **D** which is obtained from **C** by hydrolysis with KCN in methanol, is obviously a dihydroxy compound as shown by the presence of two **D**-exchangeable proton signals in the NMR spectrum. This observation indicates that both the acetoxyl and the tigloyl groups have undergone hydrolysis in this case, in contrast to the behaviour of **A** under similar conditions (see Note 8.19.3). This inference is also supported by the absence of the signals due to the tigloyl residue in the NMR spectrum. Further, the two hydroxyl proton signals appear as doublets indicating that they are secondary hydroxyls (see Note 8.19.4). The removal of the tigloyl moiety is also responsible for the disappearance of the absorption maximum at 215 nm in the UV spectrum. Therefore, **D** can be assigned the structure V.

II: R = –COCH$_3$
III: R = H

IV: R = –COCH$_3$; R' = –CO–C(H)(CH$_3$)–CH$_2$CH$_3$
V: R = R' = H

Notes

8.19.1 Alcoholic OH groups can be methylated with the help of methyl iodide or dimethyl sulphate in the presence of a strong base; NaOH is often used for this purpose, though KOH is also used, as in this case. It is worth noting that in azadirachtin A, two of the hydroxyls are tertiary.

8.19.2 The technique of liberating molecular ions from organic compounds by fast atom bombardment is 'softer' than the conventional electron-impact technique. Therefore, it is used to detect the molecular ions of sensitive compounds, like heavily hydroxylated compounds such as azadirachtin **A**, carbohydrates, and others. In this technique, the sample under examination is placed in a matrix of glycerol or *m*-nitrobenzyl alcohol and bombarded with noble gas atoms (argon is the commonly used noble gas). Both FAB positive ions as well as FAB negative ions can be detected. In a FAB positive ion spectrum, (M⁺+1) peak is detected, whereas in a FAB negative ion spectrum, (M⁺−1) peak is detected.

8.19.3 Hydrogenation of the C-22–C-23 double bond has somehow made the structure more amenable to hydrolytic cleavage, though the reason for this change is not clear.

8.19.4 Secondary and tertiary hydroxyls can be differentiated by 1H NMR spectroscopy by recording the spectra in deuterated DMSO, which, being a viscous medium, slows down the exchange of the OH protons with the matrix.

Reference

Rembold H and Kumar ChSSSR, Unpublished work; personal comunication (SSSR Kumar).

PROBLEM 8.20

From the light-grown shoots of a dwarf cultivar of the plant *Phaseolus vulgaris*, an oily compound, **A**, was isolated. It was found to be a plant growth inhibitor. High-resolution mass spectroscopy gave the molecular weight as 208.1444. Its IR spectrum showed absorption bands at 3400, 1665, 1600, 975 and 1360 cm⁻¹. Its UV spectrum had absorption maxima at 221 and 290 nm. Its 1H NMR spectrum had the following signals: δ 1.11 (3H, s), 1.12 (3H, s), 1.49 (1H, d), 1.78 (3H, s), 1.79 (1H, d), 2.09 (1H, d), 2.30 (3H, s), 2.43 (1H, d), 4.01 (1H, m), 6.11 (1H, d, J = 17 Hz) and 8.21 (1H, d, J = 17 Hz). The 13C NMR spectrum of **A** had signals at δ 21.6 (q), 27.3 (q), 28.6 (q), 30.1 (q), 36.9 (s), 42.8 (t), 48.4 (t), 64.6 (d), 132.2 (s), 132.4 (d), 135.7 (s), 142.3 (d) and 198.4 (s). The mass spectrum showed peaks at m/z 208 (M⁺), 193, 175 and 147. The compound was optically active, exhibiting a large laevo rotation. Analyse the given data and suggest a structure for **A**.

Analysis: From the high resolution mass spectral data the molecular formula can be deduced as $C_{13}H_{20}O_2$ (see Note 8.20.1). The IR spectrum indicates the presence of hydroxyl (3400 cm⁻¹), carbonyl (1665 cm⁻¹) conjugated to a double bond (1600 and 975 cm⁻¹) and geminal dimethyl (1360 cm⁻¹) groups in **A**. The UV maximum at 290 nm shows the presence of a substituted dienone moiety (see Note 8.20.2). The presence of a geminal dimethyl group is further supported by the 3H singlet signals at δ 1.11 and 1.12 in the 1H NMR spectrum. The signal at δ 1.78 can be attributed to a methyl at a double bond whereas the one at 2.30 can be assigned to a methyl attached to a carbonyl. The corresponding signals in the 13C NMR spectrum appear as quartets at δ 21.6, 27.3, 28.6 and 30.1. The presence of a double bond in conjugation with the carbonyl group is clearly and unambiguously indicated by the 1H proton doublets at δ 6.11 and 8.21, each with a coupling constant of 17 Hz; the latter shows that the double bond has the *trans* configuration (see Note 8.20.3). The other double bond of the dienone chromophore is obviously fully substituted. The molecular formula of **A** shows that it has four degrees of unsaturation of which three can be accounted for by the dienone moiety. The fourth should, therefore, be the equivalent of a ring. A structure which can satisfactorily account for all the above observations is that of a hydroxy β-ionone. The hydroxyl can be placed at position 3 as shown in the structural formula I on the basis of the following data. The two 1H proton signals at δ 2.09 and 2.43 and the

corresponding 13C triplet at δ 48.4, conclusively show the presence of a methylene group next to a double bond. In other words, position 4 of the ionone structure should be unsubstituted. Further, since these methylene proton signals appear as doublets, it is clear that position 3 has only one hydrogen and therefore the hydroxyl can be placed here. The 1H multiplet at δ 4.01 and the corresponding signal at δ 64.6 can be attributed to this part of the molecule. The remaining methylene group corresponding to position 2 is responsible for the two one-proton doublets at 1.49 and 1.79 and the 13C triplet at δ 42.8. The other 13C signal assignments are shown in the structural formula given above.

In agreement with the structure, the mass spectrum of **A** shows peaks due to the molecular ion at m/z 208, M^+–Me at m/z 193, M^+–Me –H_2O at m/z 175 and M^+–H_2O -$COCH_3$ at m/z 147. Thus, the gross structure of **A** can be deduced as 3-hydroxy-β-ionone. The configuration at the asymmetric centre has been inferred to be R from its specific rotation (see Note 8.20.4).

Notes

8.20.1 The even mass number of the molecular ion, in conjunction with other data, suggests that A is unlikely to contain nitrogen. The exact molecular formula can be determined from the molecular weight which is accurate at least to the fourth decimal point, from atomic weights of other atoms based on the carbon standard and in the present case this works out to $C_{13}H_{20}O_2$.

8.20.2 The Woodward–Fieser empirical rules based on UV absorption data have been found to be extremely useful in structural studies on polyenes and enones. The base value for an enone is assumed to be 215 nm. An increment of 30 nm is added for every additional conjugated double bond. For an α-alkyl substituent 10 nm are added whereas the corresponding increment for a β-substituent is 12 nm, and for γ- and δ-substituents, 18 nm each. Using these figures it is possible to predict that A should exhibit an absorption maximum at 299 nm which is in fair agreement with that observed.

8.20.3 Coupling constants provide valuable information regarding configurations of geometrical isomers both in aliphatic and alicyclic systems. In the case of olefins the *J* value for coupling between *trans*-hydrogen atoms is between 12 and 18 Hz (typically 17 Hz) whereas it is between

6 and 12 Hz (typically 10 Hz) for *cis*-coupling. The *J* value of 17 Hz observed in the spectrum of **A** clearly shows that the double bond has the *trans* configuration.

8.20.4 Specific rotations at a single wavelength are, generally, not as useful as ORD measurements for the determination of absolute configurations unless closely related reference compounds with known configurations are available for direct comparison. Compound A, that is, (–)R-3-hydroxy-β-ionone had earlier been isolated from the tobacco plant as well as from quince fruits.

Reference

Noguchi HK, Icosemura S, Yamamura S and Hasegawa K, *Phytochem.*, 1993, 33: 553–55.

PROBLEM 8.21

From the culture broth of the bacterium *Nocardia* sp. DSM 43130, a hitherto unknown nitrogen-containing compound was isolated. High resolution mass spectroscopy gave the molecular weight of the compound as 207.0892. The base peak in the mass spectrum had an m/z value of 164. The compound failed to give any colour with ninhydrin. Its 1H NMR spectrum had the following signals: δ 1.50 (3H, d), 2.19 (3H, s), 4.10 (1H, q), 6.49 (1H, dd), 6.68 (1H, dt), 7.38 (1H, dt) and 8.02 (1H, dd). The 13C NMR spectrum had signals at δ 17.7 (q), 24.9 (q), 58.4 (d), 111.4 (d), 115.8 (d), 132.9 (d), 135.7 (d), 150.1 (s), 172.6 (s) and 210.6 (s). Suggest a structure for the compound.

Analysis: From the accurate molecular weight given by HR-MS, the molecular formula of the compound can be worked out as $C_{11}H_{13}NO_3$. The loss of 43 mass units from the molecular ion in the mass spectrum indicates the presence of an acetyl group. The following observations show that this should be attached to a nitrogen atom.

1. The compound does not give any colour with ninhydrin, indicating the absence of a free –NH_2 group.
2. The presence of a three-proton singlet in the 1H NMR spectrum at δ 2.19.

This is further supported by the presence of the singlet signal at δ 72.6 in the 13C NMR spectrum of the compound. Thus, the compound should be an acetamide derivative. The presence of a keto carbonyl group, in addition to the amide group, is shown by the δ 210.6 signal in the 13C NMR spectrum. This should, in turn, be attached to a –CH(OH)CH$_3$ moiety as indicated by the chemical shift value of the 1H quartet at δ 4.10 (see Note 8.21.1); the methyl group of this unit appears, as expected, as a doublet at δ 1.50. The corresponding 13C signals appear at δ 58.4 and 17.7 respectively. That the compound is a disubstituted benzene derivative is clear from the 13C NMR spectrum which shows four doublet signals in the aromatic region. This inference is fully corroborated by the multiplicities exhibited by the aromatic proton signals in the 1H NMR spectrum. The one and only structure which is compatible with all the above-mentioned data is 2-(2-hydroxypropionyl) acetanilide (I) (see Note 8.21.2).

Notes

8.21.1 The position of the signal of the proton in question is quite umambiguous, and is compatible with standard data. (See, for instance, Silverstein RM, Bassler GC and Morrill TC, *Spectrometric Identification of Organic Compounds*, 5th edn, John Wiley & Sons, New York, 1991).

8.21.2 The compound discussed here was not known earlier, though 2-hydroxypropiophenone (II), also known as lactophenone, had earlier been isolated as a bacterial metabolite.

References

1.   Abraham WR and Arfmann HA, *Phytochem.*, 1993, 33: 929–30.
2.   Horio T, Yoshida K, Kikuchi H, Kawabata J and Mizutani J, *Phytochem.*, 1993, 33: 807.

PROBLEM 8.22

A methanolic extract of the roots of the plant *Chenopodium album* yielded a compound **A** whose molecular weight was found to be 343.1450 using electron-impact HR-MS. This oily compound exhibited absorption maxima in the IR region at 3300, 1650, 1580 and 1510 cm^{-1} A. Its 1H NMR spectrum had the following signals: δ 2.73 (2H, t, J = 7.3 Hz), 3.50 (2H, q, J = 7.3 Hz), 3.81 (3H, s), 3.88 (3H, s), 6.50 (1H, d, J = 15.6 Hz), 6.66 (1H, dd, J = 8.1 and 2.1 Hz), 6.74 (1H, d, J = 2.1 Hz), 6.83 (1H, d, J = 8.2 Hz), 6.85 (1H, d, J = 8.1 Hz), 7.04 (1H, dd, J = 8.2 and 1.9 Hz), 8.16 (1H, d, J = 1.9 Hz) and 7.44 (1H, d, J = 15.6 Hz). A NOE interaction was observed between the signals at δ 3.81 and 6.85. Deduce the structure of the compound and suggest a path for its synthesis.

Analysis: It is obvious that the compound is nitrogenous from the odd value of its molecular weight. The exact molecular weight also leads to the unique molecular formula, $C_{19}H_{21}NO_5$, for the compound (see Note 8.22.1). The IR absorption bands at 1650 and 1580 cm^{-1} strongly suggest that the compound is an amide. Even a cursory glance at the region between δ 6.50 and δ 7.44 in the 1H NMR spectrum reveals that **A** contains one or more aromatic units. A more careful analysis brings out the following observations. The pair of doublets, each with a J value of 15.6 Hz at δ 6.50 and δ 7.44 shows the presence of a *trans*-double bond conjugated to a carbonyl group as in a cinnamoyl or substituted cinnamoyl unit (see Note 8.22.2). The signal at δ 6.66 can be attributed to a benzene proton *ortho* to another responsible for the signal at δ 6.85 and *meta* to a third which appears as a doublet at δ 6.74. The three signals together indicate the presence of a 1,3, 4-trisubstituted benzene ring in **A**. The presence of a second 1,3, 4-trisubstituted benzene ring is shown by the following signals: the double doublet at δ 7.04 coupled *ortho* to the signal at δ 6.83 and *meta* to the one at δ 7.16. This benzene ring is, most probably, part of the substituted cinnamoyl moiety referred to earlier. The two-proton triplet signal at δ 2.73 can be assigned to a methylene group which is attached to an aromatic group on one side and another methylene on the other. The latter methylene which appears as a quartet signal can, therefore, be placed next to a NH. The presence of two methoxyls, one on each benzene ring, is shown by the signals at δ 3.81 and

3.88. That the compound contains free hydroxyls is indicated by the molecular formula as well as by the IR absorption at 3300 cm[1]. The one structure which can satisfactorily account for all of the above data is *N-trans*-feruloyl-4-*O*-methyldopamine (I).

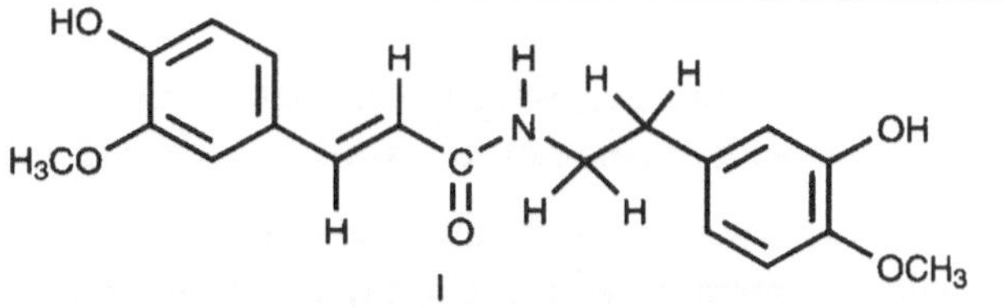

The compound has been synthesised by the following route which follows a predictable pattern. Condensation of 3-*O*-benzyl-4-methoxybenzaldehyde (II) with nitromethane gave the styrene III. The nitro group in III was reduced to an amino group with the help of LAH to obtain IV which was condensed with ferulic acid (4-hydroxy-3-methoxycinnamic acid; V) in the presence of dicyclohexyl carbodiimide (DCCI). Debenzylation of the product VI with dimethyl sulphide –BF₃ etherate yielded I (see Note 8.22.3).

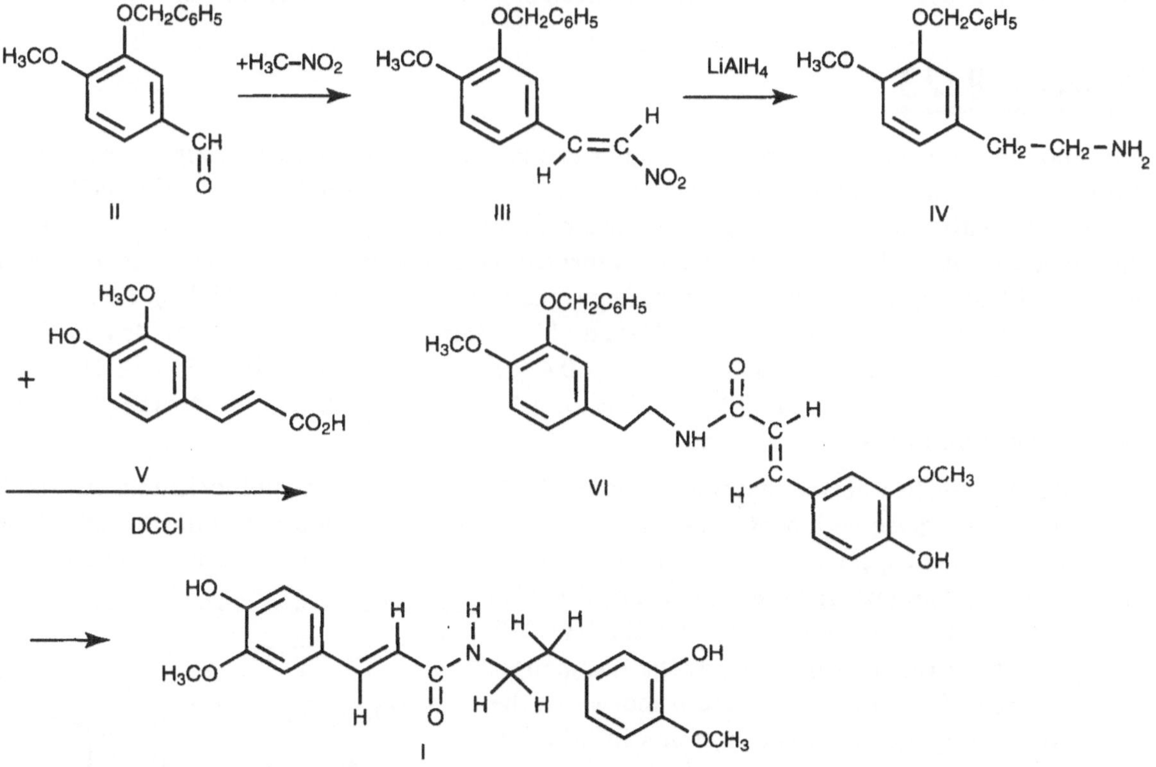

Notes

8.22.1 The molecular weight calculated for the formula $C_{19}H_{21}NO_5$ is 343.1420, which is in agreement with experimental value, namely 343.1450.

8.22.2 The chemical shift values as well as the coupling constants of the two olefinic protons are in excellent agreement with those expected for *trans*-cinnamoyl or substituted cinnamoyl moieties.

8.22.3 The benzyl group is a good protective function for hydroxyl groups in multi-step organic syntheses as it can be easily introduced and removed under mildly acidic or hydrogenolytic conditions. In the present case, a weak Lewis acid has been used to bring about the debenzylation.

The compound discussed in this problem is the first naturally occurring amide with a 4-*O*-methyldopamine moiety. The isomeric *N-trans*-feruloyl-3-*O*-methyl dopamine had earlier been isolated from plants of the families Chenopodiaceae and Lauraceae (see, for example, Suzuki T, Holden I and Cassida JE, *J. Agric. Food Chem.*, 1981, 29: 992).

Reference

Horio T, Yoshida K, Kikuchi H, Kawabata J and Mizutani J, *Phytochem.*, 1993, 33: 807–8.

PROBLEM 8.23

A compound **A** was isolated from the lipophilic exudates of crowberry leaves (*Empetrum nigrum*). The compound was found to be soluble in alkali and could be detected on TLC (silica gel) plates by spraying with $MnCl_2$ when it appeared as a reddish-brown spot. Its mass spectrum showed the molecular ion peak at m/z 260. The mass spectrum also had prominent peaks at m/z 153 and 107. Its 1H NMR spectrum had the following signals: δ 2.67 (4H, m), 3.74 (3H, s), 6.20 (2H, s), 6.57 (2H, m), 6.63 (1H, dt, *J* = 7.5 and 1.1 Hz) and δ 7.02 (1H, dd, *J* = 8.6 and 7.5 Hz). Its 13C NMR spectrum had 13 signals: δ 38.1 (t), 38.2 (t), 60.4 (q), 107.6 (d), 112.6 (d), 115.1 (d), 119.8 (d), 129.1 (d), 133.0 (s), 138.2 (s), 143.3 (s), 149.2 (s) and 156.2 (s). The compound forms a triacetate. Deduce the structure of **A**.

Analysis: It could be inferred that the compound is phenolic from its solubility in aqueous alkali and by the colour given with $MnCl_2$ (see Note 8.23.1). This is also supported by the overall profile of the 1H and NMR spectra of the compound. The formation of a triacetate shows the presence of three hydroxyl groups. The presence of a methoxyl is revealed by the 3H singlet signal at δ 3.74 in the 1H NMR spectrum. This is further supported by the 13C quartet signal at δ 60.4. The 1H NMR spectrum shows that the compound has thirteen hydrogen atoms in addition to the three hydrogen atoms of the hydroxyl groups which are not detected. Taking this into consideration, along with the fact that **A** has four oxygen atoms and a molecular weight of 260, the molecular formula can be worked out as $C_{15}H_{16}O_4$: The 1H and 13C NMR spectra together show that the compound has two benzene units, one of which carries a methoxyl group and is symmetrically substituted. The latter inference can be drawn from the fact that the carbon spectrum has only 13 signals. The two aromatic moieties are obviously connected to each other by two methylene groups as shown by the two triplet signals at δ 38.1 and 38.2 in the 13C spectrum and the 4H multiplet at δ 2.67 in the 1H spectrum. Thus, the compound is an unsymmetrical biaryl. A careful and systematic analysis of the 1H NMR spectrum clearly shows that one benzene unit is symmetrically trisubstituted and the other has a single substituent at position 3. These observations suggest two possible structures, I and II. That the correct structure is I is evident from the fact that the mass spectrum shows two fragmentation ion peaks at m/z 153 and 107. Thus, the compound can be assigned the structure, 3,5, 3'-trihydroxy-4-methoxybibenzyl (see Note 8.23.2). The complete 13C and 1H signal assignments are shown in the table given below.

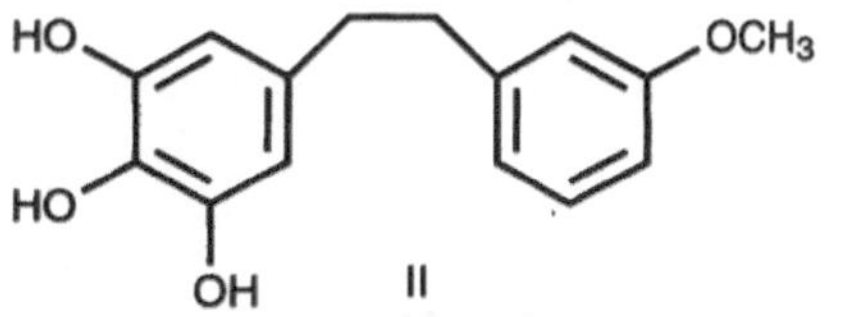

1H	δ (ppm)	13C	δ (ppm)
2 and 6	6.20	1	143.3
2' and 4'	6.57	2 and 6	107.6
5'	7.02	3 and 5	149.2
6'	6.63	4	138.2
2 × CH₂	2.67	1'	133.0
OCH₃	3.74	2'	115.1
		3'	156.2
		4'	112.6
		5'	129.1
		6'	119.8
		2 × CH₂	38.1 and 38.2
		OCH₃	60.4

Notes

8.23.1 $MnCl_2$ is an alternative to $FeCl_3$ for the detection of phenols and is particularly suitable as a spray reagent for TLC plates.

8.23.2 The bibenzyl discussed in this problem is accompanied by its monomethyl ether having the structure III and other compounds which include two chalcones, three dihydrochalcones and two dihydrophenanthrene derivatives. All these compounds are present in the lipophilic exudate which is produced in glandular structures in the small ericoid leaves of *Empetrum nigrum*.

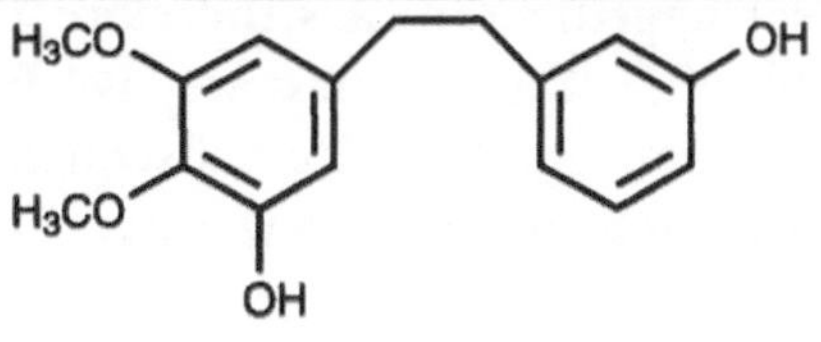

Reference

Arriaga-Giner FJ, Wollenweber E and Dorr M, *Phytochem.*, 1993, 33: 725–26.

PROBLEM 8.24

A novel phenolic compound, related to the flavonoids, was recently isolated from the roots of *Glycyrrhiza aspera* which is related to the common licorice. Its 1H and 13C NMR data are given below.

1H NMR: δ 1,63 (3H, br, s), 1.76 (3H, br, s), 3.33 (2H, br, d), 3.98(3H, s), 5.18 (1H, br, t), 6.29 (1H, d), 6.46 (1H, d), 6.68 (1H, s), 7.02 (1H, d), 8.32 (1H, br, s), 8.39 (1H, s), 8.67 (1H, br, s) and 12.39 (IH, s).13C NMR: δ 17.85 (Qm), 21.92 (Td), 25.85 (Qm), 56.79 (Q), 91.15 (D), 105.17 (Dd), 106.85 (St), 107.35 (Dd), 113.43 (Sm), 121.43 (D), 122.68 (Dm), 132.08 (Sm), 139.01 (Sm), 142.89 (Sd), 149.16 (D), 150.70 (Sm), 156.29 (Sm), 157.31 (Sdd), 158.77 (Sdt) and 178.54 (Sd).

The dimethyl ether of the compound was synthesised from 2, 4-dimethoxybenzaldehyde as per the incomplete scheme given below. Fill in the missing structures in the scheme and thereby find out the structures of the two final products, designated as **H** and **I**. Which one of them is compatible with the spectral data described above?

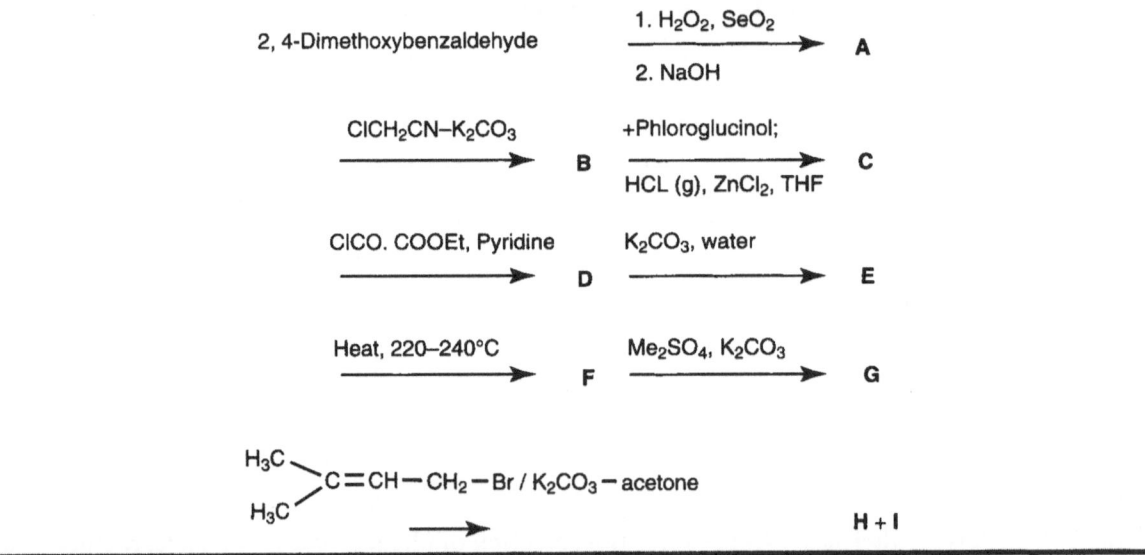

Analysis: The first step in the synthesis is a Dakin oxidation, involving a Baeyer–Villiger rearrangement and A is, therefore, 2, 4-dimethoxyphenol (I) (see Note 8.24.1). **B** is the cyanomethyl ether (II) of **A**. The next step is a Hoesch condensation and **C** can be assigned the structure III (see Note 8.24.2). The subsequent reaction is a Claisen condensation which results in the formation of a γ-pyrone ring; diethyloxalate is used for the conversion of 2-hydroxyphenyl benzyl ketones into the corresponding isoflavones (see Note 8.24.3). **D** is thus IV. Hydrolysis of the ester group in IV gives V (**E**) which on heating loses carbon dioxide to yield VI (**F**). Selective *O*-methylation of VI then leads to the formation of G (VII) (see Note 8.24.4). The final step involves *C*-isoprenylation, with the free phenolic hydroxyl directing the C-5 moiety to the *ortho* and *para* positions. Therefore, **H** and **I** can be formulated as VIII and IX (see Note 8.24.5). While the 1H NMR data are consistent with either of the structures, the multiplicities of the signals can be accounted for only by the structure VIII. The original phenolic compound itself can, therefore, be formulated as X.

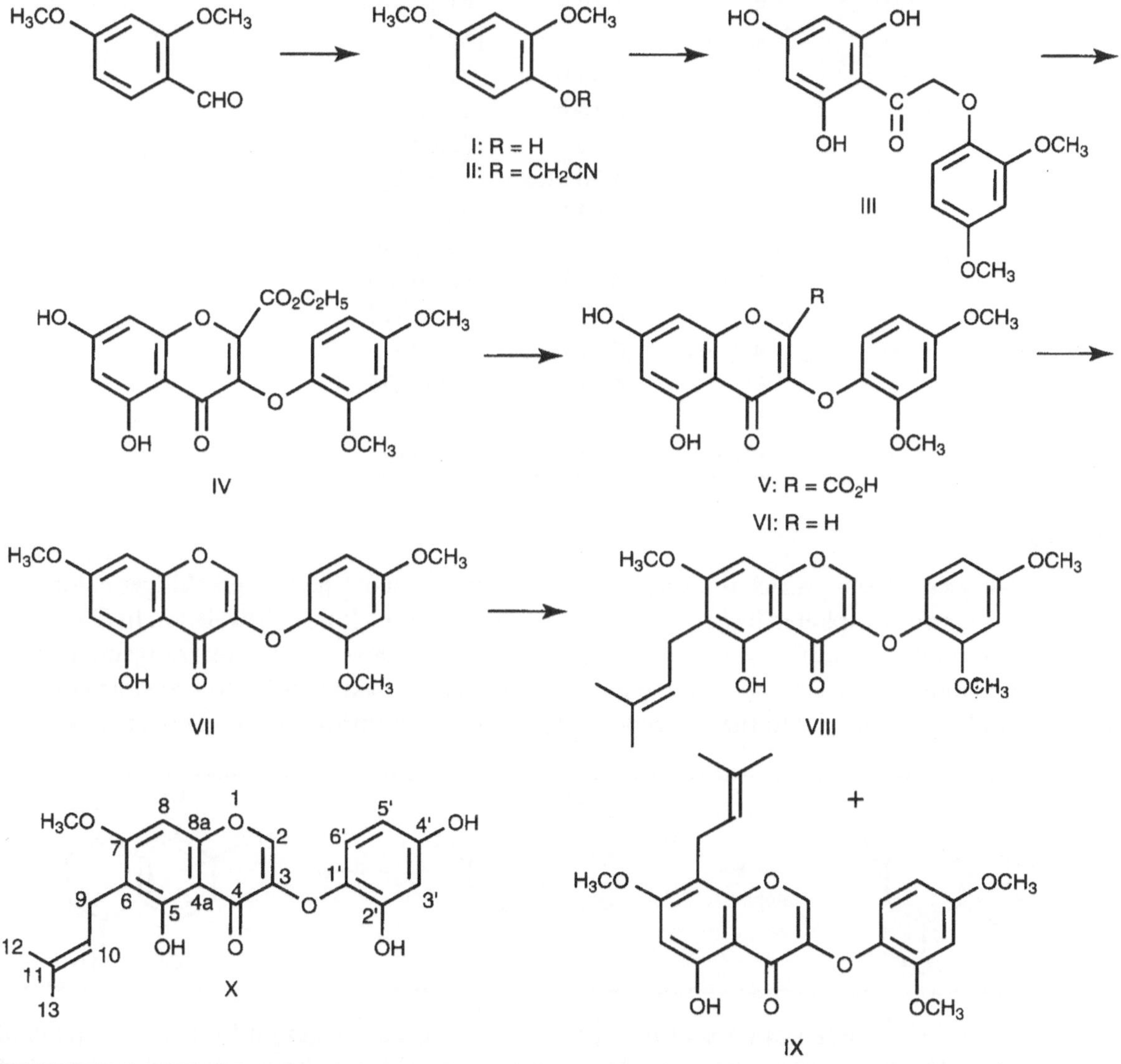

Carbon	δH	δC
2	8.39 s	149.16 D
3		142.89 Sd
4		178.54 Sd
4a		106.85 St
5 (OH)	12.35 s	158.77 Sdt
6		113.43 Sm
7		164.77 Sm
8	6.68 s	91.15 D
8a		157.31 Sd
9	3.33 br, d	21.92 Td
10	5.18 br, t	122.68 Dm
11		132.08 Sm
12	1.76 br, s	17.85 Qm
13	1.63 br, s	25.85 Qm
1'		139.01 Sm
2' (OH)	8.32 br, s	150.70 Sm
3'	6.46 d	105.17 Dd
4' (OH)	8.67 br, s	156.29 Sm
5'	6.29 d	107.35 Dd
6'	7.02 d	121.43 D
OCH$_3$	3.98 s	56.79 Q

Notes

8.24.1 Dakin's oxidation, which is a preparative reaction based on the Baeyer–Villiger oxidation, has been successfully exploited by Seshadri and co-workers for nuclear hydroxylation in flavonoids and related compounds. The strategy involved was to combine a Fries rearrangement with Dakin's oxidation to convert a phenol into its *ortho*-hydroxy derivative as in the transformation of 4-methylumbelliferone (XI) to the corresponding daphnetin derivative (XII) shown below.

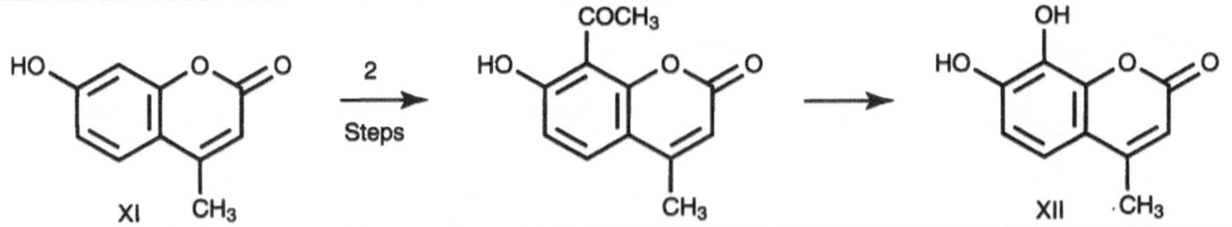

8.24.2 The Hoesch condensation is also a very good preparatory reaction for the conversion of reactive phenols into acetophenones. In this example, zinc chloride has been used as a catalyst but in a number of cases, where phloroglucinol is the substrate, the reaction works in the absence of a Lewis acid catalyst. If the reaction is meticulously carried out under perfectly anhydrous conditions, the yields of the desired products are, often, quantitative. This is an experiment which has considerable instructive value in the laboratory as its proper execution demands both patience and skill in equal measure.

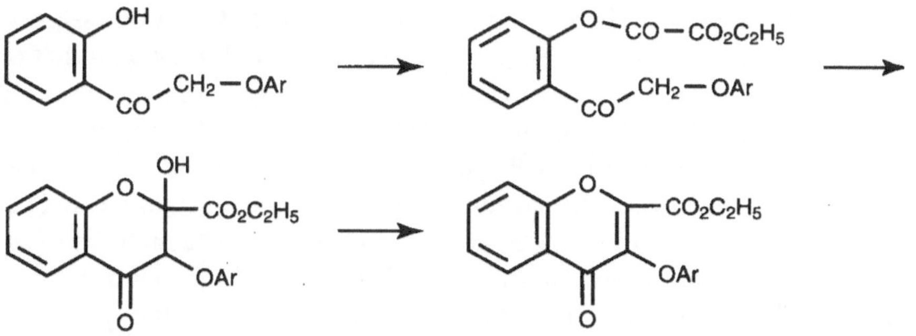

8.24.3 The course of this reaction, which involves an intramolecular Claisen condensation, may be depicted as shown above.

8.24.4 Using 1 (or 1.1) equivalent of dimethyl sulphate, selective methylation of the more reactive 7-hydroxyl could be achieved. For this purpose, the reaction is best brought about in a non-aqueous solvent using the milder base, potassium carbonate, instead of aqueous alkali.

Reference

Zeny L, Fukui T, Nomura T, Yi Zhang R and Lou ZC, *J Chem. Soc. Perkin*, 1993, 1: 1153.

PROBLEM 8.25

From the wood of *Artocarpus heterophyllus*, a pair of flavonoids with uncommon oxygenation patterns were isolated. One of them, **A**, had the molecular formula $C_{19}H_{20}O_7$ and was optically active. Its UV spectrum had maxima at 285 and 333 nm which shifted to 310 and 362 nm respectively on addition of $AlCl_3$; other shift reagents such as sodium acetate or sodium acetate–boric acid had no effect. The IR spectrum of **A** exhibited a strong absorption band at 1645 cm^{-1}. The 1H NMR spectrum of the compound had the following signals: δ 2.50 (1H, dd, $J = 17, 3$ Hz), 3.78 (3H, s), 3.80 (6H, s), 3.83 (3H, s), 3.92 (1H, dd, $J = 17, 14$ Hz), 5.98 (1H, dd, $J = 14, 3$ Hz), 6.02 (1H, d, $J = 2.2$ Hz), 6.03 (1H, d, $J = 2.2$ Hz), 6.16 (2H, s) and 12.23 (1H, s). The 13C NMR spectrum had signals at δ 40.6, 55.9, 56.0, 56.3(2C), 72.0, 91.6, 94.3, 95.2, 103.7, 106.8, 160.6, 162.6, 164.7, 164.8, 168.0 and 198.8. The mass spectrum of the compound had the molecular ion peak at m/z 360. Significant peaks were seen at m/z 194, 179 (100%), 167, 166, 151 and 135. Interpret the data and suggest an acceptable structure for the compound.

Analysis: The UV absorption spectrum (with the longest wavelength maximum below 340 nm in spite of heavy oxygenation), optical activity and the signals at δ 2.50, 3.92 and 5.98 in the

1H NMR spectrum of **A**, strongly suggest that the compound is a flavanone, with a chiral centre at C-2 (see Note 8.25.1). The 1H signal at δ 5.98 can be attributed to the hydrogen at this position. It appears as a double doublet due to coupling with the methylene protons at C-3, which are non-equivalent and which give rise to the signals at δ 2.50 and 3.92, each of which is a double doublet due to geminal (17 Hz) and vicinal couplings; the signal at δ 2.50 can be assigned to the axial hydrogen with the equatorial hydrogen resonating at a lower field. It is evident that the compound contains four methoxyl groups, from both 1H and 13C NMR spectra [signals at δ 3.78, 3.80 and 3.83, and 55.9, 56.0 and 56.3 (2C)]. The presence of an intramolecularly hydrogen-bonded hydroxyl is shown by the signal at 12.23 in the 1H NMR spectrum, the bathochromic shift caused by AlCl$_3$ in the UV maxima and the position of the carbonyl band in the IR spectrum at 1645 cm^{-1} (see Note 8.25.2). This hydroxyl can be unambiguously located at position 5. The inability of sodium acetate and sodium acetate–boric acid to bring about any shifts in the UV spectrum shows the absence of 7-hydroxyl and catechol groupings. Therefore, one of the four methoxyls can be placed at position 7. Thus, the compound contains one hydroxyl and four methoxyl groups and this accounts for all the seven oxygen atoms as the two other oxygens are part of the pyrone ring. A careful analysis of the aromatic region of the 1H NMR spectrum shows that there are two *meta*-coupled hydrogen atoms which can be assigned to positions 6 and 8 of the flavanone structure. The hydrogens of the **B** ring appear as a two-proton singlet which indicates that this ring is symmetrically substituted. The only structure which can account for all the data is 5-hydroxy-7, 2′, 4′, 6′ tetramethoxyflavanone (I) (see Note 8.25.3). The peaks at m/z 194 and 166 in the mass spectrum can be attributed to the species II and III respectively, arising from a retro-Diels–Alder fragmentation. The base peak is probably due to the loss of a methyl group from II.

Notes

8.25.1 Unlike flavones and flavonols, flavanones are optically active due to the presence of a chiral centre at C-2. They also absorb below 350 nm in the UV region. They are biosynthetically formed from the corresponding 2′-hydroxychalcones with which they often co-occur.

8.25.2 The low wave number of the carbonyl stretching vibration band shows that it is hydrogen-bonded to a hydroxyl *peri* to it. This inference is also supported by the observation that the signal due to this OH proton is shifted far downfield to 12.23 ppm.

8.25.3 A phloroglucinol-type oxygenation pattern in the **B** ring of flavonoids is very uncommon; the usual pattern is a catechol- or a pyrogallol-type. However, *meta*-hydroxylation among flavonoids of the Artocarpus species is well documented. The compound described in this problem, named as heteroflavanone-A, is accompanied by its 8-*C*-isoprenyl derivative, IV. *C*-Alkyl flavonoids commonly occur in species of the Artocarpus genus.

Reference

Lu CM and Lin CN, *Phytochem.* 1993, 33: 909–11.

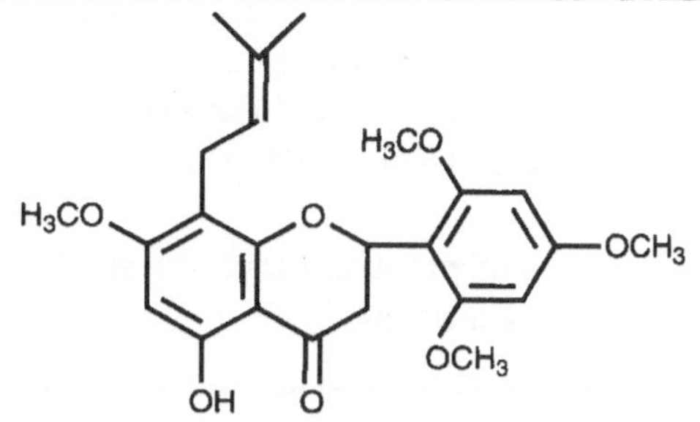

IV

PROBLEM 8.26

Betanin (!) is the violet-red pigment of beet roots (*Beta vulgaris* var. rubra). Its aglucone, betanidin (II), is obtained by acid hydrolysis of I. The 1H NMR spectrum (in trifluoroacetic acid) of betanidin shows signals at 3.68 (4H, m), 4.80 (1H, br, m), 5.55 (1H, br, m), 6.41(1H, d, *J* = 12.5 Hz), 6.74 (1H, s), 7.3 (1H, s), 7.4 (1H, s) and 8.75 (1H, d, *J* = 12.5 Hz)ppm. Betanidin, on heating with hydrochloric acid, breaks down into 4-methylpyridine-2,6-dicarboxylic acid (III) and S-cycloDOPA (IV) and formic acid. A two-step reaction involving esterification with methanolic acid followed by acetylation with acetic anhydride–pyridine, on the other hand, converts betanidin into the neobetanidin derivative (V). Assign the NMR signals to the appropriate protons in betanidin and account for the products obtained in the reactions mentioned. Which amino acid can be obtained by ozonolysis of betanidin?

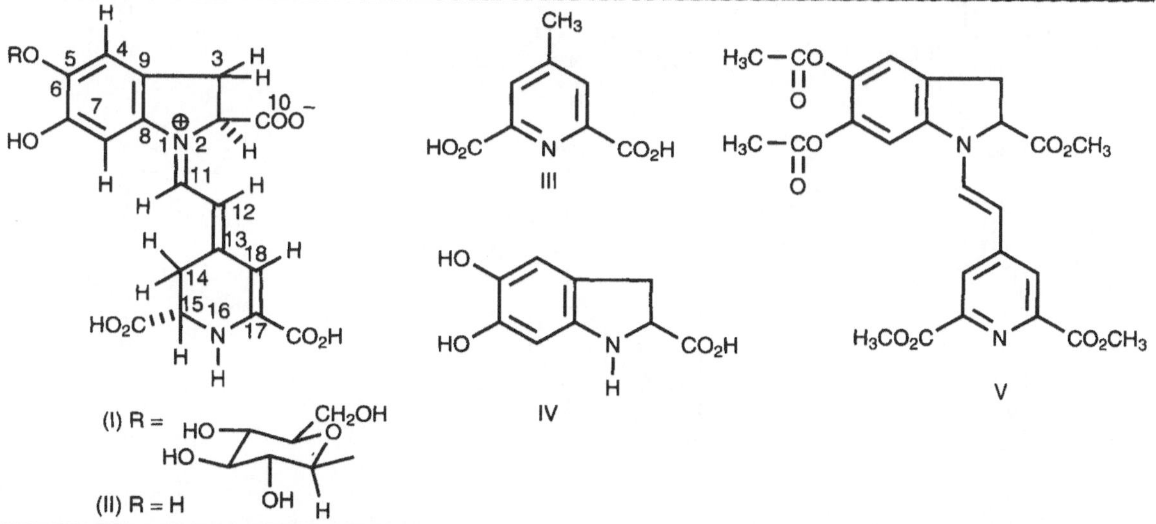

Analysis

1. 1H NMR spectrum: The two singlets at 7.3 and 7.4 ppm can be assigned to the aromatic hydrogens on C-4 and C-7 (interchangeable). The singlet at 6.74 ppm is due to the olefinic hydrogen on C-18. The doublets at 6.41and 8.75 ppm are assigned to the other two olefinic hydrogens

on C-11 and C-12, which are coupled to each other. The hydrogen atom on C-2 resonates at 5.55, whereas the one, in a similar but not identical environment at C-15, appears at 4.80 ppm. The multiplicities of these signals are due to coupling with the protons on C-3 and C-14 respectively. Incidentally, the last mentioned hydrogen atoms appear as a four-proton multiplet at 3.68 ppm.

2. The probable mechanistic course of the proton-catalysed conversion of betanidin into III and IV is shown below.

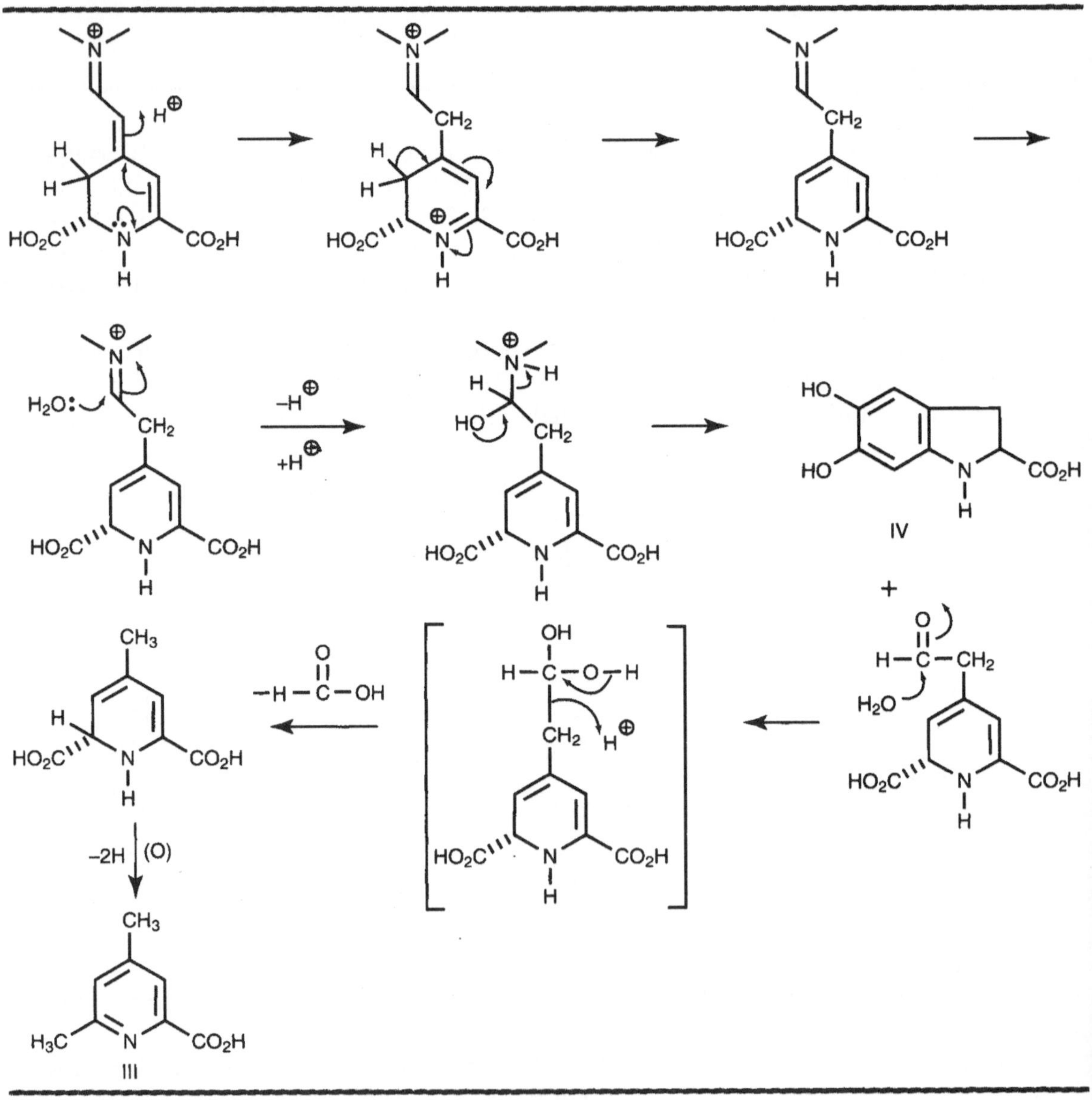

3. In the conversion of betanidin into V, the first step is the esterification of the three carboxyl groups, followed by a proton loss and aromatisation as shown below.

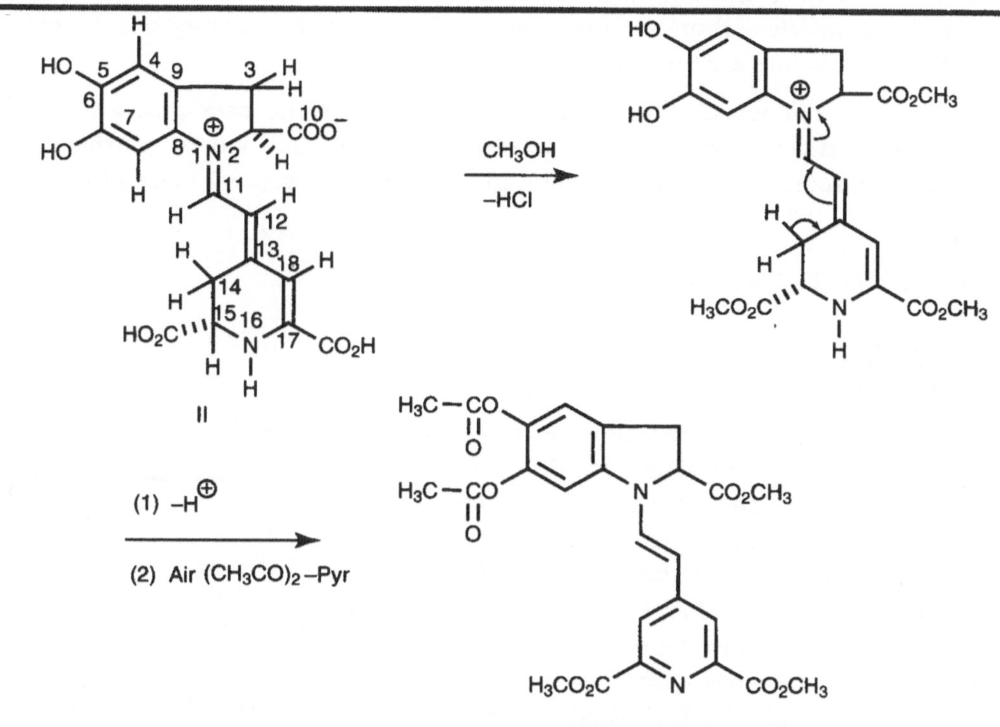

4. The amino acid obtained is L-aspartic acid (VI) as shown below.

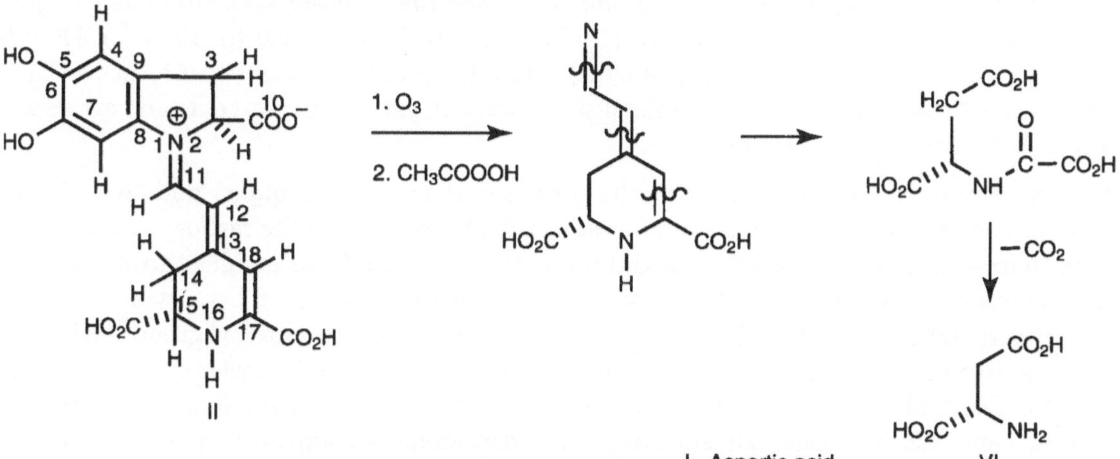

L–Aspartic acid

Notes

8.26.1 The glucose unit is attached to the hydroxyl group on C-5 through a beta linkage. The hydrogen on the anomeric carbon atom resonates at 5.26 ppm and appears as a doublet with a J value of 5 Hz. These values are typical of beta glycosides.

8.26.2 Betacyanins and betaxanthins, collectively known as the betalains, are sap soluble pigments occurring in plants of the order Centrospermae. Common sources for these pigments are, besides the beet root, bougainvillea flowers and species of the genus Amaranthus.

Reference

Mabry TJ, *J. Nat. Prod.*, 2001, **64**: 1596–1604.

PROBLEM 8.27

From a fungus found in the inner stem of a New Zealand tree, *Knightia excelsa*, a compound named as excelsione has been isolated. Its ultraviolet spectrum (absorption maxima at 213, 283 and 326 nm) resembles those of typical depsidones such as stictic acid (I). It has the molecular formula, $C_{18}H_{14}O_8$; the M^+–H peak in the HRESMS (high resolution electron spray mass spectrum) appears at 359.0704 m/z. Its infrared spectrum has absorption peaks at 3340, 3160, 1744, 1697, 1604, 1257, 779 and 725 cm^{-1}. Its 1H NMR spectrum (recorded in DMSO d6 at 500 MHz) has only five signals, all of which are singlets. The chemical shift values are 2.22 (3H, s), 2.46 (3H, s),

4.87 (2H, s), 5.32 (2H, s), and 6.79 (1H, s) ppm. However, the 13 NMR spectrum has 18 signals, with the following chemical shift values: 11.0 (q), 21.1 (q), 52.3 (t), 68.0 (t), 109.4 (s), 110.9 (s), 113.9 (s), 115.3 (s), 138.8 (s), 144.8 (s), 144.9 (s), 148.2 (s), 148.2 (s), 159.9 (s), 161.2 (s), 162.1 (s), and 168.2 (s). Assuming a depsidone skeletal structure, analyse the spectral data and arrive at an acceptable structure for excelsione.

Analysis: From the infrared spectrum the presence of hydroxyl group(s) and two carbonyl functions can be inferred. Since the compound is a depsidone, one of the carbonyl groups is that of the lactone unit. The absorption maximum at 1697 cm^{-1} could be assigned to it. The other carbonyl should be a part of a five-membered lactone as in I as 1744 cm^{-1} is a typical value for such a functionality. From the 1H NMR spectrum, the presence of two methyl groups on benzene rings, and two methylene groups each flanked by a benzene ring and a hydroxyl or oxygen can be readily deduced. The signal at 6.79 ppm should be due to an aromatic hydrogen. Therefore, it can be concluded that one benzene ring of the depsidone is completely substituted and the other penta substituted. On biosynthetic grounds, one ring can be presumed to be derived from orsellinic acid (II; see Chapter 1). If two such units combine, one of the possible structures would be III. Intramolecular oxidative coupling would then give IV. To account for the presence of a five-membered lactone, another C-methyl group and a – CH₂OH group, in exelsione, structure IV can be further modified step-wise, as shown below, to arrive at the structure V for excelsione. This is the structure given to the compound by the authors of the paper cited below. However, on the basis of the available data, other possible structures cannot be ruled out.

Notes

8.27.1 The tree, *Knightia excelsa*, belongs to the family Proteacea and is also known as New Zealand honeysuckle.

8.27.2 Preparation of a triacetate derivative would have provided additional useful information. For instance, in the triacetate, the signal at 4.87 ppm could be expected to move further downfield, whereas the other signal at 5.32 ppm would remain unaffected. The formation of such a derivative would have also confirmed the presence of three hydroxyl groups.

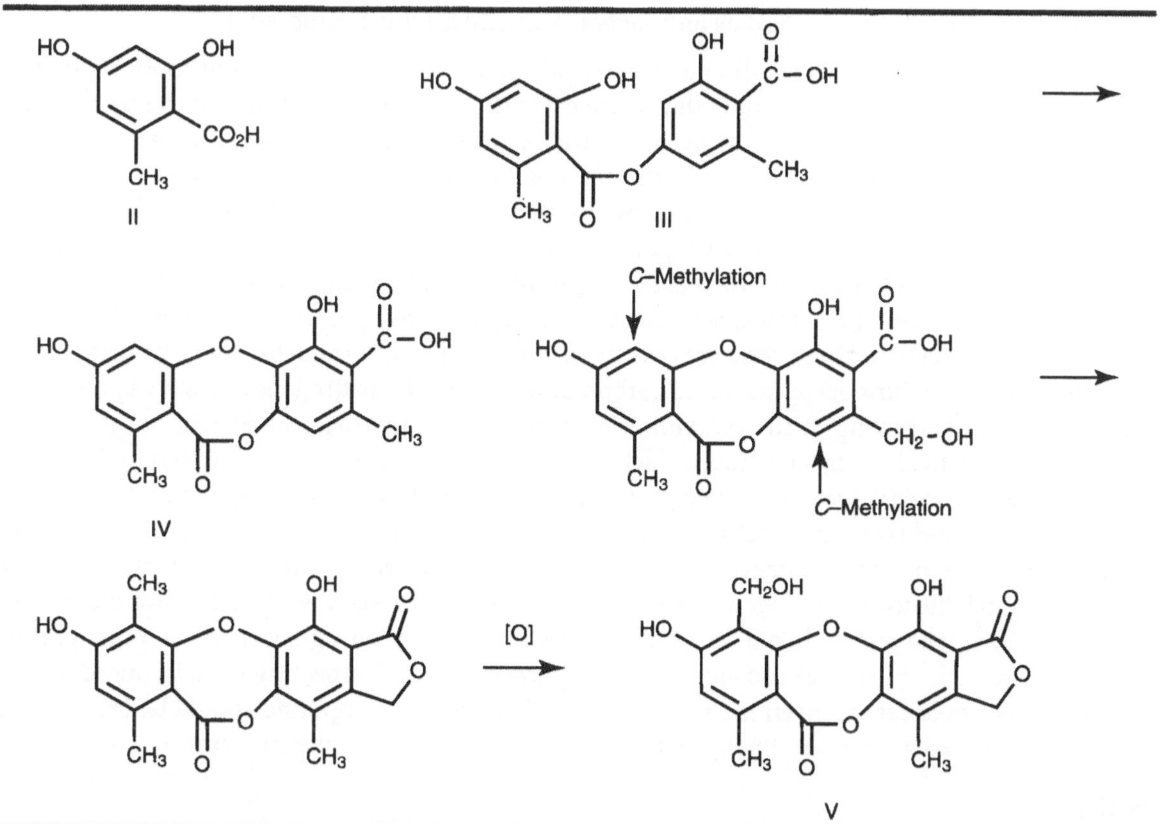

8.27.3 By the action of ethereal diazomethane, a mono methyl ether of (V) could have been obtained, as alcoholic hydroxyls and hydrogen-bonded phenolic hydroxyls do not undergo methylation by this reagent.

Reference

Lang G, Cole ALJ, Blunt JW, Robinson WT and Munro MHG, *J. Nat. Prod.*, 2007, 70: 310–11.

PROBLEM 8.28

From the plant, *Piperomia duclouxii,* several new lignans have recently been isolated. One of them is insoluble in aqueous alkali and its infrared spectrum does not show any absorption in the 3300–3600 cm^{-1} and 1650–1750 cm^{-1} regions. However, there are absorption bands at 2948 and 2892 cm^{-1}. Its molecular formula is determined as $C_{22}H_{22}O_8$ from the high resolution electron impact mass spectrum which gives a value of m/z 414.1291 for its molecular ion. The relative intensity of the molecular ion peak is 92%. The base peak occurs at 179 m/z. However, its 13C NMR spectrum has only eleven signals at 54.3 (d), 56.6 (q), 71.8 (t), 85.8 (d), 100.0 (d), 100.5 (s), 101.5 (t), 134.6 (s), 135.7 (s), 143.6 (s) and 149.0 (s). Its 1H NMR spectrum has the following signals: 3.04 (2H, m), 3.87 (2H, dd, *J* = 9.3, 3.4 Hz), 3.90 (6H, s), 4.25 (2H, dd, *J* = 9.3, 6.8 Hz), 4.69 (2H, d, *J* = 4.2 Hz), 5.95 (4H, s), 6.51 (2H, d, *J* = 2Hz) and 6.53 (2H, d, *J* = 2Hz). Interpret the data and suggest a structure for the compound.

Analysis: The insolubility in alkali and the absence of absorption bands in the 3300–3600 cm^{-1} region of the infrared spectrum show the absence of phenolic hydroxyl groups in the compound. That the compound is predominantly aromatic and symmetrically substituted is evident from the 13C NMR spectrum. The signals seen in the 1H and 13C NMR spectra are due to a pair each of eleven hydrogen and carbon atoms of which two methoxyls [3.90 ppm in the proton NMR spectrum and 56.6 (q) ppm in the carbon spectrum] and two methylenedioxy groups [5.95 ppm and 101.5 (t) ppm in the proton and carbon NMR spectra respectively] can be readily identified. The signals at 6.51 and 6.53 ppm in the 1H NMR spectrum can be assigned to two pairs of *meta*-coupled aromatic protons. Taken together, these data suggest the presence of two tetra-substituted benzene units with the structure I. This accounts for 16 carbon atoms, fourteen hydrogens and six oxygens. The connecting unit consisting of the remaining 6 carbons, eight hydrogens and two oxygen atoms contains two identical methylene groups, each attached to an oxygen atom (giving rise to the signal at 71.8 in the 13C NMR spectrum), two methine carbons, each next to an oxygen and a benzene ring (85.8 ppm) and two other methine carbons (54.3 ppm). They form a chain as shown by the coupling interactions of the hydrogen atoms attached to them. This chain can be written as II. Of these, the signal due to the hydrogen on the middle carbon is not well resolved. The two methylene hydrogens, being magnetically non-equivalent, give rise to separate signals, each as a one proton double doublet, due to geminal and vicinal coupling interactions. Connecting I and II, one identical half of the molecule can be given the structure II and hence, the compound should be IV. The ion responsible for the base peak in the mass spectrum could be assigned the structure V.

Notes

8.28.1 For the detection of the methylenedioxy group a colour test known as the Labat test is useful. In this test, the compound is treated with a solution of gallic acid in sulphuric acid, when a blue green colour develops if the compound contains a methylenedioxy functionality. Chromotropic acid can also be used for detecting this functional group.

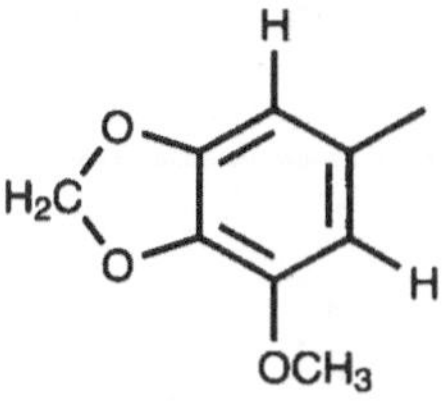

8.28.2 For a comprehensive account of lignans, see Ayres DC and Loike JD, *Lignans, Chemical, Biological and Clinical Properties*, Cambridge: Cambridge University Press, 1990.

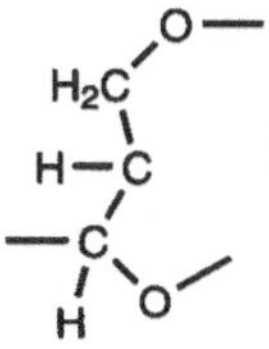

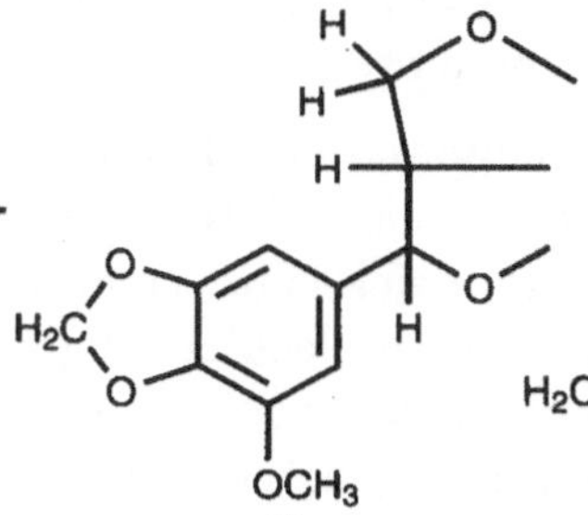

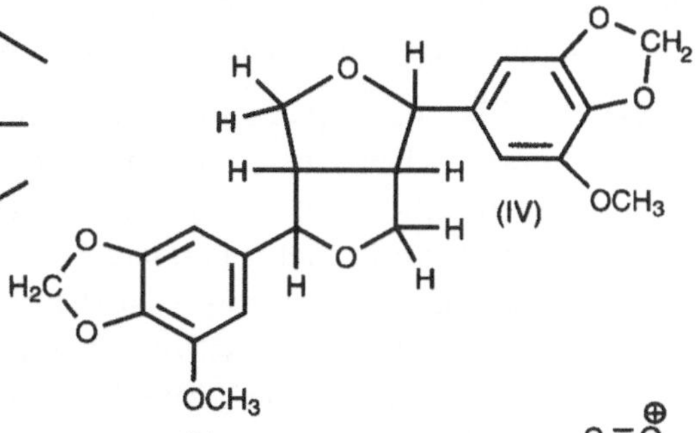

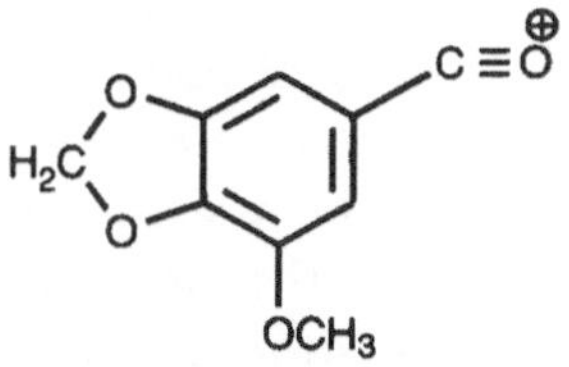

8.28.3 For the nomenclature of lignans and neolignans, refer to Moss GP, *Pure Appl. Chem.*, 2000, 72: 1493–1523.

8.28.4 For the mass spectral fragmentation of a lignan of similar structure, see Chapter 2, Section 2.1.

Reference

Li N, Wu J-L, Hasegawa T, Sakai J-i, Bai L-M, Wang L-Y, Kakuta S, Furuya Y, Ogura H, Kataoka T, Tomida A, Tsuruo T and Ando M, *J. Nat. Prod.*, 2007, 70: 544–48.

PROBLEM 8.29

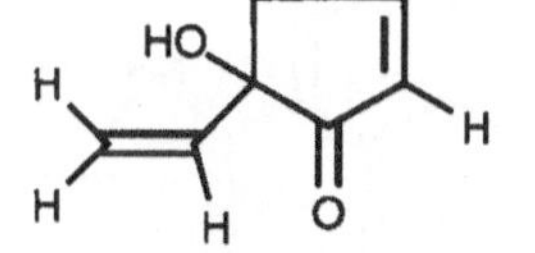

From a marine isolate of the fungus, Bortrytis, a compound named myrthenone having the structure I has been obtained along with a new compound. The latter is a red coloured oil and in its spectral and general behaviour closely resembles I. Its infrared spectrum has prominent absorption bands at 3321, 3188 and 1630 (very intense) cm^{-1}. In its mass spectrum, the molecular ion peak is seen at m/z 217 with an isotope peak at m/z of equal intensity. Other peaks are seen at m/z 138, 110, and 55 (base peak). In the 1H NMR spectrum

of myrthene (I), there is a singlet signal at 4.70 ppm which is absent in the spectrum of the new compound. Other signals in the proton NMR spectrum of the new compound recorded in DMSO d6 are seen at 2.79 (2H), 5.08 (1H, dd, J = 10.6, 1.6 Hz), 5.27 (1H, dd, J = 18.2, 1.6 Hz), 5.79 (1H,dd, J = 18.2, 10.6 Hz), 7.54 (1H, brs) and 8.00 (1H, br, s). In the 13C NMR spectrum seven signals are seen: 42.1 (CH_2), 76.5 (qC), 86.1 (qC), 113.3 (CH_2), 140.5 (CH), 169.6 (qC) and 193.3 (qC). Interpret the data and suggest a structure for the new compound.

Analysis: The mass spectrum of the compound clearly indicates the presence of a bromine atom. The odd m/z value of the molecular ion shows that it contains a nitrogen atom which is in accordance with the observation that it is related to I. The molecular weight of I is 139. Therefore, it is obvious that the new compound is a bromo-derivative of myrthenone, with the bromine replacing a hydrogen atom. That the vinyl group is intact is shown by the signals at 5.08, 5.27 and 5.79 ppm. The multiplicities and coupling constants are in perfect agreement with those expected for this functionality. The two proton signals at 2.79 ppm can be assigned to the hydrogen atoms on C-4. The amino protons resonate at 7.54 and 8.00 ppm; the OH proton signal is not seen due to deuterium exchange. That means that the bromine atom is attached to C-2.This is also supported by the observation that the signal at 4.70 ppm seen in the spectrum of I is absent in the spectrum of the new compound, which can, therefore, be assigned the structure II. It is easy to assign the 13C signals to the various carbon atoms. The carbonyl carbon resonates at 193.3 ppm. Perhaps, the only signal which requires some explanation is the one at 86.1 ppm, assigned to C-2. The somewhat highfield position for this olefinic proton can be attributed to the presence of the bromine atom on this carbon; in I this signal appears at 124.9 ppm. The 169.6 ppm signal is due to C-3 which is the beta carbon of a conjugated enone system. C-6 of the vinyl group resonates at 140.5 and C-7 at 113.3 ppm. The signal due to C-5 appears at 76.5 ppm, a value in good accordance for a carbon atom attached to an oxygen atom. In the mass spectrum, the peak at 138 m/z is due to the M^+–Br ion. Loss of a carbony group from this ion would generate the one with m/z 110.

Notes

8.29.1 Bromine-containing compounds occur fairly widespread in marine flora and fauna. For example, two bromopyrrole derivatives, orodin (III) and dispyrin (IV), have recenty been isolated from a Caribbean sponge, *Agelas dispar* (Pina IC, White KN, Cabrera G, Rivero E and Crews P, *J. Nat. Prod.*, 2007, 70: 613–617).

8.29.2 For the detection of bromine in a very small quantity of a natural product, mass spectroscopy is the most direct and reliable method.

8.29.3 In the infrared spectrum of II the keto carbonyl absorption frequency is seen at an unusually low value for a cyclopentenone carbonyl. This could be due to the electron-donating amino function at the beta carbon of the enone system. In a related structure (V), in which this amino group is absent, the carbonyl stretching frequency is seen at 1716 cm^{-1}.

Reference

Li X, Zhang D, Lee U, Li X, Cheng J, Zhu W, Jung JH, Choi HD, and Son BW, *J. Nat. Prod.*, 2007, 70: 307–9.

PROBLEM 8.30

In a recent publication, the isolation and chracterisation of two new flavanones from the bark of the leguminous tree, *Erythrina addisoniae*, are described. Herein, the reported data for one of them is given to be analysed. Its molecular formula, as determined from mass spectral data, is $C_{26}H_{30}O_6$.

In its ultraviolet spectrum in methanol the main absorption maximum is at 290 nm with minor bands at 232 and and 332 nm. Addition of anhydrous aluminium chloride is expected to bring about a bathochromic shift of the 290 nm band. In the 1H NMR spectrum, the signals are seen at 1.71 (6H, s), 1.79 (3H, s), 2.68 (1H, dd, J = 13.2, 4.7 Hz), 2.73 (1H, dd, J = 17.0, 3.2 Hz), 2.84 (1H, m), 3.15 (1H, dd, J = 17.0, 13.0 Hz), 3.39 (2H, d, J = 7.3 Hz), 3.77 (3H,s), 4.23 (1H, m), 4.73 (1H, s), 4.90 (1H, s),5.30 (1H,t, J = 7.3 Hz), 5.46 (1H, dd, J = 13.0, 3.2 Hz), 5.94 (1H, d, J = 2.2 Hz), 5.95 (1H, d, J = 2.2 Hz), 7.23 (1H, d, J = 2.1 Hz), 7.34 (1H, d, J = 2.1 Hz), 9.74 (1H, s), and 12.17 (1H,s). In the 13C NMR spectrum, the signals are at 18.1 (CH$_3$), 25.8 (CH$_3$), 29.0 (CH$_2$), 38.2 (CH$_2$), 43.5 (CH$_2$), 61.3 (CH$_3$), 76.1 (CH), 79.9 (CH), 95.8 (CH), 96.8 (CH), 103.2 (C), 110.5 (CH$_2$), 123.9 (CH), 127.3 (CH), 128.3 (CH), 132.9 (C), 133.8 (C), 135.0 (C), 135.5 (C), 149.3 (C), 157.8 (C), 164.2 (C), 165.3 (s), 167.4 (s) and 197.0 (C). Of the 25 resonances in the carbon NMR spectrum, the values of sixteen are very close to those observed in the spectrum of naringenin 4'-*O*-methyl ether (I). Interpret the data and deduce the structure of the compound.

Analysis: The UV absorption maximum at 290 nm strongly suggests a flavanone structure with hydroxyls at the 5, 7 positions. The 2, 4, 6-trihydroxyacetophenone chromophore present in the compound is responsible for this absorption band (calculated value as per rules explained in Chapter 2: 246 + 25 + 2 × 7 = 285 nm). The hydroxyl group at position 5, being involved in intramolecular hydrogen bonding with the pyrone carbonyl group, is responsible for the far downfield signal at 12.17 ppm in the 1H NMR spectrum. The other hydroxyl group gives the signal at 9.74 ppm. The use of shift reagents (sodium acetate and aluminium chloride) would have provided confirmation for these assignments. The two aromatic protons in the **A** ring, which are *meta*-coupled are responsible for the signals at 5.94 and 5.95 ppm. Two other aromatic protons, also *meta*-coupled (signals at 7.23 and 7.34 ppm) can be located in the side phenyl ring (C) which should therefore have substituents at positions 3', 4' and 5'. The lone methoxyl group (signal at 3.77 ppm) can be placed at 4' position since it is recorded that the compound under discussion bears close resemblance in its spectral characterestics to naringenin 4'-O-methyl ether (I). This accounts for 16 of the 26 carbon atoms, five of the six oxygen atoms and 12 of the 30 hydrogen atoms. The hydrogen atoms on C-2 and C-3 resonate at 5.46 and 2.73 and 3.15 ppm respectively; the methylene protons on C-3 being magnetically non-equivalent give rise to two separate signals and are coupled to each other by geminal coupling (17.0Hz). The *J* values of the other coupling interactions in this part of the molecule are readily comprehensible in terms of the principles explained in Chapter 2. The remaining ten carbon atoms, one oxygen and 18 hydrogens are constituents of two separate isoprenoid moieties both of which are on the side phenyl ring at positions 3' and 5'. One unit is γ, γ-dimethylallyl moiety (II) as can be figured out from the proton signals at 3.39 (2H, d) (a -CH$_2$ between a double bond and a benzene ring), 5.30 (1H, t) (olefinic proton coupled to the methylene) and 1.71 (6H) (the two terminal groups). Similarly, from the other proton signals, the second unit can be identified as 3-hydroxyisopentenyl moiety (III). Therefore, the new flavanone can be assigned the structure (IV). We leave the 13C NMR data to be interpreted by the readers!

Notes

8.30.1 Flavonoids containing isoprenyl groups do occur in a number of plants. Species of the genus Artocarpus have provided a number of such compounds.

8.30.2 It has been reported that the stem and root bark of *Erythrina addisoniae* (Leguminoseae) is used in some parts of Africa for the treatment of dysentery, hepatitis and rheumatic disorders.

8.30.3 In this case too, additional information, obtained particularly from the use of shift reagents in UV spectroscopy would have been useful.

8.30.4 It is mentioned in the paper that in spite of having two free phenolic hydroxyls, the compound does not exhibit any anti-oxidant property. That is intriguing as the two isoprenyl units are on a different ring and cannot, therefore, possibly interfere with the reactvity of the hydroxyl groups.

8.30.5 For HMBC (1H detected multiple-bond quantum coherence) spectral data see the original paper mentioned below. This highl;y sensitive method is useful for observing long-range coupling.

Reference

Watjen W, Suckow-Schnitker AK, Rohrig R, Kulawik A, Addae-Kyereme J, Wright CW, and Passreiter CM, *J. Nat. Prod.*, 2008, 71: 735–38.

Index